Subcellular Biochemistry

Volume 92

The book series SUBCELLULAR BIOCHEMISTRY is a renowned and well recognized forum for disseminating advances of emerging topics in Cell Biology and related subjects. All volumes are edited by established scientists and the individual chapters are written by experts on the relevant topic. The individual chapters of each volume are fully citable and indexed in Medline/Pubmed to ensure maximum visibility of the work.

More information about this series at http://www.springer.com/series/6515

Andreas Kuhn
Editor

Bacterial Cell Walls and Membranes

Springer

Editor
Andreas Kuhn
Institute of Microbiology
University of Hohenheim
Stuttgart, Germany

ISSN 0306-0225
Subcellular Biochemistry
ISBN 978-3-030-18767-5 ISBN 978-3-030-18768-2 (eBook)
https://doi.org/10.1007/978-3-030-18768-2

This Springer imprint is published by the registered company Springer Nature Switzerland AG
The registered company address is: Gewerbestrasse 11, 6330 Cham, Switzerland

Contents

Chapter 1
The Bacterial Cell Wall and Membrane—A Treasure Chest for Antibiotic Targets

Andreas Kuhn

Abstract Astonishing progress has been made in recent years to understand the structural complexity and functions of the biosynthetic pathways of the bacterial and archaeal envelopes. This progress has prompted me to assemble the present book that provides a detailed overview and the state-of-art of the respective research field. Ideally, the book will provide students and advanced scientists an up to date picture of the different parts of the bacterial and archaeal cell envelope and enable them to understand their functional roles.

Keywords Antibiotic targets · Biosynthetic pathways · Conformational cycles · Multicomponent complex assembly · Regulatory interactions · Substrate movement in transporters

Outer Membrane

We will start off in Chap. 2 with the outer membrane and its outer exposed layer, the lipopolysaccharide (LPS). Here, tremendous progress has been made in unraveling the biosynthetic pathway. In this pathway, Lipid A with core sugars is flipped across the inner membrane catalyzed by MsbA. The flipped lipid A then binds to $LptB_2FG$ and is transferred to LptA and LptC that together forms a bridge across the periplasm. At the outer membrane, LptED receives the LPS precursor and can implant it most likely through a lateral gate allowing it to integrate into the outer leaflet of the outer membrane. An atomic structure of all the participating proteins is now available. This sets the basis for the understanding the molecular events that occur along the pathway and for further investigating the required conformational movements in these proteins which are driven by ATP hydrolysis within LptB.

The next Chap. 3 is focusing on lipoproteins which are found mostly in the inner leaflet of the outer membrane but also in the outer leaflets of both membranes to fulfill their many functional roles. Lipoproteins are synthesized in the cytoplasm with

A. Kuhn (✉)
Institute of Microbiology, University of Hohenheim, Stuttgart, Germany
e-mail: Andreas.Kuhn@uni-hohenheim.de

A. Kuhn (ed.), *Bacterial Cell Walls and Membranes*, Subcellular Biochemistry 92, https://doi.org/10.1007/978-3-030-18768-2_1

a signal peptide that is cleaved off by a unique signal peptidase (termed SPase II) after its transport across the inner membrane by the Sec translocase. The cysteine residue at the N-terminus of the mature protein is modified by three fatty acids that are recognized by LolCDE complex. Lipoproteins destined to the outer membrane are released from LolCDE and transferred to LolA, a periplasmic carrier protein. LolA then moves across the periplasm and delivers the lipoprotein to LolB at the outer membrane for its insertion into the inner leaflet. Although there is a clear distinction between the biosynthetic pathways of lipoproteins and lipopolysaccharides, common principles of their transport are evident. In both pathways, ABC transporter components are involved in releasing the substrate from the inner membrane and bind their substrates in the periplasm for transfer onto carrier components.

Porins, the main proteins of the outer membrane, are discussed in Chap. 4. A major focus is on the unique regulation of the expression of the porins by stress response circuits involving two component systems and anti-sense RNA molecules. The biosynthesis and folding of the β-barrel porins by the BAM complex is discussed that is currently a topic of ongoing research. A further, very important aspect of the porins is their involvement in multidrug resistance. Interestingly, distinct transcriptional regulation systems offer possibilities to manipulate the sensitivity to antibiotics that can also lead to resistance.

Periplasm

We then move to the periplasm (Chap. 5), which is filled with a sponge-like structure of the peptidoglycan. Although the chemical structure and the major enzymes involved in peptidoglycan synthesis has been known for a long time, a more detailed picture has emerged in analyzing the peptidoglycan from different species, including Gram-positive bacteria. There is a variety among the cross-linking peptides and the modifications that occurs within the glycan strands. In addition, the usual 4-3 crosslinking can be replaced by 3-3 crosslinks. This diversity has a big impact on whether antibiotics kill certain bacteria or develop resistance. Moreover, it has been noticed that pathogenic strains are enriched in such modifications, which may contribute to the survival of the bacteria in host organisms.

The biosynthetic machineries for peptidoglycan synthesis are the elongasome and divisome that coordinate transpeptidase and glycosyltransferase activities but consist of specific components. The details of how septation is regulated and how the two new daughter sacculi that are formed are controlled by the temporal and spatial amidase (Ami ABC) activities is currently investigated.

The chaperones Skp and SurA play an essential role for escorting the outer membrane proteins to the outer membrane after they leave the Sec translocon at the periplasmic side of the inner membrane. In Chap. 6 the mechanistic details of Skp and how it binds various client proteins are discussed including the important role of the flexible arms of the Skp trimer that holds onto the client protein. SurA has, as some other periplasmic proteins do, a prolyl-isomerase activity domain. However, the functional relevance of this activity is still unknown. The SurA domain that is involved in binding to the client proteins does this in a clamp-like fashion.

Preproteins and prolipoproteins, which are processed by signal peptidases, are discussed in Chap. 7. SPase II removes the signal peptide of prolipoproteins and was detected after the discovery of globomycin from *Actinomyces* that inhibits the peptidase. In contrast, SPase I cleaves the signal peptides of periplasmic and outer membrane proteins and is important for the release from the inner membrane allowing them to reach their destination. Recent research has led to the finding that arylomycins are potent inhibitors of signal peptidases. These arylomycins are promising candidates for a new class of antibiotics.

Inner Membrane

The many different proteins of the inner membrane are represented here by transporters, proteins translocases and respiratory complexes. We start in Chap. 8 with the phosphotransferase PTS systems which are central for sugar uptake in bacteria. The predominant function the PTS systems have in bacterial metabolism is reflected by the many regulatory interactions of their components. The phosphorylation cascade and network is a beautiful example of a simple signal transduction system in cells. The coupling of the phosphorylation cascade to transport sugar across the membrane is a mechanistically challenging problem but the novel structural information of the maltose and the diacetylchitobiose transporters from *Bacillus cereus* has now shed light on this. Intriguingly, the substrates are moved through the membrane by an "elevator mechanism" involving topological rearrangements of the transmembrane helices of the transporter. This mechanism of substrate movement is distinct from the rocking switch mechanism found in LacY.

The secondary active transporters are found with 3 different protein folds and are discussed in Chap. 9. LacY and other members of the major facilitator superfamily (MFS) are organized as two symmetric 6-helical bundles that bind the substrates in a central cavity between the two bundles. The rocking movement of the bundles switches LacY from an outwards open to inwards open state, which is used to transport the lactose. Only the protonated LacY can bind the substrate sugar at the periplasmic face and is released after deprotonation at the cytoplasmic face of the membrane. In contrast to MFS, the LeuT and NhaA folds are organized pseudo-symmetrically with interwound 5-helical bundles. In addition, structural flexibility has been observed for NhaA after substrate binding, which lead the authors to propose that transport occurs also here by an elevator transport mechanism.

A wide variety of respiratory complexes present in different bacteria is covered in Chap. 10. Starting with aerobic systems in *E. coli* that operate to establish a proton gradient across the inner membrane and following with respiration of mainly marine bacteria, that generate a sodium ion gradient are discussed. The most fascinating feature of the respiratory complexes is the electron transfer among the individual modules each containing electron binding cofactors. The aerobic and anaerobic respiratory chains act as redox loops and are coupled to proton (or sodium) pumps. Bacteria show a wide variety of different respiration systems. Nevertheless, some of these systems could serve as antibiotic targets depending on their essential function in specific environments.

Newly synthesized membrane proteins are inserted by the multi-subunit Sec translocase which operates to translocate unfolded protein chains. In Chap. 11 the details of targeting the ribosome nascent chain or the preprotein chaperone complex to the Sec translocase are presented as well what is known about the movement of the protein chain through SecY and how the movement is powered by conformational cycles of the SecA ATPase. The roles of the YidC insertase alone, or of YidC together with Sec to insert proteins are discussed. Folded proteins, in particular metallo-proteins are translocated by the TAT system which might operate in a concerted action of TatA oligomers surrounding and translocating the client protein in a yet unknown mechanism.

Pili

Since pili are exposed on the bacterial cell wall they are primary antigenic targets and are also involved in biofilm formation. In addition, pili are important for the adherence of bacteria to eukaryotic cells and often involved in the invasion into these cells. Because of their importance in pathogenic infections there is a major medical interest to control the invasion process by compounds that interact with pili or even inhibit pilus formation. Chapter 12 gives an overview on the different pili systems known, starting with the chaperone-usher (CU) system of Gram-negative bacteria. Their assembly pathway at the periplasmic face of the outer membrane is an impressive example of protein folding events leading to consecutive subunit interactions. The folding mechanisms of donor-strand-exchange and the donor-strand-complementation between the subunits lead to subunit pairing and filament formation, respectively. Different from the pili that are assembled by the CU system are the type IV pili, which are involved in the twitching motility but also in cell adhesion, phage and DNA uptake. They are anchored to the inner membrane and to the outer membrane as multi-component complexes that also control filament extension and retraction reactions. New cryo-EM and crystal structures of several components of the type IV pili provide now the first structural details.

The amyloid fibrillar structure of Curli and Fab pili is another fascinating and ongoing area in protein folding research. The pili subunits are transported from the periplasm through a β barrel in the outer membrane and presumably assemble into an amyloid cross- β architecture by a mechanism supported by a chaperone on the extracellular surface.

Different from the pili involved in adhesion are the conjugative pili of the secretion system T4SS. These pili support DNA uptake and DNA delivery into recipient cells in natural competence and bacterial conjugation. The T4SS conjugative pili consist of a multi-component structure that is anchored in the outer and inner membrane forming a continuous structure from the cytoplasm to the extracellular surface. The conjugative pili are retractile and lead to an intimate donor-recipient cell contact before the DNA is transported through the membranes of the mating cells.

Actinobacteria and Archaea

The cell walls of the Gram-positive Actinobacteria and of archaea are of great interest because of their differences to the well-studied Gram-negative cell wall. In addition, research on the pathogenic Mycobacteria and Corynebacteria has made substantial advances in our knowledge of the actinobacterial cell walls during recent years. Although most of the Actinobacteria have the monodermic Gram-positive cell wall architecture, Mycobacteria and Corynebacteria have evolved a diderm cell envelope. Interestingly, the outer membrane of Mycobacteria, the mycomembrane, is rich in mycolic acids especially in the inner leaflet of the outer membrane bilayer. Another component is the mycocerosyl lipid also present in the outer leaflet which is presumably transported by the ß-barrel protein LppX during its biosynthesis. These components lead to the dense and waxy character of the mycomembrane. Together with a complex capsule structure, which is composed of secreted polysaccharides, a thick electron transparent zone surrounds the mycobacterial cells. The unique properties of the actinobacterial cell wall pose many evolutionary questions such as how they have been developed but they also provide a multitude of promising new chemotherapeutic targets.

Archaea have the capacity to develop an astonishing variety of cellular envelopes. For example, monodermic and didermic cell walls are found in archaea. The didermic cell walls are similar to those of Gram-negative although there are differences in their biochemical composition. Most common among most archaea is the existence of a surface layer (S-layer) consisting of highly glycosylated proteins arranged in 2-dimensional crystal layers. The S-layers contribute to the extreme heat and osmotic stability of the archaeal cells.

Part I
Outer Membranes

Chapter 2
Lipopolysaccharide Biosynthesis and Transport to the Outer Membrane of Gram-Negative Bacteria

Paola Sperandeo, Alessandra M. Martorana and Alessandra Polissi

Abstract Gram-negative bacteria have an outer membrane that is positioned at the frontline of the cell's interaction with the environment and that serves as a barrier against noxious molecules including many antibiotics. This protective function mainly relies on lipopolysaccharide, a complex glycolipid located in the outer leaflet of the outer membrane. In this chapter we will first summarize lipopolysaccharide structure, functions and biosynthetic pathway and then we will discuss how it is transported and assembled to the cell surface. This is a remarkably complex process, as amphipathic lipopolysaccharide molecules must traverse three different cellular compartments to reach their final destination.

Keywords Lipopolysaccharide · Outer membrane biogenesis · Lpt machinery · ABC transporters · β-Jellyroll

Introduction

Gram-negative bacteria have a multi-layered envelope consisting of two distinct membranes that define an aqueous space, termed periplasm, containing a thin layer of peptidoglycan (Silhavy et al. 2010). The cytoplasmic or inner membrane (IM) is a classical phospholipid bilayer whereas the external, outer membrane (OM) is highly asymmetric and contains phospholipids in the inner leaflet and lipopolysaccharides (LPS), in the outer leaflet (Kamio and Nikaido 1976). Because of the presence of two membranes, Gram negatives are also defined as "diderm" bacteria (Sutcliffe 2010) (Fig. 2.1a). IM and OM also differ with respect to their integral membrane proteins. Integral IM proteins span the membrane as α-helices, almost entirely composed of hydrophobic residues, while most integral outer membrane proteins (OMPs) are composed of amphipathic β-strands which adopt a β-barrel structure (Fairman et al. 2011; Schulz 2002). An important class of OMPs is constituted by porins, which form spe-

P. Sperandeo · A. M. Martorana (✉) · A. Polissi
Dipartimento di Scienze Farmacologiche e Biomolecolari, Università degli Studi di Milano, Milan, Italy
e-mail: alessandra.polissi@unimi.it

A. Kuhn (ed.), *Bacterial Cell Walls and Membranes*, Subcellular Biochemistry 92, https://doi.org/10.1007/978-3-030-18768-2_2

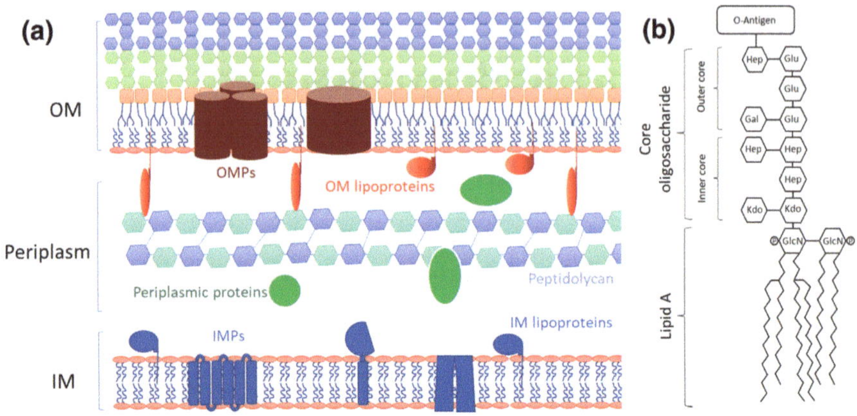

Fig. 2.1 The Gram-negative envelope and lipopolysaccharide structure. **a** The architecture of the Gram-negative envelope. The inner membrane (IM) and outer membrane (OM) are separated by an aqueous periplasm that contains the peptidoglycan cell wall. The IM is a symmetric phospholipid bilayer whereas the OM is asymmetric with LPS in the outer leaflet (for simplicity only inner and outer core are shown as colored light green and light blue hexagons, respectively). Major membrane proteins are shown and include: inner membrane proteins (IMP) and IM lipoproteins colored in blue; outer membrane proteins (OMP) colored in brown; OM lipoproteins colored in orange. Soluble proteins in the periplasm are colored in green. **b** The tripartite structure of LPS in Gram-negative bacteria. Lipid A, core oligosaccharide and O-antigen moieties are shown. Residues are indicated as hexagons: GlcN, glucosamine; Kdo, 3-deoxy-D-manno-oct-2-ulosonic acid; Hep, L-glycero-D-manno-heptose; Gal, Galactose; Glu, Glucose. Phosphate groups are indicated as circled P. See text for details

cific or non-specific channels that orchestrate the flux of hydrophilic molecules across the OM (Zeth and Thein 2010). IM and OM also contains lipoproteins anchored to the respective membranes via an N-terminal N-acyl-diacylglycerylcysteine residue (Sankaran and Wu 1994). Lipoproteins are located exclusively at the periplasmic side of the IM, whereas at the OM they can be anchored at the inner leaflet and thus extend into the periplasm (Okuda and Tokuda 2011) or be exposed at the cell surface (Konovalova and Silhavy 2015).

LPS, the major component of the OM outer leaflet, is a complex glycolipid that performs several functions. It is a fundamental structural component of the OM, essential in most Gram-negative bacteria (Moffatt et al. 2010; Peng et al. 2005; Steeghs et al. 1998). LPS is mainly responsible for the peculiar permeability barrier properties of the OM, which is effective against many toxic molecules and provides intrinsic resistance to a plethora of clinically useful antibiotics (Nikaido 2003). Finally, LPS is a major virulence factor and plays a crucial role in host-pathogen interaction by modulating the innate immune response (Kagan 2017).

In this chapter we will discuss several aspects of LPS biology. We will first review LPS structure, its barrier function and its ability to serve as signaling molecule for the immune system. Since most LPS functions stem from its localization at the OM outer leaflet, we will next revise the strategies bacteria have evolved to maintain OM asymmetry. The LPS is a large glycolipid and its transport poses several challenges to the cell as this molecule needs to cross several compartments with different physicochemical characteristics. In the last part of this chapter we will describe how bacterial cells have solved the problem by using an intermembrane transport system.

LPS Structure and Function

LPS is a unique glycolipid whose chemical structure has been fully elucidated not only in *Enterobacteria* but also in an increasing number of *Proteobacteria*, the phylum containing the majority of known Gram-negative species. LPS is organized in three structural domains: lipid A, a core oligosaccharide and a highly variable O-antigen made of repeating oligosaccharide units (Raetz and Whitfield 2002) (Fig. 2.1b). Lipid A is the hydrophobic moiety that anchors LPS to the OM outer leaflet; in *Escherichia coli* and many Enterobacteriaceae it is a glucosamine disaccharide that is phosphorylated at the 1 and 4′ position and contains six acyl chains attached via an amide linkage (2 and 2′ positions) and via an ester bond (3 and 3′ positions) (Raetz and Whitfield 2002) (Fig. 2.1b). Lipid A is the most conserved portion of the LPS molecule, although variations exist across different species with respect to the length and number of acyl chains (Raetz et al. 2007). The core oligosaccharide is covalently linked to the lipid A and is often divided in an inner and outer core. The chemical structure of the outer core is variable whereas the inner core is more conserved between different isolates of the same species. The inner core contains at least one residue of 3-deoxy-D-manno-oct-2-ulosonic acid (Kdo), that links the inner core to lipid A, and usually also L-glycero-D-manno-heptose (heptose) (Holst 2007; Raetz and Whitfield 2002) (Fig. 2.1b). The O-antigen, composed of repeating units of one to six different residues, is the distal surface-exposed LPS moiety responsible for the immunogenic properties of this macromolecule; it is the most variable portion of LPS, a characteristic that has been used as tool for strains classification based on the different serological properties (Raetz and Whitfield 2002). Notably, the O-antigen is not synthesized in the *E. coli* K12 derivatives, which are mainly used as laboratory strains, due to a mutation in the *rfb* locus where the genes responsible for the O-antigen biosynthesis are clustered (Liu and Reeves 1994; Stevenson et al. 1994). *E. coli* strains unable to produce O-antigen are typically referred as "rough" strains, as opposed to the wild-type "smooth" strains. The O-antigen is also absent in several Gram-negative species including pathogens such as *Neisseria meningitidis* and *Bordetella pertussis* (Griffiss et al. 1987; Peppler and Schrumpf 1984), whose molecules composed of only lipid A and core oligosaccharide are typically referred as lipooligosaccharides (LOS).

The structural complexity of LPS reflects the many functions this molecule performs in the bacterial cell. LPS is the major component of the OM outer leaflet and its amphipathic nature contributes to a large extent to the permeability of the OM. The hydrophobic lipid A moiety provides a barrier to the flux of hydrophilic molecules through the OM. The hydrophilic nature of the core oligosaccharide and O-antigen offers instead a barrier against the passage of hydrophobic compounds. Moreover, the chemical composition of the LPS aliphatic domain made of fully saturated fatty acyl chains is thought to create a gel-like lipid interior of very low fluidity further contributing to the low permeability of hydrophobic solutes across the OM (Carpenter et al. 2016; Nikaido 2003). The OM barrier properties are also the consequence of the very tightly packed layer of the LPS molecules bridged by the action of divalent Mg^{2+} and Ca^{2+} cations. The positive charges of Mg^{2+} and Ca^{2+} counteract repulsion between negatively charged phosphates in lipid A and in the core oligosaccharide (Nikaido 2003). Overall, the presence and the peculiar arrangement of LPS in the outer layer make the OM a formidable permeability barrier, a property that allows Gram-negative bacteria to colonize many different environments and survive in very harsh conditions.

LPS molecules cover the majority of the bacterial surface (Kamio and Nikaido 1976) and therefore play an important role in host-microbe interactions. The O-antigen, the outermost and variable portion of the molecule, protects the bacteria from phagocytosis and complement-mediated lysis (Lerouge and Vanderleyden 2002). Lipid A, the most conserved moiety of LPS and also known as endotoxin, is recognized by the innate immune system through the TLR4/MD2 (Toll-Like Receptor 4/Myeloid Differentiation factor 2) complex which initiates a robust signal cascade leading to cytokine production that is crucial for clearance of infection (Kagan 2017). Lipid A is therefore a pathogen associated molecular pattern (PAMP), namely a "pathogenic barcode", interpreted by the mammalian innate immune system as a sign of infection (Miyake 2004; Park and Lee 2013). The hexa-acylated lipid A from *E. coli* acts as a potent agonist of the TLR4/MD2 (Park et al. 2009) whereas variations in the lipid A acylation patterns alter the strength of TLR4/MD2 signaling (Needham et al. 2013). Many bacteria have evolved enzymes that alter their lipid A structure by modifying its number of acyl chains in response to environmental conditions or host interactions (Maldonado et al. 2016; Needham et al. 2013; Raetz et al. 2007). As an example, *Shigella flexneri* and *Pseudomonas aeruginosa* are able to modify the degree of acylation in their LPS during intracellular residence or in chronic lung infections, respectively (Ernst et al. 1999; Paciello et al. 2013). This is a powerful strategy to lower the sensing activity of the immune system and to escape from downstream effector mechanisms.

Maintaining the LPS Outer Layer

Maintenance of the LPS outer leaflet is critical to protect Gram-negative cells against bile salts, detergents, antimicrobial peptides and antibiotics, allowing bacterial cells to survive in harsh environments (Bishop 2008; Nikaido 2003). While OM asymmetry is primarily established by direct placement of LPS into the outer leaflet by the Lpt machinery (see section "Assembly of LPS at the OM Outer Leaflet"), several mechanisms are responsible for its disruption. Defects in LPS biosynthesis or transport, exposure to chelating agents or antimicrobial peptides all lead to perturbation of the LPS layer and to a compensatory accumulation of phospholipids in the outer leaflet (Balibar and Grabowicz 2016; Bishop et al. 2000; Nikaido 2003). Phospholipids in the outer leaflet disrupt OM asymmetry and create patches in the membrane that are more susceptible to the influx of toxic molecules (Nikaido 2005).

In *E. coli* there are several systems that monitor the presence of phospholipids in the OM outer leaflet and act to restore OM asymmetry: the OM PagP and PldA enzymes act by degrading phospholipids in the outer leaflet of the OM, whereas the Maintenance of lipid asymmetry (Mla) system removes phospholipid from the OM via a retrograde transport system.

The OM palmitoyltransferase PagP catalyzes the transfer of a palmitate chain from the *sn*-1 position of a surface exposed phospholipid to the lipid A resulting in a *sn*-1-lyso phospholipid and a hepta-acylated lipid A (Bishop et al. 2000). It has been postulated that hepta-acylated LPS molecules reduce lipid fluidity and increase lateral interactions between LPS molecules thus stabilizing the OM (Bishop 2005). Notably, in *Salmonella*, *pagP* expression is induced via the PhoPQ two component system in response of low extracellular Mg^{2+}-ion concentration (Dalebroux et al. 2014), a condition known to induce migration of phospholipids in the OM outer leaflet.

The OM phospholipase PldA catalyzes the hydrolysis of acyl ester bonds in phospholipids. The PldA enzyme resides in the OM as an inactive monomer and the formation of a catalytically active PldA dimer occurs following phospholipids migration in the outer leaflet (Dekker 2000).

The MlaA-F system is a multiprotein complex whose components are located in every cellular compartment and functions as a retrograde phospholipid transport system. The MlaA-F proteins are not essential and deletion of any Mla component leads to phospholipids accumulations at the OM outer leaflet and, as a consequence, to OM permeability defects (Malinverni and Silhavy 2009). Retrograde transport is energized by the IM MlaBDEF ABC transporter in which the hexameric MlaD protein binds phospholipids, the proposed substrates of the Mla system, and modulate the ATPase activity of the complex (Thong et al. 2016). MlaC is the periplasmic component of the system which is thought to interact with MlaD. Direct transfer of phospholipids between these proteins has been recently shown (Ercan et al. 2018). At the OM the MlaA lipoprotein has been shown to interact with the OmpC porin (Chong et al. 2015) and the complex has been proposed to remove phospholipids directly from the outer leaflet to maintain OM asymmetry.

How the OmpC-Mla pathway mediates retrograde phospholipids transport from the OM back to the IM is not fully clear yet. The architecture of the OmpC-MlaA complex has been determined by X-ray crystallography and UV-photocrosslinking experiments (Abellon-Ruiz et al. 2017; Yeow et al. 2018). Structural data suggest that a donut-shaped MlaA is embedded in the inner leaflet of the OM, where it forms an amphipathic channel to deliver the removed phospholipids from the OM outer leaflet to the periplasmic MlaC component. Although stably associated with MlaA, the OmpC porin does not appear to play an active role in phospholipid transport and is thought to serve as a scaffold for correct positioning of MlaA in the lipid bilayer

(Chong et al. 2015; Abellon-Ruiz et al. 2017; Yeow et al. 2018). A recent mutation reported in MlaA (designated *mlaA**) provided a more robust understanding of the Mla pathway. In MlaA* mutants phospholipids accumulates in the OM outer leaflet suggesting that the MlaA* variant may have the reverse activity of the wild-type protein (Sutterlin et al. 2016). The MlaA structure explains why the MlaA* variant accumulates phospholipids at the OM outer leaflet. Indeed, the *mlaA** mutation disrupts the donut shape and allows phospholipids from the inner leaflet to enter the amphipathic channel and flow in the outer leaflet in a process likely driven by mass action (Abellon-Ruiz et al. 2017).

A functional connection between the Mla pathway and the PldA system was initially proposed since PldA overexpression suppresses the OM permeability defects of Mla mutants (Malinverni and Silhavy 2009). The overlapping between the two systems has been further supported by the finding that *pldA* deletion suppresses the death phenotype of the gain-of-function *mlaA** allele in which accumulation of phospholipids at the OM outer leaflet is accompanied by a corresponding and detrimental increase in LPS production (Sutterlin et al. 2016). The *pldA* suppression phenotypes have been nicely explained by the finding that PldA, in addition to degrade mislocalized phospholipids, acts as a sensor signaling to cells that more LPS is needed to properly feed the OM outer layer (May and Silhavy 2018).

Overall, the MlaA-OmpC complex represents a novel class of lipid transport proteins and defines a novel mechanism in bacterial lipid homeostasis.

LPS Biosynthesis

The biosynthesis of LPS is a complex process that takes place in different cellular compartments and requires spatial and temporal coordination of several independent pathways that converge to produce the full-length LPS molecule. The synthesis of lipid A substituted with the Kdo residues of the inner core, the so called Kdo_2-lipid A moiety, occurs in the cytoplasm and at the inner leaflet of the IM. This pathway, which is well conserved across Gram-negative bacteria, has been extensively characterized in *E. coli* and *Salmonella* and it is referred to as the "Raetz pathway", since most of research work on it has been performed by Christian Raetz and his team (Raetz et al. 2009; Raetz and Whitfield 2002; Raetz et al. 2007).

The first step of Kdo_2-lipid A biosynthesis is the fatty acylation of *N*-acetylglucosamine linked to a nucleotide carrier (UDP-Glc*N*Ac) by LpxA (Fig. 2.2). In this reaction the thioester *R*-3-hydroxymyristoyl acyl carrier protein (ACP) is the donor substrate (Anderson et al. 1993; Anderson and Raetz 1987). The product of the reaction, UDP-3-*O*-(acyl)-Glc*N*Ac, is next deacetylated to UDP-3-*O*-(acyl)-GlcN by LpxC, a Zn^{2+}-dependent metalloenzyme (Young et al. 1995). Since the equilibrium constant for UDP-Glc*N*Ac acylation by LpxA is unfavorable (Anderson et al. 1993), the deacetylation of UDP-3-*O*-(acyl)-Glc*N*Ac by LpxC is the first committed step of Kdo_2-lipid A biosynthesis (Fig. 2.2). Indeed, LpxC represents the major control point of LPS synthesis whose turnover is controlled by the FtsH protease (Fuhrer et al. 2006; Ogura et al. 1999), the YciM/LapB heat-shock proteins (Klein et al. 2014; Mahalakshmi et al. 2014) and a not yet identified function (Emiola et al. 2016). Following deacetylation, a second *R*-3-hydroxymyristate chain is added by LpxD to make UDP-2,3-diacyl-GlcN (Kelly et al. 1993). LpxH is the pyrophosphatase that cleaves the pyrophosphate linkage of UDP-2,3-diacyl-GlcN to form 2,3-diacyl-GlcN-1-phosphate (lipid X) (Babinski et al. 2002). Formation of the $\beta,1'$-6-linked lipid A disaccharide is generated by LpxB, which condenses UDP-2,3-diacyl-GlcN with lipid X and releases UDP (Fig. 2.2) (Radika and Raetz 1988). LpxK phosphorylates the 4′-position of the disaccharide 1-phosphate produced by LpxB to generate lipid IV_A (Ray and Raetz 1987). The reaction catalyzed by LpxK precedes the addition of the Kdo residues by the bifunctional enzyme WaaA (formerly KdtA) (Fig. 2.2) (Belunis and Raetz 1992). The Kdo residue is synthesized by a separate pathway that requires four sequentially acting enzymes (Cipolla et al. 2009). The sugar nucleotide CMP-Kdo is the donor substrate for WaaA that catalyzes the sequential incorporation of two Kdo residues in lipid IV_A to generate the Kdo_2-lipid IV_A moiety. Notably, the FtsH protease also controls the turnover of WaaA (Katz and Ron 2008) to effectively control the concentrations of both, the lipid A and the sugar moiety (Kdo) precursors. The two last steps of Kdo_2-lipid A biosynthesis in *E. coli* involve the addition of the secondary lauroyl and myristoyl residues to the distal glucosamine unit by LpxL and LpxM (previously designated HtrB and MsbB, respectively), which require the Kdo disaccharide moiety in their substrates for activity (Brozek and Raetz 1990; Clementz et al. 1996, 1997). Based on their activity in the last steps of the LPS biosynthetic pathway, LpxL and LpxM are sometimes referred to as "late" acyltransferases (Fig. 2.2). However, *P. aeruginosa* LPS biosynthesis differs since fully acylated lipid A is required before Kdo residue addition (King et al. 2009). Finally, the additional sugars composing the oligosaccharide core are added to the Kdo_2-lipid A moiety by specific glycosyl-transferases (Raetz and Whitfield 2002). The complete lipid A-core moiety anchored at the IM inner leaflet is now ready to be translocated across the IM by the MsbA ABC transporter (see section "Translocation Across the IM: The ABC Transporter MsbA").

In O-antigen producing strains, residues composing the repeat units are synthesized in the cytoplasm, flipped to the periplasmic face of the IM attached to the lipid carrier undecaprenyl diphosphate and extension of the O-polysaccharide chain may occur in the periplasm via wzy- or synthase-dependent pathways (Greenfield and Whitfield 2012).

LPS Export to the Cell Surface

After completing its synthesis at the cytoplasmic side of the IM, lipid A-core is translocated across the IM to gain the periplasmic orientation necessary for the WaaL ligase-mediated O-antigen decoration (Han et al. 2012; Ruan et al. 2012) and for the subsequent transport to the cell surface. The trans-bilayer movement ("flipping") of amphipathic molecules into phospholipid membranes is energetically unfavorable

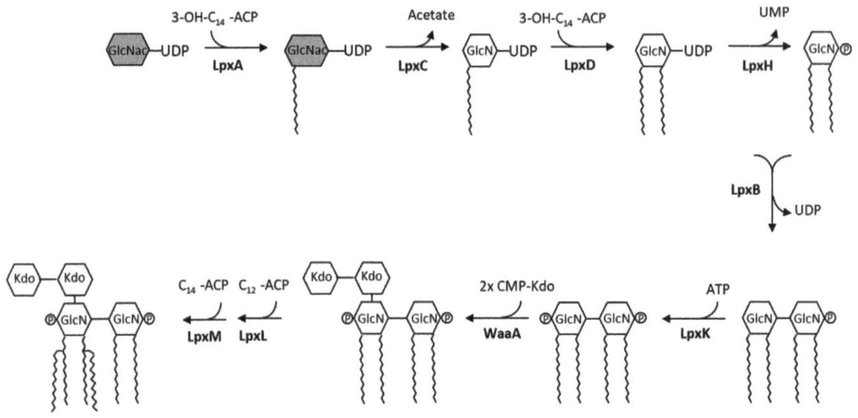

Fig. 2.2 Schematic representation of the Raetz pathway for the synthesis of Kdo_2-lipid A. Grey hexagons indicate: Glc*N*Ac, *N*-acetylglucosamine; white hexagons indicate: GlcN, glucosamine; Kdo, 3-deoxy-D-manno-oct-2-ulosonic acid. Phosphate groups are indicated as circled P. See text for details

and spontaneous translocation is very slow, therefore the cell needs a dedicated protein to carry out this function. The essential ABC transporter MsbA is the best-known bacterial lipid flippase among the ABC (ATP binding cassette) transporter superfamily and couples ATP hydrolysis to trafficking of lipid A-core across the IM (Doerrler et al. 2004).

Most of the heterogeneity displayed by the LPS molecules, not only among different Gram-negative bacteria but also within a single bacterial strain, is due to modifications of the conserved lipid A-core module (O-antigen ligation, chemical modifications on the glucosamines of lipid A, phosphate group decoration, etc.), that may take place at the periplasmic leaflet of the IM and, therefore, downstream of the MsbA-mediated lipid A-core flipping. These modifications will not be discussed in this chapter as they have been the topic of excellent reviews (Raetz et al. 2007; Whitfield 2006). The mature modified LPS molecules must then be escorted across the periplasm and assembled at the OM outer leaflet.

Due to the chemical nature of LPS and of the different cellular compartments it crosses during its journey to the OM, the transport of mature LPS molecules across the periplasm to the cell surface is a challenging process. LPS transport indeed implies the extraction of the lipid A acyl chains from the hydrophobic milieu of the IM, their shielding during transport across the aqueous environment of the periplasm and their insertion into the pre-constituted phospholipid/LPS layer of the OM. Accumulation of LPS in the IM is detrimental for the cells (Zhang et al. 2013), therefore such a demanding process must occur unidirectionally from IM and OM, against concentration gradient and without perturbing the integrity of the OM.

During the past two decades, the joint efforts of several research groups, mainly using *E. coli* as model organism, have led to the discovery of a multiprotein complex made of seven essential proteins located in every cellular compartment that assemble to span the entire cell envelope. This trans-envelope complex harnesses the energy of ATP hydrolysis in the cytoplasm to transport LPS from the IM to the OM (Braun and Silhavy 2002; Chng et al. 2010a, b; Ruiz et al. 2008; Sperandeo et al. 2007, 2008; Wu et al. 2006). These proteins named LptABCDEFG from Lipopolysaccharide transport (Sperandeo et al. 2007), physically interact with each other employing a common structural domain and function in a concerted way to coordinate the chemical energy to the mechanical work through the transmission of conformational changes in their structures (Fig. 2.3). The Lpt machinery functions as a single device and when any component is depleted LPS accumulates at the IM outer leaflet (Ruiz et al. 2008; Sperandeo et al. 2008).

The engine of the system is constituted by $LptB_2FG$, the second ABC transporter involved in LPS biogenesis after MsbA (Dong et al. 2017; Luo et al. 2017; Ruiz et al. 2008). The energy obtained by the ATP hydrolysis is then used to push LPS into the periplasmic bridge formed by LptC, LptA and the N-terminal domain of LptD

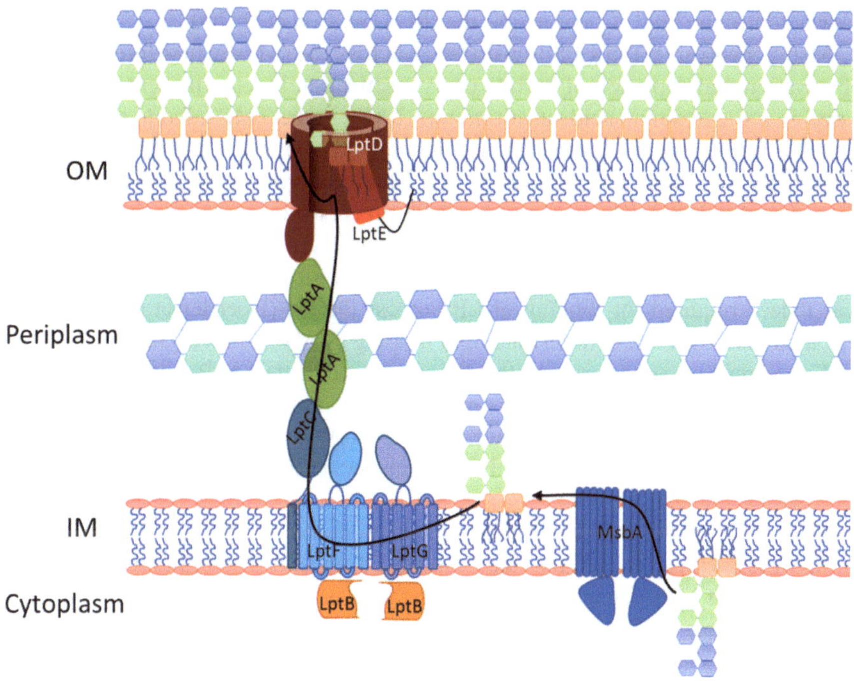

Fig. 2.3 The LPS export pathway in *E. coli*. The lipid A-core moiety synthesized at the inner leaflet of the IM is flipped across the IM by the MsbA ABC transporter. The flipped lipid A-core molecule is then extracted from the IM, escorted across the periplasm and assembled at the OM outer leaflet by the transenvelope Lpt machinery. LPS inner and outer core are shown as colored light green and light blue hexagons, respectively. See text for details. In O-antigen producing strains, the O-antigen polymer, linked to undecaprenyl phosphate carrier, is translocated across the IM by a MsbA-independent pathway and is ligated to the lipid A-core by the WaaL ligase. Here, the O-antigen domain is omitted

(Freinkman et al. 2012; Sperandeo et al. 2007, 2008, 2011). At the OM, the complex formed by the LptD β-barrel protein, and the tightly associated LptE lipoprotein, facilitates final LPS assembly at the cell surface (Chng et al. 2010a, b; Freinkman et al. 2011) (Fig. 2.3).

Translocation Across the IM: The ABC Transporter MsbA

The conserved IM protein MsbA is the first transporter working in the LPS export pathway and plays an essential role in building the LPS OM layer in LPS-producing bacteria. MsbA belongs to the ubiquitous ABC transporter superfamily (Locher 2016; Rees et al. 2009) and functions as a homodimer of two 64.3 kDa subunits, each containing one cytoplasmic nucleotide-binding domain (NBD) and one transmembrane domain (TMD), made by 6 transmembrane helices, which contains the LPS binding site (Fig. 2.4). The NBDs bind to ATP and catalyze its hydrolysis to provide the energy to flip the lipid A-core across the IM (Davidson et al. 2008).

MsbA involvement in LPS translocation across the IM was first inferred by the observation that depletion of MsbA, or its inactivation in a thermosensitive mutant carrying a single amino acid substitution (A270T) in the TMDs, led to the accumulation of LPS precursors in the IM, accompanied by extensive membrane invagination in the cytoplasm and, ultimately, cell death (Doerrler et al. 2001; Polissi and Georgopoulos 1996). The accumulated LPS precursor molecules in the $msbA_{\mathrm{A270T}}$ mutant grown at the non-permissive temperature were later shown to be inaccessible by periplasmic lipid A modification enzymes, strongly suggesting that MsbA was indeed required for flipping of LPS across the IM (Doerrler et al. 2004). Initial attempts to directly demonstrate lipid flipping by MsbA in vitro failed (Kol et al. 2003) and only several years later the direct measurement of ATP-dependent translocation of phospholipids in proteoliposomes was reported (Eckford and Sharom 2010). However, neither the substrate specificity of MsbA, nor the molecular mechanism that couples energy production with LPS translocation have emerged clearly from these initial studies.

Several structural and biochemical data produced on *E. coli* MsbA and its homologs have been carried out during the last decade to shed light on these issues. A milestone in MsbA functional characterization is represented by the X-ray crystal structure determination of three MsbA orthologs from *E. coli*, *Vibrio cholerae* and *Salmonella enterica* (serovar Typhimurium) trapped in different conformations (Ward et al. 2007). The analysis of these crystal structures revealed the existence of two different inward-facing and one outward-facing conformation, suggesting that, at least under the experimental conditions used for crystal production, the protein undergoes a large range of motions as consequence of ATP hydrolysis. These conformational changes might be required for LPS flipping. Based on these evidences, the authors suggested an "inward-outward" model for MsbA function. According to this model, when MsbA is in the apo-form (without nucleotide bound), its TMDs open creating a cavity that faces the cytoplasmic side of the IM, possibly allowing

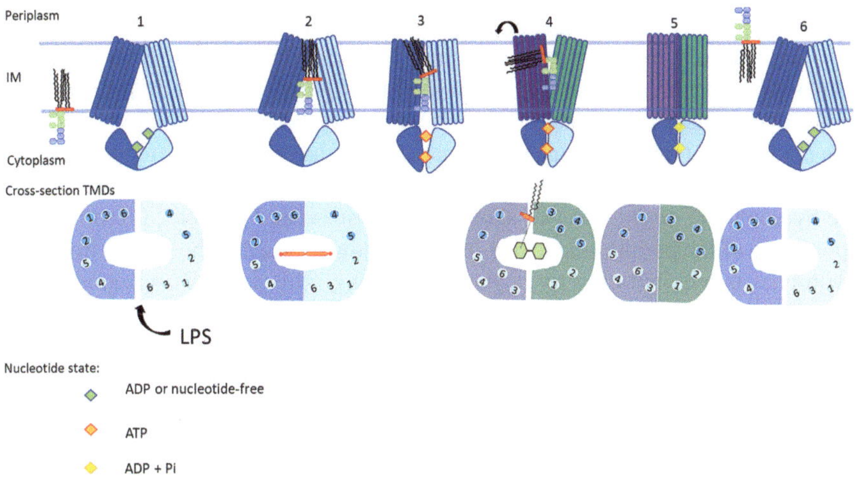

Fig. 2.4 "Trap and flip" model for MsbA-mediated lipid A-core flipping across the IM. MsbA catalyses lipid A-core flipping across the IM alternating between an inward (cytoplasmic)-facing and an outward (periplasmic)-facing conformation that allow binding and ATP hydrolysis-dependent release of lipid A-core through a six-step cycle. Upper panel: in the ADP or nucleotide-free state, MsbA dimer adopts the inward-facing conformation, opening the TMDs to allow LPS entry (step 1). LPS binding induces NBDs alignment for ATP binding (step 2). ATP binding triggers conformational changes in the TMDs that weaken lipid A-core interaction (step 3) and promote the release of the lipid A acyl chains into the periplasmic leaflet of the membrane (step 4). Lipid A-core translocation and MsbA rearrangement lead to ATP hydrolysis. After lipid A-core release, the TM helices adopt a compact assess (step 5). Finally, phosphate release induces MsbA to return in the inward-facing conformation (step 6). MsbA monomers in the resting state (inward-facing conformation) are shown in dark and pale blue. TMD rearrangement after ATP binding is represented by colour change of the monomers from dark and pale blue to violet and pale green, respectively. Lower panel: schematic representation of the cross-sections through the MsbA TMDs at the level of lipid A glucosamines for each step of the translocation process, as shown in the upper panel. Numbered circles represent the individual TM helices provided by each MsbA monomer, coloured as dark blue or pale blue according to the MsbA monomer they belong to. Glucosamine moiety of lipid A and phosphates are represented by orange ovals and red circles, respectively. The core oligosaccharide is depicted as in Fig. 2.1. Green, red and yellow diamonds represent ADP, ATP and ADP with phosphate, respectively. Adapted from Mi et al. (2017)

LPS binding from this side of the IM. On the contrary, upon ATP binding (mimicked by the binding of ADP-vanadate), MsbA TMDs open towards the periplasmic side of the IM, possibly leading to LPS release into the outer leaflet of the membrane (Ward et al. 2007). However, the major limitation of this work is that LPS was not co-crystallized with MsbA, undermining the reliability of the model.

The real turning point in the study of the molecular mechanisms of MsbA-mediated flipping of LPS across the IM has come very recently from the combined use of nanodiscs, instead of detergent, for MsbA solubilization, and cryo-electron microscopy (Cryo-EM) to image the protein in its native conformations (Mi et al. 2017; Voss and Trent 2018). Cryo-EM images of MsbA bound to LPS revealed that, under native conditions, MsbA undergoes to a smaller range of motions compared to what observed in the crystal structures. Strikingly, the higher ATPase activity displayed by MsbA in nanodiscs confirmed that the protein was in a more native conformation than in detergent. The structure of MsbA in nanodiscs showed only one predominant inward-facing conformation with the two TMDs extending parallel from the membrane and the cytoplasmic NBDs close to each other. Interestingly, the hydrophobic cavity that was shown to accommodate LPS is localized in correspondence of the periplasmic leaflet of the IM, suggesting that the energy from ATP hydrolysis is required only for acyl chain flipping. This cavity is surrounded by a ring of charged residues required to position the glucosamine-linked phosphates into the TMDs. Upon nucleotide binding, the NBDs were shown to closely associate, determining a reorganization of the TM helices that block the LPS binding site and, probably, induce the opening of the hydrophobic pocket to allow the acyl chains to insert into the periplasmic side of the IM (Mi et al. 2017).

These evidences led the authors to propose the new "trap-and-flip" model for MsbA-mediated LPS flipping. According to this model, when MsbA is bound to ADP or is in a nucleotide-free state, LPS enters MsbA internal cavity exploiting the path formed by the TM4 and TM6 from two different monomers. LPS binding aligns the NBDs for ATP hydrolysis. This agrees with previous data from fluorescence resonance energy transfer (FRET) experiments, that showed NBDs dimerization and the stimulation of ATP hydrolysis upon lipid A binding (Doshi and van Veen 2013). Upon ATP binding and hydrolysis, conformational changes occur that close the cavity and expose the acyl chains of lipid A to the periplasm. This leads LPS to slip out the protein from the path formed by TM1 and TM3 from two different monomers. Upon release of LPS, MsbA resets to the inward-facing conformation (Mi et al. 2017) (Fig. 2.4).

From structural studies and inspection of the cavity in MsbA that hosts LPS, it can be speculated that both the strong interactions with lipid A and the dimension of the cavity itself might have a role in selecting the molecule that has to be accommodated (Ho et al. 2018). This is in line with the hypothesis that MsbA plays the role of "quality control" system for LPS export to the OM. Accordingly, although several lipid A precursor molecules, in addition to LPS, can stimulate MsbA ATPase activity in vitro (Doerrler and Raetz 2002; Eckford and Sharom 2008), *E. coli* MsbA is not able to transport lipid A precursor molecules missing Kdo residues and the lauryl and myristoyl acyl chains in vivo, resulting in death of the corresponding LPS biosynthesis mutants (see above, Mamat et al. 2008; Meredith et al. 2006). However, MsbA overexpression or mutations rescue the growth phenotype of these mutants. It is worth noting that MsbA was originally identified as a multicopy suppressor of defects resulting from a mutation in *lpxL*, that encodes for one of the "late" Kdo-dependent lipid A acyltransferases (Karow and Georgopoulos 1993). This suggests that an increase in the transport rate of poor substrates, due to protein overexpression or mutations that alter the substrate selectivity, may enable MsbA to transport under-acylated LPS precursors, even in vivo. Interestingly, in a screening for suppressor mutations of lipid A biosynthesis defects, a single amino acid substitution in an IM protein with unknown function, YhjD, has been shown to allow LPS transport in the absence of MsbA (Mamat et al. 2008). The role of YhjD is still unknown and the reason why the identified mutation allows YhjD to functionally replace MsbA, albeit under extreme circumstances, is still an unsolved question.

Another open question is whether MsbA can handle other lipophilic substrates besides LPS in vivo. In *E. coli*, the thermosensitive $msbA_{\text{A270T}}$ mutant was shown to be impaired not only in LPS but also in phospholipid translocation at the non-permissive temperature, suggesting that MsbA could be the transporter of the two major lipids in the cell envelope of Gram-negative bacteria (Doerrler et al. 2001). Accordingly, the ATPase activity of MsbA reconstituted in proteoliposomes is stimulated by different lipids (Eckford and Sharom 2008). On the contrary, the deletion of *msbA* can be obtained in *N. meningitidis* and the mutant is viable and produces an OM mainly made out of phospholipids (Tefsen et al. 2005a, b). Since this organism can survive without LPS (Steeghs et al. 1998), this suggests that MsbA might not be strictly required for phospholipid translocation. Nevertheless, this evidence can simply reflect a difference between the lipid transport systems of the two organisms and suggest the existence of another pathway for phospholipid flipping across the IM, at least in *N. meningitidis*.

Finally, it has been demonstrated that MsbA can bind simultaneously lipid A and amphipathic drugs, suggesting the presence of different binding sites in the protein (Siarheyeva and Sharom 2009). The relevance of such MsbA activity in drug resistance has still to be demonstrated.

Extraction from the IM: The Atypical $LptB_2FG$-C Transporter

After MsbA-dependent translocation, LPS molecules are anchored at the periplasmic leaflet of the IM. LPS accumulation at this location is toxic for the cell and LPS molecules must be efficiently removed from the IM to reach their final destination at the cell surface. The LPS export from the IM outer leaflet to the OM is accomplished by the Lpt complex, a specialized multiprotein machinery that uses energy to move millions of LPS molecules at each cell division, in a process that begins with its extraction from the IM. The energy for the process is provided by the $LptB_2FG$ transporter. $LptB_2FG$ is a non-conventional ABC transporter since, unlike MsbA, it does not translocate its substrate across the membrane, but rather it couples the energy of the ATP hydrolysis in the cytoplasm with the detachment of LPS from the outer leaflet of the IM. Moreover, $LptB_2FG$ is a tetramer of different subunits, comprising a dimer of LptB and two polytopic proteins, LptF and LptG (Narita and Tokuda 2009), thus sharing features with the class of the ABC importers. Another unconventional trait of $LptB_2FG$ is its stable association with LptC (Chng et al. 2010a, b; Villa et al. 2013), a single pass IM protein constituted by a transmembrane helix and a large periplasmic domain folded as a β-jellyroll (Fig. 2.5) (Tran et al. 2010), whose function within the ABC transporter is still not clear (see below). For its peculiar proprieties, the $LptB_2FG$ has been recently classified as a new type of ABC transporter, the type VI exporter (Thomas and Tampe 2018).

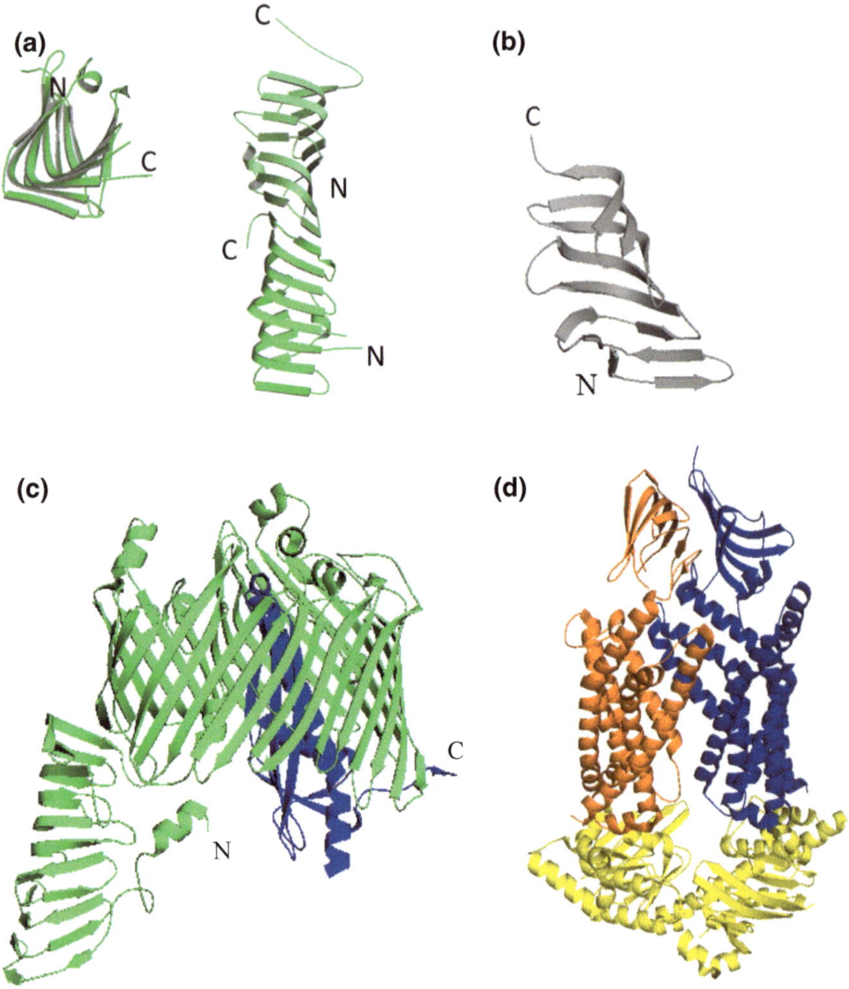

Fig. 2.5 Crystal structures of Lpt proteins. **a** Right: ribbon diagram of two molecules of *E. coli* LptA organized in head-to-tail fashion (PDB 2R19). Left: β-jellyroll structure of LptA, rotated by 90° to highlight the channel formed along the length of LptA. **b** The periplasmic domain of LptC from *E. coli* (gray, PDB 3M2Y). **c** LptDE complex from *K. pneumoniae* (PDB 5IV9). LptE, depicted in blue, resides within the LptD lumen (depicted in green). **d** $LptB_2FG$ complex from *P. aeruginosa* (PDB 5X5Y). LptB, LptF and LptG are represented in yellow, orange, and blue, respectively. N- and C-terminal end of LptA, LptC and LptD proteins are indicated

Clues on the molecular mechanism of LPS extraction from the IM by $LptB_2FG$ have recently come from the crystal structures of $LptB_2FG$ from *P. aeruginosa* and *Klebsiella pneumoniae*, obtained in the absence of nucleotides (Dong et al. 2017; Luo et al. 2017). Inspections of these structures revealed that the $LptB_2FG$ core transporter adopts a "milk jug" architecture, consisting of a pair of LptB that form the NBDs and two TM proteins, LptF and LptG, each containing 6 transmembrane helices and a large periplasmic domain between TM3 and TM4, folded as a β-jellyroll (Dong et al. 2017; Luo et al. 2017) (Fig. 2.5). LptF and LptG interact each other through their respective TM1 and TM5 helices and form an outward-facing V-shaped cavity that spans the IM and extends slightly into the periplasm. The internal surface of the IM cavity is hydrophobic, whereas the residues lining at the IM–periplasm interface are positively charged, suggesting that the region located deep inside the IM makes contact with the acyl chains of lipid A, whereas the region at the IM-periplasm interface interacts with the negatively charged phosphate groups. Interestingly, the periplasmic domains of LptF and LptG take opposite side-by-side orientations in both crystal structures and are directly connected to the hydrophobic cavity of $LptB_2FG$. This is in line with the hypothesis that LPS is accommodated within the cavity of $LptB_2FG$ before being pushed toward the OM through the periplasmic domains of LptF and/or LptG. Accordingly, in the crystal structure of $LptB_2FG$ from *K. pneumoniae*, a non-assigned extra electron density, which is compatible with a molecule of LPS, was observed within the central cavity (Dong et al. 2017). The structural data also revealed that the helices connecting TM2 and TM3 of LptF and LptG constitute the region of interaction between LptF/LptG and the LptB dimer (the so-called coupling helices), in agreement with previous predictions based on genetic and biochemical data. Indeed, residues within the coupling helices have been shown to be crucial for the assembly and the function of the IM complex by connecting LptF and LptG with a groove in LptB that experiences a large range of motions during ATP hydrolysis (Sherman et al. 2014; Simpson et al. 2016). It is likely that these interactions coordinate the ATPase activity of LptB with LPS extraction.

The interfaces between TM1 and TM5 from opposing LptF and LptG monomers have a limited number of interactions, suggesting that they may open as a consequence of the conformational changes triggered by ATP binding to LptB, thus leading to LPS entry into the cavity. Accordingly, a cluster of residues crucial for LPS interaction has been identified at the membrane-periplasm interface of the TM1 of LptG in *E. coli* and *Burkholderia coenocepacia* by a series of genetic and biochemical analyses (Bertani et al. 2018; Hamad et al. 2012). Interestingly, this cluster contains a residue (K43 in *E. coli* LptG) that is not conserved among LptG homologues and whose chemical properties correlate with the modifications occurring on the lipid A moiety in different bacterial species, suggesting that it might interact via electrostatic interactions with lipid A. The strong decrease in LPS release activity of mutant $LptG_{K34D}$ has been shown by UV-photocrosslinking assay, confirming the importance of the residue. The recognition between $LptB_2FG$ and lipid A might occur in an early step of the extraction process, driving the selection of the LPS variants to be transported to the cell surface under different conditions (Bertani et al. 2018).

The current model for $LptB_2FG$ function suggests that the transporter cycles between three conformational states: in the nucleotide-free state (represented by the published structures), LptB dimer adopts an open conformation. Upon ATP binding, LptB monomers move close together and the conformational changes are then transmitted from LptB, through the coupling helices, to LptF and LptG inducing the opening of the lateral gate between one of the TM1-TM5 interfaces of LptFG to allow the entry of LPS into the internal cavity of $LptB_2FG$. ATP hydrolysis and ADP release, then, induce the conformational switch back to the nucleotide-free state, closing the lateral gate and expelling LPS into the periplasmic domains of LptF and/or LptG (Fig. 2.6a) (Luo et al. 2017). Interestingly, the published structures of $LptB_2FG$ show a different orientation of the periplasmic domains of the transporter, with opposing opening states of the lateral gates. This suggests that $LptB_2FG$ might extract LPS from both the lateral gates, transferring the molecules alternately to the periplasmic domains of LptF or LptG.

The proposed model raises the question of whether the transporter might work in a symmetric way. It is worth noting that once extracted from the membrane by the $LptB_2FG$ complex, LPS is delivered to LptC (Okuda et al. 2012). However, since the interaction of $LptB_2FG$ with LptC has not been characterized yet, it is not possible to exclude that LptC could receive LPS alternately from LptF and LptG. Moreover, there are still no evidences of LPS interaction with the periplasmic domain of LptF and/or LptG. Nevertheless, two major observations argue against the hypothesis of a symmetric transport and suggest that LptF and LptG may have different functions within the transporter: (i) mutations in corresponding residues within the coupling helices of LptF and LptG lead to different degree of OM permeability defects (Simpson et al. 2016); (ii) mutations at a unique position in the periplasmic domain of LptF rescue the growth defect of a mutant lacking LptC, suggesting that LPS flow might have a preferential direction within the transporter (see below) (Benedet et al. 2016).

LptC stably associates to the $LptB_2FG$ complex, although its role in LPS transport is not completely clear. Several lines of evidence indicate that LptC is strictly required to funnel LPS within the periplasmic bridge towards the OM. UV-photocrosslinking experiments showed that LPS association to LptA in spheroplasts overexpressing the $LptB_2FG$ transporter depends on the ATP hydrolysis but cannot proceed in the absence of LptC (Okuda et al. 2012). Moreover, LptC itself has been demonstrated to be part of the periplasmic bridge, interacting directly with LptA (Bowyer et al. 2011; Freinkman et al. 2012; Sperandeo et al. 2011), and binding LPS (Okuda et al. 2012; Tran et al. 2010). These findings strengthen the idea that LptC might have a direct role in LPS transport. However, LptC can tolerate several mutations. A soluble version of LptC missing its hydrophobic anchor is still functional and can assemble into the complex via the N-terminal region of its soluble domain (Villa et al. 2013). Moreover, the C-terminal region of the periplasmic domain can be deleted upon overexpression of LptB, suggesting that also a truncated LptC variant can fulfill its role in the Lpt complex, albeit with less efficiency (Martorana et al. 2016). Strikingly, the complete lack of LptC can be overcome by a single amino acid substitution at residue 212 in the periplasmic domain of LptF (Benedet et al. 2016). Taken together, these observations implicate the periplasmic domain of LptF into the

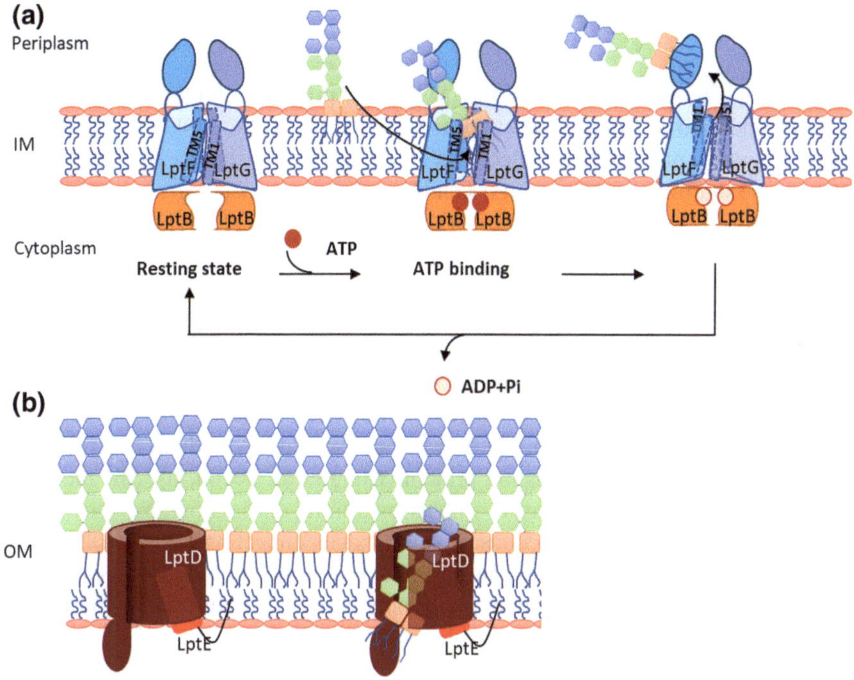

Fig. 2.6 Models for LPS extraction from the IM and assembly at the OM. **a** The $LptB_2FG$ transporter extracts and transports LPS through discrete steps of ATP hydrolysis, cycling from to the ATP-free state (resting state) to the ATP-binding state and finally ATP hydrolyzed state. Binding of LptB dimer to ATP triggers the conformational change of LptF-TM5 and LptG-TM1 resulting in the opening of a lateral gate between these two helices and the subsequent capture of LPS in the $LptB_2FG$ internal hydrophobic cavity. Upon ATP hydrolysis the lipid A is extracted from the IM and pushed to the periplasmic domain of LptF (or LptG). Finally, release of ADP induces $LptB_2FG$ to return to the ATP-free state. **b** The LPS molecule is delivered from the LptA to LptDE complex for insertion into the OM. The Lipid A portion of LPS passes from the N-terminal domain of LptD into the core of the membrane. The lateral gate formed by β1/β26 strands opens and allows the hydrophilic portion of LPS to pass through the lumen of LptD

periplasmic bridge and point to a regulative role of LptC within the transporter. The initial biochemical characterization of $LptB_2FG$-C purified in detergent showed that the ATPase activity of the transporter was not altered by the presence of LptC (Narita and Tokuda 2009), as it would be expected by a regulative subunit. However, it has been recently demonstrated that the association of LptC led to almost 40% decrease of the ATPase activity when the transporter was reconstituted in proteoliposomes, a condition that better mimics the membrane environment (Luo et al. 2017; Sherman et al. 2018).

Transport Across the Periplasm

Once LPS has been extracted from the IM by the $LptB_2FG$ complex it is transferred to LptC, and then to LptA, at the expense of ATP hydrolysis (Okuda et al. 2012). LptA is a soluble protein (Sperandeo et al. 2007) and initially proposed to function as a chaperone shuttling LPS between IM and OM (Ruiz et al. 2009) in analogy with the Lol pathway for lipoprotein transport (Okuda and Tokuda 2011). However, several lines of evidence argue against the chaperon model and support instead the view of LPS transported across the periplasm via a protein bridge connecting IM and OM: (i) LPS is not released into the medium after addition of periplasmic extract to *E. coli* spheroplasts as observed for lipoproteins, and in spheroplasts newly synthesized LPS is still transported to the OM, strongly suggesting that LPS transport could take place via contact sites between the membranes (Tefsen et al. 2005a, b); (ii) LptA fractionates with both IM and OM in sucrose density gradient centrifugation despite being a soluble periplasmic protein (Chng et al. 2010a, b); (iii) the seven Lpt proteins can be copurified as a complex by affinity chromatography, supporting the idea that they form a continuous protein bridge connecting IM and OM (Chng et al. 2010a, b); (iv) mutations impairing Lpt complex assembly lead to degradation of the periplasmic LptA component (Sperandeo et al. 2011) further supporting the biological significance of the transenvelope architecture.

LptA connects IM and OM by binding to LptC and LptD, respectively. The architecture of the protein bridge has been defined by co-purification and UV-photocrosslinking experiments. Whereas the N-terminal end of LptA interacts with the C-terminal end of LptC, the C-terminal end of LptA interacts with the N-terminal periplasmic region of LptD (Sperandeo et al. 2011; Freinkman et al. 2012). These intermolecular contacts are crucial for LPS transport as genetic disruption of these edge regions compromises the assembly and function of the Lpt bridge (Falchi et al. 2018; Freinkman et al. 2012; Sperandeo et al. 2011; Villa et al. 2013). Notably, the formation of LptC-LptA-LptD protein bridge occurs via the structurally homologous β-jellyroll domain, also termed "Lpt fold", shared by the three proteins (Fig. 2.5) (Botos et al. 2016; Qiao et al. 2014; Suits et al. 2008; Tran et al. 2010). The β-jellyroll fold represents a unique protein architecture initially identified in LptA, the first protein of the Lpt system for which a crystal structure was obtained (Suits et al. 2008), and afterwards found in all Lpt proteins with a periplasmic domain (Dong et al. 2017; Luo et al. 2017; Qiao et al. 2014; Tran et al. 2010). It consists of a variable number of consecutive, antiparallel β-strands arranged to form a semi-opened and slightly twisted structure whose interior is covered by hydrophobic residues (Fig. 2.5). The arrangement of the β-jellyroll domains of LptC, LptA and the N-terminal periplasmic region of LptD therefore forms a hydrophobic groove which is thought to accommodate the acyl chains of the lipid A moiety during LPS transport along the protein bridge, while the hydrophilic oligosaccharide portion remains exposed in the periplasm. Indeed, residues of LptA and LptC that crosslink to LPS are located in the β-jellyroll interior (Okuda et al. 2012).

LptA has a strong tendency to oligomerize in a head-to-tail orientation forming long fibrils in crystals obtained in the presence of LPS (Suits et al. 2008) and in solution in a concentration-dependent manner (Merten et al. 2012; Santambrogio et al. 2013). However, the affinity of LptA for LptC seems stronger than of LptA for itself in the oligomerization reaction (Bowyer et al. 2011; Schultz et al. 2013). The stoichiometry of LptA in vivo in the periplasmic protein bridge is still an open question. It has been recently demonstrated that an oligomeric-deficient form of LptA, lacking the last C-terminal β-sheet, partially fulfills its role in the transport of LPS to the OM (Laguri et al. 2017), suggesting that Lpt machinery with a single LptA molecule can be functional. However, it is more likely that LptA oligomerization imparts to the Lpt machinery the required flexibility to respond to variations in the periplasm width as a consequence of envelope stress or other growth conditions. This hypothesis is in line with the presence of an additional σ^E stress response promoter upstream the *lptA* gene (Martorana et al. 2011) and the recent finding that Enterobacteria are able to respond to changes of periplasm width by activating the Rcs stress response system (Asmar et al. 2017).

Assembly of LPS at the OM Outer Leaflet

The efficient insertion of LPS into the outer leaflet of the OM, avoiding a premature stop at inner leaflet, is essential to maintain the asymmetry of the OM. The step of LPS delivery and assembly at the OM surface is mediated by the OM translocon constituted by the β-barrel protein LptD and the lipoprotein LptE (Braun and Silhavy 2002; Wu et al. 2006). LptD consists of a C-terminal transmembrane β-barrel domain and a periplasmic N-terminal domain connected by two disulfide bonds between two non-consecutive cysteine residues (Botos et al. 2016; Dong et al. 2014; Qiao et al. 2014). The OM lipoprotein LptE is buried in the lumen of the barrel domain of LptD adopting a unique plug-and-barrel architecture (Fig. 2.5) (Freinkman et al. 2011).

The assembly and correct maturation of a functional LptDE complex is crucial to avoid LPS mistargeting and LptE accomplishes a non-obvious function within the translocon, assisting LptD in its folding and its assembly into the Lpt complex.

First evidence of LptE involvement in LptD assembly/maturation came from a screening for mutants able to suppress permeability defects of a two-codon *lptE* deletion that altered LptE interaction with LptD. In this screening, suppressor mutants were mapped not only in *lptD* but also in *bamA* (β-barrel protein essential for OMP assembly at the OM) (for a review see Konovalova et al. 2017; Noinaj et al. 2017), revealing that LptE association has a fundamental role in LptD assembly by the Bam (β-barrel assembly machinery) complex (Chimalakonda et al. 2011). Moreover, LptE is crucial for the functionality of LptD, since it helps the formation of the mature LptD form containing two disulfide bonds between non-consecutive cysteines (Chimalakonda et al. 2011). The plug function of LptE is critical to control the accessibility of the otherwise too large LptD lumen, as demonstrated by the isolation of a mutation in LptE that affects its interaction with LptD, altering OM permeability without impairing LptD assembly or LPS export (Freinkman et al. 2011).

In *E. coli*, LptE does not have solely a structural role in LPS biogenesis, as it is able to specifically bind LPS (Chng et al. 2010a, b) and, in vitro, to disaggregate LPS molecules through electrostatic interactions between the lipid A moiety of LPS and a patch of positively-charged residues in the exposed loop connecting the β-strands 2 and 3 of the protein. Accordingly, disruption of this LPS-binding patch increases membrane permeability but does not affect LptD assembly (Malojčić et al. 2014). Based on these observations, it has been proposed that LptE facilitates LPS transfer into the outer leaflet while preventing its insertion into the inner leaflet of the OM. Overall these evidences point toward a triple role of LptE in LPS transport: to manage the assembly of a functional LptD, to preserve membrane permeability and to assist LPS insertion into the cell surface.

The crystal structure of the LptDE complexes from several microorganisms have been recently reported (Botos et al. 2016; Dong et al. 2014; Qiao et al. 2014). Notably, the basic plug and barrel architecture of the LptDE protein complex is maintained in all solved structures. LptD consists of a large C-terminal β-barrel domain, made by 26 antiparallel β-strands embedded in the OM, and a N-terminal periplasmic β-jellyroll domain made by 11 antiparallel β-strands arranged in two sheets. The LptD β-strands are connected via periplasmic and extracellular loops; the longer loops face at the extracellular side closing off most of the β-barrel pore. Extracellular loops 4 and 8 insert inside the barrel, binding to the LptE protein which plugs the pore thus preserving membrane permeability (Botos et al. 2016; Dong et al. 2014; Qiao et al. 2014). Hydrogen bonding between the first and last β-strands of LptD β-barrel is disrupted creating a crenellation, or small gap, essential for the lateral migration of the LPS in the OM. Indeed, mutations that introduce disulfide bonds between the first and last β-strands results in lethal phenotypes (Dong et al. 2014).

A significant rotation of the N-terminal domain with respect to the β–barrel domain is also observed in the available crystal structures. The configuration of the LptD N-terminal domain with respect to the C-terminal barrel depends on the correct formation of disulfide bonds in LptD and has a fundamental role to allow the insertion of the lipid A moiety of LPS into the membrane (Dong et al. 2014), and the interaction with LptA (Okuda et al. 2016).

LptE forms a sandwich of two α-helices packed against a sheet of four β-strands. Its N-terminal lipid moiety is located outside of the barrel and inserted in the inner leaflet of the OM. In the LptDE complex, approximately 75% of LptE is located inside the β-barrel of LptD (Botos et al. 2016; Dong et al. 2014; Malojčić et al. 2014; Qiao et al. 2014).

Based on genetic and structural data, a two-portal mechanism has been proposed. According to this model the lipid A moiety of LPS moves from LptA directly into the β-jellyroll domain of LptD allowing its insertion into the membrane through a hole formed between the periplasmic and the β-barrel domain of LptD, while the hydrophilic domain of LPS flows through the lumen of the β-barrel. Then, the lateral gap of LptD may open enough to allow the passage of the hydrophilic portion of LPS through the lumen up to the OM (Gu et al. 2015; Li et al. 2015) (Fig. 2.6b). In this model, LptE helps LPS movement into the OM by providing more favourable interactions between itself and LPS rather than between aggregated LPS molecules. Finally, LptE pumps the LPS out of the LptD lumen towards its final destination (Malojčić et al. 2014).

LPS molecules must be transported to the OM to ensure cell growth and division and this process needs to be finely regulated to avoid the mislocalization of an otherwise toxic molecule. Early investigations on the LptDE complex suggested that only a correctly folded OM translocon allows the building of a functional Lpt system, enlightening a first step of regulation in the formation of the protein bridge between IM and OM, as failure of LptD folding makes impossible to form a transenvelope bridge (Chng et al. 2012; Freinkman et al. 2012; Okuda et al. 2016). As mentioned before, LptD has four cysteine residues, two located in the N-terminal periplasmic domain and the remaining two in the C-terminal domain (Ruiz et al. 2010). Functional LptD requires the formation of disulfide bonds between non-consecutive cysteines thus allowing the correct configuration of the N-terminal domain with respect to the C-terminal β-barrel (Freinkman et al. 2012; Ruiz et al. 2010). Notably, LptD mutant proteins that lack native disulfide bonds do not interact with LptA preventing the formation of the transenvelope bridge. It is worth to note that for LptD binding to LptA the formation of at least one of the two non-consecutive disulfide bonds is sufficient. These findings suggest a mechanism in which the C-terminal β-barrel domain of LptD hinders the interaction of the N-terminal periplasmic domain with LptA when LptE-plug is not properly inserted (Freinkman et al. 2012) disrupting the β-jellyroll oligomerization process.

Altogether, the previous functional data and the available crystal structures of the Lpt proteins support the so-called PEZ model, in which the energy released from ATP hydrolysis in the cytoplasm would be used to push a continuous stream of LPS along the periplasmic bridge to the cell surface (Okuda et al. 2016). Indeed, by using an in vitro system based on *right-side-out vesicles* it has been shown that LPS transits along Lpt components trough discrete steps that require energy of ATP hydrolysis to transfer LPS from $LptB_2FG$ to LptC, then from $LptB_2FG$-C to LptA (Okuda et al. 2012) and from LptA to OM-proteoliposomes containing LptDE (Sherman et al. 2018). Notably, the transfer of LPS from IM to OM proteoliposomes only occurs when LptA is added to the system further supporting the key role of LptA in connecting IM and OM (Sherman et al. 2018).

The PEZ model postulates that the transport through the channel formed by Lpt-CAD functions as a "candy dispenser" in which the "candies" LPS molecules are pushed by ATP hydrolysis in a continuous flow up to the OM LptDE translocon driven by a constant pressure from the IM $LptB_2FG$ complex. However, the recent finding that the OM translocon is able to communicate with the IM complex to arrest the LPS transport, opens new questions about the regulation and the functioning of the Lpt machinery. Using a fluorescent assay that allows to follow the fate of LPS in the passage from the IM to OM proteoliposomes, Khane and co-workers showed that when the OM proteoliposomes reach a saturating quantity of LPS, the IM complex stops to hydrolyze ATP to avoid an otherwise useless consumption of energy (Xie et al. 2018). This finding unveils a negative feedback in which the OM LptDE translocon can control the ATPase activity of the IM $LptB_2GF$ transporter. This mechanism adds a second step of regulation within the Lpt system explaining why LPS stacks at the IM when transport is impaired at the OM. How the LPS delivery at the OM and the LPS extraction from the IM could be coupled will be the next fascinating question to answer about the Lpt system.

Conclusions and Perspectives

The OM is a fundamental organelle of the Gram-negative cell envelope that protects the cell from the entry of many noxious compounds. The LPS asymmetric distribution largely contributes to the peculiar properties of the OM. Much progress has been made in the last years in understanding how LPS amphipathic molecules can be efficiently synthesized, transported and assembled at the cell surface. However, the regulation mechanisms of LPS transport during cell growth remain virtually unknown.

LPS biogenesis has been successfully exploited as a target with the potential to develop compounds acting either as new drugs or as potentiator molecules (Zabawa et al. 2016). Despite there are no molecules that have been approved for clinical use yet, there are some promising candidates under validation and development. A novel class of inhibitors of LpxC deacetylase have been recently identified and proved capable of curing infections in animal models (Lee et al. 2016; Lemaitre et al. 2017). Another promising candidate is the peptidomimetic L27-11 which specifically targets LptD from *P. aeruginosa* (Srinivas et al. 2010; Werneburg et al. 2012). Notably, POL7080, a derivative of L27-11 with improved drug properties, is currently under clinical development. These findings highlight the potential of LPS biogenesis inhibitors as novel therapeutics to treat Gram-negative infections.

References

Abellon-Ruiz J, Kaptan SS, Basle A, Claudi B, Bumann D, Kleinekathofer U (2017) Structural basis for maintenance of bacterial outer membrane lipid asymmetry. Nat Microbiol 2(12):1616–1623

Anderson MS, Bull HG, Galloway SM, Kelly TM, Mohan S, Radika K, Raetz CR (1993) UDP-N-acetylglucosamine acyltransferase of *Escherichia coli*. The first step of endotoxin biosynthesis is thermodynamically unfavorable. J Biol Chem 268(26):19858–19865

Anderson MS, Raetz CR (1987) Biosynthesis of lipid A precursors in *Escherichia coli*. A cytoplasmic acyltransferase that converts UDP-N-acetylglucosamine to UDP-3-O-(R-3-hydroxymyristoyl)-N-acetylglucosamine. J Biol Chem 262(11):5159–5169

Asmar AT, Ferreira JL, Cohen EJ, Cho SH, Beeby M, Hughes KT, Collet JF (2017) Communication across the bacterial cell envelope depends on the size of the periplasm. PLoS Biol 15(12):e2004303

Babinski KJ, Kanjilal SJ, Raetz CR (2002) Accumulation of the lipid A precursor UDP-2,3-diacylglucosamine in an *Escherichia coli* mutant lacking the *lpxH* gene. J Biol Chem 277(29):25947–25956

Balibar CJ, Grabowicz M (2016) Mutant alleles of *lptD* increase the permeability of *Pseudomonas aeruginosa* and define determinants of intrinsic resistance to antibiotics. Antimicrob Agents Chemother 60(2):845–854

Belunis CJ, Raetz CR (1992) Biosynthesis of endotoxins. Purification and catalytic properties of 3-deoxy-D-manno-octulosonic acid transferase from *Escherichia coli*. J Biol Chem 267(14):9988–9997

Benedet M, Falchi FA, Puccio S, Di Benedetto C, Peano C, Polissi A, Deho G (2016) The lack of the essential LptC protein in the trans-envelope lipopolysaccharide transport machine is circumvented by suppressor mutations in LptF, an inner membrane component of the *Escherichia coli* transporter. PLoS ONE 11(8):e0161354

Bertani BR, Taylor RJ, Nagy E, Kahne D, Ruiz N (2018) A cluster of residues in the lipopolysaccharide exporter that selects substrate variants for transport to the outer membrane. Mol Microbiol 109(4):541–554

Bishop RE (2005) The lipid A palmitoyltransferase PagP: molecular mechanisms and role in bacterial pathogenesis. Mol Microbiol 57(4):900–912

Bishop RE (2008) Structural biology of membrane-intrinsic beta-barrel enzymes: sentinels of the bacterial outer membrane. Biochim Biophys Acta 1778(9):1881–1896

Bishop RE, Gibbons H, Guina T, Trent MS, Miller SI, Raetz CR (2000) Transfer of palmitate from phospholipids to lipid A in outer membranes of gram-negative bacteria. EMBO J 19(19):5071–5080

Botos I, Majdalani N, Mayclin SJ, McCarthy JG, Lundquist K, Wojtowicz D, Barnard TJ, Gumbart JC, Buchanan SK (2016) Structural and functional characterization of the LPS transporter LptDE from gram-negative pathogens. Structure 24(6):965–976

Bowyer A, Baardsnes J, Ajamian E, Zhang L, Cygler M (2011) Characterization of interactions between LPS transport proteins of the Lpt system. Biochem Biophys Res Commun 404(4):1093–1098

Braun M, Silhavy TJ (2002) Imp/OstA is required for cell envelope biogenesis in *Escherichia coli*. Mol Microbiol 45(5):1289–1302

Brozek KA, Raetz CR (1990) Biosynthesis of lipid A in *Escherichia coli*. Acyl carrier protein-dependent incorporation of laurate and myristate. J Biol Chem 265(26):15410–15417

Carpenter TS, Parkin J, Khalid S (2016) The free energy of small solute permeation through the *Escherichia coli* outer membrane has a distinctly asymmetric profile. J Phys Chem Lett 7(17):3446–3451

Chimalakonda G, Ruiz N, Chng SS, Garner RA, Kahne D, Silhavy TJ (2011) Lipoprotein LptE is required for the assembly of LptD by the beta-barrel assembly machine in the outer membrane of *Escherichia coli*. Proc Natl Acad Sci USA 108(6):2492–2497

Chng SS, Gronenberg LS, Kahne D (2010a) Proteins required for lipopolysaccharide assembly in *Escherichia coli* form a transenvelope complex. Biochemistry 49(22):4565–4567

Chng SS, Ruiz N, Chimalakonda G, Silhavy TJ, Kahne D (2010b) Characterization of the two-protein complex in *Escherichia coli* responsible for lipopolysaccharide assembly at the outer membrane. Proc Natl Acad Sci USA 107(12):5363–5368

Chng SS, Xue M, Garner RA, Kadokura H, Boyd D, Beckwith J, Kahne D (2012) Disulfide rearrangement triggered by translocon assembly controls lipopolysaccharide export. Science 337(6102):1665–1668

Chong ZS, Woo WF, Chng SS (2015) Osmoporin OmpC forms a complex with MlaA to maintain outer membrane lipid asymmetry in *Escherichia coli*. Mol Microbiol 98(6):1133–1146

Cipolla L, Polissi A, Airoldi C, Galliani P, Sperandeo P, Nicotra F (2009) The Kdo biosynthetic pathway toward OM biogenesis as target in antibacterial drug design and development. Curr Drug Discov Technol 6(1):19–33

Clementz T, Bednarski JJ, Raetz CR (1996) Function of the *htrB* high temperature requirement gene of *Escherichia coli* in the acylation of lipid A: HtrB catalyzed incorporation of laurate. J Biol Chem 271(20):12095–12102

Clementz T, Zhou Z, Raetz CR (1997) Function of the *Escherichia coli msbB* gene, a multicopy suppressor of htrB knockouts, in the acylation of lipid A. Acylation by MsbB follows laurate incorporation by HtrB. J Biol Chem 272(16):10353–10360

Dalebroux ZD, Matamouros S, Whittington D, Bishop RE, Miller SI (2014) PhoPQ regulates acidic glycerophospholipid content of the *Salmonella Typhimurium* outer membrane. Proc Natl Acad Sci USA 111(5):1963–1968

Davidson AL, Dassa E, Orelle C, Chen J (2008) Structure, function, and evolution of bacterial ATP-binding cassette systems. Microbiol Mol Biol Rev 72(2):317–364

Dekker N (2000) Outer-membrane phospholipase A: known structure, unknown biological function. Mol Microbiol 35(4):711–717

Doerrler WT, Gibbons HS, Raetz CR (2004) MsbA-dependent translocation of lipids across the inner membrane of *Escherichia coli*. J Biol Chem 279(43):45102–45109

Doerrler WT, Raetz CR (2002) ATPase activity of the MsbA lipid flippase of *Escherichia coli*. J Biol Chem 277(39):36697–36705

Doerrler WT, Reedy MC, Raetz CR (2001) An *Escherichia coli* mutant defective in lipid export. J Biol Chem 276(15):11461–11464

Dong H, Xiang Q, Gu Y, Wang Z, Paterson NG, Stansfeld PJ, He C, Zhang Y, Wang W, Dong C (2014) Structural basis for outer membrane lipopolysaccharide insertion. Nature 511(7507):52–56

Dong H, Zhang Z, Tang X, Paterson NG (2017) Structural and functional insights into the lipopolysaccharide ABC transporter $LptB_2FG$. Nat Commun 8(1):222

Doshi R, van Veen HW (2013) Substrate binding stabilizes a pre-translocation intermediate in the ATP-binding cassette transport protein MsbA. J Biol Chem 288(30):21638–21647

Eckford PD, Sharom FJ (2008) Functional characterization of *Escherichia coli* MsbA: interaction with nucleotides and substrates. J Biol Chem 283(19):12840–12850

Eckford PD, Sharom FJ (2010) The reconstituted *Escherichia coli* MsbA protein displays lipid flippase activity. Biochem J 429(1):195–203

Emiola A, Andrews SS, Heller C, George J (2016) Crosstalk between the lipopolysaccharide and phospholipid pathways during outer membrane biogenesis in *Escherichia coli*. Proc Natl Acad Sci U S A 113(11):3108–3113

Ercan B, Low WY, Liu X, Chng SS (2018) Characterization of interactions and phospholipid transfer between substrate binding proteins of the OmpC-Mla system. Biochemistry. https://doi.org/10.1021/acs.biochem.8b00897

Ernst RK, Yi EC, Guo L, Lim KB, Burns JL, Hackett M, Miller SI (1999) Specific lipopolysaccharide found in cystic fibrosis airway *Pseudomonas aeruginosa*. Science 286(5444):1561–1565

Fairman JW, Noinaj N, Buchanan SK (2011) The structural biology of beta-barrel membrane proteins: a summary of recent reports. Curr Opin Struct Biol 21(4):523–531

Falchi FA, Maccagni EA, Puccio S, Peano C, De Castro C, Palmigiano A, Garozzo D, Martorana AM, Polissi A, Dehò G, Sperandeo P (2018) Mutation and suppressor analysis of the essential lipopolysaccharide transport protein LptA reveals strategies to overcome severe outer membrane permeability defects in *Escherichia coli*. J Bacteriol 200(2):e00487–17

Freinkman E, Chng SS, Kahne D (2011) The complex that inserts lipopolysaccharide into the bacterial outer membrane forms a two-protein plug-and-barrel. Proc Natl Acad Sci USA 108(6):2486–2491

Freinkman E, Okuda S, Ruiz N, Kahne D (2012) Regulated assembly of the transenvelope protein complex required for lipopolysaccharide export. Biochemistry 51(24):4800–4806

Fuhrer F, Langklotz S, Narberhaus F (2006) The C-terminal end of LpxC is required for degradation by the FtsH protease. Mol Microbiol 59(3):1025–1036

Greenfield LK, Whitfield C (2012) Synthesis of lipopolysaccharide O-antigens by ABC transporter-dependent pathways. Carbohydr Res 356:12–24

Griffiss JM, O'Brien JP, Yamasaki R, Williams GD, Rice PA, Schneider H (1987) Physical heterogeneity of neisserial lipooligosaccharides reflects oligosaccharides that differ in apparent molecular weight, chemical composition, and antigenic expression. Infect Immun 55(8):1792–1800

Gu Y, Stansfeld PJ, Zeng Y, Dong H, Wang W, Dong C (2015) Lipopolysaccharide is inserted into the outer membrane through an intramembrane hole, a lumen gate, and the lateral opening of LptD. Structure 23(3):496–504

Hamad MA, Di Lorenzo F, Molinaro A, Valvano MA (2012) Aminoarabinose is essential for lipopolysaccharide export and intrinsic antimicrobial peptide resistance in *Burkholderia cenocepacia*. Mol Microbiol 85(5):962–974

Han W, Wu B, Li L, Zhao G, Woodward R, Pettit N, Zhao G, Cai L, Thon V, Wang PG, Wang PG (2012) Defining function of lipopolysaccharide O-antigen ligase WaaL using chemoenzymatically synthesized substrates. J Biol Chem 287(8):5357–5365

Ho H, Miu A, Alexander MK, Garcia NK, Oh A, Zilberleyb I, Reichelt M, Austin CD, Tam C, Shriver S, Hu H, Labadie SS, Liang J, Wang L, Wang J, Lu Y, Purkey HE, Quinn J, Franke Y, Clark K, Beresini MH, Tan MW, Sellers BD, Maurer T, Koehler MFT, Wecksler AT, Kiefer JR, Verma V, Xu Y, Nishiyama M, Payandeh J, Koth CM (2018) Structural basis for dual-mode inhibition of the ABC transporter MsbA. Nature 557(7704):196–201

Holst O (2007) The structures of core regions from enterobacterial lipopolysaccharides—an update. FEMS Microbiol Lett 271(1):3–11

Kagan JC (2017) Lipopolysaccharide detection across the kingdoms of life. Trends Immunol 38(10):696–704

Kamio Y, Nikaido H (1976) Outer membrane of Salmonella typhimurium: accessibility of phospholipid head groups to phospholipase c and cyanogen bromide activated dextran in the external medium. Biochemistry 15(12):2561–2570

Karow M, Georgopoulos C (1993) The essential *Escherichia coli msbA* gene, a multicopy suppressor of null mutations in the *htrB* gene, is related to the universally conserved family of ATP-dependent translocators. Mol Microbiol 7(1):69–79

Katz C, Ron EZ (2008) Dual role of FtsH in regulating lipopolysaccharide biosynthesis in *Escherichia coli*. J Bacteriol 190(21):7117–7122

Kelly TM, Stachula SA, Raetz CR, Anderson MS (1993) The *firA* gene of *Escherichia coli* encodes UDP-3-O-(R-3-hydroxymyristoyl)-glucosamine N-acyltransferase. The third step of endotoxin biosynthesis. J Biol Chem 268(26):19866–19874

King JD, Kocincova D, Westman EL, Lam JS (2009) Review: lipopolysaccharide biosynthesis in *Pseudomonas aeruginosa*. Innate Immun 15(5):261–312

Klein G, Kobylak N, Lindner B, Stupak A, Raina S (2014) Assembly of lipopolysaccharide in *Escherichia coli* requires the essential LapB heat shock protein. J Biol Chem 289(21):14829–14853

Kol MA, van Dalen A, de Kroon AI, de Kruijff B (2003) Translocation of phospholipids is facilitated by a subset of membrane-spanning proteins of the bacterial cytoplasmic membrane. J Biol Chem 278(27):24586–24593

Konovalova A, Kahne DE, Silhavy TJ (2017) Outer membrane biogenesis. Annu Rev Microbiol 71:539–556

Konovalova A, Silhavy TJ (2015) Outer membrane lipoprotein biogenesis: Lol is not the end. Philos Trans R Soc Lond B Biol Sci 370(1679):20150030

Laguri C, Sperandeo P, Pounot K, Ayala I, Silipo A, Bougault CM, Molinaro A, Polissi A, Simorre JP (2017) Interaction of lipopolysaccharides at intermolecular sites of the periplasmic Lpt transport assembly. Sci Rep 7(1):9715

Lee CJ, Liang X, Wu Q, Najeeb J, Zhao J, Gopalaswamy R, Titecat M, Sebbane F, Lemaitre N, Toone EJ, Zhou P (2016) Drug design from the cryptic inhibitor envelope. Nat Commun 7:10638

Lemaitre N, Liang X, Najeeb J, Lee CJ, Titecat M, Leteurtre E, Simonet M, Toone EJ, Zhou P, Sebbane F (2017) Curative treatment of severe gram-negative bacterial infections by a new class of antibiotics targeting LpxC. MBio 8(4):e00674–17

Lerouge I, Vanderleyden J (2002) O-antigen structural variation: mechanisms and possible roles in animal/plant-microbe interactions. FEMS Microbiol Rev 26(1):17–47

Li X, Gu Y, Dong H, Wang W, Dong C (2015) Trapped lipopolysaccharide and LptD intermediates reveal lipopolysaccharide translocation steps across the *Escherichia coli* outer membrane. Sci Rep 5:11883
Liu D, Reeves PR (1994) Presence of different O antigen forms in three isolates of one clone of *Escherichia coli*. Genetics 138(1):6–10
Locher KP (2016) Mechanistic diversity in ATP-binding cassette (ABC) transporters. Nat Struct Mol Biol 23(6):487–493
Luo Q, Yang X, Yu S, Shi H, Wang K, Xiao L, Zhu G, Sun C, Li T, Li D, Zhang X, Zhou M, Huang Y (2017) Structural basis for lipopolysaccharide extraction by ABC transporter LptB2FG. Nat Struct Mol Biol 24(5):469–474
Mahalakshmi S, Sunayana MR, SaiSree L, Reddy M (2014) *yciM* is an essential gene required for regulation of lipopolysaccharide synthesis in *Escherichia coli*. Mol Microbiol 91(1):145–157
Maldonado RF, Sa-Correia I, Valvano MA (2016) Lipopolysaccharide modification in Gram-negative bacteria during chronic infection. FEMS Microbiol Rev 40(4):480–493
Malinverni JC, Silhavy TJ (2009) An ABC transport system that maintains lipid asymmetry in the gram-negative outer membrane. Proc Natl Acad Sci USA 106(19):8009–8014
Malojčić G, Andres D, Grabowicz M, George AH, Ruiz N, Silhavy TJ, Kahne D (2014) LptE binds to and alters the physical state of LPS to catalyze its assembly at the cell surface. Proc Natl Acad Sci USA 111(26):9467–9472
Mamat U, Meredith TC, Aggarwal P, Kuhl A, Kirchhoff P, Lindner B, Hanuszkiewicz A, Sun J, Holst O, Woodard RW (2008) Single amino acid substitutions in either YhjD or MsbA confer viability to 3-deoxy-D-manno-oct-2-ulosonic acid-depleted *Escherichia coli*. Mol Microbiol 67(3):633–648
Martorana AM, Sperandeo P, Polissi A, Deho G (2011) Complex transcriptional organization regulates an *Escherichia coli* locus implicated in lipopolysaccharide biogenesis. Res Microbiol 162(5):470–482
Martorana AM, Benedet M, Maccagni EA, Sperandeo P, Villa R, Deho G, Polissi A (2016) Functional interaction between the cytoplasmic ABC protein LptB and the inner membrane LptC protein, components of the lipopolysaccharide transport machinery in *Escherichia coli*. J Bacteriol 198(16):2192–2203
May KL, Silhavy TJ (2018) The *Escherichia coli* phospholipase PldA regulates outer membrane homeostasis via lipid signaling. MBio 9(2):e00379–18
Meredith TC, Aggarwal P, Mamat U, Lindner B, Woodard RW (2006) Redefining the requisite lipopolysaccharide structure in *Escherichia coli*. ACS Chem Biol 1(1):33–42
Merten JA, Schultz KM, Klug CS (2012) Concentration-dependent oligomerization and oligomeric arrangement of LptA. Protein Sci 21(2):211–218
Mi W, Li Y, Yoon SH, Ernst RK, Walz T, Liao M (2017) Structural basis of MsbA-mediated lipopolysaccharide transport. Nature 549(7671):233–237
Miyake K (2004) Innate recognition of lipopolysaccharide by Toll-like receptor 4-MD-2. Trends Microbiol 12(4):186–192
Moffatt JH, Harper M, Harrison P, Hale JD, Vinogradov E, Seemann T, Henry R, Crane B, St Michael F, Cox AD, Adler B, Nation RL, Li J, Boyce JD (2010) Colistin resistance in Acinetobacter baumannii is mediated by complete loss of lipopolysaccharide production. Antimicrob Agents Chemother 54(12):4971–4977
Narita S, Tokuda H (2009) Biochemical characterization of an ABC transporter LptBFGC complex required for the outer membrane sorting of lipopolysaccharides. FEBS Lett 583(13):2160–2164
Needham BD, Carroll SM, Giles DK, Georgiou G, Whiteley M, Trent MS (2013) Modulating the innate immune response by combinatorial engineering of endotoxin. Proc Natl Acad Sci USA 110(4):1464–1469
Nikaido H (2003) Molecular basis of bacterial outer membrane permeability revisited. Microbiol Mol Biol Rev 67(4):593–656
Nikaido H (2005) Restoring permeability barrier function to outer membrane. Chem Biol 12(5):507–509

Noinaj N, Gumbart JC, Buchanan SK (2017) The beta-barrel assembly machinery in motion. Nat Rev Microbiol 15(4):197–204
Ogura T, Inoue K, Tatsuta T, Suzaki T, Karata K, Young K, Su LH, Fierke CA, Jackman JE, Raetz CR, Coleman J, Tomoyasu T, Matsuzawa H (1999) Balanced biosynthesis of major membrane components through regulated degradation of the committed enzyme of lipid A biosynthesis by the AAA protease FtsH (HflB) in *Escherichia coli*. Mol Microbiol 31(3):833–844
Okuda S, Freinkman E, Kahne D (2012) Cytoplasmic ATP hydrolysis powers transport of lipopolysaccharide across the periplasm in *E. coli*. Science 338(6111):1214–1217
Okuda S, Sherman DJ, Silhavy TJ, Ruiz N, Kahne D (2016) Lipopolysaccharide transport and assembly at the outer membrane: the PEZ model. Nat Rev Microbiol 14(6):337–345
Okuda S, Tokuda H (2011) Lipoprotein sorting in bacteria. Annu Rev Microbiol 65:239–259
Paciello I, Silipo A, Lembo-Fazio L, Curcuru L, Zumsteg A, Noel G, Ciancarella V, Sturiale L, Molinaro A, Bernardini ML (2013) Intracellular *Shigella* remodels its LPS to dampen the innate immune recognition and evade inflammasome activation. Proc Natl Acad Sci USA 110(46):E4345–E4354
Park BS, Lee JO (2013) Recognition of lipopolysaccharide pattern by TLR4 complexes. Exp Mol Med 45:e66
Park BS, Song DH, Kim HM, Choi BS, Lee H, Lee JO (2009) The structural basis of lipopolysaccharide recognition by the TLR4-MD-2 complex. Nature 458(7242):1191–1195
Peng D, Hong W, Choudhury BP, Carlson RW, Gu XX (2005) *Moraxella catarrhalis* bacterium without endotoxin, a potential vaccine candidate. Infect Immun 73(11):7569–7577
Peppler MS, Schrumpf ME (1984) Phenotypic variation and modulation in *Bordetella bronchiseptica*. Infect Immun 44(3):681–687
Polissi A, Georgopoulos C (1996) Mutational analysis and properties of the msbA gene of *Escherichia coli*, coding for an essential ABC family transporter. Mol Microbiol 20(6):1221–1233
Qiao S, Luo Q, Zhao Y, Zhang XC, Huang Y (2014) Structural basis for lipopolysaccharide insertion in the bacterial outer membrane. Nature 511(7507):108–111
Radika K, Raetz CR (1988) Purification and properties of lipid A disaccharide synthase of *Escherichia coli*. J Biol Chem 263(29):14859–14867
Raetz CR, Guan Z, Ingram BO, Six DA, Song F, Wang X, Zhao J (2009) Discovery of new biosynthetic pathways: the lipid A story. J Lipid Res 50(Suppl):S103–S108
Raetz CR, Whitfield C (2002) Lipopolysaccharide endotoxins. Annu Rev Biochem 71:635–700
Raetz CRH, Reynolds CM, Trent MS, Bishop RE (2007) Lipid A modification systems in gram-negative bacteria. Annu Rev Biochem 76(1):295–329
Ray BL, Raetz CR (1987) The biosynthesis of gram-negative endotoxin. A novel kinase in *Escherichia coli* membranes that incorporates the 4′-phosphate of lipid A. J Biol Chem 262(3):1122–1128
Rees DC, Johnson E, Lewinson O (2009) ABC transporters: the power to change. Nat Rev Mol Cell Biol 10(3):218–227
Ruan X, Loyola DE, Marolda CL, Perez-Donoso JM, Valvano MA (2012) The WaaL O-antigen lipopolysaccharide ligase has features in common with metal ion-independent inverting glycosyltransferases. Glycobiology 22(2):288–299
Ruiz N, Chng SS, Hiniker A, Kahne D, Silhavy TJ (2010) Nonconsecutive disulfide bond formation in an essential integral outer membrane protein. Proc Natl Acad Sci USA 107(27):12245–12250
Ruiz N, Gronenberg LS, Kahne D, Silhavy TJ (2008) Identification of two inner-membrane proteins required for the transport of lipopolysaccharide to the outer membrane of *Escherichia coli*. Proc Natl Acad Sci USA 105(14):5537–5542
Ruiz N, Kahne D, Silhavy TJ (2009) Transport of lipopolysaccharide across the cell envelope: the long road of discovery. Nat Rev Microbiol 7(9):677–683
Sankaran K, Wu HC (1994) Lipid modification of bacterial prolipoprotein. Transfer of diacylglyceryl moiety from phosphatidylglycerol. J Biol Chem 269(31):19701–19706

Santambrogio C, Sperandeo P, Villa R, Sobott F, Polissi A, Grandori R (2013) LptA assembles into rod-like oligomers involving disorder-to-order transitions. J Am Soc Mass Spectrom 24(10):1593–1602
Schultz KM, Feix JB, Klug CS (2013) Disruption of LptA oligomerization and affinity of the LptA-LptC interaction. Protein Sci 22(11):1639–1645
Schulz GE (2002) The structure of bacterial outer membrane proteins. Biochim Biophys Acta 1565(2):308–317
Sherman DJ, Lazarus MB, Murphy L, Liu C, Walker S, Ruiz N, Kahne D (2014) Decoupling catalytic activity from biological function of the ATPase that powers lipopolysaccharide transport. Proc Natl Acad Sci USA 111(13):4982–4987
Sherman DJ, Xie R, Taylor RJ, George AH, Okuda S, Foster PJ, Needleman DJ, Kahne D (2018) Lipopolysaccharide is transported to the cell surface by a membrane-to-membrane protein bridge. Science 359(6377):798–801
Siarheyeva A, Sharom FJ (2009) The ABC transporter MsbA interacts with lipid A and amphipathic drugs at different sites. Biochem J 419(2):317–328
Silhavy TJ, Kahne D, Walker S (2010) The bacterial cell envelope. Cold Spring Harb Perspect Biol 2(5):a000414
Simpson BW, Owens TW, Orabella MJ, Davis RM, May JM, Trauger SA, Kahne D, Ruiz N (2016) Identification of residues in the lipopolysaccharide ABC transporter that coordinate ATPase activity with extractor function. MBio 7(5):e01729–16
Sperandeo P, Cescutti R, Villa R, Di Benedetto C, Candia D, Deho G, Polissi A (2007) Characterization of *lptA* and *lptB*, two essential genes implicated in lipopolysaccharide transport to the outer membrane of *Escherichia coli*. J Bacteriol 189(1):244–253
Sperandeo P, Lau FK, Carpentieri A, De Castro C, Molinaro A, Deho G, Silhavy TJ, Polissi A (2008) Functional analysis of the protein machinery required for transport of lipopolysaccharide to the outer membrane of *Escherichia coli*. J Bacteriol 190(13):4460–4469
Sperandeo P, Villa R, Martorana AM, Samalikova M, Grandori R, Deho G, Polissi A (2011) New insights into the Lpt machinery for lipopolysaccharide transport to the cell surface: LptA-LptC interaction and LptA stability as sensors of a properly assembled transenvelope complex. J Bacteriol 193(5):1042–1053
Srinivas N, Jetter P, Ueberbacher BJ, Werneburg M, Zerbe K, Steinmann J, Van der Meijden B, Bernardini F, Lederer A, Dias RL, Misson PE, Henze H, Zumbrunn J, Gombert FO, Obrecht D, Hunziker P, Schauer S, Ziegler U, Käch A, Eberl L, Riedel K, DeMarco SJ, Robinson JA (2010) Peptidomimetic antibiotics target outer-membrane biogenesis in *Pseudomonas aeruginosa*. Science 327(5968):1010–1013
Steeghs L, den Hartog R, den Boer A, Zomer B, Roholl P, van der Ley P (1998) Meningitis bacterium is viable without endotoxin. Nature 392(6675):449–450
Stevenson G, Neal B, Liu D, Hobbs M, Packer NH, Batley M, Redmond JW, Lindquist L, Reeves P (1994) Structure of the O antigen of *Escherichia coli* K-12 and the sequence of its rfb gene cluster. J Bacteriol 176(13):4144–4156
Suits MDL, Sperandeo P, Dehò G, Polissi A, Jia Z (2008) Novel structure of the conserved gram-negative lipopolysaccharide transport protein a and mutagenesis analysis. J Mol Biol 380(3):476–488
Sutcliffe IC (2010) A phylum level perspective on bacterial cell envelope architecture. Trends Microbiol 18(10):464–470
Sutterlin HA, Shi H, May KL, Miguel A, Khare S, Huang KC, Silhavy TJ (2016) Disruption of lipid homeostasis in the Gram-negative cell envelope activates a novel cell death pathway. Proc Natl Acad Sci USA 113(11):E1565–E1574
Tefsen B, Bos MP, Beckers F, Tommassen J, de Cock H (2005a) MsbA is not required for phospholipid transport in Neisseria meningitidis. J Biol Chem 280(43):35961–35966
Tefsen B, Geurtsen J, Beckers F, Tommassen J, de Cock H (2005b) Lipopolysaccharide transport to the bacterial outer membrane in spheroplasts. J Biol Chem 280(6):4504–4509

Thomas C, Tampe R (2018) Multifaceted structures and mechanisms of ABC transport systems in health and disease. Curr Opin Struct Biol 51:116–128

Thong S, Ercan B, Torta F, Fong ZY, Wong HY, Wenk MR, Chng SS (2016) Defining key roles for auxiliary proteins in an ABC transporter that maintains bacterial outer membrane lipid asymmetry. Elife 5:e19042

Tran AX, Dong C, Whitfield C (2010) Structure and functional analysis of LptC, a conserved membrane protein involved in the lipopolysaccharide export pathway in *Escherichia coli*. J Biol Chem 285(43):33529–33539

Villa R, Martorana AM, Okuda S, Gourlay LJ, Nardini M, Sperandeo P, Dehò G, Bolognesi M, Kahne D, Polissi A (2013) The *Escherichia coli* Lpt transenvelope protein complex for lipopolysaccharide export is assembled via conserved structurally homologous domains. J Bacteriol 195(5):1100–1108

Voss BJ, Trent MS (2018) LPS transport: flipping out over MsbA. Curr Biol 28(1):R30–r33

Ward A, Reyes CL, Yu J, Roth CB, Chang G (2007) Flexibility in the ABC transporter MsbA: alternating access with a twist. Proc Natl Acad Sci USA 104(48):19005–19010

Werneburg M, Zerbe K, Juhas M, Bigler L, Stalder U, Kaech A, Ziegler U, Obrecht D, Eberl L, Robinson JA (2012) Inhibition of lipopolysaccharide transport to the outer membrane in *Pseudomonas aeruginosa* by peptidomimetic antibiotics. ChemBioChem 13(12):1767–1775

Whitfield C (2006) Biosynthesis and assembly of capsular polysaccharides in *Escherichia coli*. Annu Rev Biochem 75:39–68

Wu T, McCandlish AC, Gronenberg LS, Chng SS, Silhavy TJ, Kahne D (2006) Identification of a protein complex that assembles lipopolysaccharide in the outer membrane of *Escherichia coli*. Proc Natl Acad Sci USA 103(31):11754–11759

Xie R, Taylor RJ, Kahne D (2018) Outer membrane translocon communicates with inner membrane ATPase To stop lipopolysaccharide transport. Science 140(40):12691–12694

Yeow J, Tan KW, Holdbrook DA, Chong ZS, Marzinek JK, Bond PJ, Chng SS (2018) The architecture of the OmpC-MlaA complex sheds light on the maintenance of outer membrane lipid asymmetry in *Escherichia coli*. J Biol Chem 293(29):11325–11340

Young K, Silver LL, Bramhill D, Cameron P, Eveland SS, Raetz CR, Hyland SA, Anderson MS (1995) The *envA* permeability/cell division gene of *Escherichia coli* encodes the second enzyme of lipid A biosynthesis. UDP-3-O-(R-3-hydroxymyristoyl)-N-acetylglucosamine deacetylase. J Biol Chem 270(51):30384–30391

Zabawa TP, Pucci MJ, Parr TR Jr, Lister T (2016) Treatment of Gram-negative bacterial infections by potentiation of antibiotics. Curr Opin Microbiol 33:7–12

Zeth K, Thein M (2010) Porins in prokaryotes and eukaryotes: common themes and variations. Biochem J 431(1):13–22

Zhang G, Meredith TC, Kahne D (2013) On the essentiality of lipopolysaccharide to Gram-negative bacteria. Curr Opin Microbiol 16(6):779–785

Chapter 3
Lipoproteins: Structure, Function, Biosynthesis

Volkmar Braun and Klaus Hantke

Abstract The Lpp lipoprotein of *Escherichia coli* is the first identified protein with a covalently linked lipid. It is chemically bound by its C-terminus to murein (peptidoglycan) and inserts by the lipid at the N-terminus into the outer membrane. As the most abundant protein in *E. coli* (10^6 molecules per cell) it plays an important role for the integrity of the cell envelope. Lpp represents the type protein of a large variety of lipoproteins found in Gram-negative and Gram-positive bacteria and in archaea that have in common the lipid structure for anchoring the proteins to membranes but otherwise strongly vary in sequence, structure, and function. Predicted lipoproteins in known prokaryotic genomes comprise 2.7% of all proteins. Lipoproteins are modified by a unique phospholipid pathway and transferred from the cytoplasmic membrane into the outer membrane by a special system. They are involved in protein incorporation into the outer membrane, protein secretion across the cytoplasmic membrane, periplasm and outer membrane, signal transduction, conjugation, cell wall metabolism, antibiotic resistance, biofilm formation, and adhesion to host tissues. They are only found in bacteria and function as signal molecules for the innate immune system of vertebrates, where they cause inflammation and elicit innate and adaptive immune response through Toll-like receptors. This review discusses various aspects of Lpp and other lipoproteins of Gram-negative and Gram-positive bacteria and archaea.

Keywords Lipoprotein · Structure · Function · Biosynthesis · Immune stimulation · Bacteria

V. Braun (✉)
Department of Protein Evolution, Max Planck Institute for Developmental Biology, Max Planck Ring 5, 72076 Tübingen, Germany
e-mail: volkmar.braun@tuebingen.mpg.de

K. Hantke
IMIT, University of Tuebingen, Auf der Morgenstelle 28, 72076 Tübingen, Germany
e-mail: klaus.hantke@uni-tuebingen.de

A. Kuhn (ed.), *Bacterial Cell Walls and Membranes*, Subcellular Biochemistry 92, https://doi.org/10.1007/978-3-030-18768-2_3

Introduction

The first identified and fully characterized protein with a covalently linked lipid was the murein lipoprotein Lpp of *Escherichia coli*, also called the major lipoprotein or Braun's lipoprotein. It is the type protein of a large variety of proteins in Gram-negative and Gram-positive bacteria and in archaea. These proteins are abundant in the cells and all have in common a lipid structure through which they anchor the proteins to a membrane outside of the cytoplasm. However, their sequence, structure, and function strongly vary. They are involved in membrane protein incorporation, protein secretion, signal transduction, conjugation, cell wall metabolism, antibiotic resistance, biofilm formation, adhesion to host tissues, and signaling for the innate immune system of vertebrates. This review will discuss various aspects of *E. coli* Lpp and highlight features of other lipoproteins from other bacterial and archaeal species. The early work on Lpp has been summarized in (Braun 1975; Braun and Hantke 1974; Braun and Wu 1994).

E. coli Lpp

Discovery and Components of Lpp

The Lpp lipoprotein covalently bound to murein (peptidoglycan) of *Escherichia coli* was discovered in a search for a bond that upon cleavage with trypsin decreased the absorbance of the isolated cell envelope much more rapidly than when any other protease was used (Braun and Rehn 1969). Cell envelopes treated with trypsin separated into two membranes, namely the outer membrane and cytoplasmic membrane, as observed by electron microscopy of ultra-thin sections. Samples not treated with trypsin were closely attached to each other. Since trypsin released only small amounts of protein into the medium, the trypsin cleavage site was sought in the insoluble part.

It had been shown earlier that protein was associated with isolated murein, but the protein material was not investigated further because the focus was on the chemical composition of murein (Weidel and Pelzer 1964). Treatment of cell envelopes with a hot sodium dodecyl sulfate solution left murein as the sole insoluble component. It became morphologically thinner when treated with proteases, but kept the size and rod form of the exponentially growing cells from which it was isolated. Acid hydrolysis of murein of exponentially growing cells yielded the constituents of murein and a mixture of all but six common amino acids (Cys, His, Pro, Gly, Phe, and Trp; by contrast, in stationary phase cells, the proteins associated with murein consist of all amino acids). The specific lack of these common amino acids suggested that a defined protein was associated with murein. After trypsin treatment, only lysine remained associated with murein, which indicated that it linked the protein and murein. Further analysis showed that the lysine is bound through the ε-amino group to the optical L-center of diaminopimelic acid of the murein peptide side chain, where it replaces

D-alanine (Braun and Bosch 1972). The lysine residue is at the C-terminal end of Lpp (Fig. 3.1a).

The mature Lpp protein consists of 57 residues and an additional N-terminal modified cysteine. The amino acid sequence reveals 14 repeats composed of seven residues, of which every third or fourth residue is hydrophobic (Fig. 3.1a). Such a periodicity is typical for coiled-coils of α-helices in which seven residues (heptad repeats) form two turns of a helix and that the side chains of parallel helices interlock systematically. Lpp was the first coiled-coil protein to be sequenced (Lupas et al. 2017). Its high α-helical content was deduced from circular dichroism spectra (Braun et al. 1976b). Recombinant Lpp devoid of the lipid forms a long parallel α-helical trimer. The X-ray crystal structure reveals a three-stranded coiled-coil domain

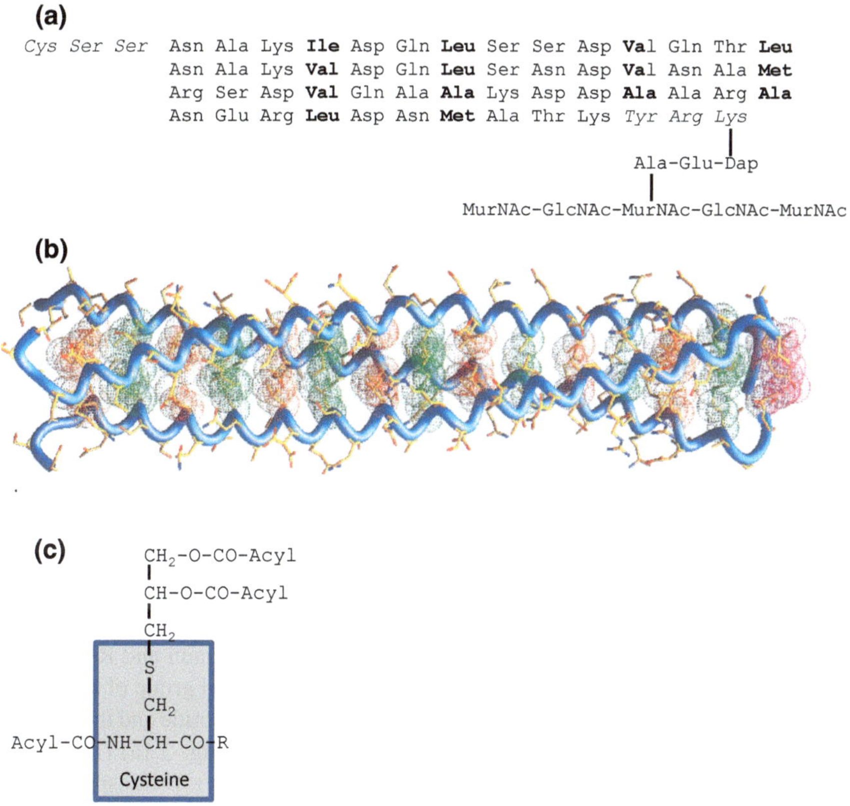

Fig. 3.1 **a** Amino acid sequence of the *E. coli* murein lipoprotein (Lpp) with bound murein (peptidoglycan) subunits. The heptad repeats (3 + 4) are in boldface. The N-terminal lipid attachment site and the C-terminal murein attachment site (in italics) are outside the heptad register. **b** High-resolution crystal structure of the Lpp trimer (residues 2–56) comprising a parallel three-stranded coiled coil and two helix capping motifs (Shu et al. 2000). **c** Chemical structure of glycerylcysteine with three bound fatty acids (mostly palmitic acid)

from residue 5 to 53, with the heptad repeat in register throughout the entire region (Fig. 3.1b) (Shu et al. 2000). The exceedingly high trypsin cleavage rate is explained by the accumulation of three trypsin cleavage sites at the very C-terminus and their exposure outside the triple helix. The sequences Tyr-Arg-Lys and Tyr-Lys-Lys are most frequently found in lipoproteins linked to murein.

Lipid Covalently Bound to the Protein

When structural work on Lpp began, lipoproteins were not defined molecules but rather mixtures of proteins with non-covalently linked mixtures of lipids. But in Lpp, lipid was associated with a protein that could not be removed with detergents or organic solvents and was therefore considered to be covalently bound. The chemical structure of the protein lipid attachment site was determined by chemically degrading isolated Lpp, incorporation of radioactively labeled suspected precursors, and finally by chemical synthesis of *S*-glycerylcysteine thioether (Hantke and Braun 1973). The structure was shown to be a glyceryl group with a fatty acid composition similar to that of phospholipids and bound as a thioether to the sulfhydryl group of cysteine, and was named glycerylcysteine (Fig. 3.1c). The amino group of the N-terminal cysteine is bound to a fatty acid, mainly palmitic acid. This unique lipid structure was later shown to be part of all bacterial lipoproteins with a covalently bound lipid. The first three-dimensional structure of triacylated cysteine in a lipid bilayer was recently found in the electron cryomicroscopy structure of a cytochrome oxidase (Sun et al. 2018). In this structure, the lipid anchors at the ActB and ActE proteins are tilted with respect to the plane of the lipid bilayer, thereby restricting the ability of other lipids to pack around them. The lipids are localized close to the entry point of menaquinol in the complex.

Linkage of Lpp to Murein (Peptidoglycan)

In *E. coli*, one-third of Lpp is covalently bound to murein (Braun and Rehn 1969; Inouye et al. 1972). The linkage is formed between the carboxyl group of the optical L-center of meso-diaminopimelate (Dpm) of the murein tetrapeptide and the ε-amino group of the carboxyl-terminal lysine (Lys) of lipoprotein (Fig. 3.1a) (Braun and Bosch 1972). The reaction is catalyzed by three very similar L-D-transpeptidases, namely ErfK, YcfS, and YbiS (Magnet et al. 2007), whereby the D-Ala linked to Dpm is replaced by Lys and the peptide energy of the Dpm-Ala bond is preserved to form the Dpm-Lys bond. The genes encoding all three enzymes must be deleted to obtain murein free of Lpp, but YbiS plays the major transpeptidation because the deletion of its gene largely prevents binding of Lpp to murein. The presence of several transpeptidases is not surprising because redundant enzymes in murein synthesis and murein modification have been frequently observed (Pazos et al. 2017). Murein is essential for the structure of cells that they cannot afford a defective synthesis. In addition, the structure of murein changes under different growth condition which requires different enzyme activities (Vollmer et al. 2008).

The C-terminal Lys in the Tyr-Arg-Lys triad plays a critical role in linkage of Lpp to murein (Zhang and Wu 1992; Zhang et al. 1992). By contrast, alterations at the N-terminus, such as lack of lipid modification, addition of a signal peptide, and fusion of the outer membrane protein OmpF, reduce but do not abolish covalent attachment of Lpp to murein (Zhang et al. 1992). The amount of Lpp bound to murein can vary; for example, in stationary phase cells, Lpp binding increases by 70% (Glauner et al. 1988).

Free Lpp

Free Lpp not attached to murein was detected when proteins of cell envelopes labeled with radioactive arginine or histidine were separated by SDS-polyacrylamide gel electrophoresis (Inouye et al. 1972). A fast migrating band, i.e., a small protein, contained labeled arginine but not histidine. The only protein known to be devoid of histidine in *E. coli* was the small protein Lpp. Its identity was confirmed by double labeling with other amino acids present and absent in Lpp. The electrophoretic mobility of free Lpp was faster than that of Lpp released from murein by lysozyme; the latter preparation also contains muropeptides. The amount of free Lpp was twice as high as that of murein-bound Lpp.

Distinct Arrangement of Bound and Free Lpp in the Outer Membrane

The early proposal (Braun and Rehn 1969) that Lpp extends from murein across the periplasm into the outer membrane was supported by the immunological determination of Lpp in separated outer and cytoplasmic membranes (Bosch and Braun 1973). Lpp is bound by the C-terminal lysine to diaminopimelic acid of murein and is fixed to the outer membrane by insertion of the lipid part into the inner lipid leaflet of the outer membrane. It was assumed that free lipoprotein occupies the same location as the bound form in the cell envelope. This assumption was supported by cross-linking the free form with the bound form and non-covalent binding of the free form by the C-terminal lysine to murein (Choi et al. 1987). However, distinct cellular locations of bound and free Lpp were discovered over two decades later (Cowles et al. 2011). Free Lpp spans the outer membrane and its C-terminus is surface-exposed, whereas bound Lpp resides in the periplasm. This conclusion was drawn from differential labeling with membrane-impermeable biotin reagents and proteolytic removal of the labels at the cell surface. The biotin label was attached to the C-terminal Lys55 and Lys58 residues. In addition, the C-terminal FLAG epitope (DYKDDDDK) of a constructed Lpp derivative could be proteolytically removed by treating cells with trypsin, which showed that the C-terminus of free Lpp is exposed at the cell surface.

The question arises how the free form inserts into and is partially translocated across the outer membrane. The parallel three-stranded coiled-coil structure of Lpp is too hydrophilic to insert spontaneously into a lipid bilayer. Since the free form is first synthesized and then converted to the bound form (Inouye et al. 1972), the two forms might transiently associate with each other, which would explain their cross-linkage. The free form is either translocated by the Bam complex, or by a specific translocase. Specific translocases were identified for various proteins, e.g., NalP or RcsF with the Bam complex, pullulanase with the type II secretion system, and in *Neisseria* Slam1 with TbPB, LbpB, fHb1, and Slam2 with HpuA (see section "Lipoproteins in Other Gram-Negative Bacteria"). The translocation of lipoproteins to the cell surface has recently been reviewed (Hooda and Moraes 2018).

Biosynthesis of Lpp

E. coli Lpp is synthesized in the cytoplasm, modified with lipids at the outer leaflet of the cytoplasmic membrane, and translocated to and inserted into the outer membrane. All these steps are mediated by dedicated proteins.

Unmodified lipoprotein can be synthesized in vitro with an mRNA of about 250 nucleotides (Hirashima et al. 1974). The mRNA has an unusually long average half-life of 11.5 min, in contrast to the half-life of 2.1 min of cytoplasmic protein mRNAs (Hirashima and Inouye 1973). The long half-life is suitable for allowing synthesis of this most abundant protein in *E. coli* (about 10^6 molecules per cell).

Lpp is synthesized in the cytoplasm with a signal sequence that serves to transfer the protein across the cytoplasmic membrane by the Sec translocon. The cysteine residue located at the border between the signal sequence and the mature protein is strictly conserved in all lipoproteins. It is surrounded by a partially conserved sequence in lipoproteins—Leu(Ala/Val)-Leu-Ala(Ser)-Gly(Ala)-Cys. This so-called lipobox is recognized by lipid-modifying enzymes. Lipid modification takes place at the periplasmic side of the cytoplasmic membrane. First, diacylglycerol is added to the sulfhydryl group of the cysteine by phosphatidylglycerol/prolipoprotein diacylglyceryl transferase (Lgt) (Fig. 3.2). Then, a special lipoprotein signal peptidase (Lsp or signal peptidase II) cleaves the signal peptide in front of the Cys residue (Innis et al. 1984). Signal peptidase II preferentially recognizes didecanoyl glycerol as the optimal lipid length and exclusively the enantio (*R*) form of the diacylglycerol (Kitamura and Wolan 2018). The protein remains attached to the cytoplasmic membrane via the diacylglycerol. Finally, a fatty acid is transferred from phospholipids to the amino group of the Cys residue by membrane-bound phospholipid/apolipoprotein transacylase (Lnt) (Tokunaga et al. 1982; Noland et al. 2017).

Genes encoding the three enzymes Lgt, Lsp, and Lnt are essential for *E. coli* and most other Gram-negative bacteria. The crystal structure of the diacylglyceryl

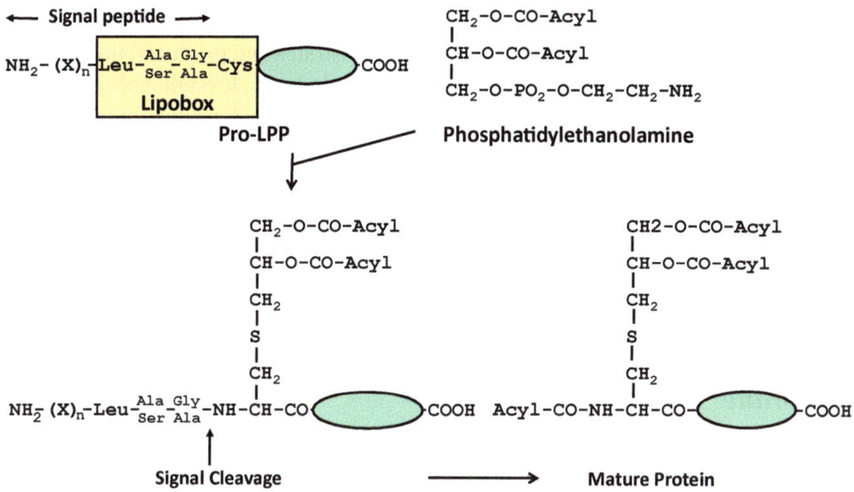

Fig. 3.2 Biosynthesis of Lpp. Refer to the text for details. The polypeptide chain of LPP is indicated in green

transferase Lgt shows a central cavity about 20 Å deep, with a major opening to the periplasmic side and two clefts exposed to the membrane lipid (Mao et al. 2016). Phosphatidylglycerol is positioned with the two acyl chains in the lower hydrophobic part of the cleft and the hydrophilic head protruding into the interface between the outer leaflet of the cytoplasmic membrane and the periplasm. It is predicted that the two clefts (6 and 10 Å wide) are entrances for phosphatidylglycerol and the lipobox of the lipoprotein. Substrates and product enter and leave laterally relative to the lipid bilayer.

The crystal structure of the Lsp signal peptidase II from *Pseudomonas aeruginosa* complexed with the inhibitor globomycin has been determined (Vogeley et al. 2016). The enzyme consists of a periplasmic domain fixed to the cytoplasmic membrane by a domain of four transmembrane helices. Globomycin acts as a noncleavable peptide that sterically blocks the active site. It mimics diglyceride Lpp in that it is composed of a 19-member cyclic depsipeptide that includes an α-methyl-β-hydroxy fatty acid. It is first partitioned in the cytoplasmic membrane and then diffuses laterally into the active site of the enzyme. The lipopeptide helix of Lpp fits into the space between membrane helices 12 and 14, which positions the Cys residue of the lipobox into the active site while leaving the protein in the periplasm. It is predicted that Lsp is an aspartyl endopeptidase with two Asp residues suitably positioned to form the catalytic diad. After release of the signal peptide, lipoprotein moves out of the active site but remains bound to the cytoplasmic membrane by the diacylglyceride moiety.

The crystal structures of the *N*-acylase Lnt of *E. coli* (Noland et al. 2017) and *P. aeruginosa* (Wiktor et al. 2017) have been determined. Both structures are composed of a large periplasmic domain that is fixed to the cytoplasmic membrane by eight transmembrane helices. The active site with the catalytic Cys387 is positioned at the base of a long channel above the outer leaflet of the cytoplasmic membrane. It contains an opening towards the membrane for substrates and products; this opening serves as the portal through which phospholipid and glycerylcysteine of lipoprotein enter the active site. The Lnt enzyme transfers a fatty acid from a phospholipid, preferentially phosphatidylethanolamine, to form *S*-acyl cysteine. The acyl chain, usually palmitate, is transferred to the α-amino group of *S*-diacyl-glyceryl lipoprotein. Triacylated lipoprotein leaves the active site and enters the membrane, and the active site is ready for the next round of lipoprotein acylation.

Transfer of Lpp from the Cytoplasmic Membrane to the Outer Membrane

Sorting of the lipoprotein from the cytoplasmic membrane to the outer membrane was studied mainly with *E. coli* Lpp and *P. aeruginosa* Pal (Okuda and Tokuda 2011; Narita and Tokuda 2017). The final destination of Lpp and Pal is the inner leaflet of the outer membrane; the protein moiety remains in the periplasm, but the N-terminal lipid anchors the lipoprotein to the outer membrane. Five proteins, designated LolA, LolB, LolC, LolD, and LolE, mediate lipoprotein sorting from the cytoplasmic membrane into the outer membrane (Fig. 3.3). LolCDE belong to the ABC transporter superfamily; LolC and LolE are the transmembrane subunits and LolD is the ATPase. One LolD subunit is bound to LolC, and one is bound to LolE. The LolCDE complex releases lipoprotein from the cytoplasmic membrane by ATP hydrolysis. Lipoprotein attaches to the soluble periplasmic protein LolA, and is then handed over to lipoprotein LolB, which is attached to the outer membrane by its lipid moiety. The affinity of LolB for lipoprotein is higher than that of LolA. The triacyl group of LolB possibly distorts the packing of the lipid bilayer, as has been found for cytochrome oxidase (Sun et al. 2018), and by this means facilitates entry of the triacyl group of lipoprotein into the inner leaflet of the outer membrane. LolE of *E. coli* binds lipoprotein and interacts with LolA, whereas LolC does neither (Okuda and Tokuda 2009).

Crystal structures of LolA and LolB have been determined and are similar despite their different amino acid sequences (Takeda et al. 2003). Both structures contain hydrophobic cavities that serve as binding sites for the acyl chains of lipoproteins. The sequence similarity between the periplasmic regions of LolC and LolE and between LolA and LolB suggest the presence of cavities in LolC and LolE similar to those in LolA and LolB. It is likely that lipoprotein is transferred from LolCDE to LolA to LolB through their hydrophobic cavities (Okuda and Tokuda 2009). Only triacylated lipoprotein is transferred across the periplasm by the Lol proteins. Deletion of any of the genes encoding the Lol proteins are not tolerated by the cells, presumably because aberrant mislocated lipoproteins accumulate.

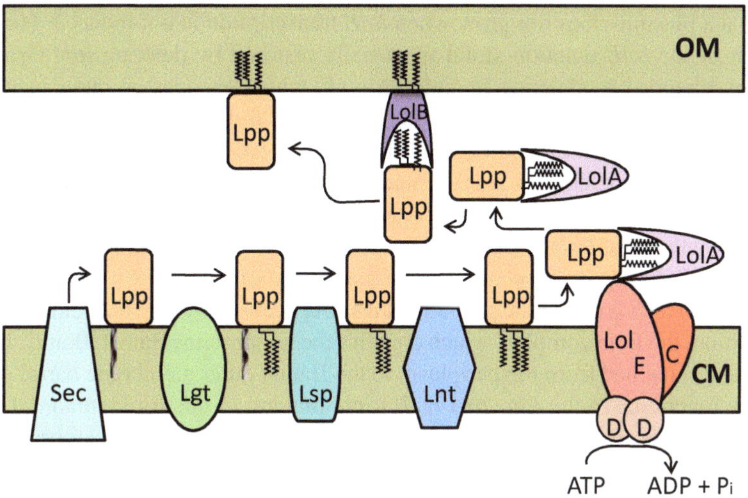

Fig. 3.3 Model of the transfer of Lpp from the cytoplasmic membrane across the periplasm to the outer membrane. See text for details

Modified Lipoproteins that Remain in the Cytoplasmic Membrane

When Ser2 of mature Lpp is replaced with Asp, Lpp is retained in the cytoplasmic membrane; other amino acids direct sorting into the outer membrane (Yamaguchi et al. 1988). The Asp-dependent retention of Lpp in the cytoplasmic membrane is influenced by the residue at position 3, and certain combinations of residues at positions 2 and 3 act as cytoplasmic membrane signals. Asp2 interferes with recognition of *N*-acyl *S*-diacylglyceryl Cys by LolCDE and therefore prevents Lpp release from the cytoplasmic membrane. Lpp with Asp at position2 also does not form a complex with LolA. In *P.* a*eruginosa*, residues at positions 3 and 4 but also Asp2 determine membrane localization (Narita and Tokuda 2007; Lewenza et al. 2008). Structural requirements around the Cys residue for the release of Lpp by LolA from the cytoplasmic membrane have been studied in detail with point mutants (summarized in Tokuda et al. 2007). The lipoprotein signal peptidase Lsp specifically recognizes the enantio (*R*) form of the diacylglycerol (Kitamura and Wolan 2018).

Alternative Lol Pathway

The essentiality of the Lol pathway in transferring lipoproteins from the cytoplasmic membrane into the outer membrane was recently questioned (Grabowicz and Silhavy 2017b). A chromosomal *lolB* deletion mutant carrying an arabinose-inducible *lolB* gene on a plasmid does not grow when *lolB* transcription is not induced. However, growth of the *lolB* deletion strain is partially restored by deleting *lpp*. Growth is further enhanced by deleting in addition *rcsF*, which encodes an outer membrane lipoprotein that controls the Rcs stress response. *lolB* deletion leads to a toxic activation of the Rcs stress response caused by accumulation of mislocalized Lpp. The *lpp rcsF* double deletion mutant is markedly more viable during LolB depletion because the Rcs regulon is not activated. Furthermore, suppressor mutations that allow growth of the *lolB* mutant are in *cpxA*. CpxA is part of the Cpx stress regulon. An *lpp rcsB* double deletion mutant that carries an activating *cpxA* mutation is still viable when *lolB* is deleted. CpxA alleviates stress caused by Lpp trafficking defects. In addition, the Bam complex, which contains the lipoproteins BamB, BamC, BamD, and BamE attached from the periplasm to the BamA outer membrane barrel protein is fully functional in the *lpp cpxA rcsB* triple mutant. In this triple mutant, LolB is not required for the delivery of lipoproteins from the cytoplasmic membrane to the outer membrane. Under conditions in which LolB is not essential, LolA is also not required. LolCDE are still essential and cannot be bypassed. These data suggest the presence of alternative LolAB-independent lipoprotein pathways with LolCDE as a constituent. Such pathways might exist in α- and ε-proteobacterial classes of the major Gram-negative phyla in which no LolB homolog is apparent.

Control of Lpp Biosynthesis

Since Lpp is by far the most abundant protein in *E. coli* (approximately 10^6 copies per cell), its synthesis must be controlled to avoid upon overexpression depletion of substrates and enzymes of RNA and protein synthesis required for other cell components and overloading of the processing and assembly machineries. Synthesis of the essential lipoproteins BamD of the Bam outer membrane protein assembly complex and LptE of LPS assembly directly competes with the lipid modification and Lol-dependent transfer across the periplasm of Lpp. Sigma E is the major regulator of outer membrane biogenesis. It controls transcription of 100 genes, including all of the machinery involved in transport and assembly of outer membrane proteins (Guo et al. 2014). Upon induction of sigma E synthesis, translation of the mRNA encoding Lpp is repressed specifically through the small sRNA MicL attached to the Hfq protein (Guo et al. 2014). Transient overproduction of Lpp decreases insertion of the BamD and LptE lipoproteins into the outer membrane and assembly of their respective machineries. This disrupts insertion of LPS and proteins into the outer membrane, and the accumulated LPS and outer membrane proteins trigger sigma E activation. Down-regulation of *lpp* mRNA translation increases synthesis of these and other outer membrane components to restore outer membrane homeostasis. Upon stress and during transition to stationary growth phase, sigma E upregulates MicL, which down-regulates *lpp* mRNA translation. Inhibition of *lpp* mRNA translation is more effective than inhibition of *lpp* transcription because the amount of *lpp* mRNA is high (10% of all cellular mRNA) (Guo et al. 2014) and the mRNA has an unusually long half-life (Hirashima and Inouye 1973; Nilsson et al. 1984).

E. coli controls extracytoplasmic stress using several systems that respond to aberrant structural and functional alterations in the cell envelope (Grabowicz and Silhavy 2017a, b). The RcsF lipoprotein functions as a sensor that recognizes outer membrane defects and transmits the information into the periplasm to activate the stress response. The N-terminal lipid of RcsF in the outer membrane is exposed to the cell surface, and the carboxyl group is in the periplasm, where it interacts with the major outer membrane proteins OmpA, OmpC, and OmpF (Konovalova et al. 2014, 2016). Inhibition of the Lol pathway activates the Rcs phosphorelay system, which results in enhanced transcription of the *lolA* gene (Tao et al. 2012). Mislocalized lipoproteins in the cytoplasmic membrane caused by Lol defects or mutations in RcsF activate the RcsF stress response.

The Cpx regulon consists of a two-component system that responds to signaling that originates, among others, from the NlpE outer membrane lipoprotein. When lipoprotein trafficking is impaired, CpxA induces a protective response through NlpE that improves cell viability. CpxA is a histidine kinase that spans the cytoplasmic membrane and activates CpxR, which up-regulates a number of genes for periplasmic chaperons and proteases. Cpx is necessary to alleviate stress caused by lipoprotein trafficking defects.

Function of Lpp in E. coli

An *lpp* deletion mutant was fortuitously isolated during construction of an F′ plasmid (Hirota et al. 1977). The mutant grows and divides normally and remains susceptible to phages. It is hypersensitive to EDTA, cationic dyes, and detergents. Enzymes leak out of the periplasm, and outer membrane vesicles (blebs) are formed and released into the medium, which suggests that the integrity of the outer membrane is disturbed. This phenomenon was originally observed when Lpp was released from murein by treating cell envelopes with trypsin, whereby the outer membrane detaches from the cytoplasmic membrane (Braun and Rehn 1969). A phenotype similar to that of the *lpp* deletion mutant is observed for mutants lacking the C-terminal lysine or that are devoid of the three L,D-transpeptidases that link Lpp to murein (Asmar et al. 2017).

Lpp plays an important structural role in that it fixes the outer membrane to murein, which also affects interaction of the cytoplasmic membrane with murein and the outer membrane. Recently, it was found that murein-bound Lpp determines the distance between the outer membrane and the cytoplasmic membrane (Asmar et al. 2017). Insertion of 14 or 21 amino acids into Lpp increases the distance between the outer membrane and cytoplasmic membrane by 3 nm and 4 nm, respectively, as revealed by electron cryomicroscopy. Response to stress elicited by β-lactams depends on the correct distance between the outer membrane and the cytoplasmic membrane, which is determined by the correct (wild-type) length of Lpp. The correct distance is required for the stress-responsive RcsF protein in the outer membrane to contact the stress-transferring IgaA protein in the cytoplasmic membrane and to confer antibiotic resistance. If signaling is interrupted, the antibiotics A22 and mecillinam cause cell lysis.

The length of the flagellum rod (25 nm) is determined by the width of the space between the outer membrane and the cytoplasmic membrane, as shown in studies of *Salmonellla typhimurium*. The length of the flagellum rod of non-motile flagellin *flgG* mutants is higher (up to 60 nm). Selection for motile suppressor mutants of *S. typhimurium* unexpectedly identified deletions in *lppA*. In five *flgG* mutants tested, deletion of *lppA* resulted in a significant increase in swim diameter on soft agar plates. An *flgG* wild-type *lppA* deletion mutant is impaired in swimming through liquid (individual cell behavior) and in swarming motility (moving of a swarm of flagellated bacteria across a hydrated surface). Further investigations revealed that the length of LppA determines the width of the periplasm; the width decreased in LppA mutants lacking 21 amino acids and increased in LppA mutants having an additional 21 amino acids (Cohen et al. 2017). Purified flagellum rods of the LppA mutant lacking 21 amino acids were shorter than those of the LppA wild-type. All LppA length mutants (+21, +42, +63, and −21) could not swarm, but three of the mutants (+21, +42, and +63) could swim. The growing flagellum senses when it hits the outer membrane and stops growing. Without LppA, the space between murein and the outer membrane is less constricted and allows FlgG distal rods to be positioned perpendicular to the outer membrane as they grow longer than 25 nm, which permits formation of the periplasmic flagellar ring-lipopolysaccharide complex.

An *lpp* deletion mutant of uropathogenic *E. coli* CFT073 is impaired in its ability to produce capsular polysaccharides. In this *lpp* mutant, the amount of KpsD—the putative outer membrane translocon for type 2 capsular polysaccharide export—is reduced. Shortage of KpsD decreases polysaccharide capsule formation and as a consequence serum resistance in vitro and complement-mediated clearance in vivo (Diao et al. 2017). Binding of Lpp to murein is important as indicated by the similar phenotypes of both an *lpp* mutant lacking the C-terminal lysine and the *lpp* deletion mutant. Detailed studies of these specific systems revealed the important function of Lpp, which was less obvious in earlier studies of *lpp* deletion mutants that multiplied normally but exhibited outer membrane defects. The findings also indicate the importance of a correct cell envelope architecture, specifically a precise distance between the outer membrane and the cytoplasmic membrane. This distance is determined by the abundant Lpp which acts as a spacer.

Rojas et al. (2018) recently showed that the outer membrane of Gram-negative bacteria is an essential load-bearing element. It, along with murein, contributes to the mechanical properties of the cell envelope. An *lpp* deletion mutant strongly contracts upon plasmolysis and lysis; the authors concluded that the coupling between murein and the outer membrane was diminished.

Lipoprotein Prediction Algorithms

Most algorithms for predicting Lpp in prokaryotes were published before 2010. T2003hey were based mainly on the increasing number of data sets obtained with experimentally proven *E. coli* lipoproteins (Juncker et al. ; Gonnet et al. 2004; Babu et al. 2006). An extended list with 90 experimentally proven lipoproteins of *E. coli* was published (Tokuda et al. 2007). Unfortunately, the proven negative results were not recorded. The website DOLOP (Babu et al. 2006) is dedicated to lipoproteins; its prediction program attempts to integrate the advances of its predecessors. However, the site has not been modified since 2005. A recent unpublished analysis (September 7, 2018) based on the prediction algorithm of Juncker et al. (2003) examined 126,697,114 gene sequences, of which 3,472,106 genes (2.7%) encode proteins with a lipobox (Jens Baßler, Max Planck Institute for Developmental Biology).

To date, most predicted lipoproteins have not been experimentally proven and a certain number of false positives is obtained. Determination of a lipid anchor is not routine in proteome analyses of cell envelopes, and special analyses are required to identify the lipid anchor and the fatty acid composition (Haake and Zückert 2017; Asanuma et al. 2011). An old but still efficient method is to use radioactively labeled precursors, e.g., glycerol or palmitate. The lipoproteins are often enriched by Triton X-114 partitioning before protease digestion and mass spectroscopy analysis (Asanuma et al. 2011). Even if some internal peptides of a predicted lipoprotein are found in a proteome analysis, this only indicates synthesis of the protein and not its lipidation, which must still be shown. Therefore, even for the well-studied *E. coli*, proof for the lipid modification of every predicted lipoprotein has not been obtained. In addition, the expression of 96 true and predicted lipoproteins was tested under three different growth conditions, aerobic and anaerobic ± nitrate. The mRNA of 21 lipoproteins was not expressed under any condition (Brokx et al. 2004). This indicates that many of these unknown predicted lipoproteins may have no discernible function for the strain.

In Gram-positive bacteria, many substrate-binding proteins are anchored by a lipid to the outer leaflet of the cytoplasmic membrane. This type of modification has also been predicted and found for the binding protein MetQ of a methionine uptake system in *E. coli* (Tokuda et al. 2007). However, when *metQ* was recently cloned and expressed in *E. coli*, the crystallized protein isolated for X-ray structure determination did not contain the lipid anchor; the signal peptide was split off after Gly at position 16 (the Cys at position 23 was not lipidated) (Nguyen et al. 2015b). Unfortunately, the authors did not discuss this discrepancy to the lipidation observed by Tokuda et al. (2007). It is possible that only part of the protein is lipidated under certain conditions. Such a case has been described by Hayashi et al. (1988) for FtsI, the penicillin-binding protein 3 (PBP3). They found that less than 15% of plasmid-encoded overproduced FtsI was lipidated. Lipidation of FtsI in wild-type *E. coli* K-12 was below the detection limit (Hayashi et al. 1988) and further studies excluded that the observed lipidation is important in vivo. In addition, no N-terminal processing was observed.

PRED-LIPO is a special Lpp prediction program for Gram-positive bacteria (Bagos et al. 2008) that was trained with a set of lipoproteins from Gram-positive bacteria. Interestingly, *Oenococcus oeni* is listed there as a bacterium without a lipoprotein. A blast search for Lgt and LspA homologues in the family Leuconostocaceae (which includes species of *Leuconostoc*, *Oenococcus*, and *Weissella*) revealed that these enzymes seem to be missing in this group of bacteria. This supports the prediction that these bacteria do not have lipoproteins in contrast to other lactic acid bacteria. How this evolved is an interesting but unsolved question. Another interesting observance is that mitochondria and plastids have lost this type of protein lipidation during evolution, which makes this structure a specific bacterial recognition signal.

Functions of *E. coli* Lipoproteins Other Than Murein Lipoprotein Lpp

The activity of lipoproteins is as multifaceted as the multitude of tasks the cell envelope has to fulfill for the cell. In nearly every task of the envelope of *E. coli*, a lipoprotein is involved (Table 3.1). Of the 116 lipoproteins listed, 38 have an unknown function or evidence of their function is inconclusive. In the following, we will discuss some of the lipoproteins with interesting functions.

In Gram-negative bacteria, lipoproteins are located either in the outer leaflet of the cytoplasmic membrane or in the inner leaflet of the outer membrane and exposed to the cell surface. The sorting processes are catalyzed by specific chaperones, e.g., LolA and LolB. LolB is an essential protein. Being itself a lipoprotein, it helps to insert other lipoproteins into the outer membrane (Szewczyk and Collet 2016).

The Bam proteins are another essential protein complex necessary for the assembly of the outer membrane. The β-barrel protein BamA and the associated proteins BamB, BamC, BamD, and BamE integrate β-barrel proteins into the outer membrane. All Bam proteins except BamA are anchored by a lipid on the periplasmic site of the outer membrane (Konovalova et al. 2017; Botos et al. 2017).

The 14 lipoproteins AmiC, Lpp, MepS, MltABCDE, NlpCDI,YgeR, YraP, LppA participate in the handling of murein. Remarkable is NlpD, which is located in the outer membrane at the divisome and activates the amidase AmiC in the process of cell division.

LPS is exported to the outer leaflet of the outer membrane, while phospholipids are mainly found in the inner leaflet of the outer membrane. This asymmetry of lipid distribution in the outer membrane is critical for resistance to detergents in the presence of EDTA (Malinverni and Silhavy 2009). LPS is integrated into the outer membrane by LptD and the lipoprotein LptE. The asymmetry of the phospholipid distribution is accomplished by the lipoprotein MlaA in collaboration with outer membrane β-barrel proteins OmpC or OmpF, MlaC in the periplasm, and the ABC transporter MlaFEDB in the inner membrane. Additional factors, such as the Tol-Pal system (Pal is an outer membrane lipoprotein) (Shrivastava et al. 2017) and

Table 3.1 *E. coli* lipoproteins (compiled by Tokuda et al. 2007) and their functions (extracted from the EcoCyc database; (Keseler et al. 2017). Additional information was obtained from the NCBI database gene

Lipoprotein	Function, location, or association
AcrA	Multidrug resistance; partner of TolC
AcrE	65% identical to AcrA; multidrug resistance
AmiD	Outer membrane *N*-acetylmuramic acid-L-alanine amidase that cleaves the bond between muramic acid and L-alanine within murein, muropeptides, and anhydro-muropeptides
ApbE	Periplasmic flavin transferase
BamB	Part of the BAM complex; assembly and insertion of β-barrel proteins into the outer membrane
BamC	Part of the BAM complex
BamD	Part of the BAM complex
Blc	Outer membrane lipocalin structure; might play a role in lipid transport
BorD	Part of the defective prophage DLP12
BsmA	Gene is upregulated during biofilm growth (?)
ChiQ	(?)
CsgG	Outer membrane; believed to form the secretion channel for curli subunits
CusC	Outer membrane; part of the CusCFBA copper/silver efflux system
CyoA	Subunit II of the cytochrome *bo* terminal oxidase complex
DcrB	Possibly involved in DNA injection by phage C1 and C6
EcnA	Entericidin A; functions as an antidote to the bacteriolytic entericidin B lipoprotein
FlgH	Basic subunit of the L ring in the outer membrane of the flagellar basal body
GfcB	Capsule polysaccharide synthesis
GfcE	Polysaccharide export
HslJ	Novobiocin resistance in a *cysB* background
LoiP	Outer membrane metalloprotease, preferentially cleaves between Phe/Phe residues
LolB	Outer membrane; Lpp required for LolA-dependent localization of Lpps
LpoB	Outer membrane; Lpp forms a complex with penicillin binding protein 1B (PBP1B)
Lpp	Murein lipoprotein
LptE	Outer membrane; LptDE complex, assembly of LPS at the cell surface
MdtE	Membrane fusion protein of the MdtEFTolC multidrug transporter
MepS	Outer membrane murein DD-endopeptidase with specificity for D-Ala-m-Dap cross links; associated with murein incorporation during cell growth
MlaA	Outer membrane; implicated in phospholipid asymmetry in the outer membrane
MltA	Outer-membrane-bound lytic murein transglycosylase
MltB	Outer-membrane-bound lytic murein transglycosylase

(continued)

Table 3.1 (continued)

Lipoprotein	Function, location, or association
MltC	Outer-membrane- bound lytic murein transglycosylase
MltD	Outer-membrane- bound lytic murein transglycosylase
MltE	Outer-membrane-bound lytic murein transglycosylase
NlpA	Inner membrane (?)
NlpC	Similarity to LytE of *Bacillus subtilis*
NlpD	Outer membrane; activates peptidoglycan hydrolase AmiC involved in septal splitting
NlpE	Outer membrane; necessary for activation of the CpxAR two-component signal transduction system in response to interactions such as adhesion
NlpI	Outer membrane; interacts with the periplasmic protease Prc and the murein endopeptidase MepS
OsmB	Inducible by both osmotic stress and growth phase signals; assumed to be in the outer membrane
OsmE	Osmotically inducible
Pal	Outer membrane component of the Tol-Pal system; integrity of the outer membrane
PgaB	Outer membrane metal-dependent de-*N*-acetylase and transport of poly-β-1,6-*N*-acetyl-D-glucosamine
QseG	Activates the QseEF two-component signal transduction system
RcsF	Surface exposed; signal transduction from the cell surface to RscC
RlpA	(?)
Slp	Possibly takes part in acid resistance
SlyB	Outer membrane; regulated by PhoPQ
Wza	Outer membrane; polysaccharide export
YaeF	Presumed peptidase
YafT	(?)
YafY	Cytoplasmic membrane; encoded by prophage CP4-6
YaiW	Outer membrane, surface exposed (?)
YajG	(?)
YajI	(?)
YbaY	(?)
YbfP	(?)
YbhC	Outer membrane; member of the carbohydrate esterase family 8
YbjP	(?)
YcaL	Periplasmic metalloprotease
yceB	Contains a tubular lipid binding domain (TULIP)
YcjN	Predicted periplasmic binding protein of a putative ATP-binding cassette transporter (YcjNOPV)
YdhY	Predicted ferredoxin-like protein. Based on sequence similarity, the *ydhYVWXUT* operon might encode an oxidoreductase system

(continued)

Table 3.1 (continued)

Lipoprotein	Function, location, or association
YeaY	Similarity to Slp
YecR	Regulated by FlhDC
YedR	(?)
YegR	(?)
YehR	(?)
YfeY	(?)
YfgH	Outer membrane integrity
YfiB	Exopolysacharide biosynthesis (?)
YfiL	(?)
YfjS	Cytoplasmic membrane; encoded on prophage CP4-57
YgdI	(?)
YgdR	(?)
YgeR	Putative peptidase containing a LysM domain
YghG	Outer membrane pilotin of a type II secretion system
YhfL	(?)
YiaD	Member of the OmpA-OmpF porin family
YidQ	(?)
YiiG	(?)
YjbF	Outer membrane; exopolysaccharide export
YjeI	(?)
YnbE	(?)
YnfC	(?)
YoaF	(?)
YqhH	(?)
YraK	Encoded in a putative chaperone-usher fimbrial operon
YraP	Outer membrane; localized at the septal ring division site. Might be involved in the NlpD-mediated activation of AmiC
Additional true or predicted Lpps not mentioned by Tokuda et al. (2007)	
BamE	Part of the BAM complex
Ccp (YhjA)	Predicted; cytochrome C quinol peroxidase
EcnB	Entericidin B, bacteriolytic toxin
LpoA	Outer membrane, stimulates the peptide crosslinking activity PBP1A
MliC	Predicted, lysozyme inhibitor
PqiC (YmbA)	OM-LPP facing the periplasm, interacts with PqiB and bridges inner membrane and outer membrane
QseG(YfhG)	Predicted, part of the QseEF two component system
RzoD	Predicted, part of the lysis cassette of the lambdoid prophage DLP12
RzoQ	Predicted, encoded on the Qin prophage

(continued)

Table 3.1 (continued)

Lipoprotein	Function, location, or association
RzoR	Predicted, encoded on the Rac prophage
YceK	Predicted, expression in biofilm repressed
YcfL	Predicted, (?)
YdbJ	Predicted, (?)
YdcL	Predicted, (?)
YddW	Predicted, (?)
YdeK	Predicted, pseudogene in *E. coli* K-12
YdjY	Predicted, 4 iron sulfur cluster ferredoxin
YedD	Predicted, (?)
YfbK	Predicted, (?)
YfhM	Predicted, protease inhibitor alpha2-macroglobulin
YfiM	Predicted, (?)
YghJ	Predicted, fragment of a TSS2 secretion system
YhdV	Predicted, (?)
YidX	Predicted, (?)
YifL	Predicted, (?)
YjaH	Predicted, (?)
YjbH	Predicted, part of a polysaccharide secretion system
YpdI	Predicted, colonic acid synthesis
YsaB	Predicted, (?)

(?) unknown function or inconclusive evidence

the phospholipase PldA (May and Silhavy 2018) also play a role. The molecular mechanisms are mostly not understood.

RcsF is one of the lipoproteins whose lipid is inserted in the outer leaflet of the outer membrane; the transmembrane part is threaded through the pore of outer membrane β-barrel proteins with the C-terminus in the periplasm. It serves as a sensor for envelope perturbations and interacts with the periplasmic domain of RcsC, a histidine kinase, and the Rcs system regulator YrfF in the inner membrane. Capsular polysaccharide synthesis, biofilm formation, and other cell surface properties depend on the Rcs regulatory system (Sato et al. 2017).

YcaL is a predicted lipoprotein with protease activity against misfolded outer membrane proteins. It possibly "cleans" the Bam complex of unsuitable substrates at a certain stage, thereby helping to avoid clogging of the Bam pathway (Soltes et al. 2017).

The Iss lipoproteins residing in the outer membrane confer serum resistance to *E. coli* and are encoded by ColV/B plasmids. Related proteins, called Bor, are encoded by lambdoid phages, e.g., on the defective DLP12 prophage in *E. coli* K-12 (Johnson et al. 2008).

The TraT lipoproteins encoded on the F plasmid and related plasmids are surface-exposed proteins responsible for surface exclusion, which prevents futile matings between bacteria with similar or identical plasmids. In addition, they might also contribute to serum resistance (Arutyunov and Frost 2013).

The group A colicins are released by concomitantly synthesized lysis lipoproteins. Some mature slowly, but all seem to be exported to the outer membrane by the Lol system. It has been shown that the release from the cell of a group B colicin lacking a specific lysis protein is stimulated by the concomitantly induced lysis protein of a lambdoid phage (Nedialkova et al. 2016). Possibly this type of colicin is dependent on the complementation by a phage lysis protein or by another colicin lysis protein.

Lipoproteins in Other Gram-Negative Bacteria

Of the great variety of lipoproteins (2.7% of all predicted proteins), only some examples can be presented in this review.

Borrelia Spp.

Lipoproteins of pathogenic bacteria have attracted special interest because they stimulate the immune system of vertebrates. For example, various species of *Borrelia*, the causative agents of Lyme disease, alter their surface proteins when they switch hosts from ticks to vertebrates. In the mammalian host, the lipoproteins OspA and OspB are repressed by BosR, a copper-dependent Fur family regulator. In this connection, it is interesting to note that ticks have copper-chelating antimicrobial peptides to control their bacterial inhabitants, which might generate surroundings with low copper availability for *Borrelia* spp. By contrast, the synthesis of the *Borrelia* spp. lipoproteins OspC and VlsE in the mammalian host are enhanced by pH, temperature, and short-chain fatty acids. In *Borrelia burgdorferi*, the lipoprotein VlsE (variable major protein-like sequence) becomes a dominating surface protein (Bankhead 2016) that is an important immune evasion factor of the spirochete. Through antigenic variation, the bacteria escape the immune defense of their host, which may lead to persistence and chronic infection. Multiple variants of VlsE are constantly produced, and it is assumed that VlsE shields dominant epitopes of other outer membrane proteins so that they cannot be recognized by protective antibodies (Batool et al. 2018).

In addition to the lipoproteins that have been studied intensely as vaccine candidates (see later), *B. burgdorferi* has a lipoproteome of at least 125 proteins; 86 are surface exposed, and only 8 seem to reside in the inner leaflet of the outer membrane and 31 in the cytoplasmic membrane (Dowdell et al. 2017).

B. burgdorferi contains several episomal prophages of the cp32 family. These plasmids contain different replication/segregation loci that allow stable maintenance of several episomes in one cell. The *erp* locus on these plasmids encodes surface-exposed lipoproteins with a conserved N-terminal peptide and very variable protein sequences attached (Brisson et al. 2013). The different *erp* loci also contribute to the variability in the appearance of *B. burgdorferi*.

Species of the Phylum Bacteroidetes

Bacteroidetes have many surface-exposed lipoproteins that are mainly engaged in the utilization of polysaccharides. One of the first systems studied in detail was the starch utilization system (Sus). Lipoproteins on the cell surface bind the polysaccharide, and a lipoprotein glycosidase sets oligosaccharides free. The oligosaccharides are transported by a TonB-dependent protein into the periplasm, where glycoside hydrolases split the oligosaccharides to disaccharides and monosaccharides. These are taken up by dedicated transport systems in the cytoplasmic membrane (Foley et al. 2016).

Genome sequences revealed that certain *Bacteroides* species have about 100 Sus-like polysaccharide utilization systems. An X-ray structure of a SusD-like substrate-binding lipoprotein in complex with the corresponding SusC-like TonB-dependent transporter has been solved. SusC is capped by the lipoprotein SusD, and the substrate is bound in the interface between the two proteins (Glenwright et al. 2017).

Results of a bioinformatic analysis led Lauber et al. (2016) to postulate a new lipoprotein export system that determines the cell-surface localization of lipoproteins in species of the phylum *Bacteroidetes*. The N-terminal consensus sequence CXK(DD or EE) was found in surface-exposed Sus-like lipoproteins of *Capnocytophaga canimorsus*. These amino acids allowed export of the periplasmic lipoprotein sialidase to the cell surface. Although the transporter-associated Sus-like proteins have a certain resemblance to neisserial outer membrane lipoproteins, such as TbpB and HupA, the lipoproteins seem to be exported by different means. In species of *Neisseria*, Slam proteins support this process (see below). Species of the phylum Bacteroidetes lack Slam-like proteins, and the system that recognizes the above-mentioned consensus sequence remains to be identified.

Neisseria Spp.

The first better-known examples of surface-exposed lipoproteins were found in *Neisseria* spp. in complex with TonB-dependent receptors, where they help in binding host iron-containing molecules (TbpB, transferring binding protein; LbpB, lactoferrin binding protein) and facilitate the uptake of iron. In a screening for *Neisseria meningitidis* mutants that do not expose the lipoprotein TbpB on the cell surface, a mutation was identified that also prevented the surface exposure of LbpB and fHbp (factor H binding protein). The encoded protein was named Slam (surface lipoprotein assembly modulator (Hooda et al. 2016).

Slam homologues were found in pathogenic species of, e.g., *Moraxella, Acinetobacter, Haemophilus, Vibrio*, and other proteobacteria but not in *E. coli*. Heterologous expression of the gene encoding Slam in *E. coli* results in exposure of TbpB, LbpB, and fHbp on the cell surface. Interestingly, Slam proteins have substrate specificity. For example, another Slam protein in *N. meningitidis* causes exposure of only HpuA (hemoglobin-haptoglobin binding protein) on the cell surface, but not of TbpB, LbpB, or fHbp (Hooda et al. 2016). The sequence identity between the two Slam proteins of *N. meningitidis* (Slam1 and Slam2) is only 26% (Hooda et al. 2017).

The function of Slam was challenged by (Fantappie et al. 2017), who found fHbp exposed on the surface of the outer membrane of *E. coli* independent of Slam. However, with Slam present, the amount of surface-exposed fHbp clearly increased, which indicates that Slam is at least a facilitator of surface exposure in *E. coli*. The authors proposed that certain uncharacterized structure determinants allow the Slam-independent surface exposure of these proteins.

Gram-Positive Bacteria and Archaea

Low-GC Gram-Positive Bacteria

Homologues of the apolipoprotein *N*-acyltransferase Lnt have not been found in low-GC Gram-positive bacteria. Therefore, it was generally assumed that these bacteria have lipoproteins with only two ester-bound fatty acids on the *S*-glycerolcysteine. However, Toll-like-receptor (TLR)-dependent stimulation of macrophages provided the first hint that *Staphylococcus aureus* might indeed have triacylated lipoproteins. Analysis of MntC (SitC), a binding protein of a manganese ABC transporter, revealed *N*-acylation of MntC in *S. aureus*. This was also shown for other *S. aureus* lipoproteins (Kurokawa et al. 2009). However, lipoproteins of *S. aureus* cells in the stationary phase grown under certain stress conditions contain only two ester-bound fatty acids and no *N*-acylation (Kurokawa et al. 2012b). Another important difference in lipidation is found between commensal *S. aureus* and non-commensal *Staphylococcus carnosus*. The lipoproteins of *S. carnosus* are *N*-acetylated and lead to a tenfold stronger immune response (Nguyen et al. 2017). It is assumed that the low immune response of the *S. aureus* lipoprotein is an evolutionary adaptation to the commensal lifestyle.

These results stimulated Kurokawa et al. (2012b) to analyze the lipid structure in other low-GC Gram-positive bacteria. As expected, they found lipoprotein with only two ester-bound fatty acids in *Listeria monocytogenes*, but to their surprise, three other lipoprotein modifications were detected in other bacteria. In *Enterococcus faecalis* and several other species (e.g., *Bacillus cereus, Lactobacillus bulgaricus, Streptococcus sanguinis*), a lysoform, i.e., an *N*-acylated lipoprotein with one fatty acid esterified to the glycerol residue, was detected. However, in several species of *Bacillus* (*B. subtilis, B. licheniformis, B. halodurans*), two fatty acids were bound to glycerol and the amino group was acetylated. The structure of some lipoproteins of *Mycoplasma fermentans* is unusual in that the signal peptidase cleaves at two residues before the *S*-diacyl-glyceryl-cysteine. However, in two other lipoproteins of *M. fermentans*, the *S*-diacyl-glycerylcysteine is the N-terminal amino acid (Kurokawa et al. 2012b). The various structures found are critically reviewed in detail in Nakayama et al. (2012).

The genesis of the unusual lyso-lipoprotein structures was explained 5 years later. In an attempt to clone the gene responsible for the unknown Lnt activity that leads to *N*-acylation in *E. faecalis*, Armbruster and Meredith (2018) devised an *lnt* complementation screening method in *E. coli*. The enzyme that they identified intramolecularly transacylated an ester-bound fatty acid to the α-amino group of the cysteine. The identified protein was called lipoprotein intramolecular transacylase (Lit) (Armbruster and Meredith 2017). Interestingly, homologues of this protein are found in species of *Bacillus* and *Lactobacillus*, but not in staphylococci. Therefore, the enzyme that acylates the amino group in *S. aureus* still remains to be identified.

High-GC Gram-Positive Bacteria

In contrast to many low-GC Gram-positive bacteria, species of the high-GC Gram-positive genera *Mycobacteria* and *Streptomyces* encode Lnt homologues and have lipoproteins with three fatty acids (Tschumi et al. 2009; Brulle et al. 2013). *Streptomyces coelicolor* encodes about 200 predicted lipoproteins; the majority are solute-binding proteins of transport systems, as generally found in Gram-positive bacteria. About a quarter of the 200 lipoproteins are not exported by the Sec system, but instead via the twin-arginine translocation (Tat) system. Another peculiarity is that two Lgt homologues are encoded, both of which are active. In addition, all species of *Streptomyces* studied contain two Lnt homologues. Both homologues in the potato pathogen *Streptomyces scabies* can be deleted without yielding a growth, differentiation, or virulence phenotype (Thompson et al. 2010).

Mycobacteria

Several lipoproteins of *Mycobacterium tuberculosis* have been studied as potential virulence factors and vaccine candidates (extensively reviewed and discussed in (Becker and Sander 2016). As in other pathogens, not all lipoproteins stimulate the immune response. Some even have immune-suppressing properties and help *M. tuberculosis* survive in the host. These latter lipoproteins are not suitable for a vaccine because they exacerbate the disease, as shown in challenged, previously vaccinated mice.

Of the 100 predicted lipoproteins, we will mention only a few well-studied cases. The LppX protein assists in localization of a group of complex lipids—the phthiocerol dimycocerosates—in the outer lipid-rich membrane of *M. tuberculosis* (Sulzenbacher et al. 2006). The crystal structure shows structural similarities to that of LolA and LolB, which also transport a lipidic substrate. Without LppX, phthiocerol dimycocerosates are not released from the cell surface into the Triton-X100-containing medium, which indicates that the lipid has not been properly inserted into the outer membrane. The correct positioning of phthiocerol dimycocerosates seems to be important, because this lipid helps in hiding pathogen-associated molecular patterns from the innate immune system of the host (Cambier et al. 2014).

The lipoprotein LprG similarly helps in positioning lipoarabinomannan on the cell surface, which inhibits phagosome-lysosome fusion, an important component in the survival strategy of *M. tuberculosis* (Shukla et al. 2014). LprG also has immune-suppressing activity.

The lipoprotein RpfB is necessary for resuscitation, which is an important pathogenicity factor of *M. tuberculosis*. The enzyme has a lysozyme fold and lytic transglycosylase activity. In a joint action of RpfB with an endopeptidase anhydromuramic acid containing muropeptides were set free from cell walls. The muropeptides stimulated resuscitation of dormant mycobacteria (Nikitushkin et al. 2015).

LpqH and LprN are lipoproteins with immune-suppressing functions. They contribute to the immune escape and late immune response of the host, which allows propagation of the infecting bacteria to high numbers (Becker and Sander 2016). Mass spectrometry of LpqH revealed that it contains, in addition to the N-terminal *N*-acyl-*S*-diacylglycerolcysteine, up to nine hexoses on the N-terminal peptide (Parra et al. 2017). Of the 41 proteins enriched from the growth medium with ConA-lectin, 34 were lipoproteins (Gonzalez-Zamorano et al. 2009). The glycosylation of lipoproteins influences envelope permeability and consequently antibiotic resistance in addition to the immunogenic properties (van Els et al. 2014).

Archaea

The first hint that archaea synthesize lipoproteins was obtained in 1994 by the analysis of halocyanin from *Natronobacterium pharaonis* of the phylum Euryarchaeota. The protein had an appropriate molecular mass and contained a lipobox and a blocked N-terminal end, which are good indicators of a lipid modification (Mattar et al. 1994).

Of the archaea, only members of the phylum Euryarchaeota seem to make lipoproteins (Storf et al. 2010). Additional experimental evidence comes mainly from the analysis of lipoproteins of *Haloferax volcanii* that are exported by the Tat system. The presence of lipoproteins was unexpected because Lgt, Lsp, and Lnt homologues have not been detected in archaeal genomes. However, an appropriate lipobox with cysteine was predicted for about 50% of the predicted Tat substrates and for a few Sec substrates in members of the class Haloarchaea (Storf et al. 2010). A cysteine-to-serine mutation in a maltose-binding protein or an iron-binding protein resulted in Tat-dependent secretion into the medium, which is good evidence for cysteine-dependent lipidation. In addition, globomycin inhibits both growth of the cells and maturation of the maltose-binding protein (Gimenez et al. 2007). However, an in-depth structural analysis of the lipid part is still lacking.

Immunological Properties of *E. coli* Lpp

Lpp as an Antigen

E. coli Lpp spontaneously coats erythrocytes, which are then lysed upon addition of complement (Bosch and Braun 1973). This method was used to determine the distribution of Lpp between the outer membrane and the cytoplasmic membrane. More than 10- to 16-fold Lpp was found in the outer membrane than in the cytoplasmic membrane. When entire cells are used as antigens, anti-Lpp titers are higher when the cells are of rough mutants lacking O-antigen and parts of the LPS core region, possibly because Lpp might be better exposed when portions of LPS are lacking. Antibodies raised against Lpp released from murein by lysozyme recognize the C-terminus with the attached murein subunits. Lpp released from murein with trypsin [trypsin-Lpp; trypsin cleaves the Lys(55)-Tyr(56) and Arg(57)-Lys(58) bonds] does not react with antiserum raised against Lpp released by lysozyme. Anti-Lpp raised against trypsin-Lpp only reacts with trypsin-Lpp. The antiserum specifically recognizes the C-terminal Lys. Removal of Lys with carboxypeptidase B inactivates the antigenicity. It is remarkable that antisera raised against lysozyme-Lpp and antisera raised against trypsin-Lpp contain no antibodies against the polypeptide chain. By contrast, antisera raised against cells react with both lysozyme-Lpp and trypsin-Lpp. In this case, antibody formation is elicited by the polypeptide integrated into the membrane. Release of the ester-linked fatty acids on glycerylcysteine does not affect antigenicity of isolated Lpp. LPS as a control does not interfere with the Lpp immunogenicity and antigenicity determinations (Braun et al. 1976a).

Lpp: A B-Lymphocyte Mitogen

Mitogens serve as tools to understand processes during proliferation and differentiation of B-cells into plasma cells. Mitogenic stimulation can be monitored by an increase in thymidine uptake, the appearance of plaque-forming cells assayed with densely coupled trinitrophenylated sheep red blood cells, and increase in synthesis and secretion of radiolabeled IgM compared to synthesis and secretion of other cellular proteins. These methods show that splenic lymphocytes of mice are activated by lipoprotein released from murein by lysozyme or trypsin. Removal of the ester-linked fatty acids by alkaline hydrolysis abolishes the mitogenic activity. Anti-immunoglobulin antibodies inhibit the mitogenic stimulation of B-cells by lipoprotein. Activation is T-cell independent and does not need serum factors or adherent cells. Small resting B-cells display the highest lipoprotein sensitivity (Melchers et al. 1975).

Lpp and Lpp-Derived Lipopeptides Act as Strong Adjuvants in the Innate Immune Response

The mitogenic activity of Lpp prompted further studies on the immunogenic activities of Lpp and the immunogenic domains of Lpp. Various lipopeptides were synthesized and tested for their B lymphocyte mitogenicity in an in vitro cell culture system (Bessler et al. 1985). The lipopeptides consisted of *S*-[2,3-bis-(palmitoleyloxy)propyl)-*N*-palmitoleylcysteine (Pam3C) with the following C-terminal extensions: cysteine methylester, cysteine-serine, cysteine-serine-serine, cysteine-serine-serine-asparagine, and cysteine-serine-serine-asparagine-alanine (see structure in Fig. 3.1a). To be fully active, the compounds had to carry at least one serine. The lipopeptide-induced stimulation of B lymphocytes is as strong as stimulation by the complete lipoprotein.

Using synthetic lipid, it has been shown that of the two possible diastereometric configurations (*R* and S) at position C2 of the propyl moiety, the *R* configuration is the most likely one in the native molecule (Wiesmüller et al. 1983). A later systematic study showed that lipopeptides with the *R* configuration activate B-cells more than the *S* forms. Induction of specific CD8(+) T-cells is significantly higher in mice injected with the *R* forms than with the *S* forms (Khan et al. 2009).

Tests of a range of peptides revealed that for full mitogenic activity and polyclonal stimulation, the hydrophilic dipeptide structure is necessary. Further investigations revealed that synthetic lipopeptides constitute potent macrophage activators. Co-administration of lipopeptides with haptens or low-molecular-weight antigens that are not immunogenic per se, or covalent coupling, result in highly immunogenic conjugates. Such conjugates might serve as synthetic vaccines and provide protection by enhancing the antibody-mediated immune response in vivo (Bessler et al. 1997). Many studies have shown that lipopeptides derived from Lpp promote immunological defense, inflammation, or sepsis. Without the use of adjuvant (self-adjuvanting activity) lipopeptides generate potent immune responses (summarized in Asmar and Collet 2018; Becker and Sander 2016; Moyle and Toth 2008; Nguyen and Götz 2016; Steinhagen et al. 2011; Zaman and Toth 2013).

Lipopeptides constitute potent immunoadjuvants in various species in parental immunizations. Lipopeptide adjuvants are comparable or even superior to Freund's adjuvant without showing side effects. Lipopeptide Pam3CSK4 also enhances serum antibody responses to various proteins when administered in combination with antigens via the nasal route in mice (Baier et al. 2000). Thus, they constitute strong adjuvants for mucosal immunizations.

Lpp and Lpp-Derived Lipopeptides Act as Ligands of Toll-like Receptors TLR2-TLR1 and TLR2-TLR6

Lpp and lipopeptides derived from Lpp are recognized by the innate immune system, which represents the first line of defense against invasive infectious pathogens. They bind to Toll-like receptor TLR2 together with TLR1 and TLR6 (Aliprantis et al. 1999). The degree of acylation at the lipid moiety discriminates the interaction of TLR2 with TLR1 and TLR-6; tri-acylated lipopeptide binds to the heterodimer TLR2/TLR1, and di-acylated lipopeptide interacts with TLR2/TLR6 (Kang et al. 2009). Signals are transduced from the lipopeptide-loaded TLR receptors to nuclear factor NF-kappaB by a cascade of phosphorylations, resulting in transcription and synthesis of proinflammatory cytokines and chemokines. Lipopeptides of various structures are important ligands for activation of the TLR receptors, which are essential for induction of the innate and adaptive immune responses.

TLRs are composed of three domains: an extracellular ligand-binding domain, a single transmembrane helix, and an intracellular signaling domain. The crystal structure of the extracellular domain of human TLR1-TLR2 heterodimer with bound tri-acylated lipopeptide Pam3CSK4 has been determined (Jin et al. 2007). The three lipid chains mediate formation of an M-shaped heterodimer. The two ester-linked lipid chains are inserted in a pocket of TLR2, while the amide-bound lipid chain is inserted in a hydrophobic channel in TLR1. Heterodimer formation of TLR6 with TLR2 is triggered by binding of diacyl-Pam2CSK4. The resulting structure is similar to that of the extracellular domain of TLR1-TLR2-Pam3CSK4 (Kang et al. 2009). The binding site of the amide-linked acyl chain is blocked by two phenylalanines, which explains the preferred binding of diacyl lipids versus triacyl lipids. The structures of Toll-like receptors with various ligands are discussed in Kang and Lee (2011).

Although low-GC Gram-positive bacteria usually contain diacylated lipoprotein, the abundant lipoprotein SidC of the Gram-positive *S. aureus* carries a triacylated lipid (Kurokawa et al. 2009, 2012a). Lipoprotein lipid structures of commensal and non-commensal staphylococci induce different immune responses (Nguyen et al. 2017). TheTLR2 response induced by commensal *S. aureus* and *Staphylococcus epidermidis* is almost ten times lower than that induced by non-commensal *S. carnosus*. The N-terminus of *S. aureus* lipoprotein and of *S. epidermidis* lipoprotein carries a long acyl chain (C17), whereas *S. carnosus* lipoprotein contains a short acetyl group. The long-chain *N*-acylated lipoprotein recognized by TLR-2/TLR-1 receptors decreases the innate and adaptive immune responses, which results in immune evasion, whereas the short chain *N*-acetylated lipoproteins recognized by TLR2/TLR6 receptor boosts IL8 production.

Low-GC Gram-positive bacteria, e.g., *Bacillus cereus*, *Enterococcus faecalis*, *Streptococcuc sanguinis*, and *Lactobacillus bulgaricus*, have a modified lipoprotein that carries the structure *N*-acyl-*S*-monoacyl-glycerylcysteine (lyso-form) (Kurokawa et al. 2012b). The purified lipoprotein induces proinflammatory cytokine production from mice peritoneal macrophages in a TLR2-dependent and TLR1-

independent manner. Lyso-lipoprotein from *B. cereus* induces TLR2-dependent and TLR1- and TLR6-independent tumor necrosis factor (TNF-α) and interleukin-6. By contrast, *E. faecalis* lyso-lipoprotein induces TLR2/TLR6-dependent and TLR1-independent cytokine secretion, as does *N*-acetyl-*S*-diacyl-glyceryl-cysteine lipoprotein from *B. subtilis*.

S. aureus lipoproteins activate Toll-like receptors when lipoproteins are released from the cytoplasmic membrane by phenol-soluble modulin (PSM) peptides (Hanzelmann et al. 2016). The different capacities of *S. aureus* clinical isolates in activating TLR2 depends on high-level production of PSM peptides. PSM peptides have a strong impact on the severity of sepsis, and TLR2 plays an important role in systemic *S. aureus* infections only for strains that produce PSM peptides.

TLR-Independent Lipopeptide Immunogenicity

Although the action of Lpp and other lipopeptides is usually mediated through interaction with TLR2 associated with either TLR6 or TLR1, lipopeptides can be immunogenic without TLR1 and TLR6 interaction. For example, diacylated lipopeptides, e.g., Pam2CSK4, induce TLR6-independent B lymphocyte proliferation and TNF-α secretion in macrophages, as determined with cells from TLR6-deficient mice (Buwitt-Beckmann et al. 2005). Diacylated Pam2CSK4 and triacylated Pam3 CGNNEDESNISFKEK stimulate murine spleen cells isolated from TLR1- and TLR6-deficient mice. Thus, distinct lipopeptides are recognized by TLR2 in a TLR1- and TLR6-independent manner (Buwitt-Beckmann et al. 2006). Even a TLR-independent enhancement of infection by lipopeptides has been observed with various virus infection models. The respiratory syncytial virus (RSV) elicits respiratory tract infections and bronchiolitis, which are associated with bacterial co-infections (Nguyen et al. 2010). Therefore, the authors examined whether bacterial Toll-like receptor agonists influence RSV infections in human primary cells or cell lines. Pam3-CSK4 enhanced RSV infection in primary epithelial, myeloid, and lymphoid cells. The lipopeptide not only enhanced RSV but also HIV-1, measles virus, and human metapneumovirus infection independent of TLR signaling.

Cytotoxic T lymphocytes (CTLs) play an important role in the immune response to viral infections. CTLs recognize peptides derived from viral proteins together with the major histocompatibility complex class I molecules at the surface of infected cells. They usually require in vivo priming with infectious virus. Synthetic viral peptides covalently linked to tripalmitoyl-*S*-glycerylcysteinyl-Ser-Ser (Pam3CSS) efficiently prime influenza-virus-specific cytotoxic T lymphocytes in vivo to a level as high as infectious virus. Lipopeptide priming is MHC class I-restricted (Deres et al. 1989).

Suppression of Immune Response by Lipopeptides

Lipopeptides not only enhance the immune response but can also suppress immune reactions. For example, skin exposed to synthetic diacylated lipopeptides suppresses the immune response by interleukin-6-dependent induction of granulocytic and monocytic myeloid-derived suppressor cells. Triacylated lipopeptides are inactive in this assay (Skabytska et al. 2014). As discussed before, the long-chain *N*-acylated lipoprotein recognized by TLR2/TLR1 receptors of *S. aureus* decreases innate and adaptive immune responses, resulting in immune evasion. Of the many surface-exposed lipoproteins of *Mycobacterium tuberculosis* (Becker and Sander 2016), LpqH promotes immune evasion by inhibiting MHC class II expression and interference with cytokine production in macrophages. LprG is immunosuppressive in that it inhibits TLR2-mediated MHC II antigen presentation.

Virulence Enhancement by Lpp

Salmonella enterica serovar Typhimurium encodes in tandem two lipoproteins of the murein lipoprotein type (Lpp). All mice orally infected with an *lppA lppB* double mutant survive. The mutant triggers robust innate and adaptive immune responses (Erova et al. 2016). Mice immunized with the mutant are completely protected against a lethal oral challenge dose of wild-type *S. typhimurium*. Since the mutant is among the most attenuated and immunogenic strains, it is considered an excellent candidate for a vaccine against *S. typhimurium* infection. Also single *lpp* mutants of *S. enterica* are highly attenuated in in vitro and in vivo models of pathogenesis (Fadl et al. 2005). The virulence potential of a double *lpp* knockout mutant was examined in cell cultures of intestinal epithelial cells, a macrophage cell line, and mice infected intraperitoneally (Sha et al. 2004). The mutant is defective in invading and inducing cytotoxic effects. Induction of proinflammatory cytokines tumor necrosis factor-α and interleukin 8 is significantly reduced. Intracellular survival and replication of *Salmonella* sp. is not affected. Similar results have been obtained with *Yersinia pestis*. Deletion of *lpp* reduces the virulence of *Y. pestis* lacking plasmid pPCP1 in a mouse model of pneumonic plague. pPCP1 encodes virulence-associated traits, such as the plasminogen-activating protease. Mice infected with an *lpp* deletion mutant survive longer than mice infected with an *lpp* wild-type strain (Agar et al. 2009). In addition, the levels of most cytokines and chemokines strongly decrease. Transfer of sera and splenocytes from mice immunized with the *lpp* mutant to naive animals provides protection against a lethal dose of the *Y. pestis* parent strain (Liu et al. 2010). Deletion of *lpp* provides a strong attenuation in a mouse model of pneumonic plague.

Species of *Staphylococcus* encode between 55 and 70 lipoproteins arranged in tandem. Deletion of the entire cluster results in a decreased TLR2-dependent stimulation of proinflammatory cytokines in human monocytes, macrophages, and keratinocytes. The lipoprotein cluster contributes to invasion of *S. aureus* into human keratinocytes and mouse skin; transfer of the gene cluster into *S. carnosus*, which is devoid of the cluster, renders *S. carnosus* invasive (Nguyen et al. 2015a).

All lipoproteins can be mutated at once by inactivation of the *lgt* gene, which attaches the diacylglyceryl group to the invariant cysteine (Stoll et al. 2005). The mutants show a decreased induction of the proinflammatory cytokines IL6 and IL8 and monocytic chemoattractant protein in human monocytic, epithelial, and endothelium cells compared to that of the wild-type. They are also affected in induction of early cytokines via TLR2 in murine peritoneal macrophages and are severely affected in pathogenicity. However, one has to bear in mind that these deletion mutants are sick even when the *lgt* mutants contain the lipid-free proteins still attached to the cytoplasmic membrane. Additional changes in the composition of the cell envelope can be predicted, and these might contribute to the immunogenic properties of the mutants. The same conclusion applies to *lspA* mutants of *Mycobacterium tuberculosis*, which do not release the signal peptides from the predicted 99 prolipoproteins. The mutants show a decreased intracellular multiplication in mouse macrophages and a decreased growth in lungs and spleens (Sander et al. 2004).

Mycoplasma synthesize a number of lipoproteins. The lipopeptide MALP-2 of *Mycoplasma fermentans* displays an extremely high immunologic activity. It stimulates macrophages at concentrations as low as 0.2 ng/ml, whereas lipoproteins of other bacterial sources require concentrations of 1 μg/ml for a half-maximal response. It consists of an N-terminal diacylglycerol cysteine linked to the amino acid sequence GNNDESNISFKEK (Mühlradt et al. 1997). Synthetic lipopeptide displays the same activity as lipopeptide isolated from *M. fermentans*. It stimulates murine macrophages to release TNF, IL-1, IL6, prostaglandins, and nitric oxide from IFN-primed murine macrophages. In addition, MALP-2 down-regulates class II MHC expression on macrophages, which results in impaired antigen presentation to T-cells.

The examples given above demonstrate that natural and synthetic lipopeptides are potent macrophage and B cell activators. Synthetic conjugates consisting of lipopeptides with T helper cell and CTL epitopes from viral and bacterial proteins are efficient low-molecular-weight vaccines with strong adjuvant activity.

Lpp-Derived Peptides as Adjuvants in Vaccines

A vaccine that induces a long-lasting high protection against foot-and-mouse disease and serotype-specific virus-neutralizing antibodies in guinea pigs after a single administration contains the lipopeptide Pam3CSS N-terminally bound to VPI and a synthetic lipopeptide of 20 residues derived from the FMDV protein, which serves as amphiphilic α-helical T-cell epitope (Wiesmüller et al. 1989). A vaccine containing Pam3C linked to the outer surface protein A (OspA) of *Borrelia burgdorferi*, the causative agent of Lyme disease, has been clinically tested in over 20,000 volunteers (Steinhagen et al. 2011). The induction of protective immunity correlated with the development of antibodies against an epitope on the C-terminus of OspA.

Three doses of vaccine induced protective immunity in over 70% of subjects. It was licensed by the FDA in 1998, but withdrawn three years later by the manufacturer because the media reported possible autoimmune side effects. However, neither the FDA nor the CDC found a connection between the vaccine and the development of autoimmunity (Nigrovic and Thompson 2007).

Pam3C has been used as adjuvant in a vaccine against malaria. The vaccine contains multiple B cell epitopes and a universal T cell epitope derived from the *Plasmodium falciparum* circumsporozoite protein CSP. After three immunizations, all ten volunteers developed peptide specific IgG1, IgG3, and IgG4 antibodies, whereas a vaccine lacking P3C induced only IgG1 and IgG3 isotypes (Nardin et al. 2001).

Mutants of *Mycobacterium tuberculosis* deficient in the surface lipoprotein LpqS are attenuated and are highly protective against a *M. tuberculosis* challenge in guinea pigs (Sakthi et al. 2016).

A totally synthetic vaccine activates TLR2 of dendritic cells (Jackson et al. 2004). It consists of a single helper T-cell epitope, a target epitope that is either recognized by B cells or CD8+ T cells, and Pam2Cys inserted between the two epitopes to form a branched configuration. The CD8+ epitopes are from influenza virus and *Listeria monocytogenes*. Each of the vaccines induces either CD8+ cell or antibody-mediated immune responses in mice. The lipidated vaccines but not the non-lipidated vaccines mediate protection against viral or bacterial infections and confer prophylactic and therapeutic anticancer activity. Additional studies have shown that a conserved minimal peptide from the M protein (J14) of group A streptococci linked to Pam2C and a universal T cell epitope administered intranasally to mice protects them from lethal respiratory challenge with group A streptococci and induces J14-specific mucosal immunoglobulins (Batzloff et al. 2006). These results demonstrate anti-disease and transmission-inhibiting activities, which could form the basis for development of an anti-streptococcal vaccine.

Lipopeptides as adjuvants were investigated in vaccines against influenza virus, group A streptococci, hepatitis C virus, leishmaniasis, *Listeria monocytogenes*, and other disease-causing microorganisms (summarized in Moyle and Toth 2008; Zaman and Toth 2013). Although strong lipopeptide-dependent immune reactions were elicited, no vaccine is in use. The induction of a lasting CD8+ cytotoxic lymphocyte response was presumably too low to elicit a satisfactory immunity.

Weak Adaptive Immune Response to Lipoproteins

The adaptive immune response to lipoproteins, in contrast to the native immune response, has been studied in healthy *S. aureus* human carriers and non-carriers (Vu et al. 2016). Specific B-cell and T-cell responses were determined. In contrast to expectations derived from the strong innate immune response described above, the titers of antibodies (IgG) binding to lipoproteins were low. Proliferation assays and cytokine profiling data revealed only subtle responses of T-cells. It remains to be determined why in this study the adaptive immune system reacts only weakly to lipoproteins (Kretschmer et al. 2016). Access of the lipoproteins in cells might not be the cause of low antigenicity, as most isolated recombinant *S. aureus* lipoproteins are also weak antigens. Impaired processing of lipoproteins by antigen-presenting cells might be a reason and opens new avenues of immune research. The results question the use of lipoprotein and its derived peptides as antigens for production of vaccines against *S. aureus*.

Acknowledgements The authors work was supported by institutional funds of the Max Planck Society (V.B.) and the German Science Foundation (V.B., K.H.). We thank Jens Baßler for his lipoprotein sequence analysis. V.B. thanks Andrei Lupas for his generous hospitality in his Department of Protein Evolution in the Max Planck Institute for Developmental Biology. K.H. thanks Karl Forchhammer and Joachim E. Schultz for generous hospitality.

References

Agar SL, Sha J, Baze WB, Erova TE, Foltz SM, Suarez G, Wang S, Chopra AK (2009) Deletion of Braun lipoprotein gene (lpp) and curing of plasmid pPCP1 dramatically alter the virulence of Yersinia pestis CO92 in a mouse model of pneumonic plague. Microbiology 155(Pt 10):3247–3259. https://doi.org/10.1099/mic.0.029124-0

Aliprantis AO, Yang RB, Mark MR, Suggett S, Devaux B, Radolf JD, Klimpel GR, Godowski P, Zychlinsky A (1999) Cell activation and apoptosis by bacterial lipoproteins through toll-like receptor-2. Science 285(5428):736–739

Armbruster KM, Meredith TC (2017) Identification of the lyso-form N-acyl intramolecular transferase in low-GC firmicutes. J Bacteriol 199(11). https://doi.org/10.1128/jb.00099-17

Armbruster KM, Meredith TC (2018) Enrichment of bacterial lipoproteins and preparation of N-terminal lipopeptides for structural determination by mass spectrometry. J Vis Exp (135). https://doi.org/10.3791/56842

Arutyunov D, Frost LS (2013) F conjugation: back to the beginning. Plasmid 70(1):18–32. https://doi.org/10.1016/j.plasmid.2013.03.010

Asanuma M, Kurokawa K, Ichikawa R, Ryu KH, Chae JH, Dohmae N, Lee BL, Nakayama H (2011) Structural evidence of alpha-aminoacylated lipoproteins of Staphylococcus aureus. FEBS J 278(5):716–728. https://doi.org/10.1111/j.1742-4658.2010.07990.x

Asmar AT, Collet JF (2018) Lpp, the Braun lipoprotein, turns 50-major achievements and remaining issues. FEMS Microbiol Lett 365(18). https://doi.org/10.1093/femsle/fny199

Asmar AT, Ferreira JL, Cohen EJ, Cho SH, Beeby M, Hughes KT, Collet JF (2017) Communication across the bacterial cell envelope depends on the size of the periplasm. PLoS Biol 15(12):e2004303. https://doi.org/10.1371/journal.pbio.2004303

Babu MM, Priya ML, Selvan AT, Madera M, Gough J, Aravind L, Sankaran K (2006) A database of bacterial lipoproteins (DOLOP) with functional assignments to predicted lipoproteins. J Bacteriol 188(8):2761–2773. https://doi.org/10.1128/JB.188.8.2761-2773.2006

Bagos PG, Tsirigos KD, Liakopoulos TD, Hamodrakas SJ (2008) Prediction of lipoprotein signal peptides in gram-positive bacteria with a Hidden Markov model. J Proteome Res 7(12):5082–5093. https://doi.org/10.1021/pr800162c

Baier W, Masihi N, Huber M, Hoffmann P, Bessler WG (2000) Lipopeptides as immunoadjuvants and immunostimulants in mucosal immunization. Immunobiol 201(3-4):391–405

Bankhead T (2016) Role of the VlsE lipoprotein in immune avoidance by the Lyme disease spirochete Borrelia burgdorferi. For Immunopathol Dis Therap 7(3–4):191–204. https://doi.org/10.1615/ForumImmunDisTher.2017019625

Batool M, Caoili SEC, Dangott LJ, Gerasimov E, Ionov Y, Piontkivska H, Zelikovsky A, Waghela SD, Rogovskyy AS (2018) Identification of surface epitopes associated with protection against highly immune-evasive VlsE-expressing Lyme disease spirochetes. Infect Immun 86(8). https://doi.org/10.1128/iai.00182-18

Batzloff MR, Hartas J, Zeng W, Jackson DC, Good MF (2006) Intranasal vaccination with a lipopeptide containing a conformationally constrained conserved minimal peptide, a universal T cell epitope, and a self-adjuvanting lipid protects mice from group A streptococcus challenge and reduces throat colonization. J Infect Dis 194(3):325–330. https://doi.org/10.1086/505146

Becker K, Sander P (2016) Mycobacterium tuberculosis lipoproteins in virulence and immunity—fighting with a double-edged sword. FEBS Lett 590(21):3800–3819. https://doi.org/10.1002/1873-3468.12273

Bessler WG, Cox M, Lex A, Suhr B, Wiesmüller KH, Jung G (1985) Synthetic lipopeptide analogs of bacterial lipoprotein are potent polyclonal activators for murine B lymphocytes. J Immunol 135(3):1900–1905

Bessler WG, Heinevetter L, Wiesmüller KH, Jung G, Baier W, Huber M, Lorenz AR, Esche UV, Mittenbühler K, Hoffmann P (1997) Bacterial cell wall components as immunomodulators–I. Lipopeptides as adjuvants for parenteral and oral immunization. Int J Immunopharmacol 19(9–10):547–550

Bosch V, Braun V (1973) Distribution of murein-lipoprotein between the cytoplasmic and outer membrane of Escherichia coli. FEBS Lett 34(2):307–310

Botos I, Noinaj N, Buchanan SK (2017) Insertion of proteins and lipopolysaccharide into the bacterial outer membrane. Philos Trans R Soc Lond B Biol Sci 372(1726). https://doi.org/10.1098/rstb.2016.0224

Braun V (1975) Covalent lipoprotein from the outer membrane of Escherichia coli. Biochim Biophys Acta 415(3):335–377

Braun V, Bosch V (1972) Repetitive sequences in the murein-lipoprotein of the cell wall of Escherichia coli. Proc Natl Acad Sci USA 69(4):970–974

Braun V, Hantke K (1974) Biochemistry of bacterial cell envelopes. Annu Rev Biochem 43:89–121. https://doi.org/10.1146/annurev.bi.43.070174.000513

Braun V, Rehn K (1969) Chemical characterization, spatial distribution and function of a lipoprotein (murein-lipoprotein) of the E. coli cell wall. The specific effect of trypsin on the membrane structure. Eur J Biochem 10(3):426–438

Braun V, Wu HC (1994) Lipoproteins, structure, function, biosynthesis and model for protein export. In: Ghuysen J-M, Hagenbeck R (eds) Bacterial cell wall. Elsevier, Amsterdam, pp 319–341

Braun V, Bosch V, Klumpp ER, Neff I, Mayer H, Schlecht S (1976a) Antigenic determinants of murein lipoprotein and its exposure at the surface of Enterobacteriaceae. Eur J Biochem 62(3):555–566

Braun V, Rotering H, Ohms JP, Hagenmaier H (1976b) Conformational studies on murein-lipoprotein from the outer membrane of Escherichia coli. Eur J Biochem 70(2):601–610

Brisson D, Zhou W, Jutras BL, Casjens S, Stevenson B (2013) Distribution of cp32 prophages among Lyme disease-causing spirochetes and natural diversity of their lipoprotein-encoding erp loci. Appl Environ Microbiol 79(13):4115–4128. https://doi.org/10.1128/AEM.00817-13

Brokx SJ, Ellison M, Locke T, Bottorff D, Frost L, Weiner JH (2004) Genome-wide analysis of lipoprotein expression in Escherichia coli MG1655. J Bacteriol 186(10):3254–3258

Brulle JK, Tschumi A, Sander P (2013) Lipoproteins of slow-growing Mycobacteria carry three fatty acids and are N-acylated by apolipoprotein N-acyltransferase BCG_2070c. BMC Microbiol 13:223. https://doi.org/10.1186/1471-2180-13-223

Buwitt-Beckmann U, Heine H, Wiesmüller KH, Jung G, Brock R, Akira S, Ulmer AJ (2005) Toll-like receptor 6-independent signaling by diacylated lipopeptides. Eur J Immunol 35(1):282–289. https://doi.org/10.1002/eji.200424955

Buwitt-Beckmann U, Heine H, Wiesmüller KH, Jung G, Brock R, Akira S, Ulmer AJ (2006) TLR1- and TLR6-independent recognition of bacterial lipopeptides. J Biol Chem 281(14):9049–9057. https://doi.org/10.1074/jbc.M512525200

Cambier CJ, Takaki KK, Larson RP, Hernandez RE, Tobin DM, Urdahl KB, Cosma CL, Ramakrishnan L (2014) Mycobacteria manipulate macrophage recruitment through coordinated use of membrane lipids. Nature 505(7482):218–222. https://doi.org/10.1038/nature12799
Choi DS, Yamada H, Mizuno T, Mizushima S (1987) Molecular assembly of the lipoprotein trimer on the peptidoglycan layer of Escherichia coli. J Biochem 102(5):975–983
Cohen EJ, Ferreira JL, Ladinsky MS, Beeby M, Hughes KT (2017) Nanoscale-length control of the flagellar driveshaft requires hitting the tethered outer membrane. Science 356(6334):197–200. https://doi.org/10.1126/science.aam6512
Cowles CE, Li Y, Semmelhack MF, Cristea IM, Silhavy TJ (2011) The free and bound forms of Lpp occupy distinct subcellular locations in Escherichia coli. Mol Microbiol 79(5):1168–1181. https://doi.org/10.1111/j.1365-2958.2011.07539.x
Deres K, Schild H, Wiesmüller KH, Jung G, Rammensee HG (1989) In vivo priming of virus-specific cytotoxic T lymphocytes with synthetic lipopeptide vaccine. Nature 342(6249):561–564. https://doi.org/10.1038/342561a0
Diao J, Bouwman C, Yan D, Kang J, Katakam AK, Liu P, Pantua H, Abbas AR, Nickerson NN, Austin C, Reichelt M, Sandoval W, Xu M, Whitfield C, Kapadia SB (2017) Peptidoglycan association of murein lipoprotein is required for KpsD-dependent group 2 capsular polysaccharide expression and serum resistance in a Uropathogenic Escherichia coli isolate. MBio 8(3). https://doi.org/10.1128/mbio.00603-17
Dowdell AS, Murphy MD, Azodi C, Swanson SK, Florens L, Chen S, Zückert WR (2017) Comprehensive spatial analysis of the Borrelia burgdorferi lipoproteome reveals a compartmentalization bias toward the bacterial surface. J Bacteriol 199(6). https://doi.org/10.1128/jb.00658-16
Erova TE, Kirtley ML, Fitts EC, Ponnusamy D, Baze WB, Andersson JA, Cong Y, Tiner BL, Sha J, Chopra AK (2016) Protective immunity elicited by oral immunization of Mice with Salmonella enterica Serovar Typhimurium Braun Lipoprotein (Lpp) and Acetyltransferase (MsbB) mutants. Front Cell Infect Microbiol 6:148. https://doi.org/10.3389/fcimb.2016.00148
Fadl AA, Sha J, Klimpel GR, Olano JP, Niesel DW, Chopra AK (2005) Murein lipoprotein is a critical outer membrane component involved in Salmonella enterica serovar typhimurium systemic infection. Infect Immun 73(2):1081–1096. https://doi.org/10.1128/IAI.73.2.1081-1096.2005
Fantappie L, Irene C, De Santis M, Armini A, Gagliardi A, Tomasi M, Parri M, Cafardi V, Bonomi S, Ganfini L, Zerbini F, Zanella I, Carnemolla C, Bini L, Grandi A, Grandi G (2017) Some gram-negative lipoproteins keep their surface topology when transplanted from one species to another and deliver foreign polypeptides to the bacterial surface. Mol Cell Proteomics 16(7):1348–1364. https://doi.org/10.1074/mcp.M116.065094
Foley MH, Cockburn DW, Koropatkin NM (2016) The Sus operon: a model system for starch uptake by the human gut Bacteroidetes. Cell Mol Life Sci 73(14):2603–2617. https://doi.org/10.1007/s00018-016-2242-x
Gimenez MI, Dilks K, Pohlschroder M (2007) Haloferax volcanii twin-arginine translocation substates include secreted soluble, C-terminally anchored and lipoproteins. Mol Microbiol 66(6):1597–1606. https://doi.org/10.1111/j.1365-2958.2007.06034.x
Glauner B, Holtje JV, Schwarz U (1988) The composition of the murein of Escherichia coli. J Biol Chem 263(21):10088–10095
Glenwright AJ, Pothula KR, Bhamidimarri SP, Chorev DS, Basle A, Firbank SJ, Zheng H, Robinson CV, Winterhalter M, Kleinekathofer U, Bolam DN, van den Berg B (2017) Structural basis for nutrient acquisition by dominant members of the human gut microbiota. Nature 541(7637):407–411. https://doi.org/10.1038/nature20828
Gonnet P, Rudd KE, Lisacek F (2004) Fine-tuning the prediction of sequences cleaved by signal peptidase II: a curated set of proven and predicted lipoproteins of Escherichia coli K-12. Proteomics 4(6):1597–1613. https://doi.org/10.1002/pmic.200300749
Gonzalez-Zamorano M, Mendoza-Hernandez G, Xolalpa W, Parada C, Vallecillo AJ, Bigi F, Espitia C (2009) Mycobacterium tuberculosis glycoproteomics based on ConA-lectin affinity capture of mannosylated proteins. J Proteome Res 8(2):721–733. https://doi.org/10.1021/pr800756a

Grabowicz M, Silhavy TJ (2017a) Envelope stress responses: an interconnected safety net. Trends Biochem Sci 42(3):232–242. https://doi.org/10.1016/j.tibs.2016.10.002
Grabowicz M, Silhavy TJ (2017b) Redefining the essential trafficking pathway for outer membrane lipoproteins. Proc Natl Acad Sci USA 114(18):4769–4774. https://doi.org/10.1073/pnas.1702248114
Guo MS, Updegrove TB, Gogol EB, Shabalina SA, Gross CA, Storz G (2014) MicL, a new sigmaE-dependent sRNA, combats envelope stress by repressing synthesis of Lpp, the major outer membrane lipoprotein. Genes Dev 28(14):1620–1634. https://doi.org/10.1101/gad.243485.114
Haake DA, Zückert WR (2017) Spirochetal lipoproteins in pathogenesis and immunity. Curr Top Microbiol Immunol. https://doi.org/10.1007/82_2017_78
Hantke K, Braun V (1973) Covalent binding of lipid to protein. Diglyceride and amide-linked fatty acid at the N-terminal end of the murein-lipoprotein of the Escherichia coli outer membrane. Eur J Biochem 34(2):284–296
Hanzelmann D, Joo HS, Franz-Wachtel M, Hertlein T, Stevanovic S, Macek B, Wolz C, Gotz F, Otto M, Kretschmer D, Peschel A (2016) Toll-like receptor 2 activation depends on lipopeptide shedding by bacterial surfactants. Nat Commun 7:12304. https://doi.org/10.1038/ncomms12304
Hayashi S, Hara H, Suzuki H, Hirota Y (1988) Lipid modification of Escherichia coli penicillin-binding protein 3. J Bacteriol 170(11):5392–5395
Hirashima A, Inouye M (1973) Specific biosynthesis of an envelope protein of Escherichia coli. Nature 242(5397):405–407
Hirashima A, Wang S, Inouye M (1974) Cell-free synthesis of a specific lipoprotein of the Escherichia coli outer membrane directed by purified messenger RNA. Proc Natl Acad Sci USA 71(10):4149–4153
Hirota Y, Suzuki H, Nishimura Y, Yasuda S (1977) On the process of cellular division in Escherichia coli: a mutant of E. coli lacking a murein-lipoprotein. Proc Natl Acad Sci USA 74(4):1417–1420
Hooda Y, Moraes TF (2018) Translocation of lipoproteins to the surface of gram negative bacteria. Curr Opin Struct Biol 51:73–79. https://doi.org/10.1016/j.sbi.2018.03.006
Hooda Y, Lai CC, Judd A, Buckwalter CM, Shin HE, Gray-Owen SD, Moraes TF (2016) Slam is an outer membrane protein that is required for the surface display of lipidated virulence factors in Neisseria. Nat Microbiol 1:16009. https://doi.org/10.1038/nmicrobiol.2016.9
Hooda Y, Shin HE, Bateman TJ, Moraes TF (2017) Neisserial surface lipoproteins: structure, function and biogenesis. Pathog Dis 75(2). https://doi.org/10.1093/femspd/ftx010
Innis MA, Tokunaga M, Williams ME, Loranger JM, Chang SY, Chang S, Wu HC (1984) Nucleotide sequence of the Escherichia coli prolipoprotein signal peptidase (lsp) gene. Proc Natl Acad Sci USA 81(12):3708–3712
Inouye M, Shaw J, Shen C (1972) The assembly of a structural lipoprotein in the envelope of Escherichia coli. J Biol Chem 247(24):8154–8159
Jackson DC, Lau YF, Le T, Suhrbier A, Deliyannis G, Cheers C, Smith C, Zeng W, Brown LE (2004) A totally synthetic vaccine of generic structure that targets Toll-like receptor 2 on dendritic cells and promotes antibody or cytotoxic T cell responses. Proc Natl Acad Sci USA 101(43):15440–15445. https://doi.org/10.1073/pnas.0406740101
Jin MS, Kim SE, Heo JY, Lee ME, Kim HM, Paik SG, Lee H, Lee JO (2007) Crystal structure of the TLR1-TLR2 heterodimer induced by binding of a tri-acylated lipopeptide. Cell 130(6):1071–1082. https://doi.org/10.1016/j.cell.2007.09.008
Johnson TJ, Wannemuehler YM, Nolan LK (2008) Evolution of the iss gene in Escherichia coli. Appl Environ Microbiol 74(8):2360–2369. https://doi.org/10.1128/AEM.02634-07
Juncker AS, Willenbrock H, Von Heijne G, Brunak S, Nielsen H, Krogh A (2003) Prediction of lipoprotein signal peptides in gram-negative bacteria. Protein Sci 12(8):1652–1662. https://doi.org/10.1110/ps.0303703
Kang JY, Lee JO (2011) Structural biology of the Toll-like receptor family. Annu Rev Biochem 80:917–941. https://doi.org/10.1146/annurev-biochem-052909-141507

Kang JY, Nan X, Jin MS, Youn SJ, Ryu YH, Mah S, Han SH, Lee H, Paik SG, Lee JO (2009) Recognition of lipopeptide patterns by Toll-like receptor 2-Toll-like receptor 6 heterodimer. Immunity 31(6):873–884. https://doi.org/10.1016/j.immuni.2009.09.018

Keseler IM, Mackie A, Santos-Zavaleta A, Billington R, Bonavides-Martinez C, Caspi R, Fulcher C, Gama-Castro S, Kothari A, Krummenacker M, Latendresse M, Muniz-Rascado L, Ong Q, Paley S, Peralta-Gil M, Subhraveti P, Velazquez-Ramirez DA, Weaver D, Collado-Vides J, Paulsen I, Karp PD (2017) The EcoCyc database: reflecting new knowledge about Escherichia coli K-12. Nucleic Acids Res 45(D1):D543–D550. https://doi.org/10.1093/nar/gkw1003

Khan S, Weterings JJ, Britten CM, de Jong AR, Graafland D, Melief CJ, van der Burg SH, van der Marel G, Overkleeft HS, Filippov DV, Ossendorp F (2009) Chirality of TLR-2 ligand Pam3CysSK4 in fully synthetic peptide conjugates critically influences the induction of specific CD8+ T-cells. Mol Immunol 46(6):1084–1091. https://doi.org/10.1016/j.molimm.2008.10.006

Kitamura S, Wolan DW (2018) Probing substrate recognition of bacterial lipoprotein signal peptidase using FRET reporters. FEBS Lett 592(13):2289–2296. https://doi.org/10.1002/1873-3468.13155

Konovalova A, Perlman DH, Cowles CE, Silhavy TJ (2014) Transmembrane domain of surface-exposed outer membrane lipoprotein RcsF is threaded through the lumen of beta-barrel proteins. Proc Natl Acad Sci USA 111(41):E4350–4358. https://doi.org/10.1073/pnas.1417138111

Konovalova A, Mitchell AM, Silhavy TJ (2016) A lipoprotein/beta-barrel complex monitors lipopolysaccharide integrity transducing information across the outer membrane. Elife 5. https://doi.org/10.7554/elife.15276

Konovalova A, Kahne DE, Silhavy TJ (2017) Outer membrane biogenesis. Annu Rev Microbiol 71:539–556. https://doi.org/10.1146/annurev-micro-090816-093754

Kretschmer D, Hanzelmann D, Peschel A (2016) Lipoprotein immunoproteomics question the potential of Staphylococcus aureus TLR2 agonists as vaccine antigens. Proteomics 16(20):2603–2604. https://doi.org/10.1002/pmic.201600351

Kurokawa K, Lee H, Roh KB, Asanuma M, Kim YS, Nakayama H, Shiratsuchi A, Choi Y, Takeuchi O, Kang HJ, Dohmae N, Nakanishi Y, Akira S, Sekimizu K, Lee BL (2009) The triacylated atp binding cluster transporter substrate-binding lipoprotein of Staphylococcus aureus functions as a native ligand for toll-like receptor 2. J Biol Chem 284(13):8406–8411. https://doi.org/10.1074/jbc.M809618200

Kurokawa K, Kim MS, Ichikawa R, Ryu KH, Dohmae N, Nakayama H, Lee BL (2012a) Environment-mediated accumulation of diacyl lipoproteins over their triacyl counterparts in Staphylococcus aureus. J Bacteriol 194(13):3299–3306. https://doi.org/10.1128/JB.00314-12

Kurokawa K, Ryu KH, Ichikawa R, Masuda A, Kim MS, Lee H, Chae JH, Shimizu T, Saitoh T, Kuwano K, Akira S, Dohmae N, Nakayama H, Lee BL (2012b) Novel bacterial lipoprotein structures conserved in low-GC content gram-positive bacteria are recognized by Toll-like receptor 2. J Biol Chem 287(16):13170–13181. https://doi.org/10.1074/jbc.M111.292235

Lauber F, Cornelis GR, Renzi F (2016) Identification of a new lipoprotein export signal in gram-negative bacteria. MBio 7(5). https://doi.org/10.1128/mbio.01232-16

Lewenza S, Mhlanga MM, Pugsley AP (2008) Novel inner membrane retention signals in Pseudomonas aeruginosa lipoproteins. J Bacteriol 190(18):6119–6125. https://doi.org/10.1128/JB.00603-08

Liu T, Agar SL, Sha J, Chopra AK (2010) Deletion of Braun lipoprotein gene (lpp) attenuates Yersinia pestis KIM/D27 strain: role of Lpp in modulating host immune response, NF-kappaB activation and cell death. Microb Pathog 48(1):42–52. https://doi.org/10.1016/j.micpath.2009.09.002

Lupas AN, Bassler J, Dunin-Horkawicz S (2017) The structure and topology of alpha-helical coiled coils. Subcell Biochem 82:95–129. https://doi.org/10.1007/978-3-319-49674-0_4

Magnet S, Bellais S, Dubost L, Fourgeaud M, Mainardi JL, Petit-Frere S, Marie A, Mengin-Lecreulx D, Arthur M, Gutmann L (2007) Identification of the L,D-transpeptidases responsible for attachment of the Braun lipoprotein to Escherichia coli peptidoglycan. J Bacteriol 189(10):3927–3931. https://doi.org/10.1128/JB.00084-07

Malinverni JC, Silhavy TJ (2009) An ABC transport system that maintains lipid asymmetry in the gram-negative outer membrane. Proc Natl Acad Sci USA 106(19):8009–8014. https://doi.org/10.1073/pnas.0903229106

Mao G, Zhao Y, Kang X, Li Z, Zhang Y, Wang X, Sun F, Sankaran K, Zhang XC (2016) Crystal structure of E. coli lipoprotein diacylglyceryl transferase. Nat Commun 7:10198. https://doi.org/10.1038/ncomms10198

Mattar S, Scharf B, Kent SB, Rodewald K, Oesterhelt D, Engelhard M (1994) The primary structure of halocyanin, an archaeal blue copper protein, predicts a lipid anchor for membrane fixation. J Biol Chem 269(21):14939–14945

May KL, Silhavy TJ (2018) The Escherichia coli phospholipase PldA regulates outer membrane homeostasis via lipid signaling. MBio 9(2). https://doi.org/10.1128/mbio.00379-18

Melchers F, Braun V, Galanos C (1975) The lipoprotein of the outer membrane of Escherichia coli: a B-lymphocyte mitogen. J Exp Med 142(2):473–482

Moyle PM, Toth I (2008) Self-adjuvanting lipopeptide vaccines. Curr Med Chem 15(5):506–516

Mühlradt PF, Kiess M, Meyer H, Süssmuth R, Jung G (1997) Isolation, structure elucidation, and synthesis of a macrophage stimulatory lipopeptide from Mycoplasma fermentans acting at picomolar concentration. J Exp Med 185(11):1951–1958

Nakayama H, Kurokawa K, Lee BL (2012) Lipoproteins in bacteria: structures and biosynthetic pathways. FEBS J 279(23):4247–4268. https://doi.org/10.1111/febs.12041

Nardin EH, Calvo-Calle JM, Oliveira GA, Nussenzweig RS, Schneider M, Tiercy JM, Loutan L, Hochstrasser D, Rose K (2001) A totally synthetic polyoxime malaria vaccine containing Plasmodium falciparum B cell and universal T cell epitopes elicits immune responses in volunteers of diverse HLA types. J Immunol 166(1):481–489

Narita S, Tokuda H (2007) Amino acids at positions 3 and 4 determine the membrane specificity of Pseudomonas aeruginosa lipoproteins. J Biol Chem 282(18):13372–13378. https://doi.org/10.1074/jbc.M611839200

Narita SI, Tokuda H (2017) Bacterial lipoproteins; biogenesis, sorting and quality control. Biochim Biophys Acta Mol Cell Biol Lipids 1862(11):1414–1423. https://doi.org/10.1016/j.bbalip.2016.11.009

Nedialkova LP, Sidstedt M, Koeppel MB, Spriewald S, Ring D, Gerlach RG, Bossi L, Stecher B (2016) Temperate phages promote colicin-dependent fitness of Salmonella enterica serovar Typhimurium. Environ Microbiol 18(5):1591–1603. https://doi.org/10.1111/1462-2920.13077

Nguyen MT, Götz F (2016) Lipoproteins of gram-positive bacteria: key players in the immune response and virulence. Microbiol Mol Biol Rev 80(3):891–903. https://doi.org/10.1128/MMBR.00028-16

Nguyen DT, de Witte L, Ludlow M, Yuksel S, Wiesmüller KH, Geijtenbeek TB, Osterhaus AD, de Swart RL (2010) The synthetic bacterial lipopeptide Pam3CSK4 modulates respiratory syncytial virus infection independent of TLR activation. PLoS Pathog 6(8):e1001049. https://doi.org/10.1371/journal.ppat.1001049

Nguyen MT, Kraft B, Yu W, Demircioglu DD, Hertlein T, Burian M, Schmaler M, Boller K, Bekeredjian-Ding I, Ohlsen K, Schittek B, Götz F (2015a) The nuSa alpha specific lipoprotein like cluster (lpl) of S. aureus USA300 contributes to immune stimulation and invasion in human cells. PLoS Pathog 11(6):e1004984. https://doi.org/10.1371/journal.ppat.1004984

Nguyen PT, Li QW, Kadaba NS, Lai JY, Yang JG, Rees DC (2015b) The contribution of methionine to the stability of the Escherichia coli MetNIQ ABC transporter-substrate binding protein complex. Biol Chem 396(9–10):1127–1134. https://doi.org/10.1515/hsz-2015-0131

Nguyen MT, Uebele J, Kumari N, Nakayama H, Peter L, Ticha O, Woischnig AK, Schmaler M, Khanna N, Dohmae N, Lee BL, Bekeredjian-Ding I, Götz F (2017) Lipid moieties on lipoproteins of commensal and non-commensal staphylococci induce differential immune responses. Nat Commun 8(1):2246. https://doi.org/10.1038/s41467-017-02234-4

Nigrovic LE, Thompson KM (2007) The Lyme vaccine: a cautionary tale. Epidemiol Infect 135(1):1–8. https://doi.org/10.1017/S0950268806007096

Nikitushkin VD, Demina GR, Shleeva MO, Guryanova SV, Ruggiero A, Berisio R, Kaprelyants AS (2015) A product of RpfB and RipA joint enzymatic action promotes the resuscitation of dormant mycobacteria. FEBS J 282(13):2500–2511. https://doi.org/10.1111/febs.13292
Nilsson G, Belasco JG, Cohen SN, von Gabain A (1984) Growth-rate dependent regulation of mRNA stability in Escherichia coli. Nature 312(5989):75–77
Noland CL, Kattke MD, Diao J, Gloor SL, Pantua H, Reichelt M, Katakam AK, Yan D, Kang J, Zilberleyb I, Xu M, Kapadia SB, Murray JM (2017) Structural insights into lipoprotein N-acylation by Escherichia coli apolipoprotein N-acyltransferase. Proc Natl Acad Sci USA 114(30):E6044–E6053. https://doi.org/10.1073/pnas.1707813114
Okuda S, Tokuda H (2009) Model of mouth-to-mouth transfer of bacterial lipoproteins through inner membrane LolC, periplasmic LolA, and outer membrane LolB. Proc Natl Acad Sci USA 106(14):5877–5882. https://doi.org/10.1073/pnas.0900896106
Okuda S, Tokuda H (2011) Lipoprotein sorting in bacteria. Annu Rev Microbiol 65:239–259. https://doi.org/10.1146/annurev-micro-090110-102859
Parra J, Marcoux J, Poncin I, Canaan S, Herrmann JL, Nigou J, Burlet-Schiltz O, Riviere M (2017) Scrutiny of Mycobacterium tuberculosis 19 kDa antigen proteoforms provides new insights in the lipoglycoprotein biogenesis paradigm. Sci Rep 7:43682. https://doi.org/10.1038/srep43682
Pazos M, Peters K, Vollmer W (2017) Robust peptidoglycan growth by dynamic and variable multi-protein complexes. Curr Opin Microbiol 36:55–61. https://doi.org/10.1016/j.mib.2017.01.006
Rojas ER, Billings G, Odermatt PD, Auer GK, Zhu L, Miguel A, Chang F, Weibel DB, Theriot JA, Huang KC (2018) The outer membrane is an essential load-bearing element in gram-negative bacteria. Nature 559(7715):617–621. https://doi.org/10.1038/s41586-018-0344-3
Sakthi S, Palaniyandi K, Gupta UD, Gupta P, Narayanan S (2016) Lipoprotein LpqS deficient M. tuberculosis mutant is attenuated for virulence in vivo and shows protective efficacy better than BCG in guinea pigs. Vaccine 34(6):735–743. https://doi.org/10.1016/j.vaccine.2015.12.059
Sander P, Rezwan M, Walker B, Rampini SK, Kroppenstedt RM, Ehlers S, Keller C, Keeble JR, Hagemeier M, Colston MJ, Springer B, Bottger EC (2004) Lipoprotein processing is required for virulence of Mycobacterium tuberculosis. Mol Microbiol 52(6):1543–1552. https://doi.org/10.1111/j.1365-2958.2004.04041.x
Sato T, Takano A, Hori N, Izawa T, Eda T, Sato K, Umekawa M, Miyagawa H, Matsumoto K, Muramatsu-Fujishiro A, Matsumoto K, Matsuoka S, Hara H (2017) Role of the inner-membrane histidine kinase RcsC and outer-membrane lipoprotein RcsF in the activation of the Rcs phosphorelay signal transduction system in Escherichia coli. Microbiology 163(7):1071–1080. https://doi.org/10.1099/mic.0.000483
Sha J, Fadl AA, Klimpel GR, Niesel DW, Popov VL, Chopra AK (2004) The two murein lipoproteins of Salmonella enterica serovar Typhimurium contribute to the virulence of the organism. Infect Immun 72(7):3987–4003. https://doi.org/10.1128/IAI.72.7.3987-4003.2004
Shrivastava R, Jiang X, Chng SS (2017) Outer membrane lipid homeostasis via retrograde phospholipid transport in Escherichia coli. Mol Microbiol 106(3):395–408. https://doi.org/10.1111/mmi.13772
Shu W, Liu J, Ji H, Lu M (2000) Core structure of the outer membrane lipoprotein from Escherichia coli at 1.9 A resolution. J Mol Biol 299(4):1101–1112. https://doi.org/10.1006/jmbi.2000.3776
Shukla S, Richardson ET, Athman JJ, Shi L, Wearsch PA, McDonald D, Banaei N, Boom WH, Jackson M, Harding CV (2014) Mycobacterium tuberculosis lipoprotein LprG binds lipoarabinomannan and determines its cell envelope localization to control phagolysosomal fusion. PLoS Pathog 10(10):e1004471. https://doi.org/10.1371/journal.ppat.1004471
Skabytska Y, Wölbing F, Günther C, Köberle M, Kaesler S, Chen KM, Guenova E, Demircioglu D, Kempf WE, Volz T, Rammensee HG, Schaller M, Röcken M, Götz F, Biedermann T (2014) Cutaneous innate immune sensing of Toll-like receptor 2-6 ligands suppresses T cell immunity by inducing myeloid-derived suppressor cells. Immunity 41(5):762–775. https://doi.org/10.1016/j.immuni.2014.10.009

Soltes GR, Martin NR, Park E, Sutterlin HA, Silhavy TJ (2017) Distinctive roles for periplasmic proteases in the maintenance of essential outer membrane protein assembly. J Bacteriol 199(20). https://doi.org/10.1128/jb.00418-17

Steinhagen F, Kinjo T, Bode C, Klinman DM (2011) TLR-based immune adjuvants. Vaccine 29(17):3341–3355. https://doi.org/10.1016/j.vaccine.2010.08.002

Stoll H, Dengjel J, Nerz C, Götz F (2005) Staphylococcus aureus deficient in lipidation of prelipoproteins is attenuated in growth and immune activation. Infect Immun 73(4):2411–2423. https://doi.org/10.1128/IAI.73.4.2411-2423.2005

Storf S, Pfeiffer F, Dilks K, Chen ZQ, Imam S, Pohlschröder M (2010) Mutational and bioinformatic analysis of haloarchaeal lipobox-containing proteins. Archaea 2010. https://doi.org/10.1155/2010/410975

Sulzenbacher G, Canaan S, Bordat Y, Neyrolles O, Stadthagen G, Roig-Zamboni V, Rauzier J, Maurin D, Laval F, Daffe M, Cambillau C, Gicquel B, Bourne Y, Jackson M (2006) LppX is a lipoprotein required for the translocation of phthiocerol dimycocerosates to the surface of Mycobacterium tuberculosis. EMBO J 25(7):1436–1444. https://doi.org/10.1038/sj.emboj.7601048

Sun C, Benlekbir S, Venkatakrishnan P, Wang Y, Hong S, Hosler J, Tajkhorshid E, Rubinstein JL, Gennis RB (2018) Structure of the alternative complex III in a supercomplex with cytochrome oxidase. Nature 557(7703):123–126. https://doi.org/10.1038/s41586-018-0061-y

Szewczyk J, Collet JF (2016) The journey of lipoproteins through the cell: one birthplace, multiple destinations. Adv Microb Physiol 69:1–50. https://doi.org/10.1016/bs.ampbs.2016.07.003

Takeda K, Miyatake H, Yokota N, Matsuyama S, Tokuda H, Miki K (2003) Crystal structures of bacterial lipoprotein localization factors. LolA and LolB. EMBO J 22(13):3199–3209. https://doi.org/10.1093/emboj/cdg324

Tao K, Narita S, Tokuda H (2012) Defective lipoprotein sorting induces lolA expression through the Rcs stress response phosphorelay system. J Bacteriol 194(14):3643–3650. https://doi.org/10.1128/JB.00553-12

Thompson BJ, Widdick DA, Hicks MG, Chandra G, Sutcliffe IC, Palmer T, Hutchings MI (2010) Investigating lipoprotein biogenesis and function in the model gram-positive bacterium Streptomyces coelicolor. Mol Microbiol 77(4):943–957. https://doi.org/10.1111/j.1365-2958.2010.07261.x

Tokuda H, Matsuyama S, Tanaka-Masuda K (2007) Structure, function, and transport of lipoproteins in Escherichia coli. In: Ehrmann M (ed) The periplasm. ASM, Washington D.C., pp 67–80

Tokunaga M, Tokunaga H, Wu HC (1982) Post-translational modification and processing of Escherichia coli prolipoprotein in vitro. Proc Natl Acad Sci USA 79(7):2255–2259

Tschumi A, Nai C, Auchli Y, Hunziker P, Gehrig P, Keller P, Grau T, Sander P (2009) Identification of apolipoprotein N-acyltransferase (Lnt) in mycobacteria. J Biol Chem 284(40):27146–27156. https://doi.org/10.1074/jbc.M109.022715

van Els CA, Corbiere V, Smits K, van Gaans-van den Brink JA, Poelen MC, Mascart F, Meiring HD, Locht C (2014) Toward understanding the essence of post-translational modifications for the mycobacterium tuberculosis immunoproteome. Front Immunol 5:361. https://doi.org/10.3389/fimmu.2014.00361

Vogeley L, El Arnaout T, Bailey J, Stansfeld PJ, Boland C, Caffrey M (2016) Structural basis of lipoprotein signal peptidase II action and inhibition by the antibiotic globomycin. Science 351(6275):876–880. https://doi.org/10.1126/science.aad3747

Vollmer W, Blanot D, de Pedro MA (2008) Peptidoglycan structure and architecture. FEMS Microbiol Rev 32(2):149–167. https://doi.org/10.1111/j.1574-6976.2007.00094.x

Vu CH, Kolata J, Stentzel S, Beyer A, Gesell Salazar M, Steil L, Pane-Farre J, Ruhmling V, Engelmann S, Gotz F, van Dijl JM, Hecker M, Mader U, Schmidt F, Volker U, Broker BM (2016) Adaptive immune response to lipoproteins of Staphylococcus aureus in healthy subjects. Proteomics 16(20):2667–2677. https://doi.org/10.1002/pmic.201600151

Weidel W, Pelzer H (1964) Bagshaped macromolecules—a new outlook on bacterial cell walls. Adv Enzymol Relat Subj Biochem 26:193–232

Wiesmüller KH, Bessler W, Jung G (1983) Synthesis of the mitogenic S-[2,3-bis(palmitoyloxy)propyl]-N-palmitoylpentapeptide from Escherichia coli lipoprotein. Hoppe Seylers Z Physiol Chem 364(5):593–606

Wiesmuller KH, Jung G, Hess G (1989) Novel low-molecular-weight synthetic vaccine against foot-and-mouth disease containing a potent B-cell and macrophage activator. Vaccine 7(1):29–33

Wiktor M, Weichert D, Howe N, Huang CY, Olieric V, Boland C, Bailey J, Vogeley L, Stansfeld PJ, Buddelmeijer N, Wang M, Caffrey M (2017) Structural insights into the mechanism of the membrane integral N-acyltransferase step in bacterial lipoprotein synthesis. Nat Commun 8:15952. https://doi.org/10.1038/ncomms15952

Yamaguchi K, Yu F, Inouye M (1988) A single amino acid determinant of the membrane localization of lipoproteins in E. coli. Cell 53(3):423–432

Zaman M, Toth I (2013) Immunostimulation by synthetic lipopeptide-based vaccine candidates: structure-activity relationships. Front Immunol 4:318. https://doi.org/10.3389/fimmu.2013.00318

Zhang WY, Wu HC (1992) Alterations of the carboxyl-terminal amino acid residues of Escherichia coli lipoprotein affect the formation of murein-bound lipoprotein. J Biol Chem 267(27):19560–19564

Zhang WY, Inouye M, Wu HC (1992) Neither lipid modification nor processing of prolipoprotein is essential for the formation of murein-bound lipoprotein in Escherichia coli. J Biol Chem 267(27):19631–19635

Chapter 4
Outer Membrane Porins

Muriel Masi, Mathias Winterhalter and Jean-Marie Pagès

Abstract The transport of small molecules across membranes is essential for the import of nutrients and other energy sources into the cell and, for the export of waste and other potentially harmful byproducts out of the cell. While hydrophobic molecules are permeable to membranes, ions and other small polar molecules require transport via specialized membrane transport proteins. The two major classes of membrane transport proteins are transporters and channels. With our focus here on porins—major class of non-specific diffusion channel proteins, we will highlight some recent structural biology reports and functional assays that have substantially contributed to our understanding of the mechanism that mediates uptake of small molecules, including antibiotics, across the outer membrane of *Enterobacteriaceae*. We will also review advances in the regulation of porin expression and porin biogenesis and discuss these pathways as new therapeutic targets.

Keywords *Enterobacteriaceae* · Porins · Outer membrane proteins · Outer membrane protein biogenesis · Porin regulation · Envelope permeability · Protein channels · Antibiotic resistance

Introduction

Antibacterial drug resistance is broadly recognized as a growing threat to human health (O'Neill 2016; WHO 2014; NIAID 2014). As such, increasing antibiotic treatment failures due to multidrug-resistant (MDR) Gram-negative bacteria have stirred an urgent need to better understand the molecular mechanisms underlying antibiotic uptake and intracellular accumulation; and promote innovation with the development of new antibiotics and alternative therapies (Fraimow and Tsigrelis

M. Masi · J.-M. Pagès (✉)
UMR_MD1, Inserm U1261, IRBA, Membranes et Cibles Thérapeutiques, Facultés de Médecine et de Pharmacie, Aix-Marseille Université, Marseille, France
e-mail: jean-marie.pages@univ-amu.fr

M. Winterhalter
Department of Life Sciences and Chemistry, Jacobs University, Bremen, Germany

A. Kuhn (ed.), *Bacterial Cell Walls and Membranes*, Subcellular Biochemistry 92, https://doi.org/10.1007/978-3-030-18768-2_4

2011; Bush et al. 2011; Boucher et al. 2013; Stavenger and Winterhalter 2014; Page and Bush 2014; Pierluigi et al. 2015; Silver 2016; Zgurskaya et al. 2015). Logically, the efficacy of antibacterial drugs depends on their capacity to reach critical intracellular concentrations for the inhibition of their respective target(s). However, this is particularly challenging for drugs directed against Gram-negative bacteria due to their complex and sophisticated two-membrane cell envelope (Stavenger and Winterhalter 2014; Silver 2016; Zgurskaya et al. 2015). The outer membrane (OM) is composed of phospholipids (PL) in the inner leaflet and lipopolysaccharides (LPS) in the outer leaflet (Nikaido 2003). The asymmetric structure of the OM bilayer together with the features of the LPS structure, such as the presence of lipid A, core and O-antigen chains, creates a formidable barrier for the translocation of most amphiphilic compounds, including antibacterial drugs (Nikaido 2003; Delcour 2009; Henderson et al. 2016). Nevertheless some access is provided by the presence of water filled β-barrel channel proteins called porins (Nikaido 2003; Delcour 2009; Pagès et al. 2008). The model Gram-negative bacterium *Escherichia coli* is known to produce three classical trimeric porins—namely OmpF, OmpC and PhoE (Fig. 4.1). Since the studies of these porins formed the basis of our current knowledge of many other porins, they (as well as their orthologues in closely related enterobacteria) are called "classical porins". Classical porins show general preferences for charge and size of the solute, with OmpF and OmpC preferring cations slightly over anions and PhoE preferring anions; and with OmpF allowing the permeation of slightly larger solutes than OmpC. Others than classical porins, which will not be described in this chapter, the OM also contains specialized protein channels and receptors used for the uptake of specific substrates (for example LamB and BtuB for maltodextrins and vitamin B12 transport, respectively), proteins involved in OM and surface appendages biogenesis (for example, Omp85/BamA for membrane protein insertion, and a large array of translocators used in the assembly of adhesins, pili and flagella), OM components of toxin secretion systems and efflux pumps (for example, secretins of Type II secretion systems and TolC as a central component of Type I secretion systems and tripartite drug efflux pumps), various enzymes (such as the *E. coli* OmpT protease) and proteins involved in LPS assembly and modifications (for example LptD for LPS transport and assembly and PagP for LPS modifications) (Fig. 4.1). The reader is referred to recent reviews for more information on these proteins (Gerlach and Hensel 2007; Koronakis et al. 2004, 2005; Botos et al. 2017; Noinaj et al. 2017). In contrast, the inner membrane (IM) is exclusively composed of PL, thus relatively permeable for most amphiphilic compounds. Despite this leakiness, the IM provides a major contribution to bacterial defenses against drugs by housing a variety of multidrug transporters (Nikaido and Pagès 2012; Li et al. 2015) (Fig. 4.1). Together, the slow influx of antibiotics across the OM, the specificity and efficiency of efflux pumps define the intracellular steady-state concentrations of antibiotics as well as the observed differences in antibiotic susceptibilities between different Gram-negative bacteria (Zgurskaya et al. 2015; Masi et al. 2017; Krishnamoorthy et al. 2016, 2017; Westfall et al. 2017; Nichols 2017; Zgurskaya et al. 2018) (Fig. 4.1). Importantly, multidrug resistant (MDR) clinical isolates of *Enterobacteriaceae* generally exhibit porin loss and/or increased efflux (Charrel et al. 1996; Mallea et al. 1998; Chevalier

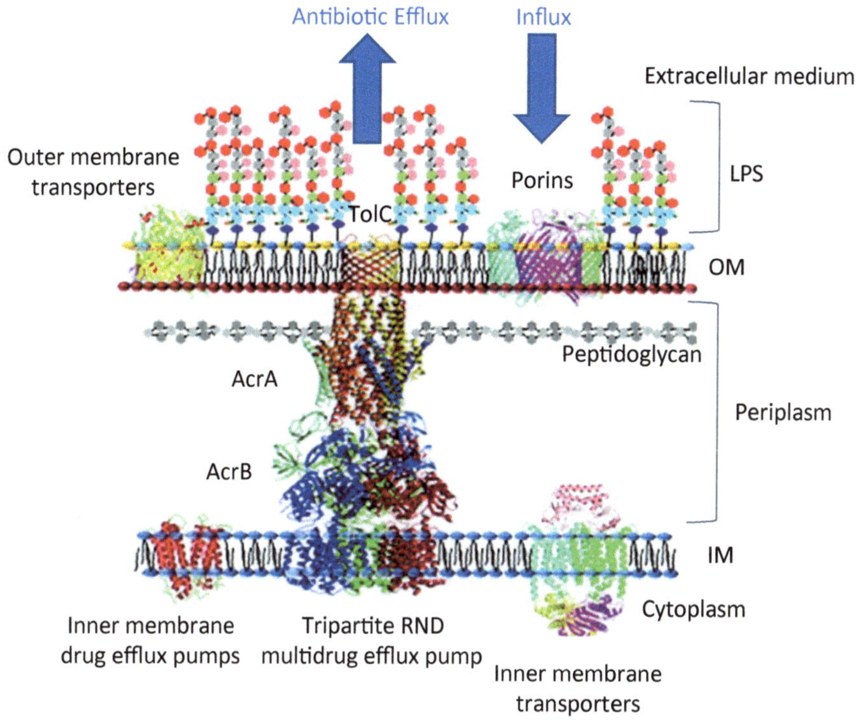

Fig. 4.1 Schematic of the Gram-negative cell envelope highlighting some of the many barriers to small molecule penetration. Adapted with permission from Manchester JI, Buurman ET, Bisacchi GS, McLaughlin RE. Molecular determinants of AcrB-mediated bacterial efflux implications for drug discovery, J. Med. Chem., 55:2532–2537. Copyright (2012) American Chemical Society. The lipopolysaccharide (LPS) chains emanating from the outer membrane (OM) present an initial and effective barrier to all but generally polar compounds small enough to pass through porin proteins. In the periplasm (Nikaido 2003), compounds too polar to diffuse rapidly through the cytoplasmic inner membrane (IM) are expelled primarily by tripartite root nodulation division (RND)-type pumps (AcrAB/TolC family) (Yu et al. 2003) and compounds in the cytoplasm are susceptible to efflux by a multitude of other pumps (e.g., drug efflux pumps). Uptake of nutrients is facilitated by specific channels in the inner and outer membranes, respectively. In addition to the more generalized mechanisms represented here, there are multiple species-specific and compound-specific mechanisms of acquired and innate resistance

et al. 1999, 2000; Dé et al. 2001; Thiolas et al. 2004, 2005; Lavigne et al. 2013; Philippe et al. 2015); see (Delcour 2009; Pagès et al. 2008; Nikaido and Pagès 2012; Li et al. 2015; Masi et al. 2017; Davin-Régli et al. 2008 for reviews).

In this chapter, we will review works by our group and others that have addressed the structural and functional characterization of classical porins. These works have led to a better understanding of the mechanism(s) of penetration of antibacterial drugs (defining routes of entry) and contributed to establishing rules for arrival in the periplasm. Some works also involve methods for measuring accumulation of compounds in cellular compartments. Focusing on the porin-mediated influx of antibiotics, we will also give a perspective on the factors, including major regulatory pathways and stress responses that control porin expression in *E. coli* and *Enterobacteriaceae*. This will be discussed with respect to clinical data, which illustrate the bacterial strategies caused by loss or structural alterations of porins to limit antibiotic entry. Finally, we will give a brief overview of recent breakthrough studies of the β-barrel assembly machinery (BAM) and periplasmic chaperones, which mediates the biogenesis of outer membrane proteins (OMPs), including that of classical porins. Notably, both regulatory and biogenesis pathways are well-conserved among γ-proteobacteria and could be considered as new therapeutic targets.

Classical Porins

Structure

X-ray crystallographic analysis showed that porins exist as trimers of transmembrane β-barrels. Until recently, crystal structures were available for only a few porins and comparison of primary sequences gave limited information. Indeed, the external loops between the β-strands are highly variable due to rapid mutational alterations as they interact with antibodies, polyamines, bacteriocins and phages present in the external medium. This creates alignment bias with gaps at improper positions. Interestingly, Ferenci proposed to first identify then compare the transmembrane β-strands, which are less variant regions (Ferenci 1994). This and other approaches revealed that porins of γ-proteobacteria share strong similarity and that porins of enterobacteria all contain a characteristic PEFGGD signature sequence in loop 3 (see below) (Ferenci 1994; Jeanteur et al. 1991). Undoubtedly, one of the most important progresses in the study of porins was the elucidation of their 3D structures by electron diffraction and X-ray crystallography. The latter approach had its first success with a *Rhodobacter capsulatus* trimeric porin (Weiss et al. 1989, 1990, 1991a, b; Weiss and Schulz 1992), which was followed by the elucidation of the structure of the *E. coli* OmpF, PhoE and OmpC porins (Cowan et al. 1992; Baslé et al. 2006). Important conclusions from these studies include the following:

(i) As predicted from earlier studies, porin monomers cross the OM bilayer as a β-barrel in a series of 16 β-strands. The strands are tilted rather strongly (by 30–60°) in relation to the barrel axis and this tilting increases the diameter of the barrel (for the concept of sheer number, which is related to the degree of tilt, (see reference Schulz 2002). The length of each transmembrane strand spans the range from only 7 (in strand 5) to 16 (in strand 1) residues in OmpF. Contact among the monomers is stabilized by hydrophobic and polar interactions, and loop 2 tends to bend over the wall of the barrel of the neighboring subunit, playing a significant role in stabilization.

(ii) The external surface of the barrel is occupied by lipophilic side chains. One striking observation about the *R. capsulatus* porin was the presence of many aromatic amino acid residues at both the outer and inner interfaces between the bilayer and the aqueous medium (Weiss et al. 1991). The presence of these "aromatic girdles" may assist in OM insertion has since been observed in the structure of porins of *E. coli* and other *Enterobacteriaceae* (Cowan et al. 1992; Baslé et al. 2006; Schulz 2002; Dutzler et al. 1999; Arunmanee et al. 2016; Acosta-Gutierrez et al. 2018) as well as in many other OM proteins. In particular, porin sequences almost invariably end with a C-terminal phenylalanine, which is located at the OM/periplasm interface (Struyve et al. 1991). Several studies subsequently demonstrated that this C-terminal residue is essential for porin biogenesis (see section "Biogenesis of β-Barrel Outer Membrane Proteins").

(iii) Transmembrane strands are connected by short "turns" on the periplasmic side, but the "loops" that connect the strands on the external sides are often long. Loop 2 folds back outward and contributes to the connection with the neighboring monomer. Loop 3 (L3), connecting strand 5 with strand 6, is especially long (33 residues in OmpF) and folds into the barrel to produce the narrowing of the channel (often called the "eyelet" or constriction region, CR) (in orange in Fig. 4.2). The size of the CR of OmpF is 7 × 11 Å (Cowan et al. 1992), very close to the estimated diameter of 12 Å from sugar diffusion studies (see section "Functional Assays"). A conserved set of charged residues decorates the CR: negatively charged residues (in red in Fig. 4.2) are typically found on the L3 itself, and positive charges (in blue in Fig. 4.2) form a cluster on the opposite barrel wall. This arrangement creates a local transversal electrostatic field that plays an important role in selectivity for the size and charge of permeating molecules (see below and section "Antibiotic Permeation Assays with New Approaches").

(iv) The nature of the residues lining the channel wall provides the explanation of the diffusion characteristics through porins. OmpF and PhoE prefer cations and anions, respectively, despite having a 72% similarity in their mature-protein sequence; this difference in charge preference was shown to be due mainly to the replacement of Gly131 in the CR of OmpF with the positively charged Lys125 in PhoE (Cowan et al. 1992). The electrostatic properties of the OmpF and PhoE channels were calculated and compared (Karshikoff et al. 1994). The OmpC channel appears to be slightly smaller than the OmpF channel on the basis of diffusion rates studies (Nikaido and Rosenberg 1981; Nikaido and Rosenberg 1983; Nikaido et al. 1983). The crystal structure of *E. coli* OmpC as well as that of the OmpC orthologues from *Klebsiella pneumoniae* has been determined more recently (Baslé et al. 2006; Schulz 2002; Dutzler et al. 1999). However, the size of the CR is almost exactly the same as in OmpF, and therefore does not account for the difference in diffusion rates. Instead, the authors pointed out that more negatively charged residues are pointing toward the pore lumen in OmpC, mainly above the CR (Baslé et al. 2006; Schulz 2002), which may decrease the functional radius of the solute diffusion pathway or

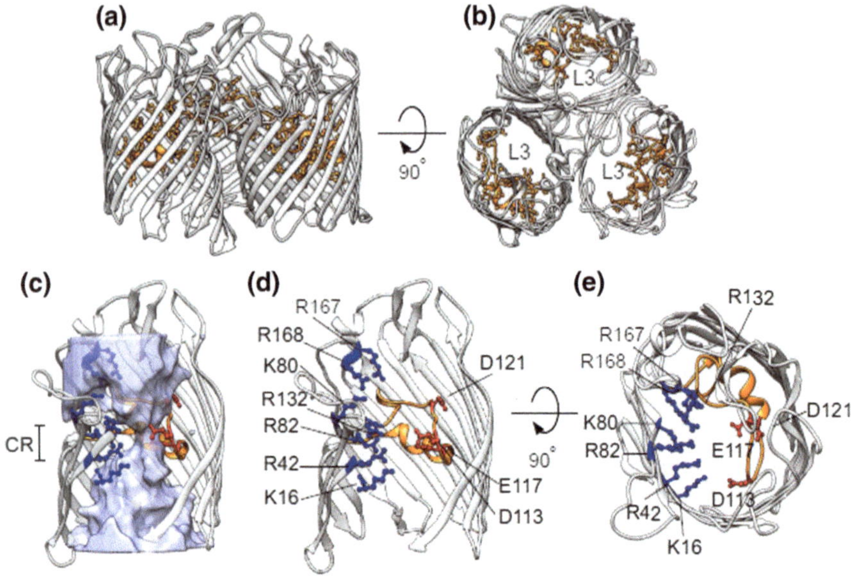

Fig. 4.2 Architecture of the major outer membrane porins OmpF and OmpC. Adapted with permission from Acosta-Guiterrez S, Ferrara L, Pathania M, Masi M, Wang J, Bodrenko I, Zahn M, Winterhalter M, Stavenger RA, Pagès JM, Naismith JH, van den Berg B, Page MGP, Ceccarelli M. 2018 Getting drugs intro Gram-negative bacteria: rational rules for permeation through general porins. ACS Infect Dis. https://doi.org/10.1021/acsinfecdis.8b00108. [Epub ahead of print]. Copyright (2018) American Chemical Society. The main features of the OmpF and OmpC orthologues are depicted for OmpF from *E. coli*: **a** trimeric arrangement depicted as ribbon; **b** top view of the trimer with the loop L3 highlighted in orange balls and sticks; **c** for one of the monomers the empty lumen (calculated from all-atom simulations) is highlighted as a purple surface and the constriction region (CR) is indicated; **d** the main charged residues are depicted as balls and sticks and colored according to their charge; **e** top-view of (**d**)

affect the transversal electric field. Molecular dynamics simulations and other permeation assays combined to porin mutagenesis have been valuable tools in solving this question (see section "Antibiotic Permeation Assays with New Approaches").

The crystal structures of OmpF and OmpC orthologues from *Enterobacter aerogenes* (Omp35 and Omp36), *Enterobacter cloacae* (OmpE35 and OmpE36) and *K. pneumoniae* (OmpK35 and OmpK36) have been solved very recently (Dutzler et al. 1999; Arunmanee et al. 2016; Acosta-Gutierrez et al. 2018). These porins share on average ~70% sequence identity. As expected, the structures are very similar and consist of trimers of 16-stranded β-barrels, each with a CR halfway down the central axis formed by L3. The acidic residues at the CR of OmpF (D113, E117, D121 in L3) are conserved in all the structures, while some differences were found for the residues forming the basic ladder on the opposing face. Overall, OmpC and orthologues Omp36, OmpE36, OmpK36 possess a smaller pore radius, a lower con-

Table 4.1 (Adapted from Acosta-Gutierrez et al. 2018): Physicochemical properties of classical porins of *Enterobacteriaceae*

Porin[a]	G (nS)[b]	P[c]	R (Å)[d]	E (mV/Å)[e]	V (kcal/mol)[f]
OmpF	4.0 ± 0.2	1.33	3.1 ± 1.1	25.2 ± 0.56	−1.3 ± 0.06
OmpC	2.9 ± 0.1	2.20	2.8 ± 1.1	15.6 ± 0.39	−0.5 ± 0.03
Omp35	3.9 ± 0.5	1.72	3.6 ± 1.7	22.8 ± 0.78	−0.5 ± 0.11
Omp36	3.0 ± 0.2	2.22	3.0 ± 1.3	14.9 ± 0.01	−0.7 ± 0.03
OmpE35	4.4 ± 0.2	1.42	3.1 ± 1.3	25.8 ± 1.00	−1.0 ± 0.08
OmpE36	3.0 ± 0.2	2.13	3.1 ± 1.6	22.3 ± 0.64	−1.5 ± 0.04
OmpK35	4.4 ± 0.4	1.36	3.6 ± 1.4	14.9 ± 0.4	−0.6 ± 0.06
OmpK36	2.5 ± 0.2	1.68	2.8 ± 1.4	19.1 ± 0.36	−1.2 ± 0.09

[a]OmpF and OmpC of *E. coli*; Omp35 and Omp36 of *E. aerogenes*; OmpE35 and OmpE36 of *E. cloacae*; OmpK35 and OmpK36 from *K. pneumoniae*
[b]Conductance of a single trimeric porin in 1 M KCl
[c]Ion selectivity P^{+}_{K}/P^{-}_{Cl} 0.1 M versus 1 M KCl
[d]Minimum radius
[e]Transversal electric field
[f]Electrostatic potential

ductance, and a lower intensity of the transversal electric field compared to OmpF and orthologues Omp35, OmpE35, OmpK35 (Table 4.1). Consequently, OmpC- are more cation-selective than OmpF-type porins (Table 4.1) and less permeable to anionic compounds. As detailed in the following sections, exploring the porin diversity using structural determinants has permitted the analysis of experimental transport data as regards of the interactions between the pore and the permeating compounds, and validated molecular mechanisms for small molecule penetration through porins (see section "Antibiotic Permeation Assays with New Approaches").

Functional Assays

The functional properties of porins have been the subject of investigation for over 40 years. For example, earlier work using radioactive tracer molecules revealed selective uptake across the so-called "outer cell wall" in Gram-negative bacteria and pointed towards the potential role of porins (Nakae and Nikaido 1975; Nakae 1976). Porins were then isolated and reconstituted into artificial planar lipid membranes (Benz et al. 1978; Schindler and Rosenbusch 1978). Conductance measurements suggested single channel pore sizes of about 1 nm, a value very close to that what high resolution X-ray structure revealed a few years later (Cowan et al. 1992). About the same time, OMPs were reconstituted into multilamellar liposomes the induced swelling by successful penetration was used to get kinetic information (Luckey and Nikaido 1980). Liposome swelling assays became an important tool to understand uptake of nutrient or antibiotics. It allows reducing the number of parameter through inserting only one porin type and single pore sizes are averaged over a larger number of liposomes. However, swelling occurs in response to the movement of all the solutes, including components of the buffer, and extreme care is needed when this method is used to study the diffusion of charged solutes (Nikaido and Rosenberg 1983). Initial works with this approach established the size exclusion cutoffs of porins by measuring the transport of various size sugars using liposome swelling assays (Nikaido and Rosenberg 1983). A value of about 550 Daltons was determined for OmpF (Nakae 1976), which implies that ions, amino acids, and small sugars use general diffusion porins for gaining access to the periplasm. Disaccharides, larger sugars and other molecules need to use dedicated pathways for outer membrane transport (Nikaido 2003). These early studies established the molecular sieving properties of porins, and provided an explanation for the high diffusion rates of these compounds through the OM (Nikaido and Rosenberg 1981, 1983; Nakae 1976). Comparison of diffusion rates of solutes of various sizes gave remarkably reliable values of the channel size with respect to the crystallographic structure (Table 4.1 and section “Antibiotic Permeation Assays with New Approaches”). However, because swelling occurs in response to the movement of any solute, including components of the buffer, extreme care is needed when this method is used to study the diffusion of charged solutes.

The gold standard approach to characterize functional properties of isolated porin channels in vitro is reconstitution in planar lipid bilayers (also known as “black lipid membranes” or “BLM”) (Fig. 4.3a). A lipid bilayer is formed over an aperture pierced through a Teflon film separating two chambers. Each chamber contains a buffered ionic solution and an electrode used to measure electric current due to the flow of ions across the bilayer and to clamp the transmembrane potential required to promote ion movement. Purified detergent-solubilized channel proteins or proteoliposomes are added to one chamber (the so-called *cis* side) and spontaneously insert in the bilayer over time. The sequential insertions of open channels in the membrane lead to discrete current jumps due to ion movement through the open channels (Fig. 4.3b). The conductance (i.e., the amount of current per unit voltage) of a channel can be obtained from measuring the size of these current jumps. In the case of classical porins (i.e., OmpF and OmpC), this would represent the trimeric conductance, since they typically purify and insert in the bilayer as trimers. Single channel insertion is achieved empirically and depends on the protein concentration, the flow profile from protein addition to the lipid membrane, the area of the lipid membrane, the lipid composition and the detergent used. Of note, 10^8–10^{15} channels in detergent are typically diluted into detergent free buffer. Diluted OMPs subsequently “lose” their detergent micelles into the solution, the most hydrophobic OMPs may aggregate, precipitate, and only few will find the correct path to insert to the lipid bilayer. After insertion, the channel activity can be studied in various conditions (e.g. in the absence or in the presence of putative permeating molecules as detailed in section “Antibiotic Permeation Assays with New Approaches”) and membrane potentials. Overall, BLM reconstitution allows the calculation of the single-channel conductance (G, nS), which is a reasonable indicator of pore size. The single-channel conductance value for OmpF, 1 nS, is used as the reference in the discussion of many such porins (Nikaido 1992).

The patch-clamp technique has also been applied to the study of purified porins reconstituted in artificial liposomes. Here, a small patch of liposome membrane is drawn at the tip of a 1 μM-diameter glass pipette, and the current flowing through this patch is recorded at a fixed membrane potential (Baslé et al. 2004). A commercial tool (Port-a-Patch) facilitates the handling as the liposome containing preparation is simply pipette on top of a hole in a glass and aspirated to form a giga-seal (Mahendran et al. 2010; Wang et al. 2018). This technique permitted the discovery that porins flicker between multiple states, whose kinetics and conductance can be affected in mutants and in the presence of small permeating molecules. Electrophysical studies performed by the Benz and the Rosenbusch groups in the 70s and 80s established some of the hallmark properties of the general diffusion porins, such as high ionic current due to the relatively large pore size, low ionic selectivity (although some porins show preference for cations (OmpC) or anions (PhoE)), and high opening probability, in standard bilayer electrophysiology conditions of low voltage, neutral pH and high ionic strength (Benz et al. 1978, 1979, 1985; Schindler and Rosenbusch 1978). The main advantage of BLM measurements is their easy access and low consumption of protein. Although a single molecule technique the insight view of molecular details has to come from all atom molecular dynamics (MD). Indeed with the upcoming numerous high resolution structure MD simulations are a perfect complementary technique. Computational modeling studies have suggested that the paths taken by anions and cations are divergent at the CR, as cations are drawn close to the negative charges of the L3 loop, and anions flow near the positively charged cluster of the opposite barrel wall (Im and Roux 2002). This type of work emphasizes the notion that the permeating ions interact with the wall of the channel and that ion movement does not follow simple diffusion. In fact, measuring the temperature dependent OmpF channel conductance and normalizing over the temperature dependent bulk ion conductances revealed a clear deviation from the bulk. An Arrhenius plot allows quantifying these ion interactions. MD modeling revealed two distinct ion pathways when ion concentration is below 150 mM KCl; whereas above this concentration, ions fill more the entire pore volume (Chimerel et al. 2008; Pezeshki et al. 2009). Interestingly the dissociation of KCl is rather temperature independent but the association-dissociation increase in a similar manner. This was also demonstrated experimentally by measuring the conductance and selectivity of various general diffusion porins in solutions of varying ionic strength or pH, and in variants with mutations at specific pore exposed residues (Saint et al. 1996; Phale et al. 2001; Bredin et al. 2002; Danelon et al. 2003; Alcaraz et al. 2004). Bezrukov's group showed that the selectivity of OmpF for cations relative to anions increases sharply in solutions of low ionic strength (Alcaraz et al. 2004). The channel reaches nearly ideal cation selectivity in solutions of <100 mM KCl. Furthermore, at pH's <4, the channel reverses its selectivity from preferring cations to preferring anions. The authors combined these experimental observations with calculations of the distribution of charged residues in the pore lumen and concluded that electrostatic

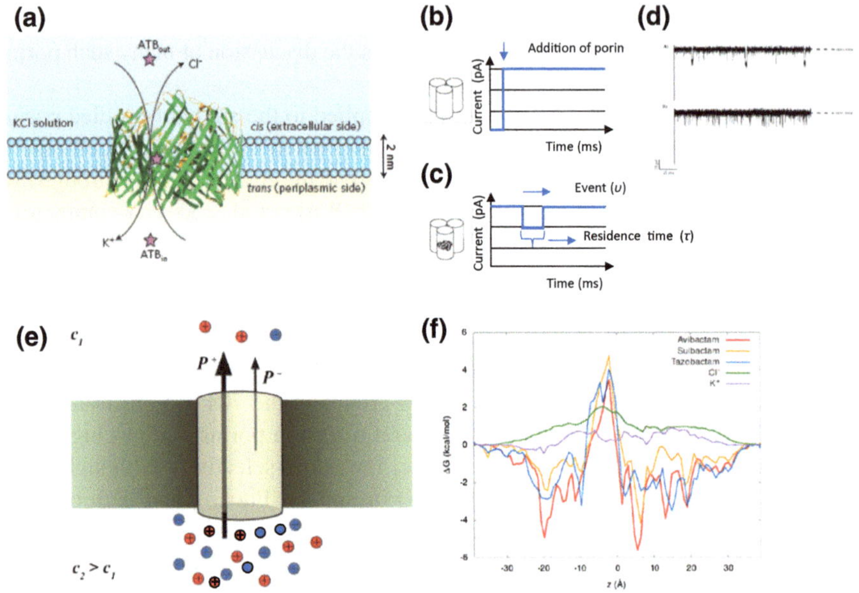

Fig. 4.3 Methods for measuring intracellular accumulation and porin-mediated uptake of antibiotics using electrophysiology. **a–c** Adapted with permission from Masi M, Réfrégiers M, Pos KM, Pagès JM (2017) Mechanisms of envelope permeability and antibiotic influx and efflux in Gram-negative bacteria. Nat Microbiol 2:17001 Copyright Nature Publishing 2017. **d** Adapted with permission from J. Phys. Chem B 116 (2012) 4433–38. Copyright 2012 ACS. **f** Reprinted (Adapted or Reprinted in part) with permission from J. Phys. Chem Lett. 8 (2017) 1295–1301. Copyright 2017 ACS. **a** A transmembrane voltage is applied across a free-standing bilayer containing a single (trimeric) porin inducing an ion current. Antibiotic molecules (ATB) will diffuse across the membrane. **b** Schematic view of a conductance recording. After addition of porins into one or both chambers within a few minutes a jump in the conductance is observed. From a statistical distribution of at least 100 of such jumps we may conclude on an average pore conductance. **c** Addition of molecules able to penetrate the porin will possibly cause blockages during the time inside the channel. To discriminate translocation from blockage only a variation of an applied external force is necessary. Increasing the transmembrane voltage increase the pulling force on charged molecules leading to shorter residence times in case of translocation whereas if the molecule get trapped stronger forces will lead longer residence times. Uncharged molecules may follow the electro-osmotic flow in a similar manner. **d** Permeation of molecules depends on the buffer: for example 100 mM enrofloxacin in presence of 5 mM Mg^{2+} (upper trace) compared to the trace in absence of Mg^{2+} (lower trace) revealed a clear difference. **e** A difference in electrophoretic mobility of charged molecules causes a diffusion potential and allows to identify translocation and to distinguish the events from sole binding. Combined with a single channel recording allows to estimate turn over numbers. **f** All atom modelling performed in the group of M. Ceccarelli indicates an energy barrier for translocation. OmpF as cation selective channel has a low and broad barrier for K^+ and a higher one for Cl^-. In contrast ß-lactam inhibitors like avibactam, sulbactam or tazobactam do have sharp energy barriers

interactions exist between the permeating ions and the charges of ionizable residues over the entire channel length (Pezeshki et al. 2009; Alcaraz et al. 2004). However, shifts in selectivity are detected upon mutations of single residues. Substitution at the pore-exposed D113 residue in OmpF (Saint et al. 1996) decreases cation-selectivity. Opposite effects are seen upon charge removal at arginines of the constriction zone (Saint et al. 1996). However, selectivity depends on the concentration and can even invert in presence of di(or tri)valent salts (Alcaraz et al. 2009; Singh et al. 2012).

Finally, diffusion rates through porin channels can be quantified in intact cells by coupling the influx of hydrophilic solutes with a measurable hydrolysis process. A convenient assay is to examine the influx of cephalosporins by coupling it to their hydrolysis by periplasmic β-lactamase (Nikaido et al. 1983; Kojima and Nikaido 2013, 2014); cephalosporins are especially useful because a diverse collection of cephalosporins has been synthesized and because the hydrolysis can be monitored easily by recording changes in the optical density at 260 nm (Zimmermann and Rosselet 1977). This approach and others as well as their application for studying porin-mediated translocation of antibiotics will be detailed in section "Antibiotic Permeation Assays with New Approaches".

Regulation of Porin Expression

Envelope Stress Responses

All living organisms have stress responses that allow them to sense and respond to environmental damaging conditions by remodeling gene expression. As such, Gram-negative bacteria possess stress responses that are uniquely targeted to the cell envelope, including membranes and cell wall. In *Enterobacteriaceae*, these envelope stress responses (ESRs) are the EnvZ/OmpR, CpxAR (Cpx), BaeRS, and Rcs phosphorelays, the stress responsive alternative sigma factor σ^E, and the phage shock response (Majdalani and Gottesman 2005; Ruiz and Silhavy 2005; Rowley et al. 2006; Raivio 2014). Each of these ESRs is activated following the perturbation of particular components of the envelope or exposure to particular environmental stresses. Although ESRs are important for reacting to damaging conditions, stress proteins also play important roles in the maintenance of basic cellular physiology (Hayden and Ades 2008; Delhaye et al. 2016). This is particularly true for the σ^E-dependent stress response in *E. coli*, as the *rpoE* gene, which encodes σ^E, is essential for viability (De Las et al. 1997). Here, we will essentially focus on ESRs that impact on antimicrobial resistance by regulating porin expression together with many other targets (regulons)—namely EnvZ/OmpR, Cpx and σ^E (see below). Additionally, with the recent highlights and advances in RNA-based techniques (Wassarman and Kiley 2010), the repertoire of small regulatory RNAs (sRNAs) has vastly increased so as their impact on the bacterial physiology (Gottesman 2004), particularly on the OM homeostasis, is continuously emerging (Guillier et al. 2006; Vogel and Papenfort 2006; Valentin-Hansen et al. 2007). sRNAs alter gene expression, allowing fast adjustment to different growth conditions (Gottesman 2004). Noteworthy, ESRs are often interconnected, regulate and are regulated by sRNAs in order to provide tied regulation of target genes both at the transcriptional and post-transcriptional levels (Guillier et al. 2006; Vogel and Papenfort 2006; Valentin-Hansen et al. 2007; Thompson et al. 2007; Vogt et al. 2014).

The expression of the two major porins, OmpF and OmpC, is exquisitely regulated. The apparent purpose of this regulation became clear when it was discovered that OmpF produces a slightly larger channel than OmpC (Nikaido and Rosenberg 1983; Nikaido et al. 1983). Thus, noxious agents such as antibiotics and bile acids diffuse far better through the larger OmpF channel, as seen clearly from the observation that low concentrations of antibiotics select for *ompF* mutants but never for *ompC* mutants (Harder et al. 1981). In its natural habitat, the intestinal tract, *E. coli* encounters 4–16 mM bile salts (Borgstrom 1974), and it is most important to minimize their influx. The conditions prevailing in the intestinal tract, high osmotic strength and high temperature, both favor the production of OmpC (with its narrower channel) and repress the production of OmpF. On the other hand, the increased production of OmpF under low-temperature, low-osmolarity conditions (for example, in lake water) will benefit *E. coli* by facilitating the influx of scarce nutrients. Although it is now admitted that the electronegativity of the pore interior rather than the pore size determines the permeability of OmpF and OmpC channels (Cowan et al. 1992; Baslé et al. 2006; Schulz 2002; Dutzler et al. 1999; Arunmanee et al. 2016; Acosta-Gutierrez et al. 2018; Kojima and Nikaido 2014), this model is still valid. The molecular mechanisms of this regulation have been studied extensively. Environmental osmolarity was one of the earliest stresses described to influence OmpF and OmpC expression via the EnvZ/OmpR two-component system (TCS) (Mizuno and Mizushima 1990; Pratt et al. 1996; Cai and Inouye 2002). EnvZ is a membrane-bound sensor kinase, and OmpR is a cytosolic response regulator, which binds to the promoter region of the porin genes. Upon activation, EnvZ autophosphorylates and the high energy phosphoryl group from EnvZ is subsequently transferred to a conserved Asp residue on OmpR. Phosphorylated OmpR (OmpR~P) serves as a transcription factor that differentially modulates the expression of the *ompF* and *ompC* porin genes. The *ompF* gene, with its high affinity OmpR-binding sites, is transcribed at low osmolarity when the phosphorylated OmpR (OmpR~P) is scarce. However, when the concentration of OmpR~P increases, additional binding of these molecules results in increased transcription of *ompC* and repression of *ompF*. High temperature and high osmolarity (through high levels of OmpR~P) have been shown to increase the transcription of the antisense RNA MicF (Andersen et al. 1987, 1989; Aiba et al. 1987; Coyer et al. 1990; Ramani et al. 1994; Delihas and Forst 2001). This RNA binds to the 5′ untranslated (UTR) region of the *ompF* mRNA and inhibits its translation (Andersen and Delihas 1990; Schmidt et al. 1995) (see section "Remodeling Porin Expression by Small Regulatory RNAs"). As described below, oxidative stress and the presence of salicylate also increase *micF* transcription and prevent the production of OmpF post-transcriptionally. The intestinal tract, the normal environment of *E. coli*, is thought to be mostly anaerobic. Interestingly, anaerobiosis was found to modify the osmoregulation of OmpF and OmpC (Matsubara et al. 2000).

Thus, under anaerobiosis, OmpC is expressed at a rather high level even in fairly low-osmolarity media, and the repression of OmpF by osmotic activity occurs more strongly than under aerobic conditions. This modification of the regulatory response, which is expected to favor the survival of *E. coli* in the intestinal tract, occurs through the cross-talk activation of OmpR by the ArcB sensor (Matsubara et al. 2000), which senses the anaerobic condition.

Accumulation of misfolded OMPs in the periplasm, presumably reflecting problems in protein assembly or transport across the IM, can be detected by regulatory sensors either comprised of the Cpx TCS, or the alternative sigma factor σ^E, the anti-sigma factor RseA, and a number of periplasmic proteases. σ^E and Cpx are the two major regulation pathways that control the envelop integrity with overlapping regulon members (Price and Raivio 2009; Raivio et al. 2013; Batchelor et al. 2005; Gerken et al. 2009; Dartigalongue et al. 2000) but responding to different inducing cues (Ruiz and Silhavy 2005). Although inducing signals may all act by causing protein misfolding, misfolded proteins are not the inducing signal per se, as some signals induce σ^E but not Cpx and vice versa. Recent studies suggest that Cpx responds to IM perturbations, while σ^E is activated by signals at the OM.

The Cpx system comprises the CpxA sensor kinase and response regulator CpxR. Envelope stresses including alkaline pH, periplasmic protein misfolding, IM abnormalities such as misfolded transporters or accumulation of the lipid II precursor, induce the dissociation of the accessory protein CpxP from CpxA, trigger CpxA-mediated phosphorylation of CpxR, which in turn alters expression of protein foldases and proteases, respiratory complexes, IM transporters, and cell wall biogenesis enzymes (Raivio 2014; Price and Raivio 2009; Raivio et al. 2013). The Cpx-mediated regulation of porins occurs at several levels. At the transcriptional level, CpxR~P has been shown to bind directly the *ompF* and *ompC* promoters (Batchelor et al. 2005). More recently, it has been found that the small IM protein MzrA connects Cpx and EnvZ/OmpR (Gerken et al. 2009). In this pathway and upon the activation of Cpx, MzrA interacts directly with EnvZ, which in turn, stabilizes OmpR~P (Gerken et al. 2009). In sensing different signals, the interconnection between Cpx and EnvZ-OmpR allows cells to adapt to diverse environmental stresses. By regulating a number of genes, Cpx has been shown to affect resistance to some but not all antibiotics (Raivio 2014; Raivio et al. 2013; Guest and Raivio 2016; Mahoney and Silhavy 2013; Moreau 2014). However, its precise role and that of other two-component systems in the development of MDR in clinical isolates are still poorly documented.

The σ^E pathway of *E. coli* was the first ESR to be identified. In the absence of inducing signals, σ^E is held at cytoplasmic side of the inner membrane by the anti-sigma factor RseA, a single-pass membrane protein (De Las et al. 1997; Missiakas et al. 1997). A periplasmic protein, RseB, binds to the periplasmic domain of RseA and enhances the inhibition of σ^E. Upon cell envelope stress, σ^E is released from RseA by a proteolytic cascade whose end result is the complete degradation of RseA and the release of σ^E to direct transcription (Ades et al. 1999). The so-called regulated intramembrane proteolysis (RIP) proteases of the cascade, DegS and RseP (formerly known as YaeL) act sequentially cleaving RseA first in the periplasmic and then in the transmembrane region (Rudner et al. 1999; Alba et al. 2002). The cytoplasmic domain of RseA ($RseA_{cyto}$) bound to σ^E is then released and degraded by cytoplasmic proteases, primarily ClpXP (Chaba et al. 2007; Flynn et al. 2003, 2004). The proteolytic cascade is induced by a conserved YxF peptide (where x is any amino acid) found at the C-terminus of OMPs (Walsh et al. 2003). This peptide is normally buried and inaccessible in folded porin trimers. When porin folding is disrupted, the peptide is exposed and binds to DegS, activating DegS to cleave RseA and initiate the response (Walsh et al. 2003). Proper porin folding and transit to the outer membrane involves a series of steps, and disruption of this pathway at any point may lead to improperly folded porins with exposed C-termini. Therefore, it has been proposed that porins provide a sensitive measure of cell envelope homeostasis (Alba and Gross 2004). This regulatory pathway is not only designed to have a sensitive trigger specifically tuned to the inducing signal, but also includes a homeostatic mechanism providing a quick and efficient method to reset the switch and deactivate the response. Once σ^E is activated it transcribes the genes in its regulon (Rhodius et al. 2006). Although the σ^E regulon includes genes that affect many aspects of the cell, a significant fraction of its known regulon members encodes chaperones required for the delivery and assembly of LPS in the OM, proteases to degrade terminally misfolded porins, and at least two small RNAs, RybB and MicA, that target mRNAs encoding porins for degradation (Thompson et al. 2007; Johansen et al. 2006; Udekwu and Wagner 2007; Rasmussen et al. 2005; Papenfort et al. 2006) and one, MicL, that target mRNA of the major OM lipoprotein Lpp (Guo et al. 2014). Therefore, the activation of the σ^E pathway increases the capacity of the cell to deliver proteins to the outer membrane, facilitates the removal of misfolded porins, and reduces new porin synthesis reducing the load on the system. Each of these systems helps to lower the level of unfolded porins, thereby reducing the inducing signal and returning DegS to the inactivated state.

Here, one has to note that the σ^E-dependent repression of porin synthesis only occurs at the post-transcriptional level, wherein base-paring sRNAs inhibits translation of *omp* mRNAs (see below section "Remodeling Porin Expression by Small Regulatory RNAs") in order to maintain the envelope homeostasis under stress conditions, as porins are abundant proteins under normal growth conditions.

Remodeling Porin Expression by Small Regulatory RNAs

The last decade has been marked by the identification and characterization of several small regulatory RNAs (sRNAs). Enterobacteria such as *E. coli* and *Salmonella* are now known to encode at least twelve OMP-regulating sRNAs—namely InvR (Pfeiffer et al. 2007), SdsR (Fröhlich et al. 2012), MicA (Udekwu and Wagner 2007; Rasmussen et al. 2005), MicL (Guo et al. 2014), MicC (Chen et al. 2004; Dam et al. 2017), MicF (Andersen et al. 1987), IpeX (Castillo-Keller et al. 2006), OmrAB (Guillier and Gottesman 2006), RseX (Douchin et al. 2006), RybB (Wassarman and Kiley 2010; Thompson et al. 2007; Papenfort et al. 2006; Bossi and Figueroa-Bossi 2007) and CyaR (Johansen et al. 2008) (see (Guillier et al. 2006; Vogel and

Papenfort 2006; Valentin-Hansen et al. 2007) for reviews). These sRNAs exert their functions under a variety of stress conditions and contribute to several stress responses including the σ^E-mediated ESR. An sRNA–OMP network is emerging in which some sRNAs act specifically on a single *omp* mRNA (i.e., *E. coli* MicC and MicF), whereas others control multiple *omp* and non-*omp* mRNA targets (i.e., *E. coli* MicA and RybB). Likewise, the same *omp* mRNA could be targeted by more than one sRNAs (i.e., *ompC* mRNA is regulated by MicC, RseX and RybB). Importantly, these sRNAs serve to provide a rapid response and can either amplify the signal or act by a negative feedback. Here, it is worth to note that all these sRNAs are *trans*-acting—functioning by imperfect base pairing with mRNA targets at their 5′ UTR that encompasses the ribosome binding site (RBS), thus inhibiting the formation of a functional translation initiation complex—and require the help of the RNA chaperone Hfq (Guillier et al. 2006; Vogel and Papenfort 2006; Valentin-Hansen et al. 2007).

(i) MicF sRNA: The 93-nucleotide MicF sRNA is located upstream the *ompC* promoter and was first shown to inhibit OmpF production through the decrease of the *ompF* mRNA (Andersen et al. 1987). The discovery of the post-transcriptional repression of OmpF by MicF revealed a direct base-pairing between MicF and a fragment of the *ompF* mRNA encompassing both the RBS and the start codon (Andersen et al. 1987; Andersen and Delihas 1990; Schmidt et al. 1995). The expression of MicF itself is subjected to multiple signals and regulatory pathways (Delihas and Forst 2001). Positive regulation occurs via the EnvZ/OmpR two-component system under high osmolarity conditions (Aiba et al. 1987; Coyer et al. 1990; Ramani et al. 1994), via SoxS in response to oxidative stress and via MarA in response to antibiotic stress (Chou et al. 1993; Cohen et al. 1988; Chubiz and Rao 2011).

(ii) MicC sRNA: The 109-nucleotide MicC sRNA has been identified more recently in a computational screen as a sRNA encoded in the *ompN-ydbK* intergenic region (Chen et al. 2004), divergent to the *ompN* gene, which encodes a quiescent porin (Prilipov et al. 1998). This study first showed that MicC represses OmpC at the post-transcriptional level by direct base-pairing to a 5′ UTR of the *ompC* mRNA, thus preventing the formation of a functional translation initiation complex (Chen et al. 2004). In addition, Northern blot analysis of MicC and MicF expression profiles from a variety of growth conditions showed the two sRNAs to accumulate in almost a mutually exclusive fashion and that they could act in conjunction with the EnvZ/OmpR two-component system to control the OmpF/C porin ratio in response to a variety of stresses, including not only the osmolarity but also the growth temperature and medium starvation. Work from our lab has further shown that *ompN* and *micC* are subjected to dual regulation upon exposure to certain antimicrobials such as β-lactams in a σ^E-dependent manner (Dam et al. 2017). This is consistent with the fact that *ompN-micC* and *ompC-micF* share similar genetic organization and previous research have shown that *ompC* and *micF* are co-induced under specific conditions too (i.e., high osmolarity via EnvZ/OmpR). The *micC* gene is well-conserved among enterobacteria and its expression is highly regulated. The physiological role of MicC is still unclear, although down-regulating OmpC at the post-transcriptional level during an antibiotic stress could be required for bacterial adaptation. Therefore, the identification of additional MicC targets as well as information of MicC expression in MDR clinical strains could help addressing this question.

(iii) σ^E-regulated sRNAs: MicA and RybB are the two major σ^E-regulated sRNAs (Johansen et al. 2006; Gogol et al. 2011). MicA was first identified in a global *E. coli* sRNA screen and observed to accumulate as a 70-nucleotide transcript when cells ceased growth (Argaman et al. 2001). Two groups have now demonstrated that MicA accounts for much of the stationary phase specific instability of *ompA* mRNA (Udekwu and Wagner 2007; Rasmussen et al. 2005). In wild type cells, MicA levels inversely correlate with *ompA* mRNA levels during growth and the stationary phase specific decrease of *ompA* mRNA levels is abrogated upon *micA* deletion. Furthermore, overexpression of MicA results in reduction of OmpA protein levels. Overall, the underlying molecular mechanism of MicA is similar to that of MicC and MicF (i.e. by masking the *ompA* RBS), MicA inhibits the formation of the translation initiation complex and accelerates the RNase E-dependent decay of this mRNA (Udekwu and Wagner 2007; Rasmussen et al. 2005). In *E. coli*, RybB is a 80-nucleotide sRNA (Wassarman and Kiley 2010; Thompson et al. 2007; Johansen et al. 2006). Transient induction of RybB was first shown to decrease the levels of the mRNAs encoding OmpC and OmpW. Like most *trans*-acting sRNAs, RybB binds to Hfq and act by imperfect base-pairing with targets mRNAs (Johansen et al. 2006; Bouvier et al. 2008; Balbontín et al. 2010). Several groups working with *E. coli* and *Salmonella* reported that both RybB and MicA require σ^E for their transcription and down-regulate σ^E activity, thereby creating an autoregulatory loop (Thompson et al. 2007; Johansen et al. 2006; Udekwu and Wagner 2007; Papenfort et al. 2006). Interestingly, these studies questioned the primary simplistic model in which the specialized function of σ^E-dependent sRNAs is to stop *de novo* synthesis of abundant OMPs upon σ^E induction. In fact, RybB and MicA together target >30 mRNAs of *E. coli*: MicA represses the synthesis of major OMPs by targeting mRNAs of *ompA*, *ompX* and *lamB* and non-OMP targets like *phoP*, *lpxT* and *htrG* (Gogol et al. 2011; Coornaert et al. 2010); RybB has even more *omp* mRNA targets and non-*omp* targets that include *waaR*, *htrG*, *fadL* and *rbsK/B* (Papenfort et al. 2006; Gogol et al. 2011; Klein and Raina 2015). Overall, the main function of MicA and RybB sRNAs seems to protect the cell from the loss of viability when σ^E activity is inadequate (Gogol et al. 2011). The connection between the σ^E regulon and sRNA down-regulation of OMPs has been further reinforced by the identification of RseX (RNA suppressor of the extracytoplasmic stress protease RseP) as another sRNA regulator of OmpA and OmpC expression. RseX was uncovered in a screen for multicopy suppressors of the growth defect associated with the depletion of one of the periplasmic proteases that degrade the RseA anti-σ factor (Douchin et al. 2006). Since RseA keeps σ^E activity low, the levels of the σ^E-regulated periplasmic chaperones required for proper outer membrane protein transport are constitutively low in this background. Several multicopy clones isolated in this screen encoded a 91-nucleotide RNA denoted RseX. The *rseX* gene is on the same strand and upstream of *yedS*, which encodes a putative homolog of OmpS1, an OMP in *Salmonella typhi*. Hfq binds the sRNA in vitro and is required for the suppression phenotype. The *ompA* and *ompC* transcripts were identified as targets of RseX because they were captured when RNA extracted from *E. coli* was incubated with synthetic RseX (Douchin et al. 2006). RseX is predicted to be capable of fairly extensive base-pairing across the ribosome-binding sites of both *ompA* and *ompC* mRNAs, and RseX was shown to interact with *ompA* in mobility shift assays. The downregulation of OmpA and OmpC expression is sufficient to explain the suppressor phenotype of the RseX clone, since the periplasmic protease deficient strain is also viable when the *ompA* and *ompC* genes are deleted (Douchin et al. 2006). Thus far, the RseX sRNA has only been detected in cells with the multicopy clone; conditions that lead to RseX expression from the chromosome have not been reported.

Chemical Stress Responses and Multidrug Resistance

Regulation of porin expression also occurs in response to the presence of chemicals, including antibiotics, by activating regulatory proteins of the XylS-AraC family. All possess helix-turn-helix (HTH) motives as DNA binding domains for remodeling target gene expression (regulons). Importantly, these pathways are considered as global transcriptional regulators of MDR in *Enterobacteriaceae* by altering the cell envelope permeability directly or indirectly (at the transcriptional and post-transcriptional levels) by simultaneously decreasing the expression of porins and increasing that of multidrug efflux pumps (Chou et al. 1993; Cohen et al. 1988; Chubiz and Rao 2011; Rosner et al. 1991; Gambino et al. 1993; Ariza et al. 1995; Tanaka et al. 1997; Rosenberg et al. 2003), and are often interconnected.

In *E. coli* and other *Enterobacteriaceae*, the *mar* (multiple antibiotic resistance) locus comprises two divergent transcriptional units: *marC* and *marRAB* separated by the operator *marO* (George and Levy 1983; Cohen et al. 1993a, b; Sulavik et al. 1997; Chollet et al. 2002). The *marRAB* operon responds to a variety of compounds (Hächler et al. 1991; Ariza et al. 1994; Cohen et al. 1993c; Miller et al. 1994) including antibiotics such as tetracycline, chloramphenicol and imipenem (Hächler et al. 1991; Bornet et al. 2000). Deletion or inactivation of the *marRAB* operon results in increased susceptibility to multiple antibiotics, a variety of oxidative stress agents, and organic solvents (George and Levy 1983; Cohen et al. 1993a, b; White et al. 1997; Sharma et al. 2017). Here, one has to note that *marR* encodes a transcriptional repressor that prevents the activation of the Mar regulatory cascade in the absence of chemical stress, and *marA* encodes the XylS-AraC transcriptional activator that controls the expression of target genes (George and Levy 1983; Cohen et al. 1993a, b; Martin et al. 1995; Seoane and Levy 1995; Alekshun and Levy 1997). Previous analysis of *marO* indicated the presence of two MarR binding sites (site I and site II) that consist of repeated inverted sequences (Martin et al. 1995). Site I is located close to the -35 and -10 boxes of the *marRAB* promoter and site II encompasses the MarR RBS; thus MarR negatively regulates the expression of the *marRAB* operon as well as it represses its own synthesis. Activation (or de-repression) of the Mar response can occur upon mutations in *marO* or *marR*, which both affect the binding of MarR to *marO*, or in the presence of inducing chemicals such as salicylate, which also acts by direct binding to MarR and inactivate MarR function. Several mutations in *marO* and *marR* have been identified among MDR clinical isolates of *Enterobacteriaceae*, leading to the constitutive activation of MarA. Remarkably, *marR* mutations, including deletions and point mutations, are found all through its sequence (Cohen et al. 1993; Martin et al. 1995; Maneewannakul and Levy 1996), and thus do not allow the identification of a minimal region necessary for its function. MarA is an important regulator in *E. coli* implicated in adaptation to the environment and protection against external aggressions, by controlling the expression more than 60 genes, directly—by binding to conserved "marbox" sequences present in the promoter region of target genes—or indirectly (Barbosa and Levy 2000). In particular, MDR in *mar* mutants occurs through a combined decrease in drug influx, via a decrease in OmpF expression (Cohen et al. 1988; Gambino et al. 1993), and increase in active drug efflux, via an overexpression of the AcrAB efflux pump (Okusu et al. 1996). With our purpose here on porin regulation, MarA negatively regulates the production of OmpF at the post-transcriptional level, by activating the expression of the sRNA MicF (Gambino et al. 1993).

SoxS is the effector of the *soxRS* global superoxide response regulon. SoxS exhibits about 50% homology with MarA and also belongs to the XylS-AraC family of transcriptional activators (Demple 1996). In the presence of oxidizing agents such as H_2O_2, NO or paraquat, SoxR shifts from a reduced (inactive) to oxidized (active) state and activates the transcription of *soxS* (Demple 1996; Martin et al. 1999). SoxS is involved in MDR in *E. coli* and *Salmonella enterica* Serovar Typhimurium and can induce the transcription of *micF* and *acrAB*. Marboxes are also target sequences for the binding of SoxS and the phenotype induced by SoxS is similar to that induced by MarA (Miller and Sulavik 1996). Interestingly, SoxS is also able to activate the expression of MarA and promote the activation of the mar regulon even in the absence of typical Mar-inducing chemicals (Martin and Rosner 1997). Mutations in *soxR* have been identified in clinical strains of *E. coli* and *S. enterica* isolated from patients undergoing quinolone treatment as well as in laboratory strains selected on high concentrations of fluoroquinolones, and shown to contribute to MDR (O'Regan et al. 2009; Bialek-Davenet et al. 2011; Kehrenberg et al. 2009).

Rob also regulates genes involved in resistance to antibiotics, organic solvents and heavy metals (Ariza et al. 1995; Nakajima et al. 1995). Over-expression of Rob in *E. coli* produces both increased organic solvent tolerance and low-level resistance to multiple antimicrobial agents, due to increased expression of AcrAB and a decreased expression of OmpF (Jair et al. 1996). In contrast to MarA and SoxS, Rob is constitutively expressed, and its precise role in the regulation of MDR is still unknown.

RamA was first described in a *K. pneumoniae* MDR mutant (George et al. 1995). It shares 45% identity with MarA with high conservation of the two DNA binding motifs essential to the regulatory function. Thus, it can be expected that RamA and MarA might share overlapping regulons. Expression of *ramA* in *E. coli* generates a high-level resistance to multiple antibiotics including chloramphenicol, tetracycline, tigecycline, fluoroquinolones, trimethoprim, and act by a decreased expression of OmpF and an active efflux (O'Regan et al. 2009; George et al. 1995; Schneiders et al. 2003). RamA has been identified in *Salmonella enterica* serovar Paratyphi B and Typhimurium, and in *E. aerogenes* and *E. cloacae* (Yassien et al. 2002; Chollet et al. 2004; van der Straaten et al. 2004; Keeney et al. 2007). In *E. aerogenes*, this regulator is involved in the modification of OM permeability and in the active extrusion of intracellular antibiotics (Chollet et al. 2004). RamA has also been demonstrated to bind to the *mar* operator in *E. coli* and enhance its transcription, suggesting an interaction between these two systems (Yassien et al. 2002; Chollet et al. 2004). The putative marbox sequence within the *ram* promoter is well conserved according to the consensus (Martin et al. 1999). This suggests that MarA could regulate the transcription of *ramA*, although constitutive expression of RamA results in a MDR phenotype even in the absence of the *mar* locus. RamA is a transcriptional activator of the Mar regulon and is also a self-governing activator of the MDR cascade. RamA plays a role in the oxidative stress response in partnership with *soxRS*, but seems more important than *marA* and *soxS* in the development of MDR in *Salmonella* spp. (van der Straaten et al. 2004; Ricci et al. 2006).

Negative regulation by repressors of porins and efflux pump also impacts MDR phenotype. OmpX is a small OMP, of which overexpression is associated with a decreased expression of Omp36 (the OmpC ortholog in *E. aerogenes*) and a decreased susceptibility to β-lactams. Studies have indicated that expression of OmpX itself is controlled by a number of environmental factors, including salicylate via MarA and paraquat via SoxS (Dupont et al. 2004). A very rapid MarA-dependent response pathway for upregulation of *ompX* has been shown to occur within 60–120 min upon cell exposure to salicylate. This work by Dupont *et al.* identified a dramatic decrease in OmpF levels, as a first line of defense together with the development of resistance to β-lactams and fluoroquinolones by altering OM permeability (Dupont et al. 2007).

Biogenesis of β-Barrel Outer Membrane Proteins

In Gram-negative bacteria, the IM contains integral membrane proteins that have an α-helical fold consisting of one or more α-helices which anchor the protein into the membrane. The OM, however, contains almost exclusively integral membrane proteins that have a β-barrel fold (the porins) consisting of 8–26 antiparallel β-strands that anchor the protein in the OM. In the last decade, studies have identified BAM (β-barrel assembly machinery) as a general pathway and conserved machinery that is responsible for the biogenesis of OMPs (Voulhoux et al. 2003; Knowles et al. 2009; Hagan et al. 2011; Ricci and Silhavy 2012; Kim et al. 2012; Noinaj et al. 2015).

OMPs are synthesized in the cytoplasm in the form of precursors with an N-terminal leader peptide, which directs them to the Sec translocon for transport across the inner membrane (Driessen and Nouwen 2008). In the periplasm, molecular chaperones such as SurA and Skp escort mature OMPs during their transit to the outer membrane (Behrens et al. 2001; Bitto and McKay 2003; Chen and Henning 1996; Walton et al. 2009; Sklar et al. 2007), where they are recognized by the multicomponent BAM complex for folding, assembly and insertion into the OM. Interestingly, genetic analyses have shown that mutants lacking SurA or Skp are viable, but cells cannot tolerate loss of both, suggesting they function in different biogenesis pathways (Sklar et al. 2007; Rizzitello et al. 2001). Indeed, while major OMPs, such as porins, show preference for the SurA pathway (Sklar et al. 2007), only LptD—an essential OMP involved by the biogenesis of lipopolysaccharide—has been shown to require the Skp pathway (Schwalm et al. 2013). Additionally, it has been proposed that the primary role of the Skp pathway, acting together with the periplasmic protease DegP (Krojer et al. 2008; Shen et al. 2009; Ge et al. 2014), may be to rescue OMPs that have fallen off the normal assembly pathway, particularly when cells encounter stressful conditions (Sklar et al. 2007). In *E. coli*, the BAM complex consists of five components called BamA (YaeT/Omp85), BamB (YfgL), BamC (NlpB), BamD (YfiO), and BamE (SmpA) (Wu et al. 2005; Malinverni et al. 2006; Sklar et al. 2007). BamA, a β-barrel OMP itself, is the central and essential component of the complex; BamB, BamC, BamD, and BamE are all lipoproteins, which are anchored to the OM via lipidation of their N-terminal cysteine residue. BamA and BamD are essential for viability; however, all components are required for efficient OMP biogenesis (Malinverni et al. 2006; Hagan et al. 2010). Studies have shown that both BamB and BamD interact directly with BamA via non-overlapping binding sites, while BamC and BamE interact directly with BamD to stabilize the complex (Malinverni et al. 2006; Sklar et al. 2007). Structures of all the BAM components have now been reported including partial complexes of BamAB and BamCD (Kim et al. 2007, 2011; Albrecht and Zeth 2011; Noinaj et al. 2011, 2013; Heuck et al. 2011; Knowles et al. 2011; Dong et al. 2012; Ni et al. 2014; Albrecht et al. 2014; Jansen et al. 2015). The full-length structure of BamA from *Neisseria gonorrhoeae* revealed a large periplasmic domain consisting of five polypeptide transport associated (POTRA) domains and a C-terminal 16-stranded β-barrel domain. Subsequent studies showed that lateral opening of the barrel domain was required for function in BamA, strengthening an existing hypothesis that the barrel domain must open laterally in the membrane to allow insertion of the substrate OMPs into the outer membrane (Noinaj et al. 2013, 2014; Albrecht et al. 2014). It has been proposed that BamB might serve as a scaffold, assisting in the handoff of OMPs by SurA or Skp to BamA, while BamC, BamD, and BamE may serve support roles in regulating the function of BamA (Noinaj et al. 2011; Jansen et al. 2012, 2015). These structures have offered clues to how each component may function within the complex; however, the lack of structural information regarding the fully assembled complex has hindered progress towards exploring the mechanism further. To address this, Backelar et al. have solved the structure of the BAM complex from *E. coli* and showed that the periplasmic domain of BamA in a closed state prevents access to the barrel lumen from the periplasm (Bakelar et al. 2016). Furthermore, binding of BamCDE to BamA causes an unprecedented conformational change, leading to opening of the top of the barrel domain along the exit pore and structural rearrangement of the lateral opening site. These structural changes suggest that the role of BamCDE may be to modulate the conformational states of BamA, thereby serving as a regulatory step in the function of the BAM complex (Bakelar et al. 2016).

In being essential for OMP biogenesis, the BAM components and periplasmic chaperones may serve as attractive targets for the development of novel antibacterial drugs in the near future. This has been recently exemplified by the discovery of a peptidomimetic antibiotic that targets BamA and disrupts the OM of *E. coli* (Urfer et al. 2016).

Porins and Antibiotic Resistance

Route for Antibiotic Uptake

Several Gram-negative bacterial species of the ESKAPE group (Boucher et al. 2009, 2013), such as *Pseudomonas aeruginosa* and *Acinetobacter baumannii*, possess an innate resistance to β-lactam antibiotics, due to their extremely low OM permeability (Vila et al. 2007; Chevalier et al. 2017). As previously mentioned, with our focus here on *Enterobacteriaceae* species (such as *Citrobacter*, *Enterobacter*, *Escherichia*, *Klebsiella*, etc.), β-lactam uptake is closely associated to the presence of general porins in the OM—*E. coli* OmpC and OmpF porins being the archetype of the general non-specific enterobacterial porins (Zgurskaya et al. 2015; Nikaido 2003; Delcour 2009; Pagès et al. 2008). By using a skillful approach, Allam et al. (2017) have recently dissected the role of porins and OM permeability in the translocation of ceftazidime, a clinically relevant cephalosporin, across the OM in *E. coli*. The data obtained provide a clear illustration of the dual contribution of OM and enzymatic barriers on translocation, as previously mentioned in the pioneer model proposed by H. Nikaido (Yoshimura and Nikaido 1985).

Several publications have described modifications of the porin expression pattern in antibiotic-resistant *Enterobacteriaceae*: some show alterations of the OmpF/C expression ratio in favor of OmpC, with lower antibiotic permeability, while others show a drastic reduction in the porin production level as a whole, or produce of a mutated porin that negatively impacts the pore properties (for reviews see Pagès et al. 2008; Masi et al. 2017). As a consequence, strains either isolated from human specimens during the course of an antibiotherapy or selected for growth in the presence of increasing antibiotic concentrations show a decrease in cephalosporin and carbapenem susceptibility (Davin-Régli et al. 2008). This altered porin phenotype is also commonly associated to the synthesis of β-lactamases, cephalosporinases, or carbapenemases, which efficiently contribute to a high level of β-lactam resistance (Davin-Régli et al. 2008; Martínez-Martínez 2008; Blair et al. 2015; Nicolas-

Chanoine et al. 2018). Here, it is also important to consider one of the clinical impacts of reducing drug influx across the OM: some β-lactams (e.g. imipenem, cefoxitine), reaching the periplasmic space at low concentrations, are potent inducers of β-lactamase production (Fisher and Mobashery 2014; Bush 2018). As such, reduced OM permeability through porin modifications is the starting point for bigger adaptive changes, leading to high resistance levels to all β-lactams (Nicolas-Chanoine et al. 2018).

By exploring each mechanism we illustrate a bacterial adaptive response to antibiotherapy that leads to MDR and preserve a relative fitness cost (Ferenci 2005; Phan and Ferenci 2017).

Antibiotic Resistance Caused by Loss or Modification of Porin

- ***Porin exchange and decreased porin production***

Today a large number of publications have reported the loss of porins in clinical isolates collected during antibiotic treatments of infected patients. Here, we will review a few well-documented cases illustrating porin loss and antibiotic resistance in *E. coli*, *E. aerogenes* and *K. pneumoniae*.

Expression of OmpC-over OmpF-type porins is usually described in *Enterobacteriaceae* including *E. coli*, *E. aerogenes*, *E. cloacae* and *K. pneumoniae* (Pagès et al. 2008; Dam et al. 2018). Indeed, in their host habitat, *Enterobacteriaceae* encounter environmental conditions (i.e., elevated osmolarity, temperature, presence of bile of salts and accessible nutrients) that favor the expression of OmpC instead of OmpF orthologues (see section "Envelope Stress Responses"). In *E. aerogenes*, sequential alterations in porin expression have been reported during antibiotic treatments of infected patients: during imipenem (IMI) therapy, diverse phenotypes successively appeared IMI intermediate-ertapenem (ERT) resistant exhibiting Omp36 production, and finally, IMI-resistant and ERT-resistant with lack of porins (Lavigne et al. 2013; Philippe et al. 2015). Similar recent observations have been reported with *K. pneumoniae* clinical strains (Nikaido 1985; Bialek et al. 2010; Pantel et al. 2016; Humphries and Hemarajata 2017; Pulzova et al. 2017; Wise et al. 2018; Ma et al. 2018), which thereby confirm previous reports (for a review see Pagès et al. 2008). Interestingly, these porin minus strains were significantly less virulent compared to others isolates indicating that this porin failure can induce some adverse fitness problems in colonization and infection models (Ferenci 2005; Phan and Ferenci 2017; Bialek et al. 2010; Pantel et al. 2016; Lavigne et al. 2012).

Several observations indicate alteration of the porin balance (OmpF/OmpC) during antibiotherapy is due to regulation or mutation events in porin gene or in a regulatory cascade that control bacterial envelope permeability (for reviews see Pagès et al. 2008; Davin-Régli et al. 2008; Dam et al. 2018). Regarding *E. aerogenes* and *K. pneumoniae*, clinical isolates that exhibit a loss of porin or a reduced level of porin expression in association with LamB overexpression have been described (Ardanuy et al. 1998; Gayet et al. 2003). Interestingly, an OmpX overexpression has been reported in these resistant isolates and this protein has been previously reported to downregulate the expression of general porins such as OmpF and OmpC (Dupont et al. 2007; Gayet et al. 2003).

Recently, three consecutive *K. pneumoniae* isolates were collected from a single patient during treatment. They were resistant to several antibiotic classes, such as extended-spectrum cephalosporins, aminoglycosides, tetracycline, chloramphenicol, and fluoroquinolones, but expressed different carbapenem-susceptibility phenotypes. The first isolate displayed non-carbapenemase-related carbapenem resistance, the second was susceptible to all carbapenems and the third exhibited heterogeneous susceptibility to carbapenems (Bialek-Davenet et al. 2017). These three isolates were isogenic strains as they all carried mutations responsible for carbapenem resistance found in the first isolate and, the second and third isolates exhibited additional mutations in the regulators of Pho regulon. These mutations induced the expression of the PhoE porin and restored susceptibility to carbapenems (Bialek-Davenet et al. 2017). The involvement of *phoE* expression in carbapenem susceptibility has first been suggested by Kaczmarek et al. following the concomitant isolation of two *K. pneumoniae* strains producing a plasmid-encoded AmpC and lacking general porins with respect to their respective phenotype towards carbapenems (Kaczmarek et al. 2006). This specific porin exchange reflects the bacterial strategy based on the regulation of porin expression that provides appropriate function for nutrient uptake as well as reduced permeability for antibiotics (Pagès et al. 2008; Knopp and Andersson 2015).

The continuous exposure to antibiotics selects for successive modifications of porin expression, resulting in further reduced influx at each step. A complete impermeability to β-lactams is achieved via total absence of porin in resistant isolates and may represent an "extreme step" in the porin adaptive response (Pagès et al. 2008; Masi et al. 2017). Porin loss can induce severe reduction of bacterial fitness due to restricted entry of nutrients and essential metabolites, but allow bacterial growth in the presence of antibiotics (Davin-Régli et al. 2008; Ferenci 2005; Phan and Ferenci 2017). Moreover, a rapid change in porin expression confers an important advantage to the pathogen compared to the susceptible commensal microflora during antibiotic treatment. Interestingly, the impact of porin expression has been also demonstrated on bacterial susceptibility towards the combination of ceftazidime-avibactam—the latter is a β-lactamase inhibitor—in *E. coli*, *E. aerogenes* and *K. pneumoniae* isolates (Pagès et al. 2015).

Another option for *Enterobacteriaceae* to maintain fitness following the loss of OmpF and OmpC orthologues involves an original porin exchange to express the quiescent porin OmpN (Doménech-Sánchez et al. 1999). Primarily identified in a resistant clinical strain of *K. pneumoniae*, the expression of OmpK37 provides low levels of antibiotic susceptibility. In *E. coli*, dual expression of OmpN and MicC sRNA occurs under specific conditions that activate the σ^E envelope stress response (Dam ct al. 2017) and alter antibiotic susceptibility (Dam et al., personal results). Although OmpN orthologues are structurally similar to OmpF and OmpC orthologues, this difference in antibiotic permeability could be due to differences in the physicochemical properties of amino acid residues near the constriction region (Pagès et al. 2008; Doménech-Sánchez et al. 1999).

- ***Porin mutations***

Clinical examples of altered antibiotic translocation across the porin channel also involve amino acid substitution at or near the constriction region of the porin channel. To date, two series of OmpC-type porin mutants have been isolated from patients under chemotherapy and characterized (Dé et al. 2001; Thiolas et al. 2004, 2005; Lou et al. 2011; Bajaj et al. 2016). Interestingly, all these mutations yield significant conformational changes in the OmpC pore lumen strongly reduced translocation of β-lactams thus bacterial susceptibilities.

In parallel, the role of specific amino acid residues has been investigated by site directed mutagenesis for many years and pointed to the importance of the constriction region of the porin channel (see section "Functional Assays").

Antibiotic Permeation Assays with New Approaches

Our limited understanding of the molecular basis for compound entry into and efflux out of Gram-negative bacteria is a key bottleneck for the rational discovery of novel antibacterial drugs. This section describes recent approaches that address this knowledge gap.

- ***Defining molecular rules of translocation***

The antibiotic class for which we understand the most with respect to OM translocation is the β-lactams. Indeed, the role of porins in β-lactam translocation and the structural features of compounds required for uptake have been studied for nearly three decades by Nikaido and co-workers. The main approaches use liposome swelling assays with purified porins (Nikaido and Rosenberg 1981, 1983; Nakae 1976; Yoshimura and Nikaido 1985) and determination of permeation rates into intact cells, which could be calculated from the degradation rates by an endogenous AmpC β-lactamase (Nikaido and Rosenberg 1981; Nikaido et al. 1983; Kojima and Nikaido 2013, 2014; Zimmermann and Rosselet 1977; Nikaido and Normark 1987; Sugawara et al. 2016). Early studies with these two approaches showed that permeation across porins is a passive diffusion mechanism with a mean of roughly 1 μm/s for zwitterionic compounds through OmpF and that: (i) there is an inverse relationship between the hydrophobicity and permeation rate through porins among monoanionic cephalosporins; (ii) zwitterionic compounds penetrate rapidly while additional negative charges retard transport; and (iii) OmpF is more permeable than OmpC. This latter observation was first attributed to the OmpC pore being slightly more constricted in this porin compared to OmpF (Cowan et al. 1992; Baslé et al. 2006; Schulz 2002; Dutzler et al. 1999; Acosta-Gutierrez et al. 2018). Although the two porins share high sequence similarity, the pore interior is more negative in

OmpC than in OmpF. This is in line with the previous observation that OmpC is more cation-selective than OmpF (Benz et al. 1985) and can also account for the low permeability of OmpC for anionic β-lactams. Moreover, the replacement of all ten titratable residues that differ between OmpC and OmpF in the pore-lining region leads to the exchange of antibiotic permeation properties (Kojima and Nikaido 2014). Together, these structural and functional data clearly demonstrate that the charge distribution at pore linings, but not pore size, is a critical parameter that physiologically distinguishes OmpC from OmpF.

The recent structural elucidation of many porins has prompted several groups to study their structural and physical properties in order to predict the translocation of small molecules. There has been great progress in this area despite the intense computational efforts that are required for these studies. Computational study of translocation is distinct from traditional structure-based drug design, as it is not simply seeking to measure binding. Instead, molecular dynamics (MD) simulations are used to explore the potential trajectories that small molecules may take to traverse porins, which could lead to a better understanding of the structural and physicochemical features that contribute to recognition and permeation. Some important observations have already emerged. Ceccarelli and co-workers have demonstrated that changes in the electric field in the constriction zone of mutated OmpC in *E. coli* impacts the translocation of imipenem and meropenem (Lou et al. 2011; Hajjar et al. 2010). Even more recently, starting from the structure of several porin orthologues and small permeating molecules (i.e., β-lactam antibiotics), we proposed a physical mechanism underlying transport and condense it in a computationally efficient scoring function (Acosta-Gutierrez et al. 2018). This scoring function is based on the structural and physicochemical properties of both the porins at the CR (i.e., the electric field, the electrostatic potential and the size) and the compounds described by their partial atomic charges (charges and dipole) and size (minimal projection area). The scoring function shows good agreement with experimental translocation data obtained by using liposome swelling assays. Thus, this should enable structure-activity relationship (SAR) studies with respect to the screening of virtual databases to identify molecules with optimal permeability through porins and the rational drug-design for the optimization of antibiotics with poor permeation (Acosta-Gutierrez et al. 2018).

Electrophysiology has also been used in combination with MD simulations approaches (Nestorovich et al. 2002; Danelon et al. 2006; Winterhalter and Ceccarelli 2015) or susceptibility assays (James et al. 2009) to elucidate drug uptake through various porins as changes in ion flux (i.e., noise analysis) can be readily measured in these systems (Fig. 4.3c, d). For example, Bezrukov and co-workers have demonstrated that ampicillin transiently reduces the ion current through the OmpF channel in a pH-dependent manner, with a maximum decrease in the pH range where ampicillin is zwitterionic (Nestorovich et al. 2002). In addition, MD simulations suggested that zwitterionic ampicillin occludes the OmpF pore by interacting simultaneously with negatively charged residues of L3 (E117) and positively charged residues of the opposite barrel wall (R132, R82 and R42) (Nestorovich et al. 2002). Such complementation between the charge distribution of the drug and the

CR of OmpF has also been found for amoxicillin, another zwitterionic β-lactam (Danelon et al. 2006). In contrast, no interaction was detected for the di-anionic carbenicillin or the mono-anionic azlocillin and piperacillin (Danelon et al. 2006). Thus, it appears that, although weak and transient, interactions at the CR correlate with enhanced diffusion of compounds through OmpF. This is consistent with previous findings that site-directed mutagenesis of key residues in the porin CR affects β-lactam uptake and susceptibility. For example, the diffusion of radiolabeled cefepime is dramatically reduced in the G119D and G119E OmpF mutants while the R132A and R132D mutants show an increased influx rate (Simonet et al. 2000). However, it is worthwhile to note that these electrophysiological data are obtained under conditions that are not physiologically relevant, with high salt concentrations and high concentrations of drugs (in the mM range). Also, changes in ion flux by the addition of small molecules may only reflect the reversible blocking at the CR without actual permeation. Recently, a number of tricks have been developed to circumvent this problem. For example, molecular binding can be distinguished from transport by applying an external force on charged molecules. As such, an electric field can drag or repel charged permeating molecules, which is then translated into modification of the residence time. Thus, measuring the average translocation time for different external electric fields should provide a clear picture (Singh et al. 2012). In the case of uncharged molecules, the molecular movement could be modulated by electro-osmotic flow. Most of the porin channels have an excess of one charge species, thus an externally applied electric field would cause a net unidirectional incoming flow; and molecules diffusing with the flow would have shorter residence times. In a series of investigations the specific permeation of α-cyclodextrin across CymA of *K. oxytoca* was quantified and the individual steps have been simulated (Bhamidimarri et al. 2016). A more recent suggestion to enhance the signal to translocation is to engineer a barrier for exit (Wang et al. 2018). To identify true permeation the residence time revealed from ion current fluctuations in absence and presence of an additional barrier at the periplasmic side of the porin to identify inasmuch the molecules reaching the exit. Starting from the well-studied porin OmpF a single point mutation at position 181 $OmpF^{E181C}$ was introduced and crosslinked with either sodium (2-sulfonatoethyl) methanethiosulfonate (MTSES) or glutathion. The modification of $OmpF^{E181C}$ by MTSES builds a barrier sufficient to block the pathway of norfloxacin. The modulation of the interaction dwell time allows us to conclude on successful permeation of norfloxacin across wild-type OmpF. This approach might allow discriminating blockage from translocation events for a wide range of substrates. A potential application could be screening for scaffolds to improve the permeability of antibiotics. This approach allows working in the μM range and allows parallelization. However the data analysis is time consuming, depending on the quality of information one to a few substrates can be characterized per day (Wang et al. 2018) A different approach is to use from an unbalanced charge accumulation (Ghai et al. 2017, 2018) (Fig. 4.3e). Adding the charged compound of interest on one side of the channel induce a concentration driven flux. In most cases one of the ion diffuses slower compared to the counter ion, the difference in flux create the so called diffusion potential. Measuring the diffusion potential as a function of the concentration gradient provides relative fluxes. In combination with a single channel conductance measurement a more quantitative analysis is possible. However, to avoid electrode effects salt bridges are needed and thus the experimental volume cannot be readily miniaturized (mM solubility of the compound, large volume). Note that the cations permeation strongly depends on that of the anions and vice-versa. The permeation of charged compounds depends on the actual solvent salt available. In case of charged antibiotic molecules their channel permeability can be estimated by application of a concentration gradient and measuring the potential built up by an unequal permeation of the cation versus the anion. This technique is similar to a selectivity measurement and a numerical solution based on the Goldman-Hodgkin-Katz ion current equation can be used. Permeation of three β-lactamase inhibitors (avibactam, sulbactam and tazobactam) through OmpF and OmpC orthologues from four enterobacterial species was recently characterized by using this approach (Ghai et al. 2017). Here again MD simulations can provide an effective energy barrier (Fig. 4.3f). In case of OmpF, the energy barrier is low and broad whereas for the β-lactam inhibitors the barrier is substantially higher and narrower. Surprisingly there is also a shallow affinity site just before the entry.

These approaches need to be further validated in vivo by using susceptibility and accumulation assays in intact bacterial cells. To date, only one study has combined whole-cell susceptibility and cell-free assays to analyze the translocation of β-lactams across Omp36 of *E. aerogenes* (James et al. 2009). In this study, high affinity constant (k_{on}) values for ertapenem and cefepime binding to Omp36 correlated well with their rapid killing. Conversely, ampicillin and ceftazidime were shown to have low k_{on} values. Although the rates of killing were not reported, other studies have demonstrated that these two antibiotics show a preference for OmpF-type channels (Acosta-Gutierrez et al. 2018; Kojima and Nikaido 2014; Pagès et al. 2015; Ziervogel and Roux 2013).

Mathematical modeling approaches represent an alternative to the computationally intensive MD study of porin permeation and/or antibiotic efflux. This approach could provide quantitative methods to compare compounds where ultimately the effects of influx and efflux could be calculated across a series of compounds. Initial approaches toward this modeling were built upon the premise that compound uptake into cells was driven by passive diffusion and thus should obey Fick's Law (Nichols 2017; Kojima and Nikaido 2013). More recently, Zgurskaya, Rybenkov and co-workers proposed a more general model, which depends on two parameters to account for the rate of uptake and efflux, and allows analysis of the nonlinearity of accumulation (Westfall et al. 2017). This is nicely exemplified by the fitting for the uptake of the bis-benzimide Hoechst 33342, a fluorescent topoisomerase inhibitor, in wild type cells as well as cells expressing a mutant hyperporinated variant of FhuA, which lacks its central plug domain (Krishnamoorthy et al. 2016).

Taken together, these learnings should dissuade us from seeking an overarching single principle based solely on physicochemical and structural properties of compounds to discover novel antibacterial drugs. Thus, we are more likely to define emerging guidelines for optimal permeation into Gram-negative bacteria, which may differ between species.

- ***Measuring antibiotic accumulation in bacteria***

Traditionally, comparative MIC assays (i.e., comparing MICs of a compound of interest against wild type versus isogenic mutant strains) have been the workhorse of assays used in the field of antibacterial research to define the molecular drivers of antibiotic uptake and efflux. Indeed, demonstration of measurable MICs against efflux-compromised or membrane-permeabilized strains is logically still among the first objectives for most of the earliest stage projects in the field. However, the possibility that a particular phenotype related to drug uptake might be masked due to either redundant or complementary functions for inactivated genes is a potential caveat to this approach. Comparative MICs is also a blunt tool when trying to evaluate kinetic effects such as rates of entry and/or efflux. An alternate approach, known as "time kill" assay, is traditionally employed to define the pharmacodynamic properties of antibiotics, can be used to perform more refined, kinetic assessments of the molecular drivers of antibiotic activity. However, the standard readout of this method, namely, determining relative bacterial CFUs at various time points after antibiotic exposure by manually counting serially diluted samples plated on agar, is very time-, labor-, and resource-intensive. We have recently addressed these limitations by adapting a previously described resazurin-based reporter system as a surrogate for bacterial viability (Fig. 4.4). The change in resorufin fluorescence in real-time qualitatively mirrors the relative changes in bacterial load and tracks drug-mediated cell killing over time. Specifically, we adapted a tightly regulated, arabinose-inducible system to a similar expression of any porin of interest in a porin-less *E. coli* mutant strain (in both efflux-plus and efflux-minus backgrounds). We found this system to be sufficiently sensitive to establish relationships between porin-mediated translocation and antibacterial activity for the passage of β-lactams through OmpF and OmpC orthologues of several *Enterobacteriaceae* (Masi et al. 2017).

Zgurskaya and co-workers have developed a different take on porin overexpression to study antibiotic permeation (Westfall et al. 2017). This cleverly engineered "hyperporination" system is based on the controlled expression of a large, nonselective variant of the FhuA transporter (without compromising the integrity of the membrane bilayer) in either wild type or efflux-deficient *E. coli* strains. The authors used this approach to determine the contributions of membrane permeability vs efflux in *E. coli* for several antibiotic classes that differed significantly in their physicochemical properties and mechanisms of action. They have since deployed the system in more problematic pathogens with distinct permeability barriers, namely, *A. baumannii*, *P. aeruginosa*, and *Burkholderia spp* (Krishnamoorthy et al. 2017; Zgurskaya et al. 2018). Doing so, they found that no universal rule of antibiotic permeation into Gram-negative bacteria exists. However, they were able to classify antibiotics into four groups according to specific biological determinants such as the presence of specific porins in the OM, targeting of the OM, or specific recognition by efflux pumps. Finally, the range of physicochemical properties for each cluster of antibiotics was found to be quite broad.

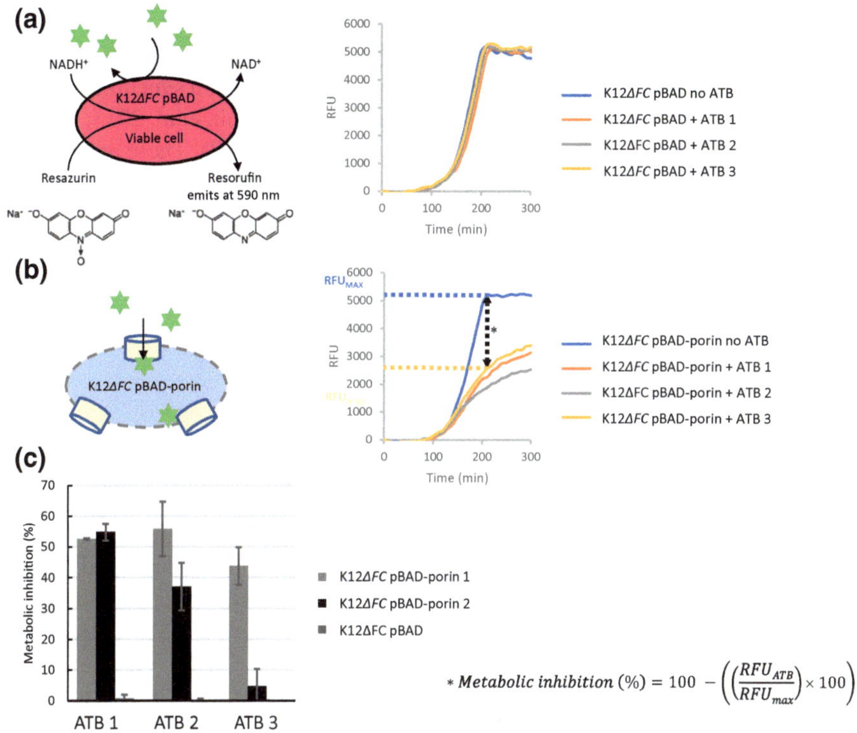

Fig. 4.4 Methods for measuring intracellular accumulation and porin-mediated uptake of antibiotics: whole cell based and real-time susceptibility assays. Actively metabolizing bacterial cells are able to reduce resazurin to resofurin, which emits fluorescence at 590 nm (**a**). Individual porins were cloned into a pBAD vector and expressed to similar levels upon the addition of arabinose in a porin-less *E. coli* strain (K12*ΔFC*) (**b**). Kinetic assays in the absence or in the presence of antibiotics (ATB 1, 2, 3, …) provide a population-based direct readout for drug antibacterial activity, which can be translated as the porin contribution to drug translocation. This assay is also useful to investigate the impact of drugs structural and physicochemical properties on their capacity of translocate through one or several types of porins. In (**a**) and (**b**), RFU stands for relative fluorescence units (RFU_{MAX} is obtained in the absence of antibiotics). **c** For clarity, percentage of metabolic inhibition is shown for each strain after exposure to a single antibiotic. As expected, the porin-less strain transformed with the empty vector showed is not affected by the presence of antibiotics. These results also indicate porin 1 allows translocation of ATB 1, 2 and 3 to a similar level as cells expressing porin 1 show similar metabolic inhibition (45–55%) in the presence of any of these three compounds. In contrast, porin 2 is permissive for translocation of ATB 1 and 2 but not ATB 3, as cells expressing porin 2 only exhibited a metabolic inhibition of 5%

Numerous biophysical methods to quantify antibiotic passage and accumulation have been described in the literature, including electrophysiological measurements on reconstituted membranes or whole cells (Tran et al. 2014) and synchrotron deep-UV (micro)spectrofluorimetry (Kaščáková et al. 2012; Cinquin et al. 2015; Vergalli et al. 2017, 2018) (Fig. 4.5) and nanofluidics (Hong et al. 2017). These approaches, while informative, are not readily amenable to development into a screening platform. In contrast, mass spectroscopy-based methods may have more potential to quantify intracellular antibiotic concentrations, as recently exemplified by the study from the Hergenrother group (Richter et al. 2017).

Concluding Remarks

It is clear that OM porins play a key role in bacterial life and adaptation by ensuring the uptake of nutrients necessary for the bacterial metabolism but also by controlling the diffusion of environmental molecules via their unique structural and functional properties. A number of well-documented reviews has previously described the role of porin channels, the constriction region and associated amino acid residues, in the diffusion rate of solutes across the bacterial envelope (Nikaido 2003; Delcour 2009; Pagès et al. 2008).

OM general porins have been investigated for many years, but structural and mechanistic insights into the process of antibiotic translocation have emerged only recently. Indeed, the molecular flux inside the porin channel ultimately determines the intracellular concentration of permeating compounds, thereby impacting on bacterial susceptibility (Masi et al. 2017). With the development of original and innovative methodologies, we can now measure in real-time the molecular flux through bacterial membranes and across individual purified porins (Masi et al. 2017).

These new studies will pave the way to identify and characterize the physicochemical properties of permeating compound as well as the channel domains that determine the OM permeability and ensure the addressing of compounds inside the bacterial cell. Moreover, the various methodological approaches using mass spectrometry, fluorimetry, microfluidic assays offer new opportunities to better understand the relationships between genetic control of the porin expression and bacterial adaptation to diverse environmental stresses. The measurement of intrabacterial accumulation of specific molecules, inducers, inhibitors, antimicrobial drugs, etc., performed in real-time represent an important breakthrough to study the early stages of envelope changes during bacterial life.

These advances will aid in efforts to understand a fundamental process in living cell: how the transport of hydrophilic molecules is organized across lipid bilayers and what are the molecular events that manage the efficiency of the diffusion rate in order to fit in with the metabolism need of a living cell.

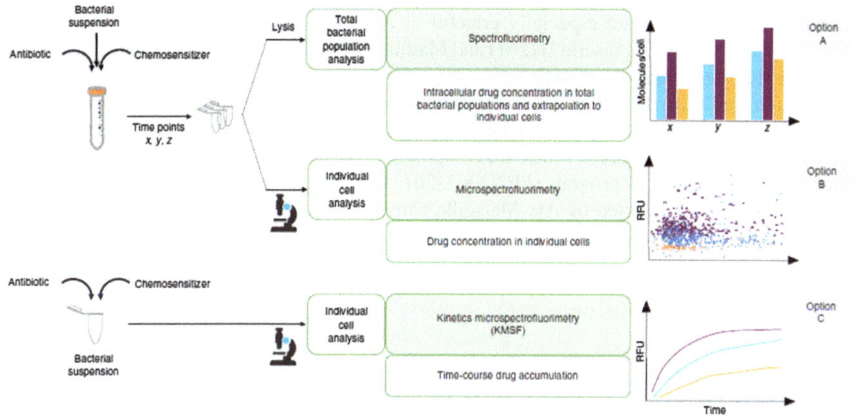

Fig. 4.5 Methods for measuring intracellular accumulation and porin-mediated uptake of antibiotics: Intracellular accumulation assays using intact cells and fluorescent drugs. Adapted with permission from Vergalli J, Dumont E, Pajović J, Cinquin B, Maigre L, Masi M, Réfrégiers M, Pagès JM (2018) Spectrofluorimetric quantification of antibiotic drug concentration in bacterial cells for the characterization of translocation across bacterial membranes. Nat Protoc 13:1348–1361 Copyright Nature Publishing 2018. In 1989, Chapman and Georgopapadakou first reported measurement of the intracellular accumulation of fleroxacin in *E. coli* based on the natural fluorescence of the fluoroquinolone (Chapman and Georgopapadakou 1989). Since then, the original protocol has been improved and adapted to other bacterial species in order to follow the intracellular accumulation of fluoroquinolones (McCaffrey et al. 1992; Mortimer and Piddock 1993; Asuquo and Piddock 1993; Piddock et al. 1999). However, a major limitation of this assay was the absence of internal standard, which limited the ability to quantify antibiotic content in single bacterial cells. A major breakthrough came in 2012, when Kašcáková et al. proposed the use of the intrinsic bacterial fluorescence (i.e., the relative fluorescence of tryptophan) as a reference to standardize the relative fluorescence of the accumulated fluoroquinolones in order to compare their respective accumulation levels in various bacterial strains and in different conditions (Kaščáková et al. 2012). Option A uses spectrofluorimetry measurements in whole cell lysates (WCLs) (Allam et al. 2017; Kaščáková et al. 2012; Cinquin et al. 2015; Vergalli et al. 2017, 2018). Typically, intact cells are incubated with the antibiotic in a series of defined conditions and centrifuged for bacteria collection. Then a cell-soft lysis procedure using glycine–HCl releases the intracellular content including the accumulated antibiotic, and the fluorescence spectra of the WCLs are read. For the creation of calibration curves, WCLs of untreated cells are mixed with increasing concentrations of antibiotics. Normalization of the antibiotic signal and calibration allow precise determination of the antibiotic concentration inside bacterial cells. Here, it is worth noting that spectrofluorimetry can be scaled up to a 96- or 384-well microplate format for high-throughput compound screening. Option B is based on microspectrofluorimetry, and the corresponding protocol follows that of spectrofluorimetry, except that the accumulated antibiotics are directly measured in individual intact cells. Here, illumination of the samples by deep ultraviolet (DUV) photons also allows standardization of the antibiotic fluorescence with respect to the intrinsic bacterial fluorescence (Allam et al. 2017; Kaščáková et al. 2012; Cinquin et al. 2015; Vergalli et al. 2018). Option C (KMSF) monitors the time-course accumulation of an antibiotic during a short period of time (2–30 min) (Vergalli et al. 2018). In this case, bacterial cells are incubated in the presence of the selected antibiotic and directly visualized under the DUV microscope, and serial measurements are made under DUV flashes. In options A, B and C data in blue, purple and orange correspond to *E. coli* AG100 (wild-type), AG100A (AG100 *acrB*::Kan, efflux minus), and AG102 (a gain-of-function *mar* mutant derivative of AG100 that overexpresses AcrAB) respectively. Although these assays have been primarily developed for studying the impact of drug efflux on antibiotic accumulation, these can now be adapted to porin-minus mutants to investigate the impact of porin-mediated influx of drugs

Acknowledgements We are especially grateful to Anne Davin-Regli, Julia Vergalli and Jean-Michel Bolla (UMR_MD1, Inserm U1261) and Matthieu Réfrégiers (Disco Beamline, Synchrotron Soleil) for sharing fruitful discussions.
The research leading to these results was conducted as part of the TRANSLOCATION consortium, and it has received support from the Innovative Medicines Initiatives Joint Undertaking under Grant Agreement n°115525, resources which are composed of financial contribution from the European Union's seventh framework program (FP7/2007–2013) and EFPIA companies in kind contribution. This work was also supported by Aix-Marseille Univ. and Service de Santé des Armées, and by Soleil programs.

References

Acosta-Gutierrez S, Ferrara L, Pathania M, Masi M, Wang J, Bodrenko I, Zahn M, Winterhalter M, Stavenger RA, Pages JM, Naismith JH, van den Berg B, Page M, Ceccarelli M (2018) Getting drugs into Gram-negative bacteria: rational rules for permeation through general porins. ACS Infect Dis. https://doi.org/10.1021/acsinfecdis.8b00108. [Epub ahead of print]

Ades SE, Connolly LE, Alba BM, Gross CA (1999) The *Escherichia coli* sigma(E)-dependent extracytoplasmic stress response is controlled by the regulated proteolysis of an anti-sigma factor. Genes Dev 13:2449–2461

Aiba H, Matsuyama S, Mizuno T, Mizushima S (1987) Function of *micF* as an antisense RNA in osmoregulatory expression of the *ompF* gene in *Escherichia coli*. J Bacteriol 169:3007–3012

Alba BM, Gross CA (2004) Regulation of the *Escherichia coli* sigma-dependent envelope stress response. Mol Microbiol 52:613–619

Alba BM, Leeds JA, Onufryk C, Lu CZ, Gross CA (2002) DegS and YaeL participate sequentially in the cleavage of RseA to activate the sigma(E)-dependent extracytoplasmic stress response. Genes Dev 16:2156–2168

Albrecht R, Zeth K (2011) Structural basis of outer membrane protein biogenesis in bacteria. J Biol Chem 286:27792–27803

Albrecht R, Schütz M, Oberhettinger P, Faulstich M, Bermejo I, Rudel T, Diederichs K, Zeth K (2014) Structure of BamA, an essential factor in outer membrane protein biogenesis. Acta Crystallogr D Biol Crystallogr 70:1779–1789

Alcaraz A, Nestorovich EM, Aguilella-Arzo M, Aguilella VM, Bezrukov SM (2004) Salting out the ionic selectivity of a wide channel: the asymmetry of OmpF. Biophys J 87:943–957

Alcaraz A, Nestorovich EM, López ML, García-Giménez E, Bezrukov SM, Aguilella VM (2009) Diffusion, exclusion, and specific binding in a large channel: a study of OmpF selectivity inversion. Biophys J 96:56–66

Alekshun MN, Levy SB (1997) Regulation of chromosomally mediated multiple antibiotic resistance: the *mar* regulon. Antimicrob Agents Chemother 41:2067–2075

Allam A, Maigre L, Vergalli J, Dumont E, Cinquin B, Alves de Sousa R, Pajovic J, Pinet E, Smith N, Herbeuval JP, Réfrégiers M, Artaud I, Pagès JM (2017) Microspectrofluorimetry to dissect the permeation of ceftazidime in Gram-negative bacteria. Sci Rep 7:986

Andersen J, Delihas N (1990) *micF* RNA binds to the 5′ end of *ompF* mRNA and to a protein from *Escherichia coli*. Biochemistry 29:9249–9256

Andersen J, Delihas N, Ikenaka K, Green PJ, Pines O, Ilercil O, Inouye M (1987) The isolation and characterization of RNA coded by the *micF* gene in *Escherichia coli*. Nucl Acids Res 15:2089–2101

Andersen J, Forst SA, Zhao K, Inouye M, Delihas N (1989) The function of *micF* RNA. *micF* RNA is a major factor in the thermal regulation of OmpF protein in *Escherichia coli*. J Biol Chem 264:17961–17970

Ardanuy C, Linares J, Dominguez MA, Hernandez-Allés S, Benedi VJ, Martinez-Martinez L (1998) Outer membrane profiles of clonally related *Klebsiella pneumoniae* isolates from clinical samples and activities of cephalosporins and carbapenems. Antimicrob Agents Chemother 42:1636–1640

Argaman L, Hershberg R, Vogel J, Bejerano G, Wagner EG, Margalit H, Altuvia S (2001) Novel small RNA-encoding genes in the intergenic regions of *Escherichia coli*. Curr Biol 11:941–950

Ariza RR, Cohen SP, Bachhawat N, Levy SB, Demple B (1994) Repressor mutations in the *marRAB* operon that activate oxidative stress genes and multiple antibiotic resistance in *Escherichia coli*. J Bacteriol 176:143–148

Ariza RR, Li Z, Ringstad N, Demple B (1995) Activation of multiple antibiotic resistance and binding of stress-inducible promoters by *Escherichia coli* Rob protein. J Bacteriol 177:1655–1661

Arunmanee W, Pathania M, Solovyova AS, Le Brun AP, Ridley H, Baslé A, van den Berg B, Lakey JH (2016) Gram-negative trimeric porins have specific LPS binding sites that are essential for porin biogenesis. Proc Natl Acad Sci USA 113:E5034–E5043

Asuquo AE, Piddock LJ (1993) Accumulation and killing kinetics of fifteen quinolones for *Escherichia coli*, *Staphylococcus aureus* and *Pseudomonas aeruginosa*. J Antimicrob Chemother 31:865–880

Bajaj H, Scorciapino MA, Moynié L, Page MG, Naismith JH, Ceccarelli M, Winterhalter M (2016) Molecular basis of filtering carbapenems by porins from β-lactam-resistant clinical strains of *Escherichia coli*. J Biol Chem 291:2837–2847

Bakelar J, Buchanan SK, Noinaj N (2016) The structure of the β-barrel assembly machinery complex. Science 351:180–186

Balbontín R, Fiorini F, Figueroa-Bossi N, Casadesús J, Bossi L (2010) Recognition of heptameric seed sequence underlies multi-target regulation by RybB small RNA in *Salmonella enterica*. Mol Microbiol 78:380–394

Barbosa TM, Levy SB (2000) Differential expression of over 60 chromosomal genes in *Escherichia coli* by constitutive expression of MarA. J Bacteriol 182:3467–3474

Baslé A, Iyer R, Delcour AH (2004) Subconductance states in OmpF gating. Biochim Biophys Acta 1664:100–107

Baslé A, Rummel G, Storici P, Rosenbusch JP, Schirmer T (2006) Crystal structure of osmoporin OmpC from *E. coli* at 2.0 A. J Mol Biol 362:933–942

Batchelor E, Walthers D, Kenney LJ, Goulian M (2005) The *Escherichia coli* CpxA-CpxR envelope stress response system regulates expression of the porins OmpF and OmpC. J Bacteriol 187:5723–5731

Behrens S, Maier R, de Cock H, Schmid FX, Gross CA (2001) The SurA periplasmic PPIase lacking its parvulin domains functions in vivo and has chaperone activity. EMBO J 20:285–294

Benz R, Janko K, Boos W, Lauger P (1978) Formation of large, ion-permeable membrane channels by the matrix protein (porin) of *Escherichia coli*. Biochim Biophys Acta 511:305–319

Benz R, Janko K, Lauger P (1979) Ionic selectivity of pores formed by the matrix protein (porin) of *Escherichia coli*. Biochim Biophys Acta 551:238–247

Benz R, Schmid A, Hancock RE (1985) Ion selectivity of gram-negative bacterial porins. J Bacteriol 162:722–727

Bhamidimarri SP, Prajapati JD, van den Berg B, Winterhalter M, Kleinekathöfer U (2016) Role of electroosmosis in the permeation of neutral molecules: CymA and cyclodextrin as an example. Biophys J 110:600–611

Bialek S, Lavigne JP, Chevalier J, Marcon E, Leflon-Guibout V, Davin A, Moreau R, Pagès JM, Nicolas-Chanoine MH (2010) Membrane efflux and influx modulate both multidrug resistance and virulence of *Klebsiella pneumoniae* in a *Caenorhabditis elegans* model. Antimicrob Agents Chemother 54:4373–4378

Bialek-Davenet S, Marcon E, Leflon-Guibout V, Lavigne JP, Bert F, Moreau R, Nicolas-Chanoine MH (2011) In vitro selection of *ramR* and *soxR* mutants overexpressing efflux systems by fluoroquinolones as well as cefoxitin in *Klebsiella pneumoniae*. Antimicrob Agents Chemother 55:2795–2802

Bialek-Davenet S, Mayer N, Vergalli J, Duprilot M, Brisse S, Pagès JM, Nicolas-Chanoine MH (2017) In-vivo loss of carbapenem resistance by extensively drug-resistant *Klebsiella pneumoniae* during treatment via porin expression modification. Sci Rep 7:6722

Bitto E, McKay DB (2003) The periplasmic molecular chaperone protein SurA binds a peptide motif that is characteristic of integral outer membrane proteins. J Biol Chem 278:49316–49322

Blair JM, Webber MA, Baylay AJ, Ogbolu DO, Piddock LJ (2015) Molecular mechanisms of antibiotic resistance. Nat Rev Microbiol 13:42–51

Borgstrom B (1974) Bile salts—their physiological functions in the gastrointestinal tract. Acta Med Scand 196:1–10

Bornet C, Davin-Regli A, Bosi C, Pages JM, Bollet C (2000) Imipenem resistance of *Enterobacter aerogenes* mediated by outer membrane permeability. J Clin Microbiol 38:1048–1052

Bossi L, Figueroa-Bossi N (2007) A small RNA downregulates LamB maltoporin in *Salmonella*. Mol Microbiol 65:799–810

Botos I, Noinaj N, Buchanan SK (2017) Insertion of proteins and lipopolysaccharide into the bacterial outer membrane. Philos Trans R Soc Lond B Biol Sci 372(1726). pii:20160224

Boucher HW, Talbot GH, Bradley JS, Edwards JE, Gilbert D, Rice LB, Scheld M, Spellberg B, Bartlett J (2009) Bad bugs, no drugs: no ESKAPE! An update from the Infectious Diseases Society of America. Clin Infect Dis 48:1–12

Boucher HW, Talbot GH, Benjamin DK Jr, Bradley J, Guidos RJ, Jones RN, Murray BE, Bonomo RA, Gilbert D (2013) 10 × '20 progress—development of new drugs active against gram-negative bacilli: an update from the Infectious Diseases Society of America. Clin Infect Dis 56:1685–1694

Bouvier M, Sharma CM, Mika F, Nierhaus KH, Vogel J (2008) Small RNA binding to 5′ mRNA coding region inhibits translational initiation. Mol Cell 32:827–837

Bredin J, Saint N, Malléa M, Dé E, Molle G, Pagès JM, Simonet V (2002) Alteration of pore properties of *Escherichia coli* OmpF induced by mutation of key residues in anti-loop 3 region. Biochem J 363:521–528

Bush K (2018) Past and present perspectives on β-lactamases. Antimicrob Agents Chemother 62:pii:e01076–18

Bush K, Courvalin P, Dantas G, Davies J, Eisenstein B, Huovinen P, Jacoby GA, Kishony R, Kreiswirth BN, Kutter E, Lerner SA, Levy S, Lewis K, Lomovskaya O, Miller JH, Mobashery S, Piddock LJ, Projan S, Thomas CM, Tomasz A, Tulkens PM, Walsh TR, Watson JD, Witkowski J, Witte W, Wright G, Yeh P, Zgurskaya HI (2011) Tackling antibiotic resistance. Nat Rev Microbiol 9:894–896

Cai SJ, Inouye M (2002) EnvZ-OmpR interaction and osmoregulation in *Escherichia coli*. J Biol Chem 277:24155–2415561

Castillo-Keller M, Vuong P, Misra R (2006) Novel mechanism of *Escherichia coli* porin regulation. J Bacteriol 188:576–586

Chaba R, Grigorova IL, Flynn JM, Baker TA, Gross CA (2007) Design principles of the proteolytic cascade governing the sigmaE-mediated envelope stress response in *Escherichia coli*: keys to graded, buffered, and rapid signal transduction. Genes Dev 21:124–136

Chapman JS, Georgopapadakou NH (1989) Fluorometric assay for fleroxacin uptake by bacterial cells. Antimicrob Agents Chemother 33:27–29

Charrel RN, Pagès JM, De Micco P, Mallea M (1996) Prevalence of outer membrane porin alteration in beta-lactam-antibiotic-resistant *Enterobacter aerogenes*. Antimicrob Agents Chemother 40:2854–2858

Chen R, Henning U (1996) A periplasmic protein (Skp) of *Escherichia coli* selectively binds a class of outer membrane proteins. Mol Microbiol 19:1287–1294

Chen S, Zhang A, Blyn LB, Storz G (2004) MicC, a second small-RNA regulator of Omp protein expression in *Escherichia coli*. J Bacteriol 186:6689–6697

Chevalier J, Pagès JM, Malléa M (1999) In vivo modification of porin activity conferring antibiotic resistance to *Enterobacter aerogenes*. Biochem Biophys Res Commun 266:248–251

Chevalier J, Pagès JM, Eyraud A, Malléa M (2000) Membrane permeability modifications are involved in antibiotic resistance in *Klebsiella pneumoniae*. Biochem Biophys Res Commun 274:496–499

Chevalier S, Bouffartigues E, Bodilis J, Maillot O, Lesouhaitier O, Feuilloley MGJ, Orange N, Dufour A, Cornelis P (2017) Structure, function and regulation of *Pseudomonas aeruginosa* porins. FEMS Microbiol Rev 41:698–722

Chimerel C, Movileanu L, Pezeshki S, Winterhalter M, Kleinekathöfer U (2008) Transport at the nanoscale: temperature dependence of ion conductance. Eur Biophys J 38:121–125

Chollet R, Bollet C, Chevalier J, Malléa M, Pagès JM, Davin-Regli A (2002) *mar* operon involved in multidrug resistance of *Enterobacter aerogenes*. Antimicrob Agents Chemother 46:1093–1097

Chollet R, Chevalier J, Bollet C, Pages JM, Davin-Regli A (2004) RamA is an alternate activator of the multidrug resistance cascade *in Enterobacter aerogenes*. Antimicrob Agents Chemother 48:2518–2523

Chou JH, Greenberg JT, Demple B (1993) Post-transcriptional repression of *Escherichia coli* OmpF protein in response to redox stress: positive control of the *micF* antisense RNA by the *soxRS* locus. J Bacteriol 175:1026–1031

Chubiz LM, Rao CV (2011) Role of the *mar-sox-rob* regulon in regulating outer membrane porin expression. J Bacteriol 193:2252–2260

Cinquin B, Maigre L, Pinet E, Chevalier J, Stavenger RA, Mills S, Réfrégiers M, Pagès JM (2015) Microspectrometric insights on the uptake of antibiotics at the single bacterial cell level. Sci Rep 5:17968

Cohen SP, McMurry LM, Levy SB (1988) *marA* locus causes decreased expression of OmpF porin in multiple-antibiotic-resistant (Mar) mutants of *Escherichia coli*. J Bacteriol 170:5416–5422

Cohen SP, Hächler H, Levy SB (1993a) Genetic and functional analysis of the multiple antibiotic resistance (*mar*) locus in *Escherichia coli*. J Bacteriol 175:1484–1492

Cohen SP, Yan W, Levy SB (1993b) A multidrug resistance regulatory chromosomal locus is widespread among enteric bacteria. J Infect Dis 168:484–488

Cohen SP, Levy SB, Foulds J, Rosner JL (1993c) Salicylate induction of antibiotic resistance in *Escherichia coli*: activation of the *mar* operon and a mar-independent pathway. J Bacteriol 175:7856–7862

Coornaert A, Lu A, Mandin P, Springer M, Gottesman S, Guillier M (2010) MicA sRNA links the PhoP regulon to cell envelope stress. Mol Microbiol 76:467–479

Cowan SW, Schirmer T, Rummel G, Steiert M, Ghosh R, Pauptit RA, Jansonius JN, Rosenbusch JP (1992) Crystal structures explain functional properties of two *E. coli* porins. Nature 358:727–733

Coyer J, Andersen J, Forst SA, Inouye M, Delihas N (1990) *micF* RNA in *ompB* mutants of *Escherichia coli*: different pathways regulate *micF* RNA levels in response to osmolarity and temperature change. J Bacteriol 172:4143–4150

Dam S, Pagès JM, Masi M (2017) Dual regulation of the small RNA MicC and the quiescent porin OmpN in response to antibiotic stress *in Escherichia coli*. Antibiotics (Basel) 6pii:E33

Dam S, Pagès JM, Masi M (2018) Stress responses, outer membrane permeability control and antimicrobial resistance in *Enterobacteriaceae*. Microbiology 164:260–267

Danelon C, Suenaga A, Winterhalter M, Yamato I (2003) Molecular origin of the cation selectivity in OmpF porin: single channel conductances vs. free energy calculation. Biophys Chem 104:591–603

Danelon C, Nestorovich EM, Winterhalter M, Ceccarelli M, Bezrukov SM (2006) Interaction of zwitterionic penicillins with the OmpF channel facilitates their translocation. Biophys J 90:1617–1627

Dartigalongue C, Missiakas D, Raina S (2000) Characterization of the *Escherichia coli* sigma E regulon. J Biol Chem 276:20866–20875

Davin-Régli A, Bolla JM, James CE, Lavigne JP, Chevalier J, Pagès JM (2008) Membrane permeability and regulation of drug "influx and efflux" in enterobacterial pathogens. Curr Drug Targets 9:750–759

De Las Penas A, Connolly L, Gross CA (1997) SigmaE is an essential sigma factor in *Escherichia coli*. J Bacteriol 179:6862–6864
Dé E, Baslé A, Jaquinod M, Saint N, Malléa M, Molle G, Pagès JM (2001) A new mechanism of antibiotic resistance in *Enterobacteriaceae* induced by a structural modification of the major porin. Mol Microbiol 41:189–198
Delcour AH (2009) Outer membrane permeability and antibiotic resistance. Biochim Biophys Acta 1794:808–816
Delhaye A, Collet JF, Laloux G (2016) Fine-Tuning of the Cpx envelope stress response is required for cell wall homeostasis in *Escherichia coli*. MBio 7:e00047-16
Delihas N, Forst S (2001) MicF: an antisense RNA gene involved in response of *Escherichia coli* to global stress factors. J Mol Biol 313:1–12
Demple B (1996) Redox signaling and gene control in the *Escherichia coli soxRS* oxidative stress regulon—a review. Gene 179:53–57
Doménech-Sánchez A, Hernández-Allés S, Martínez-Martínez L, Benedí VJ, Albertí S (1999) Identification and characterization of a new porin gene of *Klebsiella pneumoniae*: its role in beta-lactam antibiotic resistance. J Bacteriol 181(9):2726–2732
Dong C, Hou HF, Yang X, Shen YQ, Dong YH (2012) Structure of *Escherichia coli* BamD and its functional implications in outer membrane protein assembly. Acta Crystallogr D Biol Crystallogr 68:95–101
Douchin V, Bohn C, Bouloc P (2006) Down-regulation of porins by a small RNA bypasses the essentiality of the regulated intramembrane proteolysis protease RseP in *Escherichia coli*. J Biol Chem 281:12253–12259
Driessen AJ, Nouwen N (2008) Protein translocation across the bacterial cytoplasmic membrane. Annu Rev Biochem 77:643–667
Dupont M, Dé E, Chollet R, Chevalier J, Pagès JM (2004) *Enterobacter aerogenes* OmpX, a cation-selective channel mar- and osmo-regulated. FEBS Lett 569:27–30
Dupont M, James CE, Chevalier J, Pagès JM (2007) An early response to environmental stress involves regulation of OmpX and OmpF, two enterobacterial outer membrane pore-forming proteins. Antimicrob Agents Chemother 51:3190–3198
Dutzler R, Rummel G, Albertí S, Hernández-Allés S, Phale P, Rosenbusch J, Benedí V, Schirmer T (1999) Crystal structure and functional characterization of OmpK36, the osmoporin of *Klebsiella pneumoniae*. Structure 7:425–434
Ferenci T (1994) From sequence alignment to structure prediction: the case of the OmpF porin family. Mol Microbiol 14:188–189
Ferenci T (2005) Maintaining a healthy SPANC balance through regulatory and mutational adaptation. Mol Microbiol 57:1–8
Fisher JF, Mobashery S (2014) The sentinel role of peptidoglycan recycling in the β-lactam resistance of the Gram-negative *Enterobacteriaceae* and *Pseudomonas aeruginosa*. Bioorg Chem 56:41–48
Flynn JM, Neher SB, Kim YI, Sauer RT, Baker TA (2003) Proteomic discovery of cellular substrates of the ClpXP protease reveals five classes of ClpX-recognition signals. Mol Cell 11:671–683
Flynn JM, Levchenko I, Sauer RT, Baker TA (2004) Modulating substrate choice: the SspB adaptor delivers a regulator of the extracytoplasmic-stress response to the AAA+ protease ClpXP for degradation. Genes Dev 18:2292–2301
Fraimow HS, Tsigrelis C (2011) Antimicrobial resistance in the intensive care unit: mechanisms, epidemiology, and management of specific resistant pathogens. Crit Care Clin 27:163–205
Fröhlich KS, Papenfort K, Berger AA, Vogel J (2012) A conserved RpoS-dependent small RNA controls the synthesis of major porin OmpD. Nucl Acids Res 40:3623–3640
Gambino L, Gracheck SJ, Miller PF (1993) Overexpression of the MarA positive regulator is sufficient to confer multiple antibiotic resistance in *Escherichia coli*. J Bacteriol 175:2888–2894
Gayet S, Chollet R, Molle G, Pagès JM, Chevalier J (2003) Modification of outer membrane protein profile and evidence suggesting an active drug pump in *Enterobacter aerogenes* clinical strains. Antimicrob Agents Chemother 47:1555–1559

Ge X, Wang R, Ma J, Liu Y, Ezemaduka AN, Chen PR, Fu X, Chang Z (2014) DegP primarily functions as a protease for the biogenesis of β-barrel outer membrane proteins in the Gram-negative bacterium *Escherichia coli*. FEBS J 281:1226–1240

George AM, Levy SB (1983) Gene in the major cotransduction gap of the *Escherichia* coli K-12 linkage map required for the expression of chromosomal resistance to tetracycline and other antibiotics. J Bacteriol 155:541–548

George AM, Hall RM, Stokes HW (1995) Multidrug resistance in *Klebsiella pneumoniae*: a novel gene, *ramA*, confers a multidrug resistance phenotype in *Escherichia coli*. Microbiology 141:1909–1920

Gerken H, Charlson ES, Cicirelli EM, Kenney LJ, Misra R (2009) MzrA: a novel modulator of the EnvZ/OmpR two-component regulon. Mol Microbiol 72:1408–1422

Gerlach RG, Hensel M (2007) Protein secretion systems and adhesins: the molecular armory of Gram-negative pathogens. Int J Med Microbiol 297:401–415

Ghai I, Pira A, Scorciapino MA, Bodrenko I, Benier L, Ceccarelli M (2017) General method to determine the flux of charged molecules through nanopores applied to β-lactamase inhibitors and OmpF. J Phys Chem Lett 8:1295–1301

Ghai I, Bajaj H, Bafna JA, Hussein H, Winterhalter M, Wagner R (2018) Ampicillin permeation across OmpF, the major outer-membrane channel in *Escherichia coli*. J Biol Chem 293:7030–7037

Gogol EB, Rhodius VA, Papenfort K, Vogel J, Gross CA (2011) Small RNAs endow a transcriptional activator with essential repressor functions for single-tier control of a global stress regulon. Proc Natl Acad Sci USA 108:12875–12880

Gottesman S (2004) The small RNA regulators of *Escherichia coli*: roles and mechanisms. Annu Rev Microbiol 58:303–328

Guest RL, Raivio TL (2016) Role of the Gram-negative envelope stress response in the presence of antimicrobial agents. Trends Microbiol 24:377–390

Guillier M, Gottesman S (2006) Remodeling of the *Escherichia coli* outer membrane by two small regulatory RNAs. Mol Microbiol 59:231–247

Guillier M, Gottesman S, Storz G (2006) Modulating the outer membrane with small RNAs. Genes Dev 20:2338–2348

Guo MS, Updegrove TB, Gogol EB, Shabalina SA, Gross CA, Storz G (2014) MicL, a new σ^E-dependent sRNA, combats envelope stress by repressing synthesis of Lpp, the major outer membrane lipoprotein. Genes Dev 28:1620–1634

Hächler H, Cohen SP, Levy SB (1991) *marA*, a regulated locus which controls expression of chromosomal multiple antibiotic resistance in *Escherichia coli*. J Bacteriol 173:5532–5538

Hagan CL, Kim S, Daniel Kahne D (2010) Reconstitution of outer membrane protein assembly from purified components. Science 328:890–892

Hagan CL, Silhavy TJ, Kahne D (2011) β-barrel membrane protein assembly by the Bam complex. Annu Rev Biochem 80:189–210

Hajjar E, Bessonov A, Molitor A, Kumar A, Mahendran KR, Winterhalter M, Pagès JM, Ruggerone P, Ceccarelli M (2010) Toward screening for antibiotics with enhanced permeation properties through bacterial porins. Biochemistry 49:6928–6935

Harder KJ, Nikaido H, Matsuhashi M (1981) Mutants of *Escherichia coli* that are resistant to certain beta-lactam compounds lack the *ompF* porin. Antimicrob Agents Chemother 20:549–552

Hayden JD, Ades SE (2008) The extracytoplasmic stress factor, sigmaE, is required to maintain cell envelope integrity in *Escherichia coli*. PLoS ONE 3(2):e1573

Henderson JC, Zimmerman SM, Crofts AA, Boll JM, Kuhns LG, Herrera CM, Trent MS (2016) The power of asymmetry: architecture and assembly of the Gram-negative outer membrane lipid bilayer. Annu Rev Microbiol 70:255e78

Heuck A, Schleiffer A, Clausen T (2011) Augmenting β-augmentation: structural basis of how BamB binds BamA and may support folding of outer membrane proteins. J Mol Biol 406:659–666

Hong S, Moritz TJ, Rath CM, Tamrakar P, Lee P, Krucker T, Lee LP (2017) Assessing antibiotic permeability of Gram-negative Bacteria via nanofluidics. ACS Nano 11:6959–6967

Humphries RM, Hemarajata P (2017) Resistance to ceftazidime-avibactam in *Klebsiella pneumoniae* due to porin mutations and the increased expression of KPC-3. Antimicrob Agents Chemother 61:pii:e00537–17

Im W, Roux B (2002) Ions and counterions in a biological channel: a molecular dynamics simulation of OmpF porin from *Escherichia coli* in an explicit membrane with 1 M KCl aqueous salt solution. J Mol Biol 319:1177–1197

Jair KW, Yu X, Skarstad K, Thöny B, Fujita N, Ishihama A, Wolf RE Jr (1996) Transcriptional activation of promoters of the superoxide and multiple antibiotic resistance regulons by Rob, a binding protein of the *Escherichia coli* origin of chromosomal replication. J Bacteriol 178:2507–2513

James CE, Mahendran KR, Molitor A, Bolla JM, Bessonov AN, Winterhalter M, Pagès JM (2009) How beta-lactam antibiotics enter bacteria: a dialogue with the porins. PLoS ONE 4:e5453

Jansen KB, Baker SL, Sousa MC (2012) Crystal structure of BamB from *Pseudomonas aeruginosa* and functional evaluation of its conserved structural features. PLoS ONE 7:e49749

Jansen KB, Baker SL, Sousa MC (2015) Crystal structure of BamB bound to a periplasmic domain fragment of BamA, the central component of the β-barrel assembly machine. J Biol Chem 290:2126–2136

Jeanteur D, Lakey JH, Pattus F (1991) The bacterial porin superfamily: sequence alignment and structure prediction. Mol Microbiol 5:2153–2164

Johansen J, Rasmussen AA, Overgaard M, Valentin-Hansen P (2006) Conserved small non-coding RNAs that belong to the sigmaE regulon: role in down-regulation of outer membrane proteins. J Mol Biol 364:1–8

Johansen J, Eriksen M, Kallipolitis B, Valentin-Hansen P (2008) Down-regulation of outer membrane proteins by noncoding RNAs: unraveling the cAMP-CRP- and sigmaE-dependent CyaR-ompX regulatory case. J Mol Biol 383:1–9

Kaczmarek FM, Dib-Hajj F, Shang W, Gootz TD (2006) High-level carbapenem resistance in a *Klebsiella pneumoniae* clinical isolate is due to the combination of *bla*(ACT-1) beta-lactamase production, porin OmpK35/36 insertional inactivation, and down-regulation of the phosphate transport porin PhoE. Antimicrob Agents Chemother 50:3396–3406

Karshikoff A, Spassov V, Cowan SW, Ladenstein R, Schirmer T (1994) Electrostatic properties of two porin channels from *Escherichia coli*. J Mol Biol 240:372–384

Kaščáková S, Maigre L, Chevalier J, Réfrégiers M, Pagès JM. Antibiotic transport in resistant bacteria: synchrotron UV fluorescence microscopy to determine antibiotic accumulation with single cell resolution (2012) PLoS ONE 7:e38624

Keeney D, Ruzin A, Bradford PA (2007) RamA, a transcriptional regulator, and AcrAB, an RND-type efflux pump, are associated with decreased susceptibility to tigecycline in *Enterobacter cloacae*. Microb Drug Resist 3:1–6

Kehrenberg C, Cloeckaert A, Klein G, Schwarz S (2009) Decreased fluoroquinolone susceptibility in mutants of *Salmonella* serovars other than Typhimurium: detection of novel mutations involved in modulated expression of *ramA* and *soxS*. J Antimicrob Chemother 64:1175–1180

Kim S, Malinverni JC, Sliz P, Silhavy TJ, Harrison SC, Kahne D (2007) Structure and function of an essential component of the outer membrane protein assembly machine. Science 317:961–964

Kim KH, Aulack S, Paetzel M (2011) Crystal structure of β-barrel assembly machinery BamCD protein complex. J Biol Chem 286:39116–39121

Kim KH, Aulakh S, Paetzel M (2012) The bacterial outer membrane β-barrel assembly machinery. Protein Sci 21:751–768

Klein G, Raina S (2015) Regulated control of the assembly and diversity of LPS by noncoding sRNAs. Biomed Res Int 2015:153561

Knopp M, Andersson DI (2015) Amelioration of the fitness costs of antibiotic resistance due to reduced outer membrane permeability by upregulation of alternative porins. Mol Biol Evol 32:3252–3263

Knowles TJ, Scott-Tucker A, Overduin M, Henderson IR (2009) Membrane protein architects: the role of the BAM complex in outer membrane protein assembly. Nat Rev Microbiol 7:206–214

Knowles TJ, Browning DF, Jeeves M, Maderbocus R, Rajesh S, Sridhar P, Manoli E, Emery D, Sommer U, Spencer A, Leyton DL, Squire D, Chaudhuri RR, Viant MR, Cunningham AF, Henderson IR, Overduin M (2011) Structure and function of BamE within the outer membrane and the β-barrel assembly machine. EMBO Rep 12:123–128

Kobayashi Y, Takahashi I, Nakae T (1982) Diffusion of beta-lactam antibiotics through liposome membranes containing purified porins. Antimicrob Agents Chemother 22:775–780

Kojima S, Nikaido H (2013) Permeation rates of penicillins indicate that *Escherichia coli* porins function principally as nonspecific channels. Proc Natl Acad Sci USA 110:E2629–E2634

Kojima S, Nikaido H (2014) High salt concentrations increase permeability through OmpC channels of *Escherichia coli*. J Biol Chem 289:26464–26473

Koronakis V, Eswaran J, Hughes C (2004) Structure and function of TolC: the bacterial exit duct for proteins and drugs. Annu Rev Biochem 73:467–489

Kostakioti M, Newman CL, Thanassi DG, Stathopoulos C (2005) Mechanisms of protein export across the bacterial outer membrane. J Bacteriol 187:4306–4314

Krishnamoorthy G, Wolloscheck D, Weeks JW, Croft C, Rybenkov VV, Zgurskaya HI (2016) Breaking the permeability barrier of *Escherichia coli* by controlled hyperporination of the outer membrane. Antimicrob Agents Chemother 60:7372–7381

Krishnamoorthy G, Leus IV, Weeks JW, Wolloscheck D, Rybenkov VV, Zgurskaya HI (2017) Synergy between active efflux and outer membrane diffusion defines rules of antibiotic permeation into Gram-negative bacteria. MBio pii:e01172–17

Krojer T, Sawa J, Schäfer E, Saibil HR, Ehrmann M, Clausen T (2008) Structural basis for the regulated protease and chaperone function of DegP. Nature 453:885–890

Lavigne JP, Sotto A, Nicolas-Chanoine MH, Bouziges N, Bourg G, Davin-Regli A, Pagès JM (2012) Membrane permeability, a pivotal function involved in antibiotic resistance and virulence in *Enterobacter aerogenes* clinical isolates. Clin Microbiol Infect 18:539–545

Lavigne JP, Sotto A, Nicolas-Chanoine MH, Bouziges N, Pagès JM, Davin-Regli A (2013) An adaptive response of *Enterobacter aerogenes* to imipenem: regulation of porin balance in clinical isolates. Int J Antimicrob Agents 41:130–136

Li XZ, Plésiat P, Nikaido H (2015) The challenge of efflux-mediated antibiotic resistance in Gram-negative bacteria. Clin Microbiol Rev 28:337–418

Lou H, Chen M, Black SS, Bushell SR, Ceccarelli M, Mach T, Beis K, Low AS, Bamford VA, Booth IR, Bayley H, Naismith JH (2011) Altered antibiotic transport in OmpC mutants isolated from a series of clinical strains of multi-drug resistant *E. coli*. PLoS ONE 6:e25825

Luckey M, Nikaido H (1980) Specificity of diffusion channels produced by lambda phage receptor protein of *Escherichia coli*. Proc Natl Acad Sci USA 77:167–171

Ma P, Laibinis HH, Ernst CM, Hung DT (2018) Carbapenem resistance caused by high-level expression of OXA-663 β-lactamase in an OmpK36-deficient *Klebsiella pneumoniae* clinical isolate. Antimicrob Agents Chemother. pii:AAC.01281-18 [Epub ahead of print]

Mahendran KR, Kreir M, Weingart H, Fertig N, Winterhalter M (2010) Permeation of antibiotics through *Escherichia coli* OmpF and OmpC porins: screening for influx on a single-molecule level. J Biomol Screen 15:302–307

Mahoney TF, Silhavy TJ (2013) The Cpx stress response confers resistance to some, but not all, bactericidal antibiotics. J Bacteriol 195:1869–1874

Majdalani N, Gottesman S (2005) The Rcs phosphorelay: a complex signal transduction system. Annu Rev Microbiol 59:379–405

Malinverni JC, Werner J, Kim S, Sklar JG, Kahne D, Misra R, Silhavy TJ (2006) YfiO stabilizes the YaeT complex and is essential for OM protein assembly in *Escherichia coli*. Mol Microbiol 61:151–164

Mallea M, Chevalier J, Bornet C, Eyraud A, Davin-Regli A, Bollet C, Pagès JM (1998) Porin alteration and active efflux: two in vivo drug resistance strategies used by *Enterobacter aerogenes*. Microbiology 144:3003–3009

Maneewannakul K, Levy SB (1996) Identification for *mar* mutants among quinolone-resistant clinical isolates of *Escherichia coli*. Antimicrob Agents Chemother 40:1695–1698

Martin RG, Rosner JL (1997) Fis, an accessorial factor for transcriptional activation of the mar (multiple antibiotic resistance) promoter of *Escherichia coli* in the presence of the activator MarA, SoxS, or Rob. J Bacteriol 179:7410–7419
Martin RG, Nyantakyi PS, Rosner JL (1995) Regulation of the multiple antibiotic resistance (*mar*) regulon by *marORA* sequences in *Escherichia coli*. J Bacteriol 177:4176–4178
Martin RG, Gillette WK, Rhee S, Rosner JL (1999) Structural requirements for marbox function in transcriptional activation of mar/sox/rob regulon promoters I *Escherichia coli*: sequence, orientation and spatial relationship to the core promoter. Mol Microbiol 34:431–441
Martínez-Martínez L (2008) Extended-spectrum beta-lactamases and the permeability barrier. Clin Microbiol Infect 14:82–89
Masi M, Réfrégiers M, Pos KM, Pagès JM (2017) Mechanisms of envelope permeability and antibiotic influx and efflux in Gram-negative bacteria. Nat Microbiol 2:17001
Matsubara M, Kitaoka SI, Takeda SI, Mizuno T (2000) Tuning of the porin expression under anaerobic growth conditions by his-to-Asp cross-phosphorelay through both the EnvZ-osmosensor and ArcB-anaerosensor in *Escherichia coli*. Genes Cells 5:555–569
McCaffrey C, Bertasso A, Pace J, Georgopapadakou NH (1992) Quinolone accumulation in *Escherichia coli*, *Pseudomonas aeruginosa*, and *Staphylococcus aureus*. Antimicrob Agents Chemother 36:1601–1605
Miller PF, Sulavik MC (1996) Overlaps and parallels in the regulation of intrinsic multiple-antibiotic resistance in *Escherichia coli*. Mol Microbiol 21:441–448
Miller PF, Gambino LF, Sulavik MC, Gracheck SJ (1994) Genetic relationship between *soxRS* and *mar* loci in promoting multiple antibiotic resistance in *Escherichia coli*. Antimicrob Agents Chemother 38:1773–1779
Missiakas D, Mayer MP, Lemaire M, Georgopoulos C, Raina S (1997) Modulation of the *Escherichia coli* sigmaE (RpoE) heat-shock transcription-factor activity by the RseA, RseB and RseC proteins. Mol Microbiol 24:355–371
Mizuno T, Mizushima S (1990) Signal transduction and gene regulation through the phosphorylation of two regulatory components: the molecular basis for the osmotic regulation of the porin genes. Mol Microbiol 4:1077–1082
Moreau PL (2014) Protective role of the RpoE (σE) and Cpx envelope stress responses against gentamicin killing of nongrowing *Escherichia coli* incubated under aerobic, phosphate starvation conditions. FEMS Microbiol Lett 357:151–156
Mortimer PG, Piddock LJ (1993) The accumulation of five antibacterial agents in porin-deficient mutants of *Escherichia coli*. J Antimicrob Chemother 32:195–213
Nakae T (1976) Identification of outer membrane protein of *E. coli* that produces transmembrane channels in reconstituted vesicle membranes. Biochem Biophys Res Commun 71:871–884
Nakae T, Nikaido H (1975) Outer membrane as a diffusion barrier in *Salmonella typhimurium*. Penetration of oligo- and polysaccharides into isolated outer membrane vesicles and cells with degraded peptidoglycan layer. J Biol Chem 250:7359–7365
Nakajima H, Kobayashi K, Kobayashi M, Asako H, Aono R (1995) Overexpression of the *robA* gene increases organic solvent tolerance and multiple antibiotic and heavy metal ion resistance in *Escherichia coli*. Appl Environ Microbiol 61:2302–2307
Nestorovich EM, Danelon C, Winterhalter M, Bezrukov SM (2002) Designed to penetrate: time-resolved interaction of single antibiotic molecules with bacterial pores. Proc Natl Acad Sci USA 99:9789–9794
Ni D, Wang Y, Yang X, Zhou H, Hou X, Cao B, Lu Z, Zhao X, Yang K, Huang Y (2014) Structural and functional analysis of the β-barrel domain of BamA from *Escherichia coli*. FASEB J 28:2677–2685
NIAID (2014). http://www.niaid.nih.gov/topics/antimicrotopics/antimicrobialresistance/documents/arstrategicplan2014.pdf
Nichols WW (2017) Modeling the kinetics of the permeation of antibacterial agents into growing bacteria and its interplay with efflux. Antimicrob Agents Chemother 61:e02576e16

Nicolas-Chanoine MH, Mayer N, Guyot K, Dumont E, Pagès JM (2018) Interplay between membrane permeability and enzymatic barrier leads to antibiotic-dependent resistance in *Klebsiella Pneumoniae*. Front Microbiol 9:1422
Nikaido H (1985) Role of permeability barriers in resistance to beta-lactam antibiotics. Pharmacol Ther 27:197–231
Nikaido H (1992) Porins and specific channels of bacterial outer membranes. Mol Microbiol 6:435–442
Nikaido H (2003) Molecular basis of bacterial outer membrane permeability revisited. Microbiol Mol Biol Rev 67:593–656
Nikaido H, Normark S (1987) Sensitivity of *Escherichia coli* to various beta-lactams is determined by the interplay of outer membrane permeability and degradation by periplasmic beta-lactamases: a quantitative predictive treatment. Mol Microbiol 1:29–36
Nikaido H, Pagès JM (2012) Broad-specificity efflux pumps and their role in multidrug resistance of Gram-negative bacteria. FEMS Microbiol Rev 36:340–363
Nikaido H, Rosenberg EY (1981) Effect on solute size on diffusion rates through the transmembrane pores of the outer membrane of *Escherichia coli*. J Gen Physiol 77:121–135
Nikaido H, Rosenberg EY (1983) Porin channels in *Escherichia coli*: studies with liposomes reconstituted from purified proteins. J Bacteriol 153:241–252
Nikaido H, Rosenberg EY, Foulds J (1983) Porin channels in *Escherichia coli*: studies with beta-lactams in intact cells. J Bacteriol 153:232–240
Noinaj N, Fairman JW, Buchanan SK (2011) The crystal structure of BamB suggests interactions with BamA and its role within the BAM complex. J Mol Biol 407:248–260
Noinaj N, Kuszak AJ, Gumbart JC, Lukacik P, Chang H, Easley NC, Lithgow T, Buchanan SK (2013) Structural insight into the biogenesis of β-barrel membrane proteins. Nature 501:385–390
Noinaj N, Kuszak AJ, Balusek C, Gumbart JC, Buchanan SK (2014) Lateral opening and exit pore formation are required for BamA function. Structure 22:1055–1062
Noinaj N, Rollauer SE, Buchanan SK (2015) The β-barrel membrane protein insertase machinery from Gram-negative bacteria. Curr Opin Struct Biol 31:35–42
Noinaj N, Gumbart JC, Buchanan SK (2017) The β-barrel assembly machinery in motion. Nat Rev Microbiol 15:197–204
O'Neill J (2016). https://amr-review.org/sites/default/files/160525_Final%20paper_with%20cover.pdf
Okusu H, Ma D, Nikaido H (1996) AcrAB efflux pump plays a major role in the antibiotic resistance phenotype of *Escherichia coli* multiple antibiotic resistance (Mar) mutants. J Bacteriol 178:306–308
O'Regan E, Quinn T, Pagès JM, McCusker M, Piddock L, Fanning S (2009) Multiple regulatory pathways associated with high-level ciprofloxacin and multidrug resistance in *Salmonella enterica* serovar *enteritidis*: involvement of RamA and other global regulators. Antimicrob Agents Chemother 53:1080–1087
Page MG, Bush K (2014) Discovery and development of new antibacterial agents targeting Gram-negative bacteria in the era of pandrug resistance: is the future promising? Curr Opin Pharmacol 18:91–97
Pagès JM, James CE, Winterhalter M (2008) The porin and the permeating antibiotic: a selective diffusion barrier in Gram-negative bacteria. Nat Rev Microbiol 6:893–903
Pagès JM, Peslier S, Keating TA, Lavigne JP, Nichols WW (2015) The role of the outer membrane and porins in the susceptibility of β-lactamase-producing *Enterobacteriaceae* to ceftazidime-avibactam. Antimicrob Agents Chemother 60:1349–1359
Pantel A, Dunyach-Remy C, Ngba Essebe C, Mesureur J, Sotto A, Pagès JM, Nicolas-Chanoine MH, Lavigne JP (2016) Modulation of membrane influx and efflux in *Escherichia coli* sequence type 131 has an impact on bacterial motility, biofilm formation, and virulence in a *Caenorhabditis elegans* model. Antimicrob Agents Chemother 60:2901–2911

Papenfort K, Pfeiffer V, Mika F, Lucchini S, Hinton JC, Vogel J (2006) SigmaE-dependent small RNAs of *Salmonella* respond to membrane stress by accelerating global *omp* mRNA decay. Mol Microbiol 62:1674–1688
Pezeshki S, Chimerel C, Bessonov AN, Winterhalter M, Kleinekathöfer U (2009) Understanding ion conductance on a molecular level: an all-atom modeling of the bacterial porin OmpF. Biophys J 97:1898–1906
Pfeiffer V, Sittka A, Tomer R, Tedin K, Brinkmann V, Vogel J (2007) A small non-coding RNA of the invasion gene island (SPI-1) represses outer membrane protein synthesis from the *Salmonella* core genome. Mol Microbiol 66:1174–1191
Phale PS, Philippsen A, Widmer C, Phale VP, Rosenbusch JP, Schirmer T (2001) Role of charged residues at the OmpF porin channel constriction probed by mutagenesis and simulation. Biochemistry 40:6319–6325
Phan K, Ferenci T (2017) The fitness costs and trade-off shapes associated with the exclusion of nine antibiotics by OmpF porin channels. ISME J 11:1472–1482
Philippe N, Maigre L, Santini S, Pinet E, Claverie JM, Davin-Régli AV, Pagès JM, Masi M (2015) In vivo evolution of bacterial resistance in two cases of *Enterobacter aerogenes* infections during treatment with imipenem. PLoS ONE 10:e0138828
Piddock LJ, Jin YF, Ricci V, Asuquo AE (1999) Quinolone accumulation by *Pseudomonas aeruginosa*, *Staphylococcus aureus* and *Escherichia coli*. J Antimicrob Chemother 43:61–70
Pierluigi V, Maddalena G, Sara T, Russell L (2015) Treatment of MDR-Gram negative infections in the 21st century: a never ending threat for clinicians. Curr Opin Pharmacol 24:30–37
Pratt LA, Hsing W, Gibson KE, Silhavy TJ (1996) From acids to *osmZ*: multiple factors influence synthesis of the OmpF and OmpC porins in *Escherichia coli*. Mol Microbiol 20:911–917
Price NL, Raivio TL (2009) Characterization of the Cpx regulon in Escherichia coli strain MC4100. J Bacteriol 191:1798–1815
Prilipov A, Phale PS, Koebnik R, Widmer C, Rosenbusch JP (1998) Identification and characterization of two quiescent porin genes, *nmpC* and *ompN*, in *Escherichia coli* B^E. J Bacteriol 180:3388–3392
Pulzova L, Navratilova L, Comor L (2017) Alterations in outer membrane permeability favor drug-resistant phenotype of *Klebsiella pneumoniae*. Microb Drug Resist 23:413–420
Raivio TL (2014) Everything old is new again: an update on current research on the Cpx envelope stress response. Biochim Biophys Acta 1843:1529–1541
Raivio TL, Leblanc SK, Price NL (2013) The *Escherichia coli* Cpx envelope stress response regulates genes of diverse function that impact antibiotic resistance and membrane integrity. J Bacteriol 195:2755–2767
Ramani N, Hedeshian M, Freundlich M (1994) *micF* antisense RNA has a major role in osmoregulation of OmpF in *Escherichia coli*. J Bacteriol 176:5005–5010
Rasmussen AA, Eriksen M, Gilany K, Udesen C, Franch T, Petersen C, Valentin-Hansen P (2005) Regulation of *ompA* mRNA stability: the role of a small regulatory RNA in growth phase-dependent control. Mol Microbiol 58:1421–1429
Rhodius VA, Suh WC, Nonaka G, West J, Gross CA (2006) Conserved and variable functions of the sigmaE stress response in related genomes. PLoS Biol 4(1):e2
Ricci D, Silhavy TJ (2012) The Bam machine: a molecular cooper. Biochim Biophys Acta 1818:1067–1084
Ricci V, Tzakas P, Buckley A, Piddock LJ (2006) Ciprofloxacin-resistant *Salmonella enterica* serovar Typhimurium strains are difficult to select in the absence of AcrB and TolC. Antimicrob Agents Chemother 50:38–42
Richter MF, Drown BS, Riley AP, Garcia A, Shirai T, Svec RL, Hergenrother PJ (2017) Predictive compound accumulation rules yield a broad-spectrum antibiotic. Nature 545:299–304
Rizzitello AE, Harper JR, Silhavy TJ (2001) Genetic evidence for parallel pathways of chaperone activity in the periplasm of *Escherichia coli*. J Bacteriol 183:6794–6800

Rosenberg ED, Bertenthal ML, Nilles K, Bertrand P, Nikaido H (2003) Bile salts and fatty acids induce the expression of *Escherichia coli* AcrAB multidrug efflux pump through their interaction with Rob regulatory protein. Mol Microbiol 48:1609–1619

Rosner JL, Chai KJ, Foulds J (1991) Regulation of OmpF porin expression by salicylate in *Escherichia coli*. J Bacteriol 173:5631–5638

Rowley G, Spector M, Kormanec J, Roberts M (2006) Pushing the envelope: extracytoplasmic stress responses in bacterial pathogens. Nat Rev Microbiol 4:383–394

Rudner DZ, Fawcett P, Losick R (1999) A family of membrane-embedded metalloproteases involved in regulated proteolysis of membrane-associated transcription factors. Proc Natl Acad Sci USA 96:14765–14770

Ruiz N, Silhavy TJ (2005) Sensing external stress: watchdogs of the *Escherichia coli* cell envelope. Curr Opin Microbiol 8:122–126

Saint N, Lou KL, Widmer C, Luckey M, Schirmer T, Rosenbusch JP (1996) Structural and functional characterization of OmpF porin mutants selected for larger pore size. II. Functional characterization. J Biol Chem 271:20676–20680

Schindler H, Rosenbusch JP (1978) Matrix protein from *Escherichia coli* outer membranes forms voltage-controlled channels in lipid bilayers. Proc Natl Acad Sci USA 75:3751–3755

Schmidt M, Zheng P, Delihas N (1995) Secondary structures of *Escherichia coli* antisense *micF* RNA, the 5′-end of the target *ompF* mRNA, and the RNA/RNA duplex. Biochemistry 34:3621–3631

Schneiders T, Levy SB (2006) MarA-mediated transcriptional repression of the rob promoter. J Biol Chem 281:10049–10055

Schneiders T, Amyes SG, Levy SB (2003) Role of AcrR and ramA in fluoroquinolone resistance in clinical *Klebsiella pneumoniae* isolates from Singapore. Antimicrob Agents Chemother 47:2831–2837

Schulz GE (2002) The structure of bacterial outer membrane proteins. Biochim Biophys Acta 1565:308–317

Schwalm J, Mahoney TF, Soltes GR, Silhavy TJ (2013) Role for Skp in LptD assembly in *Escherichia coli*. J Bacteriol 195:3734–3742

Seoane AS, Levy SB (1995) Characterization of MarR, the repressor of the multiple antibiotic resistance (*mar*) operon of *Escherichia coli*. J Bacteriol 177:3414–3419

Sharma P, Haycocks JRJ, Middlemiss AD, Kettles RA, Sellars LE, Ricci V, Piddock LJV, Grainger DC (2017) The multiple antibiotic resistance operon of enteric bacteria controls DNA repair and outer membrane integrity. Nat Commun 8:1444

Shen QT, Bai XC, Chang LF, Wu Y, Wang HW, Sui SF (2009) Bowl-shaped oligomeric structures on membranes as DegP new functional forms in protein quality control. Proc Natl Acad Sci USA 106:4858–4863

Silver LL (2016) A Gestalt approach for Gram-negative entry. Bioorg Med Chem 24:6379–6389

Simonet V, Malléa M, Pagès JM (2000) Substitutions in the eyelet region disrupt cefepime diffusion through the *Escherichia coli* OmpF channel. Antimicrob Agents Chemother 44:311–315

Singh PR, Ceccarelli M, Lovelle M, Winterhalter M, Mahendran KR (2012) Antibiotic permeation across the OmpF channel: modulation of the affinity site in the presence of magnesium. J Phys Chem B 116:4433–4438

Sklar JG, Wu T, Kahne D, Silhavy TJ (2007a) Defining the roles of the periplasmic chaperones SurA, Skp, and DegP in *Escherichia coli*. Genes Dev 21:2473–2484

Sklar JG, Wu T, Gronenberg LS, Malinverni JC, Kahne D, Silhavy TJ (2007b) Lipoprotein SmpA is a component of the YaeT complex that assembles outer membrane proteins in *Escherichia coli*. Proc Natl Acad Sci USA 104:6400–6405

Stavenger RA, Winterhalter M (2014) TRANSLOCATION project: how to get good drugs into bad bugs. Sci Transl Med 6:228ed7

Struyve M, Moons M, Tommassen J (1991) Carboxy-terminal phenylalanine is essential for the correct assembly of a bacterial outer membrane protein. J Mol Biol 218:141–148

Sugawara E, Kojima S, Nikaido H (2016) *Klebsiella pneumoniae* major porins OmpK35 and OmpK36 allow more efficient diffusion of β-lactams than their *Escherichia coli* homologs OmpF and OmpC. J Bacteriol 198:3200–3208

Sulavik MC, Dazer M, Miller PF (1997) The *Salmonella typhimurium mar* locus: molecular and genetic analyses and assessment of its requirement for virulence. J Bacteriol 179:1857–1866

Tanaka T, Horii T, Shibayama K, Sato K, Ohsuka S, Arakawa Y, Yamaki K, Takagi K, Ohta M (1997) RobA-induced multiple antibiotic resistance largely depends on the activation of the AcrAB efflux. Microbiol Immunol 41:697–702

Thiolas A, Bornet C, Davin-Régli A, Pagès JM, Bollet C (2004) Resistance to imipenem, cefepime, and cefpirome associated with mutation in Omp36 osmoporin of *Enterobacter aerogenes*. Biochem Biophys Res Commun 317:851–856

Thiolas A, Bollet C, La Scola B, Raoult D, Pagès JM (2005) Successive emergence of *Enterobacter aerogenes* strains resistant to imipenem and colistin in a patient. Antimicrob Agents Chemother 49:1354–1358

Thompson KM, Rhodius VA, Gottesman S (2007) SigmaE regulates and is regulated by a small RNA in *Escherichia coli*. J Bacteriol 189:4243–4256

Tran QT, Pearlstein RA, Williams S, Reilly J, Krucker T, Erdemli G. (2014) Structure-kinetic relationship of carbapenem antibacterials permeating through *E. coli* OmpC porin. Proteins: Struct Funct Genet 82:2998–3012

Udekwu KI, Wagner EG (2007) Sigma E controls biogenesis of the antisense RNA MicA. Nucl Acids Res 35:1279–1288

Urfer M, Bogdanovic J, Lo Monte F, Moehle K, Zerbe K, Omasits U, Ahrens CH, Pessi G, Eberl L, Robinson JA (2016) A peptidomimetic antibiotic targets outer membrane proteins and disrupts selectively the outer membrane in *Escherichia coli*. J Biol Chem 291:1921–1932

Valentin-Hansen P, Johansen J, Rasmussen AA (2007) Small RNAs controlling outer membrane porins. Curr Opin Microbiol 10:152–155

van der Straaten T, Janssen R, Mevius DJ, van Dissel JT (2004) *Salmonella* gene *rma* (*ramA*) and multiple-drug-resistant *Salmonella enterica* serovar *typhimurium*. Antimicrob Agents Chemother 48:2292–2294

Vergalli J, Dumont E, Cinquin B, Maigre L, Pajovic J, Bacqué E, Mourez M, Réfrégiers M, Pagès JM (2017) Fluoroquinolone structure and translocation flux across bacterial membrane. Sci Rep 7:9821

Vergalli J, Dumont E, Pajović J, Cinquin B, Maigre L, Masi M, Réfrégiers M, Pagés JM (2018) Spectrofluorimetric quantification of antibiotic drug concentration in bacterial cells for the characterization of translocation across bacterial membranes. Nat Protoc 13:1348–1361

Vila J, Martí S, Sánchez-Céspedes J (2007) Porins, efflux pumps and multidrug resistance in *Acinetobacter baumannii*. J Antimicrob Chemother 59:1210–1215

Vogel J, Papenfort K (2006) Small non-coding RNAs and the bacterial outer membrane. Curr Opin Microbiol 9:605–611

Vogt SL, Evans AD, Guest RL, Raivio TL (2014) The Cpx envelope stress response regulates and is regulated by small noncoding RNAs. J Bacteriol 196:4229–4238

Voulhoux R, Bos MP, Geurtsen J, Mols M, Tommassen J (2003) Role of a highly conserved bacterial protein in outer membrane protein assembly. Science 299:262–265

Walsh NP, Alba BM, Bose B, Gross CA, Sauer RT (2003) OMP peptide signals initiate the envelope-stress response by activating DegS protease via relief of inhibition mediated by its PDZ domain. Cell 113:61–71

Walton TA, Sandoval CM, Fowler CA, Pardi A, Sousa MC (2009) The cavity-chaperone Skp protects its substrate from aggregation but allows independent folding of substrate domains. Proc Natl Acad Sci USA 106:1772–1777

Wang JJ, Benier L, Winterhalter M (2018) Quantifying permeation of small charged molecules across channels: electrophysiology in small volumes. ACS Omega 3:17481–17486

Wang J, Bafna JA, Bhamidimarri SP, Winterhalter M (2019) Small molecule permeation across membrane channels: chemical modification to quantify transport across OmpF. Angew Chem Int Ed Engl 58:4737–4741

Wassarman KM, Kiley PJ (2010) Global approaches for finding small RNA and small open reading frame functions. J Bacteriol 192:26–28

Weiss MS, Schulz GE (1992) Structure of porin refined at 1.8 Å resolution. J Mol Biol 227:493–509

Weiss MS, Wacker T, Nestel U, Woitzik D, Weckesser J, Kreutz W, Welte W, Schulz GE (1989) The structure of porin from *Rhodobacter capsulatus* at 0.6 nm resolution. FEBS Lett 256:143–146

Weiss MS, Wacker T, Weckesser J, Welte W, Schulz GE (1990) The three-dimensional structure of porin from *Rhodobacter capsulatus* at 3 Å resolution. FEBS Lett 267:268–272

Weiss MS, Abele U, Weckesser J, Welte W, Schiltz E, Schulz GE (1991a) Molecular architecture and electrostatic properties of a bacterial porin. Science 254:1627–1630

Weiss MS, Kreusch A, Schiltz E, Nestel U, Welte W, Weckesser J, Schulz GE (1991b) The structure of porin from *Rhodobacter capsulatus* at 1.8 Å resolution. FEBS Lett 280:379–382

Westfall DA, Krishnamoorthy G, Wolloscheck D, Sarkar R, Zgurskaya HI, Rybenkov VV (2017) Bifurcation kinetics of drug uptake by Gram-negative bacteria. PLoS ONE 12:e0184671

White DG, Goldman JD, Demple B, Levy SB (1997) Role of the *acrAB* locus in organic solvent tolerance mediated by expression of *marA*, *soxS*, or *robA* in *Escherichia coli*. J Bacteriol 179:6122–6126

WHO (2014). http://apps.who.int/iris/bitstream/handle/10665/112642/97892&x00A0;jsessionid=B56E948F8CBF337BBB55209EFE8AFE73?sequence=1

Winterhalter M, Ceccarelli M (2015) Physical methods to quantify small antibiotic molecules uptake into Gram-negative bacteria. Eur J Pharm Biopharm 95:63–67

Wise MG, Horvath E, Young K, Sahm DF, Kazmierczak KM (2018) Global survey of *Klebsiella pneumoniae* major porins from ertapenem non-susceptible isolates lacking carbapenemases. J Med Microbiol 67:289–295

Wu T, Malinverni J, Ruiz N, Kim S, Silhavy TJ, Kahne D (2005) Identification of a multicomponent complex required for outer membrane biogenesis in *Escherichia coli*. Cell 121:235–245

Yassien MA, Ewis HE, Lu CD, Abdelal AT (2002) Molecular cloning and characterization of the *Salmonella enterica* Serovar Paratyphi B *rma* gene, which confers multiple drug resistance in *Escherichia coli*. Antimicrob Agents Chemother 46:360–366

Yoshimura F, Nikaido H (1985) Diffusion of beta-lactam antibiotics through the porin channels of *Escherichia coli* K-12. Antimicrob Agents Chemother 27:84–92

Yu EW, McDermott G, Zgurskaya HI, Nikaido H, Koshland DE Jr (2003) Structural basis of multiple drug-binding capacity of the AcrB multidrug efflux pump. Science 300:976–980

Zgurskaya HI, Lopez CA, Gnanakaran S (2015) Permeability barrier of Gram-negative cell envelopes and approaches to bypass it. ACS Infect Dis 1:512–522

Zgurskaya HI, Rybenkov VV, Krishnamoorthy G, Leus IV (2018) Trans-envelope multidrug efflux pumps of Gram-negative bacteria and their synergism with the outer membrane barrier. Res Microbiol pii:S0923-2508(18)30022-6

Ziervogel BK, Roux B (2013) The binding of antibiotics in OmpF porin. Structure 21:76–87

Zimmermann W, Rosselet A (1977) Function of the outer membrane of *Escherichia coli* as a permeability barrier to beta-lactam antibiotics. Antimicrob Agents Chemother 12:368–372

Part II
The Periplasm

Chapter 5
Peptidoglycan

Manuel Pazos and Katharina Peters

Abstract The peptidoglycan sacculus is a net-like polymer that surrounds the cytoplasmic membrane in most bacteria. It is essential to maintain the bacterial cell shape and protect from turgor. The peptidoglycan has a basic composition, common to all bacteria, with species-specific variations that can modify its biophysical properties or the pathogenicity of the bacteria. The synthesis of peptidoglycan starts in the cytoplasm and the precursor lipid II is flipped across the cytoplasmic membrane. The new peptidoglycan strands are synthesised and incorporated into the pre-existing sacculus by the coordinated activities of peptidoglycan synthases and hydrolases. In the model organism *Escherichia coli* there are two complexes required for the elongation and division. Each of them is regulated by different proteins from both the cytoplasmic and periplasmic sides that ensure the well-coordinated synthesis of new peptidoglycan.

Keywords Peptidoglycan · Sacculus · Lipid II · PBPs · LDTs · TPase · GTase · Glycan strands · Cross-linking · Elongation · Septation

Introduction

The peptidoglycan (PG) sacculus is an elastic and net-like polymer that surrounds the cytoplasmic membrane in most bacteria. It is rigid enough to maintain the species-specific bacterial cell shape, serving as a scaffold to attach proteins and other polymers, but also porous enough to allow the diffusion of chemical signals, nutrients and virulence factors. The PG protects the cell from bursting due to its turgor, which pushes the cytoplasmic membrane towards the cell wall. In Gram-negative bacteria, a thin and single PG layer is located in the periplasm surrounding the cytoplasmic

M. Pazos (✉) · K. Peters
Centre for Bacterial Cell Biology, Institute for Cell and Molecular Biosciences, Newcastle University, Baddiley-Clark Building, Richardson Road, Newcastle upon Tyne NE2 4AX, UK
e-mail: manuel.pazos@ncl.ac.uk

K. Peters
e-mail: katharina.peters@ncl.ac.uk

A. Kuhn (ed.), *Bacterial Cell Walls and Membranes*, Subcellular Biochemistry 92, https://doi.org/10.1007/978-3-030-18768-2_5

membrane. In contrast, the PG in Gram-positive bacteria is thicker and multi-layered with covalently attached cell wall compounds like capsular polysaccharides, cell surface proteins and wall teichoic acids (Fig. 5.1). Bacteria belonging to Rickettsiaceae, Anaplasmataceae and Mycoplasmataceae families do not have PG. The presence of PG in Chlamydiaceae has been proven (Pilhofer et al. 2013; Packiam et al. 2015; Liechti et al. 2016). In order to preserve the structural integrity of the cell envelope during cell growth and division, the synthetic and hydrolytic PG enzymes must be coordinated to enlarge and divide the sacculi. According to the current model this coordination is achieved by multi-enzyme complexes that extend from the cytoplasm to the outer membrane.

This chapter is mainly focused on the model organism *Escherichia coli*, but the reader will find references to other organisms along the different sections. After describing the PG composition, modifications and its biophysical properties, the different enzymatic activities and their regulation are presented in the cellular context. The importance of the fluorescent D-amino acids for the field is highlighted, and a detailed description of the current understanding of the PG substrate flippase candidates and the unusual LD-transpeptidases (LDTs) is included.

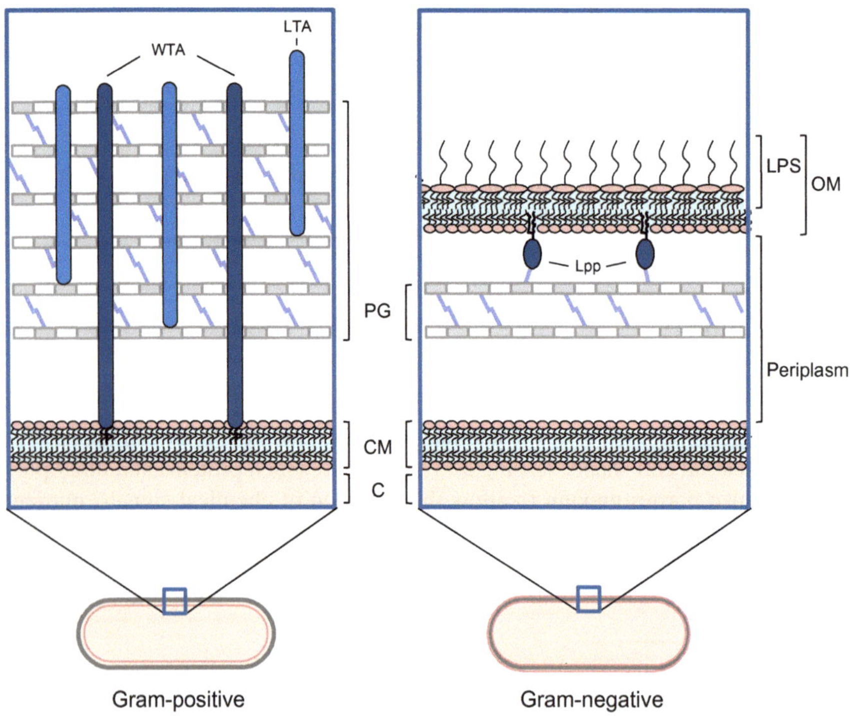

Fig. 5.1 Schematic structure of the cell wall in Gram-positive and Gram-negative bacteria. C, cytoplasm; CM, cytoplasmic membrane; PG, peptidoglycan; OM, outer membrane; LPS, lipopolysaccharide; LTA, lipoteichoic acid; WTA, wall teichoic acid; Lpp, Braun's lipoprotein

The Peptidoglycan Composition, Its Modifications and Their Function

The PG sacculus is an essential macromolecule exclusively found in bacteria. Its basic structure comprises linear glycan strands that are cross-linked by short peptides (Schleifer and Kandler 1972). The glycan strands, whose length varies between species, are made out of alternating β-1,4-connected *N*-acetylglucosamine (Glc*N*Ac) and *N*-acetylmuramic acid (Mur*N*Ac) residues. Mur*N*Ac is a lactic acid ether derivative of Glc*N*Ac, and the stem pentapeptides are attached to its D-lactyl moiety at position C3. The stem pentapeptides contain L- and D-amino acids, whereby the latter is characteristic of bacterial PG. The PG structure shows high diversity between different species, but it is well conserved in bacteria belonging to the same species. Modifications in the PG can enhance the bacterial fitness and virulence, providing resistance against environmental stresses, hydrolytic host enzymes and antimicrobial agents.

Variations in the Stem Peptide

The amino acid composition of the pentapeptide shows species-specific variations. Some examples are shown in Fig. 5.2 and described in the following lines. In Gram-negative bacteria like *E. coli* the pentapeptide in nascent PG consists of L-Ala-D-iGlu-mDAP-D-Ala-D-Ala (mDAP: *meso*-diaminopimelic acid). Two adjacent peptides are more frequently cross-linked via an amide bond between the carboxyl group of the D-Ala at position four of one peptide and the ε-amino group of mDAP at position three of another one. The D-Ala-D-Ala motif of the pentapeptide is the universal substrate of the PG cross-linking enzymes. The terminal D-Ala is released during the cross-linking reaction.

In *Mycobacterium leprae* the first amino acid of the stem peptide is Gly instead of L-Ala. This modification is suggested to be due to the growth environment (Mahapatra et al. 2000). In the Gram-positive pathogens *Streptococcus pneumoniae* and *Staphylococcus aureus*, and in *Mycobacterium tuberculosis* the α-carboxyl group of the D-iso-glutamate on position two of the stem peptide is amidated into D-iso-glutamine by the essential amidotransferase complex MurT/GatD (Zapun et al. 2013; Morlot et al. 2018; Münch et al. 2012; Figueiredo et al. 2012). In *S. pneumoniae* unamidated glutamate is predominantly found in uncross-linked PG monomers (Bui et al. 2012). In vitro assays using recombinant pneumococcal PBPs revealed that the amidation of the stem peptides is needed for an efficient cross-linking reaction (Zapun et al. 2013). In *S. aureus* the lack of amidation correlates with a decrease in cross-linking and with a higher susceptibility to antibiotics (Stranden et al. 1997; Boyle-Vavra et al. 2001; Figueiredo et al. 2012). Thus, understanding the molecular mechanism of the amidation reaction may help in the development of new therapeutics to target these important pathogens. Recent studies provide biochemical and structural

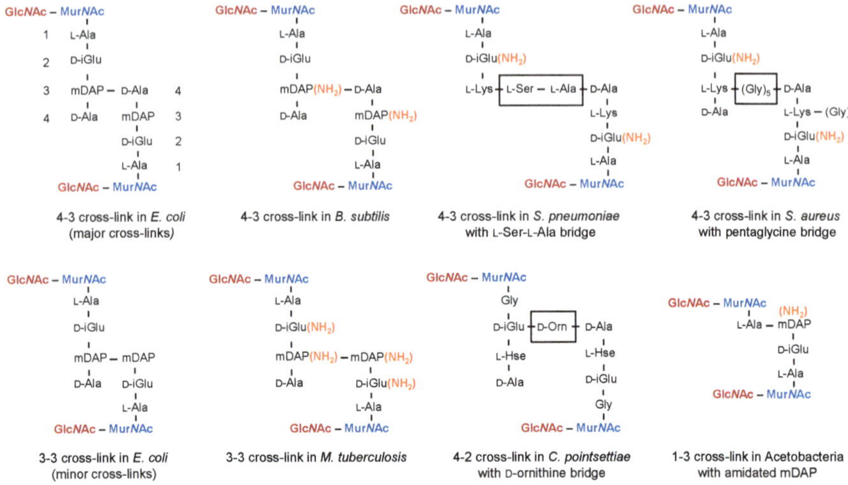

Fig. 5.2 Examples of peptides and cross-link types in the peptidoglycan of different species. Amidation of residues is depicted in orange. Interpeptide bridges are framed with a square

insights about the MurT/GatD complex of *S. pneumoniae* (Morlot et al. 2018) and *S. aureus* (Noldeke et al. 2018). The amino acid at position three of the stem peptide shows the greatest variation. Most Gram-negative bacteria, Mycobacteria and Bacilli contain an mDAP residue at this position. *Bacillus subtilis* has an amidated mDAP (Atrih et al. 1999), due to the action of the amidotransferase AsnB (Dajkovic et al. 2017). In *M. tuberculosis* AsnB also amidates the mDAP residue, and this modification is essential for cell growth (Ngadjeua et al. 2018). The amidotransferase LtsA performs the amidation of the mDAP residue in the PG of *Corynebacterium glutamicum* (Levefaudes et al. 2015). Spirochetes, such as Borrelia or Treponema, have an ornithine residue instead of mDAP (Schleifer and Kandler 1972; Yanagihara et al. 1984). Other species contain at position three different diamino acids like *meso*-lanthionine (*Fusobacterium nucleatum*) and D-Lys (*Thermatoga maritima*), or monoamino acids like L-Ala, L-Glu or L-homoserine (reviewed in Vollmer et al. 2008a). However, most Gram-positive bacteria have an L-Lys at position three, which often carries a linear peptide branch linked to its ε-amino group. These interpeptide bridges show a great diversity, with sizes varying from two to seven residues and a wide range of amino acids. In case of *S. aureus* the FemXAB peptidyltransferases catalyse the addition of a characteristic Gly_5-interpeptide bridge (Schleifer and Kandler 1972). *S. pneumoniae* contains branched stem peptides with an L-Ser-L-Ala or L-Ala-L-Ala dipeptide linked to the ε-amino group of L-Lys. This modification is added by MurM and MurN (Filipe and Tomasz 2000). The degree of branching and cross-linking of PG varies between strains. While most pneumococcal strains contain a small percentage of branched peptides, the PG of resistant strains is highly branched and cross-linked (Garcia-Bustos and Tomasz 1990; Severin et al. 1996). Most bacteria contain D-Ala-D-Ala at positions four and five of the stem peptide.

This motif is recognized by vancomycin and other glycopeptide antibiotics that form a complex with the PG precursors, preventing their incorporation into the sacculus of Gram-positive bacteria. The replacement of the D-Ala at position five by D-Lac or D-Ser prevents the binding of the antibiotic to the precursor and mediates resistance of enterococci (Arthur et al. 1993, 1996).

Variations in the Peptide Cross-Links

The type and extent of the peptide cross-links is different between species. The most abundant cross-links connect the D-Ala at position four of one stem peptide with the mDAP (or L-Lys) at position three of another one (4-3 cross-links). They are synthesized by DD-transpeptidases (DD-TPases). Less frequent are the 4-2 cross-links, found in Corynebacteria, that connect the D-Ala at position four and the D-iGlu at position two of adjacent stem peptides. *Corynebacterium pointsettiae* contains an L-homoserine at position three of the stem peptide, which is non-reactive to form cross-links. Therefore, the cross-link starts at D-iGlu and is created via a D-ornithine bridge (Fig. 5.2) (Schleifer and Kandler 1972). Some bacteria contain a small amount of 3-3 cross-links, while pathogens like *Clostridium difficile* and *M. tuberculosis* contain predominantly 3-3 cross-links in the PG. These cross-links are made by LD-transpeptidases (LDTs) that connect two mDAP residues of adjacent stem peptides (see section "Peptidoglycan Synthesis" for further details on the cross-linking reactions).

Recently, novel PG structures have been described in Acetobacteria, which proliferate at low pH and produce acetic acid. These modifications include the amidation of the α-(L)-carboxyl group of mDAP and a novel LD (1-3) cross-link, which connects the L-Ala residue of one stem peptide with the mDAP of another one (Fig. 5.2) (Espaillat et al. 2016). The enzymes catalyzing both modifications are unknown.

Modifications of the Glycan Strands

The glycan strands can be modified by *N*-deacetylation and *O*-acetylation on either one or both Glc*N*Ac and Mur*N*Ac subunits, and by *N*-glycolylation on Mur*N*Ac (Fig. 5.3) (reviewed in Vollmer 2008; Yadav et al. 2018). The *N*-glycolylation reaction occurs in the cytoplasm during the synthesis of the PG precursors. The *N*-deacetylation and *O*-acetylation reactions take place outside the cytoplasm, once the glycan strands have been synthesized. In pathogenic species these modifications contribute to the survival in the host organism by increasing the resistance to PG degrading enzymes, preventing cell lysis and the release of bacterial products that could be detected by the host immune system.

In many Gram-positive and Gram-negative bacteria the PG chains contain an extra acetyl group linked to the C6–OH group of some Mur*N*Ac (*O*-acetylation). The

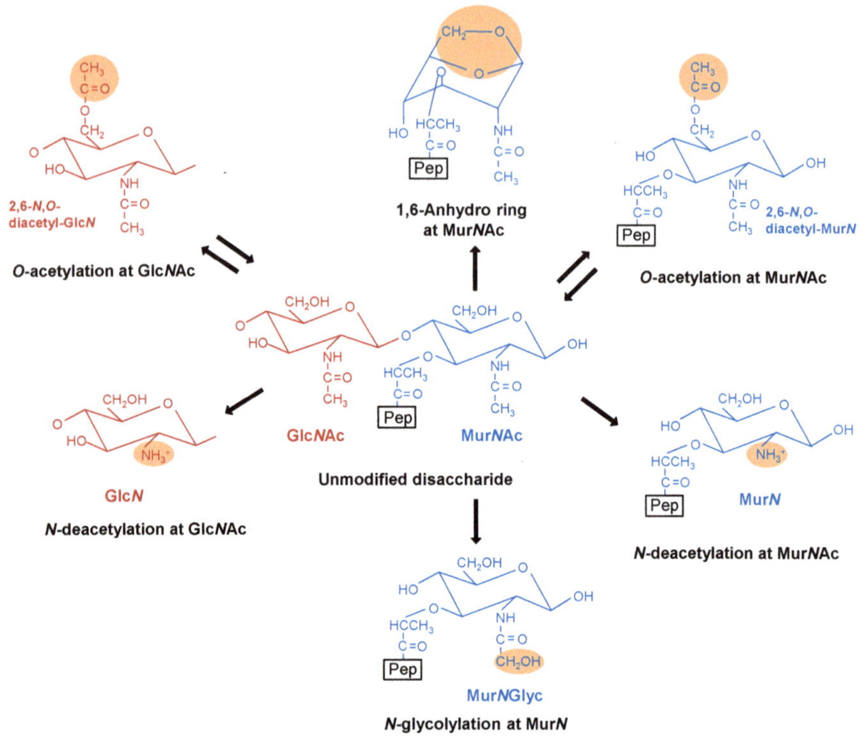

Fig. 5.3 Summary of modifications in the PG glycan strands. The structure of the unmodified GlcNAc-MurNAc disaccharide (middle) and of selected modifications in the GlcNAc (red) and MurNAc (blue) are shown. Modifications are highlighted in orange. The *O*-acetylation of GlcNAc or MurNAc is reversible. Pep, peptide linked to MurNAc. Modified from Vollmer (2008) and Yadav et al. (2018)

degree of *O*-acetylation varies between <20 and 70% depending on the species and growth conditions. The bulky acetyl group represents a steric hindrance and prevents the binding of the muramidase lysozyme (Pushkaran et al. 2015), which is secreted by the host immune system cells to hydrolyse the glycan strands between MurNAc and GlcNAc. Therefore, PG *O*-acetylation is a major virulence factor in pathogenic bacteria. For the *O*-acetylation of MurNAc, the acetyl moiety is translocated from a cytoplasmic donor molecule to the periplasm or extracellular space and transferred to the MurNAc subunit. In Gram-positive bacteria the *O*-acetyltransferases of the OatA-type perform both processes (Bera et al. 2005; Bernard et al. 2012), while Gram-negative bacteria require the coordinated action of multiple enzymes of the Pat or Pac family (Weadge et al. 2005; Moynihan and Clarke 2010; Dillard and Hackett 2005). In *S. aureus* OatA provides resistance to lysozyme (Bera et al. 2005), protects against killing by macrophages (Shimada et al. 2010), reduces the induction of pro-inflammatory cytokines and permits reinfection (Sanchez et al. 2017). OatA homologs have been identified in different Gram-positive species. In *S. pneumoniae* the OatA homolog Adr catalyses the *O*-acetylation of Mur*N*Ac and protects dividing cells from cleavage by the pneumococcal autolysin LytA. Adr localizes at the septa and shows mislocalization in some cell division mutants, suggesting an important role in pneumococcal cell division (Bonnet et al. 2017). In vancomycin-resistant *Enterococcus faecalis* the vancomycin treatment increases the levels of *O*-acetylation, which leads to lysozyme resistance and to an increase in virulence (Chang et al. 2017). *O*-acetylation of Mur*N*Ac blocks the function of the lytic transglycosylases (LTs), which have the same substrate specificity as lysozyme but catalyse a transglycosylation reaction resulting in the formation of 1,6-anhydro-Mur*N*Ac (see section "Peptidoglycan Hydrolysis and Remodelling" for further information).

In Gram-negative bacteria the level of *O*-acetylation regulates the activities of the LTs (Weadge et al. 2005; Weadge and Clarke 2006). Glycan strands can be *O*-deacetylated at the Mur*N*Ac *O*-acetyl subunit. The *O*-acetylesterase Ape reverts the *O*-acetylation of Mur*N*Ac, and Ape homologs have been found in Gram-positive and Gram-negative bacteria (Weadge et al. 2005). In *Campylobacter jejuni* the deletion of *ape1* results in the accumulation of *O*-acetylated PG, which impairs the cellular fitness and leads to defects in morphology, motility, biofilm formation and virulence (Ha et al. 2016). This highlights the importance of regulating the degree of *O*-acetylation via the activity of *O*-acetyltransferases and *O*-acetylesterases for the bacterial cell. In contrast, the *O*-acetylation of Glc*N*Ac has been only described in *Lactobacillus plantarum*, where it inhibits the major autolysin Acm2 (Bernard et al. 2011).

N-deacetylation of Glc*N*Ac occurs mostly in Gram-positive bacteria (Vollmer and Tomasz 2000; Boneca et al. 2007; Peltier et al. 2011; Benachour et al. 2012) but also in some Gram-negative like *Shigella flexneri* (Kaoukab-Raji et al. 2012). The PG deacetylase A enzyme (PgdA), identified for the first time in *S. pneumonia*, removes the acetyl group at position C2 of the Glc*N*Ac. PgdA mutants are more susceptible to lysozyme and less virulent (Vollmer and Tomasz 2000). In *Listeria monocytogenes* 50% of the Glc*N*Ac residues are deacetylated (Boneca et al. 2007) and ~23% of the Mur*N*Ac residues are *O*-acetylated at position C6 (Aubry et al. 2011), whereby both modifications enhance synergistically the resistance against lysozyme. In this organism, PgdA activity is regulated by the cell division protein GpsB and the PG synthase PBP A1 (Rismondo et al. 2018). The absence of GpsB increases the lysozyme resistance due to a rise in *N*-deacetylated muropeptides, and the absence of PBP A1 reverses this phenotype. This regulation is supported by in vitro studies suggesting that all three proteins form a complex. In *Lactococcus lactis* YvhB catalyses the *O*-acetylation of Mur*N*Ac, and XynD performs the *N*-deacetylation of Glc*N*Ac. Both modifications increase the cross-linkage and the cell wall integrity, leading to acid resistance and to the production of the polycyclic antibacterial peptide nisin (Cao et al. 2018). In the predatory bacteria *Bdellovibrio bacteriovorus*, the PgdA homologs Bd0468 and Bd3279 deacetylate the Glc*N*Ac residues of the prey PG during predation, making it more susceptible for destruction at the end of predation (Lambert et al. 2016).

The *N*-glycolylation of MurNAc subunits exists in Mycobacteria and related Actinomycetales. This modification is added by the cytoplasmic UDP-MurNAc hydroxylase NamH during the synthesis of the UDP-linked precursors, where the *N*-acetyl group at the C2 of the muramyl dipeptides is hydroxylated to an *N*-glycolyl group. The degree of *N*-glycolylation varies depending on the bacterial species, and it changes in response to antibiotics. The deletion of *namH* increases the susceptibility to lysozyme and β-lactams (Raymond et al. 2005). *N*-glycolylated MurNAc activates the innate immunity more than *N*-acetylated MurNAc, underlining its contribution to the unusual immunogenicity of Mycobacteria (Coulombe et al. 2009) but not to the pathogenicity of *M. tuberculosis* infection (Hansen et al. 2014).

Biophysical Properties of Peptidoglycan

The PG sacculus has essential stress-bearing functions in the cell, which include maintaining the cellular shape during cell growth and division within changing environmental conditions, and constraining the cell volume under turgor. In order to fulfil these functions, the PG is elastic and stiff at the same time, but also porous to allow diffusion of small proteins.

Thickness of Peptidoglycan

The osmotic pressure in Gram-negative bacteria like *E. coli* is up to 3 atm (Cayley et al. 2000) and in Gram-positive bacteria such as *B. subtilis* and *S. aureus* is significantly higher (up to 20 atm) (Whatmore and Reed 1990). The PG provides mechanical strength to counteract the osmotic pressure in the cell (Höltje 1998; Koch 1988). The thickness of the PG, which can vary depending on the species, contributes to cell stiffness. Consistent with that, Gram-positive bacteria contain a multi-layered and thick PG. The PG of Gram-negative bacteria is a thin and single layer located between the inner and the outer membranes (Fig. 5.1). A recent study suggests that mechanical loads are balanced between the outer membrane and the PG, and postulates a stress-bearing function for the outer membrane during osmotic changes (Rojas et al. 2018).

Isolated PG sacculi maintain the shape and the size of intact cells. Using atomic force microscopy (AFM) dried sacculi from *E. coli* are 3 nm thick, while rehydrated sacculi are 6 nm thick (Yao et al. 1999). In the same study rehydrated sacculi of *Pseudomonas aeruginosa* are only 3 nm thick. These data are in agreement with measurements performed by cryo-transmission electron microscopy (cryo-TEM) of frozen-hydrated sections, which preserves native structures (*E. coli*: 6.35 ± 0.53 nm; *P. aeruginosa*: 2.41 ± 0.54 nm) (Matias et al. 2003).

Cryo-TEM analysis of frozen-hydrated sections of *B. subtilis* cells revealed two distinct zones: a 22 nm thick low-density zone surrounding the cytoplasmic membrane and a 33 nm thick high-density zone surrounding the previous one. The high density zone contains PG, teichoic acids and associated proteins (Matias and Beveridge 2005). A thinner *B. subtilis* Δ*ponA* mutant (lacking the bifunctional PG synthase PBP1) shows a 30 nm thick PG (Beeby et al. 2013).

Strength, Stiffness and Elasticity of Peptidoglycan

The mechanical properties of PG have been analysed using different techniques. The cellular stiffness can be described by the Young's modulus, also known as the modulus of elasticity. To determine it, a physical force is applied to a sample (either cells or isolated sacculi) and the degree by which it can be stretched is measured. If the sample can withstand the force, it will return to the preloaded size after removing the force. If the load is too high for the elastic limits of the sample, it will irreversibly deform and rupture. The elasticity of isolated *E. coli* sacculi, measured by stretching them with the tip of the AFM, varies on the hydration state as air dried sacculi are more rigid than rehydrated ones. Differences in the elastic anisotropy between the long and the short axis of the sacculus suggests that the more flexible peptide cross-links are longitudinally orientated and the more rigid glycan strands are circumferentially orientated (Yao et al. 1999). The proposed architecture is supported by the fact that changes in the volume of osmotically shocked *E. coli* cells are mainly due to changes in cell length but not in cell diameter (van den Bogaart et al. 2007). Recent analysis of individual *E. coli* glycan chains by AFM have shown long and circumferentially orientated glycan chains (200 nm) under normal conditions and shorter and disordered glycan chains when a spheroid cell shape is induced (Turner et al. 2018).

Another technique to measure the mechanical properties of bacterial cells is the Cell Length Analysis of Mechanical Properties (CLAMP), which enables a high throughput screening. Using this technique the Young's moduli of Gram-negative and Gram-positive rod-shaped bacteria are estimated on the basis of initial growth rates of hydrogel-encapsulated living cells with varying stiffness (Tuson et al. 2012). A decrease in relative cell elongation over time is observed with increasing stiffness of the hydrogels. The estimated Young's modulus for *E. coli* is fourfold higher than the result obtained with AFM for hydrated *E. coli* sacculi (Yao et al. 1999). This difference can be explained by the fact that the outer membrane also contributes to the cell wall stiffness. Using the CLAMP technique the Young's moduli of the cell envelopes of *P. aeruginosa* and *B. subtilis* are similar to *E. coli*. Several genes affecting cell stiffness have been identified by measuring the cell growth of hydrogel-encapsulated mutant cells from the entire KEIO-collection (*E. coli* single-gene deletion mutants). These genes have roles in diverse processes, suggesting that cell stiffness is affected by different cellular functions beside the cell wall-related ones (Auer et al. 2016).

In addition to the thickness, other structural parameters as the amount of cross-links and the glycan strands length influence the mechanical properties of the PG. Compared to exponentially growing cells, stationary cells of *E. coli* contain slightly more cross-links, higher level of LD cross-links (see section "LD-Transpeptidases") and more cross-links between the outer membrane-anchored protein Lpp and the PG, which might strengthen the cell envelope and presumably enhance the cellular stiffness (Vollmer and Bertsche 2008). The cell division process of *S. aureus* changes the mechanical properties of the cell wall, as the newly formed septum is stiffer than the surrounding mature cell wall, most likely due to a higher level of cross-links (Bailey et al. 2014b). A decrease in the cross-linkage of *S. aureus* PG, by the absence of PBP4, results in a small reduction of cell wall stiffness (Loskill et al. 2014). The absence of glucosaminidases, which hydrolyse the bonds between GlcNAc and MurNAc, increases the cellular stiffness in *S. aureus*, increasing the glycan chains length (Wheeler et al. 2015).

Bacteria from diverse phyla produce non-canonical D-amino acids that are incorporated into the PG and modify its strength and flexibility. In *Vibrio cholerae* the racemase BsrV produces these amino acids (mainly D-Met and D-Leu) during stationary growth and LDTs catalyse their incorporation into the PG. The *bsrV* mutant cells have shorter glycan strands and both, *bsrV* and *ldtA ldtB* (encoding for LDTs) mutants are significantly more sensitive to osmotic shock and accumulate an excess of PG during stationary phase (Lam et al. 2009; Cava et al. 2011).

Porosity of Peptidoglycan

In early work the pore sizes in *B. subtilis* and *Bacillus licheniformis* PG were estimated (radius of 2.5 nm) by measuring the diffusion of different sized proteins out of the cells, after *n*-butanol treatment to permeabilize the lipid bilayers (Hughes et al. 1975). A more recent study using isolated *E. coli* and *B. subtilis* sacculi and fluorescein-labelled dextrans of known molecular weights, determined a similar pore radius for both sacculi (2.06 nm in *E. coli*, 2.12 nm in *B. subtilis*) (Demchick and Koch 1996). On the basis of these values it was estimated that a globular, hydrophilic protein with a molecular weight of 25 kDa can freely pass through the relaxed sacculi, while globular proteins of about 50 kDa should be able to diffuse through stretched sacculi (Demchick and Koch 1996). A homeostatic mechanism has been postulated for γ-proteobacteria like *E. coli*, in which the PG synthesis rate is coupled to the growth rate of the cell via the PG pores size (Typas et al. 2010).

Peptidoglycan Precursor Synthesis

The synthesis of the PG precursor starts with the formation of both nucleotide sugar-linked precursor UDP-GlcNAc and UDP-MurNAc in the cytoplasm (Fig. 5.4) (Barreteau et al. 2008). UDP-GlcNAc is synthesised from fructose 6-phosphate by the

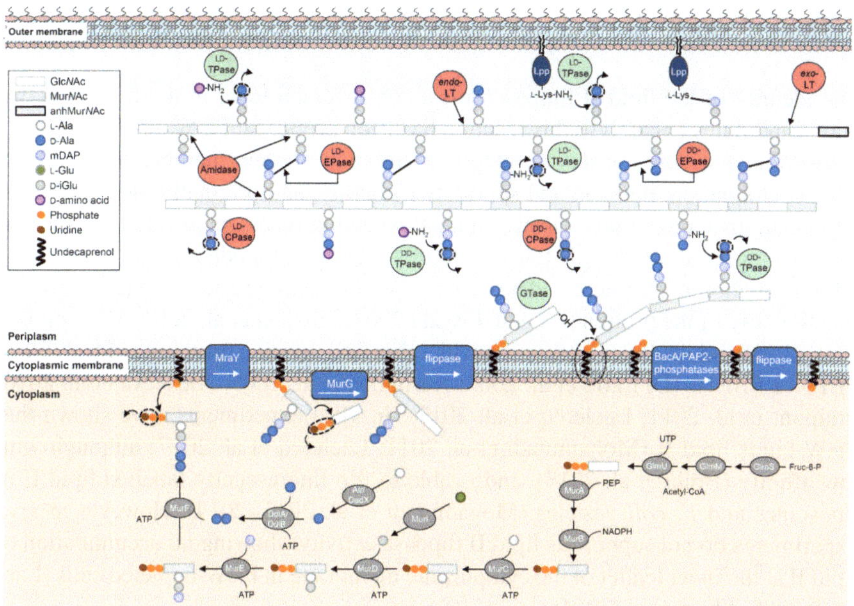

Fig. 5.4 Synthetic and hydrolytic reactions occurring during the synthesis and incorporation of new peptidoglycan in *E. coli*. Modified from Typas et al. (2011)

enzymes GlmS, GlmM and GlmU (the last reaction requires acetyl-coenzyme A and uridine triphosphate). UDP-MurNAc is synthesised by the addition of enolpyruvate to UDP-GlcNAc (by MurA) and the following reduction (by MurB). The pentapeptide moiety is added to the UDP-MurNAc by MurC, MurD, MurE and MurF enzymes in a sequence of ATP-dependent reactions that add L-Ala, D-iGlu, mDAP and D-Ala-D-Ala, respectively. The racemases generate the D-Ala (Alr or DadX) and D-iGlu (MurI) enantiomers, and the ligases DdlA and DdlB) synthesize the D-Ala-D-Ala dipeptide required. The first three amino acids (L-Ala-D-iGlu-mDAP) can also be added as a tripeptide by the ligase Mpl through the "recycling pathway". Phospho-MurNAc-pentapeptide is then transferred onto the lipid carrier undecaprenyl phosphate by MraY, which crystal structure has been recently solved (Hakulinen et al. 2017), resulting in the formation of undecaprenyl pyrophosphate MurNAc pentapeptide (lipid I) and releasing uridine monophosphate (UMP). The subsequent transfer of GlcNAc to the lipid I by MurG results in the formation of undecaprenyl pyrophosphate MurNAc(GlcNAc) pentapeptide (lipid II) and the release of uridine diphosphate (UDP). In a final step lipid II is translocated (or flipped) from the inner to the outer leaflet of the cytoplasmic membrane where it is used as substrate for the synthesis of new PG, releasing the lipid carrier undecaprenol pyrophosphate. Due to the essentiality of the process, these reaction steps are potential targets for antibiotics (Hrast et al. 2014; Kouidmi et al. 2014; Liu and Breukink 2016).

Lipid II Flippase(s)

The identity of the lipid II flippase(s) is a controversial topic in the field, and two main candidates to catalyse this reaction has been proposed: FtsW and MurJ.

FtsW is a widely conserved integral membrane protein, member of the SEDS (shape, elongation, division and sporulation) family, and originally identified from a filamentous temperature sensitive mutant, showing its essential role for cell division (Ishino et al. 1989; Khattar et al. 1994). Different studies have dissected its 10 transmembrane segments, identifying several essential residues required for the functionality of the protein (Lara and Ayala 2002; Pastoret et al. 2004; Mohammadi et al. 2014) and its direct interaction with other cell division proteins as FtsN, FtsQ, PBP3 or PBP1B (Di Lallo et al. 2003; Karimova et al. 2005; Alexeeva et al. 2010; Fraipont et al. 2011; Leclercq et al. 2017). In vitro experiments have shown that FtsW binds lipid II (Mohammadi et al. 2011; Leclercq et al. 2017) although with low affinity (Bolla et al. 2018), and is able to flip fluorescently labelled lipid II in liposomes and *E. coli* vesicles (Mohammadi et al. 2011, 2014). However in vivo experiments do not support its lipid II flippase activity, showing no accumulation of lipid II at the inner leaflet of the cytoplasmic membrane in FtsW-depleted cells (Lara et al. 2005; Sham et al. 2014).

MurJ is an integral membrane protein that belongs to the MOP (multidrug/oligosaccharidyl-lipid/polysaccharide) exporter superfamily of proteins. It was identified by bioinformatic analysis and isolated from a temperature sensitive mutant (Ruiz 2008; Inoue et al. 2008). It contains 14 transmembrane segments that adopt a V-shaped structure with a central solvent-exposed cavity (Kuk et al. 2017; Zheng et al. 2018). Several residues are necessary for the correct functionality of the protein, including some of the charged residues located in the central cavity (Butler et al. 2013, 2014; Zheng et al. 2018). The inactivation of MurJ causes cell shape defects and lysis (Ruiz 2008; Inoue et al. 2008) and, supporting its flippase activity, an accumulation of lipid II intermediates in the cell (Sham et al. 2014; Qiao et al. 2017; Chamakura et al. 2017; Rubino et al. 2018). Its essential role can be substituted by other transporters from different species as *B. subtilis* (MurJ and Amj) (Meeske et al. 2015) or *Streptococcus pyogenes* (YtgP) (Ruiz 2009), or by transporters with a more relaxed substrate specificity (Elhenawy et al. 2016; Sham et al. 2018). In vitro evidences do not support its lipid II flippase activity in proteoliposomes and *E. coli* vesicles (Mohammadi et al. 2011) and show contrary results on lipid II binding (Leclercq et al. 2017; Bolla et al. 2018).

Lipid Carrier

Undecaprenyl phosphate is the essential and unique lipid carrier used for the synthesis of the bacterial extracellular cell wall polymers as PG, teichoic acids and O-antigen. It is produced by dephosphorylation of the undecaprenyl pyrophosphate, either from de novo synthesis or from the recycling pathway, by membrane embedded phosphatases (BacA or PAP2-type phosphatases) (Manat et al. 2014). In the de novo synthesis pathway, UppS uses farnesyl pyrophosphate and eight isopentenyl pyrophosphate molecules to produce undecaprenyl pyrophosphate. In the recycling pathway, the undecaprenyl pyrophosphate is released after the glycosyltransferase reaction performed by the PG synthases to incorporate the disaccharide pentapeptide moiety of the lipid II substrate into the PG glycan chains. In both cases undecaprenyl phosphate is generated in the outer leaflet of the cytoplasmic membrane, as the active sites of BacA and the PAP2-type phosphatases PgpB, YbjG and LpxT are oriented towards the periplasm (Touze et al. 2008; Fan et al. 2014; Tatar et al. 2007; Manat et al. 2015; El Ghachi et al. 2018; Workman et al. 2018). The lipid carrier has to be flipped across the membrane, as the synthesis of the PG precursor takes place in the inner leaflet, but the required protein has not yet been identified.

Peptidoglycan Synthesis

The synthesis of new PG into the sacculus comprises the polymerization of glycan chains by glycosyltransferases (GTases) and their incorporation to new or pre-existing chains through cross-linkage of the stem peptides by transpeptidases (TPases). There are different TPase reactions depending on the donor muropeptide (Fig. 5.4). The most abundant reaction occurs when a pentapeptide donor cross-links the D-Ala at position four with the mDAP at position three of the acceptor muropeptide (DD-TPase reaction, 4-3 cross-link). The DD-TPases are named penicillin binding proteins (PBPs), as β-lactams covalently bind to the catalytic site serine residue, block the access to the donor muropeptide and therefore inhibit the enzymatic reaction (see Sauvage et al. 2008 for a detailed description of PBPs). In a less frequent reaction two muropeptides are cross-linked by their mDAP residues at position three (LD-TPase reaction, 3-3 cross-link).

The best characterised PG GTases belong to the GT51 family. They contain a conserved sequence with five motifs, a structure mainly composed of α-helices and a catalytic glutamate residue. The phosphoglycolipid antibiotic moenomycin binds to the active site, competing with the newly synthesised glycan chain for the donor site. *E. coli* contains four GT51 GTases: a monofunctional GTase (MtgA) and three bifunctional class A PBPs (encoding also DD-TPase domain) known as PBP1A, PBP1B and PBP1C (Sauvage et al. 2008). PBP1A and PBP1B are the most important class A PBPs and both contain a non-catalytic domain (ODD in PBP1A and UB2H in PBP1B) involved in the regulation of the GTase and TPase activities (Typas et al. 2010). Either PBP1A or PBP1B is required for cell viability (Yousif et al. 1985; Kato et al. 1985). PBP1A is preferentially involved in the synthesis of PG during cell elongation and PBP1B during septation. Both proteins are functionally semi-redundant, meaning that the cell can compensate the absence of each other although not under all tested conditions (Garcia del Portillo and de Pedro 1990, 1991; Denome et al. 1999; Pepper et al. 2006; Ranjit and Young 2013). Several in vitro studies have characterised the activities of both proteins either alone (Barrett et al. 2004; Bertsche et al. 2005; Born et al. 2006) or in presence of different interacting partners, as in case of PBP1A and PBP2 (Banzhaf et al. 2012) or LpoA (Typas et al. 2010; Lupoli et al. 2014), and PBP1B and FtsN (Müller et al. 2007) or LpoB (Typas et al. 2010; Egan et al. 2014, 2018; Lupoli et al. 2014). The crystal structure of *E. coli* PBP1B has been described in complex with different antibiotics (Sung et al. 2009; King et al. 2017). Crystal structures of other bifunctional PBPs from different bacteria are also available (Jeong et al. 2013; Lovering et al. 2007; Yuan et al. 2007).

E. coli has two essential monofunctional DD-TPases or class B PBPs, PBP2 and PBP3. Whereas PBP2 is required for cell elongation and its inactivation produces a spherical phenotype, PBP3 is needed for septation and its inactivation generates long filamented cells. In both cases cells eventually lyse. Crystal structures of *E. coli* PBP3 and *H. pylori* PBP2 are solved (Sauvage et al. 2014; Contreras-Martel et al. 2017) (see section "Regulation of Peptidoglycan Growth" for further details on the elongation and septation complexes).

Recently the SEDS family protein RodA has been described as a GTase. It does not belong to the known GT51 family and is not inhibited by the antibiotic moenomycin. The crystal structure of RodA from *Thermus thermophilus* has been solved (Meeske et al. 2016; Cho et al. 2016; Emami et al. 2017; Sjodt et al. 2018). Future work will show if the SEDS protein FtsW, essential for cell division, is also a GTase (Taguchi et al. 2018).

LD-*Transpeptidases*

In *E. coli* only 5–15% of the PG cross-links are 3-3 (Glauner et al. 1988). This transpeptidase reaction (LD-TPase) is catalysed by the non-essential LD-transpeptidases (LDTs) (Magnet et al. 2008) that use a disaccharide-tetrapeptide as acyl donor and form cross-links between two mDAP residues of adjacent stem peptides. LDTs contain a catalytic cysteine residue (Mainardi et al. 2005) and are insensitive to β-lactams, with the exception of carbapenems that also target PBPs (Mainardi et al. 2007). Even though the 3-3 are the minor cross-links in the PG of most bacteria, the LDTs are promising targets for new antimicrobial drugs against important pathogens, and therefore they will be discussed in more detail.

Ldt_{fm}, from an ampicillin-resistant *Enterococcus faecium* strain, was the first discovered LDT with documented LD-TPase activity. In the presence of the antibiotic *E. faecium* cells produce exclusively 3-3 cross-links (Mainardi et al. 2000, 2002, 2005). This alternative cross-linkage enables the bypass of PBPs and confers resistance to β-lactam antibiotics, although a DD-CPase reaction is needed to produce the tetrapeptides that can be used by the LDTs. Since then, Ldt homologues with a catalytic YkuD like domain (PFAM 03744) have been found and described among pathogen and non-pathogen Gram-positive and Gram-negative bacteria.

E. coli has six periplasmic LDTs (LdtA-F) with different functions. LdtC and LdtE contain an N-terminal PG-binding lysin motif (LysM) domain, which is found in many cell wall hydrolases (Buist et al. 2008), suggesting that these enzymes may be active when bound to PG. LdtA, LdtB and LdtC (ErfK, YbiS and YcfS) catalyse the covalent attachment of the outer membrane lipoprotein Lpp (Braun's lipoprotein) to PG, stabilizing the cell envelope (Magnet et al. 2007). These enzymes link the ε-amino group of the C-terminal lysine of Lpp to the α-carboxyl group of the mDAP residue in the stem peptide of the PG (Fig. 5.4). Lpp is a small α-helical protein that with 1 million molecules per cell is the most abundant protein in *E. coli* (Li et al. 2014). Lpp is anchored to the outer membrane by a lipid moiety composed of acyl chains attached to the cysteine residue at its N-terminus. Under normal growth conditions, about one third of the Lpp molecules are linked to PG (Inouye et al. 1972). LdtD and LdtE (YcbB and YnhG) form 3-3 cross-links (Fig. 5.4) (Magnet et al. 2008). The expression of *ldtD* is regulated by the Cpx pathway, which mediates adaption to cell envelope stress (Bernal-Cabas et al. 2015). In response to Cpx-activated conditions the production of LdtD is upregulated resulting in increased 3-3 cross-linkage, suggesting a role for LdtD under cellular stress conditions (Bernal-Cabas et al. 2015; Delhaye et al. 2016). Furthermore, LdtD is twice the size of the other *E. coli* LDTs and shares lower sequence similarity with them (Sanders and Pavelka 2013). LdtF (YafK) is an Ldt homolog protein with yet unknown enzymic function, although a role in biofilm formation has been suggested in pathogenic *E. coli* (Sheikh et al. 2001). The deletion of all six *ldt* genes leads to a lack of 3-3 cross-links and a reduction in Lpp-PG attachment, indicating the non-essentiality of these enzymes (Peters et al. 2018). Remarkably, the ampicillin resistant *E. coli* strain M1 shows elevated levels of the alarmone (p)ppGpp and is able to replace the DD-TPase activity of the PBPs by production of the β-lactam-insensitive LdtD (Hugonnet et al. 2016). This finding shows a new mode of PG polymerization in *E. coli* requiring the GTase activity of PBP1B, the DD-CPase activity of PBP5 and the LD-TPase activity of LdtD (Hugonnet et al. 2016).

LDTs play important roles in different pathogenic bacteria. The formation of 3-3 cross-links is essential for the virulence of the pathogen *Salmonella enterica* serovar Typhi. In this organism the secretion of the typhoid toxin depends on the activity of the *N*-acetyl-β-D-muramidase TtsA, which hydrolyses the 3-3 cross-links synthesised by the LDT YcbB (Geiger et al. 2018). LDTs from *E. coli* and *E. faecium* are inhibited by copper ions at sub-millimolar concentrations, likely through the binding to the thiol group of the catalytic cysteine residue and its activity inhibition. The resulting lack of 3-3 cross-links and the decrease in the Lpp-PG attachment impair the robustness of the cell envelope (Peters et al. 2018). Additionally this inhibition counteracts the LDT-mediated β-lactam resistance of *E. coli* and *E. faecium* strains (Peters et al. 2018). The PG of *C. difficile* and *M. tuberculosis* is predominantly 3-3 cross-linked, with several LDTs encoded in each organism, suggesting an important role for these enzymes (Peltier et al. 2011; Lavollay et al. 2008, 2011; Sutterlin et al. 2018). The deletion of two of the three LDTs in *C. difficile* leads to a significant decrease in 3-3 cross-links, partially compensated by an increase in 4-3 cross-links by PBPs, and a less cross-linked PG (Peltier et al. 2011). Attempts to inactivate the third LDT have been unsuccessful. Out of the five LDT homologues in *M. tuberculosis* (Ldt_{Mt1}–Ldt_{Mt5}), Ldt_{Mt2} is the dominant one, as its inactivation causes altered colony morphology, loss of virulence and increased susceptibility to amoxicillin (Gupta et al. 2010). The LD-TPase activity of Ldt_{Mt2} requires an amidated mDAP in the stem peptide, catalysed by the amidotransferase AsnB (Ngadjeua et al. 2018). Ldt_{Mt1}, Ldt_{Mt3}, Ldt_{Mt4} and Ldt_{Mt5} are all of similar size and share an amino acid sequence identity of 28–36% with Ldt_{Mt2}. The crystal structure of Ldt_{Mt5} shows structural and functional differences to the other LDTs, since it contains a different catalytic site and a proline rich C-terminal domain. Cells lacking a functional Ldt_{Mt5} displays aberrant growth and increased sensitivity to osmotic shock and certain carbapenems, indicating its critical role in maintaining the cell wall integrity (Brammer et al. 2015).

Peptidoglycan Hydrolysis and Remodelling

E. coli cells coordinate the synthesis of PG with the cleavage of the pre-existing PG material by dedicated hydrolases to incorporate new PG into the sacculi. PG hydrolases are also involved in autolysis, maturation, turnover and recycling of PG, showing substrate and PG-linkage specificity. According to the cleavage site they can be classified as *N*-acetylmuramoyl-L-Alanine amidases (amidases), carboxypeptidases, endopeptidases and lytic transglycosylases (van Heijenoort 2011; Vollmer et al. 2008b) (Fig. 5.4). Further information about the PG-recycling and β-lactamase induction pathways can be found in different reviews (Park and Uehara 2008; Juan et al. 2017; Dik et al. 2018).

N-Acetylmuramoyl-L-Alanine Amidases

Amidases specifically cleave the amide bond linking the L-Ala residue of the stem peptide to the Mur*N*Ac subunit of the muropeptide. *E. coli* contains five different amidases, which are grouped in two superfamilies based on the amino acid sequence similarity: AmiA, AmiB and AmiC (pfam: amidase 3) and AmpD and AmiD (pfam: amidase 2). AmiA, AmiB and AmiC exhibit specificity for Mur*N*Ac substrates, AmpD for 1,6-anhydro-Mur*N*Ac, and AmiD cleave both Mur*N*Ac and 1,6-anhydro-Mur*N*Ac substrates.

AmiA, AmiB and AmiC are exported to the periplasm via the Tat system (AmiA and AmiC) and through the Sec translocon (AmiB) (Ize et al. 2003; Bernhardt and de Boer 2003). Although AmiB and AmiC are recruited to the septal position during cell division, and AmiA remains dispersed throughout the cell periplasm (Bernhardt and de Boer 2003; Peters et al. 2011), all of them perform redundant reactions that compensate the absence of the others. The inactivation of all three amidases results in cell chains, in which daughter cells are not separated from each other. Single and double mutants have milder phenotypes, showing shorter cell chains (Heidrich et al. 2001). Genetic evidences indicate a partially redundant role of the Rcs and Cpx stress responses in supporting the growth and viability of the triple mutant

(Yakhnina et al. 2015). The activity of the amidases is dependent on the presence of two LytM-domain containing proteins, EnvC and NlpD, as the inactivation of both genes shows the same cell chain phenotype than the triple *amiABC* mutant. Their LytM-domains do not contain the residues required for the coordination of the catalytic Zn^{2+} ion and lack the ion itself (Uehara et al. 2009, 2010; Peters et al. 2013), typical from the metallo-peptidase family M23 (Pfam: Peptidase_M23). EnvC activates AmiA and AmiB in the outer-leaflet of the cytoplasmic membrane, and the outer membrane-anchored lipoprotein NlpD activates AmiC (Uehara et al. 2010). The structure of AmiC (Rocaboy et al. 2013) and studies on AmiB show the presence of a conserved α-helix blocking the access to the catalytic site (Yang et al. 2012). Conformational changes induced by the interaction with the activator enable the open access to the active site (Yang et al. 2012) (see section "Regulation of Peptidoglycan Growth" for further details on EnvC and NlpD functioning).

AmpD is a cytoplasmic Zn^{2+}-dependent amidase required for the PG-recycling process during bacterial cell growth. The specificity for 1,6-anhydro-Mur*N*Ac containing substrates (Jacobs et al. 1995) prevents AmpD of interfering with the synthesis of cell wall precursors. The inactivation of AmpD leads to the accumulation of its substrate 1,6-anhydro-Mur*N*Ac-L-Ala-D-iGlu-mDAP (Jacobs et al. 1994).

AmiD is an outer membrane-anchored lipoprotein with Zn^{2+}-dependent amidase activity. It cleaves the amide bond of intact PG or soluble fragments containing tri-, tetra- or pentapeptides, regardless of the presence or absence of 1,6-anhydro-Mur*N*Ac (Uehara and Park 2007; Pennartz et al. 2009). AmpD and AmiD are the only proteins with specificity for 1,6-anhydro-Mur*N*Ac substrates (Uehara and Park 2007). A reaction mechanism has been proposed based on the crystal structures of AmiD, alone and in complex with either the 1,6-anhydro-Mur*N*Ac-tripeptide or the tripeptide (Kerff et al. 2010). AmiD is not required for cell separation (Uehara and Park 2007).

Lytic Transglycosylases

Lytic transglycosylases (LTs) are periplasmic *N*-acetylmuramidases that catalyse the cleavage of the β-1,4 glycosidic bond between Mur*N*Ac and Glc*N*Ac, generating a 1,6-anhydro bond in the Mur*N*Ac. The resultant muropeptides containing 1,6-anhydro-Mur*N*Ac are transported from periplasm to cytoplasm, via the transmembrane protein AmpG. The cleavage reaction performed by LTs can take place at the end of the glycan chains (exolytic activity) or within them (endolytic activity). *E. coli* contains eight LTs (MltA-MltG and Slt). LTs are functionally redundant and the absence of six of them (MltA-E and Slt) does not affect cell viability although generates cell chain phenotype (Heidrich et al. 2002). RlpA, recently described as an interacting partner of the cell division protein FtsK in *E. coli* (Berezuk et al. 2018), is considered a LT based on its activity in *P. aeruginosa* (Jorgenson et al. 2014). Slt is a periplasmic LT, MltA-F are outer membrane-anchored lipoproteins, and MltG is anchored to the outer-leaflet of the cytoplasmic membrane. LTs are classified based on their domain architecture, belonging to family 1 (Slt, family 1A; MltC, family 1B; MltE, family 1C; MltD, family 1D; MltF, family 1E), family 2 (MltA), family 3 (MltB) and YceG-family (MltG). Slt is described in more detail in the following lines, and further information about LTs can be found in different reviews (van Heijenoort 2011; Dik et al. 2017; Alcorlo et al. 2017).

Slt has a main exolytic activity performed on non cross-linked muropeptides containing stem peptides, although a low endolytic activity has been described (Lee et al. 2013, 2018). It has a "doughnut-shaped" structure, as shown in complex with the specific inhibitor bulgecin or the 1,6-anhydro-muropeptide (Thunnissen et al. 1994, 1995; van Asselt et al. 1999). Recent works with *P. aeruginosa* Slt and *Neisseria meningitidis* LtgA suggest that conformational rearrangements on the active site take place during the hydrolytic reactions (Lee et al. 2018; Williams et al. 2018). In *E. coli* Slt interacts with the PG synthases PBP1B, PBP1C, PBP2 and PBP3, and with the PG endopeptidase PBP7 (Romeis and Höltje 1994b; von Rechenberg et al. 1996). In the absence of *slt*, or protein inactivation by addition of bulgecin, cells do not show different phenotype or alter the PG composition but modified sensitivity to certain β-lactams. Inactivation of PBP3 in *slt* mutant cells produces bulges and rapid lysis (Templin et al. 1992), instead of cell filamentation and lysis as in WT cells (Spratt 1975). The sensitivity to the PBP2-specific inhibitor mecillinam in *slt* cells depends on the FtsZ protein levels, being more sensitive at WT levels (Templin et al. 1992) and more resistant at high levels (Vinella et al. 1993). The hypersensitivity to mecillinam has been described as a consequence of the accumulation of newly synthesised PG glycan chains and its misincorporation into the sacculus by LDTs. In WT cells the non cross-linked new PG glycan chains are removed by Slt, contributing to the increased PG turnover by the β-lactam mecillinam stress response (Cho et al. 2014).

Endopeptidases

Endopeptidases (EPases) catalyse the cleavage of the amide bonds between amino acids from different stem peptides. According to the substrate specificity, they are classified as DD-endopeptidases (DD-EPases) and LD-endopeptidases (LD-EPases). DD-EPases cleave the bond formed by the 4-3 cross-links, in *E. coli* between D-Ala at position four from one stem peptide and mDAP at position three from a different stem peptide. LD-EPases show specificity for the bond formed by the 3-3 cross-links, in *E. coli* between the mDAP residues at position three of different stem peptides. *E. coli* contain seven proteins encoding for DD-EPase activity, three of them are sensitive to β-lactams (the class C PBPs PBP4, PBP7 and AmpH) and four are insensitive to β-lactams (MepA, MepH, MepM and MepS). The absence of each single protein does not change the cell phenotype, indicating functional redundancy between them. Recently it has been shown that the overproduction of MepA, MepM, PBP7 or MepS confers resistance to mecillinam through the stimulation of the PG synthesis by PBP1B (Lai et al. 2017).

PBP4 is a periplasmic protein that shows partial association with the cytoplasmic membrane (Korat et al. 1991; Jacoby and Young 1988; Leidenix et al. 1989). In addition to the DD-EPase activity, PBP4 also shows DD-carboxypeptidase (DD-CPase) activity (see section “Carboxypeptidases”), using either soluble muropeptides or isolated PG as substrate (Korat et al. 1991; Li et al. 2004; Clarke et al. 2009). The crystal structure of the protein alone and bound to different antibiotics shows three different domains and the formation of a soluble dimer (Kishida et al. 2006). Residues close to the catalytic serine encoded in the domain I are required for the accommodation of the mDAP residue of the stem peptide substrate (Clarke et al. 2009). In vivo studies based on the combination of gene deletions suggest a role for PBP4 in different cell processes including cell morphology, separation of daughter cells and biofilm formation (Meberg et al. 2004; Priyadarshini et al. 2006; Gallant et al. 2005).

The periplasmic protein PBP7 shows a loose association with the cytoplasmic membrane, which can be abolished by high salt treatment (Romeis and Höltje 1994a). A C-terminal truncated variant, identified as PBP8, led to confusion in the past (Henderson et al. 1994). PBP7 shows DD-EPase activity when isolated PG is used as substrate, but not in case of purified dimeric muropeptides (Romeis and Höltje 1994a). It interacts with Slt, which enhances its LTase activity (Romeis and Höltje 1994b). The inactivation or overproduction of PBP7 does not generate any cellular defect (Henderson et al. 1995). Combined inactivation of PBP7, PBP4 and PBP5 enhances the morphological defects caused by the absence of PBP5 (main DD-CPase in *E. coli*) (Meberg et al. 2004), and the additional inactivation of AmpH activates the Rcs phosphorelay and Cpx stress responses (Evans et al. 2013).

AmpH is a periplasmic protein associated to the cytoplasmic membrane, despite the lack of any membrane anchoring domain. In contrast to PBP4 and PBP7, high salt treatment does not dissociate the protein from the membrane fraction, requiring detergent for that purpose. AmpH displays DD-EPase and DD-CPase activities using both intact sacculi and isolated dimeric muropeptides as substrates, and weak β-lactamase activity (Gonzalez-Leiza et al. 2011). Inactivation of AmpH does not cause any defect in the cell, unless it is combined with other mutations producing morphological changes (Henderson et al. 1997) or stress response activation (Evans et al. 2013). A role in PG recycling and remodelling has been suggested based on the wide range of substrates used by AmpH.

MepA is a periplasmic protein belonging to the LAS metallopeptidase family (lysostaphin-type enzymes, D-alanyl-D-alanine CPase, and sonic hedgehog protein), which is characterised for the presence of a Zn^{2+}-binding site in the catalytic domain. It shows DD-EPase activity on intact sacculi and isolated muropeptides (Keck and Schwarz 1979; Firczuk and Bochtler 2007) and LD-EPase activity on intact sacculi (Engel et al. 1992). The conserved metal ligands are required for the correct folding and catalytic activity of the protein (Firczuk and Bochtler 2007), which can be inhibited by DNA, lipoteichoic acid and metal-chelating agents (Tomioka and Matsuhashi 1978; Keck and Schwarz 1979). Inactivation or overproduction of MepA does not cause any change in the PG composition or cell phenotype

(Iida et al. 1983; Keck et al. 1990). The combined inactivation of PBP4, PBP7 and MepA does not cause any defect in the cell, but enhances the cell chain phenotype due to the absence of amidases and LTs (Heidrich et al. 2002).

MepH, MepM and MepS, in contrast to the previously described EPases, are redundantly essential as their absence cause cell lysis (Singh et al. 2012). MepH is a periplasmic protein of the NlpC/P60 peptidase superfamily with the conserved Cys-His-His catalytic triad (Aramini et al. 2008). MepM is a bitopic cytoplasmic membrane protein of the metallopeptidase family M23, containing the characteristic LytM domain and the catalytic Zn^{2+}-binding site. Both proteins show DD-EPase activity against isolated muropeptides and, partially, intact sacculi (Singh et al. 2012). MepS is an outer membrane-anchored lipoprotein belonging to the NlpC/P60 peptidase superfamily, which shows DD-EPase and weak LD-CPase activity against isolated muropeptides but not against intact sacculi (Singh et al. 2012). MepS protein levels are higher during exponential phase, being modulated by the proteolytic system NlpI-Prc (Singh et al. 2015). NlpI is an outer membrane-anchored lipoprotein containing tetratricopeptide-repeats that facilitates the interaction between MepS and the periplasmic protease Prc (Singh et al. 2015; Su et al. 2017). A reaction mechanism for the degradation of MepS has been proposed based on the crystal structure and biophysical and mutational analyses of the NlpI-Prc complex (Su et al. 2017). The impairment of this proteolytic system alters the morphology of the cells at high-salinity growth conditions (Kerr et al. 2014), changes the PG dynamics in stationary phase, decreases the Lpp-PG attachment and increases the formation of outer membrane vesicles (Schwechheimer et al. 2015).

Carboxypeptidases

Carboxypeptidases (CPases) remove the terminal residues from stem peptides. *E. coli* encodes for six non-essential DD-CPases (PBP4, PBP4b, PBP5, PBP6, PBP6b and AmpH) that remove the D-Ala residue from pentapeptides, and one cytoplasmic LD-CPase (LdcA) that removes the D-Ala residue from tetrapeptides. PBP4 and AmpH also show DD-EPase activity and were described in the section "Endopeptidases".

PBP4b is a non-essential protein, showing sequence homology to AmpH and weak DD-CPase activity on artificial substrate (Vega and Ayala 2006) but not on natural isolated muropeptides (J. Ayala, unpublished results mentioned in Sauvage et al. 2008).

PBP5 is one of the best characterised and most abundant PBP in the cell. It is a membrane protein anchored to the outer leaflet of the cytoplasmic membrane by its C-terminal amphipathic helix. It shows sequence similarities with PBP4, PBP6 and certain β-lactamases, and has been recently described to dimerise (Meiresonne et al. 2017). The crystal structure of a soluble form of PBP5 shows the presence of an N-terminal domain (domain I) including the catalytic site and a C-terminal domain (domain II) containing a hydrophobic surface and the mentioned membrane-binding helix. The domain I contains a Ω-loop-like region, similar to the one found in certain β-lactamases, which is essential for keeping the β-lactam resistance and the cell shape (Dutta et al. 2015; Kar et al. 2018). The function of the domain II is not well known, although it seems to be required for the stabilization and full function of the protein. It shows DD-CPase activity against soluble muropeptides and cross-linked and uncross-linked PG. Although non-essential, the loss of PBP5 activity causes cell shape defects and an increase in the pentapeptide content in the PG. These shape defects are enhanced and lead to the formation of branches when combined with the absence of other PBPs. Variations in the level of the cell division protein FtsZ and in the formation of the FtsZ-ring structure also enhance the cell shape defects (Varma and Young 2004; Varma et al. 2007; Potluri et al. 2012a). PBP5 localizes to areas of ongoing PG synthesis, meaning along the lateral cell wall and at division sites before septation starts (preseptal sites) (Potluri et al. 2010). The absence of the membrane binding helix does not affect the catalytic activity of the protein, although the resultant soluble PBP5 does not localize at preseptal sites or restore the phenotype in mutant cells, as also observed for inactive PBP5 variants (Nelson and Young 2000; Nelson et al. 2002; Potluri et al. 2010). Overproduction of PBP5 causes non-viable spherical cells (Markiewicz et al. 1982), suggesting a role for PBP5 in regulation of the pentapeptide subunits required for the formation of cross-links by DD-TPase. The combination of the DD-CPase activity of PBP5, the LD-TPase activity of LdtD and the GTase activity of PBP1B has been described to bypass the DD-TPase activity of the PBPs (Hugonnet et al. 2016).

PBP6 and PBP6b have strong sequence and structural similarities with PBP5, including the C-terminal membrane anchor that can substitute the one from PBP5 (Nelson et al. 2002). PBP6 shows weaker activity than PBP5 against similar substrates (Amanuma and Strominger 1980). In the absence of PBP5, the main DD-CPase activity is performed by PBP4 at neutral pH, and by PBP6b at low pH. The role of PBP6b at low pH is also supported by its ability to maintain the WT cell shape in the absence of PBP5 (Peters et al. 2016).

Regulation of Peptidoglycan Growth

During the *E. coli* cell cycle the PG sacculus is enlarged and split through the enzymatic activities encoded in the protein complexes named elongasome and divisome, which facilitate cell elongation and cell division, respectively. Both, synthetic and hydrolytic activities are required for the incorporation of the new PG material, as proposed in the 3-for-1 growth model (Höltje 1998). The control of these activities, from the cytoplasm by the cytoskeletal proteins and from the periplasmic side by outer membrane and periplasmic proteins, ensures the precise and coordinated synthesis of new PG. The composition and functioning of both elongasome and divisome are largely unknown. However, the protein interactions described in the literature support the formation of these multienzymatic complexes, although it is unlikely that all the interactions take place at the same time (Fig. 5.5) (Typas et al. 2011; Egan and Vollmer 2013). The presence of large sets of seemingly redundant enzymes, under

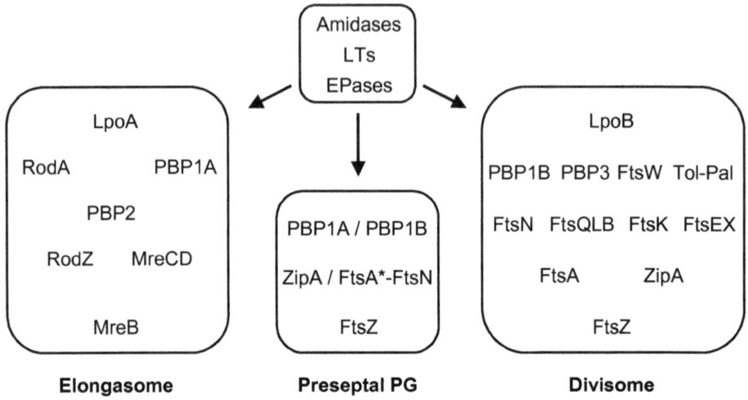

Fig. 5.5 Components of the multienzymatic complexes suggested for new PG incorporation during cell growth and division in *E. coli*. The hydrolytic enzymes (amidases, LTs and EPases) are not specified, as their coordination with the synthetic activities remains largely unknown in each case. LTs, lytic transglycosylases; EPases, endopeptidases

laboratory standard conditions, may be required to ensure a robust peptidoglycan growth under the different environmental conditions (Pazos et al. 2017).

Regulation of Peptidoglycan Synthesis by the Cytoskeletal Proteins

The elongasome is required for the elongation and shape maintenance of rod-shape cells. It contains the synthetic activities of RodA (GTase), PBP2 (TPase) and PBP1A (GTase and TPase). The activity of the elongasome is regulated by the cytoskeletal protein MreB and the membrane-associated MreC, MreD and RodZ proteins. MreB is an actin homolog that polymerizes in an ATP-dependent manner into dynamic double antiparallel filaments (van den Ent et al. 2014), which localize on the inner-leaflet of the cytoplasmic membrane in a curvature-dependent way and rotate along the short cell axis as patches (Garner et al. 2011). MreB binds directly to the membrane through its N-terminal amphiphatic helix (Salje et al. 2011). The disruption of these filaments has an impact on the cell shape, including the eventual spherical phenotype obtained after inactivation of MreB (Kruse et al. 2005). The MreB inhibitor A22 competes for the ATP binding pocket of MreB, which is not able to polymerize and remains homogeneously localized in the cytoplasm. The presence of A22-blocked MreB monomers reduces the characteristic growth heterogeneity and generates cell wall growth at the poles (Ursell et al. 2014), which were previously suggested to be inert (de Pedro et al. 1997). Cell poles are enriched in cardiolipin and phosphatidylglycerol, which prevent the localization of MreB filaments but not of MreB monomers (Kawazura et al. 2017). The fact that MreB motion requires an active elongasome, and the elongasome can function in the absence of an active MreB, suggests a passive role of MreB in guiding the elongasome (van Teeffelen et al. 2011; Ursell et al. 2014). Single particle tracking data showed that these protein complexes are very dynamic, as observed by the different motion of PBP2 and MreB (Lee et al. 2014), the presence of MreB-like slower subpopulations of RodA and PBP2 (Cho et al. 2016) or the fast and diffusive behaviour of the class A PBPs (Cho et al. 2016; Lee et al. 2016), which can be modified either by specific inhibitors or activators (Lee et al. 2016). MreB localization requires a functional Sec-translocon system for the correct insertion of RodZ into the membrane (Govindarajan and Amster-Choder 2017; Rawat et al. 2015). RodZ is required for the attachment of MreB filaments to the cytoplasmic membrane, modulating their localization in the cell (Colavin et al. 2018; Bratton et al. 2018). Further information about MreB and associated proteins can be found in reviews (Errington 2015; Shi et al. 2018).

Despite the essentiality of MreB and the elongasome, the overproduction of the tubulin homolog FtsZ is able to restore cell viability, but not rod-shape, upon deletion of any of them (Bendezu and de Boer 2008). Although the mechanism of this FtsZ-mediated rescue is not clear, it is known that both MreB and FtsZ interact directly at the cell division site and that this interaction is essential for cell division but not for rod-shape maintenance. An impaired interaction is lethal for the cell, which shows neither preseptal nor septal PG synthesis (Fenton and Gerdes 2013). Preseptal PG is synthesised at the future division site before any constriction is visible, in a PBP3-independent manner. It is considered as a transition stage between cell elongation and cell division, or an early cell division stage (Nanninga 1991; de Pedro et al. 1997). So far, in *E. coli* only the cell division proteins FtsZ and ZipA, and either PBP1A or PBP1B are described as essential proteins for preseptal PG synthesis. Several proteins from the elongasome or divisome are not required, e.g. RodA, FtsA, FtsEX, FtsK or FtsQ (Potluri et al. 2012b). ZipA links the cytosolic Z-ring and the PG synthases PBP1A or PBP1B, an essential role that can be compensated by the mutant FtsA* presumably bound to FtsN (Fig. 5.5) (Pazos et al. 2018). In the absence of both ZipA and FtsN, the cells are not viable and do not synthesise preseptal PG. Further work is required to study the essentiality of the preseptal PG synthesis for the correct cell septation.

The divisome is involved in the septation and separation of the two daughter cells. Several proteins are required for the correct assembly and functioning of the divisome and many described interactions support the hypothesis of a multiprotein complex (Fig. 5.5) (reviewed in Egan and Vollmer 2013). The synthesis of the septal PG is carried out by PBP1B (GTase and TPase) and PBP3 (TPase). Controversial results about the SEDS family protein FtsW propose that either it is a lipid II flippase or a GTase. Future work will clarify its role. The cell division protein FtsZ is the scaffold in which the divisome is build up. FtsZ polymerizes into a ring-like structure, termed the Z-ring. FtsZ polymerization and depolymerization requires the binding and hydrolysis of GTP, respectively. These filaments are bound to the inner-leaflet of the cytoplasmic membrane by the membrane-attached proteins FtsA and ZipA. The GTP hydrolysis generates conformational changes in the FtsZ filaments that adopt a curved shape (Lu et al. 2000), which has been proposed to supply the constriction force in liposomes (Osawa et al. 2008; Osawa and Erickson 2013). The cellular localization of the Z-ring is regulated by the Min system, the nucleoid occlusion, and the Ter region of the chromosome (reviewed in Schumacher 2017; Wettmann and Kruse 2018; Bailey et al. 2014a). The placement of the Z-ring at mid-cell is essential to generate two identical daughter cells after septum synthesis and cleavage. Septal PG synthesis is guided and driven by the dynamic FtsZ polymers, which move around the ring structure by treadmilling (Yang et al. 2017; Bisson-Filho et al. 2017). Remarkably, in the round-shaped bacteria *S. aureus* the recruitment of the proposed lipid II flippase MurJ to the divisome speeds up the synthesis of septal PG and makes it independent of FtsZ-treadmilling (Monteiro et al. 2018). In *E. coli*, FtsZ is not present at the division site during the whole process of septation, as it moves to the future cell division sites before constriction is finished (Soderstrom et al. 2014, 2016). Instead of that, the cell division proteins FtsN and PBP3, both spatially separated from the FtsZ ring, remain at division site until septum synthesis is completed (Soderstrom et al. 2016, 2018). PBP3 activity, but not FtsZ, has been shown to be the main regulator of septum closure (Coltharp et al. 2016), leading to different models about the source of constrictive force (Xiao and Goley 2016; Schoenemann and Margolin 2017; Holden 2018).

Regulation of Peptidoglycan Synthesis from the Periplasmic Side

E. coli contains periplasmic proteins that regulate the activities of the PG multienzymatic complexes. LpoA and LpoB are two outer membrane-anchored lipoproteins that activates the synthetic activities of PBP1A and PBP1B, respectively. The presence of one of these proteins is essential for cell viability, in a similar way to PBP1A and PBP1B, suggesting that the Lpo's are required for the proper functioning of the PBPs (Typas et al. 2010; Paradis-Bleau et al. 2010). The regions involved in the interaction with their cognate PBPs are identified (Jean et al. 2014; Egan et al. 2014; King et al. 2014; Sathiyamoorthy et al. 2017). In vitro studies have shown that LpoA stimulates the TPase activity of PBP1A, and LpoB stimulates both GTase and TPase activities of PBP1B (Lupoli et al. 2014; Egan et al. 2014, 2018). The identification of PBP1B mutants able to bypass the in vivo requirement of LpoB support the role of LpoB in PBP1B activation (Markovski et al. 2016). It has been hypothesised that the activation by Lpo's could regulate the PG enlargement depending on the sacculus pores size, which might be different depending on the cellular growth rate (Typas et al. 2010).

During cell division several proteins have been shown to regulate and coordinate the activity of the PG synthases and hydrolases. In the next lines some recent results are briefly mentioned, including the Tol-Pal complex, FtsN and PG amidases. The Tol-Pal complex is a well conserved system in Gram negative bacteria and includes cytoplasmic membrane, periplasmic and outer membrane proteins. These proteins interact with each other and, in the presence of proton motive force, maintain the stability of the outer membrane (Cascales et al. 2000, 2001, 2002). TolA and CpoB interact directly with PBP1B and regulate its synthetic activities. TolA enhances the GTase activity of PBP1B, either in presence or absence of LpoB, and CpoB decreases the TPase activity of PBP1B in presence of LpoB without interfering with the GTase activation (Gray et al. 2015; Egan et al. 2018). The CpoB effect can be reverted by TolA (Gray et al. 2015).

FtsN is considered the septation trigger, as it is mainly recruited at late stages of cell division just before constriction starts. This recruitment is due to the binding of its periplasmic SPOR (sporulation-related repeat) domain to PG glycan chains (Ursinus et al. 2004) lacking the stem peptides, suggesting that it follows the action of amidases (Yahashiri et al. 2015). The cytoplasmic region of FtsN is required for an earlier recruitment of FtsN to midcell by the interaction with FtsA (Busiek et al. 2012; Busiek and Margolin 2014; Pichoff et al. 2015). Whereas favouring the FtsA-FtsN interaction the cell bypasses different impairments in other cell division proteins (Pichoff et al. 2018), point mutations in FtsA abolish the essentiality of FtsN (Bernard et al. 2007). The essentiality of FtsN seems to be encoded in a small peptide from the periplasmic domain (Gerding et al. 2009; Liu et al. 2015), which would be involved in the eventual activation of septation. FtsN interacts with different cell division proteins including the PG synthases PBP3 and PBP1B, and the potential PG synthase FtsW (Di Lallo et al. 2003; Müller et al. 2007; Alexeeva et al. 2010), and in vitro experiments have shown that FtsN enhances both GTase and TPase activities of PBP1B (Müller et al. 2007). Due to the non-essentiality of PBP1B, the activation of PBP3 and/or FtsW is likely to be the final target of FtsN. Genetic evidences have shown that the conserved cell division proteins FtsQLB are involved in this activation pathway (Liu et al. 2015; Tsang and Bernhardt 2015).

E. coli contains three PG amidases (AmiA, AmiB and AmiC) required for splitting the PG sacculi of two daughter cells during septation, although only AmiB and AmiC are recruited to the division site. Their catalytic activities are regulated by protein activators: AmiA and AmiB by EnvC, and AmiC by NlpD. The correct localization of both activators is essential for the temporal and spatial regulation of the amidase activities. EnvC is localized at preseptal positions (Peters et al. 2011) by the cell division ATP-binding cassette complex FtsEX, in which FtsE is the cytoplasmic nucleotide-binding protein and FtsX is the integral membrane component that binds to EnvC (de Leeuw et al. 1999; Yang et al. 2011). Conformational changes in FtsX, driven by the ATPase activity of FtsE, are essential for the amidase activation by EnvC (Yang et al. 2011). NlpD, as EnvC, it is located at the division site before septal PG is synthesised (Peters et al. 2011). Additional proteins, as the Tol-Pal complex and YraP, are required for the activation of AmiC by NlpD (Tsang et al. 2017).

Fluorescent D-Amino Acids as Tools

The recent development of new fluorescent D-amino acids (FDAAs), combined with the continuous improvement of microscopic techniques, enables the visualization of PG synthesis, remodelling and dynamics in high resolution. These powerful tools are used to study PG synthesis processes during cell elongation and division in diverse Gram-positive and Gram-negative bacteria. The FDAAs designed by VanNieuwenhze's and Brun's labs mimic the acyl acceptor during the PG synthesis reaction and are thought to be incorporated into the stem peptides through a D-amino-exchange reaction performed by either DD-TPase (Lupoli et al. 2011) or DD-CPase and LD-TPase activities (Cava et al. 2011; Hsu et al. 2017) (Fig. 5.4). Based on their specificity, FDAAs efficiently label the active PG synthesis sites with minimal cell toxicities (Kuru et al. 2012). FDAAs have been successfully used in many PG studies of diverse species, providing new insights about PG synthesis, recycling and turnover (for a detailed review about FDAAs, including an updated list of studies using them (see Radkov et al. 2018)).

Concluding Remarks

In the last years we have gained insight into many different aspects of the bacterial cell wall, but we are still far from understand this complex cellular structure. The introduction of new tools and technologies (as FDAAs, high resolution microscopy or high-throughput genetic screenings), combined with the multidisciplinary contribution to the field, is currently boosting the knowledge of the field. Besides the new biophysical and structural approaches, the better understanding of the essential proteins roles might lead to the identification of new antimicrobial drug targets, for example the LDTs. New antimicrobial drugs are required to reduce the dramatic spread of multidrug resistant bacteria. In terms of basic knowledge, the formation and coordination of the elongasome and divisome complexes remain elusive. The further characterisation of the transition between both elongation and division, and the following septum formation and cell constriction will aid the comprehension of this complex but exciting macromolecule, the peptidoglycan.

Acknowledgements The authors thank Professor Waldemar Vollmer at Newcastle University for critical reading this manuscript.

References

Alcorlo M, Martinez-Caballero S, Molina R, Hermoso JA (2017) Carbohydrate recognition and lysis by bacterial peptidoglycan hydrolases. Curr Opin Struct Biol 44:87–100

Alexeeva S, Gadella Tw Jr, Verheul J, Verhoeven GS, Den Blaauwen T (2010) Direct interactions of early and late assembling division proteins in *Escherichia coli* cells resolved by FRET. Mol Microbiol 77:384–389

Amanuma H, Strominger JL (1980) Purification and properties of penicillin-binding proteins 5 and 6 from *Escherichia coli* membranes. J Biol Chem 255:11173–11180

Aramini JM, Rossi P, Huang YJ, Zhao L, Jiang M, Maglaqui M, Xiao R, Locke J, Nair R, Rost B, Acton TB, Inouye M, Montelione GT (2008) Solution NMR structure of the NlpC/P60 domain of lipoprotein Spr from *Escherichia coli*: structural evidence for a novel cysteine peptidase catalytic triad. Biochemistry 47:9715–9717

Arthur M, Molinas C, Depardieu F, Courvalin P (1993) Characterization of Tn1546, a Tn3-related transposon conferring glycopeptide resistance by synthesis of depsipeptide peptidoglycan precursors in *Enterococcus faecium* BM4147. J Bacteriol 175:117–127

Arthur M, Reynolds P, Courvalin P (1996) Glycopeptide resistance in *enterococci*. Trends Microbiol 4:401–407

Atrih A, Bacher G, Allmaier G, Williamson MP, Foster SJ (1999) Analysis of peptidoglycan structure from vegetative cells of *Bacillus subtilis* 168 and role of PBP 5 in peptidoglycan maturation. J Bacteriol 181:3956–3966

Aubry C, Goulard C, Nahori MA, Cayet N, Decalf J, Sachse M, Boneca IG, Cossart P, Dussurget O (2011) OatA, a peptidoglycan O-acetyltransferase involved in *Listeria monocytogenes* immune escape, is critical for virulence. J Infect Dis 204:731–740

Auer GK, Lee TK, Rajendram M, Cesar S, Miguel A, Huang KC, Weibel DB (2016) Mechanical genomics identifies diverse modulators of bacterial cell stiffness. Cell Syst 2:402–411

Bailey MW, Bisicchia P, Warren BT, Sherratt DJ, Mannik J (2014a) Evidence for divisome localization mechanisms independent of the min system and SlmA in *Escherichia coli*. PLoS Genet 10:e1004504

Bailey RG, Turner RD, Mullin N, Clarke N, Foster SJ, Hobbs JK (2014b) The interplay between cell wall mechanical properties and the cell cycle in *Staphylococcus aureus*. Biophys J 107:2538–2545

Banzhaf M, van den berg van saparoea B, Terrak M, Fraipont C, Egan A, Philippe J, Zapun A, Breukink E, Nguyen-Disteche M, den Blaauwen T, Vollmer W (2012) Cooperativity of peptidoglycan synthases active in bacterial cell elongation. Mol Microbiol 85:179–194

Barreteau H, Kovac A, Boniface A, Sova M, Gobec S, Blanot D (2008) Cytoplasmic steps of peptidoglycan biosynthesis. FEMS Microbiol Rev 32:168–207

Barrett DS, Chen L, Litterman NK, Walker S (2004) Expression and characterization of the isolated glycosyltransferase module of *Escherichia coli* PBP1b. Biochemistry 43:12375–12381

Beeby M, Gumbart JC, Roux B, Jensen GJ (2013) Architecture and assembly of the Gram-positive cell wall. Mol Microbiol 88:664–672

Benachour A, Ladjouzi R, le Jeune A, Hebert L, Thorpe S, Courtin P, Chapot-Chartier MP, Prajsnar TK, Foster SJ, Mesnage S (2012) The lysozyme-induced peptidoglycan *N*-acetylglucosamine deacetylase PgdA (EF1843) is required for *Enterococcus faecalis* virulence. J Bacteriol 194:6066–6073

Bendezu FO, de Boer PA (2008) Conditional lethality, division defects, membrane involution, and endocytosis in mre and mrd shape mutants of *Escherichia coli*. J Bacteriol 190:1792–1811

Bera A, Herbert S, Jakob A, Vollmer W, Gotz F (2005) Why are pathogenic staphylococci so lysozyme resistant? the peptidoglycan *O*-acetyltransferase OatA is the major determinant for lysozyme resistance of *Staphylococcus aureus*. Mol Microbiol 55:778–787

Berezuk AM, Glavota S, Roach EJ, Goodyear MC, Krieger JR, Khursigara CM (2018) Outer membrane lipoprotein RlpA is a novel periplasmic interaction partner of the cell division protein FtsK in *Escherichia coli*. Sci Rep 8:12933

Bernal-Cabas M, Ayala JA, Raivio TL (2015) The Cpx envelope stress response modifies peptidoglycan cross-linking via the LD-transpeptidase LdtD and the novel protein YgaU. J Bacteriol 197:603–614
Bernard CS, Sadasivam M, Shiomi D, Margolin W (2007) An altered FtsA can compensate for the loss of essential cell division protein FtsN in *Escherichia coli*. Mol Microbiol 64:1289–1305
Bernard E, Rolain T, Courtin P, Guillot A, Langella P, Hols P, Chapot-Chartier MP (2011) Characterization of *O*-acetylation of *N*-acetylglucosamine: a novel structural variation of bacterial peptidoglycan. J Biol Chem 286:23950–23958
Bernard E, Rolain T, David B, Andre G, Dupres V, Dufrene YF, Hallet B, Chapot-Chartier MP, Hols P (2012) Dual role for the *O*-acetyltransferase OatA in peptidoglycan modification and control of cell septation in *Lactobacillus plantarum*. PLoS ONE 7:e47893
Bernhardt TG, de Boer PA (2003) The *Escherichia coli* amidase AmiC is a periplasmic septal ring component exported via the twin-arginine transport pathway. Mol Microbiol 48:1171–1182
Bertsche U, Breukink E, Kast T, Vollmer W (2005) *In vitro* murein peptidoglycan synthesis by dimers of the bifunctional transglycosylase-transpeptidase PBP1B from *Escherichia coli*. J Biol Chem 280:38096–38101
Bisson-Filho AW, Hsu YP, Squyres GR, Kuru E, Wu F, Jukes C, Sun Y, Dekker C, Holden S, Vannieuwenhze MS, Brun YV, Garner EC (2017) Treadmilling by FtsZ filaments drives peptidoglycan synthesis and bacterial cell division. Science 355:739–743
Bolla JR, Sauer JB, Wu D, Mehmood S, Allison TM, Robinson CV (2018) Direct observation of the influence of cardiolipin and antibiotics on lipid II binding to MurJ. Nat Chem 10:363–371
Boneca IG, Dussurget O, Cabanes D, Nahori MA, Sousa S, Lecuit M, Psylinakis E, Bouriotis V, Hugot JP, Giovannini M, Coyle A, Bertin J, Namane A, Rousselle JC, Cayet N, Prevost MC, Balloy V, Chignard M, Philpott DJ, Cossart P, Girardin SE (2007) A critical role for peptidoglycan *N*-deacetylation in Listeria evasion from the host innate immune system. Proc Natl Acad Sci USA 104:997–1002
Bonnet J, Durmort C, Jacq M, Mortier-Barriere I, Campo N, Vannieuwenhze MS, Brun YV, Arthaud C, Gallet B, Moriscot C, Morlot C, Vernet T, di Guilmi AM (2017) Peptidoglycan *O*-acetylation is functionally related to cell wall biosynthesis and cell division in *Streptococcus pneumoniae*. Mol Microbiol 106:832–846
Born P, Breukink E, Vollmer W (2006) *In vitro* synthesis of cross-linked murein and its attachment to sacculi by PBP1A from *Escherichia coli*. J Biol Chem 281:26985–26993
Boyle-Vavra S, Labischinski H, Ebert CC, Ehlert K, Daum RS (2001) A spectrum of changes occurs in peptidoglycan composition of glycopeptide-intermediate clinical *Staphylococcus aureus* isolates. Antimicrob Agents Chemother 45:280–287
Brammer LAB, Ghosh A, Pan Y, Jakoncic J, Lloyd EP, Townsend CA, Lamichhane G, Bianchet MA (2015) Loss of a functionally and structurally distinct LD-transpeptidase, LdtMt5, compromises cell wall integrity in *Mycobacterium tuberculosis*. J Biol Chem 290:25670–25685
Bratton BP, Shaevitz JW, Gitai Z, Morgenstein RM (2018) MreB polymers and curvature localization are enhanced by RodZ and predict *E. coli*'s cylindrical uniformity. Nat Commun 9:2797
Bui NK, Eberhardt A, Vollmer D, Kern T, Bougault C, Tomasz A, Simorre JP, Vollmer W (2012) Isolation and analysis of cell wall components from *Streptococcus pneumoniae*. Anal Biochem 421:657–666
Buist G, Steen A, Kok J, Kuipers OP (2008) LysM, a widely distributed protein motif for binding to (peptido)glycans. Mol Microbiol 68:838–847
Busiek KK, Eraso JM, Wang Y, Margolin W (2012) The early divisome protein FtsA interacts directly through its 1c subdomain with the cytoplasmic domain of the late divisome protein FtsN. J Bacteriol 194:1989–2000
Busiek KK, Margolin W (2014) A role for FtsA in SPOR-independent localization of the essential *Escherichia coli* cell division protein FtsN. Mol Microbiol 92:1212–1226
Butler EK, Davis RM, Bari V, Nicholson PA, Ruiz N (2013) Structure-function analysis of MurJ reveals a solvent-exposed cavity containing residues essential for peptidoglycan biogenesis in *Escherichia coli*. J Bacteriol 195:4639–4649

Butler EK, Tan WB, Joseph H, Ruiz N (2014) Charge requirements of lipid II flippase activity in *Escherichia coli*. J Bacteriol 196:4111–4119

Cao L, Liang D, Hao P, Song Q, Xue E, Caiyin Q, Cheng Z, Qiao J (2018) The increase of O-acetylation and N-deacetylation in cell wall promotes acid resistance and nisin production through improving cell wall integrity in *Lactococcus lactis*. J Ind Microbiol Biotechnol

Cascales E, Bernadac A, Gavioli M, Lazzaroni JC, Lloubes R (2002) Pal lipoprotein of *Escherichia coli* plays a major role in outer membrane integrity. J Bacteriol 184:754–759

Cascales E, Gavioli M, Sturgis JN, Lloubes R (2000) Proton motive force drives the interaction of the inner membrane TolA and outer membrane pal proteins in *Escherichia coli*. Mol Microbiol 38:904–915

Cascales E, Lloubes R, Sturgis JN (2001) The TolQ-TolR proteins energize TolA and share homologies with the flagellar motor proteins MotA-MotB. Mol Microbiol 42:795–807

Cava F, de Pedro MA, Lam H, Davis BM, Waldor MK (2011) Distinct pathways for modification of the bacterial cell wall by non-canonical D-amino acids. EMBO J 30:3442–3453

Cayley DS, Guttman HJ, Record MT Jr (2000) Biophysical characterization of changes in amounts and activity of *Escherichia coli* cell and compartment water and turgor pressure in response to osmotic stress. Biophys J 78:1748–1764

Chamakura KR, Sham LT, Davis RM, Min L, Cho H, Ruiz N, Bernhardt TG, Young R (2017) A viral protein antibiotic inhibits lipid II flippase activity. Nat Microbiol 2:1480–1484

Chang JD, Foster EE, Wallace AG, Kim SJ (2017) Peptidoglycan *O*-acetylation increases in response to vancomycin treatment in vancomycin-resistant *Enterococcus faecalis*. Sci Rep 7:46500

Cho H, Uehara T, Bernhardt TG (2014) Beta-lactam antibiotics induce a lethal malfunctioning of the bacterial cell wall synthesis machinery. Cell 159:1300–1311

Cho H, Wivagg CN, Kapoor M, Barry Z, Rohs PD, Suh H, Marto JA, Garner EC, Bernhardt TG (2016). Bacterial cell wall biogenesis is mediated by SEDS and PBP polymerase families functioning semi-autonomously. Nat Microbiol 16172

Clarke TB, Kawai F, Park SY, Tame JR, Dowson CG, Roper DI (2009) Mutational analysis of the substrate specificity of *Escherichia coli* penicillin binding protein 4. Biochemistry 48:2675–2683

Colavin A, Shi H, Huang KC (2018) RodZ modulates geometric localization of the bacterial actin MreB to regulate cell shape. Nat Commun 9:1280

Coltharp C, Buss J, Plumer TM, Xiao J (2016) Defining the rate-limiting processes of bacterial cytokinesis. Proc Natl Acad Sci USA 113:E1044–E1053

Contreras-Martel C, Martins A, Ecobichon C, Trindade DM, Mattei PJ, Hicham S, Hardouin P, Ghachi ME, Boneca IG, Dessen A (2017) Molecular architecture of the PBP2-MreC core bacterial cell wall synthesis complex. Nat Commun 8:776

Coulombe F, Divangahi M, Veyrier F, de Leseleuc L, Gleason JL, Yang Y, Kelliher MA, Pandey AK, Sassetti CM, Reed MB, Behr MA (2009) Increased NOD2-mediated recognition of N-glycolyl muramyl dipeptide. J Exp Med 206:1709–1716

Dajkovic A, Tesson B, Chauhan S, Courtin P, Keary R, Flores P, Marliere C, Filipe SR, Chapot-Chartier MP, Carballido-Lopez R (2017) Hydrolysis of peptidoglycan is modulated by amidation of meso-diaminopimelic acid and Mg^{2+} in *Bacillus subtilis*. Mol Microbiol 104:972–988

de Leeuw E, Graham B, Phillips GJ, Ten Hagen-Jongman CM, Oudega B, Luirink J (1999) Molecular characterization of *Escherichia coli* FtsE and FtsX. Mol Microbiol 31:983–993

de Pedro MA, Quintela JC, Höltje JV, Schwarz H (1997) Murein segregation in *Escherichia coli*. J Bacteriol 179:2823–2834

Delhaye A, Collet JF, Laloux G (2016) Fine-tuning of the Cpx envelope stress response is required for cell wall homeostasis in *Escherichia coli*. MBio 7:e00047–16

Demchick P, Koch AL (1996) The permeability of the wall fabric of *Escherichia coli* and *Bacillus subtilis*. J Bacteriol 178:768–773

Denome SA, Elf PK, Henderson TA, Nelson DE, Young KD (1999) *Escherichia coli* mutants lacking all possible combinations of eight penicillin binding proteins: viability, characteristics, and implications for peptidoglycan synthesis. J Bacteriol 181:3981–3993 di Lallo G, Fagioli M, Barionovi D, Ghelardini P, Paolozzi L (2003) Use of a two-hybrid assay to study the assembly of a complex multicomponent protein machinery: bacterial septosome differentiation. Microbiology 149:3353–3359

Dik DA, Fisher JF, Mobashery S (2018) Cell-wall recycling of the Gram-negative bacteria and the nexus to antibiotic resistance. Chem Rev 118:5952–5984

Dik DA, Marous DR, Fisher JF, Mobashery S (2017) Lytic transglycosylases: concinnity in concision of the bacterial cell wall. Crit Rev Biochem Mol Biol 52:503–542

Dillard JP, Hackett KT (2005) Mutations affecting peptidoglycan acetylation in *Neisseria gonorrhoeae* and *Neisseria meningitidis*. Infect Immun 73:5697–5705

Dutta M, Kar D, Bansal A, Chakraborty S, Ghosh AS (2015) A single amino acid substitution in the Omega-like loop of *E. coli* PBP5 disrupts its ability to maintain cell shape and intrinsic beta-lactam resistance. Microbiology 161:895–902

Egan AJ, Jean NL, Koumoutsi A, Bougault CM, Biboy J, Sassine J, Solovyova AS, Breukink E, Typas A, Vollmer W, Simorre JP (2014) Outer-membrane lipoprotein LpoB spans the periplasm to stimulate the peptidoglycan synthase PBP1B. Proc Natl Acad Sci USA 111:8197–8202

Egan AJ, Vollmer W (2013) The physiology of bacterial cell division. Ann N Y Acad Sci 1277:8–28

Egan AJF, Maya-Martinez R, Ayala I, Bougault CM, Banzhaf M, Breukink E, Vollmer W, Simorre JP (2018) Induced conformational changes activate the peptidoglycan synthase PBP1B. Mol Microbiol 110:335–356

el Ghachi M, Howe N, Huang CY, Olieric V, Warshamanage R, Touze T, Weichert D, Stansfeld PJ, Wang M, Kerff F, Caffrey M (2018) Crystal structure of undecaprenyl-pyrophosphate phosphatase and its role in peptidoglycan biosynthesis. Nat Commun 9:1078

Elhenawy W, Davis RM, Fero J, Salama NR, Felman MF, Ruiz N (2016) The O-antigen flippase Wzk can substitute for MurJ in peptidoglycan synthesis in *Helicobacter pylori* and *Escherichia coli*. PLoS ONE 11:e0161587

Emami K, Guyet A, Kawai Y, Devi J, Wu LJ, Allenby N, Daniel RA, Errington J (2017) RodA as the missing glycosyltransferase in *Bacillus subtilis* and antibiotic discovery for the peptidoglycan polymerase pathway. Nat Microbiol 2:16253

Engel H, van Leeuwen A, Dijkstra A, Keck W (1992) Enzymatic preparation of 1,6-anhydro-muropeptides by immobilized murein hydrolases from *Escherichia coli* fused to staphylococcal protein A. Appl Microbiol Biotechnol 37:772–783

Errington J (2015) Bacterial morphogenesis and the enigmatic MreB helix. Nat Rev Microbiol 13:241–248

Espaillat A, Forsmo O, el Biari K, Bjork R, Lemaitre B, Trygg J, Canada FJ, de Pedro MA, Cava F (2016) Chemometric analysis of bacterial peptidoglycan reveals atypical modifications that empower the cell wall against predatory enzymes and fly innate immunity. J Am Chem Soc 138:9193–9204

Evans KL, Kannan S, Li G, de Pedro MA, Young KD (2013) Eliminating a set of four penicillin binding proteins triggers the Rcs phosphorelay and Cpx stress responses in *Escherichia coli*. J Bacteriol 195:4415–4424

Fan J, Jiang D, Zhao Y, Liu J, Zhang XC (2014) Crystal structure of lipid phosphatase *Escherichia coli* phosphatidylglycerophosphate phosphatase B. Proc Natl Acad Sci USA 111:7636–7640

Fenton AK, Gerdes K (2013) Direct interaction of FtsZ and MreB is required for septum synthesis and cell division in *Escherichia coli*. EMBO J 32:1953–1965

Figueiredo TA, Sobral RG, Ludovice AM, Almeida JM, Bui NK, Vollmer W, de Lencastre H, Tomasz A (2012) Identification of genetic determinants and enzymes involved with the amidation of glutamic acid residues in the peptidoglycan of *Staphylococcus aureus*. PLoS Pathog 8:e1002508

Filipe SR, Tomasz A (2000) Inhibition of the expression of penicillin resistance in *Streptococcus pneumoniae* by inactivation of cell wall muropeptide branching genes. Proc Natl Acad Sci USA 97:4891–4896

Firczuk M, Bochtler M (2007) Mutational analysis of peptidoglycan amidase MepA. Biochemistry 46:120–128

Fraipont C, Alexeeva S, Wolf B, van der Ploeg R, Schloesser M, den Blaauwen T, Nguyen-Disteche M (2011) The integral membrane FtsW protein and peptidoglycan synthase PBP3 form a subcomplex in *Escherichia coli*. Microbiology 157:251–259

Gallant CV, Daniels C, Leung JM, Ghosh AS, Young KD, Kotra LP, Burrows LL (2005) Common beta-lactamases inhibit bacterial biofilm formation. Mol Microbiol 58:1012–1024

Garcia-Bustos J, Tomasz A (1990) A biological price of antibiotic resistance: major changes in the peptidoglycan structure of penicillin-resistant pneumococci. Proc Natl Acad Sci USA 87:5415–5419

Garcia Del Portillo F, De Pedro MA (1990) Differential effect of mutational impairment of penicillin-binding proteins 1A and 1B on *Escherichia coli* strains harboring thermosensitive mutations in the cell division genes *ftsA*, *ftsQ*, *ftsZ*, and *pbpB*. J Bacteriol 172:5863–5870

Garcia Del Portillo F, De Pedro MA (1991) Penicillin-binding protein 2 is essential for the integrity of growing cells of *Escherichia coli ponB* strains. J Bacteriol 173:4530–4532

Garner EC, Bernard R, Wang W, Zhuang X, Rudner DZ, Mitchison T (2011) Coupled, circumferential motions of the cell wall synthesis machinery and MreB filaments in *B. subtilis*. Science 333:222–225

Geiger T, Pazos M, Lara-Tejero M, Vollmer W, Galan JE (2018). Peptidoglycan editing by a specific LD-transpeptidase controls the muramidase-dependent secretion of typhoid toxin. Nat Microbiol

Gerding MA, Liu B, Bendezu FO, Hale CA, Bernhardt TG, de Boer PA (2009) Self-enhanced accumulation of FtsN at division sites and roles for other proteins with a SPOR domain (DamX, DedD, and RlpA) in *Escherichia coli* cell constriction. J Bacteriol 191:7383–7401

Glauner B, Höltje JV, Schwarz U (1988) The composition of the murein of *Escherichia coli*. J Biol Chem 263:10088–10095

Gonzalez-Leiza SM, de Pedro MA, Ayala JA (2011) AmpH, a bifunctional DD-endopeptidase and DD-carboxypeptidase of *Escherichia coli*. J Bacteriol 193:6887–6894

Govindarajan S, Amster-Choder O (2017) The bacterial Sec system is required for the organization and function of the MreB cytoskeleton. PLoS Genet 13:e1007017

Gray AN, Egan AJ, Van't Veer IL, Verheul J, Colavin A, Koumoutsi A, Biboy J, Altelaar AF, Damen MJ, Huang KC, Simorre JP, Breukink E, Den Blaauwen T, Typas A, Gross CA, Vollmer W (2015) Coordination of peptidoglycan synthesis and outer membrane constriction during *Escherichia coli* cell division. Elife 4

Gupta R, Lavollay M, Mainardi JL, Arthur M, Bishai WR, Lamichhane G (2010) The *Mycobacterium tuberculosis* protein LdtMt2 is a nonclassical transpeptidase required for virulence and resistance to amoxicillin. Nat Med 16:466–469

Ha R, Frirdich E, Sychantha D, Biboy J, Taveirne ME, Johnson JG, Dirita VJ, Vollmer W, Clarke AJ, Gaynor EC (2016) Accumulation of peptidoglycan *O*-acetylation leads to altered cell wall biochemistry and negatively impacts pathogenesis factors of *Campylobacter jejuni*. J Biol Chem 291:22686–22702

Hakulinen JK, Hering J, Branden G, Chen H, Snijder A, Ek M, Johansson P (2017) MraY-antibiotic complex reveals details of tunicamycin mode of action. Nat Chem Biol 13:265–267

Hansen JM, Golchin SA, Veyrier FJ, Domenech P, Boneca IG, Azad AK, Rajaram MV, Schlesinger LS, Divangahi M, Reed MB, Behr MA (2014) *N*-glycolylated peptidoglycan contributes to the immunogenicity but not pathogenicity of *Mycobacterium tuberculosis*. J Infect Dis 209:1045–1054

Heidrich C, Templin MF, Ursinus A, Merdanovic M, Berger J, Schwarz H, de Pedro MA, Höltje JV (2001) Involvement of *N*-acetylmuramyl-L-alanine amidases in cell separation and antibiotic-induced autolysis of *Escherichia coli*. Mol Microbiol 41:167–178

Heidrich C, Ursinus A, Berger J, Schwarz H, Höltje JV (2002) Effects of multiple deletions of murein hydrolases on viability, septum cleavage, and sensitivity to large toxic molecules in *Escherichia coli*. J Bacteriol 184:6093–6099

Henderson TA, Dombrosky PM, Young KD (1994) Artifactual processing of penicillin-binding proteins 7 and 1b by the OmpT protease of *Escherichia coli*. J Bacteriol 176:256–259

Henderson TA, Templin M, Young KD (1995) Identification and cloning of the gene encoding penicillin-binding protein 7 of *Escherichia coli*. J Bacteriol 177:2074–2079

Henderson TA, Young KD, Denome SA, Elf PK (1997) AmpC and AmpH, proteins related to the class C beta-lactamases, bind penicillin and contribute to the normal morphology of *Escherichia coli*. J Bacteriol 179:6112–6121

Holden S (2018) Probing the mechanistic principles of bacterial cell division with super-resolution microscopy. Curr Opin Microbiol 43:84–91

Höltje JV (1998) Growth of the stress-bearing and shape-maintaining murein sacculus of *Escherichia coli*. Microbiol Mol Biol Rev 62:181–203

Hrast M, Sosic I, Sink R, Gobec S (2014) Inhibitors of the peptidoglycan biosynthesis enzymes MurA-F. Bioorg Chem 55:2–15

Hsu YP, Rittichier J, Kuru E, Yablonowski J, Pasciak E, Tekkam S, Hall E, Murphy B, Lee TK, Garner EC, Huang KC, Brun YV, Vannieuwenhze MS (2017) Full color palette of fluorescent D-amino acids for *in situ* labeling of bacterial cell walls. Chem Sci 8:6313–6321

Hughes RC, Thurman PF, Stokes E (1975) Estimates of the porosity of *Bacillus licheniformis* and *Bacillus subtilis* cell walls. Z Immunitatsforsch Exp Klin Immunol 149:126–135

Hugonnet JE, Mengin-Lecreulx D, Monton A, Den Blaauwen T, Carbonnelle E, Veckerle C, Brun YV, Van Nieuwenhze M, Bouchier C, Tu K, Rice LB, Arthur M (2016). Factors essential for LD-transpeptidase-mediated peptidoglycan cross-linking and beta-lactam resistance in *Escherichia coli*. Elife 5

Iida K, Hirota Y, Schwarz U (1983) Mutants of *Escherichia coli* defective in penicillin-insensitive murein DD-endopeptidase. Mol Gen Genet 189:215–221

Inoue A, Murata Y, Takahashi H, Tsuji N, Fujisaki S, Kato J (2008) Involvement of an essential gene, *mviN*, in murein synthesis in *Escherichia coli*. J Bacteriol 190:7298–7301

Inouye M, Shaw J, Shen C (1972) The assembly of a structural lipoprotein in the envelope of *Escherichia coli*. J Biol Chem 247:8154–8159

Ishino F, Jung HK, Ikeda M, Doi M, Wachi M, Matsuhashi M (1989) New mutations *fts-36*, *lts-33*, and *ftsW* clustered in the *mra* region of the *Escherichia coli* chromosome induce thermosensitive cell growth and division. J Bacteriol 171:5523–5530

Ize B, Stanley NR, Buchanan G, Palmer T (2003) Role of the *Escherichia coli* Tat pathway in outer membrane integrity. Mol Microbiol 48:1183–1193

Jacobs C, Huang LJ, Bartowsky E, Normark S, Park JT (1994) Bacterial cell wall recycling provides cytosolic muropeptides as effectors for beta-lactamase induction. EMBO J 13:4684–4694

Jacobs C, Joris B, Jamin M, Klarsov K, van Beeumen J, Mengin-Lecreulx D, van Heijenoort J, Park JT, Normark S, Frere JM (1995) AmpD, essential for both beta-lactamase regulation and cell wall recycling, is a novel cytosolic *N*-acetylmuramyl-L-alanine amidase. Mol Microbiol 15:553–559

Jacoby GH, Young KD (1988) Unequal distribution of penicillin-binding proteins among inner membrane vesicles of *Escherichia coli*. J Bacteriol 170:3660–3667

Jean NL, Bougault CM, Lodge A, Derouaux A, Callens G, Egan AJ, Ayala I, Lewis RJ, Vollmer W, Simorre JP (2014) Elongated structure of the outer-membrane activator of peptidoglycan synthesis LpoA: implications for PBP1A stimulation. Structure 22:1047–1054

Jeong JH, Kim YS, Rojviriya C, Ha SC, Kang BS, Kim YG (2013) Crystal structures of bifunctional penicillin-binding protein 4 from *Listeria monocytogenes*. Antimicrob Agents Chemother 57:3507–3512

Jorgenson MA, Chen Y, Yahashiri A, Popham DL, Weiss DS (2014) The bacterial septal ring protein RlpA is a lytic transglycosylase that contributes to rod shape and daughter cell separation in *Pseudomonas aeruginosa*. Mol Microbiol 93:113–128

Juan C, Torrens G, Gonzalez-Nicolau M, Oliver A (2017) Diversity and regulation of intrinsic beta-lactamases from non-fermenting and other Gram-negative opportunistic pathogens. FEMS Microbiol Rev 41:781–815

Kaoukab-Raji A, Biskri L, Bernardini ML, Allaoui A (2012) Characterization of SfPgdA, a *Shigella flexneri* peptidoglycan deacetylase required for bacterial persistence within polymorphonuclear neutrophils. Microbes Infect 14:619–627

Kar D, Pandey SD, Mallick S, Dutta M, Ghosh AS (2018) Substitution of alanine at position 184 with glutamic acid in *Escherichia coli* PBP5 omega-like loop introduces a moderate cephalosporinase activity. Protein J 37:122–131

Karimova G, Dautin N, Ladant D (2005) Interaction network among *Escherichia coli* membrane proteins involved in cell division as revealed by bacterial two-hybrid analysis. J Bacteriol 187:2233–2243

Kato J, Suzuki H, Hirota Y (1985) Dispensability of either penicillin-binding protein-1a or -1b involved in the essential process for cell elongation in *Escherichia coli.* Mol Gen Genet 200:272–277

Kawazura T, Matsumoto K, Kojima K, Kato F, Kanai T, Niki H, Shiomi D (2017) Exclusion of assembled MreB by anionic phospholipids at cell poles confers cell polarity for bidirectional growth. Mol Microbiol 104:472–486

Keck W, Schwarz U (1979) *Escherichia coli* murein-DD-endopeptidase insensitive to beta-lactam antibiotics. J Bacteriol 139:770–774

Keck W, van Leeuwen AM, Huber M, Goodell EW (1990) Cloning and characterization of *mepA*, the structural gene of the penicillin-insensitive murein endopeptidase from *Escherichia coli.* Mol Microbiol 4:209–219

Kerff F, Petrella S, Mercier F, Sauvage E, Herman R, Pennartz A, Zervosen A, Luxen A, Frere JM, Joris B, Charlier P (2010) Specific structural features of the *N*-acetylmuramoyl-L-alanine amidase AmiD from *Escherichia coli* and mechanistic implications for enzymes of this family. J Mol Biol 397:249–259

Kerr CH, Culham DE, Marom D, Wood JM (2014) Salinity-dependent impacts of ProQ, Prc, and Spr deficiencies on *Escherichia coli* cell structure. J Bacteriol 196:1286–1296

Khattar MM, Begg KJ, Donachie WD (1994) Identification of FtsW and characterization of a new *ftsW* division mutant of *Escherichia coli.* J Bacteriol 176:7140–7147

King DT, Lameignere E, Strynadka NC (2014) Structural insights into the lipoprotein outer membrane regulator of penicillin-binding protein 1B. J Biol Chem 289:19245–19253

King DT, Wasney GA, Nosella M, Fong A, Strynadka NC (2017) Structural insights into inhibition of *Escherichia coli* penicillin-binding protein 1B. J Biol Chem 292:979–993

Kishida H, Unzai S, Roper DI, Lloyd A, Park SY, Tame JR (2006) Crystal structure of penicillin binding protein 4 (dacB) from *Escherichia coli*, both in the native form and covalently linked to various antibiotics. Biochemistry 45:783–792

Koch AL (1988) Biophysics of bacterial walls viewed as stress-bearing fabric. Microbiol Rev 52:337–353

Korat B, Mottl H, Keck W (1991) Penicillin-binding protein 4 of *Escherichia coli*: molecular cloning of the *dacB* gene, controlled overexpression, and alterations in murein composition. Mol Microbiol 5:675–684

Kouidmi I, Levesque RC, Paradis-Bleau C (2014) The biology of Mur ligases as an antibacterial target. Mol Microbiol 94:242–253

Kruse T, Bork-Jensen J, Gerdes K (2005) The morphogenetic MreBCD proteins of *Escherichia coli* form an essential membrane-bound complex. Mol Microbiol 55:78–89

Kuk AC, Mashalidis EH, Lee SY (2017) Crystal structure of the MOP flippase MurJ in an inward-facing conformation. Nat Struct Mol Biol 24:171–176

Kuru E, Hughes HV, Brown PJ, Hall E, Tekkam S, Cava F, de Pedro MA, Brun YV, Vannieuwenhze MS (2012) *In situ* probing of newly synthesized peptidoglycan in live bacteria with fluorescent D-amino acids. Angew Chem Int Ed Engl 51:12519–12523

Lai GC, Cho H, Bernhardt TG (2017) The mecillinam resistome reveals a role for peptidoglycan endopeptidases in stimulating cell wall synthesis in *Escherichia coli.* PLoS Genet 13:e1006934

Lam H, Oh DC, Cava F, Takacs CN, Clardy J, de Pedro MA, Waldor MK (2009) D-amino acids govern stationary phase cell wall remodeling in bacteria. Science 325:1552–1555

Lambert C, Lerner TR, Bui NK, Somers H, Aizawa S, Liddell S, Clark A, Vollmer W, Lovering AL, Sockett RE (2016) Interrupting peptidoglycan deacetylation during Bdellovibrio predator-

prey interaction prevents ultimate destruction of prey wall, liberating bacterial-ghosts. Sci Rep 6:26010
Lara B, Ayala JA (2002) Topological characterization of the essential *Escherichia coli* cell division protein FtsW. FEMS Microbiol Lett 216:23–32
Lara B, Mengin-Lecreulx D, Ayala JA, van Heijenoort J (2005) Peptidoglycan precursor pools associated with MraY and FtsW deficiencies or antibiotic treatments. FEMS Microbiol Lett 250:195–200
Lavollay M, Arthur M, Fourgeaud M, Dubost L, Marie A, Veziris N, Blanot D, Gutmann L, Mainardi JL (2008) The peptidoglycan of stationary-phase *Mycobacterium tuberculosis* predominantly contains cross-links generated by LD-transpeptidation. J Bacteriol 190:4360–4366
Lavollay M, Fourgeaud M, Herrmann JL, Dubost L, Marie A, Gutmann L, Arthur M, Mainardi JL (2011) The peptidoglycan of *Mycobacterium abscessus* is predominantly cross-linked by LD-transpeptidases. J Bacteriol 193:778–782
Leclercq S, Derouaux A, Olatunji S, Fraipont C, Egan AJ, Vollmer W, Breukink E, Terrak M (2017) Interplay between penicillin-binding proteins and SEDS proteins promotes bacterial cell wall synthesis. Sci Rep 7:43306
Lee M, Batuecas MT, Tomoshige S, Dominguez-Gil T, Mahasenan KV, Dik DA, Hesek D, Millan C, Uson I, Lastochkin E, Hermoso JA, Mobashery S (2018) Exolytic and endolytic turnover of peptidoglycan by lytic transglycosylase Slt of *Pseudomonas aeruginosa*. Proc Natl Acad Sci USA 115:4393–4398
Lee M, Hesek D, Llarrull LI, Lastochkin E, Pi H, Boggess B, Mobashery S (2013) Reactions of all *Escherichia coli* lytic transglycosylases with bacterial cell wall. J Am Chem Soc 135:3311–3314
Lee TK, Meng K, Shi H, Huang KC (2016) Single-molecule imaging reveals modulation of cell wall synthesis dynamics in live bacterial cells. Nat Commun 7:13170
Lee TK, Tropini C, Hsin J, Desmarais SM, Ursell TS, Gong E, Gitai Z, Monds RD, Huang KC (2014) A dynamically assembled cell wall synthesis machinery buffers cell growth. Proc Natl Acad Sci USA 111:4554–4559
Leidenix MJ, Jacoby GH, Henderson TA, Young KD (1989) Separation of *Escherichia coli* penicillin-binding proteins into different membrane vesicles by agarose electrophoresis and sizing chromatography. J Bacteriol 171:5680–5686
Levefaudes M, Patin D, De Sousa-D'auria C, Chami M, Blanot D, Herve M, Arthur M, Houssin C, Mengin-Lecreulx D (2015) Diaminopimelic acid amidation in corynebacteriales: new insights into the role of LtsA in peptidoglycan modification. J Biol Chem 290:13079–13094
Li GW, Burkhardt D, Gross C, Weissman JS (2014) Quantifying absolute protein synthesis rates reveals principles underlying allocation of cellular resources. Cell 157:624–635
Li SY, Höltje JV, Young KD (2004) Comparison of high-performance liquid chromatography and fluorophore-assisted carbohydrate electrophoresis methods for analyzing peptidoglycan composition of *Escherichia coli*. Anal Biochem 326:1–12
Liechti G, Kuru E, Packiam M, Hsu YP, Tekkam S, Hall E, Rittichier JT, Vannieuwenhze M, Brun YV, Maurelli AT (2016) Pathogenic chlamydia lack a classical sacculus but synthesize a narrow, mid-cell peptidoglycan ring, regulated by MreB, for cell division. PLoS Pathog 12:e1005590
Liu B, Persons L, Lee L, de Boer PA (2015) Roles for both FtsA and the FtsBLQ subcomplex in FtsN-stimulated cell constriction in *Escherichia coli*. Mol Microbiol 95:945–970
Liu Y, Breukink E (2016) The membrane steps of bacterial cell wall synthesis as antibiotic targets. Antibiotics (Basel) 5
Loskill P, Pereira PM, Jung P, Bischoff M, Herrmann M, Pinho MG, Jacobs K (2014) Reduction of the peptidoglycan crosslinking causes a decrease in stiffness of the *Staphylococcus aureus* cell envelope. Biophys J 107:1082–1089
Lovering AL, de Castro LH, Lim D, Strynadka NC (2007) Structural insight into the transglycosylation step of bacterial cell-wall biosynthesis. Science 315:1402–1405
Lu C, Reedy M, Erickson HP (2000) Straight and curved conformations of FtsZ are regulated by GTP hydrolysis. J Bacteriol 182:164–170

Lupoli TJ, Lebar MD, Markovski M, Bernhardt T, Kahne D, Walker S (2014) Lipoprotein activators stimulate *Escherichia coli* penicillin-binding proteins by different mechanisms. J Am Chem Soc 136:52–55
Lupoli TJ, Tsukamoto H, Doud EH, Wang TS, Walker S, Kahne D (2011) Transpeptidase-mediated incorporation of D-amino acids into bacterial peptidoglycan. J Am Chem Soc 133:10748–10751
Magnet S, Bellais S, Dubost L, Fourgeaud M, Mainardi JL, Petit-Frere S, Marie A, Mengin-Lecreulx D, Arthur M, Gutmann L (2007) Identification of the LD-transpeptidases responsible for attachment of the Braun lipoprotein to *Escherichia coli* peptidoglycan. J Bacteriol 189:3927–3931
Magnet S, Dubost L, Marie A, Arthur M, Gutmann L (2008) Identification of the LD-transpeptidases for peptidoglycan cross-linking in *Escherichia coli*. J Bacteriol 190:4782–4785
Mahapatra S, Crick DC, Brennan PJ (2000) Comparison of the UDP-*N*-acetylmuramate:L-alanine ligase enzymes from *Mycobacterium tuberculosis* and *Mycobacterium leprae*. J Bacteriol 182:6827–6830
Mainardi JL, Fourgeaud M, Hugonnet JE, Dubost L, Brouard JP, Ouazzani J, Rice LB, Gutmann L, Arthur M (2005) A novel peptidoglycan cross-linking enzyme for a beta-lactam-resistant transpeptidation pathway. J Biol Chem 280:38146–38152
Mainardi JL, Hugonnet JE, Rusconi F, Fourgeaud M, Dubost L, Moumi AN, Delfosse V, Mayer C, Gutmann L, Rice LB, Arthur M (2007) Unexpected inhibition of peptidoglycan LD-transpeptidase from *Enterococcus faecium* by the beta-lactam imipenem. J Biol Chem 282:30414–30422
Mainardi JL, Legrand R, Arthur M, Schoot B, van Heijenoort J, Gutmann L (2000) Novel mechanism of beta-lactam resistance due to bypass of DD-transpeptidation in *Enterococcus faecium*. J Biol Chem 275:16490–16496
Mainardi JL, Morel V, Fourgeaud M, Cremniter J, Blanot D, Legrand R, Frehel C, Arthur M, van Heijenoort J, Gutmann L (2002) Balance between two transpeptidation mechanisms determines the expression of beta-lactam resistance in *Enterococcus faecium*. J Biol Chem 277:35801–35807
Manat G, el Ghachi M, Auger R, Baouche K, Olatunji S, Kerff F, Touze T, Mengin-Lecreulx D, Bouhss A (2015) Membrane topology and biochemical characterization of the *Escherichia coli* BacA undecaprenyl-pyrophosphate phosphatase. PLoS ONE 10:e0142870
Manat G, Roure S, Auger R, Bouhss A, Barreteau H, Mengin-Lecreulx D, Touze T (2014) Deciphering the metabolism of undecaprenyl-phosphate: the bacterial cell-wall unit carrier at the membrane frontier. Microb Drug Resist 20:199–214
Markiewicz Z, Broome-Smith JK, Schwarz U, Spratt BG (1982) Spherical *E. coli* due to elevated levels of D-alanine carboxypeptidase. Nature 297:702–704
Markovski M, Bohrhunter JL, Lupoli TJ, Uehara T, Walker S, Kahne DE, Bernhardt TG (2016) Cofactor bypass variants reveal a conformational control mechanism governing cell wall polymerase activity. Proc Natl Acad Sci USA 113:4788–4793
Matias VR, Al-Amoudi A, Dubochet J, Beveridge TJ (2003) Cryo-transmission electron microscopy of frozen-hydrated sections of *Escherichia coli* and *Pseudomonas aeruginosa*. J Bacteriol 185:6112–6118
Matias VR, Beveridge TJ (2005) Cryo-electron microscopy reveals native polymeric cell wall structure in *Bacillus subtilis* 168 and the existence of a periplasmic space. Mol Microbiol 56:240–251
Meberg BM, Paulson AL, Priyadarshini R, Young KD (2004) Endopeptidase penicillin-binding proteins 4 and 7 play auxiliary roles in determining uniform morphology of *Escherichia coli*. J Bacteriol 186:8326–8336
Meeske AJ, Riley EP, Robins WP, Uehara T, Mekalanos JJ, Kahne D, Walker S, Kruse AC, Bernhardt TG, Rudner DZ (2016) SEDS proteins are a widespread family of bacterial cell wall polymerases. Nature 537:634–638
Meeske AJ, Sham LT, Kimsey H, Koo BM, Gross CA, Bernhardt TG, Rudner DZ (2015) MurJ and a novel lipid II flippase are required for cell wall biogenesis in *Bacillus subtilis*. Proc Natl Acad Sci USA 112:6437–6442
Meiresonne NY, Van Der Ploeg R, Hink MA, Den Blaauwen (2017) Activity-related conformational changes in DD-carboxypeptidases revealed by *in vivo* periplasmic forster resonance energy transfer assay in *Escherichia coli*. MBio 8

Mohammadi T, Sijbrandi R, Lutters M, Verheul J, Martin NI, den Blaauwen T, de Kruijff B, Breukink E (2014) Specificity of the transport of lipid II by FtsW in *Escherichia coli*. J Biol Chem 289:14707–14718

Mohammadi T, Van Dam V, Sijbrandi R, Vernet T, Zapun A, Bouhss A, Diepeveen-De Bruin M, Nguyen-Disteche M, De Kruijff B, Breukink E (2011) Identification of FtsW as a transporter of lipid-linked cell wall precursors across the membrane. EMBO J 30:1425–1432

Monteiro JM, Pereira AR, Reichmann NT, Saraiva BM, Fernandes PB, Veiga H, Tavares AC, Santos M, Ferreira MT, Macario V, Vannieuwenhze MS, Filipe SR, Pinho MG (2018) Peptidoglycan synthesis drives an FtsZ-treadmilling-independent step of cytokinesis. Nature 554:528–532

Morlot C, Straume D, Peters K, Hegnar OA, Simon N, Villard AM, Contreras-Martel C, Leisico F, Breukink E, Gravier-Pelletier C, le Corre L, Vollmer W, Pietrancosta N, Havarstein LS, Zapun A (2018) Structure of the essential peptidoglycan amidotransferase MurT/GatD complex from *Streptococcus pneumoniae*. Nat Commun 9:3180

Moynihan PJ, Clarke AJ (2010) *O*-acetylation of peptidoglycan in gram-negative bacteria: identification and characterization of peptidoglycan *O*-acetyltransferase in *Neisseria gonorrhoeae*. J Biol Chem 285:13264–13273

Müller P, Ewers C, Bertsche U, Anstett M, Kallis T, Breukink E, Fraipont C, Terrak M, Nguyen-Disteche M, Vollmer W (2007) The essential cell division protein FtsN interacts with the murein (peptidoglycan) synthase PBP1B in *Escherichia coli*. J Biol Chem 282:36394–36402

Münch D, Roemer T, Lee SH, Engeser M, Sahl HG, Schneider T (2012) Identification and *in vitro* analysis of the GatD/MurT enzyme-complex catalyzing lipid II amidation in *Staphylococcus aureus*. PLoS Pathog 8:e1002509

Nanninga N (1991) Cell division and peptidoglycan assembly in *Escherichia coli*. Mol Microbiol 5:791–795

Nelson DE, Ghosh AS, Paulson AL, Young KD (2002) Contribution of membrane-binding and enzymatic domains of penicillin binding protein 5 to maintenance of uniform cellular morphology of *Escherichia coli*. J Bacteriol 184:3630–3639

Nelson DE, Young KD (2000) Penicillin binding protein 5 affects cell diameter, contour, and morphology of *Escherichia coli*. J Bacteriol 182:1714–1721

Ngadjeua F, Braud E, Saidjalolov S, Iannazzo L, Schnappinger D, Ehrt S, Hugonnet JE, Mengin-Lecreulx D, Patin D, Etheve-Quelquejeu M, Fonvielle M, Arthur M (2018) Critical impact of peptidoglycan precursor amidation on the activity of LD-transpeptidases from *Enterococcus faecium* and *Mycobacterium tuberculosis*. Chemistry 24:5743–5747

Noldeke ER, Muckenfuss LM, Niemann V, Müller A, Stork E, Zocher G, Schneider T, Stehle T (2018) Structural basis of cell wall peptidoglycan amidation by the GatD/MurT complex of *Staphylococcus aureus*. Sci Rep 8:12953

Osawa M, Anderson DE, Erickson HP (2008) Reconstitution of contractile FtsZ rings in liposomes. Science 320:792–794

Osawa M, Erickson HP (2013) Liposome division by a simple bacterial division machinery. Proc Natl Acad Sci USA 110:11000–11004

Packiam M, Weinrick B, Jacobs WR Jr, Maurelli AT (2015) Structural characterization of muropeptides from *Chlamydia trachomatis* peptidoglycan by mass spectrometry resolves "chlamydial anomaly". Proc Natl Acad Sci USA 112:11660–11665

Paradis-Bleau C, Markovski M, Uehara T, Lupoli TJ, Walker S, Kahne DE, Bernhardt TG (2010) Lipoprotein cofactors located in the outer membrane activate bacterial cell wall polymerases. Cell 143:1110–1120

Park JT, Uehara T (2008) How bacteria consume their own exoskeletons (turnover and recycling of cell wall peptidoglycan). Microbiol Mol Biol Rev 72:211–227 (table of contents)

Pastoret S, Fraipont C, den Blaauwen T, Wolf B, Aarsman ME, Piette A, Thomas A, Brasseur R, Nguyen-Disteche M (2004) Functional analysis of the cell division protein FtsW of *Escherichia coli*. J Bacteriol 186:8370–8379

Pazos M, Peters K, Casanova M, Palacios P, Vannieuwenhze M, Breukink E, Vicente M, Vollmer W (2018) Z-ring membrane anchors associate with cell wall synthases to initiate bacterial cell division. Nat Commun 9:5090

Pazos M, Peters K, Vollmer W (2017) Robust peptidoglycan growth by dynamic and variable multi-protein complexes. Curr Opin Microbiol 36:55–61

Peltier J, Courtin P, el Meouche I, Lemee L, Chapot-Chartier MP, Pons JL (2011) *Clostridium difficile* has an original peptidoglycan structure with a high level of *N*-acetylglucosamine deacetylation and mainly 3-3 cross-links. J Biol Chem 286:29053–29062

Pennartz A, Genereux C, Parquet C, Mengin-Lecreulx D, Joris B (2009) Substrate-induced inactivation of the *Escherichia coli* AmiD *N*-acetylmuramoyl-L-alanine amidase highlights a new strategy to inhibit this class of enzyme. Antimicrob Agents Chemother 53:2991–2997

Pepper ED, Farrell MJ, Finkel SE (2006) Role of penicillin-binding protein 1b in competitive stationary-phase survival of *Escherichia coli*. FEMS Microbiol Lett 263:61–67

Peters K, Kannan S, Rao VA, Biboy J, Vollmer D, Erickson SW, Lewis RJ, Young KD, Vollmer W (2016) The redundancy of peptidoglycan carboxypeptidases ensures robust cell shape maintenance in *Escherichia coli*. MBio 7:e00819-16

Peters K, Pazos M, Edoo Z, Hugonnet JE, Martorana AM, Polissi A, Vannieuwenhze MS, Arthur M, Vollmer W (2018) Copper inhibits peptidoglycan LD-transpeptidases suppressing beta-lactam resistance due to bypass of penicillin-binding proteins. Proc Natl Acad Sci USA 115:10786–10791

Peters NT, Dinh T, Bernhardt TG (2011) A fail-safe mechanism in the septal ring assembly pathway generated by the sequential recruitment of cell separation amidases and their activators. J Bacteriol 193:4973–4983

Peters NT, Morlot C, Yang DC, Uehara T, Vernet T, Bernhardt TG (2013) Structure-function analysis of the LytM domain of EnvC, an activator of cell wall remodelling at the *Escherichia coli* division site. Mol Microbiol 89:690–701

Pichoff S, Du S, Lutkenhaus J (2015) The bypass of ZipA by overexpression of FtsN requires a previously unknown conserved FtsN motif essential for FtsA-FtsN interaction supporting a model in which FtsA monomers recruit late cell division proteins to the Z ring. Mol Microbiol 95:971–987

Pichoff S, Du S, Lutkenhaus J (2018) Disruption of divisome assembly rescued by FtsN-FtsA interaction in *Escherichia coli*. Proc Natl Acad Sci USA 115:E6855–E6862

Pilhofer M, Aistleitner K, Biboy J, Gray J, Kuru E, Hall E, Brun YV, Vannieuwenhze MS, Vollmer W, Horn M, Jensen GJ (2013) Discovery of chlamydial peptidoglycan reveals bacteria with murein sacculi but without FtsZ. Nat Commun 4:2856

Potluri L, Karczmarek A, Verheul J, Piette A, Wilkin JM, Werth N, Banzhaf M, Vollmer W, Young KD, Nguyen-Disteche M, den Blaauwen T (2010) Septal and lateral wall localization of PBP5, the major DD-carboxypeptidase of *Escherichia coli*, requires substrate recognition and membrane attachment. Mol Microbiol 77:300–323

Potluri LP, de Pedro MA, Young KD (2012a) *Escherichia coli* low-molecular-weight penicillin-binding proteins help orient septal FtsZ, and their absence leads to asymmetric cell division and branching. Mol Microbiol 84:203–224

Potluri LP, Kannan S, Young KD (2012b) ZipA is required for FtsZ-dependent preseptal peptidoglycan synthesis prior to invagination during cell division. J Bacteriol 194:5334–5342

Priyadarshini R, Popham DL, Young KD (2006) Daughter cell separation by penicillin-binding proteins and peptidoglycan amidases in *Escherichia coli*. J Bacteriol 188:5345–5355

Pushkaran AC, Nataraj N, Nair N, Gotz F, Biswas R, Mohan CG (2015) Understanding the structure-function relationship of lysozyme resistance in *Staphylococcus aureus* by peptidoglycan *O*-acetylation using molecular docking, dynamics, and lysis assay. J Chem Inf Model 55:760–770

Qiao Y, Srisuknimit V, Rubino F, Schaefer K, Ruiz N, Walker S, Kahne D (2017) Lipid II overproduction allows direct assay of transpeptidase inhibition by beta-lactams. Nat Chem Biol 13:793–798

Radkov AD, Hsu YP, Booher G, Vannieuwenhze MS (2018) Imaging bacterial cell wall biosynthesis. Annu Rev Biochem 87:991–1014

Ranjit DK, Young KD (2013) The Rcs stress response and accessory envelope proteins are required for *de novo* generation of cell shape in *Escherichia coli*. J Bacteriol 195:2452–2462
Rawat S, Zhu L, Lindner E, Dalbey RE, White SH (2015) SecA drives transmembrane insertion of RodZ, an unusual single-span membrane protein. J Mol Biol 427:1023–1037
Raymond JB, Mahapatra S, Crick DC, Pavelka MS Jr (2005) Identification of the *namH* gene, encoding the hydroxylase responsible for the *N*-glycolylation of the mycobacterial peptidoglycan. J Biol Chem 280:326–333
Rismondo J, Wamp S, Aldridge C, Vollmer W, Halbedel S (2018) Stimulation of PgdA-dependent peptidoglycan *N*-deacetylation by GpsB-PBP A1 in *Listeria monocytogenes*. Mol Microbiol 107:472–487
Rocaboy M, Herman R, Sauvage E, Remaut H, Moonens K, Terrak M, Charlier P, Kerff F (2013) The crystal structure of the cell division amidase AmiC reveals the fold of the AMIN domain, a new peptidoglycan binding domain. Mol Microbiol 90:267–277
Rojas ER, Billings G, Odermatt PD, Auer GK, Zhu L, Miguel A, Chang F, Weibel DB, Theriot JA, Huang KC (2018) The outer membrane is an essential load-bearing element in Gram-negative bacteria. Nature 559:617–621
Romeis T, Höltje JV (1994a) Penicillin-binding protein 7/8 of *Escherichia coli* is a DD-endopeptidase. Eur J Biochem 224:597–604
Romeis T, Höltje JV (1994b) Specific interaction of penicillin-binding proteins 3 and 7/8 with soluble lytic transglycosylase in *Escherichia coli*. J Biol Chem 269:21603–21607
Rubino FA, Kumar S, Ruiz N, Walker S, Kahne DE (2018) Membrane potential is required for MurJ function. J Am Chem Soc 140:4481–4484
Ruiz N (2008) Bioinformatics identification of MurJ (MviN) as the peptidoglycan lipid II flippase in *Escherichia coli*. Proc Natl Acad Sci USA 105:15553–15557
Ruiz N (2009) *Streptococcus pyogenes* YtgP (Spy_0390) complements *Escherichia coli* strains depleted of the putative peptidoglycan flippase MurJ. Antimicrob Agents Chemother 53:3604–3605
Salje J, van den Ent F, de Boer P, Lowe J (2011) Direct membrane binding by bacterial actin MreB. Mol Cell 43:478–487
Sanchez M, Kolar SL, Müller S, Reyes CN, Wolf AJ, Ogawa C, Singhania R, de Carvalho DD, Arditi M, Underhill DM, Martins GA, Liu GY (2017) *O*-acetylation of peptidoglycan limits helper T cell priming and permits *Staphylococcus aureus* reinfection. Cell Host Microbe 22(543–551):e4
Sanders AN, Pavelka MS (2013) Phenotypic analysis of *Eschericia coli* mutants lacking LD-transpeptidases. Microbiology 159:1842–1852
Sathiyamoorthy K, Vijayalakshmi J, Tirupati B, Fan L, Saper MA (2017) Structural analyses of the *Haemophilus influenzae* peptidoglycan synthase activator LpoA suggest multiple conformations in solution. J Biol Chem 292:17626–17642
Sauvage E, Derouaux A, Fraipont C, Joris M, Herman R, Rocaboy M, Schloesser M, Dumas J, Kerff F, Nguyen-Disteche M, Charlier P (2014) Crystal structure of penicillin-binding protein 3 (PBP3) from *Escherichia coli*. PLoS ONE 9:e98042
Sauvage E, Kerff F, Terrak M, Ayala JA, Charlier P (2008) The penicillin-binding proteins: structure and role in peptidoglycan biosynthesis. FEMS Microbiol Rev 32:234–258
Schleifer KH, Kandler O (1972) Peptidoglycan types of bacterial cell walls and their taxonomic implications. Bacteriol Rev 36:407–477
Schoenemann KM, Margolin W (2017) Bacterial division: FtsZ treadmills to build a beautiful wall. Curr Biol 27:R301–R303
Schumacher MA (2017) Bacterial nucleoid occlusion: multiple mechanisms for preventing chromosome bisection during cell division. Subcell Biochem 84:267–298
Schwechheimer C, Rodriguez DL, Kuehn MJ (2015) NlpI-mediated modulation of outer membrane vesicle production through peptidoglycan dynamics in *Escherichia coli*. Microbiologyopen 4:375–389

Severin A, Figueiredo AM, Tomasz A (1996) Separation of abnormal cell wall composition from penicillin resistance through genetic transformation of *Streptococcus pneumoniae*. J Bacteriol 178:1788–1792

Sham LT, Butler EK, Lebar MD, Kahne D, Bernhardt TG, Ruiz N (2014) Bacterial cell wall. MurJ is the flippase of lipid-linked precursors for peptidoglycan biogenesis. Science 345:220–222

Sham LT, Zheng S, Yakhnina AA, Kruse AC, Bernhardt TG (2018) Loss of specificity variants of WzxC suggest that substrate recognition is coupled with transporter opening in MOP-family flippases. Mol Microbiol 109:633–641

Sheikh J, Hicks S, Dall'Agnol M, Phillips AD, Nataro JP (2001) Roles for Fis and YafK in biofilm formation by enteroaggregative *Escherichia coli*. Mol Microbiol 41:983–997

Shi H, Bratton BP, Gitai Z, Huang KC (2018) How to build a bacterial cell: MreB as the foreman of *E. coli* construction. Cell 172:1294–1305

Shimada T, Park BG, Wolf AJ, Brikos C, Goodridge HS, Becker CA, Reyes CN, Miao EA, Aderem A, Gotz F, Liu GY, Underhill DM (2010) *Staphylococcus aureus* evades lysozyme-based peptidoglycan digestion that links phagocytosis, inflammasome activation, and IL-1beta secretion. Cell Host Microbe 7:38–49

Singh SK, Parveen S, Saisree L, Reddy M (2015) Regulated proteolysis of a cross-link-specific peptidoglycan hydrolase contributes to bacterial morphogenesis. Proc Natl Acad Sci USA 112:10956–10961

Singh SK, Saisree L, Amrutha RN, Reddy M (2012) Three redundant murein endopeptidases catalyse an essential cleavage step in peptidoglycan synthesis of *Escherichia coli* K12. Mol Microbiol 86:1036–1051

Sjodt M, Brock K, Dobihal G, Rohs PDA, Green AG, Hopf TA, Meeske AJ, Srisuknimit V, Kahne D, Walker S, Marks DS, Bernhardt TG, Rudner DZ, Kruse AC (2018) Structure of the peptidoglycan polymerase RodA resolved by evolutionary coupling analysis. Nature 556:118–121

Soderstrom B, Chan H, Shilling PJ, Skoglund U, Daley DO (2018) Spatial separation of FtsZ and FtsN during cell division. Mol Microbiol 107:387–401

Soderstrom B, Mirzadeh K, Toddo S, von Heijne G, Skoglund U, Daley DO (2016) Coordinated disassembly of the divisome complex in *Escherichia coli*. Mol Microbiol 101:425–438

Soderstrom B, Skoog K, Blom H, Weiss DS, von Heijne G, Daley DO (2014) Disassembly of the divisome in *Escherichia coli*: evidence that FtsZ dissociates before compartmentalization. Mol Microbiol 92:1–9

Spratt BG (1975) Distinct penicillin binding proteins involved in the division, elongation, and shape of *Escherichia coli* K12. Proc Natl Acad Sci USA 72:2999–3003

Stranden AM, Ehlert K, Labischinski H, Berger-Bachi B (1997) Cell wall monoglycine cross-bridges and methicillin hypersusceptibility in a *femAB* null mutant of methicillin-resistant *Staphylococcus aureus*. J Bacteriol 179:9–16

Su MY, Som N, Wu CY, Su SC, Kuo YT, Ke LC, Ho MR, Tzeng SR, Teng CH, Mengin-Lecreulx D, Reddy M, Chang CI (2017) Structural basis of adaptor-mediated protein degradation by the tail-specific PDZ-protease Prc. Nat Commun 8:1516

Sung MT, Lai YT, Huang CY, Chou LY, Shih HW, Cheng WC, Wong CH, Ma C (2009) Crystal structure of the membrane-bound bifunctional transglycosylase PBP1b from *Escherichia coli*. Proc Natl Acad Sci USA 106:8824–8829

Sutterlin L, Edoo Z, Hugonnet JE, Mainardi JL, Arthur M (2018) Peptidoglycan cross-linking activity of LD-transpeptidases from *Clostridium difficile* and inactivation of these enzymes by beta-lactams. Antimicrob Agents Chemother 62

Taguchi A, Welsh MA, Marmont LS, Lee W, Kahne D, Bernhardt TG, Walker S (2018) FtsW is a peptidoglycan polymerase that is activated by its cognate penicillin-binding protein. Preprint at. https://www.biorxiv.org/content/early/2018/06/29/358663

Tatar LD, Marolda CL, Polischuk AN, van Leeuwen D, Valvano MA (2007) An *Escherichia coli* undecaprenyl-pyrophosphate phosphatase implicated in undecaprenyl phosphate recycling. Microbiology 153:2518–2529

Templin MF, Edwards DH, Höltje JV (1992) A murein hydrolase is the specific target of bulgecin in *Escherichia coli*. J Biol Chem 267:20039–20043

Thunnissen AM, Dijkstra AJ, Kalk KH, Rozeboom HJ, Engel H, Keck W, Dijkstra BW (1994) Doughnut-shaped structure of a bacterial muramidase revealed by X-ray crystallography. Nature 367:750–753

Thunnissen AM, Rozeboom HJ, Kalk KH, Dijkstra BW (1995) Structure of the 70-kDa soluble lytic transglycosylase complexed with bulgecin A. Implications for the enzymatic mechanism. Biochemistry 34:12729–12737

Tomioka S, Matsuhashi M (1978) Purification of penicillin-insensitive DD-endopeptidase, a new cell wall peptidoglycan-hydrolyzing enzyme in *Escherichia coli*, and its inhibition by deoxyribonucleic acids. Biochem Biophys Res Commun 84:978–984

Touze T, Blanot D, Mengin-Lecreulx D (2008) Substrate specificity and membrane topology of *Escherichia coli* PgpB, an undecaprenyl pyrophosphate phosphatase. J Biol Chem 283:16573–16583

Tsang MJ, Bernhardt TG (2015) A role for the FtsQLB complex in cytokinetic ring activation revealed by an *ftsL* allele that accelerates division. Mol Microbiol 95:925–944

Tsang MJ, Yakhnina AA, Bernhardt TG (2017) NlpD links cell wall remodeling and outer membrane invagination during cytokinesis in *Escherichia coli*. PLoS Genet 13:e1006888

Turner RD, Mesnage S, Hobbs JK, Foster SJ (2018) Molecular imaging of glycan chains couples cell-wall polysaccharide architecture to bacterial cell morphology. Nat Commun 9:1263

Tuson HH, Auer GK, Renner LD, Hasebe M, Tropini C, Salick M, Crone WC, Gopinathan A, Huang KC, Weibel DB (2012) Measuring the stiffness of bacterial cells from growth rates in hydrogels of tunable elasticity. Mol Microbiol 84:874–891

Typas A, Banzhaf M, Gross CA, Vollmer W (2011) From the regulation of peptidoglycan synthesis to bacterial growth and morphology. Nat Rev Microbiol 10:123–136

Typas A, Banzhaf M, Van Den Berg Van Saparoea B, Verheul J, Biboy J, Nichols RJ, Zietek M, Beilharz K, Kannenberg K, Von Rechenberg M, Breukink E, Den Blaauwen T, Gross CA, Vollmer W (2010). Regulation of peptidoglycan synthesis by outer-membrane proteins. Cell 143:1097–1109

Uehara T, Dinh T, Bernhardt TG (2009) LytM-domain factors are required for daughter cell separation and rapid ampicillin-induced lysis in *Escherichia coli*. J Bacteriol 191:5094–5107

Uehara T, Park JT (2007) An anhydro-*N*-acetylmuramyl-L-alanine amidase with broad specificity tethered to the outer membrane of *Escherichia coli*. J Bacteriol 189:5634–5641

Uehara T, Parzych KR, Dinh T, Bernhardt TG (2010) Daughter cell separation is controlled by cytokinetic ring-activated cell wall hydrolysis. EMBO J 29:1412–1422

Ursell TS, Nguyen J, Monds RD, Colavin A, Billings G, Ouzounov N, Gitai Z, Shaevitz JW, Huang KC (2014) Rod-like bacterial shape is maintained by feedback between cell curvature and cytoskeletal localization. Proc Natl Acad Sci USA 111:E1025–E1034

Ursinus A, van den Ent F, Brechtel S, de Pedro M, Höltje JV, Lowe J, Vollmer W (2004) Murein (peptidoglycan) binding property of the essential cell division protein FtsN from *Escherichia coli*. J Bacteriol 186:6728–6737

van Asselt EJ, Thunnissen AM, Dijkstra BW (1999) High resolution crystal structures of the *Escherichia coli* lytic transglycosylase Slt70 and its complex with a peptidoglycan fragment. J Mol Biol 291:877–898

van den Bogaart G, Hermans N, Krasnikov V, Poolman B (2007) Protein mobility and diffusive barriers in *Escherichia coli*: consequences of osmotic stress. Mol Microbiol 64:858–871

van den Ent F, Izore T, Bharat TA, Johnson CM, Lowe J (2014) Bacterial actin MreB forms antiparallel double filaments. Elife 3:e02634

van Heijenoort J (2011) Peptidoglycan hydrolases of *Escherichia coli*. Microbiol Mol Biol Rev 75:636–663

van Teeffelen S, Wang S, Furchtgott L, Huang KC, Wingreen NS, Shaevitz JW, Gitai Z (2011) The bacterial actin MreB rotates, and rotation depends on cell-wall assembly. Proc Natl Acad Sci USA 108:15822–15827

Varma A, de Pedro MA, Young KD (2007) FtsZ directs a second mode of peptidoglycan synthesis in *Escherichia coli*. J Bacteriol 189:5692–5704

Varma A, Young KD (2004) FtsZ collaborates with penicillin binding proteins to generate bacterial cell shape in *Escherichia coli*. J Bacteriol 186:6768–6774

Vega D, Ayala JA (2006) The DD-carboxypeptidase activity encoded by *pbp4B* is not essential for the cell growth of *Escherichia coli*. Arch Microbiol 185:23–27

Vinella D, Joseleau-Petit D, Thevenet D, Bouloc P, D'Ari R (1993) Penicillin-binding protein 2 inactivation in *Escherichia coli* results in cell division inhibition, which is relieved by FtsZ overexpression. J Bacteriol 175:6704–6710

Vollmer W (2008) Structural variation in the glycan strands of bacterial peptidoglycan. FEMS Microbiol Rev 32:287–306

Vollmer W, Bertsche U (2008) Murein (peptidoglycan) structure, architecture and biosynthesis in *Escherichia coli*. Biochim Biophys Acta 1778:1714–1734

Vollmer W, Blanot D, de Pedro MA (2008a) Peptidoglycan structure and architecture. FEMS Microbiol Rev 32:149–167

Vollmer W, Joris B, Charlier P, Foster S (2008b) Bacterial peptidoglycan (murein) hydrolases. FEMS Microbiol Rev 32:259–286

Vollmer W, Tomasz A (2000) The *pgdA* gene encodes for a peptidoglycan *N*-acetylglucosamine deacetylase in *Streptococcus pneumoniae*. J Biol Chem 275:20496–20501

von Rechenberg M, Ursinus A, Höltje JV (1996) Affinity chromatography as a means to study multienzyme complexes involved in murein synthesis. Microb Drug Resist 2:155–157

Weadge JT, Clarke AJ (2006) Identification and characterization of *O*-acetylpeptidoglycan esterase: a novel enzyme discovered in *Neisseria gonorrhoeae*. Biochemistry 45:839–851

Weadge JT, Pfeffer JM, Clarke AJ (2005) Identification of a new family of enzymes with potential *O*-acetylpeptidoglycan esterase activity in both Gram-positive and Gram-negative bacteria. BMC Microbiol 5:49

Wettmann L, Kruse K (2018) The Min-protein oscillations in *Escherichia coli*: an example of self-organized cellular protein waves. Philos Trans R Soc Lond B Biol Sci 373

Whatmore AM, Reed RH (1990) Determination of turgor pressure in *Bacillus subtilis*: a possible role for K+ in turgor regulation. J Gen Microbiol 136:2521–2526

Wheeler R, Turner RD, Bailey RG, Salamaga B, Mesnage S, Mohamad SA, Hayhurst EJ, Horsburgh M, Hobbs JK, Foster SJ (2015) Bacterial cell enlargement requires control of cell wall stiffness mediated by peptidoglycan hydrolases. MBio 6:e00660

Williams AH, Wheeler R, Rateau L, Malosse C, Chamot-Rooke J, Haouz A, Taha MK, Boneca IG (2018) A step-by-step in crystallo guide to bond cleavage and 1,6-anhydro-sugar product synthesis by a peptidoglycan-degrading lytic transglycosylase. J Biol Chem 293:6000–6010

Workman SD, Worrall LJ, Strynadka NCJ (2018) Crystal structure of an intramembranal phosphatase central to bacterial cell-wall peptidoglycan biosynthesis and lipid recycling. Nat Commun 9:1159

Xiao J, Goley ED (2016) Redefining the roles of the FtsZ-ring in bacterial cytokinesis. Curr Opin Microbiol 34:90–96

Yadav AK, Espaillat A, Cava F (2018) Bacterial strategies to preserve cell wall integrity against environmental threats. Front Microbiol 9:2064

Yahashiri A, Jorgenson MA, Weiss DS (2015) Bacterial SPOR domains are recruited to septal peptidoglycan by binding to glycan strands that lack stem peptides. Proc Natl Acad Sci USA 112:11347–11352

Yakhnina AA, McManus HR, Bernhardt TG (2015) The cell wall amidase AmiB is essential for *Pseudomonas aeruginosa* cell division, drug resistance and viability. Mol Microbiol 97:957–973

Yanagihara Y, Kamisango K, Yasuda S, Kobayashi S, Mifuchi I, Azuma I, Yamamura Y, Johnson RC (1984) Chemical compositions of cell walls and polysaccharide fractions of spirochetes. Microbiol Immunol 28:535–544

Yang DC, Peters NT, Parzych KR, Uehara T, Markovski M, Bernhardt TG (2011) An ATP-binding cassette transporter-like complex governs cell-wall hydrolysis at the bacterial cytokinetic ring. Proc Natl Acad Sci USA 108:E1052–E1060

Yang DC, Tan K, Joachimiak A, Bernhardt TG (2012) A conformational switch controls cell wall-remodelling enzymes required for bacterial cell division. Mol Microbiol 85:768–781

Yang X, Lyu Z, Miguel A, McQuillen R, Huang KC, Xiao J (2017) GTPase activity-coupled treadmilling of the bacterial tubulin FtsZ organizes septal cell wall synthesis. Science 355:744–747

Yao X, Jericho M, Pink D, Beveridge T (1999) Thickness and elasticity of gram-negative murein sacculi measured by atomic force microscopy. J Bacteriol 181:6865–6875

Yousif SY, Broome-Smith JK, Spratt BG (1985) Lysis of *Escherichia coli* by beta-lactam antibiotics: deletion analysis of the role of penicillin-binding proteins 1A and 1B. J Gen Microbiol 131:2839–2845

Yuan Y, Barrett D, Zhang Y, Kahne D, Sliz P, Walker S (2007) Crystal structure of a peptidoglycan glycosyltransferase suggests a model for processive glycan chain synthesis. Proc Natl Acad Sci USA 104:5348–5353

Zapun A, Philippe J, Abrahams KA, Signor L, Roper DI, Breukink E, Vernet T (2013) *In vitro* reconstitution of peptidoglycan assembly from the Gram-positive pathogen *Streptococcus pneumoniae*. ACS Chem Biol 8:2688–2696

Zheng S, Sham LT, Rubino FA, Brock KP, Robins WP, Mekalanos JJ, Marks DS, Bernhardt TG, Kruse AC (2018) Structure and mutagenic analysis of the lipid II flippase MurJ from *Escherichia coli*. Proc Natl Acad Sci USA 115:6709–6714

Chapter 6
The Periplasmic Chaperones Skp and SurA

Guillaume Mas, Johannes Thoma and Sebastian Hiller

Abstract The periplasm of Gram-negative bacteria contains a specialized chaperone network that facilitates the transport of unfolded membrane proteins to the outer membrane as its primary functional role. The network, involving the chaperones Skp and SurA as key players and potentially additional chaperones, is indispensable for the survival of the cell. Structural descriptions of the apo forms of these molecular chaperones were initially provided by X-ray crystallography. Subsequently, a combination of experimental biophysical methods including solution NMR spectroscopy provided a detailed understanding of full-length chaperone–client complexes. The data showed that conformational changes and dynamic re-organization of the chaperones upon client binding, as well as client dynamics on the chaperone surface are crucial for function. This chapter gives an overview of the structure-function relationship of the dynamic conformational rearrangements that regulate the functional cycles of the periplasmic molecular chaperones Skp and SurA.

Keywords Molecular chaperones · Protein folding · Periplasm · Gram-negative bacteria · Protein structure · Protein dynamics

Abbreviations

ATP	Adenosine 5′-triphosphate
BAM	β-barrel assembly machinery
Hsp60	Heat shock protein 60
NMR	Nuclear magnetic resonance
NOE	Nuclear Overhauser effect

G. Mas · S. Hiller (✉)
Biozentrum, University of Basel, Klingelbergstrasse 70, Basel 4056, Switzerland
e-mail: sebastian.hiller@unibas.ch

J. Thoma
Department of Chemistry and Molecular Biology, Wallenberg Centre for Molecular and Translational Medicine, University of Gothenburg, Medicinaregatan 9c, 405 30 Gothenburg, Sweden

A. Kuhn (ed.), *Bacterial Cell Walls and Membranes*, Subcellular Biochemistry 92, https://doi.org/10.1007/978-3-030-18768-2_6

OMP	Outer membrane protein
PPIase	Parvulin-like peptidyl-prolyl isomerase
PRE	Paramagnetic relaxation enhancement
SAXS	Small angle X-ray scattering
Skp	Seventeen Kilodalton Protein
SurA	Survival factor A

Introduction

For Gram-negative bacteria, such as *E. coli*, around 35% of the proteome is directed to the bacterial cell envelope (Krogh et al. 2001; Orfanoudaki and Economou 2014; De Geyter et al. 2016; Tsirigotaki et al. 2017). The cell envelope is composed of two membranes: the inner membrane and the outer membrane, which are separated by the aqueous periplasm (Van Wielink and Duine 1990). The periplasm is a viscous and oxidizing compartment that contains a thin layer of peptidoglycan playing a structural role as a bacterial cell wall. Periplasmic proteins are synthesized as pre-proteins in the cytoplasm and are subsequently translocated across the inner membrane by either of two secretory pathways, depending on the signal sequence they carry, the Sec machinery and the twin arginine translocation (TAT) system (Wickner et al. 1991; Driessen et al. 2001; Natale et al. 2008; Lycklama et al. 2012; Tsirigotaki et al. 2017).

Thereby, the majority of proteins are processed by the Sec machinery, which translocates its clients in an unfolded conformation. As proteins exit the Sec machinery and enter the periplasm, a complex chaperone network in charge of protein quality control ensures their folding and integrity. The network comprises two main pathways for clients, the protection of protein folding under stress conditions and the transport of client proteins towards the outer membrane. The first pathway includes the chaperones HdeA/HdeB, which are activated by a pH-controlled mechanism that helps to prevent protein aggregation in acidic environment (Hong et al. 2005; Kern et al. 2007; Ding et al. 2015; Zhang et al. 2016; Yu et al. 2017; Salmon et al. 2018; Yu et al. 2018), and the chaperone Spy (Quan et al. 2011; Stull et al. 2016; He et al. 2016). The second pathway, the transport of outer membrane proteins (OMPs) through the periplasm is controlled by the network of two chaperones SurA and Skp (Bitto and McKay 2002; Walton and Sousa 2004; Korndörfer et al. 2004). Additionally, the protease DegP is responsible for the degradation of off-pathway OMPs (Jiang et al. 2008; Clausen et al. 2011; Ge et al. 2014a, b) and has also been shown to display chaperone activity under low temperature conditions (Spiess et al. 1999). Under heat shock conditions, in turn, the chaperone FkpA has been demonstrated to play a key role in OMP biogenesis (Ge et al. 2014a, b). In addition, the biogenesis of OM lipoproteins requires the dedicated chaperone LolA (Tajima et al. 1998; Miyamoto et al. 2001; Taniguchi et al. 2005; Okuda and Tokuda 2009).

Besides these two pathways, the periplasm contains several folding catalysts that are speeding up two slow reaction steps, the cis–trans isomerization of peptidyl-prolyl bonds and the formation of disulfide bonds. Cis–trans isomerization is generally a slow process that can often be rate-limiting for protein folding and that can be accelerated by enzymes with peptidyl-prolyl cis-trans isomerases (PPIase) activity (Liu and Walsh 1990; Bardwell et al. 1991; Hayano et al. 1991; Scholz et al. 1998). Four proteins with PPIase activity have been identified in the *E. coli* periplasm, FkpA (Arie et al. 2001; Saul et al. 2004; Helbig et al. 2011), PpiA (Liu and Walsh 1990; Hayano et al. 1991), PpiD (Antonoaea et al. 2008; Matern et al. 2010) and SurA (Bitto and McKay 2002), among which SurA and FkpA are the main players. A second slow process in protein folding is the formation of disulfide bonds. In the periplasm, the formation of disulfide bond is catalyzed by the Dsb redox system that comprises the oxidation system DsbA and DsbB and the isomerization system DsbC and DsbD (Joly and Swartz 1997; Messens and Collet 2006; Vertommen et al. 2008; Ito and Inaba 2008; Heras et al. 2009; Denoncin and Collet 2013). In this chapter, we focus on the recent advances of the structural characterization of the chaperones Skp and SurA that transport the OMPs through the periplasm and promote their insertion into the outer membrane by the BAM complex.

The biogenesis of OMPs is a challenging task for the cell, since the proteins have to cross two biophysical barriers to reach the outer membrane: The inner membrane and the aqueous periplasm. On this pathway, the translocation of the OMPs across the inner membrane, from the cytoplasm into the periplasm, is facilitated by the SEC translocase (Driessen et al. 2001; van den Berg et al. 2004; Chatzi et al. 2014; Tsirigotaki et al. 2017). The second biophysical barrier for the insoluble OMPs, the aqueous periplasm, is then overcome by a unique network of molecular chaperones. This network is organized around the chaperones Skp (Seventeen Kilodalton Protein) and SurA (survival factor A), which work on two parallel pathways to target the OMP clients to the β-barrel assembly machinery (BAM) complex, which eventually facilitates their folding and insertion (Fig. 6.1) (Rizzitello et al. 2001; De Geyter et al. 2016).

Understanding the structural biology of these two chaperones is of key scientific interest due to their fundamental role in polypeptide transport and OMP biogenesis. A first milestone has been the determination of the crystal structures of the client-free (apo) forms of SurA and Skp (Bitto and McKay 2002; Walton and Sousa 2004; Korndörfer et al. 2004). These structures provided valuable information about the functional mechanisms of the individual components of the chaperone system in single, individually stabilized states. Subsequently, new developments of biophysical techniques including nuclear magnetic resonance (NMR) spectroscopy allowed the observation of these chaperone systems in aqueous solution and in complex with client proteins, thus providing complete descriptions of the chaperones dynamic functional cycles. These findings have established how structural plasticity allows these chaperones to undergo conformational rearrangements and populate conformational ensemble states that overall regulate their functional cycles. Such conformational flexibility enables the chaperones to adapt to their highly dynamic client proteins as well as to fulfil their roles in a complex interaction network in the periplasm.

The Periplasmic Chaperone Skp

The crystal structure of Skp shows a homo-trimeric assembly where the three subunits are connected by a β-barrel trimerization domain that extends into three α-helical "tentacles" or "arms" that give rise to Skp's characteristic jelly fish-like shape (Fig. 6.2a, b) (Walton and Sousa 2004; Korndörfer et al. 2004). Each monomer contributes three β-strands to the trimerization interface. Importantly, these strands are however not connected within one protomer subunit but form a small β-barrel across the trimer by connecting β-strands intermolecularly from two adjacent subunits (Fig. 6.2a). Each arm is constituted by two long α-helixes in coiled-coil arrangement, extending from the oligomerization domain and forming a central cavity. The presence of a central cavity bears some resemblance to the well-characterized Hsp60 (GroEL, CTT) chaperone, where a central cavity serves as a protective environment for the client protein (Braig et al. 1995; Xu et al. 1997; Ditzel et al. 1998; Horwich et al. 2007). The overall organization and shape of Skp bears further similarity to the ATP-independent holdase chaperone prefoldin, with the difference that the N and C-termini of Skp are located at the β-barrel trimerization domain, while for prefoldin

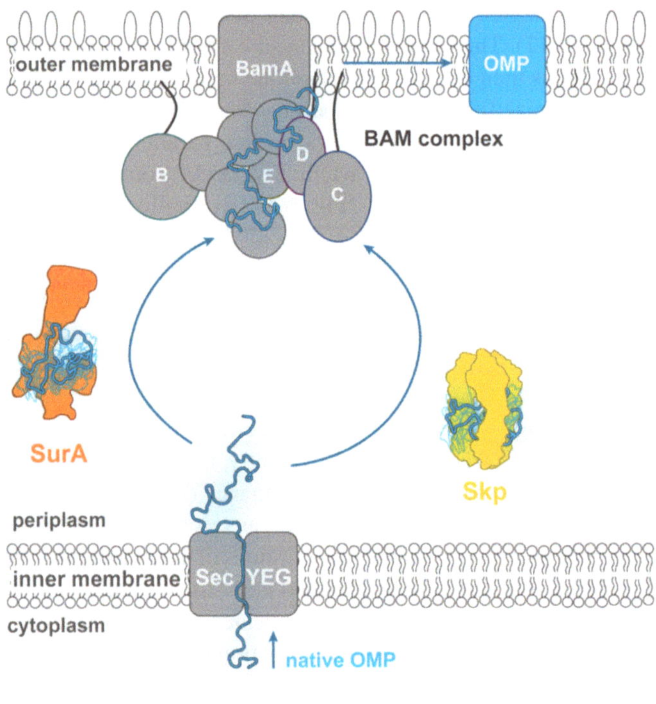

Fig. 6.1 The periplasmic chaperone network in outer membrane biogenesis. After their synthesis in the cytoplasm, unfolded native OMPs are translocated by the SecYEG machinery in a linear fashion across the inner membrane. Subsequently, they are transported by the chaperones Skp or SurA on parallel pathways to the BAM complex for subsequent insertion into the outer membrane

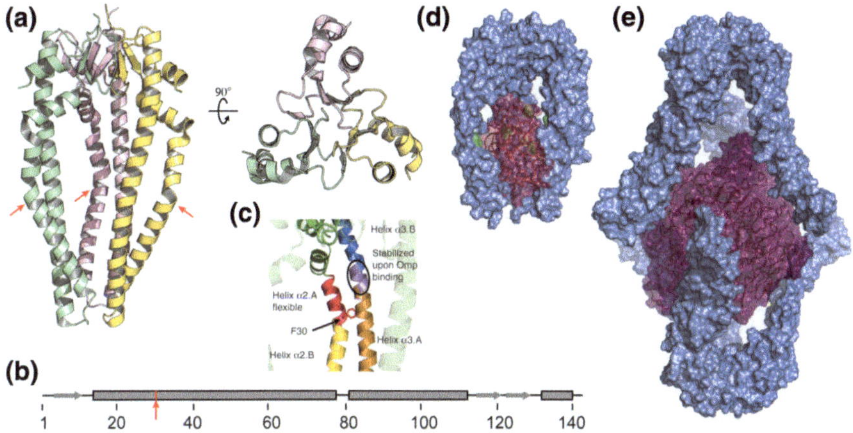

Fig. 6.2 Dynamic adaptation of the periplasmic chaperone Skp and its client protein. **a** Structure of Skp trimer [PDB code 1SG2 (Korndörfer et al. 2004)] with the individual protomers coloured in green, pink and yellow. The position of phenylalanine 30 is highlighted by red arrows. **b** Skp amino acid sequence and secondary structure. The β-sheets and α-helixes are represented by grey arrows and rectangular shapes, respectively. Phenylalanine 30 is highlighted as in (**a**). **c** Close-up on Skp arms highlighting the pivot element in helices α2 and α3. The flexible helix α2.A (red) is linked by the hinge at phenylalanine 30 to the more rigid helix α2.B (yellow). The helicity of segment Ser89–Arg93 (purple) of helix α3.A (orange) is stabilized upon client binding [adapted from (Burmann et al. 2013)]. **d** Structural model of a single conformation from the dynamic Skp(blue)–tOmpA (purple) ensemble. The residues highlighted in green correspond to the alanine identified as interacting with the unfolded tOmpA [adapted from (Callon et al. 2014)]. **e** Model of the multivalent binding of OMPs that are too large to fit in the cavity of one trimeric Skp. The arms are in an open conformation to expand the size of the cavity [adapted from (Holdbrook et al. 2017)]

they are located at the extremity of the arms (Siegert et al. 2000; Martín-Benito et al. 2002; Ohtaki et al. 2008). Just like prefoldin, Skp is classified as an ATP-independent holdase chaperone.

The protein Skp was functionally identified as a binder of OMPs in a pull-down experiment with unfolded OmpF covalently linked to Sepharose beads (Chen and Henning 1996). OmpF is however not the only client protein of Skp, as shown in a subsequent study where the Skp clientome was mapped by a combined strep-tag affinity and proteomic approach (Jarchow et al. 2008). Overall, more than 30 proteins were identified in the Skp clientome, including the OMPs OmpA and LamB, but also several soluble periplasmic proteins, such as MalE and OppA, demonstrating Skp's functional importance for a broad spectrum of clients (Table 6.1). The functional relevance of the interaction of Skp with soluble proteins was additionally highlighted by studies showing that Skp chaperone activity can improve the expression yields of recombinant proteins not only in the periplasm, but also in the cytosol when a Skp variant devoid of its signal-sequence is used (Bothmann and Plückthun 1998; Levy et al. 2001).

Table 6.1 List of unambiguously identified wild-type Skp client proteins with outer membrane localization. Data from Jarchow et al. (2008)

Number	Uniprot entry	Protein name	Function
1	P06129	BtuB	Vitamin B12 transport
2	P17315	CirA	Putative iron transport, colicin IA and IB receptor
3	P13036	FecA	Fe(III) dicitrate transport
4	P05825	FepA	Ferrienterobactin receptor
5	P02943	LamB	Maltose and maltodextrin transport
6	P0A910	OmpA	Outer membrane porin
7	P06996	OmpC	Outer membrane porin
8	P02931	OmpF	Outer membrane porin
9	P0A917	OmpX	Putative defense
10	P75780	Fiu	Catecholate siderophore transport
11	P10384	FadL	Long-chain fatty acid transport
12	P0A921	PldA	Phospholipase
13	P0A927	Tsx	Nucleoside porin
14	P76045	OmpG	Outer membrane porin
15	P02930	TolC	Multidrug exporter
16	P76115	YncD	Putative iron/siderophore receptor

The complex of an OMP client protein and Skp was initially characterized by NMR spectroscopy and fluorescence spectroscopy (Walton et al. 2009; Qu et al. 2009). A full understanding of the dynamic nature of the complexes at atomic resolution was however only obtained subsequently with a complete description of structure and dynamics of Skp–OmpX and Skp–tOmpA (transmembrane domain of OmpA) complexes determined by solution NMR spectroscopy (Burmann et al. 2013; Callon et al. 2014). The OMP polypeptide was found to adopt a disordered conformation that interacts within Skp's central cavity by both hydrophobic contacts and interactions between the charged loops of the OMP and charged parts of the cavity. Notably, these works exploited state-of-the-art NMR experiments to determine for the first time the dynamic properties and structure of a chaperone–client complex with a full-length native client protein with atomic resolution. Moreover, these studies demonstrated the conformational flexibility of Skp in its apo and in its client-bound holo state in solution. In the apo form, a hinge element located at the conserved residue phenylalanine 30 allows the Skp arms to explore a large degree of conformational flexibility (Fig. 6.2a–c). Upon OMP binding, the flexibility of the arms is decreased, leading to a stabilization of the cavity around the bound client protein. While bound, the OMP client protein explores a dynamic landscape of disordered states within the cavity of Skp with no polypeptide segment of OMP stably bound to the chaperone. Thereby, the OMP encapsulated in the cavity has a reduced compactness by approximately a factor of two in comparison to the disordered OMP in denaturant solution. The local lifetime of the individual conformations of at most 1 ms is in stark contrast to the long global lifetime of the complex in the range of hours. This discrepancy arises by avidity from a combination of multiple weak local interactions with a short lifetime. A detailed analysis of a combination of distance-based NMR experiments (NOE and PRE) confirmed the contact interface between Skp and the bound OMP in the cavity (Fig. 6.2d).

The observed position of the OMP client protein in the Skp cavity also raised the question by which mechanism Skp adapts to a wide range of different clients ranging in size from 150 residues (OmpX) to more than 700 residues (FhuA). A study combining molecular modelling, small angle X-ray scattering (SAXS) and NMR spectroscopy confirmed that the hinges on the arms permit to dramatically extend the size of the cavity in an ATP-independent manner to adapt to the size of client proteins (Fig. 6.2e) (Holdbrook et al. 2017). A complementary model was established by systematically analysing the binding capacity of a large variety of OMPs with a size range from 8 to 16 strands to Skp (Schiffrin et al. 2017). These data demonstrated that a higher Skp:OMP ratio is required to bind larger OMPs (16-stranded), while for smaller OMPs the Skp cavity can be expanded to adapt to the size of the client proteins. One common characteristic of Skp is that it exists only as a multiple of the trimeric state, indicating that this stoichiometry corresponds to the active form of the chaperone. Thereby, a recent study established that at physiological concentration Skp exists in an equilibrium between a trimeric and a monomeric state (Sandlin et al. 2015). However, the conformation of the monomeric state is still controversial as Sandlin et al. hypothesized a folded conformation by the mean of a van't Hoff analysis, while Burmann et al. (2013) observed a disordered conformation. To date no study has been able to shed light on characteristics and functionality of the Skp monomeric state.

While structure and dynamics of Skp in apo and holo states have been extensively characterized, the mechanism allowing the release of OMPs from Skp into the outer membrane is less clear. Burmann et al. (2013) have proposed a mechanism hypothesizing that in a ternary Skp–OMP–BAM transition complex the weak local affinity and the rapid structural interconversion of the OMP within the Skp cavity would allow client release by dynamic rearrangements of the polypeptide while it remains bound to Skp. According to this model, the OMP bound to Skp might be inserted sequentially one β-hairpin at a time into the lipid membrane. Furthermore, the BAM complex could guide the unfolded OMP, by transient contacts with BamA POTRA domains to the membrane entry point.

Taken together, the key element regulating Skp's functional cycle is its dynamic nature, which allows both to operate large motion of its arms and to adapt to highly dynamic client proteins. The fast dynamics of the encapsulated OMP client permit its release in an ATP-independent manner. The recent demonstration of the biological relevance of a second non-characterized monomeric state could lead to additional critical insights into the functional cycle of Skp.

The Periplasmic Chaperone SurA

The periplasmic chaperone SurA is generally considered the main pathway for the transport of OMPs to the BAM complex (Sklar et al. 2007; Lazar and Kolter 1996; Rouvière and Gross 1996). Deletion of surA results in a lower OMP density and accumulation in the periplasm of the major OMPs OmpA, OmpF, and LamB in an unfolded form (Rouvière and Gross 1996). The expression of these proteins is down-regulated upon induction of the σ^E-dependent cytoplasmic stress response, which is activated when mis- or unfolded proteins are accumulating in the periplasm (Lazar and Kolter 1996; Missiakas et al. 1996; Rouvière and Gross 1996). The drastically decreased levels of outer membrane-inserted OMPs in SurA mutant strains result in phenotypes characterized by a defective cell envelope resulting in an increased sensitivity to hydrophobic agents such as SDS-EDTA and to antibiotics such as novobiocin (Lazar and Kolter 1996).

SurA is composed of four distinct domains: a large N-terminal domain, two parvulin-like peptidyl-prolyl isomerase (PPIase) domains and a short C-terminal helix followed by a short β-strand (Fig. 6.3a, b) (Bitto and McKay 2002). The combination of the PPIase 1 domain (P1) with the N and C-terminal domains form the core of the protein and the PPIase 2 domain (P2) is flexibly connected to this core by two linkers. The C-terminal 10 last residues form a short β-strand that forms an anti-parallel β-sheet with an N-terminal β-hairpin. While the exact amino acid sequence of the C-terminus is not essential, its shortening drastically affects chaperone activity (Chai et al. 2014). A minimal construct of SurA, composed of the N-terminal domain directly linked to the C-terminal domain, displays activity in vitro and can complement the SurA deletion phenotype in vivo (Behrens et al. 2001; Webb et al. 2001). This suggests that the N/C-domain pair is responsible for the chaperone activity and corroborates the initial observation of a large cavity on the N-terminal domain that could be a binding site for unfolded clients (Fig. 6.3a, b) (Bitto and McKay 2002).

How the PPIase domains are relevant for the function of SurA is still unclear, especially as several SurA homologues have only one or no PPIase domain (Alcock et al. 2008). In general, parvulin-like domains feature enzymatic activity to catalyze the cis–trans conversion of peptide bonds preceding prolyl residues (Rahfeld et al. 1994; Rudd et al. 1995). For SurA only the P2 domain exhibits significant PPIase activity and both PPIase domains can simultaneously be deleted without significant loss of function (Rouvière and Gross 1996; Behrens et al. 2001). Isolated PPIase domains 1 and 2 do not exhibit chaperone activity and do not complement activity in a surA depletion mutant (Behrens et al. 2001). However, it cannot be excluded that their PPIase activity is important for specific client proteins and/or under certain conditions. In this context, it has recently been demonstrated that while the deletion of the P2 domain does not completely abolish the chaperoning activity it is significantly decreasing it (Soltes et al. 2016). This discovery could potentially be explained by a large dynamic conformational rearrangement of the P2 domain that allows closure of the N-terminal domain by a clamp-like mechanism. Soltes et al. also showed that the binding of the P1 domain to the core stabilizes the protein but inhibits the access to the

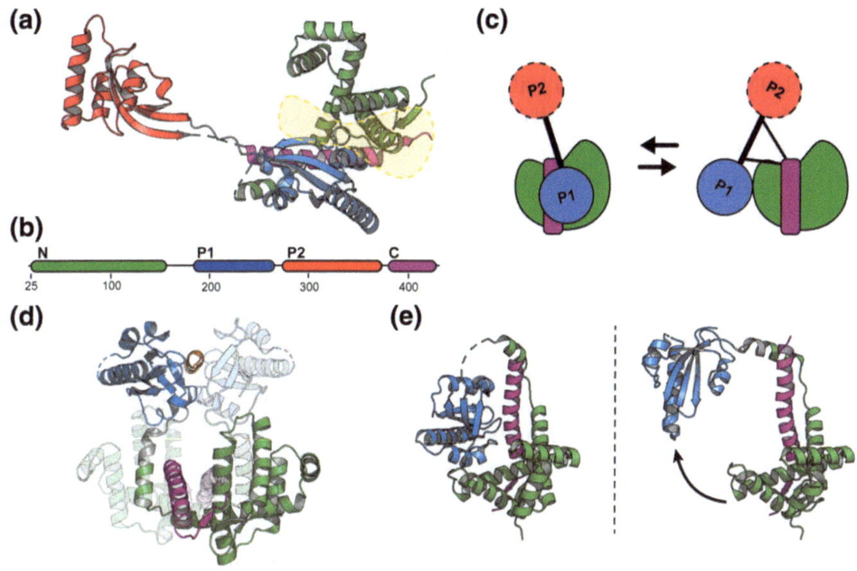

Fig. 6.3 Conformational rearrangements of the periplasmic chaperone SurA regulate its function. **a** Crystal structure of apo SurA (PDB 1M5Y). The core of the protein is formed by the N-terminal domain (green) together with the PPIase domain I (P1) (blue) and the C-terminal helix (magenta). The PPIase domain II (P2) (red) is connected by two flexible linkers and is thus located 30 Å away from the core. The hypothesized client binding site on the N-terminal domain is highlighted in yellow. **b** SurA domain composition coloured as in (**a**). **c** Schematic illustration of the regulation of SurA functional cycle by conformational dynamics of the two PPIase domains relatively to the core, coloured are as in (**a**) [adapted from (Soltes et al. 2016)]. **d** Structure of the dimeric complex of SurAΔP2 (SurA lacking the domain P2) bound to a peptide mimicking a client protein (PDB 2PV3), coloured as in (**a**). The peptide is coloured in orange. **e** Different conformations of the P1 domain in the apo state (left) and in the monomer extracted from the dimeric complex with a peptide (right). The formation of the complex triggers a major conformational change that leads to the dissociation of the P1 domain from the core domain

binding site, while the flexible unbound state increases the chaperone activity at the cost of a decreased stability (Fig. 6.3c). The combination of the studies above suggests that the N/C-domain could represent a platform providing the basal chaperone activity that then has been evolutionarily diversified by addition of accessory domains.

The involvement of SurA in the assembly and transport of β-barrel OMPs to the outer membrane is clearly demonstrated by multiple studies (Rizzitello et al. 2001; Sklar et al. 2007). A specific set of OMPs that require SurA for their proper maturation has been revealed by a differential proteomics experiment including the formerly known SurA client proteins OmpF, OmpA, FhuA, and LamB as well as newly discovered client proteins such as LptD (Vertommen et al. 2009). Some motifs determining client specificity of SurA were identified using phage display experiments (Bitto and McKay 2003; Xu et al. 2007). These studies have shown that SurA recognizes peptide sequences rich in aromatic amino acids and arranged in the specific pattern

Ar–X–Ar, where Ar is an aromatic residue and X can be any residue. The Ar–X–Ar sequence is found with increased abundance at the C-terminus of transmembrane β-barrel OMPs, indicating a possible role as recognition signal for SurA. This study lead to the development of a high affinity peptide mimicking the C-terminus of OMPs and the crystal structure of SurA with this peptide bound could be solved (Xu et al. 2007) (Fig. 6.3d). The structure shows a hydrophobic binding site on the P1 domain, however, not only one but two P1 domains are binding to one peptide, thus triggering a 2:1 SurA:peptide stoichiometry. Thereby, the P1 domain undergoes large conformational changes, which result in its separation from the core domain, while the structures of the N- and C-terminal domains remain unaffected compared to the apo form (Fig. 6.3e). Overall, the biological relevance of this structure remains unclear, since the binding of a short peptide is not necessarily representative for a full-length client protein. Furthermore, the position of the binding site contradicts in vivo data showing that the chaperoning activity is not affected by P1 deletion. This discrepancy could perhaps be explained by a decoupling of the sites of substrate recognition and chaperone activity. In the structure by Xu et al. the peptide bound to the P1 domain is oriented with its N-terminus towards the SurA N/C—domain. A full-length client protein could then bind the N/C-domain via a non-specific interaction.

Overall, the structural properties of SurA–OMP complexes are by far not as well characterized as the Skp–OMP complex, as no atomic resolution information is available. To obtain a complete picture it will be necessary to determine, among others, which functional role the conformational rearrangements of the domains have. A complete characterization of the structure and dynamics of SurA in its apo and holo states, is highly desired in order to draw a complete the picture of the molecular mechanism underlying the function of SurA.

Skp- and SurA-Mediated Stabilization and Folding of Client Proteins

The increasingly detailed structural knowledge of the periplasmic molecular chaperone network is accompanied by several studies characterizing how these chaperones interact with and influence the folding behaviour of client OMPs in vitro. Trimeric Skp has been found to form stable 1:1 (Skp_3:OMP) complexes with small OMP clients such as unfolded OmpA, but also complexes in 2:1 (Skp_3:OMP) ratio with larger OMPs such as unfolded BamA and FhuA (Schiffrin et al. 2016; Thoma et al. 2015). Thereby, Skp can bind unfolded client OMPs with high affinities, corresponding to dissociation constants in the low nM range, as determined for multiple unfolded OMP clients, such as OmpA, OmpG and BamA as well as OmpLA, OmpW and PagP (Qu et al. 2007; Moon et al. 2013). While the stoichiometry of SurA:OMP complexes has not been studied in detail, it has been proposed to bind OmpC as well as FhuA in a 2:1 ratio (Thoma et al. 2015; Li et al. 2018). The affinity of SurA to OMP clients is generally much weaker than of Skp. While SurA's affinity to full-length nascent client proteins has been determined to a lesser extent, it has been found to bind unfolded OmpG and OmpF with affinities in the low μM range (Bitto and McKay 2004). Similar values were found for its affinity to heptapeptide WEYIPNV, which contains the Ar–X–Ar motif frequently found in OMPs (Bitto and McKay 2003).

A higher client affinity of Skp compared to SurA is also consistent with experiments showing Skp binding to unfolded OmpC at faster rates than SurA as well as 6-fold longer lifetimes of chaperone-client complexes formed by Skp with unfolded FhuA compared to SurA-FhuA complexes (Wu et al. 2011; Thoma et al. 2015). Moreover, Skp has been shown to outcompete SurA in equimolar mixtures of both chaperones when binding unfolded FhuA or OmpA (Thoma et al. 2015; Schiffrin et al. 2017). The client OMP either remains bound to Skp when preassembled Skp—client complexes are added to apo SurA, or transfers to Skp when preassembled SurA–client complexes are added to apo Skp. Differences in client binding were further elucidated by a recent study reporting a mild disaggregase function for Skp, which could break down scant aggregates containing up to 5 OmpC monomers, whereas such behavior could not be observed for SurA (Li et al. 2018). These data are also in full agreement with the notion that Skp binds its clients in a tighter and more compact conformation than SurA.

While chaperones can maintain client OMPs dynamically unfolded in solution over prolonged time periods, proper folding of the client requires the presence of a lipid membrane which provides the thermodynamic stabilization of the native form (Moon et al. 2013; Burmann et al. 2013; Thoma et al. 2015). Thereby, the transition from a chaperone-stabilized state to a folded, membrane-embedded state depends on the biophysical properties of the lipid bilayer. For example, OmpA and PagP fold into negatively charged lipid bilayers from solutions containing Skp, but not into neutral bilayers (Bulieris et al. 2003; Patel et al. 2009; Patel and Kleinschmidt 2013; McMorran et al. 2013). Moreover, the presence of additional factors such as LPS, transmembrane proteins in general and in particular the presence of BamA appears to influence the transition from a chaperone-stabilized state into the lipid bilayer (Bulieris et al. 2003; Patel et al. 2009; Patel and Kleinschmidt 2013; Schiffrin et al. 2017). Especially client OMPs folding from complexes with SurA appear to profit greatly from the presence of membrane-embedded BAM, but show relatively little folding into empty lipid bilayers (Schiffrin et al. 2017). Despite the influence of different membrane composition, it evolves as a common theme from these studies that elevated concentrations of Skp prevent folding and membrane insertion of client OMPs (Bulieris et al. 2003; Schiffrin et al. 2016). The inhibitory effect of Skp was also observed when looking at chaperone assisted OMP folding on the single molecule level. The large OMP FhuA has a strong tendency to misfold in the absence of periplasmic chaperones, whereas in the presence of either of the chaperones Skp or SurA, misfolding is prevented. In agreement with the bulk measurements, folding of FhuA is largely abolished in the presence of Skp, however, FhuA can reinsert into the lipid bilayer in a series of folding steps shaped by individual beta-hairpins in the presence of SurA (Fig. 6.4a) (Thoma et al. 2012, 2015). The sequence in which individual β-hairpins insert into the membrane can be random, while a certain tendency is observed to progress from the C-terminal toward the N-terminal end of the

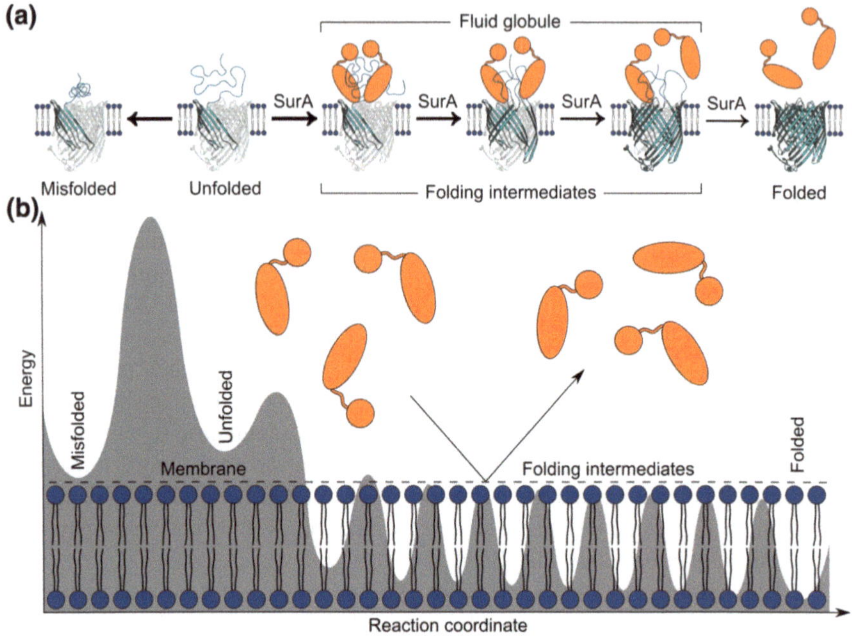

Fig. 6.4 Folding and membrane insertion model of client OMP FhuA from a SurA-FhuA complex. **a** In the absence of chaperones, the folding pathway of unfolded FhuA (cyan) is directed towards a misfolded state. The presence of SurA (orange) stabilizes the unfolded state of FhuA and promotes stepwise insertion and folding of β-hairpins into the lipid membrane until folding is completed, upon which SurA is released. **b** The hypothetical folding free-energy landscape of FhuA in the presence of SurA is characterized by a series of free-energy wells corresponding to β-hairpins stably inserted in the lipid membrane. SurA (orange) prevents misfolding of FhuA in solution but is spatially excluded from the lipid membrane (blue) and thus cannot interact with folded intermediates [adapted from (Thoma et al. 2015)]

protein. Thereby, the yet unfolded segments of the nascent FhuA polypeptide remain protected from misfolding over the entire time of the folding process. In agreement with the higher binding affinity of Skp, in an equimolar mixture of both chaperones, the inhibitory effect Skp exerted on the folding behaviour dominates over the folding promoting effect of SurA (Thoma et al. 2015). Consequentially, these experiments indicate that both chaperones could interact effectively with partially inserted client OMPs, strengthening the concept of the lipid membrane acting not only as a free-energy sink but also as a physical barrier separating partially folded segments from the chaperones (Fig. 6.4b).

In summary, these in vitro findings indicate that Skp binds clients with high affinity and a strong tendency to prevent their folding whereas SurA binds unfolded OMPs with lower affinity while promoting their folding. Together with in vivo findings, which indicate severely reduced OMP levels in SurA-depleted strains, but only mild phenotypes upon Skp depletion, the available data indicate a role of SurA acting as the main periplasmic pathway for OMP folding, while Skp forms a secondary backup pathway (Silhavy et al. 2010; Sklar et al. 2007; Wu et al. 2011). The concept of SurA acting as the primary pathway is further substantiated by the observation that the delivery of major OMPs, as well as certain large OMPs such as LptD and FhuA seem to rely particularly on the presence of SurA (Vertommen et al. 2009). However, it remains to be elucidated precisely how the differential effects of both chaperones observed in vitro are modulated in vivo to promote efficient delivery of client OMPs to the OM.

Concluding Remarks

During the last decade, ground-breaking works have drastically improved our understanding of the molecular mechanism regulating the periplasmic chaperones Skp and SurA and highlighted the major importance of conformational flexibility in their mechanisms of action. These latest advancements have been made possible by the technical development of SAXS, molecular dynamics and NMR spectroscopy methods, in addition to the ground-laying crystal structures, highlighting the critical importance of the integration of the different methods. The structure and dynamics of the periplasmic chaperone Skp in its apo and holo states have shown that large conformational rearrangements of its cavity are the key factors of its functional cycle, giving Skp the capacity to adapt in response to the properties of its client proteins. Similarly, preliminary results on the chaperone SurA highlight the role played by the dynamic rearrangements of its domains to regulate its functional cycle.

Acknowledgements This work was supported by the Swiss National Science Foundation and the NFP 72 (grants 31003A_166426 and 407240_167125 to S.H.) and by the Swiss Nanoscience institute. The authors declare no conflict of interest.

References

Alcock FH, Grossmann JG, Gentle IE, Likić VA, Lithgow T, Tokatlidis K (2008) Conserved substrate binding by chaperones in the bacterial periplasm and the mitochondrial intermembrane space. Biochem J 409:377–387. https://doi.org/10.1042/BJ20070877

Antonoaea R, Furst M, Nishiyama K, Muller M (2008) The periplasmic chaperone PpiD interacts with secretory proteins exiting from the SecYEG translocon. Biochemistry 47:5649–5656. https://doi.org/10.1021/bi800233w

Arie JP, Sassoon N, Betton JM (2001) Chaperone function of FkpA, a heat shock prolyl isomerase, in the periplasm of Escherichia coli. Mol Microbiol 39:199–210. https://doi.org/10.1046/j.1365-2958.2001.02250.x

Bardwell JC, McGovern K, Beckwith J (1991) Identification of a protein required for disulfide bond formation in vivo. Cell 67:581–589. https://doi.org/10.1016/0092-8674(91)90532-4

Behrens S, Maier R, de Cock H, Schmid FX, Gross CA (2001) The SurA periplasmic PPIase lacking its parvulin domains functions in vivo and has chaperone activity. EMBO J 20:285–294. https://doi.org/10.1093/emboj/20.1.285

Bitto E, McKay DB (2002) Crystallographic structure of SurA, a molecular chaperone that facilitates folding of outer membrane porins. Structure 10:1489–1498. https://doi.org/10.1016/s0969-2126(02)00877-8
Bitto E, McKay DB (2003) The periplasmic molecular chaperone protein SurA binds a peptide motif that is characteristic of integral outer membrane proteins. J Biol Chem 278:49316–49322. https://doi.org/10.1074/jbc.M308853200
Bitto E, McKay DB (2004) Binding of phage-display-selected peptides to the periplasmic chaperone protein SurA mimics binding of unfolded outer membrane proteins. FEBS Lett 568:94–98. https://doi.org/10.1016/j.febslet.2004.05.014
Bothmann H, Plückthun A (1998) Selection for a periplasmic factor improving phage display and functional periplasmic expression. Nat Biotechnol 16:376–380. https://doi.org/10.1038/nbt0498-376
Braig K, Adams PD, Brünger AT (1995) Conformational variability in the refined structure of the chaperonin GroEL at 2.8 Å resolution. Nat Struct Mol Biol 2:1083. https://doi.org/10.1038/nsb1295-1083
Bulieris PV, Behrens S, Holst O, Kleinschmidt JH (2003) Folding and insertion of the outer membrane protein OmpA is assisted by the chaperone Skp and by lipopolysaccharide. J Biol Chem 278:9092–9099. https://doi.org/10.1074/jbc.M211177200
Burmann BM, Wang C, Hiller S (2013) Conformation and dynamics of the periplasmic membrane-protein-chaperone complexes OmpX-Skp and tOmpA-Skp. Nat Struct Mol Biol 20:1265–1272. https://doi.org/10.1038/nsmb.2677
Callon M, Burmann BM, Hiller S (2014) Structural mapping of a chaperone-substrate interaction surface. Angew Chem Int Edit 53:5069–5072. https://doi.org/10.1002/anie.201310963
Chai Q, Ferrell B, Zhong M, Zhang X, Ye C, Wei Y (2014) Diverse sequences are functional at the C-terminus of the E. coli periplasmic chaperone SurA. Protein Eng Des Sel 27:111–116. https://doi.org/10.1093/protein/gzu003
Chatzi KE, Sardis MF, Economou A, Karamanou S (2014) SecA-mediated targeting and translocation of secretory proteins. Biochim et Biophys Acta (BBA) 1843:1466–1474. https://doi.org/10.1016/j.bbamcr.2014.02.014
Chen R, Henning U (1996) A periplasmic protein (Skp) of Escherichia coli selectively binds a class of outer membrane proteins. Mol Microbiol 19:1287–1294. https://doi.org/10.1111/j.1365-2958.1996.tb02473.x
Clausen T, Kaiser M, Huber R, Ehrmann M (2011) HTRA proteases: regulated proteolysis in protein quality control. Nat Rev Mol Cell Biol 12:152–162. https://doi.org/10.1038/nrm3065
De Geyter J, Tsirigotaki A, Orfanoudaki G, Zorzini V, Economou A, Karamanou S (2016) Protein folding in the cell envelope of Escherichia coli. Nat Microbiol 1:16107. https://doi.org/10.1038/nmicrobiol.2016.107
Denoncin K, Collet J-F (2013) Disulfide bond formation in the bacterial periplasm: major achievements and challenges ahead. Antioxid Redox Signal 19:63–71. https://doi.org/10.1089/ars.2012.4864
Ding J, Yang C, Niu X, Hu Y, Jin C (2015) HdeB chaperone activity is coupled to its intrinsic dynamic properties. Sci Rep 5:1292. https://doi.org/10.1038/srep16856
Ditzel L, Löwe J, Stock D, Stetter K-O, Huber H, Huber R, Steinbacher S (1998) Crystal structure of the thermosome, the archaeal chaperonin and homolog of CCT. Cell 93:125–138. https://doi.org/10.1016/S0092-8674(00)81152-6
Driessen AJM, Manting EH, van der Does C (2001) The structural basis of protein targeting and translocation in bacteria. Nat Struct Mol Biol 8:492–498. https://doi.org/10.1038/88549
Ge X, Wang R, Ma J, Liu Y, Ezemaduka AN, Chen PR, Fu X, Chang Z (2014a) DegP primarily functions as a protease for the biogenesis of β-barrel outer membrane proteins in the Gram-negative bacterium Escherichia coli. FEBS J 281:1226–1240. https://doi.org/10.1111/febs.12701
Ge X, Lyu Z-X, Liu Y, Wang R, Zhao XS, Fu X, Chang Z (2014b) Identification of FkpA as a key quality control factor for the biogenesis of outer membrane proteins under heat shock conditions. J Bacteriol 196:672–680. https://doi.org/10.1128/jb.01069-13

Hayano T, Takahashi N, Kato S, Maki N, Suzuki M (1991) Two distinct forms of peptidylprolyl-cis-trans-isomerase are expressed separately in periplasmic and cytoplasmic compartments of Escherichia coli cells. Biochem 30:3041–3048. https://doi.org/10.1021/bi00226a009
He L, Sharpe T, Mazur A, Hiller S (2016) A molecular mechanism of chaperone-client recognition. Sci Adv 2–11:e1601625. https://doi.org/10.1126/sciadv.1601625
Helbig S, Patzer SI, Schiene-Fischer C, Zeth K, Braun V (2011) Activation of colicin M by the FkpA prolyl cis-trans isomerase/chaperone. J Biol Chem 286:6280–6290. https://doi.org/10.1074/jbc.m110.165274
Heras B, Shouldice SR, Totsika M, Scanlon MJ, Schembri MA, Martin JL (2009) DSB proteins and bacterial pathogenicity. Nat Rev Microbiol 7:215–225. https://doi.org/10.1038/nrmicro2087
Holdbrook DA, Burmann BM, Huber RG, Petoukhov MV, Svergun DI, Hiller S, Bond PJ (2017) A spring-loaded mechanism governs the clamp-like dynamics of the Skp chaperone. Structure 25:1079–1088.e3. https://doi.org/10.1016/j.str.2017.05.018
Hong W, Jiao W, Hu J, Zhang J, Liu C, Fu X, Shen D, Xia B, Chang Z (2005) Periplasmic protein HdeA exhibits chaperone-like activity exclusively within stomach pH range by transforming into disordered conformation. J Biol Chem 280:27029–27034. https://doi.org/10.1074/jbc.m503934200
Horwich AL, Fenton WA, Chapman E, Farr GW (2007) Two families of chaperonin: physiology and mechanism. Annu Rev Cell Dev Bi 23:115–145. https://doi.org/10.1146/annurev.cellbio.23.090506.123555
Ito K, Inaba K (2008) The disulfide bond formation (Dsb) system. Curr Opin Struct Biol 18:450–458. https://doi.org/10.1016/j.sbi.2008.02.002
Jarchow S, Lück C, Görg A, Skerra A (2008) Identification of potential substrate proteins for the periplasmic Escherichia coli chaperone Skp. Proteomics 8:4987–4994. https://doi.org/10.1002/pmic.200800288
Jiang J, Zhang X, Chen Y, Wu Y, Zhou ZH, Chang Z, Sui S-F (2008) Activation of DegP chaperone-protease via formation of large cage-like oligomers upon binding to substrate proteins. Proc Natl Acad Sci USA 105:11939–11944. https://doi.org/10.1073/pnas.0805464105
Joly JC, Swartz JR (1997) In vitro and in vivo redox states of the Escherichia coli periplasmic oxidoreductases DsbA and DsbC. Biochemistry 36:10067–10072. https://doi.org/10.1021/bi9707739
Kern R, Malki A, Abdallah J, Tagourti J, Richarme G (2007) Escherichia coli HdeB is an acid stress chaperone. J Bacteriol 189:603–610. https://doi.org/10.1128/jb.01522-06
Korndörfer IP, Dommel MK, Skerra A (2004) Structure of the periplasmic chaperone Skp suggests functional similarity with cytosolic chaperones despite differing architecture. Nat Struct Mol Biol 11:1015–1020. https://doi.org/10.1038/nsmb828
Krogh A, Larsson B, von Heijne G, Sonnhammer ELL (2001) Predicting transmembrane protein topology with a hidden markov model: application to complete genomes. J Mol Biol 305:567–580. https://doi.org/10.1006/jmbi.2000.4315
Lazar SW, Kolter R (1996) SurA assists the folding of Escherichia coli outer membrane proteins. J Bacteriol 178:1770–1773. https://doi.org/10.1128/jb.178.6.1770-1773.1996
Levy R, Weiss R, Chen G, Iverson BL, Georgiou G (2001) Production of correctly folded Fab antibody fragment in the cytoplasm of Escherichia coli trxB gor mutants via the coexpression of molecular chaperones. Prot Exp Purif 23:338–347. https://doi.org/10.1006/prep.2001.1520
Li G, He C, Bu P, Bi H, Pan S, Sun R, Zhao XS (2018) Single–molecule detection reveals different roles of Skp and SurA as chaperones. ACS Chem Biol. https://doi.org/10.1021/acschembio.8b00097
Liu J, Walsh CT (1990) Peptidyl-prolyl cis-trans-isomerase from Escherichia coli: a periplasmic homolog of cyclophilin that is not inhibited by cyclosporin A. Proc Natl Acad Sci USA 87:4028–4032. https://doi.org/10.1073/pnas.87.11.4028
Lycklama A, Nijeholt JA, Driessen AJ (2012) The bacterial Sec-translocase: structure and mechanism. Philos Trans R Soc B Biol Sci 367:1016–1028. https://doi.org/10.1098/rstb.2011.0201

Martín-Benito J, Boskovic J, Gómez-Puertas P, Carrascosa JL, Simons CT, Lewis SA, Bartolini F, Cowan NJ, Valpuesta JM (2002) Structure of eukaryotic prefoldin and of its complexes with unfolded actin and the cytosolic chaperonin CCT. EMBO J 21:6377–6386. https://doi.org/10.1093/emboj/cdf640

Matern Y, Barion B, Behrens-Kneip S (2010) PpiD is a player in the network of periplasmic chaperones in Escherichia coli. BMC Microbiol 10:251. https://doi.org/10.1186/1471-2180-10-251

McMorran LM, Bartlett AI, Huysmans GHM, Radford SE, Brockwell DJ (2013) Dissecting the effects of periplasmic chaperones on the in vitro folding of the outer membrane protein PagP. J Mol Biol 425:3178–3191. https://doi.org/10.1016/j.jmb.2013.06.017

Messens J, Collet JF (2006) Pathways of disulfide bond formation in Escherichia coli. Int J Biochem Cell Biol 38:1050–1062. https://doi.org/10.1016/j.biocel.2005.12.011

Missiakas D, Betton JM, Raina S (1996) New components of protein folding in extracytoplasmic compartments of Escherichia coli SurA, FkpA and Skp/OmpH. Mol Microbiol 21:871–884. https://doi.org/10.1046/j.1365-2958.1996.561412.x

Miyamoto A, Matsuyama S, Tokuda H (2001) Mutant of LolA, a lipoprotein-specific molecular chaperone of Escherichia coli, defective in the transfer of lipoproteins to LolB. Biochem Biophys Res Commun 287:1125–1128. https://doi.org/10.1006/bbrc.2001.5705

Moon CP, Zaccai NR, Fleming PJ, Gessmann D, Fleming KG (2013) Membrane protein thermodynamic stability may serve as the energy sink for sorting in the periplasm. Proc Natl Acad Sci USA 110:4285–4290. https://doi.org/10.1073/pnas.1212527110

Natale P, Bruser T, Driessen AJ (2008) Sec- and Tat-mediated protein secretion across the bacterial cytoplasmic membrane-distinct translocases and mechanisms. Biochim Biophys Acta 1778:1735–1756. https://doi.org/10.1016/j.bbamem.2007.07.015

Okuda S, Tokuda H (2009) Model of mouth-to-mouth transfer of bacterial lipoproteins through inner membrane LolC, periplasmic LolA, and outer membrane LolB. Proc Natl Acad Sci USA 106:5877–5882. https://doi.org/10.1073/pnas.0900896106

Ohtaki A, Kida H, Miyata Y, Ide N, Yonezawa A, Arakawa T, Iizuka R, Noguchi K, Kita A, Odaka M, Miki K, Yohda M (2008) Structure and molecular dynamics simulation of archaeal prefoldin: the molecular mechanism for binding and recognition of nonnative substrate proteins. J Mol Biol 376:1130–1141. https://doi.org/10.1016/j.jmb.2007.12.010

Orfanoudaki G, Economou A (2014) Proteome-wide subcellular topologies of E. coli polypeptides database (STEPdb). Mol Cell Proteomics 13:3674–3687. https://doi.org/10.1074/mcp.O114.041137

Patel GJ, Behrens-Kneip S, Holst O, Kleinschmidt JH (2009) The periplasmic chaperone Skp facilitates targeting, insertion, and folding of OmpA into lipid membranes with a negative membrane surface potential. Biochemistry 48:10235–10245. https://doi.org/10.1021/bi901403c

Patel GJ, Kleinschmidt JH (2013) The lipid bilayer-inserted membrane protein BamA of Escherichia coli facilitates insertion and folding of outer membrane protein A from its complex with Skp. Biochemistry 52:3974–3986. https://doi.org/10.1021/bi400103t

Qu J, Behrens S, Holst O, Kleinschmidt JH (2009) Binding sites of outer membrane protein A (OmpA) in the complex with the periplasmic chaperone Skp from E. Coli. A site-directed fluorescence study. Biophys J 96:449a–450a. https://doi.org/10.1016/j.bpj.2008.12.2310

Qu J, Mayer C, Behrens S, Holst O, Kleinschmidt JH (2007) The trimeric periplasmic chaperone Skp of Escherichia coli forms 1:1 complexes with outer membrane proteins via hydrophobic and electrostatic interactions. J Mol Biol 374:91–105. https://doi.org/10.1016/j.jmb.2007.09.020

Quan S, Koldewey P, Tapley T, Kirsch N, Ruane KM, Pfizenmaier J, Shi R, Hofmann S, Foit L, Ren G, Jakob U, Xu Z, Cygler M, Bardwell JC (2011) Genetic selection designed to stabilize proteins uncovers a chaperone called Spy. Nat Struct Mol Biol 18:262–269. https://doi.org/10.1038/nsmb.2016

Rahfeld J-U, Schierhorn A, Mann K, Fischer G (1994) A novel peptidyl-prolyl cis/trans isomerase from Escherichia coli. FEBS Lett 343:65–69. https://doi.org/10.1016/0014-5793(94)80608-X

Rizzitello AE, Harper JR, Silhavy TJ (2001) Genetic evidence for parallel pathways of chaperone activity in the periplasm of Escherichia coli. J Bacteriol 183:6794–6800. https://doi.org/10.1128/JB.183.23.6794-6800.2001

Rouvière PE, Gross CA (1996) SurA, a periplasmic protein with peptidyl-prolyl isomerase activity, participates in the assembly of outer membrane porins. Genes Dev 10:3170–3182. https://doi.org/10.1101/gad.10.24.3170

Rudd KE, Sofia HJ, Koonin EV, Plunkett G, Lazar S, Rouviere PE (1995) A new family of peptidyl-prolyl isomerases. Trends Biochem Sci 20:12–14. https://doi.org/10.1016/S0968-0004(00)88940-9

Salmon L, Stull F, Sayle S, Cato C, Akgül Ş, Foit L, Ahlstrom LS, Eisenmesser EZ, Al-Hashimi HM, Bardwell JCA, Horowitz S (2018) The mechanism of HdeA unfolding and chaperone activation. J Mol Biol 430:33–40. https://doi.org/10.1016/j.jmb.2017.11.002

Sandlin CW, Zaccai NR, Fleming KG (2015) Skp trimer formation is insensitive to salts in the physiological range. Biochemistry 54:7059–7062. https://doi.org/10.1021/acs.biochem.5b00806

Saul FA, Arie JP, Vulliez-le Normand B, Kahn R, Betton JM, Bentley GA (2004) Structural and functional studies of FkpA from Escherichia coli, a cis/trans peptidyl-prolyl isomerase with chaperone activity. J Mol Biol 335:595–608. https://doi.org/10.1016/j.jmb.2003.10.056

Schiffrin B, Calabrese AN, Devine PWA, Harris SA, Ashcroft AE, Brockwell DJ, Radford SE (2016) Skp is a multivalent chaperone of outer-membrane proteins. Nat Struct Mol Biol 1–11. https://doi.org/10.1038/nsmb.3266

Schiffrin B, Calabrese AN, Higgins AJ, Humes JR, Ashcroft AE, Kalli AC, Brockwell DJ, Radford SE (2017) Effects of periplasmic chaperones and membrane thickness on BamA-catalyzed outer-membrane protein folding. J Mol Biol 429:3776–3792. https://doi.org/10.1016/j.jmb.2017.09.008

Scholz C, Scherer G, Mayr LM, Schindler T, Fischer G, Schmid FX (1998) Prolyl isomerases do not catalyze isomerization of non-prolyl peptide bonds. Biol Chem 379:361–365. https://doi.org/10.1016/s0014-5793(98)00871-0

Siegert R, Leroux MR, Scheufler C, Hartl FU, Moarefi I (2000) Structure of the molecular chaperone prefoldin: unique interaction of multiple coiled coil tentacles with unfolded proteins. Cell 103:621–632. https://doi.org/10.1016/S0092-8674(00)00165-3

Silhavy TJ, Kahne D, Walker S (2010) The bacterial cell envelope. CSH Perspect Biol 2:a000414–a000414. https://doi.org/10.1101/cshperspect.a000414

Sklar JG, Wu T, Kahne D, Silhavy TJ (2007) Defining the roles of the periplasmic chaperones SurA, Skp, and DegP in Escherichia coli. Genes Dev 21:2473–2484. https://doi.org/10.1101/gad.1581007

Spiess C, Beil A, Ehrmann M (1999) A temperature-dependent switch from chaperone to protease in a widely conserved heat shock protein. Cell 97:339–347. https://doi.org/10.1016/s0092-8674(00)80743-6

Soltes GR, Schwalm J, Ricci DP, Silhavy TJ (2016) The activity of Escherichia coli chaperone SurA is regulated by conformational changes involving a parvulin domain. J Bacteriol 198:921–929. https://doi.org/10.1128/JB.00889-15

Stull F, Koldewey P, Humes JR, Radford SE, Bardwell JCA (2016) Substrate protein folds while it is bound to the ATP-independent chaperone Spy. Nat Struct Mol Biol 23:53–58. https://doi.org/10.1038/nsmb.3133

Tajima T, Yokota N, Matsuyama S, Tokuda H (1998) Genetic analyses of the in vivo function of LolA, a periplasmic chaperone involved in the outer membrane localization of Escherichia coli lipoproteins. FEBS Lett 439:51–54. https://doi.org/10.1016/s0014-5793(98)01334-9

Taniguchi N, Matsuyama S, Tokuda H (2005) Mechanisms underlying energy- independent transfer of lipoproteins from LolA to LolB, which have similar unclosed beta-barrel structures. J Biol Chem 280:34481–34488. https://doi.org/10.1074/jbc.m507388200

Thoma J, Bosshart P, Pfreundschuh M, Müller DJ (2012) Out but not in: the large transmembrane β-barrel protein FhuA unfolds but cannot refold via β-hairpins. Structure 20:2185–2190

Thoma J, Burmann BM, Hiller S, Müller DJ (2015) Impact of holdase chaperones Skp and SurA on the folding of β-barrel outer-membrane proteins. Nat Struct Mol Biol 22:795–802. https://doi.org/10.1038/nsmb.3087

Tsirigotaki A, De Geyter J, Šoštaric N, Economou A, Karamanou S (2017) Protein export through the bacterial Sec pathway. Nat Rev Microbiol 15:4, 15:21–36. https://doi.org/10.1038/nrmicro.2016.161

Van den Berg B, Clemons WM Jr, Collinson I, Modis Y, Hartmann E, Harrison SC, Rapoport TA (2004) X-ray structure of a protein-conducting channel. Nature 427:36–44. https://doi.org/10.1038/nature02218

Van Wielink JE, Duine JA (1990) How big is the periplasmic space? Trends Biochem Sci 15:136–137. https://doi.org/10.1016/0968-0004(90)90208-S

Vertommen D, Depuydt M, Pan J, Leverrier P, Knoops L, Szikora JP, Messens J, Bardwell JC, Collet J-F (2008) The disulphide isomerase DsbC cooperates with the oxidase DsbA in a DsbD-independent manner. Mol Microbiol 67:336–349. https://doi.org/10.1111/j.1365-2958.2007.06030.x

Vertommen D, Ruiz N, Leverrier P, Silhavy TJ, Collet J-F (2009) Characterization of the role of the Escherichia coli periplasmic chaperone SurA using differential proteomics. Proteomics 9:2432–2443. https://doi.org/10.1002/pmic.200800794

Walton TA, Sandoval CM, Fowler CA, Pardi A, Sousa MC (2009) The cavity-chaperone Skp protects its substrate from aggregation but allows independent folding of substrate domains. Proc Natl Acad Sci USA 106:1772–1777. https://doi.org/10.1073/pnas.0809275106

Walton TA, Sousa MC (2004) Crystal structure of Skp, a prefoldin-like chaperone that protects soluble and membrane proteins from aggregation. Mol Cell 15:367–374. https://doi.org/10.1016/j.molcel.2004.07.023

Webb HM, Ruddock LW, Marchant RJ, Jonas K, Klappa P (2001) Interaction of the periplasmic peptidylprolyl cis-transisomerase SurA with model peptides. J Biol Chem 276:45622–45627. https://doi.org/10.1074/jbc.M107508200

Wickner W, Driessen AJ, Hartl FU (1991) The enzymology of protein translocation across the Escherichia coli plasma membrane. Annu Rev Plant Physio. Plant Mol Biol 60:101–124. https://doi.org/10.1146/annurev.bi.60.070191.000533

Wu S, Ge X, Lv Z, Zhi Z, Chang Z, Zhao XS (2011) Interaction between bacterial outer membrane proteins and periplasmic quality control factors: a kinetic partitioning mechanism. Biochem J 438:505–511. https://doi.org/10.1042/BJ20110264

Xu X, Wang S, Hu Y-X, McKay DB (2007) The periplasmic bacterial molecular chaperone SurA adapts its structure to bind peptides in different conformations to assert a sequence preference for aromatic residues. J Mol Biol 373:367–381. https://doi.org/10.1016/j.jmb.2007.07.069

Xu Z, Horwich AL, Sigler PB (1997) The crystal structure of the asymmetric GroEL–GroES–$(ADP)_7$ chaperonin complex. Nature 388:741–750. https://doi.org/10.1038/41944

Yu X-C, Yang C, Ding J, Niu X, Hu Y, Jin C (2017) Characterizations of the interactions between Escherichia coli periplasmic chaperone HdeA and its native substrates during acid stress. Biochem 56:5748–5757. https://doi.org/10.1021/acs.biochem.7b00724

Yu X-C, Hu Y, Ding J, Li H, Jin C (2018) Structural basis and mechanism of the unfolding-induced activation of HdeA, a bacterial acid response chaperone. J Biol Chem. https://doi.org/10.1101/390104

Zhang S, He D, Yang Y, Lin S, Zhang M, Dai S, Chen PR (2016) Comparative proteomics reveal distinct chaperone–client interactions in supporting bacterial acid resistance. Proc Natl Acad Sci USA 113:10872–10877. https://doi.org/10.1073/pnas.1606360113

Chapter 7
Bacterial Signal Peptidases

Mark Paetzel

Abstract Signal peptidases are the membrane bound enzymes that cleave off the amino-terminal signal peptide from secretory preproteins. There are two types of bacterial signal peptidases. Type I signal peptidase utilizes a serine/lysine catalytic dyad mechanism and is the major signal peptidase in most bacteria. Type II signal peptidase is an aspartic protease specific for prolipoproteins. This chapter will review what is known about the structure, function and mechanism of these unique enzymes.

Keywords Protein secretion · Signal peptide · Signal peptidase · Bacterial lipoprotein · Periplasmic protease

Introduction

The basic elements of the general protein secretion system are conserved in all cells; archaeal (Pohlschroder et al. 2018), bacterial (Chatzi et al. 2013; Driessen and Nouwen 2008; Tsirigotaki et al. 2017) and eukaryotic (Nyathi et al. 2013; Voorhees and Hegde 2016). A key feature of this system is the signal peptide. The signal peptide or signal sequence is approximately 20–30 residues in length and located at the amino-terminus of secretory proteins (preproteins). The signal peptide is essential for targeting to the cytoplasmic membrane or endoplasmic reticulum membrane and also translocation across the membrane. Upon translocation of secretory proteins, the signal peptide is cleaved off.

The protein machinery of the general secretion system helps catalyze the targeting and translocation events. Bacterial protein secretion occurs predominantly as a post-translational translocation event. In other words, preproteins are fully synthesized and released from the ribosome within the cytoplasm, and then targeted to and translocated across the cytoplasmic membrane. Typically, the preprotein first encounters the protein SecB that functions as a holdase, keeping the preprotein sol-

M. Paetzel (✉)
Department of Molecular Biology and Biochemistry, Simon Fraser University, South Science Building 8888 University Drive, Burnaby, BC V5A 1S6, Canada
e-mail: mpaetzel@sfu.ca

A. Kuhn (ed.), *Bacterial Cell Walls and Membranes*, Subcellular Biochemistry 92, https://doi.org/10.1007/978-3-030-18768-2_7

uble and in a translocationally competent form that can fit through the translocation channel SecYEG. The ATPase SecA feeds the preprotein through SecYEG stepwise. The energy for translocation across the membrane comes from ATP hydrolysis and the proton motive force. Once the carboxy-terminal region of the preprotein has been pushed through the translocase, the signal peptide is thought to partition into the lipid bilayer via a lateral gate in the SecYEG channel. The preprotein is then tethered to the membrane via the signal peptide. This membrane-tethered preprotein is the natural substrate of signal peptidase (SPase). It has been shown that removal of the signal peptide is essential in order for secretory proteins to be released from the cytoplasmic membrane (Dalbey and Wickner 1985). Limited proteolysis (Randall 1983) and deletion studies (Kaderbhai et al. 2008) are consistent with the carboxy-terminal region of preproteins being nearly or completely translocated across the cytoplasmic membrane before the signal peptide is cleaved off.

There are two types of bacterial SPase referred to as type I and type II. SPase I cleaves off the signal peptide from the majority of secreted proteins. It is an endopeptidase that utilizes a serine nucleophile and has the enzyme commission number EC 3.4.21.89. In the MEROPS protease database it is in clan SF and the family S26. Archaea also have an SPase I, but their sequences reveal that they maybe more related to a component of the eukaryotic signal peptidase complex. SPase II cleaves off the signal peptide from bacterial lipoproteins, proteins that associate with the membrane via an amino-terminal lipid modification. SPase II is an endopeptidase that utilizes aspartate catalytic residues and has the enzyme commission number EC 3.4.23.36. In the MEROPS protease database, it is in clan AC and the family A8. No homolog of SPase II has been found in an archaeal genome. Before discussing the research related to the structure, function and mechanism of the bacterial and archaeal SPase, the preprotein signal peptide will be briefly described.

The signal hypothesis states that secretory proteins contain an amino-terminal sequence (signal sequence) which helps the target secretory protein to the membrane and that after translocation across the membrane, the signal sequence is cleaved off by a specific peptidase (signal peptidase, SPase) (Blobel and Dobberstein 1975a, b; Milstein et al. 1972; Sabatini et al. 1971). Since this groundbreaking discovery, a great deal has been learned about the structure and function of signal peptides.

The general features of signal peptides are conserved across evolution. They have an amino-terminal region of sequence (1–5 residues in length) called the n-region that has a net-positive charge. The n-region is followed by a stretch of 7–15 hydrophobic residues (often leucines) labeled the h-region. Just beyond the h-region are 3–7 residues termed the c-region. This c-region contains the specificity residues for SPase recognition and cleavage. The consensus sequence for the cleavage site consists of small aliphatic residues at the −3 and −1 positions relative to the cleavage site (or P1 and P3 positions in Schechter and Berger nomenclature (Berger and Schechter 1970). This is often called the Ala-X-Ala rule because of the preference for alanine at the −3 and −1 position (Perlman and Halvorson 1983; von Heijne 1983, 1985), while there is a lack of preference for specific residues at the −2 position, thus the X designation (Fig. 7.1a). Signal peptide length is conserved as well. The average eukaryotic signal peptide is approximately 23 residues in length, while the average

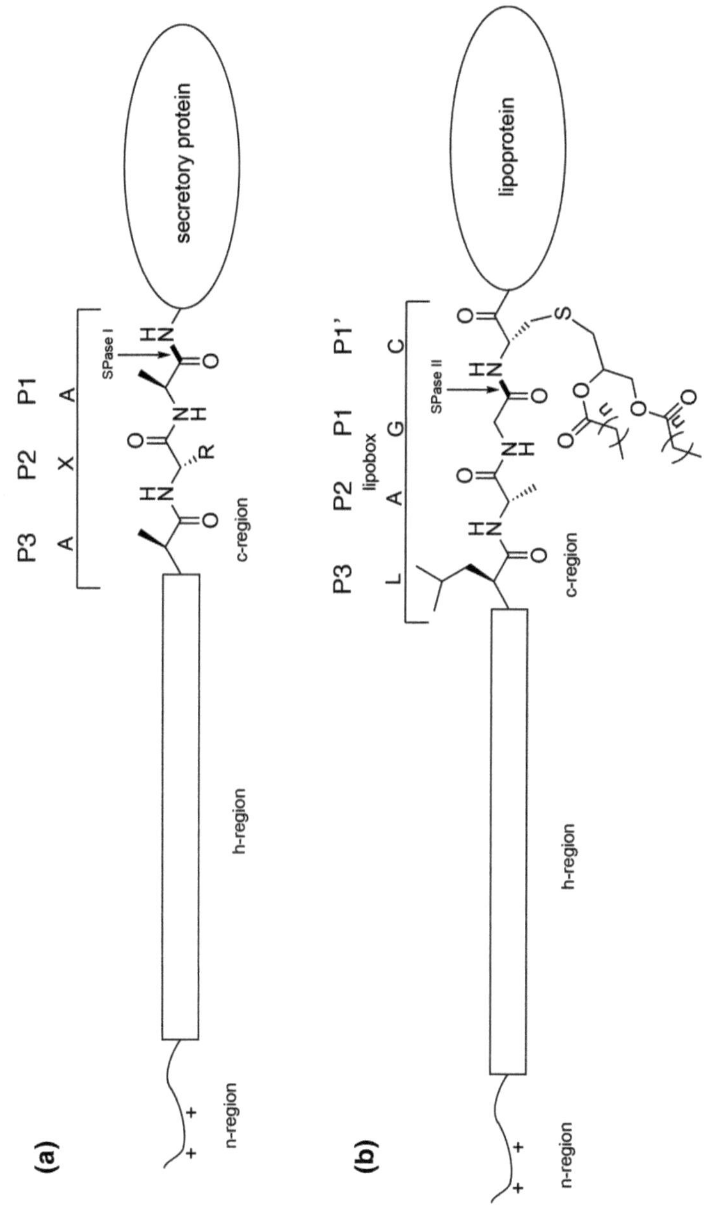
(a)
P3 P2 P1
A X A
SPase I
secretory protein
n-region
h-region
c-region
(b)
P3 P2 P1 P1'
L A G C
lipobox
SPase II
lipoprotein
n-region
h-region
c-region

◀**Fig. 7.1** Bacterial secretory preproteins. **a** Secretory proteins have three regions within their signal peptide: the n-region has a net-positive charge, the h-region contains predominantly aliphatic residues such as leucine, and the c-region contains the specificity cleavage site for signal peptidase. Small aliphatic residues such as alanine are preferred at the P1 and P3 positions (also referred to as the −1 and −3 positions). The molecular structure of the last three residues of the c-region are drawn in an extended conformation in order to show the alternating up and down topology that leads to the "Ala-X-Ala" SPase preference. The h-region is shown as a rectangle to symbolize an α-helical conformation. **b** Bacterial prolipoproteins are synthesized with a signal peptide that has similarities to secretory proteins sequences with the exception that the c-region contains the so-called "lipobox" with the consensus sequence leucine(P3)-alanine(P2)-glycine(P1)-cysteine(P1^/). The P1^/ cysteine is covalently modified by diacylglycerol. This is the substrate of SPase II

Gram-negative bacterial signal peptide is approximately 25 residues in length and the average Gram-positive bacterial signal peptide is approximately 32 residues in length (Nielsen et al. 1997a, b). It has been hypothesized that the differences observed in signal peptide length may reflect the differences in the thickness of the lipid bilayer in the different membranes.

Researchers continue to develop increasingly more accurate signal peptide prediction servers (Choo et al. 2009; Leversen et al. 2009; Petersen et al. 2011; Savojardo et al. 2018; Wang et al. 2018). The availability of genome sequences (UniProt 2013) and improved proteomic methods have vastly increased the number of predicted and experimentally verified signal peptide sequences. Biophysical and modeling studies are consistent with the h-region of the signal peptide being helical within the hydrophobic environments of detergent micelles or membrane (Bechinger et al. 1996; McKnight et al. 1991a, b; Rizo et al. 1993; Sankaram et al. 1994; Wang et al. 1993). Modeling and structural analysis is consistent with the c-region of the signal peptide being in the extended β-strand conformation (Paetzel et al. 1998, 2002; Ting et al. 2016). The substrate cleavage-site being in a β-conformation is the classical conformation within protease binding sites (Tyndall et al. 2005). The structural and biophysical information now available regarding the conformation of signal peptides is in agreement with the prediction that was made in 1983 (von Heijne 1983).

Occasionally, unusual signal peptide variants are recognized. For example, it has been observed that in *Porphyromonas gingivalis* and other *Bacteroidetes* species >50% of the preproteins have a glutamine at the P1′ position (Bochtler et al. 2018). This generates an amino-terminal glutamine after SPase I processing. The newly formed amino-terminal glutamines are cyclized to pyroglutamate by glutaminyl cyclase. These pyroglutamate containing proteins do not appear to be specific to inner membrane proteins, periplasmic proteins, outer membrane proteins or extracellular proteins. It is not yet clear what role the amino-terminal pyroglutamate plays in the *Bacteroidetes* physiology but it has been shown that glutaminyl cyclase is essential in these species.

There has been significant effort put into the optimization of signal peptides for the secretion of industrially important recombinant proteins (Freudl 2018; Ghahremanifard et al. 2018; Low et al. 2013; Molino et al. 2018; Selas Castineiras et al.

2018; Ujiie et al. 2016; Zhang et al. 2018). A recent comprehensive review on signal peptides has been published (Owji et al. 2018).

Bacterial secretory proteins that utilize bound cofactors often need to fold within the cytoplasm in order to assemble properly and bind their cofactor. These folded proteins use the twin arginine translocation (Tat) secretion system rather that the general secretion system. The name comes from the sequence within the signal peptide. These signal peptides have a so-called twin-arginine motif (SRRxFLK) located between the n- and h-regions. There are 27 known Tat substrates in *E. coli* (Tullman-Ercek et al. 2007) and it has been shown that SPase I cleaves off Tat signal peptides (Luke et al. 2009).

Archaea have an evolutionarily similar general secretion system to that seen in bacteria and eukaryotes which includes similar signal peptide characteristics (Bardy et al. 2003). The similarity allows for bacterial SPase I to recognize and cleave archaeal preproteins. *E. coli* SPase I is capable of cleaving pre-α-amylase from *Thermococcus kodakarensis* (Muhammad et al. 2017). Genomic analysis suggests that protein secretion in archaea may be more similar to that of eukaryotes than to prokaryotes. For example, archaea lack a SecA (ATPase) homolog, therefore archaeal translocation maybe more energetically similar to the eukaryotic system where co-translational translocation is the main mechanism of transport across the membrane. Consistent with this idea, archaea also appear to lack a post-translational holdase type of targeting chaperone like SecB (Pohlschroder et al. 2018). Subtle differences in archaeal signal peptides have led researchers to develop a signal peptide prediction server trained on a set of characterized archaeal signal peptides (Bagos et al. 2009).

Bacterial lipoproteins are soluble hydrophilic proteins located in the bacterial cell wall. These proteins are tethered or anchored to the inner or outer membrane by permanent covalent lipid modifications. They have a diacylglycerol linked to an amino-terminal cysteine via a thioether linkage. Many of the bacterial lipoproteins perform functions that are essential to cellular viability. They have roles in: membrane and cell shape maintenance, outer membrane protein assembly, outer membrane stabilization, transport of molecules, energy production, signal transduction, virulence mechanisms, adhesion, digestion, sensing, growth, cell motility and many more cellular functions.

Bacterial lipoproteins are initially synthesized as preprolipoproteins that have a signal sequence for targeting and translocation via the general secretion system pathway (Driessen and Nouwen 2008) (Fig. 7.1b). The signal peptides of lipoproteins are very similar to those for secretory proteins in that they are approximately 20 residues in length and have a net positively charged region (n-region) followed by a stretch of hydrophobic residues (h-region) and then a recognition sequence region (c-region). Within the c-region is the SPase II recognition sequence called the "lipobox". The consensus sequence for the lipobox is leucine most often at the −3 (P3) position relative to the cleavage site, alanine most often at the −2 (P2) position, glycine most often at the −1 (P1) position and cysteine always at the +1 (P1′) position. The sequence variations that are observed in the lipobox are [**LVI**] [**A**STVI][**G**AS]**C** (Chimalapati et al. 2013; Kovacs-Simon et al. 2011; LoVullo et al. 2015; Nakayama et al. 2012; Narita and Tokuda 2017; Zuckert 2014). The +1 (P1′) cysteine is modified with diacylglycerol (DAG) via a thioether linkage. This reaction is catalyzed by lipoprotein diacylglyceryl transferase (Lgt) and occurs in the inner membrane using the phospholipid within the inner membrane as substrate (Mao et al. 2016; Sankaran et al. 1997). The DAG-modified protein is then referred to as a prolipoprotein and is the substrate for type II signal peptide (SPase II) which cleaves off the signal peptide (Chimalapati et al. 2013). The scissile bond is the peptide bond preceding the invariant cysteine within the lipobox. Most SPase II will not cleave preprolipoproteins (protein without DAG-modification) (Inouye et al. 1983). For many Gram-positive bacterial species this is the end of lipoprotein synthesis; these mature lipoproteins are tethered to the outer leaflet of the cytoplasmic membrane, preventing them from diffusing away from the surface of the bacteria (Nguyen and Gotz 2016). For Gram-negative species another modification can occur. A fatty acid is attached, via an amide linkage, to the α-amino group of amino-terminal cysteine. This reaction is catalyzed by the enzyme lipoprotein N-acyl transferase (Lnt) (Buddelmeijer and Young 2010; Hillmann et al. 2011; Noland et al. 2017). The mature lipoprotein with three acyl-chains can then be shuttled to the outer membrane via the lipoprotein outer membrane localization (LOL) machinery (Konovalova and Silhavy 2015; Tokuda and Matsuyama 2004), or stay within the inner membrane if there is a retention signal which often consists of an aspartic acid at the +2 (P2′) position (Seydel et al. 1999; Terada et al. 2001).

The database DOLOP lists and categorizes both predicted and experimentally verified bacterial lipoproteins (Babu et al. 2006; Madan Babu and Sankaran 2002), but this database has not been updated since 2005. A video procedure for the purification of lipoproteins via Triton X-114 detergent extraction and electrophoresis, followed by chemical characterization by proteolysis, solvent extraction and mass-spectrometry is available (Armbruster and Meredith 2018). There are many recent reviews on bacterial lipoproteins (Hutchings et al. 2009; Konovalova and Silhavy 2015; Kovacs-Simon et al. 2011; LoVullo et al. 2015; Narita and Tokuda 2017; Zuckert 2014). For a review on the chemical characterization of both Gram-negative and Gram-positive bacterial lipoproteins by mass-spectrometry see the review by Nakayama et al. (2012). A lipoprotein signal peptide prediction server called LipoP is available (Juncker et al. 2003; Rahman et al. 2008). Great progress has been made in understanding the enzymes in the bacterial lipoprotein synthetic pathway and crystal structures have been solved for each of the enzymes involved: Lgt (Mao et al. 2016), Lnt (Noland et al. 2017) and SPase II (Vogeley et al. 2016). Only SPase II will be discussed in this chapter.

Gram-Negative Type 1 Signal Peptidase

Escherichia coli SPase I was the first bacterial SPase I to be studied and is the most thoroughly characterized SPase to date. The first observation of bacteria SPase I activity was in *E. coli* membranes using a nascent bacteriophage f1 pre-coat protein as the substrate (Chang et al. 1978). *E. coli* SPase I was also the first bacterial SPase I to be purified. Bacteriophage M13 procoat protein was used as the substrate to assess the 6,000 fold purification (Zwizinski and Wickner 1980). A proteoliposome assay showed that only purified *E. coli* SPase I and phospholipid were needed for preprotein processing (Watts et al. 1981). The pH optimum of *E. coli* SPase I is 8.5–9.0 and it is inhibited by NaCl concentrations above 160 mM and $MgCl_2$ concentrations above 1 mM. The enzyme has been shown to be essential for cell growth (Date 1983). *E. coli* SPase I is the product of the *lepB* gene, is 324 residues in length, and has a calculated molecular mass of 35,960 Da and a theoretical isoelectric point of 6.9. Quantitative western blot analysis is consistent with approximately 1000 SPase I molecules per cell (van Klompenburg et al. 1995). The current residue numbering system for *E. coli* SPase I is different by one residue from the numbering system used in earlier work. This is due to an error in the originally reported sequence of the *E. coli* enzyme (Wolfe et al. 1983); Arg42 in the originally reported sequence is actually Ala42. In addition there is glycine residue (Gly43) just after the Ala42 that was not in the originally reported sequence. The insertion moves the sequence up one number after that point. Therefore, in the current system, residues 1–41 are consistent with the old numbering and residues 44–324 are correct in sequence but the numbering is one residue different from the old numbering. The sequencing error corresponds to residues within the cytoplasmic region, between the two transmembrane segments. The numbering system used in this chapter matches that in the UniProt sequence database (accession number: P00803).

Protease accessibility assays show that *E. coli* SPase I is an integral membrane protein and the majority of the protein chain is on the outside of the cytoplasmic membrane (Moore and Miura 1987; Wolfe et al. 1983). Enzyme-fusion assays are consistent with both the amino- and carboxy-termini facing the periplasm (San Millan et al. 1989). The interface between the two transmembrane segments in *E. coli* SPase I has been modeled based on disulfide mapping (Whitley et al. 1993). It was found that residue pairs 3–76, 4–76, 4–77, –4–80, and 7–76 are likely in close proximity. Modeling is consistent with the two transmembrane helices packing against each other in a left-handed supercoil with an i, i + 3/i, i + 4 grooves-into-ridges arrangement. Site-directed mutagenesis and deletion studies show that the first transmembrane segment (residues 4–28) and cytoplasmic region (residues 29–57) are not essential for catalytic activity. The second transmembrane segment (residues 58–76) is essential for activity and for initiating the translocation of the periplasmic region (residues 77–324) where the protease active site resides (Bilgin et al. 1990) (Fig. 7.2). It has been shown that highly purified and concentrated *E. coli* SPase I will cleave itself in a intermolecular fashion within its cytoplasmic region (Talarico et al. 1991). Since this cleavage site region is on the opposite side of the cytoplasmic membrane from the SPase I active site, it is not likely this is a common occurrence in vivo, although disruption of the membrane or mistakes in SPase I assembly could allow for this type of cleavage.

Early studies tried to classify the mechanism of *E. coli* SPase I based on sensitivity to standard protease inhibitors (Zwizinski et al. 1981). These experiments were inconclusive and suggested that this enzyme may utilize a unique mechanism. Site-directed mutagenesis of conserved ionizable residues was consistent with Ser91

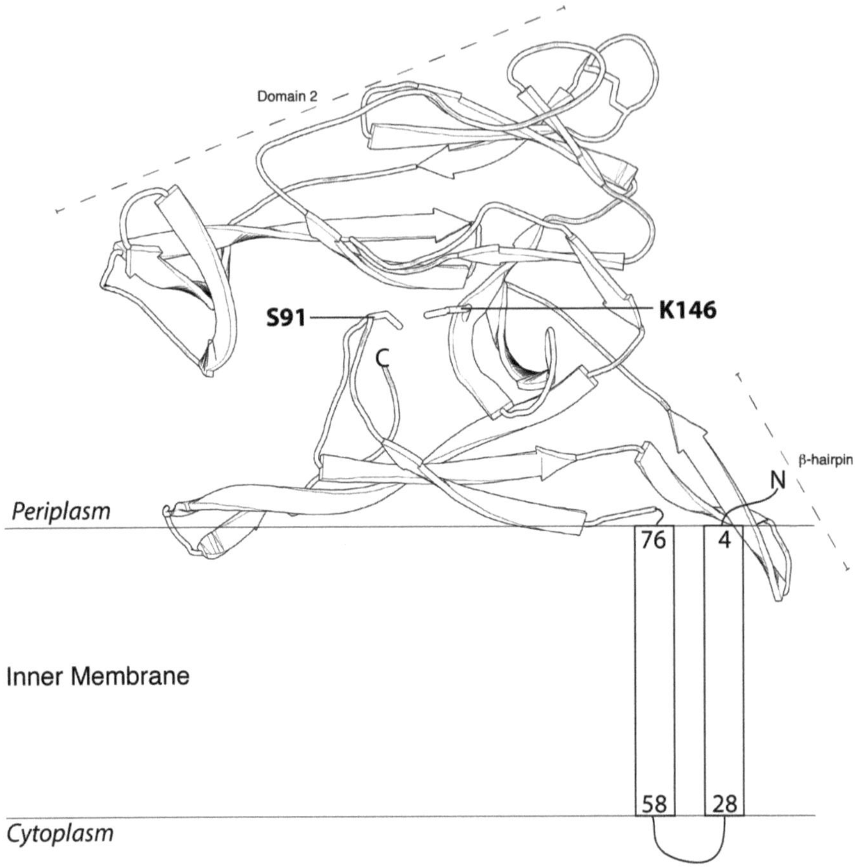

Fig. 7.2 The membrane topology and protein fold of *E. coli* SPase I. The transmembrane segments, not part of the crystal structure, are represented by rectangles. The nucleophilic Ser91 and general base Lys146 are labeled. The β-ribbon and domain 2 region not normally observed in the Gram-positive SPase I enzymes are labeled (PDB: 1B12)

and Lys146 being essential for catalytic activity (Black 1993; Black et al. 1992; Paetzel et al. 1997; Sung and Dalbey 1992; Tschantz et al. 1993). Therefore, a Ser/Lys catalytic dyad mechanism was proposed.

The first three-dimensional structure of a SPase I was that of *E. coli* SPase I (Paetzel et al. 1998). A soluble catalytically active enzyme was produced, lacking the two amino-terminal transmembrane segments (Δ2-76) (Tschantz et al. 1995). This active periplasmic region of *E. coli* SPase I produced ordered diffraction quality crystals (Paetzel et al. 1995). Crystal structures of this enzyme are available in complex with a β-lactam type of inhibitor (Paetzel et al. 1998), in complex with a lipopeptide base inhibitor called arylomycin (Paetzel et al. 2004), in complex with arylomycin and β-sultam inhibitor (Luo et al. 2009), and in complex with a lipoglycopeptide version of arylomycin (Liu et al. 2011a). In addition, a structure of *E. coli* SPase I with a free active site is available (Paetzel et al. 2002).

The crystal structures reveal that SPase I is primarily constructed of β-sheets. There are two antiparallel β-sheet domains (domain I and domain II) and an extended β-hairpin that extends from domain I (Fig. 7.2). Domain I contains the catalytic residues and has a similar protein fold to that of the proteinase domain in the UmuD protein of the SOS mutagenesis system (Paetzel and Strynadka 1999). Sequence alignments reveal that domain II is much smaller or missing in Gram-positive SPase I. A disulfide bond (Cys171–Cys177) resides within domain II of *E. coli* SPase I but mutagenesis has shown it to be nonessential (Sung and Dalbey 1992). Almost all of the conserved SPase I residues are located near the active site in domain I.

The catalytic center of SPase I resides at the end of a substrate binding groove that is provided by a missing β-stand of a would-be β-barrel within domain I. The strand and the loop region that makes up the edges of the binding groove run parallel to one another. Modeling studies based on the inhibitor complexes and the constraints related to the lipid bilayer of the inner membrane suggest that the c-region of the signal peptide within the preprotein substrate binds in a parallel β-strand conformation (Paetzel et al. 1998, 2002). This binding mode is in agreement with previous predictions (von Heijne 1983). The nucleophile Ser91 resides on a loop following the first β-stand. The general-base Lys146 resides on a β-strand with its Nζ atom within hydrogen bonding distance to the Ser91Oγ (Fig. 7.3). Lys146 Nζ is also within hydrogen bonding distance to Ser279Oγ. This is a conserved residue and site-directed mutagenesis experiments show that it is important for optimal activity (Klenotic et al. 2000). The oxyanion hole is constructed from the mainchain NH group of Ser91 and the side chain hydroxyl group of the Ser89. In vivo and in vitro analysis of site-directed mutants at this position are consistent with Ser89 playing a role in transition state stabilization (Carlos et al. 2000).

The substrate specificity binding pockets that lead to the active site explain the preference for alanine at the −1 (P1) and −3 (P3) position in preproteins (Fig. 7.3). The shallow and hydrophobic S1 substrate specificity binding pocket is made of atoms from the residues: Met92, Ile145, Leu96 and Ile87. The shallow and hydrophobic S3 substrate specificity binding pocket is constructed with atoms from residues: Phe85, Ile87, Ile102, Val133, Ile145 and Asp143. Site-directed mutagenesis, mass-spectrometry and molecular modeling were used to probe the importance of residues in the SPase I binding pockets. It was observed that Ile87 and Ile145, residues that divide the S1 and S3 binding pockets (Fig. 7.3), are important for specificity as well as cleavage site fidelity (Ekici et al. 2007; Karla et al. 2005). NMR analysis has recently been employed to look at the details of the interactions between SPase I and the signal peptide (De Bona et al. 2012; Musial-Siwek et al. 2008a, b).

E. coli SPase I has an extensive hydrophobic surface that runs along domain I and leads to the active site. This hydrophobic surface and the orientation of active site residues provide clues as to how the catalytic domain lays on the membrane surface. Chemical modification and site-directed mutagenesis experiments are consistent with Trp301 and Trp311 being important for activity and these residues reside along the hydrophobic surface (Kim et al. 1995a, b). Biophysical analysis has shown

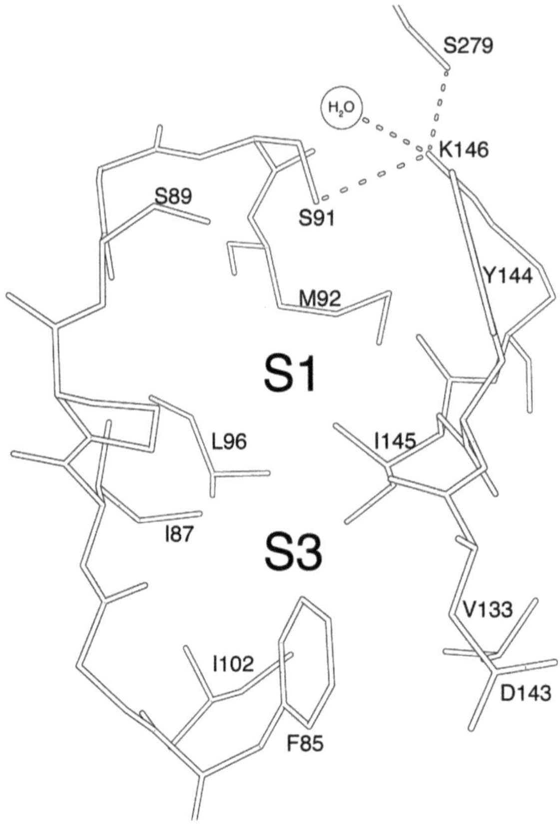

Fig. 7.3 The active site region of *E. coli* SPase I. The regions corresponding to the S1 and S3 substrate specificity pockets are labeled. Dashed lines correspond to important hydrogen bonds within the active site. The coordinates used are from a structure with a free active site. An ordered, and potentially catalytic, water at the correct distance and angle to the general base and modeled scissile carbonyl is labeled (PDB: 1KN9)

that the catalytic domain of *E. coli* SPase I penetrates into the membrane and that phosphatidylethanolamine, the most abundant phospholipid in the *E. coli* inner membrane, mediates the association (van Klompenburg et al. 1997, 1998). Kinetic analysis shows that detergent or lipid is required for optimal activity of the soluble catalytic domain of *E. coli* SPase I even though it lacks the transmembrane domains (Tschantz et al. 1995).

Significant effort has been invested into SPase I inhibitor development. The first reported inhibitor of SPase I was a β-lactam compound (Kuo et al. 1994), and a number of penem type inhibitors have been developed (Allsop et al. 1995, 1996; Barbrook et al. 1996; Black and Bruton 1998; Perry et al. 1995). The first crystal structure of *E. coli* SPase I was solved with the compound allyl (5S,6S)-6-[(R)-acetoxyethyl]-penem-3-carboxylate (Fig. 7.4a) covalently bound to the Oγ of Ser91, thus proving directly for the first time its role as the nucleophile (Paetzel et al. 1998). The acyl-enzyme inhibitor complex also revealed that the nucleophile of *E. coli* SPase I attacks from the *si*-face of the scissile bond rather than the more commonly observed *re*-face attack. Work has continued on the development of the penem SPase I inhibitors (Harris et al. 2009; Yeh et al. 2018). Lipohexapeptides called arylomycins, which have antibiotic properties, have been discovered to inhibit SPase I (Fig. 7.4b). These compounds were first isolated from extracts of *Streptomyces sp.* Tu 6075 and have the sequence: D-MeSer, D-Ala, Gly, L-MeHpG, L-Ala, and L-Tyr. The amino acid L-MeHpG is N-methyl-4-hydroxy-phenylglycine (Holtzel et al. 2002; Schimana et al. 2002). The aromatic ring of L-MeHpG is covalently bonded to the aromatic ring of L-Tyr to form a three-residue macrocycle via a 3,3-biaryl bridge. A fatty acid is attached to the amino-terminus. Crystallographic and biophysical analysis reveals that arylomycin binds to *E. coli* SPase I in a non-covalent fashion (Fig. 7.4b) (Paetzel et al. 2004). Analogs of arylomycin inhibitors have also been developed (Dufour et al. 2010; Liu et al. 2011b; Roberts et al. 2007, 2011a, b; Smith et al. 2010, 2011, 2018; Smith and Romesberg 2012). Other reported inhibitors of SPase I include boronic ester-linked macrocyclic lipopeptides (Szalaj et al. 2018) and aldehyde containing lipopeptides (Buzder-Lantos et al. 2009; De Rosa et al. 2017).

The crystal structures, site-directed mutagenesis and kinetic analysis are all consistent with the following proposed catalytic mechanism for *E. coli* SPase I. The c-region of the signal peptide binds within the SPase I substrate binding groove (Michaelis complex) which buries the catalytic residues and likely lowers the pK_a of the lysine general base, Lys146. Ser91 serves as the nucleophile and is activated by the abstraction of its hydroxyl hydrogen by the deprotonated Nζ of the general base Lys146. The Ser279Oγ is within hydrogen bonding distance and helps orient the Nζ of Lys146 towards Ser91 (Fig. 7.3). The nucleophilic attack from the *si*-face of the scissile carbonyl by the activated Ser91 Oγ results in the tetrahedral oxyanion transition state 1. The oxyanion is stabilized via hydrogen bonds to the oxyanion hole (Ser91NH and Ser89OH). The protonated Lys146 Nζ donates a proton to the main chain nitrogen of the leaving group (P′ side of the scissile bond, the mature region of the secretory protein, product 1). The main chain carbonyl carbon of the P1 residue of the signal peptide is then covalently attached via an ester bond to the Ser91 Oγ (acyl-enzyme intermediate). A nucleophilic water (a.k.a. deacylating or catalytic water) is activated via the general base Lys146. The structure of *E. coli* SPase I with a free active site revealed a likely candidate for this water (Paetzel et al. 2002) (Fig. 7.3). The hydroxyl is in position to attack the carbonyl carbon of the ester bond. This forms the tetrahedral oxyanion transition state 2, stabilized via hydrogen bonds to the oxyanion hole. This state then leads to the regeneration of the enzyme and release of product 2.

Although *E. coli* SPase I is the most characterized signal peptidase there has been a significant amount of research on other Gram-negative type 1 signal peptidases. Most Gram-negative species have a single SPase I gene that codes for a membrane bound enzyme with two transmembrane segments but there are some exceptions. Photosynthetic bacteria have two types of membranes; the cytoplasmic membrane and the thylakoid membrane. Genomic analysis of the photosynthetic bacteria *Synechocystis sp.* strain PCC 6803 has revealed two SPase I like genes (sll0716 or lepB1 and slr1377 or lepB2) (Zhbanko et al. 2005). Gene knockout analysis was used to show that lepB2 is essential for cell viability and LepB1 is essential for photoautotrophic growth. It will be interesting to look at the relative distribution of these enzymes in the two different membrane systems of this cyanobacterium given that

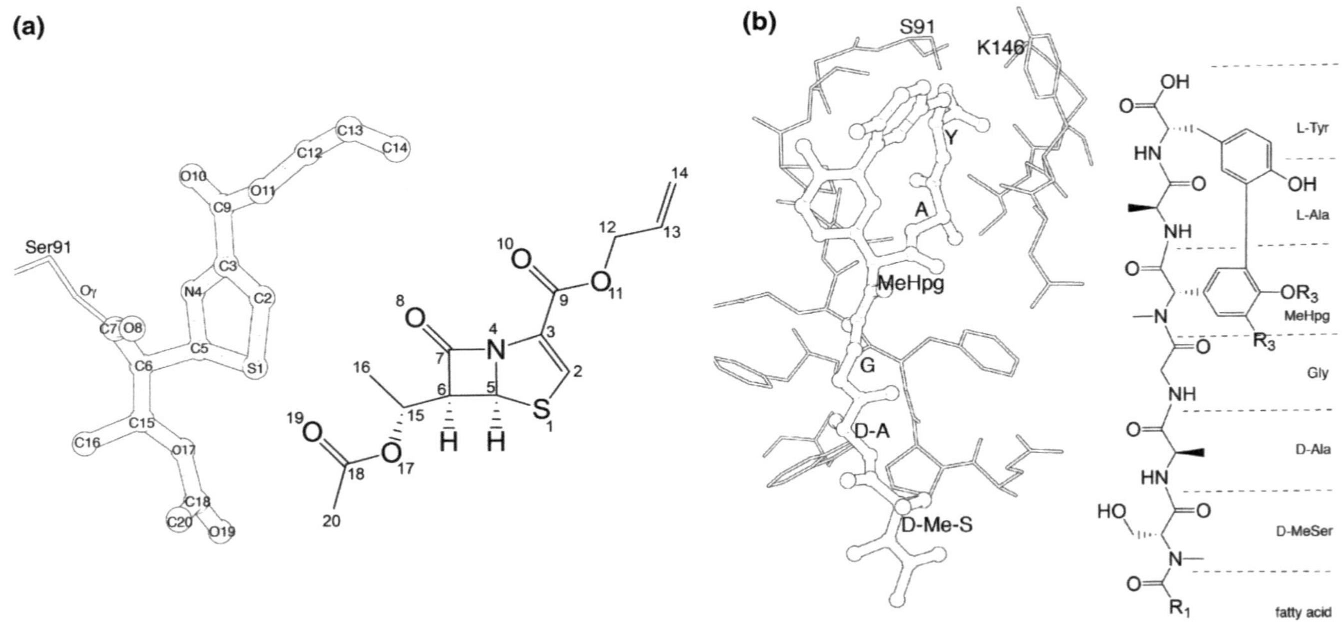

Fig. 7.4 Inhibitors of SPase I. **a** The β-lactam type inhibitor, allyl (5S,6S)-6-[(R)-acetoxyethyl]-penem-3-carboxylate (PDB: 1B12). **b** The lipopeptide based inhibitor arylomycin (PDB: 1T7D)

proteomic analysis observes some potential crossover in activity and leaves open the possibility that there maybe other SPase I enzymes. SPase I activity has also been observed from solubilized membranes of the phototrophic α-purple bacterium *Rhodobacter capsulatus* (Wieseler et al. 1992).

The *Pseudomonas aeruginosa* genome contains the SPase I genes PA0768 and PA1303 (Waite et al. 2012). In vitro assays using FRET-peptide substrates showed that both enzymes are active. PA1303 is smaller than a typical Gram-negative SPase I and is non-essential. PA0768 has more of a typical length for a Gram-negative SPase I and is essential for cell growth. Site-directed mutagenesis suggests that PA1303 plays a role in the quorum-sensing cascade and that PA0768 is the main SPase I in *P. aeruginosa*. *Bradyrhizobium japonicum* is a soil bacterium that fixes nitrogen and is yet another Gram-negative bacterium with two SPase I genes (SipS and SipF) (Bairl and Muller 1998). These enzymes are also different in that they are predicted to have a single amino-terminal transmembrane segment, similar to the Gram-positive SPase I enzymes. The thermophilic cyanobacterium *Phormidium laminosum* also has a SPase I with a predicted single amino-terminal transmembrane segment (Packer et al. 1995). *Legionella pneumophila*, the facultative intracellular Gram-negative bacterium that causes Legionnaires' disease, has a SPase I with a unique sequence feature (Lammertyn et al. 2004). The conserved methionine that immediately follows the nucleophilic serine is a leucine. The structure of *E. coli* SPase showed that this residue is located right behind the Ser/Lys catalytic dyad (Paetzel et al. 1998) (Fig. 7.3).

Other Gram-negative species whose type 1 signal peptidase activity has been confirmed by in vivo complementation assays includes: *Azotobacter vinelandii* (Jock et al. 1997), *Bordetella pertussis* (Smith et al. 2000), *Pseudomonas fluorescens* (Black et al. 1992), *Salmonella typhimurium* (van Dijl et al. 1990), *Rickettsia rickettsii* and *Rickettsia typhi* (Rahman et al. 2003).

Gram-Positive Type 1 Signal Peptidase

Unlike most Gram-negative species that possess a single gene for SPase I, Gram-positive bacteria tend to have multiple genes for SPase I. Sequence alignments reveal that Gram-positive SPase I are shorter in length than the Gram-negative SPase I. They are typically missing part of domain II in their extracytoplasmic region and they tend to have a single transmembrane segment at their amino-terminus rather than the two transmembrane segments normally observed in the Gram-negative SPase I. The best characterized Gram-positive SPase I are those from *Bacillus subtilis* and *Staphylococcus aureus*.

Bacillus subtilis SipS was the first Gram-positive SPase I to be characterized in detail (van Dijl et al. 1992). All together, *Bacillus subtilis* has five chromosomally expressed SPase I enzymes (SipS, SipT, SipU, SipV, and SipW) (Bolhuis et al. 1996; Tjalsma et al. 1997) and two plasmid expressed SPase I enzymes (SipP) (Meijer et al. 1995; Tjalsma et al. 1999b). Site-directed mutagenesis experiments are consistent with the SipS utilizing a Ser/Lys catalytic dyad mechanism where Ser43 is the nucleophile and Lys83 serves as the general base. Interestingly it was also found that the mutant M44A showed increased activity (van Dijl et al. 1995). It has been shown that all of the *sip* gene products function as type 1 signal peptidases but only cells lacking both SipS and SipT are not viable (Tjalsma et al. 1998). SipW is very unique in this group of SPase I enzymes in that it has a histidine in place of the general base lysine (Tjalsma et al. 2000). This suggests it is more like the Sec11-like component of the eukaryotic ER signal peptidase complex or the archaeal SPase I (discussed below). SipW is required for processing of the spore-associated protein TasA (Tjalsma et al. 2000). SPase I genes from other *Bacillus* species have been characterized including: *Bacillus amyloliquefaciens* (Chu et al. 2002; Hoang and Hofemeister 1995; van Roosmalen et al. 2001), *Bacillus licheniformis* (Cai et al. 2016) and *Bacillus megaterium* (Malten et al. 2005; Nahrstedt et al. 2004).

Staphylococcus aureus has two SPase I-like genes (Cregg et al. 1996). The main SPase I is coded for by the *spsB* gene and is essential. Immediately preceding the *spsB* gene is another SPase I-like gene *spsA* but it lacks the catalytic residues. The nucleophilic serine is replaced by an aspartic acid and the lysine general-base is replaced by a serine. SpsB is involved in quorum sensing. It is responsible for the amino-terminal processing of the autoinducing peptide (AIP) molecule precursor AgrD (Kavanaugh et al. 2007). Kinetic analysis using both preprotein and peptide cleavage assays shows SpsB has a pH optimum around 8 and apparent pK_a values of 6.6 and 8.7. Curiously, it has been shown that SpsB is not absolutely essential. *S. aureus* is able to survive without SpsB by over-expressing a native gene cassette that encodes a putative ABC transporter. Apparently it compensates for the lack of a SPase I by cleaving a subset of secretory proteins at a site distinct from the SpsB-cleavage site (Craney and Romesberg 2017; Hazenbos et al. 2017; Morisaki et al. 2016).

SpsB is the only SPase I besides *E. coli* SPase I to be crystallographically characterized (Ting et al. 2016). Protein engineering was used to make a crystallizable construct with a peptide substrate tethered for presentation to its binding site (Fig. 7.5a). The catalytic domain of SpsB was fused with *E. coli* maltose binding protein (MBP) to promote solubility, stability, purification and crystallization. The construct included a poly-histidine-affinity tag followed by MBP (residues 33–393), followed by a three-residue linker (Ala-Gly-Ala) and then residues 26–191 of SpsB. The carboxy-terminal residues of SpsB (176–191) that are disordered in the free active site structure were replaced with a Strep-tag II (WSHPQFEK). Four different structures of SpsB were determined; a free active site structure (PDB: 4wvg) with the nucleophilic serine mutated to an alanine (S36A) and three structures with peptides bound and with the wild type residue at position 36. The peptides were tethered covalently to the MBP at an engineered cysteine residue (Q78C) via a thioether linkage. The covalent tether increased the local effective concentration of the peptide and led to high occupancy of the peptide within the binding site and clear electron density for the bound peptide. The peptides (peptide 1: GGGGADHDAHA↓SET, peptide 2: GGGGAVPTAKA↓ASK, and peptide 3: GGGGGAPTAKAPSK) were synthesized with an amino-terminal N-bromo-acetyl moiety that reacted with the Sγ of the engineered cysteine. The structures of peptide 1 and peptide 2 bound to SpsB (PDB: 4wvh and 4wvi respectively) reveal that the substrate-like peptides have been cleaved at the scissile bond between the P1 and P1′ residues resulting in product complexes. The third peptide bound structure (PDB: 4wvj) has a noncleavable P1′ proline peptide (peptide 3) showing possible contacts that could be made by the P1′–P3′ residues of the substrate in a Michaelis complex. All of these complex structures reveal the methyl group of the P1 alanine pointing into the S1 binding pocket and the methyl group of the P3 alanine pointing into the S3 binding pocket. These structures are consistent with the binding mode previously hypothesized based on *E. coli* SPase I inhibitor complex structures (Liu et al. 2011a; Luo et al. 2009; Paetzel et al. 2002; Paetzel et al. 1998, 2004). SpsB is very similar to domain I of *E. coli* SPase I and the residues that makeup the active site and specificity binding pockets superimpose well. If the engineering of this construct does not interfere with the natural substrate cleavage site presentation, the SPase I-peptide complexes reveal the path that preprotein residues P6 to P3′ take along the enzyme surface (Fig. 7.5b).

There have been a number of other characterized Gram-positive type 1 signal peptidase including some from species that are associated with biofilm formation. *Staphylococcus epidermidis* has three SPase I genes (Sip1, Sip2, Sip3). The catalytic activity of Sip2 and Sip3 has been measured using both preprotein and synthetic peptide substrates. Sequence analysis shows that Sip1 does not have a lysine general base (Bockstael et al. 2009). The presence of SPase I paralogs without the full catalytic dyad has been observed in a number of species, such as *S. aureus* SpsA that was mentioned above, but it is not yet clear what role(s) these genes perform. *Streptococcus sanguinis*, contains two SPase I genes; SSA_0351 and SSA_0849. SSA_0351 has been shown to be essential for biofilm formation (Aynapudi et al. 2017). *Actinomyces oris* expresses two SPase I enzymes (LepB1 and LepB2). Mutational analysis is consistent with LepB2 being responsible for processing fimbrial proteins (Siegel et al. 2016).

Streptococcus pneumoniae SPase I (the *spi* gene product) has been expressed in *E. coli*, purified, and its activity characterized (Zhang et al. 1997). Site-directed mutagenesis and preprotein processing assays reveal the Ser38 and Lys76 of *S. pneumoniae* SPase I are essential for catalytic activity (Peng et al. 2001). Interestingly, biochemical studies provide evidence that this enzyme undergoes intermolecular self-cleavage that results in loss of activity and that this may play a role in regulation (Zheng et al. 2002).

Listeria monocytogenes contains three contiguous SPase I genes in its genome (SipX, SipY and SipZ). Gene knockout analysis has shown that SipZ is the major SPase I involved in the processing of most secretory proteins and that SipX and SipZ are specific for processing proteins related to pathogenicity (Bonnemain et al. 2004; Raynaud and Charbit 2005).

Streptomyces lividans contains four different chromosomally encoded SPase I genes (*sipW*, *sipX*, *sipY* and *sipZ*) (Parro et al. 1999). Sequence alignments and functional analysis are consistent with all four of the enzymes having a full set of catalytic residues and all four being able to process preproteins (Geukens et al. 2001b). The first of these genes to be cloned was that of sipZ (Parro and Mellado

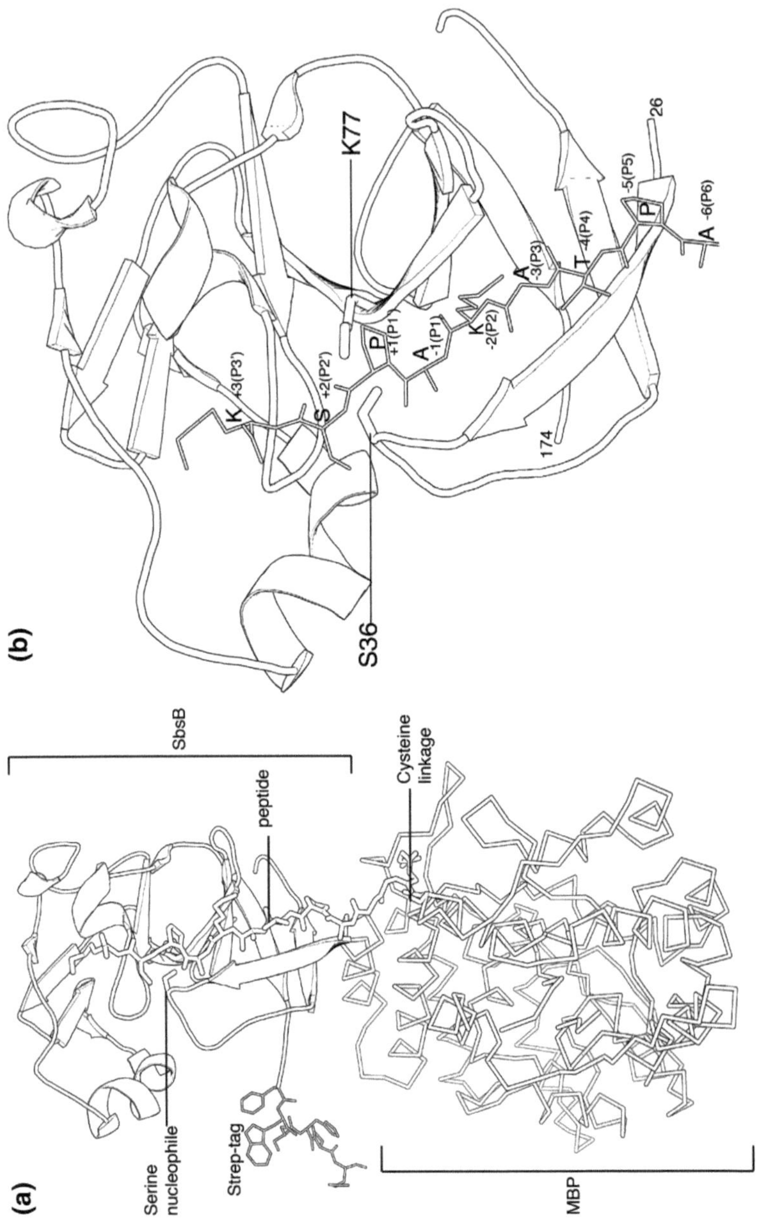
(a)
SbsB
peptide
Cysteine linkage
Serine nucleophile
Strep-tag
MBP
(b)
K77
S36
26
174
K +3(P3')
S +2(P2')
P +1(P1')
A -1(P1)
K -2(P2)
A -3(P3)
T -4(P4)
P -5(P5)
A -6(P6)

◀**Fig. 7.5** The crystal structure of the catalytic domain of *S. aureus* SPase I (SpsB). **a** Protein engineering was used to make a crystallizable construct of SpsB. Maltose Binding Protein (MBP) was engineered at the amino terminus of SpsB. A strep-affinity-tag replaced the C-terminal residues of SpsB. A series of peptides corresponding to substrate and inhibitor sequences were covalently linked to the MBP via an engineered cysteine, such that they presented the specificity residues to the SpsB substrate-binding groove. **b** A close up view of the SpsB binding groove with the P1′ proline peptide (PDB: 4wvj)

1998). Membrane topology analysis shows that SipY has one amino-terminal transmembrane segment and one carboxy-terminal transmembrane segment. Sequence analysis suggests that SipX and SipZ have a similar topology (Geukens et al. 2001a). Kinetic analysis shows that SipW and SipY have a pH optimum of 8–9 whereas SipX and SipZ have a pH optimum of 10–11. All four enzymes show improved activity in the presence of phospholipids (Geukens et al. 2002). Mutagenesis and proteomic analysis is consistent with SipY being the main SPase I for general protein secretion in *S. lividans* (Escutia et al. 2006; Palacin et al. 2002).

Archaeal Type 1 Signal Peptidase

Genome analysis has shown that archaea possess a gene with sequence similarity to bacterial SPase I, but the lysine general-base that is observed in bacterial SPase I is replaced by a histidine in archaeal SPase I (Eichler 2002). In addition, most archaeal SPase I lack a large region of domain II (Eichler 2002). This suggests that archaeal SPase I maybe similar to the Sec11 component of the signal peptidase complex (SPC) that is present in eukaryotic species (Ng et al. 2007). It is interesting that the bacterial genus *Bacillus* contain SPase I genes that are similar to the eukaryotic or archaeal SPase I homologs. As mentioned above, the *B. subtilis* SipW is a characterized SPase I whose sequence suggests that it utilizes a histidine rather that lysine as its general base (Tjalsma et al. 1998, 2000).

The first characterization of an archaeal type I signal peptidase was that of the methanogenic archaeon *Methanococcus voltae* (Ng and Jarrell 2003). The gene was cloned and expressed in *E. coli* and its activity analyzed using an in vitro assay with a truncated form of *M. voltae* S-layer protein as the substrate. This assay demonstrates that the archaeal SPase I gene product can function without the help of other components. Membrane topology prediction analysis suggests that this enzyme contains an amino-terminal and a carboxy-terminal transmembrane segment. Site directed mutagenesis studies on *M. voltae* SPase I show that Ser52 and His122 are essential residues and that Asp148 may also be essential (Bardy et al. 2005).

The halophilic archaeon *Haloferax volcanii* contains two SPase I genes in its genome (*sec11a* and *sec11b*). Gene deletion studies show that both of these genes are expressed but that only Sec11b is essential. Both enzymes were able to cleave preproteins but with different levels of efficiency (Fine et al. 2006). It is not yet clear why some arachaea have multiple SPase I paralogs. Of note, genomic analysis suggests that haloarchaea use the Tat pathway rather than the general secretion system for nearly half of their secreted proteins. Since the Tat pathway specializes in the translocation of fully folded proteins, this maybe an adaption to the extreme salt concentrations outside the cell that could prevent proper protein folding. Interestingly, both SPase I enzymes (Sec11a and Sec11b) in *H. volcanii* have an acidic isoelectric point (4.5) consistent with the negative charge seen on most of the *H. volcanii* cell surface proteins. The calculated isoelectric point for most archaeal SPase I are significantly more basic. For example, the isoelectric point of *M. voltae* SPase I is 9.0. Site-directed mutagenesis and in vitro preprotein cleavage assays reveal that the residues Ser72, His137 and Asp187 in *H. volcanii* Sec11b are essential (Fink-Lavi and Eichler 2008).

Type II Signal Peptidases

SPase II (also called type 2 signal peptidase, prolipoprotein signal peptidase, and LspA) catalyzes the removal of the amino-terminal signal peptide from bacterial proteins that are tethered to the inner membrane by the signal peptide and a diacylglycerol moiety. Many of the insights into the discovery and characterization of SPase II were facilitated by the cyclic pentapeptide (19-membered cyclic depsipeptide) globomycin (Fig. 7.6c). This natural product has four natural amino acid (glycine, L-serine, L-*allo*-isoleucine (2S,3R), L-*allo*-threonine (2S,3S)), N-methyl-L-leucine, and β-hydroxy-α-methyl carboxylic acid (α-methyl-β-hydroxynonanoyl). It is a product of non-ribosomal peptide synthesis in the Gram-positive mycelial bacteria *Actinomycetes*, first isolated in 1978, and shown to have antibacterial activity (Inukai et al. 1978a, b; Nakajima et al. 1978). It was revealed that globomycin's antimicrobial activity is the result of the inhibition of a prolipoprotein processing enzyme which converts prolipoprotein to lipoprotein (Inukai et al. 1978c). This was the first observation of SPase II activity. Globomycin was instrumental in the discovery that the cleavage of the signal peptide by SPase II is essential for the transport of lipoprotein to the outer membrane of Gram-negative bacteria (Hussain et al. 1980). Inhibition of SPase II results in the accumulation of prolipoprotein in the inner membrane which then leads to bacterial cell death (Hussain et al. 1980). This also helped confirm that modification with diacylglycerol occurs before cleavage of the signal peptide. Globomycin is a reversible noncompetitive inhibitor with a K_i of 36 nM (Dev et al. 1985). Total synthesis and crystallographic analysis led to the relative and absolute configurations of globomycin (Kogen et al. 2000). NMR analysis suggests that globomycin exists as a mixture of two rotational isomers in solution. Synthetic globomycin has the same antimicrobial activity as the naturally purified globomycin and changes to the serine position decreases antimicrobial activity. Solid-phase methods of globomycin synthesis has allowed for further optimization of this inhibitor (Sarabia et al. 2011). Intriguingly, another cyclic secondary metabolite from the Myxobacteria called TA (myxovirescin) also inhibits SPase II (Xiao et al. 2012).

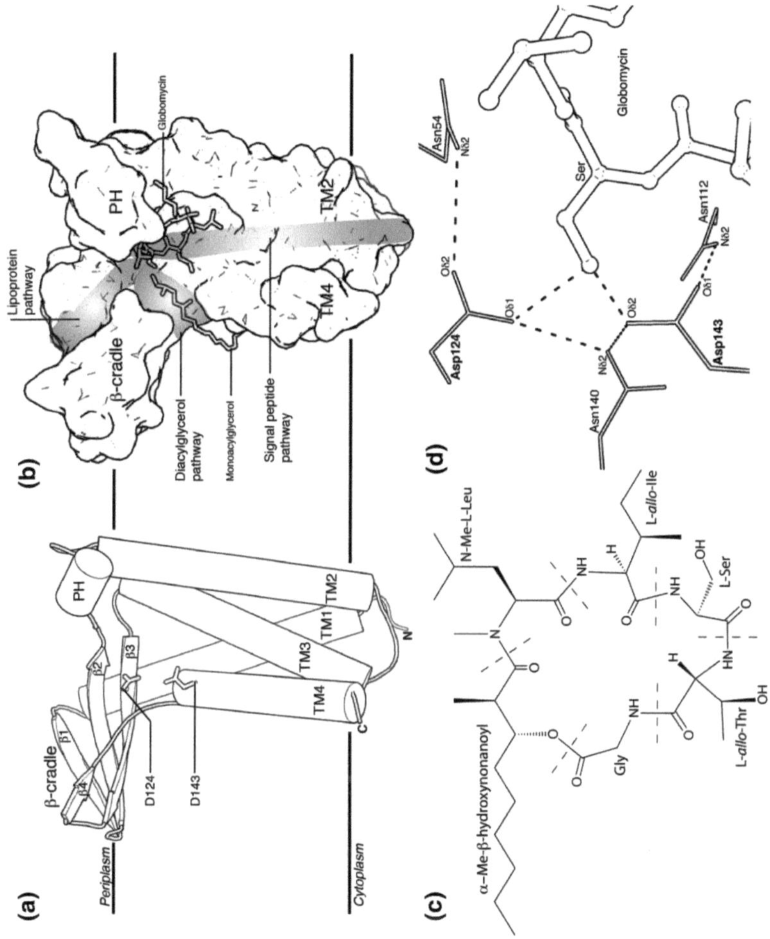
(a)
Periplasm
Cytoplasm
β-cradle
PH
TM1
TM2
TM3
TM4
D124
D143
(b)
β-cradle
PH
TM2
TM4
Lipoprotein pathway
Globomycin
Diacylglycerol pathway
Monoacylglycerol
Signal peptide pathway
(c)
α–Me-β-hydroxynonanoyl
N-Me-L-Leu
L-allo-Ile
L-Ser
L-allo-Thr
Gly
(d)
Asn54
Asp124
Asn140
Asp143
Asn112
Ser
Globomycin

◂**Fig. 7.6** The crystal structure of *Pseudomonas aeruginosa* (strain PAO1) SPase II (PDB: 5DIR). **a** The protein fold of SPase II showing the transmembrane segments 1–4 (TM1-4). The β-cradle domain and the PH (periplasmic helix) region are labeled. The proposed catalytic residues Asp124 and Asp143 are labeled. **b** The surface of SPase II reveals three pathways that meet at the proposed position of the P1′ cysteine of the prolipoprotein. The proposed pathway for the lipoprotein (mature region of the prolipoprotein) is shaded and labeled, as are the proposed pathways for the signal peptide and the diacylglycerol. The bound lipid used in the crystallization is labeled. The position of the c-region of the signal peptide is proposed based on the position of bound inhibitor globomycin. **c** The molecular structure of SPase II inhibitor globomycin. **d** The active site of SPase II. The proposed catalytic residues Asp124 and Asp143 are within hydrogen bonding distance (dashed lines) to coordinating asparagine residues. The atoms involved in the hydrogen bonding interactions are labeled. The serine residue of globomycin, which mimics the P1 residue of the substrate, is shown in ball-and-stick

SPase II activity has been measured using both in vivo and in vitro assays (Dev and Ray 1984; Kitamura and Wolan 2018; Yamagata 1983). Recently a high-throughput in vitro SPase II activity assay has been designed, allowing for screening of chemical libraries. This has led to the development of inhibitors with nanomolar half-maximal inhibitory concentration values (Kitamura et al. 2018).

The first SPase II to be investigated was the SPase II from *E. coli*. Evidence clearly showed that there existed in *E. coli* a distinct signal peptidase for the processing of lipoproteins (Tokunaga et al. 1982, 1984). Its gene was mapped and sequenced (Regue et al. 1984; Tokunaga et al. 1984) and the enzyme-fusion technique was used to determine its membrane topology. This analysis suggested that *E. coli* SPase II has four transmembrane segments with both the amino-terminus and the carboxy-terminus residing in the cytoplasm (Munoa et al. 1991). SPase II genes from other Gram-negative bacterial have been characterized including: *Enterobacter aerogenes* (Isaki et al. 1990), *Legionella pneumophila* (Geukens et al. 2006), *Myxococcus xanthus* (Paitan et al. 1999), and *Rickettsia typhi* (Rahman et al. 2007), and *Pseudomonas aeruginosa* (Vogeley et al. 2016).

Gram-positive SPase II enzymes that have been characterized include: *Bacillus subtilis* (Pragai et al. 1997; Tjalsma et al. 1999a, c), *Staphylococcus carnosus* (Witke and Gotz 1995), *Streptococcus pneumonia* (Khandavilli et al. 2008), *Streptococcus suis* (De Greeff et al. 2003), *Streptomyces coelicolor* (Munnoch et al. 2016), and *Streptomyces lividans* (Gullon et al. 2013). Sequence alignments and site-directed mutagenesis in *B. subtilis* SPase II suggest that SPase II is a unique aspartic protease (Tjalsma et al. 1999c).

One structure of a SPase II enzyme is available (Vogeley et al. 2016). LspA from *Pseudomonas aeruginosa* (strain PAO1) was crystallized using the *in meso* (lipid cubic phase) method (Caffrey 2015). The structure is refined to 2.8 Å resolution. The crystallization conditions were at pH5.6–6.0 and included MES buffer, PEG400, ammonium phosphate, and monoolein (monoacylglycerol, 9.9 MAG, 1-(9Z-octadecenoyl)-rac-glycerol). The enzyme is 169 residues in length (UniProt: Q9HVM5, molecular mass 18,997 Da, isoelectric point 8.0). The crystal structure has clear electron density for residues 2–158 (for chain A, the most ordered of the four molecules in the asymmetry unit). Structural analysis suggests SPase II is a monomer.

The crystal structure of SPase II from *Pseudomonas aeruginosa* showed there are four transmembrane segments with a topology that places both the amino- and carboxy-termini in the cytoplasm (TM1: residues 9–34, TM2: residues 68–90, TM3: residues 95–119, and TM4: residues 140–158) (Fig. 7.6a). This topology is consistent with the gene-fusion topology analysis performed on the *E. coli* SPase II homolog (Munoa et al. 1991). The first three residues of TM1 are in the 3_{10}-helical conformation. There is also a small periplasmic domain that is made up of a four-stranded antiparallel β-sheet (β1: residues 40–43, β2: residues 46–54, β3: residues 122–130, β4: residues 134–136). Because of its shape, this domain is referred to as the β-cradle. There is also a loop region with a single 3_{10}-helical turn (residues 60–62), which is referred to as the periplasmic helix (PH). The β-cradle and PH are amphipathic in nature and reside on the periplasmic region of the protein, along the surface of the periplasmic side of the inner membrane lipid bilayer, approximately perpendicular to the orientation of the TM helices. The first two strands of the β-sheet and the PH are an insertion between TM1 and 2. The last 2 β-stands within the β-sheet are an insertion between TM3 and 4. The secondary structure assignments discussed here were assigned using the program PROMOTIF(Hutchinson and Thornton 1996).

The SPase II structure was cocrystallized with globomycin. This allowed for the clear interpretation of the substrate binding groove and the catalytic active site. The substrate-binding groove is located at the periplasmic end of TM2, 3 and 4. The residues that have direct interactions with the inhibitor, and therefore likely with the prolipoprotein substrate as well, include residues that reside on TM3 such as Arg116 and Asn112. Interactions also come from Asn140 and Asp143, residing on TM4. Contacts are made from Asp124 found on the third β-strand within the β-sheet. The residues Leu-Ile-Ser of globomycin likely fit into the substrate-binding groove of SPase II in a similar fashion to the lipobox residues in the prolipoprotein. The residues surrounding the bound inhibitor correspond to the residues that are most conserved in SPase II.

Modeling and molecular dynamics simulations suggests that the Cα atom of the P1′ cysteine within the prolipoprotein substrate sits at a three-way crossroad of three substrate binding surfaces on the surface of LspA (Fig. 7.6b). The periplasmic β-cradle and the PH-loop region form a groove where the mature region of the lipoprotein would be directed. A second groove is formed by the periplasmic β-cradle and TM4. The side chain of the P1′ cysteine could fit in this groove and the thioether linkage to diacylglycerol would direct the fatty acids to lay across the surface of TM4. A binding surface is formed from TM helices 2, 3 and 4. This creates a long surface for the signal peptide to bind. The surface complementarity between this enzyme and its unique substrate allows for LspA to recognize and cleave a broad range of prolipoproteins.

The clues provided by the inhibitor complex allowed for a hypothetical Michaelis complex to be modeled. This model in turn suggested the identity of the residues involved in the catalytic mechanism. The most likely residues are Asp124 and Asp143 (Fig. 7.6a, d). Asp124 resides on the amino-terminal end of β-strand 3 (β3) and Asp143 resides on the amino-terminal end of TM4. In the globomycin-SPase II complex the globomycin serine hydroxyl group hydrogen bonds to both of the proposed catalytic residues. Sequence alignments show 14 strictly conserved residues: Asp23, Lys27, Asn54, Gly56, Gly108, Ala109, Asn112, Arg116, Val122, Asp124, Phe139, Asn140, Ala142, and Asp143. Most of these residues are located near the active site. Site-directed mutagenesis was consistent with Asp124 and Asp143 being essential residues. This suggests that these residues likely play a direct role in catalysis similar to the classic aspartic protease mechanism (Rawlings and Barrett 2013; Wlodawer et al. 2013). A constellation of asparagine residues helps to coordinate and orient the proposed catalytic residues (Fig. 7.6d). Asp143 Oδ1 is hydrogen bonded to Asn112 Nδ2 (3.0 Å). Asp124 Oδ2 is hydrogen bonded to Asn54 Nδ2 (3.4 Å) and Asp124 Oδ1 is hydrogen bonded to Asn140 Nδ2 (3.2 Å). In the future, it will be helpful to see a structure of this enzyme with a free active site; this may reveal the coordination geometry for the catalytic water. It is likely that the bound globomycin has displaced the catalytic water. If the c-region of the prolipoprotein substrate lays within the binding site of SPase II in a similar orientation to that of the cyclic peptide globomycin, then the structure suggests that the nucleophilic water attacks from the *si*-face of the substrate's scissile carbonyl. The hydroxyl oxygen of the serine within the bound globomycin sits adjacent to the two proposed catalytic carboxylate groups of Asp124 and Asp143. The hydroxyl oxygen is closer to Asp143 Oδ2 (2.4 Å) than it is to Asp 124 Oδ1 (2.9 Å). If the globomycin serine hydroxyl is mimicking one of the hydroxyls of the *gem*-diol transition state, then it is possible that the Asp143 is stabilizing the transition state via a low barrier hydrogen bond as has been proposed in some aspartic proteases (Wlodawer et al. 2013). A full stereo-electronic understanding of the SPase II mechanism awaits structural analysis at higher resolution.

Other Signal Peptide Cleaving Enzymes

Although they have not been covered in this chapter, there are a number of other bacterial peptidases associated with protein secretion such as those involved in prepilin processing (Dupuy et al. 2013) and hydrolysis of signal peptides (Dalbey and Wang 2013).

Acknowledgements The author would like to thank Dr. Andreas Kuhn for the opportunity to contribute to this book.

References

Allsop AE, Brooks G, Bruton G, Coulton S, Edwards PD, Hatton IK, Kaura AC, McLean SD, Pearson ND, Smale TC, Southgate R (1995) Penem inhibitors of bacterial signal peptidase. Biorg Medic Chem Lett 5:443–448

Allsop A, Brooks G, Edwards PD, Kaura AC, Southgate R (1996) Inhibitors of bacterial signal peptidase: a series of 6-(substituted oxyethyl)penems. J Antibiot (Tokyo) 49:921–928

Armbruster KM, Meredith TC (2018) Enrichment of bacterial lipoproteins and preparation of N-terminal lipopeptides for structural determination by mass spectrometry. J Vis Exp: JoVE

Aynapudi J, El-Rami F, Ge X, Stone V, Zhu B, Kitten T, Xu P (2017) Involvement of signal peptidase I in Streptococcus sanguinis biofilm formation. Microbiology 163:1306–1318

Babu MM, Priya ML, Selvan AT, Madera M, Gough J, Aravind L, Sankaran K (2006) A database of bacterial lipoproteins (DOLOP) with functional assignments to predicted lipoproteins. J Bacteriol 188:2761–2773

Bagos PG, Tsirigos KD, Plessas SK, Liakopoulos TD, Hamodrakas SJ (2009) Prediction of signal peptides in archaea. Protein Eng Des Sel 22:27–35

Bairl A, Muller P (1998) A second gene for type I signal peptidase in Bradyrhizobium japonicum, sipF, is located near genes involved in RNA processing and cell division. Mol Gen Genet 260:346–356

Barbrook AC, Packer JC, Howe CJ (1996) Inhibition by penem of processing peptidases from cyanobacteria and chloroplast thylakoids. FEBS Lett 398:198–200

Bardy SL, Eichler J, Jarrell KF (2003) Archaeal signal peptides—a comparative survey at the genome level. Protein Sci 12:1833–1843

Bardy SL, Ng SY, Carnegie DS, Jarrell KF (2005) Site-directed mutagenesis analysis of amino acids critical for activity of the type I signal peptidase of the archaeon Methanococcus voltae. J Bacteriol 187:1188–1191

Bechinger B, Gierasch LM, Montal M, Zasloff M, Opella SJ (1996) Orientations of helical peptides in membrane bilayers by solid state NMR spectroscopy. Solid State Nucl Magn Reson 7:185–191

Berger A, Schechter I (1970) Mapping the active site of papain with the aid of peptide substrates and inhibitors. Philos Trans R Soc Lond B Biol Sci 257:249–264

Bilgin N, Lee JI, Zhu HY, Dalbey R, von Heijne G (1990) Mapping of catalytically important domains in Escherichia coli leader peptidase. EMBO J 9:2717–2722

Black MT (1993) Evidence that the catalytic activity of prokaryote leader peptidase depends upon the operation of a serine-lysine catalytic dyad. J Bacteriol 175:4957–4961

Black MT, Bruton G (1998) Inhibitors of bacterial signal peptidases. Curr Pharm Des 4:133–154

Black MT, Munn JG, Allsop AE (1992) On the catalytic mechanism of prokaryotic leader peptidase 1. Biochem J 282:539–543

Blobel G, Dobberstein B (1975a) Transfer of proteins across membranes. I. Presence of proteolytically processed and unprocessed nascent immunoglobulin light chains on membrane-bound ribosomes of murine myeloma. J Cell Biol 67:835–851

Blobel G, Dobberstein B (1975b) Transfer of proteins across membranes. II. Reconstitution of functional rough microsomes from heterologous components. J Cell Biol 67:852–862

Bochtler M, Mizgalska D, Veillard F, Nowak ML, Houston J, Veith P, Reynolds EC, Potempa J (2018) The Bacteroidetes Q-Rule: pyroglutamate in signal peptidase I substrates. Front Microbiol 9:230

Bockstael K, Geukens N, Rao CV, Herdewijn P, Anne J, Van Aerschot A (2009) An easy and fast method for the evaluation of Staphylococcus epidermidis type I signal peptidase inhibitors. J Microbiol Methods 78:231–237

Bolhuis A, Sorokin A, Azevedo V, Ehrlich SD, Braun PG, de Jong A, Venema G, Bron S, van Dijl JM (1996) Bacillus subtilis can modulate its capacity and specificity for protein secretion through temporally controlled expression of the sipS gene for signal peptidase I. Mol Microbiol 22:605–618

Bonnemain C, Raynaud C, Reglier-Poupet H, Dubail I, Frehel C, Lety MA, Berche P, Charbit A (2004) Differential roles of multiple signal peptidases in the virulence of Listeria monocytogenes. Mol Microbiol 51:1251–1266

Buddelmeijer N, Young R (2010) The essential Escherichia coli apolipoprotein N-acyltransferase (Lnt) exists as an extracytoplasmic thioester acyl-enzyme intermediate. Biochemistry 49:341–346

Buzder-Lantos P, Bockstael K, Anne J, Herdewijn P (2009) Substrate based peptide aldehyde inhibits bacterial type I signal peptidase. Bioorg Med Chem Lett 19:2880–2883

Caffrey M (2015) A comprehensive review of the lipid cubic phase or in meso method for crystallizing membrane and soluble proteins and complexes. Acta Crystallogr Sect F, Struct Biol Commun 71:3–18

Cai D, Wei X, Qiu Y, Chen Y, Chen J, Wen Z, Chen S (2016) High-level expression of nattokinase in Bacillus licheniformis by manipulating signal peptide and signal peptidase. J Appl Microbiol 121:704–712

Carlos JL, Klenotic PA, Paetzel M, Strynadka NC, Dalbey RE (2000) Mutational evidence of transition state stabilization by serine 88 in Escherichia coli type I signal peptidase. Biochemistry 39:7276–7283

Chang CN, Blobel G, Model P (1978) Detection of prokaryotic signal peptidase in an Escherichia coli membrane fraction: endoproteolytic cleavage of nascent f1 pre-coat protein. Proc Natl Acad Sci USA 75:361–365

Chatzi KE, Sardis MF, Karamanou S, Economou A (2013) Breaking on through to the other side: protein export through the bacterial Sec system. Biochem J 449:25–37

Chimalapati S, Sankaran K, Brown JS (2013) Chapter 62—signal peptidase II. In: Handbook of proteolytic enzymes. Academic Press, pp 258–261

Choo KH, Tan TW, Ranganathan S (2009) A comprehensive assessment of N-terminal signal peptides prediction methods. BMC Bioinform 10(Suppl 15):S2

Chu HH, Hoang V, Kreutzmann P, Hofemeister B, Melzer M, Hofemeister J (2002) Identification and properties of type I-signal peptidases of Bacillus amyloliquefaciens. Eur J Biochem 269:458–469

Craney A, Romesberg FE (2017) Stable signal peptides and the response to secretion stress in staphylococcus aureus. mBio 8

Cregg KM, Wilding I, Black MT (1996) Molecular cloning and expression of the spsB gene encoding an essential type I signal peptidase from Staphylococcus aureus. J Bacteriol 178:5712–5718

Dalbey RE, Wang P (2013) Chapter 798—signal peptide peptidase A (Prokaryotes). In: Handbook of proteolytic enzymes. Academic Press, pp 3611–3613

Dalbey RE, Wickner W (1985) Leader peptidase catalyzes the release of exported proteins from the outer surface of the Escherichia coli plasma membrane. J Biol Chem 260:15925–15931

Date T (1983) Demonstration by a novel genetic technique that leader peptidase is an essential enzyme of Escherichia coli. J Bacteriol 154:76–83

De Bona P, Deshmukh L, Gorbatyuk V, Vinogradova O, Kendall DA (2012) Structural studies of a signal peptide in complex with signal peptidase I cytoplasmic domain: the stabilizing effect of membrane-mimetics on the acquired fold. Proteins 80:807–817

De Greeff A, Hamilton A, Sutcliffe IC, Buys H, Van Alphen L, Smith HE (2003) Lipoprotein signal peptidase of Streptococcus suis serotype 2. Microbiology 149:1399–1407

De Rosa M, Lu L, Zamaratski E, Szalaj N, Cao S, Wadensten H, Lenhammar L, Gising J, Roos AK, Huseby DL et al (2017) Design, synthesis and in vitro biological evaluation of oligopeptides targeting E. coli type I signal peptidase (LepB). Bioorg Med Chem 25:897–911

Dev IK, Ray PH (1984) Rapid assay and purification of a unique signal peptidase that processes the prolipoprotein from Escherichia coli B. J Biol Chem 259:11114–11120

Dev IK, Harvey RJ, Ray PH (1985) Inhibition of prolipoprotein signal peptidase by globomycin. J Biol Chem 260:5891–5894

Driessen AJ, Nouwen N (2008) Protein translocation across the bacterial cytoplasmic membrane. Annu Rev Biochem 77:643–667

Dufour J, Neuville L, Zhu J (2010) Intramolecular Suzuki-Miyaura reaction for the total synthesis of signal peptidase inhibitors, arylomycins A(2) and B(2). Chemistry 16:10523–10534

Dupuy B, Deghmane A-E, Taha M-K (2013) Chapter 63—type IV prepilin peptidase. In: Handbook of proteolytic enzymes. Academic Press, pp 261–265
Eichler J (2002) Archaeal signal peptidases from the genus Thermoplasma: structural and mechanistic hybrids of the bacterial and eukaryal enzymes. J Mol Evol 54:411–415
Ekici OD, Karla A, Paetzel M, Lively MO, Pei D, Dalbey RE (2007) Altered-3 substrate specificity of Escherichia coli signal peptidase 1 mutants as revealed by screening a combinatorial peptide library. J Biol Chem 282:417–425
Escutia MR, Val G, Palacin A, Geukens N, Anne J, Mellado RP (2006) Compensatory effect of the minor Streptomyces lividans type I signal peptidases on the SipY major signal peptidase deficiency as determined by extracellular proteome analysis. Proteomics 6:4137–4146
Fine A, Irihimovitch V, Dahan I, Konrad Z, Eichler J (2006) Cloning, expression, and purification of functional Sec11a and Sec11b, type I signal peptidases of the archaeon Haloferax volcanii. J Bacteriol 188:1911–1919
Fink-Lavi E, Eichler J (2008) Identification of residues essential for the catalytic activity of Sec11b, one of the two type I signal peptidases of Haloferax volcanii. FEMS Microbiol Lett 278:257–260
Freudl R (2018) Signal peptides for recombinant protein secretion in bacterial expression systems. Microb Cell Fact 17:52
Geukens N, Lammertyn E, Van Mellaert L, Schacht S, Schaerlaekens K, Parro V, Bron S, Engelborghs Y, Mellado RP, Anne J (2001a) Membrane topology of the Streptomyces lividans type I signal peptidases. J Bacteriol 183:4752–4760
Geukens N, Parro V, Rivas LA, Mellado RP, Anne J (2001b) Functional analysis of the Streptomyces lividans type I signal peptidases. Arch Microbiol 176:377–380
Geukens N, Lammertyn E, Van Mellaert L, Engelborghs Y, Mellado RP, Anne J (2002) Physical requirements for in vitro processing of the Streptomyces lividans signal peptidases. J Biotechnol 96:79–91
Geukens N, De Buck E, Meyen E, Maes L, Vranckx L, Van Mellaert L, Anne J, Lammertyn E (2006) The type II signal peptidase of Legionella pneumophila. Res Microbiol 157:836–841
Ghahremanifard P, Rezaeinezhad N, Rigi G, Ramezani F, Ahmadian G (2018) Designing a novel signal sequence for efficient secretion of Candida antarctica lipase B in E. coli: the molecular dynamic simulation, codon optimization and statistical analysis approach. Int J Biol Macromol 119:291–305
Gullon S, Arranz EI, Mellado RP (2013) Transcriptional characterisation of the negative effect exerted by a deficiency in type II signal peptidase on extracellular protein secretion in Streptomyces lividans. Appl Microbiol Biotechnol 97:10069–10080
Harris DA, Powers ME, Romesberg FE (2009) Synthesis and biological evaluation of penem inhibitors of bacterial signal peptidase. Bioorg Med Chem Lett 19:3787–3790
Hazenbos WL, Skippington E, Tan MW (2017) Staphylococcus aureus type I signal peptidase: essential or not essential, that's the question. Microbial cell 4:108–111
Hillmann F, Argentini M, Buddelmeijer N (2011) Kinetics and phospholipid specificity of apolipoprotein N-acyltransferase. J Biol Chem 286:27936–27946
Hoang V, Hofemeister J (1995) Bacillus amyloliquefaciens possesses a second type I signal peptidase with extensive sequence similarity to other Bacillus SPases. Biochim Biophys Acta 1269:64–68
Holtzel A, Schmid DG, Nicholson GJ, Stevanovic S, Schimana J, Gebhardt K, Fiedler HP, Jung G (2002) Arylomycins A and B, new biaryl-bridged lipopeptide antibiotics produced by Streptomyces sp. Tu 6075. II. Structure elucidation. J Antibiot (Tokyo) 55:571–577
Hussain M, Ichihara S, Mizushima S (1980) Accumulation of glyceride-containing precursor of the outer membrane lipoprotein in the cytoplasmic membrane of Escherichia coli treated with globomycin. J Biol Chem 255:3707–3712
Hutchings MI, Palmer T, Harrington DJ, Sutcliffe IC (2009) Lipoprotein biogenesis in Gram-positive bacteria: knowing when to hold 'em, knowing when to fold 'em. Trends Microbiol 17:13–21
Hutchinson EG, Thornton JM (1996) PROMOTIF—a program to identify and analyze structural motifs in proteins. Protein Sci 5:212–220

Inouye S, Franceschini T, Sato M, Itakura K, Inouye M (1983) Prolipoprotein signal peptidase of Escherichia coli requires a cysteine residue at the cleavage site. EMBO J 2:87–91
Inukai M, Enokita R, Torikata A, Nakahara M, Iwado S, Arai M (1978a) Globomycin, a new peptide antibiotic with spheroplast-forming activity. I. Taxonomy of producing organisms and fermentation. J Antibiot (Tokyo) 31:410–420
Inukai M, Nakajima M, Osawa M, Haneishi T, Arai M (1978b) Globomycin, a new peptide antibiotic with spheroplast-forming activity. II. Isolation and physico-chemical and biological characterization. J Antibiot (Tokyo) 31:421–425
Inukai M, Takeuchi M, Shimizu K, Arai M (1978c) Mechanism of action of globomycin. J Antibiot (Tokyo) 31:1203–1205
Isaki L, Kawakami M, Beers R, Hom R, Wu HC (1990) Cloning and nucleotide sequence of the Enterobacter aerogenes signal peptidase II (lsp) gene. J Bacteriol 172:469–472
Jock CA, Pulakat L, Lee S, Gavini N (1997) Nucleotide sequence and genetic complementation analysis of lep from Azotobacter vinelandii. Biochem Biophys Res Commun 239:393–400
Juncker AS, Willenbrock H, Von Heijne G, Brunak S, Nielsen H, Krogh A (2003) Prediction of lipoprotein signal peptides in Gram-negative bacteria. Protein Sci 12:1652–1662
Kaderbhai NN, Harding V, Kaderbhai MA (2008) Signal peptidase I-mediated processing of an engineered mammalian cytochrome b(5) precursor is an exocytoplasmic post-translocational event in Escherichia coli. Mol Membr Biol 25:388–399
Karla A, Lively MO, Paetzel M, Dalbey R (2005) The identification of residues that control signal peptidase cleavage fidelity and substrate specificity. J Biol Chem 280:6731–6741
Kavanaugh JS, Thoendel M, Horswill AR (2007) A role for type I signal peptidase in Staphylococcus aureus quorum sensing. Mol Microbiol 65:780–798
Khandavilli S, Homer KA, Yuste J, Basavanna S, Mitchell T, Brown JS (2008) Maturation of Streptococcus pneumoniae lipoproteins by a type II signal peptidase is required for ABC transporter function and full virulence. Mol Microbiol 67:541–557
Kim YT, Muramatsu T, Takahashi K (1995a) Identification of Trp300 as an important residue for Escherichia coli leader peptidase activity. Eur J Biochem 234:358–362
Kim YT, Muramatsu T, Takahashi K (1995b) Leader peptidase from Escherichia coli: overexpression, characterization, and inactivation by modification of tryptophan residues 300 and 310 with N-bromosuccinimide. J Biochem 117:535–544
Kitamura S, Wolan DW (2018) Probing substrate recognition of bacterial lipoprotein signal peptidase using FRET reporters. FEBS Lett 592:2289–2296
Kitamura S, Owensby A, Wall D, Wolan DW (2018) Lipoprotein signal peptidase inhibitors with antibiotic properties identified through design of a robust in vitro HT platform. Cell Chem Biol 25(301–308):e312
Klenotic PA, Carlos JL, Samuelson JC, Schuenemann TA, Tschantz WR, Paetzel M, Strynadka NC, Dalbey RE (2000) The role of the conserved box E residues in the active site of the Escherichia coli type I signal peptidase. J Biol Chem 275:6490–6498
Kogen H, Kiho T, Nakayama M, Furukawa Y, Kinoshita T, Inukai M (2000) Crystal structure and total synthesis of globomycin: establishment of relative and absolute configurations. J Am Chem Soc 122:10214–10215
Konovalova A, Silhavy TJ (2015) Outer membrane lipoprotein biogenesis: lol is not the end. Philos Trans R Soc Lond B Biol Sci 370
Kovacs-Simon A, Titball RW, Michell SL (2011) Lipoproteins of bacterial pathogens. Infect Immun 79:548–561
Kuo D, Weidner J, Griffin P, Shah SK, Knight WB (1994) Determination of the kinetic parameters of Escherichia coli leader peptidase activity using a continuous assay: the pH dependence and time-dependent inhibition by beta-lactams are consistent with a novel serine protease mechanism. Biochemistry 33:8347–8354
Lammertyn E, Van Mellaert L, Meyen E, Lebeau I, De Buck E, Anne J, Geukens N (2004) Molecular and functional characterization of type I signal peptidase from Legionella pneumophila. Microbiology 150:1475–1483

Leversen NA, de Souza GA, Malen H, Prasad S, Jonassen I, Wiker HG (2009) Evaluation of signal peptide prediction algorithms for identification of mycobacterial signal peptides using sequence data from proteomic methods. Microbiology 155:2375–2383

Liu J, Luo C, Smith PA, Chin JK, Page MG, Paetzel M, Romesberg FE (2011a) Synthesis and characterization of the arylomycin lipoglycopeptide antibiotics and the crystallographic analysis of their complex with signal peptidase. J Am Chem Soc 133:17869–17877

Liu WT, Kersten RD, Yang YL, Moore BS, Dorrestein PC (2011b) Imaging mass spectrometry and genome mining via short sequence tagging identified the anti-infective agent arylomycin in Streptomyces roseosporus. J Am Chem Soc 133:18010–18013

LoVullo ED, Wright LF, Isabella V, Huntley JF, Pavelka MS Jr (2015) Revisiting the Gram-negative lipoprotein paradigm. J Bacteriol 197:1705–1715

Low KO, Muhammad Mahadi N, Md Illias R (2013) Optimisation of signal peptide for recombinant protein secretion in bacterial hosts. Appl Microbiol Biotechnol 97:3811–3826

Luke I, Handford JI, Palmer T, Sargent F (2009) Proteolytic processing of Escherichia coli twin-arginine signal peptides by LepB. Arch Microbiol 191:919–925

Luo C, Roussel P, Dreier J, Page MG, Paetzel M (2009) Crystallographic analysis of bacterial signal peptidase in ternary complex with arylomycin A2 and a beta-sultam inhibitor. Biochemistry 48:8976–8984

Madan Babu M, Sankaran K (2002) DOLOP–database of bacterial lipoproteins. Bioinformatics 18:641–643

Malten M, Nahrstedt H, Meinhardt F, Jahn D (2005) Coexpression of the type I signal peptidase gene sipM increases recombinant protein production and export in Bacillus megaterium MS941. Biotechnol Bioeng 91:616–621

Mao G, Zhao Y, Kang X, Li Z, Zhang Y, Wang X, Sun F, Sankaran K, Zhang XC (2016) Crystal structure of E. coli lipoprotein diacylglyceryl transferase. Nat Commun 7:10198

McKnight CJ, Rafalski M, Gierasch LM (1991a) Fluorescence analysis of tryptophan-containing variants of the LamB signal sequence upon insertion into a lipid bilayer. Biochemistry 30:6241–6246

McKnight CJ, Stradley SJ, Jones JD, Gierasch LM (1991b) Conformational and membrane-binding properties of a signal sequence are largely unaltered by its adjacent mature region. Proc Natl Acad Sci USA 88:5799–5803

Meijer WJ, de Jong A, Bea G, Wisman A, Tjalsma H, Venema G, Bron S, van Dijl JM (1995) The endogenous Bacillus subtilis (natto) plasmids pTA1015 and pTA1040 contain signal peptidase-encoding genes: identification of a new structural module on cryptic plasmids. Mol Microbiol 17:621–631

Milstein C, Brownlee GG, Harrison TM, Mathews MB (1972) A possible precursor of immunoglobulin light chains. Nat New Biol 239:117–120

Molino JVD, de Carvalho JCM, Mayfield SP (2018) Comparison of secretory signal peptides for heterologous protein expression in microalgae: expanding the secretion portfolio for Chlamydomonas reinhardtii. PLoS ONE 13:e0192433

Moore KE, Miura S (1987) A small hydrophobic domain anchors leader peptidase to the cytoplasmic membrane of Escherichia coli. J Biol Chem 262:8806–8813

Morisaki JH, Smith PA, Date SV, Kajihara KK, Truong CL, Modrusan Z, Yan D, Kang J, Xu M, Shah IM et al (2016) A putative bacterial ABC transporter circumvents the essentiality of signal peptidase. mBio 7

Muhammad MA, Falak S, Rashid N, Gardner QA, Ahmad N, Imanaka T, Akhtar M (2017) Escherichia coli signal peptidase recognizes and cleaves archaeal signal sequence. Biochemistry (Mosc) 82:821–825

Munnoch JT, Widdick DA, Chandra G, Sutcliffe IC, Palmer T, Hutchings MI (2016) Cosmid based mutagenesis causes genetic instability in Streptomyces coelicolor, as shown by targeting of the lipoprotein signal peptidase gene. Sci Rep 6:29495

Munoa FJ, Miller KW, Beers R, Graham M, Wu HC (1991) Membrane topology of Escherichia coli prolipoprotein signal peptidase (signal peptidase II). J Biol Chem 266:17667–17672

Musial-Siwek M, Kendall DA, Yeagle PL (2008a) Solution NMR of signal peptidase, a membrane protein. Biochim Biophys Acta 1778:937–944
Musial-Siwek M, Yeagle PL, Kendall DA (2008b) A small subset of signal peptidase residues are perturbed by signal peptide binding. Chem Biol Drug Des 72:140–146
Nahrstedt H, Wittchen K, Rachman MA, Meinhardt F (2004) Identification and functional characterization of a type I signal peptidase gene of Bacillus megaterium DSM319. Appl Microbiol Biotechnol 64:243–249
Nakajima M, Inukai M, Haneishi T, Terahara A, Arai M, Kinoshita T, Tamura C (1978) Globomycin, a new peptide antibiotic with spheroplast-forming activity. III. Structural determination of globomycin. J Antibiot (Tokyo) 31:426–432
Nakayama H, Kurokawa K, Lee BL (2012) Lipoproteins in bacteria: structures and biosynthetic pathways. FEBS J 279:4247–4268
Narita SI, Tokuda H (2017) Bacterial lipoproteins; biogenesis, sorting and quality control. Biochim Biophys Acta 1862:1414–1423
Ng SY, Jarrell KF (2003) Cloning and characterization of archaeal type I signal peptidase from Methanococcus voltae. J Bacteriol 185:5936–5942
Ng SY, Chaban B, VanDyke DJ, Jarrell KF (2007) Archaeal signal peptidases. Microbiology 153:305–314
Nguyen MT, Gotz F (2016) Lipoproteins of Gram-positive bacteria: Key players in the immune response and virulence. Microbiol Mol Biol Rev 80:891–903
Nielsen H, Engelbrecht J, Brunak S, von Heijne G (1997a) Identification of prokaryotic and eukaryotic signal peptides and prediction of their cleavage sites. Protein Eng 10:1–6
Nielsen H, Engelbrecht J, Brunak S, von Heijne G (1997b) A neural network method for identification of prokaryotic and eukaryotic signal peptides and prediction of their cleavage sites. Int J Neural Syst 8:581–599
Noland CL, Kattke MD, Diao J, Gloor SL, Pantua H, Reichelt M, Katakam AK, Yan D, Kang J, Zilberleyb I et al (2017) Structural insights into lipoprotein N-acylation by Escherichia coli apolipoprotein N-acyltransferase. Proc Natl Acad Sci USA 114:E6044–E6053
Nyathi Y, Wilkinson BM, Pool MR (2013) Co-translational targeting and translocation of proteins to the endoplasmic reticulum. Biochim Biophys Acta 1833:2392–2402
Owji H, Nezafat N, Negahdaripour M, Hajiebrahimi A, Ghasemi Y (2018) A comprehensive review of signal peptides: structure, roles, and applications. Eur J Cell Biol 97:422–441
Packer JC, Andre D, Howe CJ (1995) Cloning and sequence analysis of a signal peptidase I from the thermophilic cyanobacterium Phormidium laminosum. Plant Mol Biol 27:199–204
Paetzel M, Strynadka NC (1999) Common protein architecture and binding sites in proteases utilizing a Ser/Lys dyad mechanism. Protein Sci 8:2533–2536
Paetzel M, Chernaia M, Strynadka N, Tschantz W, Cao G, Dalbey RE, James MN (1995) Crystallization of a soluble, catalytically active form of Escherichia coli leader peptidase. Proteins 23:122–125
Paetzel M, Strynadka NC, Tschantz WR, Casareno R, Bullinger PR, Dalbey RE (1997) Use of site-directed chemical modification to study an essential lysine in Escherichia coli leader peptidase. J Biol Chem 272:9994–10003
Paetzel M, Dalbey RE, Strynadka NC (1998) Crystal structure of a bacterial signal peptidase in complex with a beta-lactam inhibitor. Nature 396:186–190
Paetzel M, Dalbey RE, Strynadka NC (2002) Crystal structure of a bacterial signal peptidase apoenzyme: implications for signal peptide binding and the Ser-Lys dyad mechanism. J Biol Chem 277:9512–9519
Paetzel M, Goodall JJ, Kania M, Dalbey RE, Page MG (2004) Crystallographic and biophysical analysis of a bacterial signal peptidase in complex with a lipopeptide-based inhibitor. J Biol Chem 279:30781–30790
Paitan Y, Orr E, Ron EZ, Rosenberg E (1999) A nonessential signal peptidase II (Lsp) of Myxococcus xanthus might be involved in biosynthesis of the polyketide antibiotic TA. J Bacteriol 181:5644–5651

Palacin A, Parro V, Geukens N, Anne J, Mellado RP (2002) SipY is the Streptomyces lividans type I signal peptidase exerting a major effect on protein secretion. J Bacteriol 184:4875–4880
Parro V, Mellado RP (1998) A new signal peptidase gene from Streptomyces lividans TK21. DNA Seq 9:71–77
Parro V, Schacht S, Anne J, Mellado RP (1999) Four genes encoding different type I signal peptidases are organized in a cluster in Streptomyces lividans TK21. Microbiology 145(Pt 9):2255–2263
Peng SB, Wang L, Moomaw J, Peery RB, Sun PM, Johnson RB, Lu J, Treadway P, Skatrud PL, Wang QM (2001) Biochemical characterization of signal peptidase I from gram-positive Streptococcus pneumoniae. J Bacteriol 183:621–627
Perlman D, Halvorson HO (1983) A putative signal peptidase recognition site and sequence in eukaryotic and prokaryotic signal peptides. J Mol Biol 167:391–409
Perry CR, Ashby MJ, Elsmere SA (1995) Penems as research tools to investigate the activity of E.coli leader peptidase. Biochem Soc Trans 23:548S
Petersen TN, Brunak S, von Heijne G, Nielsen H (2011) SignalP 4.0: discriminating signal peptides from transmembrane regions. Nat Methods 8:785–786
Pohlschroder M, Pfeiffer F, Schulze S, Halim MFA (2018) Archaeal cell surface biogenesis. FEMS Microbiol Rev 42:694–717
Pragai Z, Tjalsma H, Bolhuis A, van Dijl JM, Venema G, Bron S (1997) The signal peptidase II (Isp) gene of Bacillus subtilis. Microbiology 143(Pt 4):1327–1333
Rahman MS, Simser JA, Macaluso KR, Azad AF (2003) Molecular and functional analysis of the lepB gene, encoding a type I signal peptidase from Rickettsia rickettsii and Rickettsia typhi. J Bacteriol 185:4578–4584
Rahman MS, Ceraul SM, Dreher-Lesnick SM, Beier MS, Azad AF (2007) The lspA gene, encoding the type II signal peptidase of Rickettsia typhi: transcriptional and functional analysis. J Bacteriol 189:336–341
Rahman O, Cummings SP, Harrington DJ, Sutcliffe IC (2008) Methods for the bioinformatic identification of bacterial lipoproteins encoded in the genomes of Gram-positive bacteria. World J Microbiol Biotechnol 24:2377
Randall LL (1983) Translocation of domains of nascent periplasmic proteins across the cytoplasmic membrane is independent of elongation. Cell 33:231–240
Rawlings ND, Barrett AJ (2013) Chapter 1—introduction: aspartic and glutamic peptidases and their clans. In: Handbook of proteolytic enzymes. Academic Press, pp 3–19
Raynaud C, Charbit A (2005) Regulation of expression of type I signal peptidases in Listeria monocytogenes. Microbiology 151:3769–3776
Regue M, Remenick J, Tokunaga M, Mackie GA, Wu HC (1984) Mapping of the lipoprotein signal peptidase gene (lsp). J Bacteriol 158:632–635
Rizo J, Blanco FJ, Kobe B, Bruch MD, Gierasch LM (1993) Conformational behavior of Escherichia coli OmpA signal peptides in membrane mimetic environments. Biochemistry 32:4881–4894
Roberts TC, Smith PA, Cirz RT, Romesberg FE (2007) Structural and initial biological analysis of synthetic arylomycin A2. J Am Chem Soc 129:15830–15838
Roberts TC, Schallenberger MA, Liu J, Smith PA, Romesberg FE (2011a) Initial efforts toward the optimization of arylomycins for antibiotic activity. J Med Chem 54:4954–4963
Roberts TC, Smith PA, Romesberg FE (2011b) Synthesis and biological characterization of arylomycin B antibiotics. J Nat Prod 74:956–961
Sabatini DD, Blobel G, Nonomura Y, Adelman MR (1971) Ribosome-membrane interaction: structural aspects and functional implications. Adv Cytopharmacol 1:119–129
San Millan JL, Boyd D, Dalbey R, Wickner W, Beckwith J (1989) Use of phoA fusions to study the topology of the Escherichia coli inner membrane protein leader peptidase. J Bacteriol 171:5536–5541
Sankaram MB, Marsh D, Gierasch LM, Thompson TE (1994) Reorganization of lipid domain structure in membranes by a transmembrane peptide: an ESR spin label study on the effect of the Escherichia coli outer membrane protein a signal peptide on the fluid lipid domain connectivity in

binary mixtures of dimyristoyl phosphatidylcholine and distearoyl phosphatidylcholine. Biophys J 66:1959–1968

Sankaran K, Gan K, Rash B, Qi HY, Wu HC, Rick PD (1997) Roles of histidine-103 and tyrosine-235 in the function of the prolipoprotein diacylglyceryl transferase of Escherichia coli. J Bacteriol 179:2944–2948

Sarabia F, Chammaa S, Garcia-Ruiz C (2011) Solid phase synthesis of globomycin and SF-1902 A5. J Org Chem 76:2132–2144

Savojardo C, Martelli PL, Fariselli P, Casadio R (2018) DeepSig: deep learning improves signal peptide detection in proteins. Bioinformatics 34:1690–1696

Schimana J, Gebhardt K, Holtzel A, Schmid DG, Sussmuth R, Muller J, Pukall R, Fiedler HP (2002) Arylomycins A and B, new biaryl-bridged lipopeptide antibiotics produced by Streptomyces sp. Tu 6075. I. Taxonomy, fermentation, isolation and biological activities. J Antibiot (Tokyo) 55:565–570

Selas Castineiras T, Williams SG, Hitchcock A, Cole JA, Smith DC, Overton TW (2018) Development of a generic beta-lactamase screening system for improved signal peptides for periplasmic targeting of recombinant proteins in Escherichia coli. Sci Rep 8:6986

Seydel A, Gounon P, Pugsley AP (1999) Testing the ' + 2 rule' for lipoprotein sorting in the Escherichia coli cell envelope with a new genetic selection. Mol Microbiol 34:810–821

Siegel SD, Wu C, Ton-That H (2016) A type I signal peptidase is required for pilus assembly in the Gram-positive, biofilm-forming bacterium Actinomyces oris. J Bacteriol 198:2064–2073

Smith PA, Romesberg FE (2012) Mechanism of action of the arylomycin antibiotics and effects of signal peptidase I inhibition. Antimicrob Agents Chemother 56:5054–5060

Smith AM, Yan H, Groves N, Dalla Pozza T, Walker MJ (2000) Co-expression of the Bordetella pertussis leader peptidase I results in enhanced processing and expression of the pertussis toxin S1 subunit in Escherichia coli. FEMS Microbiol Lett 191:177–182

Smith PA, Roberts TC, Romesberg FE (2010) Broad-spectrum antibiotic activity of the arylomycin natural products is masked by natural target mutations. Chem Biol 17:1223–1231

Smith PA, Powers ME, Roberts TC, Romesberg FE (2011) In vitro activities of arylomycin natural-product antibiotics against Staphylococcus epidermidis and other coagulase-negative staphylococci. Antimicrob Agents Chemother 55:1130–1134

Smith PA, Koehler MFT, Girgis HS, Yan D, Chen Y, Chen Y, Crawford JJ, Durk MR, Higuchi RI, Kang J et al (2018) Optimized arylomycins are a new class of Gram-negative antibiotics. Nature 561:189–194

Sung M, Dalbey RE (1992) Identification of potential active-site residues in the Escherichia coli leader peptidase. J Biol Chem 267:13154–13159

Szalaj N, Lu L, Benediktsdottir A, Zamaratski E, Cao S, Olanders G, Hedgecock C, Karlen A, Erdelyi M, Hughes D et al (2018) Boronic ester-linked macrocyclic lipopeptides as serine protease inhibitors targeting Escherichia coli type I signal peptidase. Eur J Med Chem 157:1346–1360

Talarico TL, Dev IK, Bassford PJ Jr, Ray PH (1991) Inter-molecular degradation of signal peptidase I in vitro. Biochem Biophys Res Commun 181:650–656

Terada M, Kuroda T, Matsuyama SI, Tokuda H (2001) Lipoprotein sorting signals evaluated as the LolA-dependent release of lipoproteins from the cytoplasmic membrane of Escherichia coli. J Biol Chem 276:47690–47694

Ting YT, Harris PW, Batot G, Brimble MA, Baker EN, Young PG (2016) Peptide binding to a bacterial signal peptidase visualized by peptide tethering and carrier-driven crystallization. IUCrJ 3:10–19

Tjalsma H, Noback MA, Bron S, Venema G, Yamane K, van Dijl JM (1997) Bacillus subtilis contains four closely related type I signal peptidases with overlapping substrate specificities. Constitutive and temporally controlled expression of different sip genes. J Biol Chem 272:25983–25992

Tjalsma H, Bolhuis A, van Roosmalen ML, Wiegert T, Schumann W, Broekhuizen CP, Quax WJ, Venema G, Bron S, van Dijl JM (1998) Functional analysis of the secretory precursor processing machinery of Bacillus subtilis: identification of a eubacterial homolog of archaeal and eukaryotic signal peptidases. Genes Dev 12:2318–2331

Tjalsma H, Kontinen VP, Pragai Z, Wu H, Meima R, Venema G, Bron S, Sarvas M, van Dijl JM (1999a) The role of lipoprotein processing by signal peptidase II in the Gram-positive eubacterium bacillus subtilis. Signal peptidase II is required for the efficient secretion of alpha-amylase, a non-lipoprotein. J Biol Chem 274:1698–1707

Tjalsma H, van den Dolder J, Meijer WJ, Venema G, Bron S, van Dijl JM (1999b) The plasmid-encoded signal peptidase SipP can functionally replace the major signal peptidases SipS and SipT of Bacillus subtilis. J Bacteriol 181:2448–2454

Tjalsma H, Zanen G, Venema G, Bron S, van Dijl JM (1999c) The potential active site of the lipoprotein-specific (type II) signal peptidase of Bacillus subtilis. J Biol Chem 274:28191–28197

Tjalsma H, Stover AG, Driks A, Venema G, Bron S, van Dijl JM (2000) Conserved serine and histidine residues are critical for activity of the ER-type signal peptidase SipW of Bacillus subtilis. J Biol Chem 275:25102–25108

Tokuda H, Matsuyama S (2004) Sorting of lipoproteins to the outer membrane in E. coli. Biochim Biophys Acta 1693:5–13

Tokunaga M, Loranger JM, Wolfe PB, Wu HC (1982) Prolipoprotein signal peptidase in Escherichia coli is distinct from the M13 procoat protein signal peptidase. J Biol Chem 257:9922–9925

Tokunaga M, Loranger JM, Wu HC (1984) A distinct signal peptidase for prolipoprotein in Escherichia coli. J Cell Biochem 24:113–120

Tschantz WR, Sung M, Delgado-Partin VM, Dalbey RE (1993) A serine and a lysine residue implicated in the catalytic mechanism of the Escherichia coli leader peptidase. J Biol Chem 268:27349–27354

Tschantz WR, Paetzel M, Cao G, Suciu D, Inouye M, Dalbey RE (1995) Characterization of a soluble, catalytically active form of Escherichia coli leader peptidase: requirement of detergent or phospholipid for optimal activity. Biochemistry 34:3935–3941

Tsirigotaki A, De Geyter J, Sostaric N, Economou A, Karamanou S (2017) Protein export through the bacterial Sec pathway. Nat Rev Microbiol 15:21–36

Tullman-Ercek D, DeLisa MP, Kawarasaki Y, Iranpour P, Ribnicky B, Palmer T, Georgiou G (2007) Export pathway selectivity of Escherichia coli twin arginine translocation signal peptides. J Biol Chem 282:8309–8316

Tyndall JD, Nall T, Fairlie DP (2005) Proteases universally recognize beta strands in their active sites. Chem Rev 105:973–999

Ujiie A, Nakano H, Iwasaki Y (2016) Extracellular production of Pseudozyma (Candida) antarctica lipase B with genuine primary sequence in recombinant Escherichia coli. J Biosci Bioeng 121:303–309

UniProt C (2013) Update on activities at the Universal Protein Resource (UniProt) in 2013. Nucleic Acids Res 41:D43–D47

van Dijl JM, van den Bergh R, Reversma T, Smith H, Bron S, Venema G (1990) Molecular cloning of the Salmonella typhimurium lep gene in Escherichia coli. Mol Gen Genet 223:233–240

van Dijl JM, de Jong A, Vehmaanpera J, Venema G, Bron S (1992) Signal peptidase I of Bacillus subtilis: patterns of conserved amino acids in prokaryotic and eukaryotic type I signal peptidases. EMBO J 11:2819–2828

van Dijl JM, de Jong A, Venema G, Bron S (1995) Identification of the potential active site of the signal peptidase SipS of Bacillus subtilis. Structural and functional similarities with LexA-like proteases. J Biol Chem 270:3611–3618

van Klompenburg W, Whitley P, Diemel R, von Heijne G, de Kruijff B (1995) A quantitative assay to determine the amount of signal peptidase I in E. coli and the orientation of membrane vesicles. Mol Membr Biol 12:349–353

van Klompenburg W, Ridder AN, van Raalte AL, Killian AJ, von Heijne G, de Kruijff B (1997) In vitro membrane integration of leader peptidase depends on the Sec machinery and anionic phospholipids and can occur post-translationally. FEBS Lett 413:109–114

van Klompenburg W, Paetzel M, de Jong JM, Dalbey RE, Demel RA, von Heijne G, de Kruijff B (1998) Phosphatidylethanolamine mediates insertion of the catalytic domain of leader peptidase in membranes. FEBS Lett 431:75–79

van Roosmalen ML, Jongbloed JD, de Jonf A, van Eerden J, Venema G, Bron S, van Dijl JM (2001) Detergent-independent in vitro activity of a truncated Bacillus signal peptidase. Microbiology 147:909–917
Vogeley L, El Arnaout T, Bailey J, Stansfeld PJ, Boland C, Caffrey M (2016) Structural basis of lipoprotein signal peptidase II action and inhibition by the antibiotic globomycin. Science 351:876–880
von Heijne G (1983) Patterns of amino acids near signal-sequence cleavage sites. Eur J Biochem 133:17–21
von Heijne G (1985) Signal sequences. The limits of variation. J Mol Biol 184:99–105
Voorhees RM, Hegde RS (2016) Toward a structural understanding of co-translational protein translocation. Curr Opin Cell Biol 41:91–99
Waite RD, Rose RS, Rangarajan M, Aduse-Opoku J, Hashim A, Curtis MA (2012) Pseudomonas aeruginosa possesses two putative type I signal peptidases, LepB and PA1303, each with distinct roles in physiology and virulence. J Bacteriol 194:4521–4536
Wang Z, Jones JD, Rizo J, Gierasch LM (1993) Membrane-bound conformation of a signal peptide: a transferred nuclear Overhauser effect analysis. Biochemistry 32:13991–13999
Wang S, Wang D, Li J, Huang T, Cai YD (2018) Identification and analysis of the cleavage site in a signal peptide using SMOTE, dagging, and feature selection methods. Mol Omics 14:64–73
Watts C, Silver P, Wickner W (1981) Membrane assembly from purified components. II. Assembly of M13 procoat into liposomes reconstituted with purified leader peptidase. Cell 25:347–353
Whitley P, Nilsson L, von Heijne G (1993) Three-dimensional model for the membrane domain of Escherichia coli leader peptidase based on disulfide mapping. Biochemistry 32:8534–8539
Wieseler B, Schiltz E, Muller M (1992) Identification and solubilization of a signal peptidase from the phototrophic bacterium Rhodobacter capsulatus. FEBS Lett 298:273–276
Witke C, Gotz F (1995) Cloning and nucleotide sequence of the signal peptidase II (lsp)-gene from Staphylococcus carnosus. FEMS Microbiol Lett 126:233–239
Wlodawer A, Gustchina A, James MNG (2013) Chapter 2—catalytic pathways of aspartic peptidases. In: Handbook of proteolytic enzymes. Academic Press, pp 19–26
Wolfe PB, Wickner W, Goodman JM (1983) Sequence of the leader peptidase gene of Escherichia coli and the orientation of leader peptidase in the bacterial envelope. J Biol Chem 258:12073–12080
Xiao Y, Gerth K, Muller R, Wall D (2012) Myxobacterium-produced antibiotic TA (myxovirescin) inhibits type II signal peptidase. Antimicrob Agents Chemother 56:2014–2021
Yamagata H (1983) Temperature-sensitive prolipoprotein signal peptidase in an Escherichia coli mutant: use of the mutant for an efficient and convenient assay system. J Biochem 93:1509–1515
Yeh CH, Walsh SI, Craney A, Tabor MG, Voica AF, Adhikary R, Morris SE, Romesberg FE (2018) Optimization of a beta-lactam scaffold for antibacterial activity via the inhibition of bacterial type I signal peptidase. ACS Med Chem Lett 9:376–380
Zhang YB, Greenberg B, Lacks SA (1997) Analysis of a Streptococcus pneumoniae gene encoding signal peptidase I and overproduction of the enzyme. Gene 194:249–255
Zhang W, Lu J, Zhang S, Liu L, Pang X, Lv J (2018) Development an effective system to expression recombinant protein in E. coli via comparison and optimization of signal peptides: expression of Pseudomonas fluorescens BJ-10 thermostable lipase as case study. Microb Cell Fact 17:50
Zhbanko M, Zinchenko V, Gutensohn M, Schierhorn A, Klosgen RB (2005) Inactivation of a predicted leader peptidase prevents photoautotrophic growth of Synechocystis sp. strain PCC 6803. J Bacteriol 187:3071–3078
Zheng F, Angleton EL, Lu J, Peng SB (2002) In vitro and in vivo self-cleavage of Streptococcus pneumoniae signal peptidase I. Eur J Biochem 269:3969–3977
Zuckert WR (2014) Secretion of bacterial lipoproteins: through the cytoplasmic membrane, the periplasm and beyond. Biochim Biophys Acta 1843:1509–1516

Zwizinski C, Wickner W (1980) Purification and characterization of leader (signal) peptidase from Escherichia coli. J Biol Chem 255:7973–7977
Zwizinski C, Date T, Wickner W (1981) Leader peptidase is found in both the inner and outer membranes of Escherichia coli. J Biol Chem 256:3593–3597

Part III
Inner Membrane

Chapter 8
Carbohydrate Transport by Group Translocation: The Bacterial Phosphoenolpyruvate: Sugar Phosphotransferase System

Jean-Marc Jeckelmann and Bernhard Erni

Abstract The Bacterial Phosphoenolpyruvate (PEP): Sugar Phosphotransferase System (PTS) mediates the uptake and phosphorylation of carbohydrates, and controls the carbon- and nitrogen metabolism in response to the availability of sugars. PTS occur in eubacteria and in a few archaebacteria but not in animals and plants. All PTS comprise two cytoplasmic phosphotransferase proteins (EI and HPr) and a species-dependent, variable number of sugar-specific enzyme II complexes (IIA, IIB, IIC, IID). EI and HPr transfer phosphorylgroups from PEP to the IIA units. Cytoplasmic IIA and IIB units sequentially transfer phosphates to the sugar, which is transported by the IIC and IICIID integral membrane protein complexes. Phosphorylation by IIB and translocation by IIC(IID) are tightly coupled. The IIC(IID) sugar transporters of the PTS are in the focus of this review. There are four structurally different PTS transporter superfamilies (glucose, glucitol, ascorbate, mannose). Crystal structures are available for transporters of two superfamilies: *bc*IIC^{mal} (MalT, 5IWS, 6BVG) and *bc*IIC^{chb} (ChbC, 3QNQ) of *B. subtilis* from the glucose family, and IIC^{asc} (UlaA, 4RP9, 5ZOV) of *E. coli* from the ascorbate superfamily. They are homodimers and each protomer has an independent transport pathway which functions by an elevator-type alternating-access mechanism. *bc*IIC^{mal} and *bc*IIC^{chb} have the same fold, IIC^{asc} has a completely different fold. Biochemical and biophysical data accumulated in the past with the transporters for mannitol (IICBA^{mtl}) and glucose (IICB^{glc}) are reviewed and discussed in the context of the *bc*IIC^{mal} crystal structures. The transporters of the mannose superfamily are dimers of protomers consisting of a IIC and a IID protein chain. The crystal structure is not known and the topology difficult to predict. Biochemical data indicate that the IICIID complex employs a different transport mechanism. Species specific IICIID serve as a gateway for the penetration of bacteriophage lambda DNA across, and insertion of class IIa bacteriocins into the inner membrane. PTS transporters are inserted into the membrane by SecYEG translocon and have specific lipid requirements. Immunoelectron- and fluorescence

J.-M. Jeckelmann (✉) · B. Erni
Institute of Biochemistry and Molecular Medicine, University of Bern, Bühlstrasse 28, 3012 Bern, Switzerland
e-mail: jean-marc.jeckelmann@ibmm.unibe.ch

B. Erni
e-mail: bernhard.erni@dcb.unibe.ch

A. Kuhn (ed.), *Bacterial Cell Walls and Membranes*, Subcellular Biochemistry 92, https://doi.org/10.1007/978-3-030-18768-2_8

microscopy indicate a non-random distribution and supramolecular complexes of PTS proteins.

Keywords Bacteriocin · Bacteriophage lambda · EIIC component · EIID component · Elevator mechanism · Energy coupling · Glucose · Mannitol · Mannose · PTS · Subcellular localization · Sugar transport

Introduction

Bacteria are single cell organisms that can form communities of cells expressing different and complementary metabolic pathways. Communities display metabolic diversity exceeding the metabolic potential encoded by the genome of a single bacterium. Changes of nutrient composition and community structure require a continuous adaptation of nutrient uptake and metabolism by intracellular signal transduction pathways and by intercellular communication between different community members. The phosphoenolpyruvate (PEP)-dependent sugar phosphotransferase system (PTS) couples carbohydrate uptake with the control of carbohydrate metabolism, chemotaxis, virulence and other functions. Carbohydrates are important nutrients for the majority of heterotrophic bacteria. Glucose is the primary product of photosynthesis and the building block of cellulose, the globally most abundant biopolymer. N-acetyl glucosamine is the building block of chitin, the second most abundant biopolymer (organic matrix of fungal cell walls and arthropodal exoskeletons).

Uptake of polar nutrients requires specific membrane transport proteins, which utilize different energy coupling mechanisms: (i) primary active transporters use ATP as energy source to pump solutes against a concentration gradient; (ii) secondary active transporters couple the uphill transport of a solute with the downhill transport of an ion along the electric and/or concentration potential; (iii) group translocation transporters couple solute transport with the chemical modification of the solute.

The sugar transporters of the phosphoenolpyruvate (PEP)-dependent sugar phosphotransferase system (PTS) mediate substrate uptake coupled with phosphorylation of the substrate (Fig. 8.1). Substrates are mono- and disaccharides, sugar alcohols (mannitol) and sugar acids (ascorbic acids). The phosphoryldonor is PEP. Phosphorylgroups are sequentially transferred from PEP to the incoming substrate by two "general" phosphotransferase proteins, EI and HPr, and the IIA and IIB domains of the sugar-specific enzyme II complexes (IIA, IIB, IIC, IID). IIC and IICIID are integral membrane transporters of different and sometimes overlapping sugar specificity, which are not phosphorylated (for reviews see Postma and Roseman 1976; Meadow et al. 1990; Postma et al. 1993; Saier et al. 1995; Robillard and Broos 1999; Erni 2013; Gabor et al. 2011; Roseman 1969). The IIC domain contains the substrate binding site.

Vectorial solute transport with chemical modification of the transported solute is termed group translocation (Mitchell and Moyle 1958). It exists mainly in prokaryotes unlike ion-gradient dependent primary, and ATP-dependent secondary transport, which occur in all kingdoms of life. Two examples of putative group translocation different from the PTS are uptake of pyrroline-5-carboxylate (P5C) coupled with reduction to proline, and uptake of fatty acids coupled with acylation by coenzyme A (Zou et al. 2003; Lepore et al. 2011; Mixson and Phang 1988). Group translocation allows the nutrient uptake under conditions where the membrane is deenergized and ion-coupled sugar transport is failing. It allows reinitiation of growth and survival in a hostile, the membrane integrity compromising, environment (Wells and Russell 1996).

As indicated above, the PTS plays a central role in the regulation of carbon and nitrogen metabolism in response to the availability of PTS-sugars (comprehensively reviewed by Josef Deutscher and colleagues; Saier et al. 1995; Deutscher et al. 2006; Deutscher et al. 2014; Galinier and Deutscher 2017). Input for this control is the phosphorylation state of the phosphotransferase proteins, which varies with sugar availability, uptake activity, and the production of PEP. In the limiting case of the complete absence of PTS sugars, all PTS proteins will be in the phosphorylated state. In the limiting case of the complete absence of PEP (or the interruption of the phosphoryltransfer chain by a mutation), all PTS proteins (downstream of the mutated protein) will be in the dephosphorylated state. Within these extremes the phosphorylation state of the PTS proteins varies with the supply of PEP and the demand by sugars, which are transported by the PTS. The ratio of the dephosphorylated to the phosphorylated forms of IIA or IIB components increases when sugars are translocated by IICs faster than IIA and IIB are rephosphorylated PEP via EI and HPr (Hogema et al. 1998). Phosphorylated and non-phosphorylated forms of certain (species-dependent) HPr, IIA and IIB control uptake of non PTS sugars, such as lactose or maltose (catabolite repression), intermediate metabolism (carbon, nitrogen balance, glycogen synthesis), gene expression and chemotaxis (Somavanshi et al. 2016).

The structure and the function of the different transmembrane IICs of the PTS are the main topic of this chapter. Secondary topics are: Biogenesis of IICs and IIC supramolecular complexes in the bacterial inner membrane.

The Discovery of the Phosphotransferase System (PTS)

The bacterial PEP: sugar phosphotransferase system (PTS) was discovered by Kundig, Gosh and Roseman during their investigation of the sialic acid metabolism (Roseman 1989; Basu 2003; Gracy 2003). N-acetyl-mannosamine (ManNAc), an intermediate of this pathway, was phosphorylated by an ATP dependent kinase isolated from liver. Bacteria were expected to contain a similar kinase because they were known to utilize ManNAc as the sole carbon source for growth. But no such activity was found in extracts from bacteria. On the consideration that ATP in a crude extract might be hydrolysed faster than utilized by the searched-for kinase, the extract was supplemented with an ATP-regenerating system consisting of ADP, PEP and nucleoside diphosphokinase. And this worked—but unexpectedly with PEP alone! Subsequently, the PEP dependent "kinase" was purified and shown to comprise three protein fractions. The Enzyme I (EI), HPr (heat stable, histidine containing protein) fractions were cytoplasmic soluble. The third fraction Enzyme II was membrane bound (Kundig et al. 1964; Kundig and Roseman 1971a). The activity of the Enzyme II fraction towards different sugars varied depending on which sugar was present in the growth medium, pointing to the existence of different sugar-specific and inducible enzymes II. The subsequent biochemical and genetic dissection of the PTS in *E. coli* and *S. typhimurium* revealed that enzymes II consisted of two genetic and biochemical units termed Enzyme IIA and Enzyme IIB (today's IIA^{sugar} and $IICB^{sugar}$) and an ever increasing number of PTS gene products, in particular of several enzymes II (abbreviated II^{sugar}) of different specificity for hexoses, hexitols, disaccharides and ascorbic acid (Danchin 1989; Saier 2015).

Phylogeny and Evolution of the PTS

Milton Saier and collaborators analysed the phylogenetic relationships between the PTS in hundreds of bacterial genomes (Saier et al. 2005; Nguyen et al. 2006; Chen et al. 2011). According to these analyses the IICs of the PTS originate from at least five independent ancestors, giving rise to five *superfamilies*: (1) Glucose (Glc)/Fructose (Fru)/Lactose (Lac) (GFL) superfamily, (2) Glucitol, (3) Ascorbate (Asc)/Galactitol (Gat) (AG) superfamily, (4) Mannose (Man)/Sorbose (Sor)/Fructose (MSF) superfamily, and (5) Dihydroxyacetone (Dha) superfamily (Fig. 8.1). The Dha superfamily does not comprise integral membrane proteins and therefore will not be described in more detail than this: DhaL and DhaK are a pair of soluble subunits with the same fold as the ATP- and Dha-binding domains of the ATP-dependent Dha kinase. DhaM, the third subunit, has the same fold as the IIA of the Man superfamily. DhaK contains the Dha binding site. DhaL a quasi-irreversibly bound ADP as cofactor, which is rephosphorylated in situ by DhaM (for reviews see Erni 2013; Erni et al. 2006.

High resolution X-ray structures of six IICs belonging to the GFL and AG superfamilies are known (McCoy et al. 2016; Luo et al. 2015; Cao et al. 2011) (see section "Structure and Function of Glucose (GFL) and Ascorbate (AG) Superfamily Transporters"), and the structures of the other members can be predicted/modelled with high confidence by the known structures as template (The Phyre2 web portal for protein modeling, prediction and analysis; Kelley et al. 2015). The glucitol transporters consist of two membrane spanning subunits (IIC'^{gut} and $IIBC''^{gut}$). The IICIID complex of the MSF superfamily also consists of two subunits. The structures of $IICIID^{man}$ and $IIC'IIBC''^{gut}$ are not known. Residues 176–268 of the IIBC″ subunit, corresponding to the 110 N-terminal residues (out of a 140) of the integral membrane domain IIC″ are predicted with limited confidence (Kelley et al. 2015) to assume a fold similar to the N-terminus of a dicarboxylate/sodium symporter (Mancusso et al. 2012). The $IICIID^{man}$ defied attempts to predict the membrane topology (Möller et al. 2001).

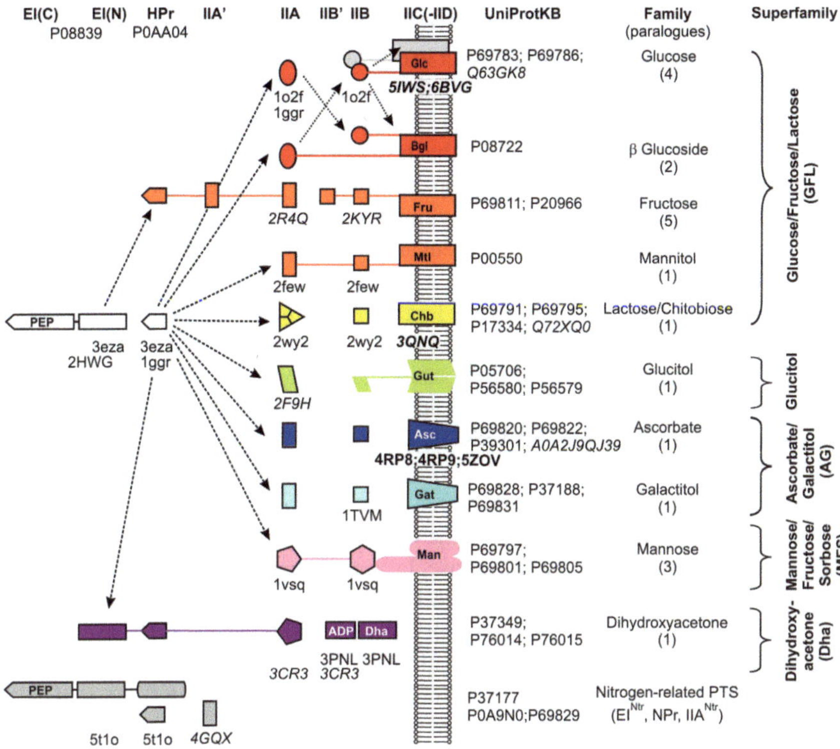

Fig. 8.1 The bacterial phosphotransferase system. Representative sugar transporters from each superfamily occurring in *Escherichia coli* are shown in different colors. The different structural folds are indicated by different shapes. Representative PDB entries (www.rcsb.org) are indicated below each shape as follows: Structures of *E. coli* proteins normal, of orthologs from other bacteria *italic*. PDB entries of intersubunit complexes are given, rather than the entries of the single subunits. X-ray structures (upper case) and NMR structures (lower case). Solid outlined shapes indicate that a structure is known or can be predicted with high confidence by known structures as template. Shapes without outlines indicate proteins of unknown tertiary structure. Interdomain linkers are indicated by straight lines (not to scale). EI(EIC) and all transporters (IIC(-IID)) are dimers (explicitly shown in red/grey only for $IICB^{glc}$). IIA^{man} and DhaM (IIA^{dha}) are homodimers, IIA^{chb} is a homotrimer. All other IIA and IIB are monomeric. *Broken arrows* indicate the phosphoryltransfer from HPr to the different IIA domains or subunits. EI, HPr, all IIA are phosphorylated at a histidine, IIB of the GFL, glucitol, and ascorbate/galactitol superfamily at a cysteine. IIB of the mannose superfamily are phosphorylated at a histidine. DhaL of the dihydroxyacetone kinase is phosphorylated by DhaM at a quasi irreversibly bound ADP cofactor. Additional *arrows* depicting the phosphoryl flow from IIA to IIB and hence to the sugar transported by IIC are not shown. One example of crosstalk occurring between structurally related enzyme II complexes of the glucose and β-glucoside family is indicated at the top. The nitrogen-related PTS (grey, PtsP, PtsO, PtsN) does not crosstalk with EI and HPr of the sugar PTS (white) (Rabus et al. 1999). The numbers of paralogues transport systems in *Escherichia coli* are indicated in parenthesis (Tchieu et al. 2001)

The structural heterogeneity of the five PTS superfamilies strongly indicates that they have different ancestors, which later in evolution were integrated in a primordial PTS. The ancestor of the GFL family appears to be a protein of the IIC^{fru} type. The ancestor of the AG family could be a protein related to the major facilitator superfamily (MFS), viz. membrane proteins that transport solutes driven by chemiosmotic ion gradients. An example of obvious ancestorship is the Dha kinases of the PTS, which share structure and mechanism with the contemporary ATP dependent dihydroxyacetone kinases of bacteria and eukaryotes (Saier et al. 2005; Barabote and Saier 2005).

The GFL *superfamily* (1) is by far the largest. 212 of its members from 34 bacterial species could be further divided into five equally divergent *families,* and each *family* again into several *subfamilies* (SF, clusters) as follows: (1) Glc family (6 SF, 41 members); (2) β-glucoside family (6 SF, 56 members); (3) Fru family (6 SF, 40 members); (4) Mannitol (Mtl) family (5 SF, 33 members); (5) Lactose-diacetylchitobiose family (6 SF, 42 members) (Chen et al. 2011). Whereas the IIC domains of the GFL superfamily share the same fold this does not hold for their cognate IIA and IIB domains/subunits. IIA and IIB of the Glc and β-glucoside families (1, 2) share one common fold. The mannitol/fructose, lactose-diacetylchitobiose families (3–5) share another common fold (Fig. 8.1) (Erni 2013; Clore and Venditti 2013). The only shared feature is that all IIA are phosphorylated at histidine and all IIB at cysteine.

Whether the glucitol transporter constitutes a superfamily of its own or belongs to the GFL superfamily is not clear. It consists of two subunits (IIC′ and $IIBC''^{gut}$), and the IIA fold (1F9H) is different from the GFL family IIA folds.

The mannose superfamily (MSF) of transporters was phylogenetically grouped in seven families (groups 1–7) of which the putative metabolic function and substrate specificity was inferred from their respective genome-context (Zuniga et al. 2005). Alternatively, the IICIID complexes were phylogenetically grouped into three families (groups I–III) (Kjos et al. 2009). The correspondence between these two groupings is not entirely clear. Of interest is group I which contains the IICIID complexes conferring sensitivity to class IIA bacteriocins (section "IICIID Mannose Transporters as Targets of Bacteriocins"). Three peptide sequences distinguish the group I IIC and IID proteins with bacteriocin receptor function from those with poor or none (Kjos et al. 2009).

Do the phylogenetic trees obtained for the PTS of different species correspond with the 16S ribosomal RNA trees (Woese and Fox 1977), which represent the evolutionary relationships of these bacteria? A comparison shows that the members of the GFL superfamily diversified in parallel with speciation (vertical transmission during evolution) (Nguyen et al. 2006; Chen et al. 2011). In contrast the phylogenetic tree of the MSF superfamily often does not agree with the order inferred from ribosomal RNAs indicating that they spread between species by horizontal gene transfer.

Distribution and Composition of PTS in Eubacteria and Archaebacteria

There are more than 10,000 complete bacterial and 300 archaeal genomes listed in the NCBI Microbial genome databank in 2018. Barabote and Saier in 2005 screened 174 bacterial genomes of 136 different bacterial species for their content of PTS genes. The results of this comprehensive analyses (Barabote and Saier 2005) can be summarized as follows.

Thirty out of the 136 species do not possess PTS-encoding genes, and 29 have an incomplete PTS, consisting of EI, HPr and IIA but lacking IIC/IIB transporters. 77 species (55%) have at least one complete PTS (EI, HPr, IIA, IIB, IIC), of which 18 have only one IIA, IIB, IIC system, 13 have only two systems and 46 have between three and 30 different systems. *Listeria monocytogenes* has 30 different systems with IIA, IIB, IIC subunits encoded by 91 genes accounting for 3.2% of all genes. *Escherichia coli* contains representatives (Fig. 8.1) from all five superfamilies (Barabote and Saier 2005; Tchieu et al. 2001).

Where only one or two systems are present they always are of the IIC^{glc} and/or IIC^{fru} type. Genomes with more than three IIC systems always include at least one IIC^{fru} and/or one IIC^{glc} system. But the number of IIC^{glc} can be as high as 14 and IIC^{fru} as high as 6. Some bacterial genomes contain as much as 8 IIC^{lac} systems, others as much as 12 $IICIID^{man}$ systems. Some of these systems may be cryptic, viz. they are not expressed unless a promoter is inserted in front of the operon, e.g. by transposition. A well studied cryptic system is the β-glucoside (*bgl*) operon of *E. coli* (Reynolds et al. 1981; Schnetz and Rak 1988; Harwani 2014).

The PTS content generally correlates with the potential for facultative anaerobic and anaerobic growth. 100% of the species with 10–30 transport systems, 86% of species with 3–9, and 39% of species with 1–3 systems are capable of anaerobic growth. Only 27% of species with incomplete PTS and only 20% of bacteria without PTS can grow anaerobically.

Multiple PTS systems occur in bacteria growing in the rumen, the oxygen-free and carbohydrate rich section of the stomach of ruminant animals, multiple $IICIID^{man}$ systems in bacteria from the intestinal microbiota (Zuniga et al. 2005).

The IIC of the GFL superfamily and of the AG superfamily (section "Phylogeny and Evolution of the PTS") can occur as single-domain proteins or in fusion proteins with IIB and IIA domains (IICBA), or with IIB alone (IICB) but never with IIA (no IICA) alone. Fusions can occur to the N-terminus ($IIBC^{fru}$), the C-terminus (*bc*$IICB^{mal}$) or to both ($IIBCA^{bgl}$). IIC and IID of the mannose superfamily occur either as tightly associated subunits (IICIID) or as fusion proteins IICD (Q8RD53). Fusions of integral membrane IIC domains with non-PTS domains (putative topoisomerases, DNA repair proteins, guanylate phosphodiesterases) are rare, but occur in *Mycoplasma, Mesoplasma* and *Vibrio* species. For comparison, fusions between EI, HPr, IIA and IIB domains and also between these domains and non-PTS domains, e.g. transcription regulators are quite common (Saier et al. 1995; Deutscher et al. 2006; Galinier and Deutscher 2017).

In archaea homologues of PTS proteins were discovered for the first time in 2006. The square halophilic archaeon *Haloquadratum walsby* has a PTS involved in the phosphorylation of dihydroxyacetone (Bolhuis et al. 2006) and a similar PTS was soon also found in *Haloferax volcanii* (Kirkland et al. 2008). *H. volcanii* in addition possesses a functional fructose (Fru) PTS comprising the IIC, IIA, EI, HPr and IIB proteins encoded in a single operon (Pickl et al. 2012). A similar Fru PTS with about 70% sequence identity also occurs in *Halalkalicoccus jeotgali* and *Haloarcula marismortui*. The haloarchaeal IIC^{fru} shows about 45% sequence identity with a IIC^{fru} homologue from *Clostridia*, indicating that it might have been acquired via horizontal gene transfer (Pickl et al. 2012). *Thermofilum pendens*, isolated from a solfatara in Iceland, has a complete Man PTS (HPr, EI, IIB, IIA, $IICIID^{man}$). Animals, plants and other eukaryotes (protists) do not have PTS.

Methods to Assay Transport and Phosphorylation Activity of IIC

Sugar uptake and fermentation by Gram-negative, enteric bacteria such as *Escherichia coli* can be detected on MacConkey indicator plates supplemented with the sugar of interest and peptides. Colonies of sugar fermenting bacteria turn red because they secrete acids, which can be detected with the pH indicator neutral red. Bacteria that cannot transport a given sugar utilize the peptides and their colonies remain pale yellow. Rate and extent of sugar uptake by intact bacteria can be measured with radioactive sugars. Exponentially growing bacteria are washed and preincubated before the radioactive sugar is added. Samples are withdrawn in short intervals, diluted into buffer and immediately collected by Millipore filtration. Uptake is fast and reaches saturation within less than 60 s. Non-metabolizable sugar analogs, such as α-methyl-glucopyranoside or 2-deoxyglucose are utilized to measure net uptake undisturbed by metabolic degradation of the transported substrate.

Phosphotransferase activity (PT activity) of cell-free systems and of the purified proteins is assayed in vitro (for standard methods' protocol see Mao et al. 1995; Navdaeva et al. 2011). Cells are ruptured mechanically by shear stress and decompression in a French pressure cell or by sonication. Cell debris are removed by low speed centrifugation, and the cytoplasmic and membrane fractions separated by high speed centrifugation. The PEP dependent PT activity of the combined fractions is assayed with radioactive sugars (Fig. 8.2a, b). The negatively charged $[^{14}C]$sugar phosphate is separated from the unreacted neutral $[^{14}C]$sugar by adsorption to an ion-exchange resin and released with dilute HCl. The activity of purified protein fractions can be monitored spectrophotometrically in a coupled assay (Fig. 8.2b): (i) PEP consumed (pyruvate) is detected with NADH and lactate dehydrogenase, or (ii) sugar phosphate produced is detected with NAD^+ and a matching sugar phosphate

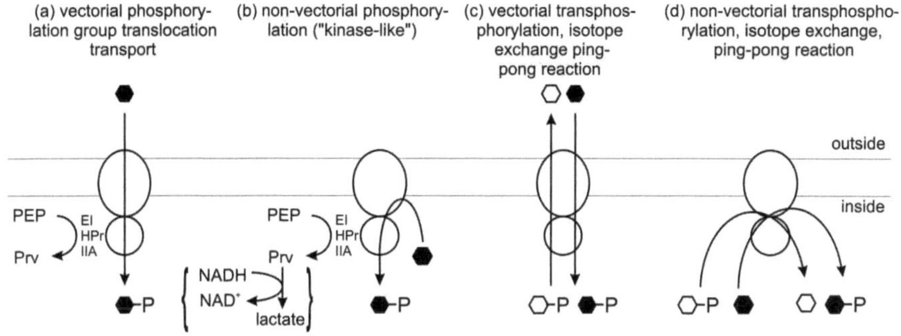

Fig. 8.2 Nomenclature of and assay methods for transport and phosphorylation reactions catalyzed by IICB complexes (sections "Methods to Assay Transport and Phosphorylation Activity of IIC" and "Membrane Transport Reconstitution"). **a** Transport with phosphorylation is assayed with intact bacteria or after reconstitution of the PTS in proteoliposomes. **b** Non-vectorial phosphorylation is assayed in cell lysates and with detergent solublized (and purified) IICB complexes. **c** The double displacement (ping-pong) reaction is assayed with proteoliposomes, and **d** with solubilized IICB. The reactions are conveniently measured with radioactive (filled hexagon) and non-radioactive (open hexagon) sugars. The phosphorylated products are separated from the non-phosphorylated educts by ion-exchange chromatography. Alternatively the phosphorylation reactions can be assayed spectrophotometrically in a coupled dehydrogenase assay

dehydrogenase. The sugar phosphorylation activity of IICB and IIC IIB complexes proceeds by a ping-pong (double displacement) mechanism. It can be measured in the absence of EI, HPr and IIA as phosphate exchange between radiolabelled sugar and unlabelled sugar phosphate (transphosphorylation reaction, Fig. 8.2c, d). PEP-dependent phosphorylation and sugar-phosphate dependent transphosphorylation can be assayed either as non-vectorial (trans)phosphorylation in homogeneous solution (for instance when IIC is solubilized in a detergent micelle) (Fig. 8.2b, d), or as vectorial (trans)phosphorylation with sealed IIC containing proteoliposomes (Fig. 8.2a, c). Proteoliposomes are either loaded with PEP, EI, HPr, IIA and IIB and sugar is added to the outside ("right-side" out orientation), or they are loaded with the sugar and EI, HPr, IIA and IIB are added to the outside ("inside-out" orientation) (Fig. 8.2a and section "Membrane Transport Reconstitution").

IIC dependent facilitated diffusion can be measured as uptake of radiolabelled sugars into proteoliposomes (Cao et al. 2011). Facilitated diffusion is not very efficient with wild-type proteins, but convenient to measure. Sugar binding to surface immobilized IIC is most conveniently assayed with a scintillation proximity assay (Jeckelmann et al. 2011).

Structure and Function of Glucose (GFL) and Ascorbate (AG) Superfamily Transporters

Molecular Structures of GFL and AG PTS Transporters

PTS transporters consist of three functional units IIA, IIB and IIC. The structures of representative IIA and IIB cytosolic subunits and domains and of some of their transient complexes have been determined by NMR and X-ray diffraction (for reviews see Robillard and Broos 1999; Erni 2013; Clore and Venditti 2013). First insights into the structure of the IIC integral membrane domains were obtained by transmission electron microscopy (TEM) of 2D crystals of the Mtl and Glc specific transporters (Jeckelmann et al. 2011; Koning et al. 1999). Crystals of *ec*IIC^{glc} prepared by limited trypsin digestion of IICB^{glc} diffracted X-rays to 4.5 Å (Zurbriggen et al. 2010).

The breakthrough was the high resolution X-ray structure of the diacetylchitobiose specific *bc*IIC^{chb} of *Bacillus cereus*, determined by Ming Zhou and collaborators (Cao et al. 2011; McCoy et al. 2015). Soon afterwards the same team solved the structures of the maltose specific *bc*IIC^{mal} of *Bacillus cereus* in two different conformations (McCoy et al. 2016; Ren et al. 2018; Lee et al. 2017). *bc*IIC^{mal} (*bc*MalT, PDB: 5IWS, 6BVG, UniProtKB: Q63GK8) and *bc*II^{chb} (*bc*ChbC, 3QNQ, Q72XQ0) belonging to the GFL superfamily have the same fold. *bc*IIC^{mal} is a representative of the biochemically well characterized Glc/β-glucoside family and will be used below as the reference for the description of structure function relationships (Fig. 8.3a–c). *bc*IIC^{chb} is a member of the lactose-chitobiose family for which biochemical data are scarce (Fig. 8.3d). Completely different from the *bc*IIC^{mal} structure is the fold of the ascorbate (vitamin C) specific transporter IIC^{asc} of *E. coli* (Fig. 8.3e–g) (Luo et al. 2015, 2018). IIC^{asc} (UlaA, SgaT; 4RP8, 4RP9 of P39301; 5ZOV of A0A2J9QJ39) is a member of the AG superfamily. The regulation of IIC^{asc}, IIB^{asc}, IIA^{asc} expression, the uptake of ascorbate by intact cells and non-vectorial phosphorylation in cell lysates have been characterized (Zhang et al. 2003), but nothing is known about the molecular function. Unlike IIC^{asc} which has a novel and unique fold, the associated IIA^{asc} and IIB^{asc} units have the same fold as the IIA and IIB of the mannitol/fructose families (Fig. 8.1).

Different Folds of *bc*IIC^{MAL} and IIC^{asc}, and Domain Organization

The six IIC structures known so far show dimers of protomers related by a two-fold rotational symmetry axis, and they all contain one substrate moiety per protomer (Fig. 8.3). Each protomer consists of a scaffold domain (SD) and a transport domain (TD). SD and TD of *bc*IIC^{mal} and *bc*IIC^{chb} each comprise five consecutive transmembrane helices, TM1–5 and TM6–10 (Fig. 8.3a). In contrast SD of IIC^{asc} consists of two non-consecutive pairs of helices, TM1, 2 and TM6, 7 and TD consists of two non-consecutive triplets TM3–5 and TM8–10 (Fig. 8.3e). Different folds notwithstanding, the same mechanism of substrate transport, namely a so-called elevator type

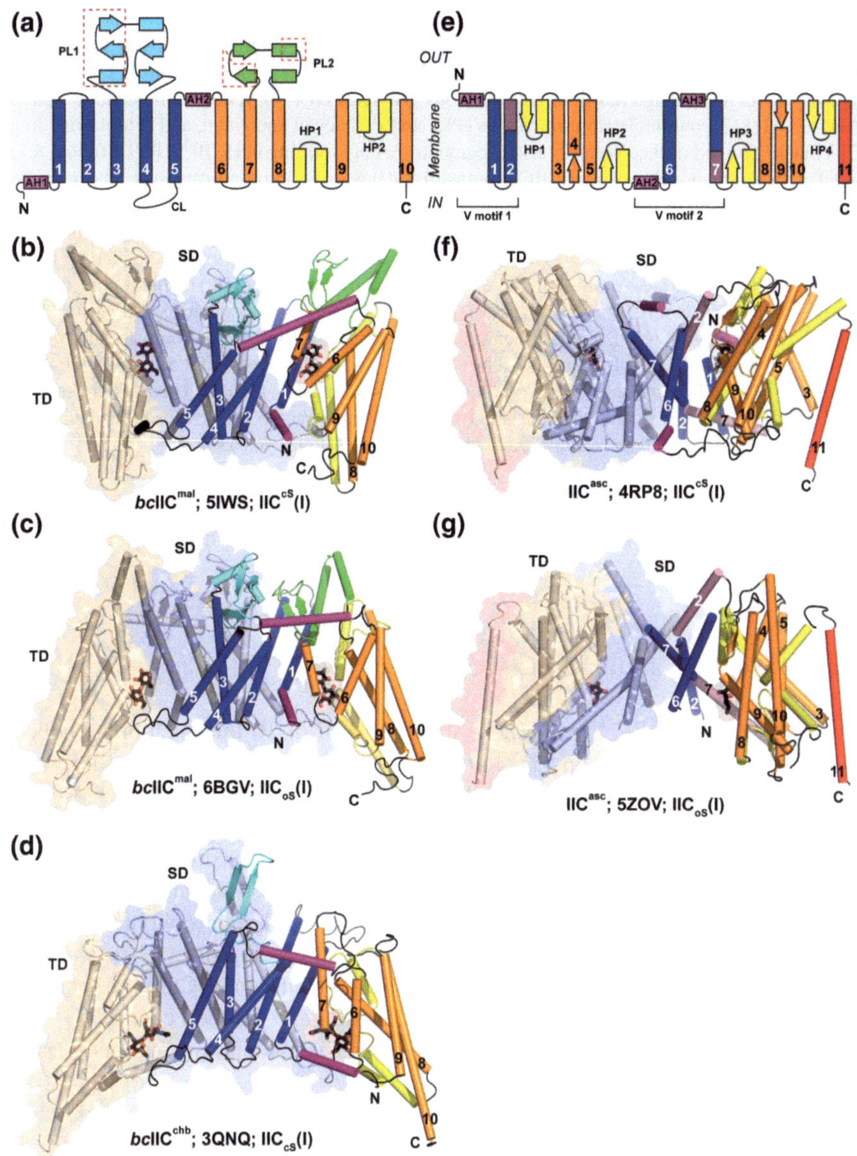
(a)
PL1
PL2
AH1
AH2
HP1
HP2
N
CL
C
(e)
OUT
Membrane
IN
AH1
AH2
AH3
HP1
HP2
HP3
HP4
V motif 1
V motif 2
(b)
SD
TD
bcIIC^mal; 5IWS; IIC^cS(I)
(c)
SD
TD
bcIIC^mal; 6BGV; IIC_oS(I)
(d)
SD
TD
bcIIC^chb; 3QNQ; IIC_cS(I)
(f)
TD
SD
IIC^asc; 4RP8; IIC^cS(I)
(g)
TD
SD
IIC^asc; 5ZOV; IIC_oS(I)

◂**Fig. 8.3** Topology and structure models of PTS transporters of the GFL and ascorbate superfamily, *bc*IIC^{mal} (**a–c**), *bc*IIC^{chb} (**a, d**), IIC^{asc} (**e–g**). (**a, e**) Topology diagrams; Scaffold/dimerization domain (SD)/(TM1–5; V motives 1 and 2, TM1, 2, 6, 7) blue; Transport domain (TD; TM6–10; TM3–5, 8–10, 11) orange/red with reentry loops (HP1, 2; HP1–4) yellow; amphipathic helices (AH1, 2; AH1–3), purple; Periplasmic loops (PL1 and PL2) cyan and green, and cytoplasmic loop (CL) black. The red dashed boxes bracket segments that are missing in IICB^{glc} (P69786, Fig. 8.1). **b–d, f, g** Structure models of dimeric IIC in outward and inward facing conformations. One protomer is colored according to the topology diagram. The other protomer is shown in transparent surface representation with SD in faint blue and TD in beige/faint red. Note in **b–d** the cytoplasmic loop (CL, black or faint blue) that reaches from TM4, 5 of one protomer to the exit of the binding cavity of the other. The substrates (maltose, N,N′-diacetylchitobiose and ascorbate) in the binding cavities are shown as sticks. **b** *bc*IIC^{mal} in the outward (periplasmic facing conformation), **c, d** *bc*IIC^{mal} and *bc*IIC^{chb} in the inward (cytoplasmic) facing conformation. **f, g** IIC^{asc} in the outward and inward facing conformation. The fold consists of the two inverted repeats (AH1–HP2 and AH2–HP4) and the extra TM11. SD consists of the two non contiguous V-motives 1 and 2. The kinked TM2 and TM7 in SD are colored in blue/brown, TM11 in red. The abbreviations below each structure refer to in-text usage, PDB databank entries, and states of the transport cycle (Fig. 8.4)

alternating-access mechanism has been proposed for both PTS transporters (McCoy et al. 2016; Luo et al. 2015; Ren et al. 2018).

The GFL Superfamily Transporters *bc*IIC^{mal} and *bc*IIC^{chb}

The *bc*IIC^{mal} and *bc*IIC^{chb} protomers are composed of 10 transmembrane helices (TM1–10) two reentrant loops (HP1 and HP2) between TM8/9 and TM9/10 respectively, two amphipathic helices (AH1 and AH2) preceding TM1 and TM6, two periplasmic loops (PL1 and PL2) between TM3/4 and TM7/8 and one cytoplasmic loop (CL) between TM4/5 (Fig. 8.3a). The ten TM constitute two structural domains: SD comprising TM1–5 and TD comprising TM6–10 and the two reentrant loops (Fig. 8.3a–d). SD contains the large and mostly hydrophobic protomer-protomer interface of 2700 Å^2. TD contains the sugar binding site and translocation pathway. The substrates, maltose (Mal) and diacetylchitobiose (Chb), respectively, are coordinated by the reentrant loops HP1 and HP2 and the loop connecting TM6 and TM7 (Fig. 8.3b–d). AH1 and AH2 lie flat on the cytoplasmic and periplasmic surface of the membrane.

The periplasmic loops (PL1 and PL2) are the two segments of strongest diversity in amino acids sequence and length. Their function is not known. They are 30–50 residues long in *bc*IIC^{mal}, 20–30 residues in the *E. coli* IIC^{glc}, and short in the Mtl-specific *E. coli* IIC^{mtl}. PL1 of IIC^{glc} lacks α1, β1 and β2, PL2 lacks β5, and β6 and PH1 are truncated (Fig. 8.3a). IIC^{glc} tolerates insertions of up to 20 residues and splitting of the polypeptide chain in PL1, retaining >30% of wild-type PT activity. No mutants with a detectable phenotype could be isolated in PL2 (Beutler et al. 2000a, b). In *Mycoplasma genitalium* IIC^{glc} (P47315; Reizer et al. 1996) the two loops are approximately 40–200 residues long. Not accounting of the 200 amino acids PL2, *mg*IIC^{glc} and *ec*IIC^{glc} share 36% sequence identity!

The cytoplasmic loop (CL) reaches from TM4/5 of one protomer to the inward oriented binding site of the other. It is the only structure, which connects the SD of one protomer with TD of the other (Fig. 8.3b–d). In inward oriented *bc*IIC^{chb} the CL occludes the binding cavity (Fig. 8.2d) (Cao et al. 2011). In inward oriented *bc*IIC^{mal} the CL leaves the access open for phosphate transfer from phospho-IIB^{mal} to the sugar (Fig. 8.2c) (Ren et al. 2018), but assumes a rigid, partially α-helical structure in outward oriented *bc*IIC^{mal} (Fig. 8.2b, 5IWS) (McCoy et al. 2016). Cysteine inserted at residue 124 in the putative CL of IICBA^{mtl} (Fig. 8.6) formed interprotomer disulfide bonds, indicating that two residues 40 Å apart in the crystal structure could approach to 4 Å in solution. The same cysteine could also be crosslinked with the active site C384 of the IIB^{mtl} domain (Van Montfort et al. 2001). The two findings suggest that the CL loop may be flexible in solution, and participate in IIC/IIB docking.

The AG Superfamily Transporter IIC^{asc}

The vitamin C specific IIC^{asc} protomer contains 11 transmembrane helices (TM1–11), four reentrant loops (HP1–4) between TM2/3, TM5/6, TM7/8 and TM10/11, and three amphipathic helices (AH1–3) preceeding TM1, TM6 and TM7 (Luo et al. 2015). Each protomer can be divided in two inverted structural repeats, which are related by a two-fold pseudosymmetry axis parallel to the membrane (Fig. 8.3e–g). The N-terminal repeat comprises AH1, TM1–5, HP1 and HP2, the C-terminal repeat AH2, TM6–10, HP3 and HP4. TM11 extends the periplasmic end of the second repeat back to the cytoplasmic face of the membrane for linkage with the IIB domain (for instance in *Pasteurella multocida* IICB^{asc}, A0A2X4V5W7). AH1, TM1 and TM2 together with the symmetry related AH2, TM6 and TM7 (V motives 1 and 2) form the scaffold domain. HP1, TM3–5 and HP2 together with the symmetry related HP3, TM6–10 and HP4 constitute the transport domain (cores 1 and 2). The substrate, ascorbate, is coordinated by the four hairpin loops of the reentrant loops HP1-HP4, TM4 and TM8. The inward facing binding site of IIC^{asc} (5ZOV) is open, and there is no structure comparable with CL of *bc*IIC^{mal} (Luo et al. 2018).

Structure of the Sugar Binding Cavities of *bc*IIC^{mal} and IIC^{asc}

The X-ray structures of *bc*IIC^{chb} and *bc*IIC^{mal} contain electron density consistent with a disaccharide in the binding site (Fig. 8.5a, b). *bc*IIC^{chb} was crystallized in the presence of N,N′-diacetylchitobiose (Cao et al. 2011). *bc*IIC^{mal} was solubilized with the detergent dodecylmaltopyranoside. Its headgroup (or contaminating maltose?) is trapped in the substrate binding cavity (McCoy et al. 2016). In both structures the C6-OH of the non-reducing hexose is oriented towards the cytoplasmic exit where it can be phosphorylated by the IIB domain. In *bc*IIC^{mal} faint electron density near the periplasmic entry site could be caused by the disordered acyl tail of the detergent.

The substrate binding cavity of *bc*IIC^{mal} is located in TD next to the TD/SD interface (Fig. 8.3). All the residues forming hydrogen bonds to the sugar come from

TD, and these H-bonding interactions (Fig. 8.5a, b) are preserved during translocation from the outward to the inward facing conformation. TM1 closes the binding cavity from the SD side, but no residue of SD appears to directionally interact with the sugar. The hydroxylgroups of the non-reducing Glc moiety interact with the following residues of the binding cavity: OH-6 with H240 (TM6/7) and E355 (HP1a/b); OH-4 with E355 (HP1a/b); OH-3 with T354 (HP1a/b), Q310 (TM8a) and R232 (TM6); OH-2 with E231 (TM6) and Q310 (TM8a). The OH* groups of the reducing Glc moiety* of the disaccharide interact: OH-6* with H241 (TM6/7); OH-3* and OH-2* with K307 (PH2/TM8a) (Fig. 8.5a, b). It is noteworthy, that the glyosidic O-1 and OH-1* are not coordinated. The residues coordinating the non-reducing Glc moiety are conserved (Fig. 8.6), and mutagenesis had indicated in the past that they are essential for binding and transport (section "Substrate Binding, Transport and Phosphorylation: Important Residues and Point Mutations"; Otte et al. 2003; Weng and Jacobson 1993; Saraceni-Richards and Jacobson 1997a, b; Lanz and Erni 1998; Boer et al. 1996; Opacic et al. 2012).

SD borders on the binding cavity by TM1. V17, V19, V20 and M23 of TM1 are within 3.5–4.5 Å from the substrate. Mutations in and next to TM1 of $IICB^{glc}$ affect substrate specificity, and the fluorescence of a Trp-30 in the putative TM1 of $IICBA^{mtl}$ is highly sensitive to Mtl binding (Fig. 8.6 and section "Substrate Binding, Transport and Phosphorylation: Important Residues and Point Mutations"; Dijkstra et al. 1996, 1997; Broos et al. 2000).

In IIC^{asc} the binding cavity is formed by elements of TM4, TM9 and the reentrant loops 1–4 (Fig. 8.3e–g). They provide hydrogen bonding contacts to the OH-2,3,5 and 6 of ascorbate. Similar to *bc*IIC^{mal}, one side of the cavity is delimitated by SD, which forms one hydrogen bond contact: in the outward facing conformation between OH-3 and S59 of TM2; in the inward facing conformation between OH-3 and R288 of TM7 (Luo et al. 2015, 2018).

Mechanism of Substrate Binding and Transport

The alternating-access mechanism is a concept of substrate transport proposed by Mitchell in 1957. It has been reviewed in the light of known membrane protein structures in 2016 by Drew and Boudker (Mitchell 1957; Drew and Boudker 2016). Three types of alternating-access mechanisms are described: the rocker switch, the rocking bundle and the elevator type mechanism. For the former two, a movement of two domains around the substrate binding site provides alternate access for the substrate from either side of the membrane. The elevator mechanism, in contrast, implies a translational rigid-body motion of TD against the immobile SD (see in this volume Chap. 9, *Permeases and Secondary Active Transporters*). The putative cycle of substrate translocation by a PTS transporter is depicted in Fig. 8.4. It begins with substrate (S) *binding* to the transporter (IIC) in the outward-facing open conformation ($IIC^{o} \rightarrow IIC^{oS}$) followed by substrate occlusion (IIC^{cS} (I) and (II)), *translocation* across the membrane (IIC_{cS}(II)) by a rigid body motion, isomerization to the inward-facing open conformation (IIC_{oS}(II)), and *release* of the substrate into the cytoplasm.

The transporter then cycles back from the inward- to the outward-facing open conformation ($IIC_o \rightarrow IIC^o$) which might be a stochastic process (Drew and Boudker 2016).

Three of the aforementioned structures (Fig. 8.3) are in an outward-facing and three in an inward-facing conformation allowing to trace the structural transitions accompanying substrate binding, translocation and release (Fig. 8.4).

The process of *substrate binding and closure of the binding site* was inferred from a molecular dynamic simulation starting with the outward facing closed conformation of *bc*IIC^{mal} (5IWS) (McCoy et al. 2016) on one hand, and by comparison of the IIC^{asc} structures 4RP9 and 4RP8 (Luo et al. 2015) on the other. The simulation showed that the *bc*IIC^{mal} binding cavity can be opened ($IIC^{cS} \rightarrow IIC^{oS}$) by a slight tilt of TM7 towards AH2 and pulling away of Y249 from the sugar. The comparison of 4RP9 and 4RP8 suggested how IIC^{asc} switches from an open (IIC^{oS}, 4RP9, C2 symmetry) to an occluded conformation (IIC^{cS}, 4RP8, P21 symmetry) by a rigid-body rotation of TD relative to SD. In both cases the outward-facing binding site can be converted from open to close by small conformational movements.

Translocation of the sugar binding cavity across the membrane ($IIC^{cS} \rightarrow IIC_{oS}$) can be traced by comparing the outward and inward oriented conformations of *bc*IIC^{mal} (5IWS $\rightarrow$ 6BVG; McCoy et al. 2016; Ren et al. 2018) and of IIC^{asc} (4RP8 $\rightarrow$ 5ZOV; Luo et al. 2015, 2018). Translocation in *bc*IIC^{mal} is accomplished by a 9 Å vertical translation and a 44° rotation of TD relative to SD. Translocation in IIC^{asc} is accomplished by vertical and horizontal translocations of 14 and 7 Å and a rotation of 11° of TD relative to the SD. The protein-sugar H-bonding interactions in the binding cavities are preserved in the outward and the inward facing conformation (Fig. 8.5a, b). Collective variable-based steered molecular dynamics (CVSMD) simulations further indicated that the H-bonding interactions are preserved also during the translocation process and that translocation is spontaneous (Lee et al. 2017).

Substrate release into the cytoplasm is the final step (Fig. 8.4 $IIC_{cS}(II) \rightarrow IIC_o$) entailing sugar phosphorylation by phospho-IIB. A comparison of electron crystallographically obtained projection maps of 2D crystals from IIC^{glc} embedded in D-Glc and L-Glc (binding and non-binding enantiomers) with the calculated projection map of the *bc*IIC^{chb} (3QNQ) indicated that Glc binding/release is correlated with changes at the reentrant loop HP1 (Kalbermatter et al. 2017). Once dephospho-IIB has dissociated from the exit of the binding site (IIC_{oS}), electrostatic repulsion of E355 might expel the negatively charged sugar-phosphate ($IIC_{oS} \rightarrow IIC_o$). Glc-6-phosphate indeed is a poor substrate of IIC^{glc} as shown in the scintillation proximity assay (Jeckelmann et al. 2011).

Single-molecule fluorescence resonance energy transfer (smFRET) between donor acceptor fluorophores at residue 288 in the PL2 and at residues 340 in HP1 of *bc*IIC^{mal}, indicated that in solution and in the absence of maltose the dimer existed in two states (Ren et al. 2018). The donor acceptor FRET distances in solution correlated with the distances between the corresponding residues in the crystals of outward and inward facing *bc*IIC^{mal}. The equilibrium ratio between the two population in solution was 3.2, indicating that the outward orientation is ca. 3 kJ/mol more stable. Only two populations of *bc*IIC^{mal} could be resolved by FRET, indicating that dimeric *bc*IIC^{mal}

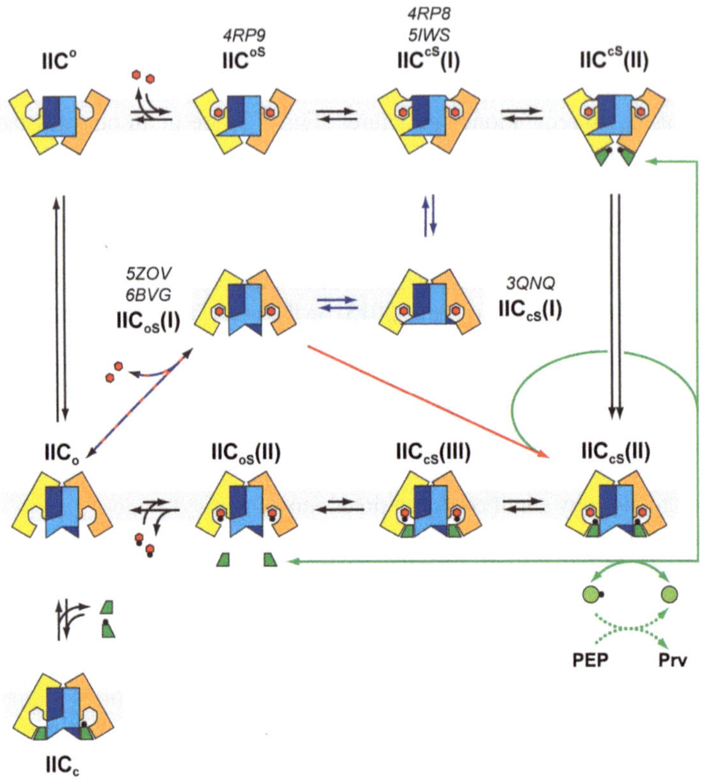

Fig. 8.4 Model of the transport and phosphorylation cycle of PTS transporters. An elevator-type transport is coupled with substrate phosphorylation from the cytoplasmic side (group translocation) . Four states are represented by crystal structures (PDB labeled), seven hypothetical states are postulated based on kinetic, biophysical and mutant experiments. The transport domains (orange and yellow) slide along the immobile scaffold/dimerization domains (blue and cyan) and thereby move the binding cavity with the substrate (red hexagon) between an outward (IIC^{oS}) and an inward facing (open) state (IIC_{oS}). After closure of the binding cavity (IIC^{cS}), the transporter isomerizes to the inward facing closed state (IIC_{cS}). It is not known whether the two protomers isomerize independently or in synchrony. In vectorial phosphorylation (group translocation) the isomerization rate is increased 100–1000 fold by the activity of the phosphorylated IIB domain (dark green) and rendered unidirectional by the concomitant phosphoryl transfer from phospho-IIB to the substrate in the binding cavity and/or by the subsequent release of the phosphorylated substrate ($IIC^{cS}(II) \rightarrow IIC_{oS}(II)$). Facilitated diffusion (blue arrows), that is isomerization without phosphorylation is a slow reaction ($IIC^{cS}(I) \rightarrow IIC_{oS}(I)$). Isomerization is not affected by non-phosphorylated IIB, but mutations mainly in the transport domain can increase the rate of facilitated diffusion. Non-vectorial phosphorylation (red arrow) of intracellular substrates ($IIC_{oS}(I) \rightarrow IIC_{cS}(II) \rightarrow IIC_{oS}(II)$) is a rapid reaction. Green arrows depict the continuous rephosphorylation of IIB by PEP mediated by EI, HPr (not shown) and IIA (faint green circle). The isomerization of substrate-free $IIC_o \rightarrow IIC^o$ must be a fast process. It is not known whether it is spontaneous or whether the rate is increased by IIB (state IIC_c). Superscripts indicate outward facing, subscripts an inward facing conformations, o and c refer to an open and closed substrate binding cavity, S indicates substrate is in the binding cavity

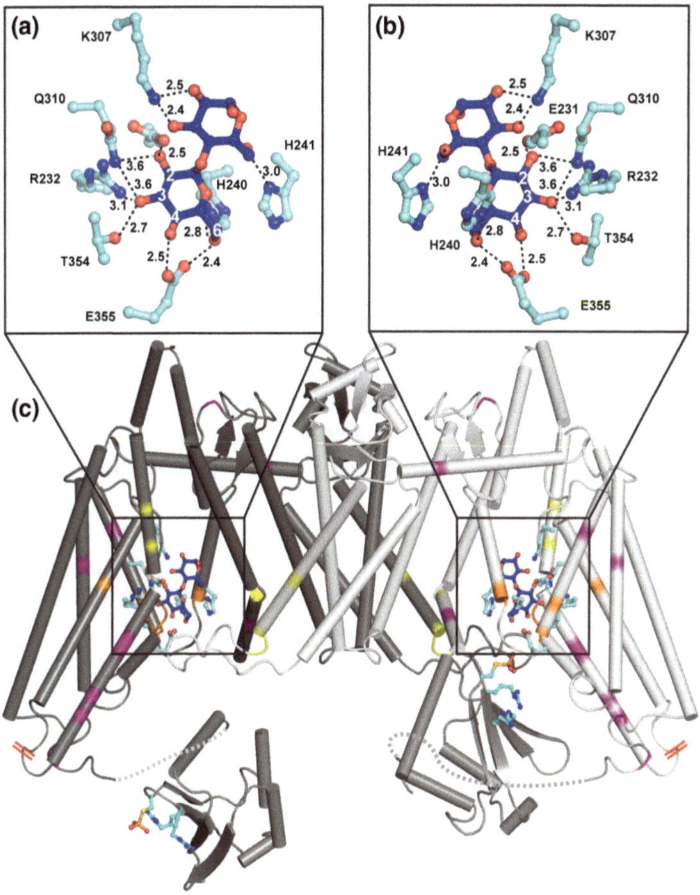

Fig. 8.5 The substrate binding cavity of *bc*IIC^{mal} and mutant sites of IICB^{glc} projected onto the structure model of *bc*IIC^{mal}-IIB^{mal}. **a**, **b** Substrate binding sites in two orientations related by 180° rotation. Shown are invariant residues in hydrogen bonding distance to the sugar hydroxyl groups. **c** The positions of the IICB^{glc} mutant sites in the *bc*IIC^{mal} model are inferred from the structural alignment generated by the Phyre2 web portal for protein modeling, prediction and analysis (Kelley et al. 2015) as shown in Fig. 8.6. The mutant sites are color-coded according to phenotype: Facilitated diffusion and uncoupling of transport and phosphorylation, *orange*; relaxed substrate specificity, *purple*; attenuated uptake but normal non-vectorial phosphorylation activity, *yellow*; point mutations that reduce activity to <50% of control, *grey*; site ‖ (*red*) of two-fragment split (IIC′ and IIC″B) with 30% activity, and C/N-termini of circularly permuted variant (IIC″BC′^{glc}) with 24% uptake and 70% non-vectorial phosphorylation activity (Beutler et al. 2000a, b). Models of the IIB domain are shown docked (DeLuca et al. 2015) to the open binding cavity and in an unrestricted orientation. Active site phosphocysteine and two nearby invariant arginines, which are essential for the phosphoryltransfer from IIB to the sugar (Lanz and Erni 1998), are shown as sticks. The linkers (*grey dotted lines*) are not to scale and shown only to illustrate the connectivity. They are of different length in *bc*IICB^{mal} and IICB^{glc}

existed in two different, symmetrical conformations. It appears from this that the protomers in the dimer are conformationally coupled and not independent of each other. Whether addition of substrate influenced the equilibrium between the two states is not reported.

Oligomeric Structure and Protomer Interactions

The oligomeric structure and function of PTS transporters was a major subject of biochemical and biophysical studies soon after the first PTS transporters were purified. Questions were: (i) Do inactive protomers functionally complement each other (interallelic complementation)? (ii) How fast do protomers exchange? (iii) Do heterodimers between subunits of different PTS transporters exist and how do they function? (iv) Do monomeric subunits function and are they physiologically relevant?

The PTS transporter of Mtl, $IICBA^{mtl}$ was the first to be purified (Jacobson et al. 1979; Robillard and Blaauw 1987), and $IICB^{glc}$ followed soon afterwards (Navdaeva et al. 2011; Erni et al. 1982; Erni and Zanolari 1986; Bouma et al. 1987; Waeber et al. 1993). The mechanism of transport and phosphorylation, the oligomeric structure, and domain interactions of these two transporters were comprehensively reviewed in 1999 by Robillard and Broos (1999) and over 50 $IICB^{glc}$ mutants and their properties have been tabulated (Erni 2001). The integral membrane domains IIC^{glc} (P69786) and IIC^{mtl} (P00550) share 32 and 19% sequence identity with *bc*IIC^{mal}, and 18% identity with *bc*IIC^{chb}. With the X-ray structure of *bc*IIC^{mal} as reference biochemical and biophysical results can be interpreted more conclusively.

Structure of the Dimer

The X-ray structures depict the IIC units as a dimer of protomers related by a twofold axis normal to the plane of the membrane. The buried surface area between protomers is 2700 Å^2, and ~75% of the interface residues are nonpolar, characteristic of obligate homodimeric proteins (Ali and Imperiali 2005). The two protomers simultaneously bind one substrate per protomer and the transport pathways along the two protomers are 50 Å apart and separated by the two SD.

Gelfiltration, ultracentrifugation (Meins et al. 1988; Erni 1986), chemical cross-linking (Meins et al. 1988; Van Montfort et al. 2002), radiation inactivation (Pas et al. 1987), fluorescence cross-correlation spectroscopy (Veldhuis et al. 2006), interallelic complementation, blue-native PAGE analysis, 2D-crystal and single particle analysis by TEM and atomic force microscopy of detergent purified $IICBA^{mtl}$ and IIC^{glc} provided strong evidence that PTS transporters are dimeric (Jeckelmann et al. 2011; Koning et al. 1999).

Interallelic Complementation

Interallelic complementation between inactive mutants was demonstrated for $IICBA^{mtl}$, $IIBCA^{bgl}$ and $IICB^{glc}$. Transport and phosphorylation activity could be detected when the complementing proteins were either coexpressed, or separate membrane preparations were detergent solubilized and then mixed to allow subunit exchange (Chen and Amster-Choder 1998; Aboulwafa and Saier 2011). The rate of $IICBA^{mtl}$ subunit exchange was slow (40 min to hours), and it depended on the size of the detergent micelle aggregates, on the phosphorylation state of the subunits, and on substrate binding. $IICBA^{mtl}$ remains dimeric down to a concentration of 1 nM, and monomers form only transiently (Veldhuis et al. 2006).

IIA/IIB complementation was demonstrated many times e.g. between the inactive phosphorylation site mutants H554A, and C384H of $IICBA^{mtl}$ (Weng et al. 1992) and between the C24S and H547R mutants of $IIBCA^{bgl}$ (Chen and Amster-Choder 1998). IIA/IIB complementation convincingly demonstrates that phosphorylgroups are transferred not only between IIA and IIB domains on the same protomer, but also between different protomers (for review see Robillard and Broos 1999).

IIB/IIC complementation occurs between protomers with inactive IIC and IIB. $IICBA^{mtl}$ heterodimers G196D/C384S, E257A/C384H, E257D/C384H and $IICB^{glc}$ heterodimers H211N/R424K, H211N/R426K had 10% of wild-type activity, indicating that B/C interprotomer phosphate transfer is less efficient than intraprotomer transfer (Lanz and Erni 1998; Boer et al. 1996). Recombinant $IIBC^{glc}$ with the B domain at the N- instead of the C-terminal end had between 20% and 70% activity depending on the length of the peptide linker. Inactive $IICBA^{mtl}$(C384H), inactive $IICB^{glc}$(C421S), a recombinant IIC^{mtl} domain, and a recombinant IIC^{glc} domain could all be complemented (inefficiently) with an excess of soluble recombinant $IIBA^{mtl}$ and IIB^{glc}, respectively (Boer et al. 1994, 1996; Buhr et al. 1994).

IIC/IIC interallelic complementation has been reported twice, between inactive $IICBA^{mtl}$ protomers H195A and E257D (Saraceni-Richards and Jacobson 1997b), and between $IICBA^{mtl}$ single Trp mutants F97W, F126W and F133W which bind Mtl but lack PT activity (Vos et al. 2009a). These results cannot be rationalized on the evidence of crystal structures (Fig. 8.6).

Negative dominance, that is inactivation of the active protomer by an inactive one, has not been observed. Although the H196D/C384S double mutant of $IICBA^{mtl}$ can neither be phosphorylated nor bind substrate, heterodimers with wild-type (wt) subunits have the same specific activity per wt monomer as wt homodimers (Boer et al. 1996). This result can be expected when the inactive protomer remains flexible to switch in concert with the active protomer between the inward and outward conformations or when the promoters are structurally uncoupled and functionally independent.

Intergenic complementation, that is phosphoryltransfer between domains of different transporters was reported, too (Fig. 8.1): For instance between $IICBA^{glcnac}$ and $IICB^{glc}$ ($IIA^{glcnac} \rightarrow IIB^{glc}$) and between $IIBCA^{bgl}$ and $IICB^{glc}$ ($IIA^{bgl} \rightarrow IIB^{glc}$ and $IIB^{Glc} \rightarrow IIC^{Bgl}$) (Schnetz et al. 1990; Vogler et al. 1988). Whether transfer occurs within transient clusters of homodimeric transporters or within transient heterodimers is not known.

Monomeric IIC was detected in detergent micelles and lipid bilayers where the apolar dimer interface may be stabilized by contacts with the aliphatic chains (Robillard and Broos 1999; Robillard and Blaauw 1987; Leonard and Saier 1983; Stephan and Jacobson 1986; Khandekar and Jacobson 1989; Lolkema and Robillard 1990; Lolkema 1993). It was suggested that monomers retain PT activity. The soluble high-speed ultracentrifugation supernatant of a cell lysate obtained by sonic disruption had Glc, Man and Mtl sugar PT activity. The active fraction had an apparent molecular mass of 40–50 kDa in gel-filtration corresponding to the monomer (Aboulwafa and Saier 2003). Transphosphorylation activity was recorded concurrent with sedimentation of "monomeric" $IICB^{glc}$ through a glycerol gradient containing [^{14}C]Glc and Glc6P (Meins et al. 1988). However, it cannot be excluded that in these and similar experiments the "monomer activity" is due to a small percentage of dimers in rapid equilibrium with the monomer under the assay conditions (temperature, pH, ionic strength).

In summary, in line with the X-ray structures, the biophysical, biochemical and genetic evidence is overwhelming that functionally active PTS transporters are dimeric (Robillard and Broos 1999). It is however not known, whether the two protomers function independently of each other or are allosterically coupled.

Substrate Binding, Transport and Phosphorylation: Important Residues and Point Mutations

The main objective of past mutant studies was generating the following phenotypes: (i) Uncoupling of transport and phosphorylation resulting and facilitated diffusion; (ii) Attenuation of transport activity with retention of non-vectorial phosphorylation activity (Fig. 8.2a, b); (iii) Relaxing substrate specificity—a complete change of substrate specificity was never found. The majority of 24 $IICB^{glc}$ mutants with these phenotypes were located in TD, and only 7 in the SD (Figs. 8.5c and 8.6; summarized in Erni 2001).

Facilitated Diffusion, Transport Without Phosphorylation

IIC mediated facilitated diffusion (Fig. 8.4, blue arrows) was measured with purified $IICBA^{mtl}$ and IIC^{mtl} in proteoliposomes (Elferink et al. 1990). With a rate of 2.5 nmol Mtl/nmol IIC^{mtl}/min compared to 250 nmol/nmol IIC^{mtl}/min of vectorial phosphorylation, facilitated diffusion does not play a significant role in transport. But it proved useful to conveniently ascertain the activity of purified transporters before crystallization (McCoy et al. 2016; Cao et al. 2011; Ren et al. 2018). IIC mutants with a higher rate of facilitated diffusion, so-called uncoupled mutants, were identified in

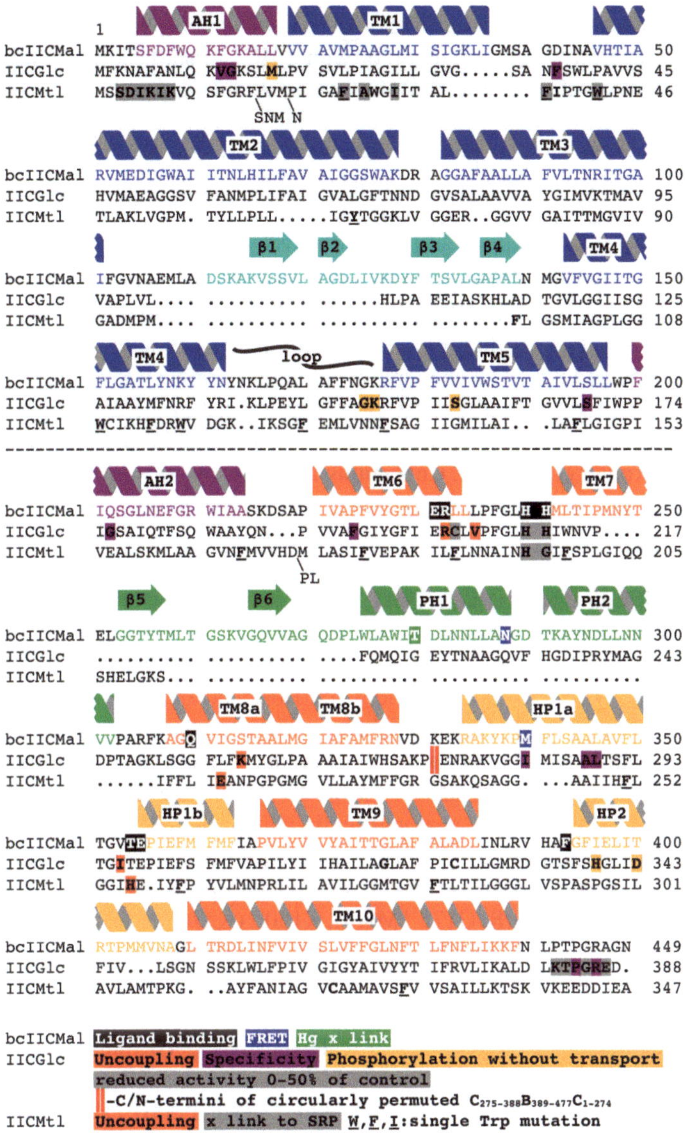

Fig. 8.6 Structural sequence alignment of IIC^{glc} and IIC^{mtl} with *bc*IIC^{mal}. The *bc*IIC^{mal} secondary structure above its sequence is shown in the same color code as in Fig. 8.3. *bc*IIC^{mal} residues in hydrogen bonding distance of the substrate, *white*; modified for FRET studies, *blue*; and crosslinked with mercury (Ren et al. 2018), *green.* The phenotypes and properties of IIC^{glc} and II^{mtl} mutations are highlighted and color–coded separately for each sequence, as indicated at the bottom. IIC^{mtl} residues cross-linked to the signal recognition particle (SRP), *grey* (exception H195, G196 are *shaded* as active site residues, not crosslinked). The *dashed black line* marks the border between scaffold and transport domains

bacteria growing on a PTS sugar without a functional phosphoryl transfer chain, for instance in an enzyme I/HPr mutant, or with a IIC mutant lacking (functional) IIA and IIB domains/subunits (Fig. 8.1). In most cases uncoupling was conditional, the rate of facilitated diffusion was increased, but the capability to phosphorylate was not lost when a functional phosphoryltransfer chain was restored. The best uncoupled mutants attained growth rates of between 20 and 50% of the control (Otte et al. 2003; Postma 1981; Ruijter et al. 1991, 1992).

The uncoupling mutations of $IICB^{glc}$, R203S (R232 TM6) and V206A (L235 TM6) are located in a conserved sequence next to the invariant, substrate binding H211/H212 pair (H240/H241 TM6/7) (indicated in parentheses are the structurally corresponding residues of *bc*IIC^{mal} and their location in the *bc*IIC^{mal} structure; Figs. 8.5c and 8.6). The uncoupling mutations I296N (V353 HP1b) of $IICB^{glc}$, and H256P/Y (T354 HP1a/b) of $IICBA^{mtl}$ are located in the conserved GI(T/H)E motif of the sugar binding site, which is directly involved in substrate binding (Otte et al. 2003; Ruijter et al. 1992). Uncoupling mutations K257N (S314) of $IICB^{glc}$ and E218A/V (I312) of $IICBA^{mtl}$ are located in TM8a (Fig. 8.6).

Phosphorylation Without Transport

Vectorial (transport) und non-vectorial PT activity (Figs. 8.2 and 8.4) of PTS transporters are not strictly coupled (Kundig and Roseman 1971b). $IICB^{glc}$ can replace glucokinase to (non-vectorially) phosphorylate intracellular Glc (Nuoffer et al. 1988). Mutants capable of phosphorylating intracellular Glc but unable to transport Glc were selected for growth on maltose (which is metabolized to intracellular Glc) in the presence of the toxic Glc analog α-methylglucoside (αMG) (Buhr et al. 1992). The following mutations resulted in a strong impairment of αMG uptake by intact cells with only minimal decrease of non-vectorial phosphotransferase activity in vitro (Figs. 8.5c and 8.6): G149S (G175 TM4/5), K150E (K176 TM4/5) and S157F (V183 TM4/5) are near the conserved loop (CL in Fig. 8.3a), which blocks the cytoplasmic access to the substrate-binding cavity on the neighboring protomer (Figs. 8.5c and 8.6). A mutation in or near this loop could facilitate the access to the binding site from the cytoplasmic side at the expense of a slower uptake, viz. prevent the transition of $IICB^{glc}$ from the inward to the outward facing conformation. How the remaining mutations M17T/I (L17 AH1) in the SD and H339Y (I396 HP2) and D343G (T400 HP2) favoured non-vectorial phosphorylation over uptake is not clear. And it is also not clear, whether resistance to the toxic αMG (the counterselection agent) is caused by a defective transport mechanism or by a change of selectivity for Glc and against αMG. Some mutants adjacent to M17 indeed affect substrate specificity (see next section).

Sugar Specificity

In keeping with the lock and key analogy, protein-ligand interactions can be characterized with protein mutations (section "Relaxed Sugar Specificity of IICB^{glc}") and with substrate analogues (section "Sugar Analogues: Inhibitors and Pseudosubstrates of IICB^{glc}").

Relaxed Sugar Specificity of IICB^{glc}

The specificity of PTS transporters for their substrate is not absolute. IICB^{glc} for instance tolerates modification at C-1, the Man transporter (IICIID^{man}) at C-2 (Garcia-Alles et al. 2002a, b) and the binding cavity of *bc*IIC^{mal} is large enough to accommodate not only maltose but also maltotriose (McCoy et al. 2016). Mutants with a relaxed sugar specificity were isolated by enrichment and selection of bacteria on a metabolizable sugar which is not the (main) substrate of the transporter under investigation (Zeppenfeld et al. 2000; Oh et al. 1999; Notley-McRobb and Ferenci 2000; Begley et al. 1996; Kornberg et al. 2000). "Relaxed sugar specificity" can be caused either by a structural change in the ligand binding site, or by overexpression of a transporter and hence increased rate of uptake of a low affinity substrate (regulatory instead of structural mutation).

Most IICB^{glc} (*bc*IIC^{mal}) mutations were detected in TD near the sugar binding site: F195L (next to F225 in TM6), G281D (K338 HP1a), I283T (M340 HP1a), A288V (A345 HP1a), L289Q (L346, HP1a), G320V/S (G377, TM9) (Figs. 8.5c and 8.6). Further away are G176D (Q202 AH2) in the periplasmic amphipathic helix that connects the scaffold with TD. P384R and E387G are in the linker connecting the IIC with the IIB domain (C-terminus of *bc*IIC^{mal}). This linker is strongly conserved, and certain mutations in the linker compromised PT activity of IICB^{glc} more than the complete absence of the linker (Lanz and Erni 1998). The E387G mutation in the linker resulted in a 7-fold overexpression of IICB^{glc}, which may explain the "relaxed" specificity. The linker sequence was shown to be part of or next to the binding site for the transcription repressor Mlc. Mlc is kept away from the operator DNA by binding to IICB^{glc} in the non-phosphorylated state, which predominates when Glc is transported and phosphorylated. In the absence of Glc, IICB^{glc} remains phosphorylated, Mlc is released to repress *ptsG* (IICB^{glc}) transcription (Lee et al. 2000; Seitz et al. 2003). Similarly the G176D mutation may be a regulatory rather than substrate specificity mutant (Plumbridge 2000).

Specificity mutations in SD were less frequent: V12F/G (F12 AH1), G13C (G13 AH1), F37Y (D42 TM1/2). Val12 and Gly13 were isolated repeatedly in different laboratories (Notley-McRobb and Ferenci 2000; Kornberg et al. 2000; Manché et al. 1999). These mutants displayed mixed phenotypes of relaxed substrate specificity, facilitated diffusion of fructose, and upregulated IICB^{glc} expression (Aboulwafa et al. 2003; Crigler et al. 2018). They are located in AH1 and TM1, the helix which lines the sugar binding cavity (Fig. 8.5c). It is thus likely that residues along TM1 exert a subtle influence on sugar specificity although none appears to directly interact with the sugar (Fig. 8.5 and section "Structure of the Sugar Binding Cavities of *bc*IIC^{mal} and IIC^{asc}"). This proximity between TM1 and ligand binding site was noticed in IICBA^{mtl}, too. For instance, the fluorescence of W30 in TM1 was the most sensitive of all single tryptophan mutants to Mtl binding (Dijkstra et al. 1996, 1997; Broos et al. 2000).

How the remaining four mutations in the SD, S169F/P (S195 TM5) and G176D (Q202 AH2) affected substrate specificity is not understood.

Sugar Analogues: Inhibitors and Pseudosubstrates of IICB^{glc}

Nineteen glucose analogues were characterized as pseudosubstrates and inhibitors of Glc uptake and non-vectorial phosphorylation by IICB^{glc} (Garcia-Alles et al. 2002a, b).

Results can be summarized as follows: (i) Non-vectorial phosphorylation of Glc was biphasic (concave Eadie-Hofstee plots) indicative of two binding sites, one of high (ligand dissociation constant K_{d1}) and one of low affinity (K_{d2}). The biphasic kinetics were observed with IICB^{glc} in crude membranes as well as highly purified IICB^{glc}, indicating that they were not caused by the simultaneous presence of IICB^{glc} and a contamination with Glc PT acitivity. (ii) Uptake of Glc in proteoliposomes, in contrast, was linear indicative of only one high affinity binding site. (iii) The pseudosubstrates and inhibitors differed in their preference for high or low affinity binding sites, and whether they preferentially inhibited non-vectorial phosphorylation or uptake (vectorial phosphorylation) .

The Glc analogue modified at C-1 (1F-Glc) behaved like Glc, it was taken up by intact cells, biphasically phosphorylated, and it competed for high and low affinity sites. The C-1 epoxide (2′,3′-epoxypropyl β-D-glucopyranoside) also behaved like Glc but competed against Glc only for the high affinity binding site ($K_{dInh1}/K_{dGlc1} =$ 3 μM/4 μM and $K_{dInh2}/K_{dGlc2} = 140$ μM/700 μM).

Glc analogues modified at C-2 (including OH epimers) were phosphorylated inefficiently (20% k_{cat}/K_m), competed only for non-vectorial phosphorylation of Glc but did not inhibit uptake (high affinity binding). Glc analogues modified at C-3 and C-4 were neither phosphorylatable substrates nor inhibitors of uptake. The exception is 3F-Glc, a pseudosubstrate which was phosphorylated at 50% efficiency in the low affinity site, inhibited uptake of Glc and was taken up. A OH-3 that can be replaced by fluorine most likely is a hydrogen acceptor (Dalvit and Vulpetti 2016). The remaining OH groups are acceptors, donors or both, in agreement with the protein-ligand hydrogen bonding network depicted in the *bc*IIC^{mal} structure (Fig. 8.5c) (McCoy et al. 2016; Ren et al. 2018).

Glc analogues modified at C-6 could not be phosphorylated. Some were strong inhibitors of non-vectorial phosphorylation but only weak inhibitors of uptake. The C-6 epoxide (6,7-anhydro-D,L-glycero-α-D-glucoheptopyranoside) showed biphasic inhibition. It had a 5 times lower affinity than Glc for the high affinity site ($K_{dInh1}/K_{dGlc1} = 19$ μM/4 μM). For the low affinity site to the contrary it had a 24 times higher affinity ($K_{dInh2/dGlc2} = 8$ μM/190 μM). The C-6 aldehydes (D-

Gluco-hexodialdo-1,5-pyranose, and the corresponding methyl-α-glucopyranoside) were the strongest inhibitors of both high (K_{dInh1}/K_{dGlc1} = 5 μM/15 μM and low affinity (K_{dInh2}/K_{dGlc2} = 3 μM/370 μM) sites. Inhibition was fully reversible and the aldehydes did not form Schiff bases with amino groups of the protein. The C-6 aldehyde exists as diol in water (–CHO + H_2O → –$CH(OH)_2$). They were synthesised to test whether the diol could be phosphorylated, and whether the resulting phosphorylhemiacetal (–CH(OH)(O–PO_3H_2) would hydrolyse spontaneously (Haddad et al. 1999), resulting in uncoupled consumption of PEP. This was not the case.

Ten Glc analogues contained chemically reactive groups (epoxi-, α-halocarbonyl-, isothiocyanyl and allylgroups). 6BrAc-Glc (6-O-Bromoacetyl-D-glucopyranose) inactivated $IICB^{glc}$ at the highest rate. $IICB^{glc}$ could be protected by phosphorylation, indicating that the active site Cys421 on the IIB domain was the target of alkylation. Inactivation by 6BrAc-Glc was 2.5 times faster than by the more reactive, unspecific iodoacetamide, indicating that 6BrAc-Glc binding was selective. Preincubation of bacteria with 6BrAc-Glc irreversibly blocked uptake of Glc. The simultaneous presence of Glc did not protect but to the contrary increased the rate of inactivation 20-fold. It is possible that if only one protomer in the dimer were occupied by 6BrAc-Glc the other could transport and phosphorylate Glc, and that the (transiently) dephosphorylated cysteine of the IIB domain then could react with 6BrAc-Glc.

In summary: Inhibitors of uptake that bind to the periplasmic site must have a free OH-6 whereas inhibitors of non-vectorial phosphorylation that bind to the cytoplasmic site may or may not have one. Glc enters the outward facing site, OH-6 first, through a narrow opening that excludes bulkier substituents. It enters the inward facing site, OH-1 first, through an opening that must be wider in order to allow access to the substrate for the IIB phosphoryldonor domain.

Systematic Mutations Without a Strong Phenotype

Thirteen single Phe to Trp and Tyr to Trp mutants in Trp less $IICBA^{mtl}$ were constructed (Fig. 8.6) for spectroscopic studies ((Opacic et al. 2012), and references there). They retained between 15 and 65% of the wild-type non-vectorial PT activity, none was inactive or unstable.

Kinetics and Mechanism of Sugar Transport and Phosphorylation

All X-ray structures solved to date show symmetric homodimeric transporters in either and outward or an inward facing conformation (Fig. 8.3). The two protomers are exactly superimposable, suggesting that they are conformationally coupled, not independent and acting in concert. Single molecule FRET experiments with *bc*IIC^{mal}

in detergent solution showed two discrete populations of molecules with FRET distances in good agreement with the distances measured in the inward and outward facing X-ray structures (Ren et al. 2018; see section "Mechanism of Substrate Binding and Transport"). No third population corresponding to a dimer with one inward and one outward oriented protomer could be discerned.

The Conformation of the Protomers in Dimeric IICBA^{mtl}

A large body of enzyme kinetic, ligand binding and Trp fluorescence studies with IICBA^{mtl} was provided by G.T. Robillard, J. Broos, B. Poolman and collaborators at Groningen University. Their studies indicated that the IICBA^{mtl} dimers existed in different states. In their interpretation these states were determined by unequal structures of the protomers in the (asymmetrical) dimer (AB $\leftrightarrow$ BA) rather than by two (or more) populations of symmetrical dimers (AA $\leftrightarrow$ BB) (Opacic et al. 2012). The experimental evidence for different structures of the protomers can be summarized as follows.

Time resolved Trp fluorescence and unequal iodide quenching of fluorescence indicated that identical Trp in the two protomers were in different microenvironments and experienced different solvent accessibility (Opacic et al. 2010, 2012; Vos et al. 2009b). Similarly FRET distance measurements between azi-mannitol and single-Trp IICBA^{mtl} indicated that the distances between azi-Mtl and the two identical Trp in the protomers were not equal, and that the Trp residues were positioned non-symmetrically relative the Mtl binding site(s) (Opacic et al. 2010). This result can be expected if only one of the two binding sites in the dimer was occupied by Mtl, or if both binding sites were occupied but in protomers of different conformation.

Mannitol Binding of IICBA^{mtl}

Mtl binding and enzyme kinetic studies consistently pointed to the existence of a high affinity (K_d 45–600 nM) and a low affinity binding site (K_d 5–10 μM). The stoichiometry was only one high affinity binding site per dimer (Pas et al. 1988; Meijberg et al. 1998a). The low affinity binding site, in contrast, could not always be detected, possibly depending on the sensitivity of the methods and experimental conditions employed. High affinity was assigned to the periplasmically oriented outward facing binding site, low affinity to the cytoplasmically oriented inward facing site. The two different Mtl dissociation constants qualitatively agreed with the Michaelis Menten constants of uptake (vectorial phosphorylation, $K_m < 10\ \mu M$, V_{max} 2.1 nmol min^{-1} mg lipid^{-1}) and non-vectorial phosphorylation (K_m 66 μM, V_{max} 22 nmol min^{-1} mg lipid^{-1}) of purified IICBA^{mtl} reconstituted in phospholipid vesicles (Elferink et al. 1990).

Mtl binding modulated the Trp fluorescence and phosphorescence of single Trp mutants (Veldhuis et al. 2005). Residue 97 is close to or may be part of the short β-sheet rich periplasmic loop 1 (PL1) according the structural alignment (Figs. 8.3

a and 8.6). The phosphorescence spectrum of Trp 97 was characteristic of a β-sheet rich structure, and Mtl binding induced its partial unfolding. Residues 126, 133 are in CL, the loop between TM4 and TM5, which reaches from the SD of one protomer to the cytoplasmic exit of TD of the other (Figs. 8.3b–d and 8.6). Mtl binding induced a more ordered structure around these residues. This is the first indication that PL1, and additional evidence that CL respond to sugar binding, although their exact function remains to be determined. In this section examples are given, how Mtl binding affects the IIC domain, in the next how the IIB domain affects IIC.

C/B Domain Interactions of IICBA^{mtl}

The IIB domain and its phosphorylation state affected both, Mtl binding and the conformation of the IIC domain. (i) The IIB domain had a moderate effect on the strength of the Mtl::IICBA^{mtl} high affinity complex (K_d 45 nM). In the absence of IIB the K_d was increased three fold. IIB phosphorylation had no effect on the affinity for perseitol, a non phosphorylatable analog of Mtl (Lolkema et al. 1993). (ii) IIB with a mutated Cys384 phosphorylation site increased the K_d three to six fold. In parallel the dissociation rate constant increased three to ten fold ($t_{1/2}$ of exchange between 45 and 5 s). These variations were interpreted in terms of two binding sites per dimer, a slow exchanging occluded one and a fast exchanging initial binding one. The C384 mutants would either enhance the rate of isomerization between the two states or completely block it, such that in either case only a fast exchange reaction was observed (Boer et al. 1995). The C384D mutation, mimicking a negative charge of the phosphothioester, was inactive and had only a small effect on sugar binding (Boer et al. 1995). Similarly phosphorylation of the IICBA^{mtl} mutant C384S, which could be stably phosphorylated at serine, did not detectably affect the environment of the Mtl binding site, as concluded from Trp fluorescence (Opacic et al. 2010). (iii) Although not affecting Mtl binding, phosphorylation of IIB did induce significant conformational changes in the IIC domain. It increased the accessibility of engineered Cys at residues 85, 90–110 to sulfhydryl reagents (Vervoort et al. 2005). These residues were accessible from the periplamic side, and according to the structural alignment are located in or next to PL1 (Figs. 8.3 and 8.6).

The C/B domain interaction of IICBA^{mtl} was probed by differential scanning calorimetry (DSC) and isothermal titration calorimetry (ITC) in the presence and absence of Mtl (Meijberg et al. 1998a, b). The comparison of the thermodynamic parameters derived from DSC of IIC^{mtl} and IICB^{mtl} suggested: (i) that Mtl binding induced a conformational change in the IIC domain that was propagated to the B domain, and (ii) that solvent-exposed parts of the proteins were removed from the surrounding water as a result of docking of preexisting surfaces or folding of unstructured parts induced by the domain interactions.

Removal of the IIB domain with trypsin destablized the IIC domain, reducing its temperature of unfolding by 3 °C. Destabilization was stronger in the presence of Mtl than in its absence. Heat capacity increments of Mtl binding to non-phosphorylated IICBA^{mtl} determined by ITC indicated that approximately 50 residues were involved (directly or indirectly) in the binding reaction. Phosphorylation of $IICBA^{mtl}$ and removal of the B domain both reduced this number. In this section examples are given how the IIB domain affects physical properties of the IIC domain, in the next how IIB affects sugar binding and transport by IIC.

Coupling of Protein and Substrate Phosphorylation with Transport

The standard free energy of phosphoryltransfer from PEP to a sugar is 48 kJ mol^{-1}. 80% are released in the last step, the transfer from phospho-Cys of IIB to the primary hydroxyl group of the sugar (Erni 2001; Rohwer et al. 2000). In theory this is enough to generate a concentration gradient of 10^5 to 10^6. In fact, the steady-state intracellular concentration of PEP is between 0.5 and 3 mM and of Glc6P ca. 50 μM. Two molecules of PEP are formed per Glc taken up by the PTS. One PEP is utilized to take up the next Glc, the second is withdrawn from the circle to generate ATP, serve as biosynthetic precursor, or for the uptake of a second Glc molecule. It is not understood how bacteria gear down the uptake to prevent being jammed by an overshoot of sugar phosphates.

Phosphorylation of the sugar empties the binding site and maintains the concentration gradient of the sugar. But this alone does not explain the 100 to 1000 fold rate increase of vectorial phosphorylation over facilitated diffusion (Lolkema et al. 1991). Facilitated diffusion follows Michaelis Menten kinetics, and the maximum rate of transport at saturating solute concentration is determined solely by the isomerization rate of the binding site between inward and outward orientation (Fig. 8.4 $IIC^{cS} \leftrightarrow IIC_{cS}$). For a 100 to 1000 fold rate increase the activation energy of isomerization must be reduced by 10–15 kJ mol^{-1} (equivalent to 1–2 hydrogen bonds). How this is achieved, how translocation and phosphorylation are coupled is not understood. The cartoon model (Fig. 8.4) suggests that it is the binding of phospho-IIB to the IIC::substrate complex, which induces a conformation change raising the complex from a minimum energy (high affinity) state to a higher energy (low affinity) state (Fig. 8.4 $IIC^{cS}(II) \rightarrow IIC_{cS}(II)$) from which the phosphorylated sugar can escape. The rate of non-vectorial sugar phosphorylation in the cytoplasmic low affinity site ($IIC_{cS}(II) \rightarrow IIC_{cS}(III) \rightarrow IIC_{oS}(II)$) must be faster than isomerization back to the minimum energy high affinity state ($IIC_{cS}(II) \rightarrow IIC^{cS}(II)$).

The model shown in Fig. 8.4 presumes that the two protomers operate in symmetric synchrony as suggested by the X-ray structure. Yet additional single molecule FRET experiments are required to elucidate the dynamics of the transport reaction.

Structure and Function of Mannose (MFS) Superfamily Transporters

The Man transporter of *E. coli* was one of the first PTS transporters to be discovered and characterized (Kundig and Roseman 1971a; Curtis and Epstein 1975). It had a broad substrate specificity, including Glc (!), mannosamine, N-acetylglucosamine, and it was the only transporter for Man, hence its name. It is believed to be a scavenger of carbohydrates released during cell wall remodeling. The transporters of the Man superfamily consist of three or four subunits: IIA, IIB, (or IIAB), IIC and IID (Erni and Zanolari 1985; Williams et al. 1986; Erni et al. 1987, 1989; Bouma and Roseman 1996). IIA^{man} and IIB^{man} are soluble proteins, each containing a histidine phosphorylation site. In *E. coli* the IIA and IIB domains are linked by a 20 residue long Ala-Pro-rich linker (Erni et al. 1989; Stolz et al. 1993; Markovic-Housley et al. 1994; Wu et al. 1990; Perham 1991). IIA forms a stable homodimer, IIB is monomeric (Nunn et al. 1996; Schauder et al. 1998). IIC and IID are integral membrane subunits, which form dimers of two IICIID protomers. IIC and IID are fused in the transporter (Q8RD53) of *Thermoanaerobacter tengcongensis* (Navdaeva et al. 2011).

Subunit Composition, Kinetics and Mechanism of the E. coli Mannose Transporter

The atomic structure of the IICIID complex is not known. IIC^{man} and IID^{man} cannot be dissociated without denaturation of the subunits. The $IIAB_2^{man}$:$(IIC^{man}IID^{man})_2$ complex can be purified intact due to strong binding between IIB and IICIID. It is controversial, whether IIC^{man} and IID^{man} can be expressed separately, and how stable they are (Williams et al. 1986; Erni et al. 1987).

According to 80 experimental protein fusions of IIC and IID with alkaline phosphatase and β-galactosidase IIC^{man} was predicted to contain six transmembrane segments, and IID^{man} to consist of a large N-terminal cytoplasmic domain anchored in the membrane by a C-terminal transmembrane region (Huber and Erni 1996). This topology is not consistent with any of over 10 in silico predictions (Dobson et al. 2015; Wehmeier et al. 1995; Tymoszewska et al. 2017) which, however, themselves are not consistent either. IIC is more resistant to proteases than IID (Mao et al. 1995). A functional histidine 6-tag could be inserted only at the N-terminus of IID^{man}. The His 6-tags were poorly accessible at the N-terminus of IIC^{man} and the C-terminus of IID^{man}, and inactivating at the C-terminus of IIC^{man}. Combining all evidence provided by experiments and prediction algorithms we propose the following speculative consensus topology: IIC, 8 (or 6) TM with N- and C-termini on the cytoplasmic side; IID, a N-terminal cytoplasmic domain of 130 residues followed by 3 (or 1) TM, a periplasmic domain followed by 2 TM and the C-terminus on the periplasmic side.

The purified Man transporter complex was reconstituted in proteoliposomes (Mao et al. 1995). The K_m and k_{cat} were 30 μM and 1.2 s^{-1} for vectorial phosphorylation and 0.1 mM and 3 s^{-1} for non-vectorial phosphorylation. Interestingly and in contrast to $IICB^{glc}$ (Mukhija and Erni 1996), non-vectorial phosphorylation of external Glc did not compete with export (vectorial-phosphorylation) of trapped [^{14}C]Glc in proteoliposomes. Noncompetition indicates that the transported Glc does not equilibrate with Glc in the bulk phase, and that transport and phosphorylation are either mechanistically coupled or that the rate of phosphorylation of bound Glc is much faster than the rate by which unphosphorylated Glc can dissociate.

E. coli $IICIID^{man}$ tolerates as substrate glucose with fluorine instead of hydroxyl at C-1 and C-3. Deletion of OH-1 is tolerated, but not alkylation. Substitutions at C-4 and C-6 are poorly tolerated (Garcia-Alles et al. 2002a, b). The least sensitive to modifications appears to be C-2. $IICIID^{man}$ transports and phosphorylates 2-deoxyglucose, N-acetylglucosamine, 2F-Glc. Large alterations at C-2 are tolerated by a IICIID complex of *S. typhimurium* (A0A0F6BB79/8). This complex took up zwitterionic glucoselysine and fructoselysine (KEGG compounds C20978 and C16488) which could be utilized as carbon and nitrogen sources (Miller et al. 2015).

Bacteriophage Lambda DNA Injection

Integral membrane proteins of the cytoplasmic inner membrane are essential for DNA injection by different bacteriophages (Garcia-Doval and van Raaij 2013). Of the PTS transporters, $IICIID^{man}$ is essential for certain bacteriophage lambda and $IICB^{glc}$ for phage HK97 entry (Cumby et al. 2015). Elliott and Arber (1978) observed *E. coli* mutants *(pel)*, which were resistant against bacteriophage lambda because they lacked a gene product required for DNA entry across the inner membrane. Genetic analysis revealed that (i) the *pel* locus was genetically linked with *ptsM* (today *manXYZ*), (ii) *pel* mutants were unable to grow on Man, and (iii) 30% of the Man (*ptsM*) mutants were *pel.* The genetic linkage between *pel* and *ptsM* suggested that the phage might utilize the *PtsM* proteins as a gateway for transfer of DNA across the inner membrane (Palva et al. 1985). The resistance of the *pel/ptsM* mutants was leaky, and stronger against phages with smaller than wild-type genomes. It could be overcome by phages with genomes larger than wild-type, with mutations in the phage tail proteins pV and pH, and also with high doses (multiplicity of infection) of phages (Emmons et al. 1975; Katsura 1983; Roessner and Ihler 1984; Scandella and Arber 1976).

IIC^{man} turned out to be the major specificity determinant for lambda infection, but not sufficient, most likely because IIC is stably expressed only in a complex with IID. The cytoplasmic $IIAB^{man}$ subunit was not required (Williams et al. 1986; Erni et al. 1987). To identify structural determinants important for penetration of lambda DNA, the homologous IICIID complexes of *E. coli (ec), K. pneumoniae (kp)* (P37082, P37083), *B. subtilis (bs)* (P26381, P26382), and chimeric complexes were characterized (Esquinas-Rychen and Erni 2001). All three supported sugar uptake in *E. coli*, but only *ec*$IICIID^{man}$ and *bs*$IICIID^{fru}$ also supported phage lambda infection. None of the six chimeric IICIID complexes supported sugar uptake, and only three (*ec/bs, ec/kp* and *bs/ec*) supported infection. Replacement with alkaline phosphatase of the putative periplasmic loop and the two C-terminal TMs of *ec*IID^{man} did not compromise infection but abolished sugar uptake. Truncation of *ec*IIC^{man} rendered the complex unstable (Esquinas-Rychen and Erni 2001).

IICIID Mannose Transporters as Targets of Bacteriocins

Bacteriocins are antimicrobial peptides produced mostly by lactic acid bacteria (for reviews see Drider et al. 2006; Duquesne et al. 2007; Hassan et al. 2012; Cotter 2014; Rios Colombo et al. 2018). They are of interest as agents for food preservation and against food born pathogens. Class II bacteriocins are cleaved from ribosomally synthesized precursors, consist of one or two peptides (20–80 amino acids), are cationic and not further modified (except disulfide bridges and circularization) . They form ion selective pores in the inner membrane resulting in membrane permeabilization, depolarization and killing of the target bacteria. The producer bacteria are protected against the lethal action by immunity proteins that are coexpressed with the bacteriocin. Bacteriocins require outer and inner membrane proteins for uptake and pore formation. Outer membrane receptors are FepA, Fiu, or Cir, which are involved in the uptake of siderophore-bound iron (Strahsburger et al. 2005; Patzer et al. 2003). Inner membrane scaffolds for insertion and pore formation are certain IICIID of the MSF superfamily (Hechard et al. 2001; Ramnath et al. 2000), IICB of the GFL superfamily (Garcia De Gonzalo et al. 2015; Swe et al. 2009), the Fo complex of the F1Fo ATPase (Rodriguez and Lavina 2003), a serine transporter (Gerard et al. 2005), or an ATP-dependent maltose transporter (Gabrielsen et al. 2012). The interactions are specific, and dependency on a cognate inner membrane protein restricts the bacteriocidal spectrum of each bacteriocin to a narrow range of target bacteria, often to species closely related to the producer. Mutations in the scaffold protein attenuate bacteriocin sensitivity up to 1000 fold. The IIA and IIB subunits are not required for bacteriocin action.

The *E. coli* IICIID^{man} complex in particular serves as a gate for the bacteriocin MccE492 (Bieler et al. 2006; Bieler et al. 2010). MccE492 is produced in *Klebsiella* as a 103 residue precursor and cleaved to the 84 residues mature peptide. In the producer bacterium MccE492 and the immunity protein MceB formed a tight complex with IICIID^{man} that could be copurified with hexahistidine tagged IID^{man} (Bieler et al. 2010). MccE492 caused loss of proton motive force in a wild-type but not in a ΔmanXYZ strain, indicating that the IICIID^{man} was essential for MccE492 toxicity. Binding of MccE492 compromised uptake of Man (a catalytic multiturnover process) but did not confer increased resistance against bacteriophage lambda (a single molecule event). In the case of lactococcin A, a bacteriocin produced by *Lactococcus*, the immunity protein LciA is copurified with the IICIID complex, but only in the presence of the bacteriocin, pointing to the existence in the producer strain of a ternary complex between, the bacteriocin, the immunity protein and the IICIID scaffold (Diep et al. 2007).

The subclass IIa and IId bacteriocin of Gram-postive bacteria (*Bacillus, Listeria, Lactococcus*) act by a similar mechanism (Hassan et al. 2012; Kjos et al. 2011; Nissen-Meyer et al. 2010). They specifically target the phyolgenetic group I of the MSF superfamily (Kjos et al. 2009) (section "Phylogeny and Evolution of the PTS"). The IIC components confer specific recognition whereas the IID subunits are essential for the stability of the complex. It is controversial, whether IICIID participates in pore formation with bacteriocins, or whether it only assists membrane penetration and pore assembly. On one hand pore formation was achieved in the absence of IICIID with a fusion protein between bacteriocin and a bitopic membrane protein. And this bacteriocin complex was neutralized by its cognate immunity protein in the absence of the IICIID complex (Barraza et al. 2017). On the other hand the MceA bacteriocin and the MceB immunity protein together inhibited sugar uptake which points to a permanent rather than transient association between bacteriocin, immunity protein and $IICIID^{man}$ (Bieler et al. 2010). Similarly dysgalacticin, a heat-labile, large bacteriocin (21.5 kDa) produced by *Streptococcus* inhibited uptake of Glc by direct interference with components of the Glc- and/or Man-PTS (Swe et al. 2009). It is not clear whether both IIC and IID, or only one subunit is required. IIC and IID each conferred some bacteriocin sensitivity but less than the IICIID complex (Diep et al. 2007; Kjos et al. 2010; Geldart and Kaznessis 2017; Zhou et al. 2016). The expression of IIC (A2RJ81) alone was sufficient to confer class IIA bacteriocin sensitivity in *Lactococcus lactis* (Ramnath et al. 2004). Hybrid complexes between IIC and IID of different *Lactobacillus* species were recognized by pediocin PA-1 but not by the cognate immunity protein PedB (Zhou et al. 2016).

Membrane Insertion and Lipid Dependencies $IICBA^{mtl}$ and $IICB^{glc}$

The N-Terminal Amphipathic Helix (AH1)

Polytopic inner membrane proteins contain hydrophobic N-terminal signal sequences, which are not cleaved off after membrane insertion. They are recognized by the signal recognition particle (SRP), conducted to the membrane and transferred from SRP to the SecYEG insertion/secretion machinery (translocon) (Fekkes and Driessen 1999; Heijne 2003; Kiefer and Kuhn 2007). The N-terminal 20 residues of $IICB^{glc}$ are amphipathic similar to the amphipathic leader peptides of imported mitochondrial proteins (Erni and Zanolari 1986; Roise et al. 1986) and different from bacterial signal sequences, which exhibit end-to-end rather than side-to-side amphiphilicity (Jones et al. 1990). Eleven out of 13 N-terminal IIC sequences were amphipathic, whereas the N-termini of 18 non-PTS membrane proteins had no or at best a weak hydrophobic moment (Saier et al. 1988). But unlike mitochondrial leader sequences, which contain a positive net charge and an excess of arginines, the N-termini of PTS transporters have positive or negative net charges and do not share sequence similarity with each other. It was proposed that they are surface seeking structures that precede the first membrane spanning helix, and target the proteins to the membrane (Saier et al. 1989). The X-ray structures of *bc*IIC^{mal} and *bc*IIC^{chb} confirm this prediction (McCoy et al. 2016; Cao et al. 2011). IIC^{asc} contains three amphipathic helices, two at the N-terminus of the structural repeats and one in the second repeat between TM6 and TM7 (Luo et al. 2015) (Fig. 8.3e). Multidomain PTS transporters with a soluble IIB domain preceding the IIC domain, do not have or at most have a weak amphipathic sequence between the domains, for instance in IIBCA^{bgl} (P08722) IIB′BC^{fru} (P20966) and IIBC^{suc} (P08470). The N-terminal sequences of PTS transporters, reminiscent of mitochondrial leader sequences (Saier et al. 1988), attracted attention and prompted studies of their function in membrane insertion and transport activity.

AH1 and Membrane Insertion of IICB^{glc} and IICBA^{mtl}

Fusion proteins between residues 1–13 and 1–53 of IICBA^{mtl} and alkaline phosphatase were translocated across the cytoplasmic membrane indicating that the N-terminus of IICBA^{mtl} can initiate insertion/secretion of this periplasmic protein across the cytoplasmic membrane. IICBA^{mtl} with the 10 N-terminal residues replaced by the N-terminus of β-galactosidase, and IICBA^{mtl} with amino acid substitutions S3P, D4P, D4L, D4R, D4H (Fig. 8.6) were completely cleaved by endogenous proteases, indicating that these proteins were not inserted into the membrane. The mutants I5N, K6P, K8P were partially cleaved, with the uncleaved forms retaining reduced PT activity (Yamada et al. 1991). 22-mer peptides of the N-terminus of IICBA^{mtl} with the above indicated amino acid substitutions were chemically synthesized and biophysically characterized. The α-helix content in trifluoroethanol was between 60 and 80% and not affected by the mutations. The peptides penetrated phospholipid monolayers against high lateral surface pressure, and only the mutants containing α-helix breaking proline had a 2–4 times smaller partition coefficient (Tamm et al. 1989; Portlock et al. 1992). Polarized attenuated total reflection infrared spectroscopy indicated that the helix axis of the peptide was aligned parallel to the membrane surface (Tamm and Tatulian 1993).

A fusion protein between residues 1–19 of IICB^{glc} and β-galactosidase could be washed off the membrane with sodium carbonate alone with no need for detergent solubilization, indicating that AH1 was apposed to the membrane surface but not membrane spanning. Fusion proteins between residues 1–41 or longer and alkaline phosphatase were successfully translocated across the cytoplasmic membrane indicating that this segment comprised a membrane spanning segment in addition to AH1 (Buhr and Erni 1993). IICB^{glc} with circularly permuted amino acid sequences ($II_{390}BC_{386}$, $II_{275}CBC_{274}$, numbers indicate the residues at the N- and C-termini of the permuted sequence, Figs. 8.5c and 8.6) retained between 25% to 70% activity although the amphipathic helix (residues 1–17) was now sandwiched between the

C-terminus of the $B_{390\text{-}477}$ and the N-terminus of the $C_{18\text{-}386}$ domain (Beutler et al. 2000b; Gutknecht et al. 1998).

Membrane protein insertion depends not only on signal sequences but also on the insertion/translocation machinery by which these sequences are recognized and processed. Membrane insertion of $IICBA^{mtl}$ was characterized in a cell-free protein synthesis system in the presence of inside-out cytoplasmic membrane vesicles (Werner et al. 1992). $IICBA^{mtl}$ inserted into membrane vesicles and became urea and protease resistant only cotranslationally but not when protein synthesis was completed before membranes were added. Insertion required the SecY(EG) translocon and the soluble signal recognition particle SRP, but it was independent of the soluble SecA and SecB chaperones (Koch et al. 1999). In vitro site-specific cross-linking between the residues 27, 29, 32, 37 and 42 of the 189 amino acid N-terminal fragment of $IICBA^{mtl}$ and the Ffh subunit of SRP revealed the hydrophobic signal anchor sequence of $IICBA^{mtl}$ (residues 1–44) as a recognition site for Ffh (Beck et al. 2000).

The fact that $IICBA^{mtl}$ inserted by the mechanism common for polytopic inner membrane proteins (Facey and Kuhn 2004) suggests that the amphipathic N-terminus is not a special signal sequence but may have another function. As described above (section "Substrate Binding, Transport and Phosphorylation: Important Residues and Point Mutations"), mutations in AH1 and the adjacent TM1 relax sugar specificity and compromise uptake but not non-vectorial phosphorylation (Buhr et al. 1992; Notley-McRobb and Ferenci 2000; Kornberg et al. 2000; Manché et al. 1999). TM1 contributes to the wall of the binding cavity and AH1 is near its cytoplasmic exit. AH1 may be a part of the binding site for the IIB phosphoryldonor domain as suggested by the in silico IIC:IIB model (Fig. 8.5c, right monomer).

Phospholipid Requirement

The phospholipid bilayer of the membrane is the solvent for membrane proteins. The hydrophobic surface of the membrane spanning helices are in intimate and sometimes highly specific contact with the fatty acid chains (Yeagle 2014; Lee 2003). Strongly bound lipids, the so-called annular lipids, do not readily exchange with bulk lipids. They sometimes copurify with or are only slowly lost during the purification of membrane proteins (Palsdottir and Hunte 2004). The changes in the lipid gel to liquid-crystalline transition indicated that $IICBA^{mtl}$ interacted with approximately 40 lipid molecules (Meijberg et al. 1998b).

It was discovered very early (Kundig and Roseman 1971a) that membrane phospholipids, in particular negatively charged phosphatidylglycerol (PG), were required for in vitro reconstitution of Glc- and Man- PT activity from deoxycholate solubilized membrane extracts ("IIB") and the soluble PTS proteins (EI, HPr and "IIA") . Phosphatidylethanolamine (PE) and PG strongly stimulated PT activity of detergent solubilized and purified $IICBA^{mtl}$ (Jacobson et al. 1983a, b). Phospholipids also induced dimerization with concomitant activation of $IICBA^{mtl}$. Both, head groups and fatty acid chain length were important, with choline being more efficient than glycerol as headgroup, and myristoyl (C14) more efficient than oleoyl (monounsaturated C18) as acyl chains (Robillard and Blaauw 1987). The PT activity of $IICB^{glc}$, purified from membranes solubilized with alkaline octyl-polyethyleneglycol, isoelectric focusing in a sucrose gradient and gel filtration was stimulated 20-fold with PG from egg yolk, but not with pure DL-dipalmitoyl-PG (Erni et al. 1982). Highly purified, delipidated PTS transporters in detergent solution had little to no PT activity. Activity could however be recovered by diluting the mixed micelles in an aqueous suspension of phospholipid vesicles. It is likely that the detergent solubilized protein partitioned into lipid vesicles upon dilution of the detergent to below the critical micellar concentration.

The phospholipid requirement of the PTS transporters for Mtl, Glc and Man was further studied in vivo in *E. coli* mutants with altered phospholipid composition (Aboulwafa and Saier 2002; Aboulwafa et al. 2004). The *E. coli psgA* mutant lacks detectable levels of the negatively charged PG (<0.1%), but has a compensating increase of negatively charged phosphatidic acid (PA, 4%) and a bulk-lipid compensating increase of PE from 80 to 90% (Kikuchi et al. 2000). The total amount of PTS transport proteins in the membranes was not affected by lipid composition. However, PTS sugar uptake by the *pgsA* mutant was reduced to 20–30% of the control, and the in vitro non-vectorial PT activity to 40–60%. IIC activity in the *pgsA* background was more sensitive to thermal inactivation and inhibition by non-ionic detergents. Man PT activity (by $IICIID^{man}$) was the most sensitive in the above conditions, Mtl PT activity (by $IICBA^{mtl}$) the least.

Membrane Transport Reconstitution

The sugar-transporters of the PTS have the advantage over other transport proteins, that their (PEP-dependent PT) activity can be assayed easily in a homogeneous solution without the need to measure vectorial solute transport between two separate compartments (Kundig and Roseman 1971a). This activity is convenient to follow activity during protein purification process, and to characterize substrate specificity and other properties of mutant proteins (Fig. 8.2b). In order to measure transport activity, the detergent solubilized membrane proteins must be reconstituted into liposomes (Seddon et al. 2004) and the proteoliposomes loaded with either the sugar or with the soluble PTS-proteins and PEP. Many of the drastic methods of loading liposome with ions and small molecules are not applicable with enzymatically active proteins.

$IICBA^{mtl}$ solubilized with β-octylglucoside was reconstituted in unilamellar liposomes of soybean lecithin and *E. coli* phospholipids by the detergent dilution method (Leonard and Saier 1983). 1 mM mannitol 1-phosphate (Mtl1P) was included in the reconstitution buffer. Mtl1P trapped in the proteoliposomes served as phosphoryl-donor for the uptake of [^{14}C]Mtl by vectorial transphosphorylation. The reaction rates increased sigmoidally with $IICBA^{mtl}$ concentration suggesting that dimeric $IICBA^{ntl}$ was the more active form. The overall reaction rates, however, were low.

IICBA^{mtl} solubilized in decyl-polyethyleneglycol was reconstituted with PE by detergent dialysis in the presence deoxycholate. Active efflux was measured with proteoliposomes preloaded with [^{14}C]Mtl to which EI, HPr and PEP were added on the outside (Fig. 8.2a). Trypsin completely abolished PEP dependent active efflux, but not slow facilitated diffusion. Active uptake was measured with liposomes preloaded with PEP, EI and HPr to which [^{14}C]Mtl was added on the outside. In order to prevent non-vectorial phosphorylation, external EI, HPr and inside-out oriented IICBA^{mtl} were first inactivated with trypsin, and PEP removed with pyruvate kinase and ADP, before [^{14}C]Mtl was added to start the uptake reaction (Fig. 8.2a, b). The turnover numbers were 4 s^{-1} for active uptake compared with 40 s^{-1} for non-vectorial phosphorylation. The K_m-values were <10 μM for uptake and 66 μM for non-vectorial phosphorylation. These experiments showed: (i) transport is coupled with phosphorylation of the transported sugar; (ii) transport has a small K_m, allowing tight substrate binding on the outside; (iii) transport has a 10 times lower turnover than non-vectorial phosphorylation. This is in line with the large amplitude rigid-body displacement of the sugar binding site from the outside to the inside (Fig. 8.3b, c, as shown by X-ray crystallography (McCoy et al. 2016). Non-vectorial phosphorylation in contrast is likely a simple hit-and-run process.

IICBA^{glcnac} (the N-acetylglucosamine transporter) was reconstituted in PE by detergent dialysis in the presence of biobeads (Mukhija and Erni 1996). Enantiomeric [^{3}H]L-Glc was used as a marker to measure the included volume and the negligible non-specific efflux by diffusion. The final concentration of GlcNAc-6P in the proteoliposome reached 2 mM at 10 mM internal PEP and 50 μM external GlcNAc concentrations. The turnover number for uptake of GlcNAc was 6 s^{-1}, the K_m was 67 μM. The turnover for non-vectorial phosphorylation was 16 s^{-1} and the K_m was 750 μM. Transport and non-vectorial phosphorylation are two competing reactions. The presence of unlabelled GlcNAc on the outside of the proteoliposomes slows down the export of radiolabelled GlcNAc in a concentration dependent way.

The Man transporter IICIID^{man} was reconstituted with PE by β-octylglucoside detergent dilution (Mao et al. 1995). The proteoliposomes were loaded by freeze thaw sonication with PEP and soluble PTS proteins, or with radiolabelled Glc. Extravesicular Glc and PEP were removed with hexokinase/ATP and pyruvate kinase/ADP, respectively and the loaded proteoliposomes were purified by spin-column gel filtration. The turnover numbers and K_m-values were 1.2 s^{-1} and 30 μM for active uptake and 3 s^{-1} and 200 μM for non-vectorial phosphorylation. Although sugar transport coupled with phosphorylation (vectorial phosphorylation) and non-vectorial phosphorylation are mediated by the same IICIID^{man}::IIAB^{man} complex, no cross-inhibition between the two reactions could be observed (unlike with IICBA^{glcnac}). Neither the translocation nor the phosphorylation of [^{14}C]Glc out of proteoliposomes into the extravesicular medium could be inhibited by unlabelled Glc added to the outside of the proteoliposomes.

The difference in competition between transport and non-vectorial phosphorylation suggests that the PTS transporters of the GFL and the MFS superfamily not only have different structures but also different mechanisms of transport and phosphorylation.

Cellular Distribution of PTS Proteins: Immunoelectron- and Fluorescence Microscopy

The protein components of functional units can transiently associate, occur as noncovalent multisubunit complexes or as covalent multdidomain proteins. Non-transient complex formation can enhance the overall catalytic activity by shortening the diffusion pathway of intermediates between the active sites, and by channelling reactive intermediates (Castellana et al. 2014; Kholodenko et al. 1998; Rohwer et al. 1998). If one of the units of a complex is a membrane protein, the soluble subunits will colocalize at the membrane surface. The prototype of cluster formation is the supramolecular complex of chemoreceptors and cytoplasmic chemotaxis proteins, which localize at cell poles and division sites (Thiem et al. 2007; Sourjik and Berg 2002)

It has been suggested quite early, that PTS subunits may occur in membrane associated supramolecular complexes. Membrane vesicles of osmotically shocked bacteria catalyzed PEP dependent sugar phosphorylation more efficiently than a mixture of the individual enzymes (Saier et al. 1982). 40% of the PTS activity remained associated with the membrane fraction after high speed centrifugation but could be completely dissociated by resuspension of the membrane pellet in water and a second round of centrifugation (Brouwer et al. 1982). The distribution of EI activity between the membrane and the cytoplasmic fraction was 77 and 13%, respectively in wild-type *Streptococcus mutans*, but inversed, namely 5% and 95% in a $IICB^{glc}$ deletion mutant, suggestive of a specific association between cytoplasmic EI and membrane spanning $IICB^{glc}$ (Liberman and Bleiweis 1984).

When cryosections of cells harvested in exponential growth were stained with anti-EI antiserum and protein A-gold (Ghosh et al. 1989), one third of the EI (2 $\times$ 63 kDa) gold particles were associated with the surface of the inner membrane. For comparison and as a control, β-galactosidase (116 kDa), a *bona fide* cytoplasmic enzyme, was uniformly distributed in the cytoplasm under the same conditions. When EI was overexpressed from a plasmid, the membrane became saturated with EI and more EI accumulated in the vicinity of the inner membrane.

GFP and its mutants were used to study the cellular localization of EI, HPr and IIA^{glc} (Patel et al. 2004; Lopian et al. 2010). Yellow fluorescent protein tagged EI (YFP-EI) was uniformly distributed in bacteria exponentially growing on PTS sugars as carbon source, but assumed a punctate or (bi)polar distribution in: (i) non-growing cells, (ii) in cells growing on an non-PTS substrate such as lactate, and (iii) in cells overexpressing YFP-EI (Patel et al. 2004). The transition between diffuse to punctate distribution was reversible switching from diffuse to punctate when cells entered stationary phase and back to diffuse when they resumed to grow exponentially. Only the non-phosphorylatable YFP-EI (His189Q) mutant remained uniformly distributed in all conditions. EI-mCherry behaved similarly, forming dense clusters (*foci*) at the cell poles which did not overlap with the nucleoid (Lopian et al. 2010; Govindarajan et al. 2018). Of the two EI domains only the recombinant C-terminal (PEP binding) EIC-mCherry domain formed clusters, whereas the N-terminal (HPr binding) EIN-mCherry remained uniformly distributed under all conditions. HPr-mCherry formed polar clusters like EI-mCherry but was partly released from the poles in cells growing on PTS sugars. Overexpression of EI-mCherry and HPr-mCherry increased the size and number of polar clusters. IIA^{glc}-GFP, in contrast, was uniformly distributed in the cytoplasm (Lopian et al. 2010), though IIA^{glc} physically interacts with several integral membrane proteins: e.g. with $IICB^{glc}$ and with the non-PTS transporters for lactose (LacY) and maltose (MalFGK), when it allosterically inhibits their transport activity (Hariharan et al. 2015; Wuttge et al. 2016). The GFP-tagged β-glucoside transporter ($IIBCA^{bgl}$-GFP), a *bona fide* representative of a PTS-transporter was uniformly distributed throughout the cell membrane, with at best a slight accumulation at the cell poles (Lopian et al. 2010). It was suggested, that the clusters form a reservoir of inactive EI and HPr from which the proteins can be mobilized on demand when cells grow exponentially and sugars are transported (Govindarajan et al. 2018). However, the EI-FP and HPr-FP foci near the cell poles also resemble aggregates of misfolded proteins which are driven to the cell poles by nucleoid occlusion (Winkler et al. 2010). The argument against foci being artefacts is that sugar-uptake and growth of cells expressing FP-tagged proteins is normal, and that fusion proteins could be purified and were fully active in vitro.

Conclusion and Outlook

The phosphotransferase system acts as an interface between energy- and signal transduction. Its main components are sugar transporters of the inner membrane, phosphoryltransfer proteins and transcription factors. Its main functions are uptake and phosphorylation of carbohydrates, expression control of PTS and non-PTS genes, allosteric control of catabolic enzymes, non-PTS transporters and chemotaxis (Erni 2013; Galinier and Deutscher 2017; Somavanshi et al. 2016; Sourjik and Berg 2004; Niwano and Taylor 1982; Adler and Epstein 1974; Lux et al. 1995). The integral membrane proteins of the PTS and their cellular context are in the focus of this review. For the other aspects the reader is referred to previous reviews (Erni 2013; Deutscher et al. 2006; Galinier and Deutscher 2017).

Five decades of biochemical and biophysical PTS research (Danchin 1989; Saier 2015) reached an apogee in 2011 with the determination of the first crystal structures of a PTS sugar transporters (Cao et al. 2011). Remaining open questions are: Do the two protomers work concertedly or independently? How do IIC and IIB domains interact during transport and phosphorylation? Still outstanding is the crystal structure of a IICIID transporter of the MSF superfamily. It most likely will reveal a novel fold unseen before. IICIID complexes are of particular interest because of their multiple functions: broad substrate specificity (Sundar et al. 2017); gateway for bactriophage DNA penetration; receptor for pore forming bacteriocins; component of the pore; interfacing sugar availability with virulence (Abranches et al. 2006). None of these functions are yet understood at the molecular mechanistic level.

In the recent years of PTS research priorities have shifted from the characterization of the molecular parts to the understanding of cellular, physiological and biotechnological aspects. Attempts to improve the fermentative production of natural compounds have placed the PTS under the spotlight of metabolic engineering and systems biology (Erni 2013; Gosset 2005; Luo et al. 2014; McCloskey et al. 2018). The unmodified PTS prevents simultaneous uptake of Glc and other carbohydrates by Glc-induced catabolite repression of non-PTS metabolic pathways. It also consumes a significant proportion of PEP creating a bottleneck for the synthesis of aromatic amine intermediates (Carmona et al. 2015). Replacement of the PTS by PEP-independent uptake mechanisms and inactivation of PTS components involved in catabolite repression have significantly improved the industrial production of aromatic metabolites, biofuels and organic acids (Balderas-Hernandez et al. 2009; Zhu et al. 2018).

The antibiotic crisis looming ahead has prompted diverse attempts to identify virulence factors within the PTS or under its control (Erni 2013; Poncet et al. 2009; Le Bouguenec and Schouler 2011). The PTS is not essential for survival of bacteria in a rich medium containing nutrients such as amino acids and non-PTS sugars, but inactivation may indirectly affect functions that are vital in an adverse host environment. It is likely that PTS transporters function as sensors for sugars, cues in the host environment that can trigger a virulence response. *PtsP* (encoding EI^{ntr} Fig. 8.1) was identified as a virulence gene in *Pseudomonas aeruginosa* by screening of transposon mutants. *PtsP* mutants were non toxic for *Caenorhabditis elegans*, grew 100 times slower on plants and were non-lethal for mice (Tan et al. 1999). *PtsP* was also identified as a virulence gene in *Legionella pneumophila* by signature tagged mutagenesis. PtsP mutants colonized the spleen and lung 100 times less efficiently than the wild-type control (Higa and Edelstein 2001). Knocking out *ptsI* (encoding EI, Fig. 8.1) in a mouse-virulent *Salmonella typhimurium* strain increased the LD_{50} (number of bacteria that kill 50% of infected mice) by almost a factor 1000 (Kok et al. 2003). Knocking out *ptsI* (M5005_Spy1120) or the Man transporter (*ManLMN*, M5005_Spy1479-81) in Group A *Streptococcus pyogenes* had the opposite effect. These mutant bacteria increased the size and the severity of lesions at the site of infection due to an early onset of expression and activity of Streptolysin S (Sundar et al. 2017; Gera et al. 2014).

The PTS still poses intriguing questions. Molecular dissection and systems biological synthesis of this important bacterial system will provide further insights into the mechanism of energy- and signal transduction.

Abbreviation and nomenclature: Four character codes are Research Collaboratory for Structural Bioinformatics (RCSB) protein data bank (PDB) identifiers; Six and more character codes are UniProtKB entry identifiers. asc, ascorbate, vitamin C; AH, amphipathic helix; *bc*, *Bacillus cereus*; bgl, β-glucoside; chb, N,N′-diacetlychitobiose; CL, cytoplasmic loop; EI, enzyme I of the PTS; FP, fluorescent protein; FRET, fluorescence resonance energy transfer; fru, fructose; glc, glucose; glcnac, N-acetylglucosamine; HP, re-entrant hairpin loop; HPr, heat stable protein of the PTS; IIA^{sugar} and IIB^{sugar}, phosphotransferase domains/subunits of sugar-specific PTS transporters ($IICBA^{sugar}$); IIC^{sugar}: integral membrane domain/subunit of a sugar-specific PTS transporter. For consistency the following abbreviation are used for IIC subunits irrespective of the designations used in the primary literature: IIC^{asc}

instead of UlaA; *bc*IICmal instead of *bc*MalT; *bc*IIC^{chb} instead *bc*ChbC. IICBA, IICIIBIIA, three-domain, three-subunit protein; $IIC_{o/c}$ in the inward-facing (cytoplasmic) open/closedconformation; $IIC^{o/c}$, IIC in the outward-facing (periplasmic) open/closed conformation; *ec, Escherichia coli*; mal, maltose; man, mannose; mtl, mannitol; PEP, phosphoenolpyruvate; PH, periplasmic helix; PL, periplasmic loop; PT, phosphotransferase; PTS, phosphotransferase system; SD, scaffold/dimerization domain; TD, transport domain; TM, transmembrane helix; wt, wild-type

Acknowledgements J.-M. J. was supported by the University of Bern, the National Centre of Competence in Research (NCCR) TransCure and the Swiss National Science Foundation (grants to Prof. Dimitrios Fotiadis, University of Bern, Switzerland). BE would like to thank his former collaborators for their commitment, and the Universities of Basel, Marburg and Bern, the DFG and the SNF for their generous support.

References

Aboulwafa M, Saier MH Jr (2002) Dependency of sugar transport and phosphorylation by the phosphoenolpyruvate-dependent phosphotransferase system on membranous phosphatidyl glycerol in Escherichia coli: studies with a pgsA mutant lacking phosphatidyl glycerophosphate synthase. Res Microbiol 153:667–677

Aboulwafa M, Saier H (2003) Soluble sugar permeases of the phosphotransferase system in Escherichia coli: evidence for two physically distinct forms of the proteins in vivo. Mol Microbiol 48:131–141

Aboulwafa M, Saier MH Jr (2011) Biophysical studies of the membrane-embedded and cytoplasmic forms of the glucose-specific enzyme II of the E. coli phosphotransferase system (PTS). PLoS One 6:e24088. https://doi.org/10.1371/journal.pone.0024088

Aboulwafa M, Chung YJ, Wai HH, Saier MH Jr (2003) Studies on the Escherichia coli glucose-specific permease, PtsG, with a point mutation in its N-terminal amphipathic leader sequence. Microbiology 149:763–771

Aboulwafa M, Hvorup R, Saier MH Jr (2004) Dependency of sugar transport and phosphorylation by the phosphoenolpyruvate-dependent phosphotransferase system on membranous phosphatidylethanolamine in Escherichia coli: studies with a pssA mutant lacking phosphatidylserine synthase. Arch Microbiol 181:26–34

Abranches J, Candella MM, Wen ZT, Baker HV, Burne RA (2006) Different roles of EIIABMan and EIIGlc in regulation of energy metabolism, biofilm development, and competence in Streptococcus mutans. J Bacteriol 188:3748–3756

Adler J, Epstein W (1974) Phosphotransferase-system enzymes as chemoreceptors for certain sugars in Escherichia coli chemotaxis. Proc Natl Acad Sci USA 71:2895–2899

Ali MH, Imperiali B (2005) Protein oligomerization: how and why. Bioorg Med Chem 13:5013–5020

Balderas-Hernandez VE, Sabido-Ramos A, Silva P, Cabrera-Valladares N, Hernandez-Chavez G, Baez-Viveros JL, Martinez A, Bolivar F, Gosset G (2009) Metabolic engineering for improving anthranilate synthesis from glucose in Escherichia coli. Microb Cell Fact 8:19

Barabote RD, Saier MH Jr (2005) Comparative genomic analyses of the bacterial phosphotransferase system. Microbiol Mol Biol Rev 69:608–634

Barraza DE, Rios Colombo NS, Galvan AE, Acuna L, Minahk CJ, Bellomio A, Chalon MC (2017) New insights into enterocin CRL35; mechanism of action and immunity revealed by heterologous expression in Escherichia coli. Mol Microbiol 105:922–933

Basu S (2003) Biography of Professor Dr. Saul Roseman. Glycoconjugate J 20:7

Beck K, Wu LF, Brunner J, Müller M (2000) Discrimination between SRP-and SecA/SecB-dependent substrates involves selective recognition of nascent chains by SRP and trigger factor. EMBO J 19:134–143

Begley GS, Warner KA, Arents JC, Postma PW, Jacobson GR (1996) Isolation and characterization of a mutation that alters the substrate specificity of the *Escherichia coli* glucose permease. J Bacteriol 178:940–942

Beutler R, Kaufmann M, Ruggiero F, Erni B (2000a) The glucose transporter of the escherichia coli phosphotransferase system: linker insertion mutants and split variants. Biochemistry 39:3745–3750

Beutler R, Ruggiero F, Erni B (2000b) Folding and activity of circularly permuted forms of a polytopic membrane protein. Proc Natl Acad Sci USA 97:1477–1482

Bieler S, Silva F, Soto C, Belin D (2006) Bactericidal activity of both secreted and nonsecreted microcin E492 requires the mannose permease. J Bacteriol 188:7049–7061

Bieler S, Silva F, Belin D (2010) The polypeptide core of Microcin E492 stably associates with the mannose permease and interferes with mannose metabolism. Res Microbiol 161:706–710

Boer H, Ten Hoeve-Duurkens RH, Schuurman-Wolters GK, Dijkstra A, Robillard GT (1994) Expression, purification, and kinetic characterization of the mannitol transport domain of the phosphoenolpyruvate-dependent mannitol phosphotransferase system of Escherichia coli—kinetic evidence that the E-coli mannitol transport protein is a functional dimer. J Biol Chem 269:17863–17871

Boer H, Ten Hoeve-Duurkens RH, Lolkema JS, Robillard GT (1995) Phosphorylation site mutants of the mannitol transport protein enzyme II^{mtl} of *Escherichia coli*: studies on the interaction between the mannitol translocating C-domain and the phosphorylation site on the energy-coupling B-domain. Biochemistry 34:3239–3247

Boer H, Ten Hoeve-Duurkens RH, Robillard GT (1996) Relation between the oligomerization state and the transport and phosphorylation function of the *Escherichia coli* mannitol transport protein: interaction between mannitol-specific enzyme II monomers studied by complementation of inactive site-directed mutants. Biochemistry 35:12901–12908

Bolhuis HH, Palm PP, Wende AA, Falb MM, Rampp MM, Rodriguez-Valera FF, Pfeiffer FF, Oesterhelt DD (2006) The genome of the square archaeon Haloquadratum walsbyi: life at the limits of water activity. BMC Genomics 7:169. https://doi.org/10.1186/1471-2164-7-169

Bouma CL, Roseman S (1996) Sugar transport by the marine chitinolytic bacterium Vibrio furnissii. Molecular cloning and analysis of the mannose/glucose permease. J Biol Chem 271:33468–33475

Bouma CL, Meadow ND, Stover EW, Roseman S (1987) II-BGlc, a glucose receptor of the bacterial phosphotransferase system: molecular cloning of ptsG and purification of the receptor from an overproducing strain of Escherichia coli. Proc Natl Acad Sci USA 84:930–934

Broos J, Strambini GB, Gonnelli M, Vos EP, Koolhof M, Robillard GT (2000) Sensitive monitoring of the dynamics of a membrane-bound transport protein by tryptophan phosphorescence spectroscopy. Biochemistry 39:10877–10883

Brouwer M, Elferink MGL, Robillard GT (1982) Phosphoenolpyruvate-dependent fructose phosphotransferase system of Rhodopseudomonas sphaeroides: purification and physicochemical and immunochemical characterization of a membrane-associated enzyme I. Biochemistry 21:82–88

Buhr A, Erni B (1993) Membrane topology of the glucose transporter of Escherichia coli. J Biol Chem 268:11599–11603

Buhr A, Daniels GA, Erni B (1992) The glucose transporter of Escherichia coli. Mutants with impaired translocation activity that retain phosphorylation activity. J Biol Chem 267:3847–3851

Buhr A, Flükiger K, Erni B (1994) The glucose transporter of *Escherichia coli*. Overexpression, purification, and characterization of functional domains. J Biol Chem 269:23437–23443

Cao Y, Jin X, Levin EJ, Huang H, Zong Y, Quick M, Weng J, Pan Y, Love J, Punta M, Rost B, Hendrickson WA, Javitch JA, Rajashankar KR, Zhou M (2011) Crystal structure of a phosphorylation-coupled saccharide transporter. Nature 473:50–54

Carmona SB, Moreno FM, Bolivar F, Gosset G, Escalante A (2015) Inactivation of the PTS as a strategy to engineer the production of aromatic metabolites in Escherichia coli. J Mol Microbiol Biotechnol 25:195–208

Castellana M, Wilson MZ, Xu Y, Joshi P, Cristea IM, Rabinowitz JD, Gitai Z, Wingreen NS (2014) Enzyme clustering accelerates processing of intermediates through metabolic channeling. Nat Biotechnol 32:1011–1018

Chen Q, Amster-Choder O (1998) BglF, the sensor of the *bgl* system and the β-glucosides permease of *Escherichia coli*: evidence for dimerization and intersubunit phosphotransfer. Biochemistry 37:8714–8723

Chen JS, Reddy V, Chen JH, Shlykov MA, Zheng WH, Cho J, Yen MR, Saier MH Jr (2011) Phylogenetic characterization of transport protein superfamilies: Superiority of superfamilytree programs over those based on multiple alignments. J Mol Microbiol Biotechnol 21:83–96

Clore GM, Venditti V (2013) Structure, dynamics and biophysics of the cytoplasmic protein-protein complexes of the bacterial phosphoenolpyruvate: sugar phosphotransferase system. Trends Biochem Sci 38:515–530

Cotter PD (2014) An 'Upp'-turn in bacteriocin receptor identification. Mol Microbiol 92:1159–1163

Crigler J, Bannerman-Akwei L, Cole AE, Eiteman MA, Altman E (2018) Glucose can be transported and utilized in Escherichia coli by an altered or overproduced N-acetylglucosamine phosphotransferase system (PTS). Microbiology 164:163–172

Cumby N, Reimer K, Mengin-Lecreulx D, Davidson AR, Maxwell KL (2015) The phage tail tape measure protein, an inner membrane protein and a periplasmic chaperone play connected roles in the genome injection process of E. coli phage HK97. Mol Microbiol 96(3):437–447. https://doi.org/10.1111/mmi.12918

Curtis SJ, Epstein W (1975) Phosphorylation of D-glucose in Escherichia coli mutants defective in glucosephosphotransferase, mannosephosphotransferase, and glucokinase. J Bacteriol 122:1189–1199

Dalvit C, Vulpetti A (2016) Weak intermolecular hydrogen bonds with fluorine: detection and implications for enzymatic/chemical reactions, chemical properties, and ligand/protein fluorine nmr screening. Chem Eur J 22:7592–7601

Danchin A (1989) The PTS after 25 Years. FEMS Microbiol Rev 5:1–200

Poncet S, Milohanic E, Maze A, Abdallah JN, Ake F, Larribe M, Deghmane AE, Taha MK, Dozot M, De B, X, Letesson JJ, Deutscher J (2009) Correlations between carbon metabolism and virulence in bacteria. Contrib Microbiol 16:88–102

DeLuca S, Khar K, Meiler J (2015) Fully flexible docking of medium sized ligand libraries with RosettaLigand. PLoS One 10:e0132508. https://doi.org/10.1371/journal.pone.0132508

Deutscher J, Francke C, Postma PW (2006) How phosphotransferase system-related protein phosphorylation regulates carbohydrate metabolism in bacteria. Microbiol Mol Biol Rev 70:939–1031

Deutscher J, Aké FMD, Derkaoui M, Zébré AC, Cao TN, Bouraoui H, Kentache T, Mokhtari A, Milohanic E, Joyet P (2014) The bacterial phosphoenolpyruvate: carbohydrate phosphotransferase system: regulation by protein phosphorylation and phosphorylation-dependent protein-protein interactions. Microbiol Mol Biol Rev 78:231–256

Diep DB, Skaugen M, Salehian Z, Holo H, Nes IF (2007) Common mechanisms of target cell recognition and immunity for class II bacteriocins. Proc Natl Acad Sci USA 104:2384–2389

Dijkstra DS, Broos J, Lolkema JS, Enequist H, Minke W, Robillard GT (1996) A fluorescence study of single tryptophan-containing mutants of enzyme II^{mtl} of the *Escherichia coli* phosphoenolpyruvate-dependent mannitol transport system. Biochemistry 35:6628–6634

Dijkstra DS, Broos J, Visser AJWG, Van Hoek A, Robillard GT (1997) Dynamic fluorescence spectroscopy on single tryptophan mutants of EII^{mtl} in detergent micelles. Effects of substrate binding and phosphorylation on the fluorescence and anisotropy decay. Biochemistry 36:4860–4866

Dobson L, Reményi I, Tusnady GE (2015) CCTOP: a Consensus Constrained TOPology prediction web server. Nucleic Acids Res 43:W408–W412

Drew D, Boudker O (2016) Shared molecular mechanisms of membrane transporters. Annu Rev Biochem 85:543–572

Drider D, Fimland G, Hechard Y, McMullen LM, Prevost H (2006) The continuing story of class IIa bacteriocins. Microbiol Mol Biol Rev 70:564–582

Duquesne S, Destoumieux-Garzon D, Peduzzi J, Rebuffat S (2007) Microcins, gene-encoded antibacterial peptides from enterobacteria. Nat Prod Rep 24:708–734

Elferink MG, Driessen AJ, Robillard GT (1990) Functional reconstitution of the purified phosphoenolpyruvate-dependent mannitol-specific transport system of Escherichia coli in phospholipid vesicles: coupling between transport and phosphorylation. J Bacteriol 172:7119–7125

Elliott J, Arber W (1978) Escherichia coli mutants which block phage lambda DNA injection coincide with ptsM which determines a component of a sugar transport system. Mol Gen Genet 161:1–8

Emmons SW, MacCosham V, Baldwin RL (1975) Tandem genetic duplications in phage lambda. III. The frequency of duplication mutants in two derivatives of phage lambda is independent of known recombination systems. J Mol Biol 91:133–146

Erni B (1986) Glucose-specific permease of the bacterial phosphotransferase system: phosphorylation and oligomeric structure of the glucose-specific IIGlc-IIIGlc complex of Salmonella typhimurium. Biochemistry 25:305–312

Erni B (2001) Glucose transport by the bacterial phosphotransferase system (PTS): an interface between energy- and signal transduction. In: Winkelmann G (ed) Microbial transport systems. Wiley-VCH, Weinheim Germany, pp 115–138

Erni B (2013) The bacterial phosphoenolpyruvate: sugar phosphotransferase system (PTS): an interface between energy and signal transduction. J Iran Chem Soc 10:593–630

Erni B, Zanolari B (1985) The mannose-permease of the bacterial phosphotransferase system. Gene cloning and purification of the enzyme IIMan/IIIMan complex of Escherichia coli. J Biol Chem 260:15495–15503

Erni B, Zanolari B (1986) Glucose-permease of the bacterial phosphotransferase system. Gene cloning, overproduction, and amino acid sequence of enzyme IIGlc. J Biol Chem 261:16398–16403

Erni B, Trachsel H, Postma PW, Rosenbusch JP (1982) Bacterial phosphotransferase system. Solubilization and purification of the glucose-specific enzyme II from membranes of Salmonella typhimurium. J Biol Chem 257:13726–13730

Erni B, Zanolari B, Kocher HP (1987) The mannose permease of Escherichia coli consists of three different proteins. Amino acid sequence and function in sugar transport, sugar phosphorylation, and penetration of phage lambda DNA. J Biol Chem 262:5238–5247

Erni B, Zanolari B, Graff P, Kocher HP (1989) Mannose permease of Escherichia coli. Domain structure and function of the phosphorylating subunit. J Biol Chem 264:18733–18741

Erni B, Siebold C, Christen S, Srinivas A, Oberholzer A, Baumann U (2006) Small substrate, big surprise: fold, function and phylogeny of dihydroxyacetone kinases. Cell Mol Life Sci 63:890–900

Esquinas-Rychen M, Erni B (2001) Facilitation of bacteriophage lambda DNA injection by inner membrane proteins of the bacterial phosphoenolpyruvate: carbohydrate phosphotransferase system (PTS). J Mol Micro Biotechnol 3:361–370

Facey SJ, Kuhn A (2004) Membrane integration of E. coli model membrane proteins. Biochim Biophys Acta 1694:55–66

Fekkes P, Driessen AJM (1999) Protein targeting to the bacterial cytoplasmic membrane. Microbiol Mol Biol Rev 63:161–173

Gabor E, Gohler AK, Kosfeld A, Staab A, Kremling A, Jahreis K (2011) The phosphoenolpyruvate-dependent glucose-phosphotransferase system from Escherichia coli K-12 as the center of a network regulating carbohydrate flux in the cell. Eur J Cell Biol 90:711–720

Gabrielsen C, Brede DA, Hernandez PE, Nes IF, Diep DB (2012) The maltose ABC transporter in Lactococcus lactis facilitates high-level sensitivity to the circular bacteriocin garvicin ML. Antimicrob Agents Chemother 56:2908–2915

Galinier A, Deutscher J (2017) Sophisticated regulation of transcriptional factors by the bacterial phosphoenolpyruvate: sugar phosphotransferase system. J Mol Biol 429:773–789

Garcia De Gonzalo CV, Denham EL, Mars RA, Stulke J, van der Donk WA, Van Dijl JM (2015) The phosphoenolpyruvate: sugar phosphotransferase system is involved in sensitivity to the glucosylated bacteriocin sublancin. LID—AAC.01519-15 [pii]. Antimicrob Agents Chemother 59:6844–6854
Garcia-Alles LF, Zahn A, Erni B (2002a) Sugar recognition by the glucose and mannose permeases of Escherichia coli. Steady-state kinetics and inhibition studies. Biochemistry 41:10077–10086
Garcia-Alles LF, Navdaeva V, Haenni S, Erni B (2002b) The glucose-specific carrier of the Escherichia coli phosphotransferase system. Eur J Biochem 269:4969–4980
Garcia-Doval C, van Raaij MJ (2013) Bacteriophage receptor recognition and nucleic acid transfer 489–518
Geldart K, Kaznessis YN (2017) Characterization of class IIa Bacteriocin resistance in Enterococcus faecium. Antimicrob Agents Chemother
Gera K, Le T, Jamin R, Eichenbaum Z, McIver KS (2014) The phosphoenolpyruvate phosphotransferase system in group A Streptococcus acts to reduce streptolysin S activity and lesion severity during soft tissue infection. Infect Immun 82:1192–1204
Gerard F, Pradel N, Wu LF (2005) Bactericidal activity of colicin V is mediated by an inner membrane protein, SdaC, of Escherichia coli. J Bacteriol 187:1945–1950
Ghosh BK, Owens K, Pietri R, Peterkofsky A (1989) Localization to the inner surface of the cytoplasmic membrane by immunoelectron microscopy of enzyme I of the phosphoenolpyruvate:sugar phosphotransferase system of Escherichia coli. Proc Natl Acad Sci USA 86:849–853
Gosset G (2005) Improvement of Escherichia coli production strains by modification of the phosphoenolpyruvate: sugar phosphotransferase system. Microb Cell Fact 4:14
Govindarajan S, Albocher N, Szoke T, Nussbaum-Shochat A, Amster-Choder O (2018) Phenotypic heterogeneity in sugar utilization by E. coli is generated by stochastic dispersal of the general PTS protein EI from polar clusters. Front Microbiol 8:2695
Gracy KN (2003) A conversation with Saul Roseman. Biochem Biophys Res Commun 300:5–8
Gutknecht R, Manni M, Mao QC, Erni B (1998) The glucose transporter of *Escherichia coli* with circularly permuted domains is active *in vivo* and *in vitro*. J Biol Chem 273:25745–25750
Haddad J, Vakulenko S, Mobashery S (1999) An antibiotic cloaked by its own resistance enzyme. J Am Chem Soc 121:11922–11923
Hariharan P, Balasubramaniam D, Peterkofsky A, Kaback HR, Guan L (2015) Thermodynamic mechanism for inhibition of lactose permease by the phosphotransferase protein IIAGlc. Proc Natl Acad Sci USA 112:2407–2412
Harwani D (2014) Regulation of gene expression: cryptic beta-glucoside (bgl) operon of Escherichia coli as a paradigm. Braz J Microbiol 45:1139–1144
Hassan M, Kjos M, Nes IF, Diep DB, Lotfipour F (2012) Natural antimicrobial peptides from bacteria: characteristics and potential applications to fight against antibiotic resistance. J Appl Microbiol 113:723–736
Hechard Y, Pelletier C, Cenatiempo Y, Frere J (2001) Analysis of sigma(54)-dependent genes in Enterococcus faecalis: a mannose PTS permease (EII(Man)) is involved in sensitivity to a bacteriocin, mesentericin Y105. Microbiology 147:1575–1580
Heijne GV (2003) Membrane protein assembly in vivo. Adv Protein Chem 63:1–18
Higa F, Edelstein PH (2001) Potential virulence role of the Legionella pneumophila ptsP ortholog. Infect Immun 69:4782–4789
Hogema BM, Arents JC, Bader R, Eijkemans K, Yoshida H, Takahashi H, Alba H, Postma PW (1998) Inducer exclusion in *Escherichia coli* by non-PTS substrates: the role of the PEP to pyruvate ratio in determining the phosphorylation state of enzyme IIA^{Glc}. Mol Microbiol 30:487–498
Huber F, Erni B (1996) Membrane topology of the mannose transporter of *Escherichia coli* K12. Eur J Biochem 239:810–817
Jacobson GR, Lee CA, Saier MH Jr (1979) Purification of the mannitol-specific enzyme II of the Escherichia coli phosphoenolpyruvate: sugar phosphotransferase system. J Biol Chem 254:249–252

Jacobson GR, Tanney LE, Kelly DM, Palman KB, Corn SB (1983a) Substrate and phospholipid specificity of the purified mannitol permease of Escherichia coli. J Cell Biochem 23:231–240
Jacobson GR, Lee CA, Leonard JE, Saier MH (1983b) Mannitol-specific enzyme II of the bacterial phosphotransferase system. I. Properties of the purified permease. J Biol Chem 258:10748–10756
Jeckelmann JM, Harder D, Mari SA, Meury M, Ucurum Z, Muller DJ, Erni B, Fotiadis D (2011) Structure and function of the glucose PTS transporter from Escherichia coli. J Struct Biol 176:395–403
Jones JD, McKnight CJ, Gierasch LM (1990) Biophysical studies of signal peptides: implications for signal sequence functions and the involvement of lipid in protein export. J Bioenerg Biomembr 22:213–232
Kalbermatter D, Chiu PL, Jeckelmann JM, Ucurum Z, Walz T, Fotiadis D (2017) Electron crystallography reveals that substrate release from the PTS IIC glucose transporter is coupled to a subtle conformational change. J Struct Biol 199:39–45
Katsura I (1983) Tail assembly and injection. In: Hendrix RW (ed) Lambda II. Cold Spring Harbor, New York, pp 331–346
Kelley LA, Mezulis S, Yates CM, Wass MN, Sternberg MJE (2015) The Phyre2 web portal for protein modeling, prediction and analysis. Nat Protoc 10:845–858
Khandekar SS, Jacobson GR (1989) Evidence for two distinct conformations of the Escherichia coli mannitol permease that are important for its transport and phosphorylation functions. J Cell Biochem 39:207–216
Kholodenko BN, Rohwer JM, Cascante M, Westerhoff HV (1998) Subtleties in control by metabolic channelling and enzyme organization. Mol Cell Biochem 184:311–320
Kiefer D, Kuhn A (2007) YidC as an essential and multifunctional component in membrane protein assembly. Int Rev Cytol 259:113–138
Kikuchi S, Shibuya I, Matsumoto K (2000) Viability of an Escherichia coli pgsANull mutant lacking detectable phosphatidylglycerol and cardiolipin. J Bacteriol 182:371–376
Kirkland PA, Gil MA, Karadzic IM, Maupin-Furlow JA (2008) Genetic and proteomic analyses of a proteasome-activating nucleotidase a mutant of the haloarchaeon Haloferax volcanii. J Bacteriol 190:193–205
Kjos M, Nes IF, Diep DB (2009) Class II one-peptide bacteriocins target a phylogenetically defined subgroup of mannose phosphotransferase systems on sensitive cells. Microbiology 155:2949–2961
Kjos M, Salehian Z, Nes IF, Diep DB (2010) An extracellular loop of the mannose phosphotransferase system component IIC is responsible for specific targeting by class IIa bacteriocins. J Bacteriol 192:5906–5913
Kjos M, Nes IF, Diep DB (2011) Mechanisms of resistance to bacteriocins targeting the mannose phosphotransferase system. Appl Environ Microbiol 77:3335–3342
Koch HG, Hengelage T, Neumann-Haefelin C, MacFarlane J, Hoffschulte HK, Schimz KL, Mechler B, Müller M, Walter P (1999) In vitro studies with purified components reveal signal recognition particle (SRP) and SecA/SecB as constituents of two independent protein-targeting pathways of Escherichia coli. Mol Biol Cell 10:2163–2173
Kok M, Bron G, Erni B, Mukhija S (2003) Effect of enzyme I of the bacterial phosphoenolpyruvate: sugar phosphotransferase system (PTS) on virulence in a murine model. Microbiology 149:2645–2652
Koning RI, Keegstra W, Oostergetel GT, Schuurman-Wolters G, Robillard GT, Brisson A (1999) The 5 A projection structure of the transmembrane domain of the mannitol transporter enzyme II. J Mol Biol 287:845–851
Kornberg HL, Lambourne LT, Sproul AA (2000) Facilitated diffusion of fructose via the phosphoenolpyruvate/glucose phosphotransferase system of Escherichia coli. Proc Natl Acad Sci USA 97:1808–1812
Kundig W, Roseman S (1971a) Sugar transport. II. Characterization of constitutive membrane-bound enzymes II of the Escherichia coli phosphotransferase system. J Biol Chem 246:1407–1418

Kundig W, Roseman S (1971b) Sugar Transport: I. Isolation of a phosphotransferase system from Escherichia coli. J Biol Chem 246:1393–1406
Kundig W, Ghosh S, Roseman S (1964) Phosphate bound to histidine in a protein as an intermediate in a novel phospho-transferase system. Proc Natl Acad Sci USA 52:1067–1074
Lanz R, Erni B (1998) The glucose transporter of the *Escherichia coli* phosphotransferase system–mutant analysis of the invariant arginines, histidines, and domain linker. J Biol Chem 273:12239–12243
Le Bouguenec C, Schouler C (2011) Sugar metabolism, an additional virulence factor in enterobacteria. Int J Med Microbiol 301:1–6
Lee AG (2003) Lipid-protein interactions in biological membranes: a structural perspective. Biochim Biophys Acta 1612:1–40
Lee SJ, Boos W, Bouche JP, Plumbridge J (2000) Signal transduction between a membrane-bound transporter, PtsG, and a soluble transcription factor, Mlc, of Escherichia coli. EMBO J 19:5353–5361
Lee J, Ren Z, Zhou M, Im W (2017) Molecular simulation and biochemical studies support an elevator-type transport mechanism in EIIC. Biophys J 112:2249–2252
Leonard JE, Saier MH (1983) Mannitol-specific enzyme II of the bacterial phosphotransferase system. II. Reconstitution of vectorial transphosphorylation in phospholipid vesicles. J Biol Chem 258:10757–10760
Lepore BW, Indic M, Pham H, Hearn EM, Patel DR, van den Berg B (2011) Ligand-gated diffusion across the bacterial outer membrane. Proc Natl Acad Sci USA 108:10121–10126
Liberman ES, Bleiweis AS (1984) Glucose phosphoenolpyruvate-dependent phosphotransferase system of Streptococcus mutans GS5 studied by using cell-free extracts. Infect Immun 44:486–492
Lolkema JS (1993) A method to study complex enzyme kinetics involving numerical analysis of enzymatic schemes. The mannitol permease of Escherichia coli as an example. J Biol Chem 268:17850–17860
Lolkema JS, Robillard GT (1990) Subunit structure and activity of the mannitol-specific enzyme II of the Escherichia coli phosphoenolypyruvate-dependent phosphotransferase system solubilized in detergent. Biochemistry 29:10120–10125
Lolkema JS, Hoeve-Duurkens RH, Dijkstra DS, Robillard GT (1991) Mechanistic coupling of transport and phosphorylation activity by enzyme IImtl of the Escherichia coli phosphoenolpyruvate-dependent phosphotransferase system. Biochemistry 30:6716–6721
Lolkema JS, Wartna ES, Robillard GT (1993) Binding of the substrate analogue perseitol to phosphorylated and unphosphorylated enzyme IImtl of the phosphoenolpyruvate-dependent phosphotransferase system of Escherichia coli. Biochemistry 32:5848–5854
Lopian L, Elisha Y, Nussbaum-Shochat A, Amster-Choder O (2010) Spatial and temporal organization of the E. coli PTS components. EMBO J 29:3630–3645
Luo Y, Zhang T, Wu H (2014) The transport and mediation mechanisms of the common sugars in Escherichia coli. Biotechnol Adv 32:905–919
Luo P, Yu X, Wang W, Fan S, Li X, Wang J (2015) Crystal structure of a phosphorylation-coupled vitamin C transporter. Nat Struct Mol Biol 22:238–241
Luo P, Dai S, Zeng J, Duan J, Shi H, Wang J (2018) Inward-facing conformation of l-ascorbate transporter suggests an elevator mechanism. Cell Discov 4:35. https://doi.org/10.1038/s41421-018-0037-y
Lux R, Jahreis K, Bettenbrock K, Parkinson JS, Lengeler JW (1995) Coupling the phosphotransferase system and the methyl-accepting chemotaxis protein-dependent chemotaxis signaling pathways of Escherichia coli. Proc Natl Acad Sci USA 92:11583–11587
Manché K, Notley-McRobb L, Ferenci T (1999) Mutational adaptation of Escherichia coli to glucose limitation involves distinct evolutionary pathways in aerobic and oxygen-limited environments. Genetics 153:5
Mancusso R, Gregorio GG, Liu Q, Wang DN (2012) Structure and mechanism of a bacterial sodium-dependent dicarboxylate transporter. Nature 491:622–626

Mao Q, Schunk T, Flükiger K, Erni B (1995) Functional reconstitution of the purified mannose phosphotransferase system of *Escherichia coli* into phospholipid vesicles. J Biol Chem 270:5258–5265
Markovic-Housley Z, Cooper A, Lustig A, Flukiger K, Stolz B, Erni B (1994) Independent folding of the domains in the hydrophilic subunit IIABman of the mannose transporter of Escherichia coli. Biochemistry 33:10977–10984
McCloskey D, Xu S, Sandberg TE, Brunk E, Hefner Y, Szubin R, Feist AM, Palsson BO (2018) Adaptive laboratory evolution resolves energy depletion to maintain high aromatic metabolite phenotypes in Escherichia coli strains lacking the phosphotransferase system. Metab Eng 48:233–242
McCoy JG, Levin EJ, Zhou M (2015) Structural insight into the PTS sugar transporter EIIC. Biochim Biophys Acta 1850:577–585
McCoy JG, Ren Z, Stanevich V, Lee J, Mitra S, Levin EJ, Poget S, Quick M, Im W, Zhou M (2016) The structure of a sugar transporter of the glucose EIIC superfamily provides insight into the elevator mechanism of membrane transport. Structure 24:956–964
Meadow ND, Fox DK, Roseman S (1990) The bacterial phosphoenolpyruvate: glycose phosphotransferase system. Annu Rev Biochem 59:497–542
Meijberg W, Schuurman-Wolters GK, Robillard GT (1998a) Thermodynamic evidence for conformational coupling between the B and C domains of the mannitol transporter of escherichia coli, enzyme IImtl. J Biol Chem 273:7949–7956
Meijberg W, Schuurman-Wolters GK, Boer H, Scheek RM, Robillard GT (1998b) The thermal stability and domain interactions of the mannitol permease of Escherichia coli. A differential scanning calorimetry study. J Biol Chem 273:20785–20794
Meins M, Zanolari B, Rosenbusch JP, Erni B (1988) Glucose permease of Escherichia coli. Purification of the IIGlc subunit and functional characterization of its oligomeric forms. J Biol Chem 263:12986–12993
Miller KA, Phillips RS, Kilgore PB, Smith GL, Hoover TR (2015) A mannose family phosphotransferase system permease and associated enzymes are required for utilization of fructoselysine and glucoselysine in Salmonella enterica serovar Typhimurium. J Bacteriol 197:2831–2839
Mitchell P (1957) A general theory of membrane transport from studies of bacteria. Nature 180:134
Mitchell P, Moyle J (1958) Nature 182:372–373
Mixson AJ, Phang JM (1988) The uptake of pyrroline 5-carboxylate. Group translocation mediating the transfer of reducing-oxidizing potential. J Biol Chem 263:10720–10724
Möller S, Croning MDR, Apweiler R (2001) Evaluation of methods for the prediction of membrane spanning regions. Bioinformatics 17:646–653
Mukhija S, Erni B (1996) Purification by Ni^{2+} affinity chromatography, and functional reconstitution of the transporter for N-acetylglucosamine of *Escherichia coli*. J Biol Chem 271:14819–14824
Navdaeva V, Zurbriggen A, Waltersperger S, Schneider P, Oberholzer AE, Bahler P, Bachler C, Grieder A, Baumann U, Erni B (2011) Phosphoenolpyruvate: sugar phosphotransferase system from the hyperthermophilic Thermoanaerobacter tengcongensis. Biochemistry 50:1184–1193
Nguyen TX, Yen MR, Barabote RD, Saier MH Jr (2006) Topological predictions for integral membrane permeases of the phosphoenolpyruvate: sugar phosphotransferase system. J Mol Microbiol Biotechnol 11:345–360
Nissen-Meyer J, Oppegard C, Rogne P, Haugen HS, Kristiansen PE (2010) Structure and mode-of-action of the two-peptide (class-IIb) bacteriocins. Probiotics Antimicrob Proteins 2:52–60
Niwano M, Taylor BL (1982) Novel sensory adaptation mechanism in bacterial chemotaxis to oxygen and phosphotransferase substrates. Proc Natl Acad Sci USA 79:11–15
Notley-McRobb L, Ferenci T (2000) Substrate specificity and signal transduction pathways in the glucose-specific enzyme II (EII(Glc)) component of the Escherichia coli phosphotransferase system. J Bacteriol 182:4437–4442
Nunn RS, Markovic-Housley Z, Génovésio-Taverne JC, Flükiger K, Rizkallah PJ, Jansonius JN, Schirmer T, Erni B (1996) Structure of the IIA domain of the mannose transporter from *Escherichia coli* at 1.7 Å resolution. J Mol Biol 259:502–511

Nuoffer C, Zanolari B, Erni B (1988) Glucose permease of Escherichia coli. The effect of cysteine to serine mutations on the function, stability, and regulation of transport and phosphorylation. J Biol Chem 263:6647–6655
Oh H, Park Y, Park C (1999) A mutated PtsG, the glucose transporter, allows uptake of D-ribose. J Biol Chem 274:14006–14011
Opacic M, Vos EP, Hesp BH, Broos J (2010) Localization of the substrate-binding site in the homodimeric mannitol transporter, EIImtl, of Escherichia coli. J Biol Chem 285:25324–25331
Opacic M, Hesp BH, Fusetti F, Dijkstra BW, Broos J (2012) Structural investigation of the transmembrane C domain of the mannitol permease from Escherichia coli using 5-FTrp fluorescence spectroscopy. Biochim Biophys Acta 1818:861–868
Otte S, Scholle A, Turgut S, Lengeler JW (2003) Mutations which uncouple transport and phosphorylation in the D-mannitol phosphotransferase system of Escherichia coli K-12 and Klebsiella pneumoniae 1033-5P14. J Bacteriol 185:2267–2276
Palsdottir H, Hunte C (2004) Lipids in membrane protein structures. Biochim Biophys Acta 1666:2–18
Palva ET, Saris P, Silhavy TJ (1985) Gene fusions to the ptsM/pel locus of Escherichia coli. Mol Gen Genet MGG 199:427–433
Pas HH, Ellory JC, Robillard GT (1987) Bacterial phosphoenolpyruvate-dependent phosphotransferase system: association state of membrane-bound mannitol-specific enzyme II demonstrated by radiation inactivation. Biochemistry 26:6689–6696
Pas HH, Hoeve-Duurkens RH, Robillard GT (1988) Bacterial phosphoenolpyruvate-dependent phosphotransferase system: mannitol-specific EII contains two phosphoryl binding sites per monomer and one high-affinity mannitol binding site per dimer. Biochemistry 27:5520–5525
Patel HV, Vyas KA, Li X, Savtchenko R, Roseman S (2004) Subcellular distribution of enzyme I of the Escherichia coli phosphoenolpyruvate: glycose phosphotransferase system depends on growth conditions. Proc Natl Acad Sci USA 101:17486–17491
Patzer SI, Baquero MR, Bravo D, Moreno F, Hantke K (2003) The colicin G, H and X determinants encode microcins M and H47, which might utilize the catecholate siderophore receptors FepA, Cir, Fiu and IroN. Microbiology 149:2557–2570
Perham RN (1991) Domains, motifs, and linkers in 2-oxo acid dehydrogenase multienzyme complexes: a paradigm in the design of a multifunctional protein. Biochemistry 30:8501–8512
Pickl A, Johnsen U, Schönheit P (2012) Fructose degradation in the haloarchaeon Haloferax volcanii involves a bacterial type phosphoenolpyruvate-dependent phosphotransferase system, fructose-1-phosphate kinase, and class II fructose-1,6-bisphosphate aldolase. J Bacteriol 194:3088–3097
Plumbridge J (2000) A mutation which affects both the specificity of PtsG sugar transport and the regulation of ptsG expression by Mlc in Escherichia coli. Microbiology 146:2655–2663
Portlock SH, Lee Y, Tomich JM, Tamm LK (1992) Insertion and folding of the amino-terminal amphiphilic signal sequences of the mannitol and glucitol permeases of Escherichia coli. J Biol Chem 267:11017–11022
Postma PW (1981) Defective enzyme II-BGlc of the phosphoenolpyruvate:sugar phosphotransferase system leading to uncoupling of transport and phosphorylation in Salmonella typhimurium. J Bacteriol 147:382–389
Postma PW, Roseman S (1976) The bacterial phosphoenolpyruvate: sugar phosphotransferase system. Biochim Biophys Acta 457:213–257
Postma PW, Lengeler JW, Jacobson GR (1993) Phosphoenolpyruvate—carbohydrate phosphotransferase systems of bacteria. Microbiol Rev 57:543–594
Rabus R, Reizer J, Paulsen I, Saier MH (1999) Enzyme I(Ntr) from Escherichia coli. A novel enzyme of the phosphoenolpyruvate-dependent phosphotransferase system exhibiting strict specificity for its phosphoryl acceptor, NPr. J Biol Chem 274:26185–26191
Ramnath M, Beukes M, Tamura K, Hastings JW (2000) Absence of a putative mannose-specific phosphotransferase system enzyme IIAB component in a leucocin a-resistant strain of Listeria monocytogenes, as shown by two-dimensional sodium dodecyl sulfate-polyacrylamide gel electrophoresis. Appl Environ Microbiol 66:3098–3101

Ramnath M, Arous S, Gravesen A, Hastings JW, Hechard Y (2004) Expression of mptC of Listeria monocytogenes induces sensitivity to class IIa bacteriocins in Lactococcus lactis. Microbiology 150:2663–2668

Reizer J, Paulsen IT, Reizer A, Titgemeyer F, Saier MH (1996) Novel phosphotransferase system genes revealed by bacterial genome analysis: the complete complement of pts genes in Mycoplasma genitalium. Microb Comp Genomics 1:151–164

Ren Z, Lee J, Moosa MM, Nian Y, Hu L, Xu Z, McCoy JG, Ferreon AC, Im W, Zhou M (2018) Structure of an EIIC sugar transporter trapped in an inward-facing conformation. Proc Natl Acad Sci USA 115:5962–5967

Reynolds AE, Felton J, Wright A (1981) Insertion of DNA activates the cryptic bgl operon in E. coli K12. Nature 293:625

Rios Colombo NS, Chalon MC, Navarro SA, Bellomio A (2018) Pediocin-like bacteriocins: new perspectives on mechanism of action and immunity. Curr Genet 64:345–351

Robillard GT, Blaauw M (1987) Enzyme II of the Escherichia coli phosphoenolpyruvate-dependent phosphotransferase system: protein-protein and protein-phospholipid interactions. Biochemistry 26:5796–5803

Robillard GT, Broos J (1999) Structure/function studies on the bacterial carbohydrate transporters, enzymes II, of the phosphoenolpyruvate-dependent phosphotransferase system. Biochim Biophys Acta 1422:73–104

Rodriguez E, Lavina M (2003) The proton channel is the minimal structure of ATP synthase necessary and sufficient for microcin h47 antibiotic action. Antimicrob Agents Chemother 47:181–187

Roessner CA, Ihler GM (1984) Proteinase sensitivity of bacteriophage lambda tail proteins gpJ and pH in complexes with the lambda receptor. J Bacteriol 157:165–170

Rohwer JM, Postma PW, Kholodenko BN, Westerhoff HV (1998) Implications of macromolecular crowding for signal transduction and metabolite channeling. Proc Natl Acad Sci USA 95:10547–10552

Rohwer JM, Meadow ND, Roseman S, Westerhoff HV, Postma PW (2000) Understanding glucose transport by the bacterial phosphoenolpyruvate: glycose phosphotransferase system on the basis of kinetic measurements in vitro. J Biol Chem 275:34909–34921

Roise D, Horvath SJ, Tomich JM, Richards JH, Schatz G (1986) A chemically synthesized presequence of an imported mitochondrial protein can form an amphiphilic helix and perturb natural and artificial phospholipid bilayers. EMBO J 5:1327–1334

Roseman S (1969) The transport of carbohydrates by a bacterial phosphotransferase system. J Gen Physiol 54:138

Roseman S (1989) Sialic acid, serendipity, and sugar transport: discovery of the bacterial phosphotransferase system. FEMS Microbiol Rev 5:3–11

Ruijter GJG, Postma PW, Van Dam K (1991) Energetics of glucose uptake in a Salmonella typhimurium mutant containing uncoupled enzyme IIGlc. Arch Microbiol 155:234–237

Ruijter GJG, van Meurs G, Verwey MA, Postma PW, Van Dam K (1992) Analysis of mutations that uncouple transport from phosphorylation in enzyme-IIGlc of the *Escherichia coli* phosphoenolpyruvate-dependent phosphotransferase system. J Bacteriol 174:2843–2850

Saier MH (2015) PTS50—the prokaryotic phosphoenolpyruvate: sugar phosphotransferase system, 50 years after its discovery. J Mol Microbiol Biotechnol 25:69–236

Saier MH, Cox DF, Feucht BU, Novotny MJ (1982) Evidence for the functional association of enzyme I and HPr of the phosphoenolpyruvate-sugar phosphotransferase system with the membrane in sealed vesicles of Escherichia coli. J Cell Biochem 18:231–238

Saier MH, Yamada M, Suda K, Erni B, Rak B, Lengeler J, Stewart GC, Waygood EB, Rapoport G (1988) Bacterial proteins with N-terminal leader sequences resembling mitochondrial targeting sequences of eukaryotes. Biochimie 70:1743–1748

Saier MH, Werner PK, Müller M (1989) Insertion of proteins into bacterial membranes: mechanism, characteristics, and comparisons with the eucaryotic process. Microbiol Rev 53:333–366

Saier MH, Chauvaux S, Deutscher J, Reizer J, Ye J-J (1995) Protein phosphorylation and regulation of carbon metabolism in Gram-negative versus Gram-positive bacteria. Trends Biochem Sci 20:267–271

Saier MH, Hvorup RN, Barabote RD (2005) Evolution of the bacterial phosphotransferase system: from carriers and enzymes to group translocators. Biochem Soc Trans 33:220

Saraceni-Richards CA, Jacobson GR (1997a) A conserved glutamate residue, Glu-257, is important for substrate binding and transport by the Escherichia coli mannitol permease. J Bacteriol 179:1135–1142

Saraceni-Richards CA, Jacobson GR (1997b) Subunit and amino acid interactions in the Escherichia coli mannitol permease: a functional complementation study of coexpressed mutant permease proteins. J Bacteriol 179:5171–5177

Scandella D, Arber W (1976) Phage lambda DNA injection into Escherichia coli pel mutants is restored by mutations in phage genes V and H. Virology 69:206–215

Schauder S, Nunn RS, Lanz R, Erni B, Schirmer T (1998) Crystal structure of the IIB subunit of a fructose permease (IIB^{Lev}) from *Bacillus subtilis*. J Mol Biol 276:591–602

Schnetz K, Rak B (1988) Regulation of the bgl operon of Escherichia coli by transcriptional antitermination. EMBO J 7:3271–3277

Schnetz K, Sutrina SL, Saier MH Jr, Rak B (1990) Identification of catalytic residues in the beta-glucoside permease of Escherichia coli by site-specific mutagenesis and demonstration of interdomain cross-reactivity between the beta-glucoside and glucose systems. J Biol Chem 265:13464–13471

Seddon AM, Curnow P, Booth PJ (2004) Membrane proteins, lipids and detergents: not just a soap opera. Biochim Biophys Acta 1666:105–117

Seitz S, Lee SJ, Pennetier C, Boos W, Plumbridge J (2003) Analysis of the interaction between the global regulator Mlc and EIIBGlc of the glucose-specific phosphotransferase system in Escherichia coli. J Biol Chem 278:10744–10751

Somavanshi R, Ghosh B, Sourjik V (2016) Sugar influx sensing by the phosphotransferase system of Escherichia coli. PLoS Biol 14(8):e2000074. https://doi.org/10.1371/journal.pbio.2000074

Sourjik V, Berg HC (2002) Localization of components of the chemotaxis machinery of Escherichia coli using fluorescent protein fusions. Mol Microbiol 37:740–751

Sourjik V, Berg HC (2004) Functional interactions between receptors in bacterial chemotaxis. Nature 428:437–441

Stephan MM, Jacobson GR (1986) Subunit interactions of the Escherichia coli mannitol permease: correlation with enzymic activities. Biochemistry 25:4046–4051

Stolz B, Huber M, Markovic-Housley Z, Erni B (1993) The mannose transporter of Escherichia-coli—structure and function of the IIAB(Man)-subunit. J Biol Chem 268:27094–27099

Strahsburger E, Baeza M, Monasterio O, Lagos R (2005) Cooperative uptake of microcin E492 by receptors FepA, Fiu, and Cir and inhibition by the siderophore enterochelin and its dimeric and trimeric hydrolysis products. Antimicrob Agents Chemother 49:3083–3086

Sundar GS, Islam E, Gera K, Le Breton Y, McIver KS (2017) A PTS EII mutant library in group A Streptococcus identifies a promiscuous man-family PTS transporter influencing SLS-mediated hemolysis. Mol Microbiol 103:518–533

Swe PM, Cook GM, Tagg JR, Jack RW (2009) Mode of action of dysgalacticin: a large heat-labile bacteriocin. J Antimicrob Chemother 63:679–686

Tamm LK, Tatulian SA (1993) Orientation of functional and nonfunctional PTS permease signal sequences in lipid bilayers. A polarized attenuated total reflection infrared study. Biochemistry 32:7720–7726

Tamm LK, Tomich JM, Saier MH (1989) Membrane incorporation and induction of secondary structure of synthetic peptides corresponding to the N-terminal signal sequences of the glucitol and mannitol permeases of Escherichia coli. J Biol Chem 264:2587–2592

Tan MW, Rahme LG, Sternberg JA, Tompkins RG, Ausubel FM (1999) Pseudomonas aeruginosa killing of Caenorhabditis elegans used to identify P. aeruginosa virulence factors. Proc Natl Acad Sci USA 96:2408–2413

Tchieu JH, Norris V, Edwards JS, Saier MH Jr (2001) The complete phosphotranferase system in Escherichia coli. J Mol Microbiol Biotechnol 3:329–346

Thiem S, Kentner D, Sourjik V (2007) Positioning of chemosensory clusters in E. coli and its relation to cell division. The EMBO J 26:1615

Tymoszewska A, Diep DB, Wirtek P, Aleksandrzak-Piekarczyk TA (2017) The non-lantibiotic bacteriocin garvicin Q targets man-PTS in a broad spectrum of sensitive bacterial genera. Sci Rep 7:8359. https://doi.org/10.1038/s41598-017-09102-7

Van Montfort BA, Schuurman-Wolters GK, Duurkens RH, Mensen R, Poolman B, Robillard GT (2001) Cysteine cross-linking defines part of the dimer and B/C domain interface of the Escherichia coli mannitol permease. J Biol Chem 276:12756–12763

Van Montfort BA, Schuurman-Wolters GK, Wind J, Broos J, Robillard GT, Poolman B (2002) Mapping of the dimer interface of the Escherichia coli mannitol permease by cysteine cross-linking. J Biol Chem 277:14717–14723

Veldhuis G, Gabellieri E, Vos EP, Poolman B, Strambini GB, Broos J (2005) Substrate-induced conformational changes in the membrane-embedded IIC(mtl)-domain of the mannitol permease from Escherichia coli, EnzymeII(mtl), probed by tryptophan phosphorescence spectroscopy. J Biol Chem 280:35148–35156

Veldhuis G, Hink M, Krasnikov V, van den Bogaart G, Hoeboer J, Visser AJ, Broos J, Poolman B (2006) The oligomeric state and stability of the mannitol transporter, EnzymeII(mtl), from Escherichia coli: a fluorescence correlation spectroscopy study. Protein Sci 15:1977–1986

Vervoort EB, Bultema JB, Schuurman-Wolters GK, Geertsma ER, Broos J, Poolman B (2005) The first cytoplasmic loop of the mannitol permease from Escherichia coli is accessible for sulfhydryl reagents from the periplasmic side of the membrane. J Mol Biol 346:733–743

Vogler AP, Broekhuizen CP, Schuitema A, Lengeler JW, Postma PW (1988) Suppression of IIIGlc-defects by enzymes IINag and IIBgl of the PEP: carbohydrate phosphotransferase system. Mol Microbiol 2:719–726

Vos EP, ter Horst R, Poolman B, Broos J (2009a) Domain complementation studies reveal residues critical for the activity of the mannitol permease from Escherichia coli. Biochim Biophys Acta 1788:581–586

Vos EPP, Bokhove M, Hesp BH, Broos J (2009b) Structure of the cytoplasmic loop between putative helices II and III of the mannitol permease of Escherichia coli: a tryptophan and 5-fluorotryptophan spectroscopy study. Biochemistry 48:5284–5290

Waeber U, Buhr A, Schunk T, Erni B (1993) The glucose transporter of Escherichia coli. Purification and characterization by Ni + chelate affinity chromatography of the IIBCGlc subunit. FEBS Lett 324:109–112

Wehmeier UF, Wöhrl BM, Lengeler JW (1995) Molecular analysis of the phosphoenolpyruvate-dependent L-sorbose: phosphotransferase system from *Klebsiella pneumoniae* and of its multidomain structure. Mol Gen Genet 246:610–618

Wells JE, Russell JB (1996) Why do many ruminal bacteria die and lyse so quickly? J Dairy Sci 79:1487–1495

Weng QP, Jacobson GR (1993) Role of a conserved histidine residue, His-195, in the activities of the Escherichia coli mannitol permease. Biochemistry 32:11211–11216

Weng QP, Elder J, Jacobson GR (1992) Site-specific mutagenesis of residues in the Escherichia coli mannitol permease that have been suggested to be important for its phosphorylation and chemoreception functions. J Biol Chem 267:19529–19535

Werner PK, Saier MH, Müller M (1992) Membrane insertion of the mannitol permease of Escherichia coli occurs under conditions of impaired SecA function. J Biol Chem 267:24523–24532

Williams N, Fox DK, Shea C, Roseman S (1986) Pel, the protein that permits lambda DNA penetration of Escherichia coli, is encoded by a gene in ptsM and is required for mannose utilization by the phosphotransferase system. Proc Natl Acad Sci USA 83:8934–8938

Winkler J, Seybert A, Konig L, Pruggnaller S, Haselmann U, Sourjik V, Weiss M, Frangakis AS, Mogk A, Bukau B (2010) Quantitative and spatio-temporal features of protein aggregation in

Escherichia coli and consequences on protein quality control and cellular ageing. EMBO J 29:910–923
Woese CR, Fox GE (1977) Phylogenetic structure of the prokaryotic domain: the primary kingdoms. Proc Natl Acad Sci USA 74:5088–5090
Wu LF, Tomich JM, Saier MH (1990) Structure and evolution of a multidomain multiphosphoryl transfer protein: nucleotide sequence of the fruB(HI) gene in Rhodobacter capsulatus and comparisons with homologous genes from other organisms. J Mol Biol 213:687–703
Wuttge S, Licht A, Timachi MH, Bordignon E, Schneider E (2016) Mode of interaction of the signal-transducing protein EIIAGlc with the maltose ABC transporter in the process of inducer exclusion. Biochemistry 55:5442–5452
Yamada Y, Chang YY, Daniels GA, Wu LF, Tomich JM, Yamada M, Saier MH Jr (1991) Insertion of the mannitol permease into the membrane of Escherichia coli. Possible involvement of an N-terminal amphiphilic sequence. J Biol Chem 266:17863–17871
Yeagle PL (2014) Non-covalent binding of membrane lipids to membrane proteins. Biochim Biophys Acta 1838:1548–1559
Zeppenfeld T, Larisch C, Lengeler JW, Jahreis K (2000) Glucose transporter mutants of *Escherichia coli* K-12 with changes in substrate recognition of IICB(Glc) and induction behavior of the *ptsG* gene. J Bacteriol 182:4443–4452
Zhang Z, Aboulwafa M, Smith MH, Saier MH Jr (2003) The ascorbate transporter of Escherichia coli. J Bacteriol 185:2243–2250
Zhou W, Wang G, Wang C, Ren F, Hao Y (2016) Both IIC and IID components of mannose phosphotransferase system are involved in the specific recognition between immunity protein PedB and bacteriocin-receptor complex. PLoS One 11(10):e0164973. https://doi.org/10.1371/journal.pone.0164973
Zhu F, Wang Y, San K, Bennett GN (2018) Metabolic engineering of Escherichia coli to produce succinate from soybean hydrolysate under anaerobic conditions. Biotechnol Bioeng 115:1743–1754
Zou Z, Tong F, Faergeman NJ, Borsting C, Black PN, DiRusso CC (2003) Vectorial acylation in Saccharomyces cerevisiae: Fat1p and fatty acyl-CoA synthetase are interacting components of a fatty acid import complex. J Biol Chem 278:16414–16422
Zuniga M, Comas I, Linaje R, Monedero V, Yebra MJ, Esteban CD, Deutscher J, Perez-Martinez G, Gonzalez-Candelas F (2005) Horizontal gene transfer in the molecular evolution of mannose PTS transporters. Mol Biol Evol 22:1673–1685
Zurbriggen A, Schneider P, Bahler P, Baumann U, Erni B (2010) Expression, purification, crystallization and preliminary X-ray analysis of the EIICGlc domain of the Escherichia coli glucose transporter. Acta Crystallogr Sect F: Struct Biol Cryst Commun 66:684–688

Chapter 9
Secondary Active Transporters

Patrick D. Bosshart and Dimitrios Fotiadis

Abstract Transport of solutes across biological membranes is essential for cellular life. This process is mediated by membrane transport proteins which move nutrients, waste products, certain drugs and ions into and out of cells. Secondary active transporters couple the transport of substrates against their concentration gradients with the transport of other solutes down their concentration gradients. The alternating access model of membrane transporters and the coupling mechanism of secondary active transporters are introduced in this book chapter. Structural studies have identified typical protein folds for transporters that we exemplify by the major facilitator superfamily (MFS) and LeuT folds. Finally, substrate binding and substrate translocation of the transporters LacY of the MFS and AdiC of the amino acid-polyamine-organocation (APC) superfamily are described.

Keywords Alternating access model · Amino acid-polyamine-organocation superfamily · Major facilitator superfamily · Membrane transporter · Secondary active transporter · Transport mechanism · Transporter fold

Introduction

Biological membranes constitute physical barriers of eukaryotic cells, organelles, nuclei, some viruses and bacteria. They are composed of an asymmetric bilayer of amphiphilic phospholipid molecules, which consist of two hydrophobic tails and a hydrophilic head. Such lipid membranes are impermeable for large polar or charged molecules, while small polar and uncharged molecules can equilibrate across the membrane via passive diffusion. This impermeability for biologically relevant molecules and ions allows separating chemically and physically distinct

P. D. Bosshart · D. Fotiadis (✉)
Swiss National Centre of Competence in Research (NCCR) TransCure, Institute of Biochemistry and Molecular Medicine, University of Bern, Bühlstrasse 28, 3012 Bern, Switzerland
e-mail: dimitrios.fotiadis@ibmm.unibe.ch

P. D. Bosshart
e-mail: patrick.bosshart@ibmm.unibe.ch

A. Kuhn (ed.), *Bacterial Cell Walls and Membranes*, Subcellular Biochemistry 92, https://doi.org/10.1007/978-3-030-18768-2_9

compartments. Consequently, electrochemical gradients are established across the membranes whose free energy is available to drive biophysical and biochemical processes. However, exchange of material across otherwise impermeable lipid bilayers is essential for maintaining vital biochemical processes. This is mediated by the incorporation of transport proteins into cell membranes, which transport cellular nutrients, waste products, certain drugs and ions into and out of cells.

In bacteria, there are three different types of membrane transporters that mediate the transport of solutes and ions across lipid bilayers (Fig. 9.1a). Primary active transporters use the energy released by ATP hydrolysis to shuttle solutes (i.e., uncharged and charged organic molecules, and inorganic ions) across membranes against their concentration gradients. This results in building up electrochemical gradients that can be used to drive other processes. Channels (e.g., the bacterial potassium-specific channel KcsA), which are open to both sides of the membrane, catalyze the highly specific movement of molecules, ions or water across a biological membrane down their concentration gradients. Secondary transporters facilitate the transfer of specific solutes down or against a concentration gradient. This class of membrane proteins can be subdivided into three different types (Fig. 9.1b): uniporters, symporters and antiporters. Uniporters transport in general one type of substrate and are driven by the energy stored in the substrate's concentration gradient across the membrane. Symporters and antiporters (secondary active transporters) catalyze the transfer of one solute against its concentration gradient, driven by the movement of co-transported solutes(s) down their electrochemical gradients. These two different types of co-transport differ in the relative direction of movement of the transported and co-transported substrates (Fig. 9.1b). The ubiquitously distributed secondary active transporters utilize the free energy stored in a concentration gradient of solutes across a membrane to facilitate the uphill transport of other substrates. The directionality of the transport process is given by the combined electrochemical potentials of the involved substrates. A fourth type of membrane transporters (i.e., group translocation systems, see Chap. 8) couples the translocation of a substrate to its chemical modification (e.g., phosphorylation) releasing a chemically modified substrate (Erni 2013).

Up to 16% of bacterial genomes encode for membrane transporters (Blattner 1997; Ren and Paulsen 2007). In *Escherichia coli* more than 500 different membrane transporters have been annotated (Busch and Saier 2008) of which ~25% can be assigned to the major facilitator superfamily (MFS; http://www.uniprot.org). The MFS superfamily is the largest evolutionary related and the most diverse superfamily of secondary active transporters. This superfamily (TCID 2.A.1) contains more than 80 distinct families (http://www.tcdb.org) and it is present in all kingdoms of life. The amino acid-polyamine-organocation (APC) superfamily (TCID 2.A.3) contains 18 families and is the second largest superfamily of secondary active transporters. The APC superfamily plays an important role in the transport of biochemically relevant molecules, e.g., amino acids and their derivatives (Jack et al. 2000; Wong et al. 2012; Vastermark et al. 2014). Many members of this superfamily function as solute:cation symporters or as solute:solute antiporters, and fulfill very different functions in bacteria. For example, specific antiporters of the APC family (e.g., AdiC

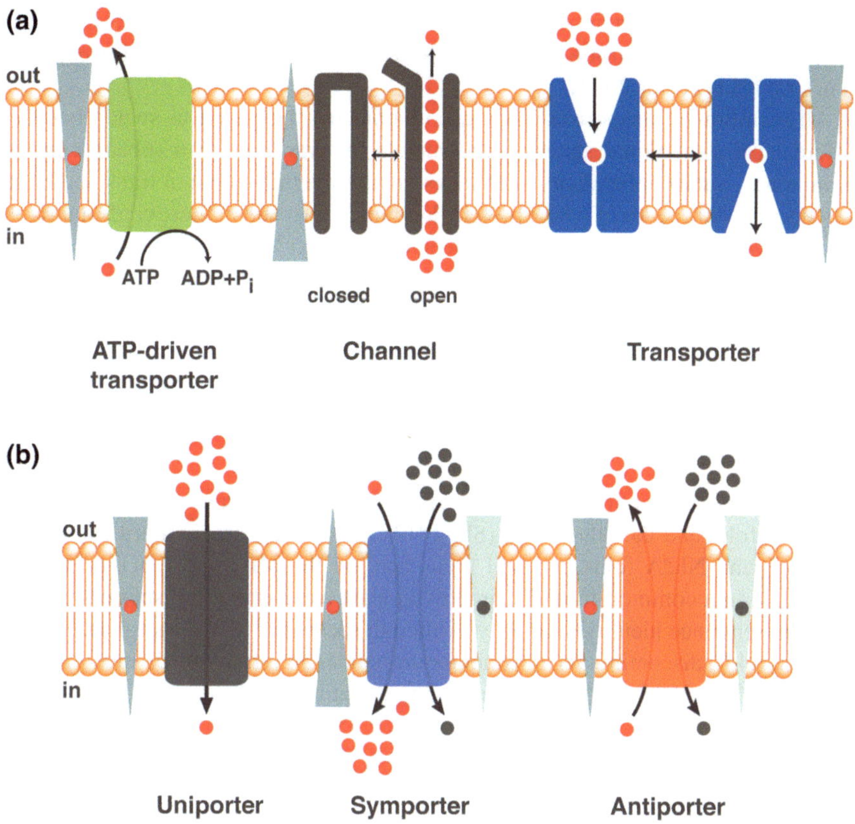

Fig. 9.1 Schematic illustration of membrane transport proteins. **a** Three major types of membrane transport proteins are shown: ATP-driven transporters (~10^0–10^3 substrates/second) that use the energy released from ATP hydrolysis to shuttle substrates (i.e., solutes) across membranes against their concentration gradients ("uphill transport"). Solutes include uncharged and charged organic molecules as well as inorganic ions. Channels (~10^7–10^8 substrates/second) catalyze the highly specific movement of specific substrates (e.g., molecules, ions or water) across a biological membrane down their concentration gradients ("downhill transport"). Transporters (~10^2–10^4 substrates/second) facilitate the transfer of specific molecules or ions down, or against a concentration gradient. **b** Three groups of transporters are shown: Uniporters move one type of solute across the membrane down the concentration gradient. Symporters and antiporters transport solutes down their concentration gradients and are coupled with the active transport of other solutes against their concentration gradients in same or opposite directions, respectively. Concentration gradients are indicated by the triangles. The tips of the triangles point towards the side of the membrane with the lower solute concentration

and GadC) play important roles in extreme acid-resistance mechanisms of *E. coli*, where they virtually extrude protons that leaked into the cytoplasm during exposure to gastric acid (Foster 2004).

All membrane transporters have in common that they need to switch between different conformational states to accomplish the transport of a substrate across membranes. Such conformational changes involve the movement of transmembrane α-helices (TMs) or even of whole protein domains. As a result of these conformational changes, which are required to complete a transport cycle, the turnover rate of most membrane transporters is relatively low (~10^2–10^4 substrates/second) in contrast to channels (~10^7–10^8 substrates/second).

Transport Protein Folds

Secondary active membrane transporters have unique structural folds: the MFS fold, the LeuT fold, and the NhaA fold (Shi 2013). Since there are similarities between the LeuT and NhaA folds, we will focus here on the MFS and LeuT folds only. By definition a common fold is shared by at least two distinct transporters with no apparent sequence identity/similarity. Although secondary active transporters adopt different folds, discontinuous transmembrane helices and structural elements that are connected by pseudo-symmetries are common features.

MFS Fold

X-ray structures of membrane transporters of the MFS (http://blanco.biomol.uci.edu/mpstruc/) with relatively low amino acid sequence identities/similarities and different substrate specificities share the common canonical MFS fold. The core consists of twelve TMs with TM1–6 and TM7–12 arranged in an N- and C-terminal six-helix bundle (Fig. 9.2 a). There is a central cavity between the two bundles, which hosts the substrate-binding site and serves as the substrate translocation pathway. The bundles are related to each other by a pseudo-twofold symmetry axis that is perpendicular to the membrane plane and is located in the substrate translocation pathway. The symmetry-related N- and C-terminal six-helix bundles stack against each other through a flat interface and are physically connected through a long intracellular loop between TM6 and TM7. Internally, the two bundles consist of a pair of two inverted-topology repeats, i.e., TM1–3 and TM4–6 for the N-terminal bundle, and TM7–9 and TM10–12 for the C-terminal bundle. The two inverted repeats in each bundle can be superimposed on each other by a ~180° rotation around an axis, which is parallel to the membrane plane. These repeats form stable and tightly packed entities due to the intertwinement of the TMs. The central cavity is surrounded by TM1, TM2, TM4 and TM5 of the N-terminal bundle, and TM7, TM8, TM10 and TM11 of the C-terminal bundle. At the ends of the longest dimension of MFS transporters

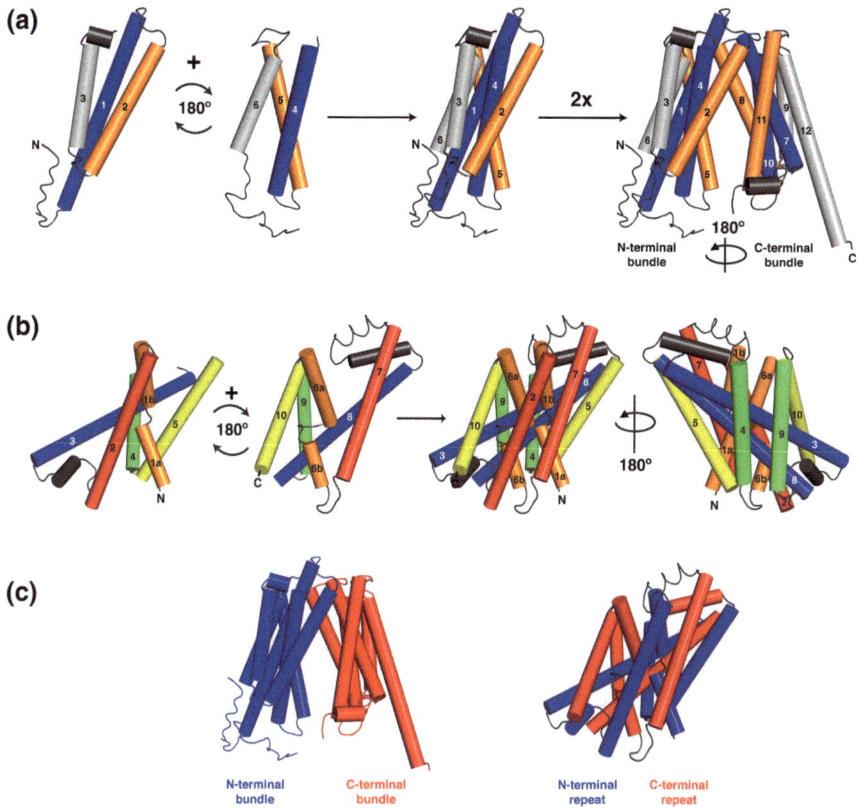

Fig. 9.2 Common folds of secondary active transporters. **a** Members of the Major Facilitator Superfamily (MFS) share a common fold that consists of 12 transmembrane α-helices (TMs) organized into an N- and a C-terminal bundle (TMs 1–6 and TMs 7–12). The bundles are related to each other by a pseudo-twofold symmetry axis that is perpendicular to the membrane plane. Each bundle consists of symmetry related inverted repeats being composed of three TMs. **b** Membrane transporters that share the LeuT fold consist of 10 TMs assembled into two inverted structural repeats (N-terminal repeat TMs 1–5, C-terminal repeat TMs 6–10—also known as the "5 + 5 inverted repeats") . TMs 1 and 6 are discontinuous. The two short α-helices in TM1 and TM6 are linked by a conserved unwound polypeptide stretch. **c** The TMs of the N- and C-terminal bundle of secondary active transporters that adopt the MFS fold are intertwined such that there is a stable TM packing within the bundle. In contrast, in the LeuT fold the TMs of one repeat are intertwined with the TMs of the other repeat through extensive interfaces

TM3, TM6, TM9 and TM12 are in contact with the lipid bilayer, not being parts of the central cavity, and are thus involved in hydrophobic mismatch interactions with the surrounding lipid bilayer. It has been postulated that proteins of the MFS resulted from internal gene duplication events of structural repeats that occurred before the MFS family diverged (Pao et al. 1998; Madej et al. 2013; Madej and Kaback 2013; Madej 2014). The gene duplication is also reflected in a specific sequence motif: GXXX(D/E)(R/K)XG[X](R/K)(R/K), which is located in the intracellular loops connecting TM2 with TM3, and TM8 with TM9 of many MFS transporters (Henderson and Maiden 1990; Henderson 1990; Griffith et al. 1992). It was also proposed that the six-helix bundles might originate from triplication of a 2-TMs hairpin structure (Reddy et al. 2012). In summary, MFS transporters consist of four structural repeats each consisting of three TMs. However, there are MFS transporters that comprise additional TMs (e.g., peptide transporters (Newstead et al. 2011)), which contribute to the stabilization of the respective membrane transporter.

LeuT Fold

A novel fold has been unveiled by the X-ray structure of a bacterial leucine-transporting homologue of the Na^+/Cl^--dependent neurotransmitter transporter (Yamashita et al. 2005) that is a member of the neurotransmitter sodium symporter (NSS) family (TCID 2.A.22). The LeuT fold is characterized by a protein core consisting of ten TMs, which are organized into two repeats, each containing five TMs (TM1–5 and TM6–10), with the termini located on the cytoplasmic side (Fig. 9.2b). The two repeats are related to each other by a pseudo-twofold symmetry axis that is parallel to the membrane plane. In contrast to the MFS fold, TMs are not tightly packed within one repeat. The TMs of one repeat are intertwined with the TMs of the other repeat through extensive interfaces (Fig. 9.2c). Similar to the MFS fold, it has been suggested that the LeuT fold has evolved from gene-duplication and gene fusion events resulting in the two inverted repeats (Khafizov et al. 2010). The first TMs of each inverted repeat (i.e., TM1 and TM6) are antiparallel and discontinuous. Therefore, they contain two short α-helices, which are linked by a conserved unwound amino acid stretch (Abramson and Wright 2009). Besides the protein core, which consists of ten TMs, many membrane transporters contain additional peripheral TMs, at the N-terminal and/or C-terminal ends. For example, two additional C-terminal TMs, i.e., TM11 and TM12, are found in the APC family member AdiC (TCID 2.A.3.2.5) to form dimers (Fang et al. 2009; Gao et al. 2009, 2010; Kowalczyk et al. 2011; Ilgü et al. 2016). The classical LeuT fold is adopted by membrane proteins of different families, e.g., APC (TCID 2.A.3), BCCT (TCID 2.A.15), NCS1 (TCID 2.A.39), NSS (TCID 2.A.22), SSS (TCID 2.A.21) families of the APC superfamily (TCID 2.A.3), which have relatively low amino acid sequence identities/similarities and have different functionalities.

Transport Mechanism of Secondary Active Transporters

Alternating Access Model

In order to accomplish the transport of a substrate across a biological membrane, all membrane transporters must switch between at least two different conformations (Fig. 9.3a). In this model, the substrate-binding site needs to be accessible to either side of the membrane at different time points during the transport cycle. This alternating accessibility, which was already proposed by Jardetzky in 1966 (Jardetzky 1966), can only be accomplished through an allosteric coupling of extra- and intracellular gates of a transporter. In contrast to a channel, a secondary transporter is not open simultaneously to both sides of the membrane. Therefore, global structural changes are required to mediate substrate transport across a membrane. In a simplified model, a membrane transporter must adopt outward-open and inward-open conformations, which are connected by an occluded state, where the substrate binding site is not accessible from either side of the membrane. Theoretically, structures of multiple distinct states of a particular transporter should be available to elucidate its transport mechanism. This has proven to be a challenging endeavor and has not been successfully accomplished. However, snapshots of distinct conformational states of different MFS transporters have been successfully captured in X-ray structures and can be used to assemble and understand the transport cycle: inward-open (Abramson et al. 2003; Huang et al. 2003, 2015; Mirza et al. 2006; Guan et al. 2007; Chaptal et al. 2011; Solcan et al. 2012; Doki et al. 2013; Guettou et al. 2013, 2014; Zheng et al. 2013; Quistgaard et al. 2013; Iancu et al. 2013; Wisedchaisri et al. 2014; Zhao et al. 2014; Lyons et al. 2014; Boggavarapu et al. 2015; Fowler et al. 2015; Heng et al. 2015; Taniguchi et al. 2015; Fukuda et al. 2015; Ma et al. 2017; Quistgaard et al. 2017; Parker et al. 2017; Martinez Molledo et al. 2018; Minhas et al. 2018), partially inward-open (Yan et al. 2013), inward-facing and partially occluded (Quistgaard et al. 2013; Martinez Molledo et al. 2018), occluded (Yin et al. 2006; Newstead et al. 2011; Yan et al. 2013; Fukuda et al. 2015; Martinez Molledo et al. 2018), almost occluded but partly open to the periplasmic space (Kumar et al. 2014, 2015), outward-facing and partly occluded (Sun et al. 2012; Ethayathulla et al. 2014), slightly occluded and outward-open (Kumar et al. 2018), and outward-open (Dang et al. 2010; Jiang et al. 2013, 2016; Taniguchi et al. 2015; Nagarathinam et al. 2018) (Table 9.1).

Within the alternating access model (Fig. 9.3a), three different mechanisms have been identified based on crystal structures: rocker-switch (Fig. 9.3b), rocking-bundle (Fig. 9.3c) and elevator (Fig. 9.3d) mechanisms. The substrate-binding site of transporters, which function according to the rocker-switch or rocking-bundle mechanisms, is located approximately in the middle of the membrane at the interface of two domains (MFS fold transporters: N- and C-terminal six helix bundles; LeuT fold transporters: "5 + 5 inverted repeats"). When the barrier between the two domains is broken on one side and reforms on the other side of the membrane (i.e., moving barrier), the alternating access to the substrate-binding site is possible. In the rocker-switch mechanism, substrate binding and the movement of the two symmetry-related

Table 9.1 Structures of bacterial members of the major facilitator superfamily

Protein	Mutation	Conformation	Ligand	Resolution (Å)[a]	PDB	Reference	Year
LacY	C154G	Inward-open	TDG	3.6	1PV7	Abramson et al. (2003)	2003
LacY	C154G	Inward-open	None	3.5	1PV6	Abramson et al. (2003)	2003
LacY	C154G	Inward-open	None/pH 6.5	2.95	2CFQ	Mirza et al. (2006)	2006
LacY	C154G	Inward-open	None/pH 5.6	3.3	2CFP	Mirza et al. (2006)	2006
LacY	WT	Inward-open	None	3.6	2V8N	Guan et al. (2007)	2007
LacY	Cys-mut[b]	Inward-open	MTS-gal	3.38	2Y5Y	Chaptal et al. (2011)	2011
LacY	G46W-G262W	Almost occluded, partly open to the periplasmic side	TDG	3.5	4OAA	Kumar et al. (2014)	2014
LacY	G46W-G262W	Almost occluded, partly open to the periplasmic side	α-NPG	3.31	4ZYR	Kumar et al. (2015)	2015
LacY	G46W-G262W	Outward-open	Nanobody	3.3	5GXB	Jiang et al. (2016)	2016
LacY	G46W-G262W	Slightly occluded, outward-open	Nanobody/α-NPG	3.0	6C9W	Kumar et al. (2018)	2018
FucP	WT	Outward-open	β-NG	3.14	3O7Q	Dang et al. (2010)	2010
FucP	N162A	Outward-open	β-NG	3.2	3O7P	Dang et al. (2010)	2010
MelB	L5M	Outward partially occluded	None	3.35	4M64	Ethayathulla et al. (2014)	2014
MelB	L5M	Outward inactive	None	3.35	4M64	Ethayathulla et al. (2014)	2014
XylE	WT	Outward-facing, partly occluded	D-Xylose	2.81	4GBY	Sun et al. (2012)	2012

(continued)

Table 9.1 (continued)

Protein	Mutation	Conformation	Ligand	Resolution (Å)[a]	PDB	Reference	Year
XylE	WT	Outward-facing, partly occluded	D-Glucose	2.90	4GBZ	Sun et al. (2012)	2012
XylE	WT	Outward-facing, partly occluded	6-Bromo-6-deoxy-D-glucose	2.6	4GC0	Sun et al. (2012)	2012
XylE	WT	Inward-open, partially occluded	None	3.8	4JA3	Quistgaard et al. (2013)	2013
XylE	WT	Inward-open	None	4.2	4JA4	Quistgaard et al. (2013)	2013
XylE	WT	Inward-open	None	3.51	4QIQ	Wisedchaisri et al. (2014)	2014
GlcP	WT	Inward-open	None	3.2	4LDS	Iancu et al. (2013)	2013
GlpT	WT	Inward-open	None	3.3	1PW4	Huang et al. (2003)	2003
EmrD	WT	Occluded	None	3.5	2GFP	Yin et al. (2006)	2006
$PepT_{So}$	T2N-T3S	Occluded	None	3.62	2XUT	Newstead et al. (2011)	2011
$PepT_{So2}$	WT	Inward-open	Ala-Tyr(L-3,5-di-Br)	3.16	4TPH	Guettou et al. (2014)	2014
$PepT_{So2}$	WT	Inward-open	Ala-Tyr(L-3,5-di-Br)-Ala	3.91	4TPG	Guettou et al. (2014)	2014
$PepT_{So2}$	WT	Inward-open	Ala-Ala-Ala	3.2	4TPJ	Guettou et al. (2014)	2014
$PepT_{So2}$	WT	Inward-open	Alafosfalin	3.2	4LEP	Guettou et al. (2013)	2013
$PepT_{So}$	WT	Inward-open	None	3.0	4UVM	Fowler et al. (2015)	2015
$PepT_{St}$	WT	Inward-open	None	3.3	4APS	Solcan et al. (2012)	2012
$PepT_{St}$	WT	Inward-open	None	2.35	4D2B	Lyons et al. (2014)	2014
$PepT_{St}$	WT	Inward-open	Ala-Phe	2.47	4D2C	Lyons et al. (2014)	2014

(continued)

Table 9.1 (continued)

Protein	Mutation	Conformation	Ligand	Resolution (Å)[a]	PDB	Reference	Year
$PepT_{St}$	WT	Inward-open	Ala-Ala-Ala	2.52	4D2D	Lyons et al. (2014)	2014
$PepT_{St}$	WT	Inward-open	None	2.3	4XNJ	Huang et al. (2015)	2015
$PepT_{St}$	WT	Inward-open	None	2.8	4XNI	Huang et al. (2015)	2015
$PepT_{St}$	WT	Inward-open	None	3.4	5MMT	Quistgaard et al. (2017)	2017
$PepT_{St}$	WT	Inward-open	None	2.4	5D6K	Ma et al. (2017)	2017
$PepT_{St}$	WT	Inward-facing partially occluded	Ala-Leu	2.66	5OXL	Martinez Molledo et al. (2018)	2018
$PepT_{St}$	WT	Inward-facing partially occluded	Ala-Gln	2.38	5OXK	Martinez Molledo et al. (2018)	2018
$PepT_{St}$	WT	Inward-open	Asp-Glu	2.3	5OXM	Martinez Molledo et al. (2018)	2018
$PepT_{St}$	WT	Inward-open	Phe-Ala	2.2	5OXN	Martinez Molledo et al. (2018)	2018
$PepT_{St}$	WT	Inward-open	HEPES	2.0	6EIA	Martinez Molledo et al. (2018)	2018
$PepT_{St}$	WT	Inward-open	HEPES	2.19	5OXQ	Martinez Molledo et al. (2018)	2018
$PepT_{St}$	WT	Occluded	PO_4^{3-}	2.37	5OXP	Martinez Molledo et al. (2018)	2018
$PepT_{St}$	WT	Inward-open	None	1.95	5OXO	Martinez Molledo et al. (2018)	2018
GkPOT	WT	Inward-open	None	1.90	4IKV	Doki et al. (2013)	2013

(continued)

Table 9.1 (continued)

Protein	Mutation	Conformation	Ligand	Resolution (Å)[a]	PDB	Reference	Year
GkPOT	WT	Inward-open	SO_4^{2-}	2.00	4IKW	Doki et al. (2013)	2013
GkPOT	E310Q	Inward-open	None	2.30	4IKX	Doki et al. (2013)	2013
GkPOT	E310Q	Inward-open	sulfate	2.1	4IKY	Doki et al. (2013)	2013
GkPOT	E310Q	Inward-open	Alafosfalin	2.40	4IKZ	Doki et al. (2013)	2013
YbgH	WT	Inward-open	None	3.4	4Q65	Zhao et al. (2014)	2014
YePEPT	WT	Inward-open	None	3.02	4W6V	Boggavarapu et al. (2015)	2015
$PepT_{Xc}$	WT	Inward-open	None	2.1	6EI3	Parker et al. (2017)	2017
$PepT_{Sh}$	WT	Inward-open	S-Cys-Gly-3M3SH	2.5	6EXS	Minhas et al. (2018)	2018
NarU	WT	Unclear	None	3.00	4IU9	Yan et al. (2013)	2013
NarU	WT	Partially inward-open	None	3.00	4IU9	Yan et al. (2013)	2013
NarU	SeMet-derivatized WT	Occluded	NO_3^-	3.11	4IU8	Yan et al. (2013)	2013
NarU	SeMet-derivatized WT	Partially inward-open	None	3.11	4IU8	Yan et al. (2013)	2013
NarK	WT	Inward-open	None	2.60	4JR9	Zheng et al. (2013)	2013
NarK	WT	Inward-open	NO_2^-	2.80	4JRE	Zheng et al. (2013)	2013
NarK	WT	Inward-open	None	2.35	4U4V	Fukuda et al. (2015)	2015
NarK	WT	Inward-open	NO_3^-	2.40	4U4T	Fukuda et al. (2015)	2015
NarK	WT	Occluded	NO_3^-	2.40	4U4W	Fukuda et al. (2015)	2015

(continued)

Table 9.1 (continued)

Protein	Mutation	Conformation	Ligand	Resolution (Å)[a]	PDB	Reference	Year
YajR	WT	Outward-open	None	3.15	3WDO	Jiang et al. (2013)	2013
MdfA	Q131R	Inward-open	Deoxycholate	2.0	4ZP0	Heng et al. (2015)	2015
MdfA	Q131R	Inward-open	LDAO	2.2	4ZP2	Heng et al. (2015)	2015
MdfA	Q131R	Inward-open	Chloramphenicol	2.45	4ZOW	Heng et al. (2015)	2015
MdfA	WT	Outward-open	None	3.4	6GV1	Nagarathinam et al. (2018)	2018
FPN	WT	Outward-open	None	2.20	5AYN	Taniguchi et al. (2015)	2015
FPN	WT	Outward-open	Fe^{2+}	3.0	5AYM	Taniguchi et al. (2015)	2015
FPN	WT	Inward-open	None	3.3	5AYO	Taniguchi et al. (2015)	2015

Abbreviations α-NPG *p*-nitrophenyl-α-D-galactopyranoside, TDG β-D-galactopyranosyl-1-thio-β-D-galactopyranoside, MTS-gal methanethiosulfonyl-galactopyranoside, β-NG n-nonyl-β-D-glucopyranoside, LDAO lauryldimethylamine *N*-oxide, WT wild-type

[a]Resolution as deposited in the PDB database (http://www.rcsb.org)

[b]C117S-A122C-C148A-C154V-C176M-C234S-C333S-C353A-C355A

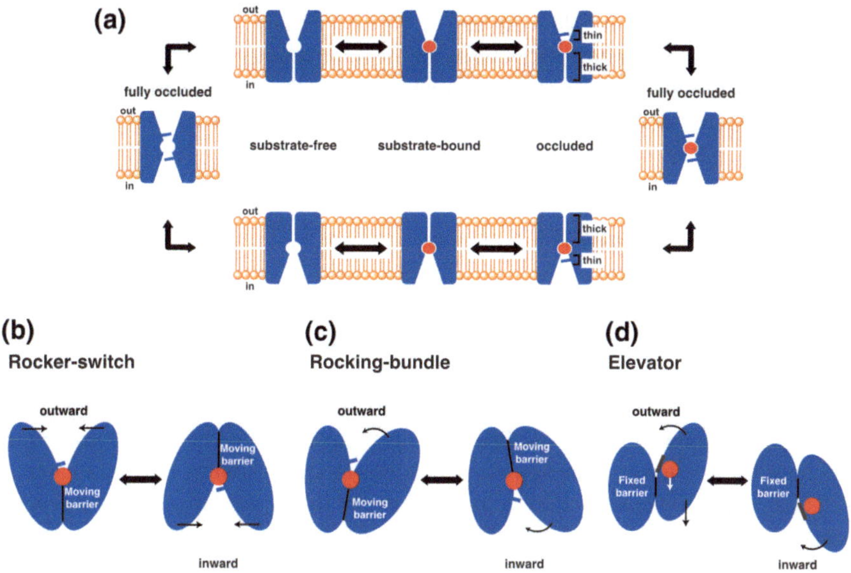

Fig. 9.3 The alternating access mechanism. **a** The major conformational changes and states, which are necessary to allow alternating access to the substrate binding site, are shown. **b** In secondary active transporters that function according to the rocker-switch principle two domains (e.g., N- and C-terminal bundle of MFS transporters) "rock" against each other to allow access to the binding site from both sides of the membrane. **c** In the rocking-bundle mechanism, the two structural domains (e.g., N- and C-terminal repeat of LeuT fold) are dissimilar. One domain rearranges against the other domain to expose the substrate binding site alternatingly to both sides of the membrane. Substrate transport requires subtle, substrate-induced structural rearrangement by amino acid side chains and helices in only one or both domains/repeats, which is highlighted by the *bar* that covers the substrate. In the rocker-switch and in the rocking-bundle mechanism there is a moving barrier (*black line*) that seals the protein on either the intra- or extracellular side. **d** In the elevator mechanism, there is a fixed barrier. The substrate binding site is located in one domain, which moves against the other immobile domain that is structurally dissimilar. This mechanism allows the transport of the substrate to the other side of the fixed barrier. Panel **a** was made and kindly provided by Dr. Hüseyin Ilgü (University of Bern)

bundles, which maintain their structures as rigid bodies, is nearly symmetrical around a central substrate-binding site (Karpowich and Wang 2008; Yan 2013). This means that the protein moves around the bound substrate. In the rocking-bundle mechanism, substrate binding occurs between two structurally distinct domains and the binding catalyzes the coupled movement of outside and inside barriers around a central substrate-binding site (Forrest and Rudnick 2009). The movement of both domains is not symmetrical since one structurally distinct domain rearranges against a more static domain. Besides the large rigid body movements of the two domains, there are also further subtle and local substrate-induced gating rearrangements of helices in either one or both domains. In the elevator mechanism (Fig. 9.3d) the substrate-binding site is located in only one domain (transport domain), which traverses the membrane against an immobile domain (scaffold domain). The two domains, which are part of the same polypeptide chain, are distinct bundles. A detailed description of this transport mechanism, which is used by the glutamate transporter homologue Glt_{Ph}, can be found elsewhere (Drew and Boudker 2016).

Coupling Mechanism

In contrast to permeases that catalyze facilitated diffusion of specific molecules down their concentration gradients, secondary active transporters utilize the electrochemical gradient of a co-substrate (e.g., an ion) to catalyze the transport of a substrate against its concentration gradient. In symporters and antiporters, transport of the substrate is only allowed if the co-substrate is also transported in the same (symporter) or opposite direction (antiporter), a concept described as coupling. Substrates and co-substrates bind to the secondary active transporter in an interdependent and ordered manner. In the case of MFS transporters, the fundamental question that is associated with their transport mechanism is how binding of substrate and co-substrate (e.g., proton or Na^+) is coupled to the conformational switch between the outward- and inward-open conformations. To understand the former, researchers introduced the "counterflow" assay (Xie 2008). In this assay, vesicles containing the protein of interest (e.g., right-side-out vesicles) are loaded with a high concentration of unlabeled substrate and are then diluted into a buffer that contains a low concentration of radiolabeled substrate. This results in an outwardly directed concentration gradient across the membrane. The buffers on both sides of the vesicle membrane have the same pH such that there is no proton gradient. Transport of substrate in and out leads to the exchange of the internal unlabeled substrate with the external labeled substrate and thus to the accumulation of radioactivity inside the vesicles. This accumulation of radiolabeled substrate ceases at the point of equilibration, where the exit of cold and hot substrates becomes equal. If a particular mutation leads to an inactive transporter in a bacteria-based uptake assay (i.e., in the presence of a transmembrane proton gradient), but reveals activity in the counterflow assay (i.e., in the absence of a transmembrane proton gradient) , it can be assumed that this mutation does not impair substrate binding, but coupling.

External energy is required if a substrate is to be transported against its concentration gradient (i.e., $\Delta\mu_{[S]} = 2.3\ RT/F\ \log([S]_{release}/[S]_{loading}) > 0$). In bacteria, many transmembrane transport processes are driven by an electrochemical proton gradient ($\Delta\mu_{[H+]}$), which consists of an electrical component ($\Delta\mu_{\Psi}$, negative inside the bacterium) and/or a pH gradient (ΔpH, alkaline inside the bacterium): $\Delta\mu_{[H+]} = \Delta\mu_{\Psi} - 2.3\ RT/F\ \Delta pH$ ($2.3\ RT/F = {\sim}60$ mV at 25 °C). The bacterial proton-coupled lactose transporter LacY is a well-studied MFS symporter, which serves as model system to investigate many aspects of secondary active symport. LacY catalyzes the strictly coupled stoichiometric transport of one lactose and one proton to drive the uphill accumulation of lactose against its concentration gradient. The free energy that is released in the energetically downhill movement of the proton is consumed in this process. Accordingly, an electrochemical proton gradient of − 60 mV in the form of $\Delta\mu_{\Psi}$ and/or ΔpH will lead to a tenfold concentration gradient of lactose. The conformational switch of LacY from outward- to inward-open is only allowed in the ternary form (i.e., substrate (lactose) and co-substrate (proton) are bound). Under physiological conditions, any binary complexes (i.e., substrate (lactose) or co-substrate (proton) are bound) are locked, which prevents the system from uncoupling. However, unloaded LacY switches back to its initial state after substrate release on the cytosolic side of the membrane. For LacY, the substrate binds to a protonated transporter while dissociation occurs in the opposite sequence (Sahin-Tóth et al. 2000). Kaback and colleagues studied the effect of bulk-phase pH on the apparent affinity K_D of purified LacY lactose permease (LacY) for sugars (Smirnova et al. 2008). Apparent K_D values increased with increasing pH with an apparent pK_a of ~10.5, which was defined as the bulk pH value leading to a half-maximal K_D. However, mutational studies have highlighted that this extraordinary high pK_a cannot be assigned to a single amino acid. Under physiological conditions (e.g., pH 7) LacY can be assumed to be in the protonated state, which is a prerequisite for substrate binding, due to the high apparent pK_a value of substrate binding. It has been postulated that translocation of protons requires the protonation and deprotonation of certain residues: glutamate, aspartate, histidine, and to some minor extent lysine, arginine and tyrosine. This fact is nicely illustrated in the case of LacY, where proton translocation involves residues that are mainly located in the C-terminal six-helix bundle. Tyr236 (TM7), Asp240 (TM7), Arg302 (TM9), Lys319 (TM10), His322 (TM10), and Glu325 (TM10) form a hydrogen-bond and salt bridge network between TMs 7, 9, and 10 (Mirza et al. 2006) (Fig. 9.4a). Charge pairs between Asp240 (TM7) and Arg302 (TM9) as well as Asp240 (TM7) and Lys319 (TM10) were described, although the corresponding functional groups are not within the expected distances in all LacY structures (Madej et al. 2014). His322 (TM10) and Glu269 (TM8), which are part of the ligand-binding site, might be involved in proton translocation and substrate binding, and are thus important for the coupling between substrate (lactose) and co-substrate (proton). Glu325 (TM10) is most likely exclusively involved in proton translocation (Carrasco et al. 1986a, b; Sahin-Tóth and Kaback 2001). Arg302 (TM9) might facilitate its deprotonation by decreasing the pKa of the protonated carboxyl group of Glu325 (TM10) (Sahin-Tóth and Kaback 2001; Kaback et al. 2001; Andersson et al. 2012). However, in X-ray structures of LacY the distance between the side chains of Arg302 (TM9) and Glu325 (TM10) is ~7 Å, and they are separated by Tyr236 (TM7) (Fig. 9.4a). A conformational change that reduces the distance between the oppositely charged residues must occur to allow efficient deprotonation (Madej et al. 2012). Reprotonation of the apo intermediate might occur stochastically from the side of the membrane with the higher proton concentration. Apparently, a negative charge at position 325 (i.e., unprotonated glutamate) counteracts substrate binding and has to be neutralized to lower the energy barrier for substrate binding. A substrate-bound structure of LacY has revealed that Tyr236 (TM7), Glu269 (TM8) and His322 (TM10), which have been previously considered to be exclusively involved in coupled proton translocation (Guan and Kaback 2006), are clearly involved in substrate binding (Kumar et al.

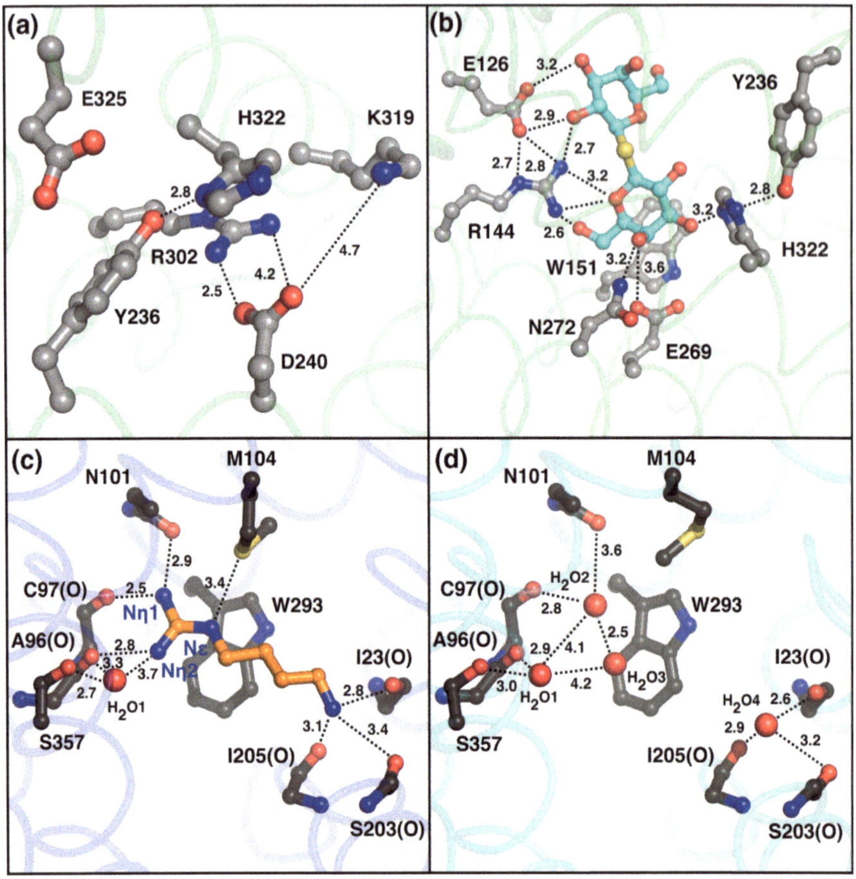

Fig. 9.4 Ligand and substrate binding mechanism in LacY and AdiC. **a** Residues in LacY that are supposed to form a hydrogen-bond and salt bridge network between TMs 7, 9, and 10 (Mirza et al. 2006). **b** Substrate binding site of LacY with bound β-D-galactopyranosyl-1-thio-β-D-galactopyranoside (TDG, *cyan*). **c** Substrate binding site of AdiC with bound agmatine. The three guanidinium group nitrogen atoms of agmatine are labeled Nε, Nη1 and Nη2. **d** Substrate binding site of AdiC in its apo state. Functionally important residues are labeled using the single letter code. Substrate interactions with main-chain carbonyl oxygen atoms of particular residues are labeled with (O). Crystallographic water molecules are highlighted as red balls. Potential hydrogen bonds are indicated as dashed lines and interatomic distances are given in Ångström, respectively

2014). However, these residues might play a dual role in substrate and co-substrate binding. Unfortunately, there is no experimental evidence so far about what triggers the conformational change from the outside- to the inside-open conformation after the co-substrate (proton) and substrate (sugar) were bound.

LacY can accumulate lactose against a ~100-fold concentration gradient in the presence of an electrochemical proton gradient (Kaback et al. 2001; Guan and Kaback 2006). The main effect of the electrochemical proton gradient on active transport by LacY seems to be kinetic (i.e., lowering of the K_m, small influence on V_{max}) (Robertson et al. 1980). It does not affect the binding affinity of LacY for sugar (i.e., K_D) (Guan and Kaback 2004), but it accelerates deprotonation at the cytoplasmic side (Viitanen et al. 1983; Guan and Kaback 2006). In the presence of an electrochemical proton gradient, substrate release seems to be the rate-limiting step (Guan and Kaback 2006). This increases the population of unloaded, inward-open LacY molecules, which can then return back to the outward-open conformation. Presumably, unloaded LacY carries a net negative charge, which senses the membrane potential (i.e., outside positive, inside negative) (Forrest et al. 2011). It can be hypothesized that the electrophoretic effect (i.e., the movement of charged objects in an electrical field) of the membrane potential might facilitate the conformational change of unloaded LacY. It has also been speculated that the required energy for this conformational change might be provided from stored energy originating from the ligand-binding step that can be released in this part of the transport cycle (Forrest et al. 2011). It has to be mentioned that so far there is no structural evidence supporting this mechanistic hypothesis.

In contrast to the hydrogen-bond and salt bridge network of LacY, the proton-coupled fucose transporter FucP has only two negatively charged residues (i.e., Asp46 (TM1) and Glu135 (TM4)) along its substrate translocation path, which can undergo protonation and deprotonation. Asp46 is essential for proton coupling but it is not involved in substrate binding, while Glu135 is important for substrate binding as well as transport (Dang et al. 2010). Interestingly, the proton-dependent sugar transporter XylE does not have any charged residues in the vicinity of its substrate-binding site in contrast to LacY and FucP (Sun et al. 2012). A detailed mechanism describing the translocation of protons in XylE is currently unclear, although charged and polar residues on the cytoplasmic side of the protein have been identified, which abrogate transport function (Sun et al. 2012).

Substrate Binding

For a detailed understanding of the transport mechanism of a particular transporter, it is important to characterize the substrate-binding site. Residues involved in substrate binding can be identified by site-directed mutagenesis in combination with an appropriate functional or spectroscopic assay (e.g., double electron-electron resonance, single molecule fluorescence resonance energy transfer, and tryptophan fluorescence). If the structure of the protein of interest in complex with its substrate is available at a sufficient resolution, the molecular binding interactions can be directly visualized. Substrate-bound structures of different secondary active transporters unveil common features in the substrate-binding site.

Substrate Binding and Translocation—The MFS Transporter LacY

For LacY, several structures with bound substrates and substrate analogues in different conformational states have been solved (Abramson et al. 2003; Guan et al. 2007; Chaptal et al. 2011; Kumar et al. 2014, 2015, 2018). A recent structure of a substrate-bound LacY mutant (G46W-G262W) in an almost occluded conformation confirms biochemical and spectroscopic findings (Kumar et al. 2014). Functional studies have shown that there is a single substrate-binding site in LacY that is enclosed by the N- and C-terminal six-helix bundles (Abramson et al. 2003). The substrate-binding site is located midway of the membrane part near the apex of the central cavity in the inward-open LacY structure (Abramson et al. 2003). There are fewer than ten residues that are irreplaceable for active lactose transport. Several structures have shown that the stoichiometry of substrate binding is one galactoside per LacY. The substrate is mainly bound to residues located in the N-terminal bundle. Glu126 (TM4), Arg144 (TM5), and Trp151 (TM5) are directly involved in the binding of the galactopyranoside ring of lactose (Fig. 9.4b). A conservative replacement of Arg144 (TM5) to lysine abolished substrate binding, and exchanging Glu126 (TM4) for an aspartate reduces substrate affinity (Kaback et al. 2001). The guanidinium group of Arg144 (TM5) and the carboxylate group of Glu126 (TM4) seem to interact in the inward-open, occluded and outward-open conformations (Abramson et al. 2003; Mirza et al. 2006; Guan et al. 2007; Chaptal et al. 2011; Kumar et al. 2014, 2015, 2018; Jiang et al. 2016). An aromatic residue at position 151 is essential for substrate binding (Guan et al. 2003) and the two charged residues are irreplaceable for substrate binding (Frillingos et al. 1998; Venkatesan and Kaback 1998; Sahin-Tóth et al. 1999). Furthermore, Glu269 (TM8) that is also essential for substrate binding, accepts hydrogen bonds from the hydroxyl-groups of the galactopyranoside ring. This residue is assumed to be involved in substrate specificity. Asn272 might also be hydrogen bonded to one of the hydroxyl-groups of the substrate and its amide group is important for substrate affinity (Jiang et al. 2014). Interestingly, His322 (TM10), which is positioned by Tyr236 (TM7), has long been assumed to be mainly responsible for proton translocation. However, a substrate-bound LacY structure shows that the irreplaceable His322 (Padan et al. 1985; Püttner et al. 2002; Smirnova et al. 2009) can act as a hydrogen bond donor between the εNH of the imidazole ring and the C3 hydroxyl group of the substrate, which are separated by 3.2 Å (Fig. 9.4b) (Kumar et al. 2014). In contrast to the occluded conformation of LacY with a bound, high-affinity lactose analogue, the structure of LacY was in an inward-open conformation if a substrate analogue (i.e., MTS-galactoside) was covalently bound to the transporter (Chaptal et al. 2011). In this case, only Trp151 (TM5) and Glu269 interact with the galactoside (Chaptal et al. 2011). Therefore, the transporter can only be shifted into its occluded conformation if the sugar is fully ligated by the binding site residues, as observed in the occluded structure (Kumar et al. 2014). Substrate translocation might start with the hydrophobic interaction between the indole ring of Trp151 (TM5) and the galactopyranoside ring. Sugar binding is postulated to lower the activation energy barrier for the transition between the outward- and inward-open LacY conformations (Madej et al. 2012). Therefore, the energy that is required for the conformational change might originate from the binding energy of increased interactions between LacY and its substrate in the occluded state. Consequently, substrate-binding allows LacY to overcome the energy barriers that separate the outward- and inward-open conformations. Substrate binding to LacY seems to be mainly entropic, which indicates the existence of different ligand-bound conformational states (Nie et al. 2006). Therefore, substrate binding involves an induced fit mechanism that results in an expanded conformational space. Binding and dissociation of the substrate seem to be the driving force for the alternating access mechanism of the secondary active transporter LacY. A new transport cycle might be initiated after deprotonation where Arg302 (TM10) approaches the deprotonated Glu325 (TM10).

Combining biochemical, spectroscopic and structure analysis data supports the existence of an induced-fit mechanism for substrate-binding in LacY (Vázquez-Ibar et al. 2004; Mirza et al. 2006). A double tryptophan mutant (G46W-G262W), which has completely lost its transport function, has been shown to behave as if it is trapped in an outward-open conformation with diffusion-limited access to the binding site (Smirnova et al. 2013). However, the structure exhibits an almost occluded LacY molecule with bound substrate. The periplasmic entrance is too narrow for the bound substrate to leave the binding site (Kumar et al. 2014). It can be hypothesized that the double tryptophan mutant of LacY is in an outward-open conformation in the absence of substrate to facilitate substrate entry to the binding-site. After substrate-binding, LacY might shift to an occluded state. However, this transition cannot be completely accomplished due to the sterically demanding tryptophan residues. Therefore, the double tryptophan mutant is able to bind substrate but a fully occluded state cannot be achieved.

It is important to mention that besides LacY there are other bacterial MFS transporters with solved structures that show proteins in ligand bound states (Table 9.1).

Substrate Binding and Translocation—The APC Transporter AdiC

Several structures of the bacterial APC transporter AdiC that catalyzes the electrogenic export of agmatine in exchange of L-arginine have been reported (Fang et al. 2009; Gao et al. 2009, 2010; Kowalczyk et al. 2011; Ilgü et al. 2016). A structure of wild-type AdiC in complex with agmatine highlights the molecular interactions that are required for substrate binding in the outward-open conformation (Ilgü et al. 2016). One molecule of agmatine is bound by several amino acids from TM1, TM3, TM6, TM8, and TM10 halfway up the AdiC transport path (Fig. 9.4c). Agmatine, which consists of two hydrophilic groups (i.e., the primary amino group and the guanidinium group) separated by one hydrophobic group (i.e., the aliphatic region connecting the previous two), is mainly recognized by hydrogen bonds (H-bonds) and one cation–π interaction. In the AdiC structure with bound agmatine, the primary amino group donates three hydrogen bonds to the backbone carbonyl oxygen atoms of Ile23 (TM1), Ser203 (TM6), and Ile205 (TM6). The three nitrogen atoms of the guanidinium group of agmatine interact with backbone and side chain atoms of residues on TM3: Nη1 with the carbonyl oxygen atom of Cys97 (TM3) and the amide group oxygen atom of Asn101 (TM3), Nη2 with the carbonyl oxygen atom of Ala96 (TM3), and Nε with the sulfur atom of Met104 (TM3). Due to the remarkable resolution of this structure even water molecules could be resolved. A water molecule (H_2O1) is hydrogen bonded to Nη2, the carbonyl oxygen atom of Ala96 (TM3) and the hydroxyl group of Ser357 (TM10). Therefore, this water molecule (H_2O1) is involved in substrate binding and in shaping the binding pocket by stabilizing interactions between TM3 and TM10.

A cation–π interaction, which is often found in protein-ligand interactions (Dougherty 1996; Gallivan and Dougherty 1999; Zacharias and Dougherty 2002; Dougherty 2013), is responsible for the binding of the guanidinium group of agmatine by interacting with Trp293 (TM8). This interaction is crucial for substrate binding to AdiC (Casagrande et al. 2008; Fang et al. 2009). The aliphatic moiety of agmatine is bound via van der Waals interactions with the side chains of Ile205 (TM6) and Trp293 (TM8). A structure of AdiC in its outward-open conformation in the absence of substrate (i.e., apo state) is also available (Fig. 9.4d) (Ilgü et al. 2016). The crystallographic water molecule H_2O_1 was found in the agmatine-bound and in the apo structures. However, three additional water molecules (i.e., H_2O2, H_2O3, and H_2O4) were present in the apo AdiC structure, which mimic most of the hydrophilic portions of the agmatine molecule. Based on the available AdiC structures, a model could be generated that describes the conformational changes that are involved in the release of agmatine into the periplasm of *E. coli* (Ilgü et al. 2016). Under extreme acidic conditions, the uptake of L-arginine into the bacterial cytoplasm is driven by the concentration gradient of agmatine across the membrane.

References

Abramson J, Wright EM (2009) Structure and function of Na^+-symporters with inverted repeats. Curr Opin Struct Biol 19:425–432

Abramson J, Smirnova I, Kasho V, Verner G, Kaback HR, Iwata S (2003) Structure and mechanism of the lactose permease of *Escherichia coli*. Science 301:610–615

Andersson M, Bondar A-N, Freites JA, Tobias DJ, Kaback HR, White SH (2012) Proton-coupled dynamics in lactose permease. Structure 20:1893–1904

Blattner FR (1997) The complete genome sequence of *Escherichia coli* K-12. Science 277:1453–1462

Boggavarapu R, Jeckelmann J-M, Harder D, Ucurum Z, Fotiadis D (2015) Role of electrostatic interactions for ligand recognition and specificity of peptide transporters. BMC Biol 13:58

Busch W, Saier MH (2008) The transporter classification (TC) system, 2002. Crit Rev Biochem Mol Biol 37:287–337

Carrasco N, Antes LM, Poonian MS, Kaback HR (1986a) *Lac* permease of *Escherichia coli*: histidine-322 and glutamic acid-325 may be components of a charge-relay system. Biochemistry 25:4486–4488
Carrasco N, Puttner IB, Antes LM, Lee JA, Larigan JD, Lolkema JS, Roepe PD, Kaback HR (1986b) Characterization of site-directed mutants in the *lac* permease of *Escherichia coli*. 2. Glutamate-325 replacements. Biochemistry 28:2533–2539
Casagrande F, Ratera M, Schenk AD, Chami M, Valencia E, Lopez JM, Torrents D, Engel A, Palacin M, Fotiadis D (2008) Projection structure of a member of the amino acid/polyamine/organocation transporter superfamily. J Biol Chem 283:33240–33248
Chaptal V, Kwon S, Sawaya MR, Guan L, Kaback HR, Abramson J (2011) Crystal structure of lactose permease in complex with an affinity inactivator yields unique insight into sugar recognition. Proc Natl Acad Sci USA 108:9361–9366
Dang S, Sun L, Huang Y, Lu F, Liu Y, Gong H, Wang J, Yan N (2010) Structure of a fucose transporter in an outward-open conformation. Nature 467:734–738
Doki S, Kato HE, Solcan N, Iwaki M, Koyama M, Hattori M, Iwase N, Tsukazaki T, Sugita Y, Kandori H, Newstead S, Ishitani R, Nureki O (2013) Structural basis for dynamic mechanism of proton-coupled symport by the peptide transporter POT. Proc Natl Acad Sci USA 110:11343–11348
Dougherty DA (1996) Cation–π interactions in chemistry and biology: a new view of benzene, Phe, Tyr, and Trp. Science 271:163–168
Dougherty DA (2013) The Cation–π interaction. Acc Chem Res 46:885–893
Drew D, Boudker O (2016) Shared molecular mechanisms of membrane transporters. Annu Rev Biochem 85:543–572
Erni B (2013) The bacterial phosphoenolpyruvate: sugar phosphotransferase system (PTS): an interface between energy and signal transduction. J Iran Chem Soc 10:593–630
Ethayathulla AS, Yousef MS, Amin A, Leblanc G, Kaback HR, Guan L (2014) Structure-based mechanism for Na^+/melibiose symport by MelB. Nat Commun 5:3009
Fang Y, Jayaram H, Shane T, Kolmakova-Partensky L, Wu F, Williams C, Xiong Y, Miller C (2009) Structure of a prokaryotic virtual proton pump at 3.2 Å resolution. Nature 460:1040–1043
Forrest LR, Rudnick G (2009) The rocking bundle: a mechanism for ion-coupled solute flux by symmetrical transporters. Physiology 24:377–386
Forrest LR, Krämer R, Ziegler C (2011) The structural basis of secondary active transport mechanisms. Biochim Biophys Acta 1807:167–188
Foster JW (2004) *Escherichia coli* acid resistance: tales of an amateur acidophile. Nat Rev Micro 2:898–907
Fowler PW, Orwick-Rydmark M, Radestock S, Solcan N, Dijkman PM, Lyons JA, Kwok J, Caffrey M, Watts A, Forrest LR, Newstead S (2015) Gating topology of the proton-coupled oligopeptide symporters. Structure 23:290–301
Frillingos S, Sahin-Tóth M, Wu J, Kaback HR (1998) Cys-scanning mutagenesis: a novel approach to structure–function relationships in polytopic membrane proteins. FASEB J 12:1281–1299
Fukuda M, Takeda H, Kato HE, Doki S, Ito K, Maturana AD, Ishitani R, Nureki O (2015) Structural basis for dynamic mechanism of nitrate/nitrite antiport by NarK. Nat Commun 6:190
Gallivan JP, Dougherty DA (1999) Cation–π interactions in structural biology. Proc Natl Acad Sci USA 96:9459–9464
Gao X, Lu F, Zhou L, Dang S, Sun L, Li X, Wang J, Shi Y (2009) Structure and mechanism of an amino acid antiporter. Science 324:1565–1568
Gao X, Zhou L, Jiao X, Lu F, Yan C, Zeng X, Wang J, Shi Y (2010) Mechanism of substrate recognition and transport by an amino acid antiporter. Nature 463:828–832
Griffith JK, Baker ME, Rouch DA, Page MGP, Skurray RA, Paulsen IT, Chater KF, Baldwin SA, Henderson PJF (1992) Membrane transport proteins: implications of sequence comparisons. Curr Opin Cell Biol 4:684–695
Guan L, Kaback HR (2004) Binding affinity of lactose permease is not altered by the H^+ electrochemical gradient. Proc Natl Acad Sci USA 101:12148–12152

Guan L, Kaback HR (2006) Lessons from lactose permease. Annu Rev Biophys Biomol Struct 35:67–91
Guan L, Hu Y, Kaback HR (2003) Aromatic stacking in the sugar binding site of the lactose permease. Biochemistry 42:1377–1382
Guan L, Mirza O, Verner G, Iwata S, Kaback HR (2007) Structural determination of wild-type lactose permease. Proc Natl Acad Sci USA 104:15294–15298
Guettou F, Quistgaard EM, Tresaugues L, Moberg P, Jegerschold C, Zhu L, Jong AJO, Nordlund P, Löw C (2013) Structural insights into substrate recognition in proton-dependent oligopeptide transporters. EMBO Rep 14:804–810
Guettou F, Quistgaard EM, Raba M, Moberg P, Löw C, Nordlund P (2014) Selectivity mechanism of a bacterial homolog of the human drug-peptide transporters PepT1 and PepT2. Nat Struct Mol Biol 21:728–731
Henderson PJF (1990) Proton-linked sugar transport systems in bacteria. J Bioenerg Biomembr 22:525–569
Henderson PJF, Maiden MCJ (1990) Homologous sugar transport proteins in *Escherichia coli* and their relatives in both prokaryotes and eukaryotes. Philos Trans R Soc Lond B Biol Sci 326:391–410
Heng J, Zhao Y, Liu M, Liu Y, Fan J, Wang X, Zhao Y, Zhang XC (2015) Substrate-bound structure of the *E. coli* multidrug resistance transporter MdfA. Cell Res 25:1060–1073
Huang Y, Lemieux MJ, Song J, Auer M, Wang D-N (2003) Structure and mechanism of the glycerol-3-phosphate transporter from *Escherichia coli*. Science 301:616–620
Huang C-Y, Olieric V, Ma P, Panepucci E, Diederichs K, Wang M, Caffrey M (2015) *In meso in situ* serial X-ray crystallography of soluble and membrane proteins. Acta Crystallogr D Biol Crystallogr 71:1238–1256
Iancu CV, Zamoon J, Woo SB, Aleshin A, Choe JY (2013) Crystal structure of a glucose/H^+ symporter and its mechanism of action. Proc Natl Acad Sci USA 110:17862–17867
Ilgü H, Jeckelmann J-M, Gapsys V, Ucurum Z, de Groot BL, Fotiadis D (2016) Insights into the molecular basis for substrate binding and specificity of the wild-type L-arginine/agmatine antiporter AdiC. Proc Natl Acad Sci USA 113:10358–10363
Jack DL, Paulsen IT, Saier MH (2000) The amino acid/polyamine/organocation (APC) superfamily of transporters specific for amino acids, polyamines and organocations. Microbiology 146:1797–1814
Jardetzky O (1966) Simple allosteric model for membrane pumps. Nature 211:969–970
Jiang D, Zhao Y, Wang X, Fan J, Heng J, Liu X, Feng W, Kang X, Huang B, Liu J, Zhang XC (2013) Structure of the YajR transporter suggests a transport mechanism based on the conserved motif A. Proc Natl Acad Sci USA 110:14664–14669
Jiang X, Villafuerte MKR, Andersson M, White SH, Kaback HR (2014) Galactoside-binding site in LacY. Biochemistry 53:1536–1543
Jiang X, Smirnova I, Kasho V, Wu J, Hirata K, Ke M, Pardon E, Steyaert J, Yan N, Kaback HR (2016) Crystal structure of a LacY-nanobody complex in a periplasmic-open conformation. Proc Natl Acad Sci USA 113:12420–12425
Kaback HR, Sahin-Tóth M, Weinglass AB (2001) The kamikaze approach to membrane transport. Nat Rev Mol Cell Biol 2:610–620
Karpowich NK, Wang D-N (2008) Symmetric transporters for asymmetric transport. Science 321:781–782
Khafizov K, Staritzbichler R, Stamm M, Forrest LR (2010) A study of the evolution of inverted-topology repeats from LeuT-fold transporters using AlignMe. Biochemistry 49:10702–10713
Kowalczyk L, Ratera M, Paladino A, Bartoccioni P, Errasti-Murugarren E, Valencia E, Portella G, Bial S, Zorzano A, Fita I, Orozco M, Carpena X, Vázquez-Ibar J, Palacín M (2011) Molecular basis of substrate-induced permeation by an amino acid antiporter. Proc Natl Acad Sci USA 108:3935–3940
Kumar H, Kasho V, Smirnova I, Finer-Moore JS, Kaback HR, Stroud RM (2014) Structure of sugar-bound LacY. Proc Natl Acad Sci USA 111:1784–1788

Kumar H, Finer-Moore JS, Kaback HR, Stroud RM (2015) Structure of LacY with an α-substituted galactoside: connecting the binding site to the protonation site. Proc Natl Acad Sci USA 112:9004–9009
Kumar H, Finer-Moore JS, Jiang X, Smirnova I, Kasho V, Pardon E, Steyaert J, Kaback HR, Stroud RM (2018) Crystal Structure of a ligand-bound LacY-nanobody complex. Proc Natl Acad Sci USA 115:8769–8774
Lyons JA, Parker JL, Solcan N, Brinth A, Li D, Shah ST, Caffrey M, Newstead S (2014) Structural basis for polyspecificity in the POT family of proton-coupled oligopeptide transporters. EMBO Rep 15:886–893
Ma P, Weichert D, Aleksandrov LA, Jensen TJ, Riordan JR, Liu X, Kobilka BK, Caffrey M (2017) The cubicon method for concentrating membrane proteins in the cubic mesophase. Nat Protoc 12:1745–1762
Madej MG (2014) Function, structure, and evolution of the major facilitator superfamily: the LacY manifesto. Adv Biol 2014:20
Madej MG, Kaback HR (2013) Evolutionary mix-and-match with MFS transporters II. Proc Natl Acad Sci USA 110:4831–4838
Madej MG, Soro SN, Kaback HR (2012) Apo-intermediate in the transport cycle of lactose permease (LacY). Proc Natl Acad Sci USA 109:2970–2978
Madej MG, Dang S, Yan N, Kaback HR (2013) Evolutionary mix-and-match with MFS transporters. Proc Nat Acad Sci USA 110:5870–5874
Madej MG, Sun L, Yan N, Kaback HR (2014) Functional architecture of MFS D-glucose transporters. Proc Natl Acad Sci USA 111:719–727
Martinez Molledo M, Quistgaard EM, Flayhan A, Pieprzyk J, Löw C (2018) Multispecific substrate recognition in a proton-dependent oligopeptide transporter. Structure 26:467–476
Minhas GS, Bawdon D, Herman R, Rudden M, Stone AP, James AG, Thomas GH, Newstead S (2018) Structural basis of malodour precursor transport in the human axilla. eLife 7:e34995
Mirza O, Guan L, Verner G, Iwata S, Kaback HR (2006) Structural evidence for induced fit and a mechanism for sugar/H^+ symport in LacY. EMBO J 25:1177–1183
Nagarathinam K, Nakada-Nakura Y, Parthier C, Terada T, Juge N, Jaenecke F, Liu K, Hotta Y, Miyaji T, Omote H, Iwata S, Nomura N, Stubbs MT, Tanabe M (2018) Outward open conformation of a Major Facilitator Superfamily multidrug/H^+ antiporter provides insights into switching mechanism. Nat Commun 9:4005
Newstead S, Drew D, Cameron AD, Postis VLG, Xia X, Fowler PW, Ingram JC, Carpenter EP, Sansom MSP, McPherson MJ, Baldwin SA, Iwata S (2011) Crystal structure of a prokaryotic homologue of the mammalian oligopeptide-proton symporters, PepT1 and PepT2. EMBO J 30:417–426
Nie Y, Smirnova I, Kasho V, Kaback HR (2006) Energetics of ligand-induced conformational flexibility in the lactose permease of *Escherichia coli*. J Biol Chem 281:35779–35784
Padan E, Sarkar HK, Viitanen PV, Poonian MS, Kaback HR (1985) Site-specific mutagenesis of histidine residues in the lac permease of *Escherichia coli*. Proc Natl Acad Sci USA 82:6765–6768
Pao SS, Paulsen IT, Saier MH (1998) Major facilitator superfamily. Microbiol Mol Biol Rev 62:1–34
Parker JL, Li C, Brinth A, Wang Z, Vogeley L, Solcan N, Ledderboge-Vucinic G, Swanson JMJ, Caffrey M, Voth GA, Newstead S (2017) Proton movement and coupling in the POT family of peptide transporters. Proc Natl Acad Sci USA 114:13182–13187
Püttner IB, Sarkar HK, Poonian MS, Kaback HR (2002) *Lac* permease of *Escherichia coli*: histidine-205 and histidine-322 play different roles in lactose/protein symport. Biochemistry 25:4483–4485
Quistgaard EM, Löw C, Moberg P, Tresaugues L, Nordlund P (2013) Structural basis for substrate transport in the GLUT-homology family of monosaccharide transporters. Nat Struct Mol Biol 20:766–768
Quistgaard EM, Martinez Molledo M, Löw C (2017) Structure determination of a major facilitator peptide transporter: Inward facing $PepT_{St}$ from *Streptococcus thermophilus* crystallized in space group $P3_121$. PLoS ONE 12:e0173126
Reddy VS, Shlykov MA, Castillo R, Sun EI, Saier MH (2012) The major facilitator superfamily (MFS) revisited. FEBS J 279:2022–2035

Ren Q, Paulsen IT (2007) Large-scale comparative genomic analyses of cytoplasmic membrane transport systems in prokaryotes. J Mol Microbiol Biotechnol 12:165–179
Robertson DE, Kaczorowski GJ, Garcia ML, Kaback HR (1980) Active transport in membrane vesicles from *Escherichia coli*: the electrochemical proton gradient alters the distribution of the lac carrier between two different kinetic states. Biochemistry 19:5692–5702
Sahin-Tóth M, Kaback HR (2001) Arg-302 facilitates deprotonation of Glu-325 in the transport mechanism of the lactose permease from *Escherichia coli*. Proc Natl Acad Sci USA 98:6068–6073
Sahin-Tóth M, le Coutre J, Kharabi D, le Maire G, Lee JC, Kaback HR (1999) Characterization of Glu126 and Arg144, two residues that are indispensable for substrate binding in the lactose permease of *Escherichia coli*. Biochemistry 38:813–819
Sahin-Tóth M, Karlin A, Kaback HR (2000) Unraveling the mechanism of the lactose permease of *Escherichia coli*. Proc Natl Acad Sci USA 97:10729–10732
Shi Y (2013) Common folds and transport mechanisms of secondary active transporters. Annu Rev Biophys 42:51–72
Smirnova I, Kasho V, Kaback HR (2008) Protonation and sugar binding to LacY. Proc Natl Acad Sci USA 105:8896–8901
Smirnova I, Kasho V, Sugihara J, Choe J-Y, Kaback HR (2009) Residues in the H^+ translocation site define the pK_a for sugar binding to LacY. Biochemistry 48:8852–8860
Smirnova I, Kasho V, Sugihara J, Kaback HR (2013) Trp replacements for tightly interacting Gly–Gly pairs in LacY stabilize an outward-facing conformation. Proc Natl Acad Sci USA 110:8876–8881
Solcan N, Kwok J, Fowler PW, Cameron AD, Drew D, Iwata S, Newstead S (2012) Alternating access mechanism in the POT family of oligopeptide transporters. EMBO J 31:3411–3421
Sun L, Zeng X, Yan C, Sun X, Gong X, Rao Y, Yan N (2012) Crystal structure of a bacterial homologue of glucose transporters GLUT1-4. Nature 490:361–366
Taniguchi R, Kato HE, Font J, Deshpande CN, Wada M, Ito K, Ishitani R, Jormakka M, Nureki O (2015) Outward- and inward-facing structures of a putative bacterial transition-metal transporter with homology to ferroportin. Nat Commun 6:8545
Vastermark A, Wollwage S, Houle ME, Rio R, Saier MH (2014) Expansion of the APC superfamily of secondary carriers. Proteins 82:2797–2811
Vázquez-Ibar JL, Guan L, Weinglass AB, Verner G, Gordillo R, Kaback HR (2004) Sugar recognition by the lactose permease of *Escherichia coli*. J Biol Chem 279:49214–49221
Venkatesan P, Kaback HR (1998) The substrate-binding site in the lactose permease of *Escherichia coli*. Proc Natl Acad Sci USA 95:9802–9807
Viitanen P, Garcia ML, Foster DL, Kaczorowski GJ, Kaback HR (1983) Mechanism of lactose translocation in proteoliposomes reconstituted with lac carrier protein purified from *Escherichia coli*. II. Deuterium solvent isotope effects. Biochemistry 22:2531–2536
Wisedchaisri G, Park M-S, Iadanza MG, Zheng H, Gonen T (2014) Proton-coupled sugar transport in the prototypical major facilitator superfamily protein XylE. Nat Commun 5:4521
Wong FH, Chen JS, Reddy V, Day JL, Shlykov MA, Wakabayashi ST, Saier MH (2012) The amino acid–polyamine–organocation superfamily. J Mol Microbiol Biotechnol 22:105–113
Xie H (2008) Activity assay of membrane transport proteins. Acta Biochim Biophys Sin 40:269–277
Yamashita A, Singh SK, Kawate T, Jin Y, Gouaux E (2005) Crystal structure of a bacterial homologue of Na^+/Cl^--dependent neurotransmitter transporters. Nature 437:215–223
Yan N (2013) Structural advances for the major facilitator superfamily (MFS) transporters. Trends Biochem Sci 38:151–159
Yan H, Huang W, Yan C, Gong X, Jiang S, Zhao Y, Wang J, Shi Y (2013) Structure and mechanism of a nitrate transporter. Cell Rep 3:716–723
Yin Y, He X, Szewczyk P, Nguyen T, Chang G (2006) Structure of the multidrug transporter EmrD from *Escherichia coli*. Science 312:741–744
Zacharias N, Dougherty DA (2002) Cation–π interactions in ligand recognition and catalysis. Trends Pharmacol Sci 23:281–287

Zhao Y, Mao G, Liu M, Zhang L, Wang X, Zhang XC (2014) Crystal structure of the *E. coli* peptide transporter YbgH. Structure 22:1152–1160

Zheng H, Wisedchaisri G, Gonen T (2013) Crystal structure of a nitrate/nitrite exchanger. Nature 497:647–651

Chapter 10
Respiratory Membrane Protein Complexes Convert Chemical Energy

Valentin Muras, Charlotte Toulouse, Günter Fritz and Julia Steuber

Abstract The invention of a biological membrane which is used as energy storage system to drive the metabolism of a primordial, unicellular organism represents a key event in the evolution of life. The innovative, underlying principle of this key event is respiration. In respiration, a lipid bilayer with insulating properties is chosen as the site for catalysis of an exergonic redox reaction converting substrates offered from the environment, using the liberated Gibbs free energy (ΔG) for the build-up of an electrochemical H^+ (proton motive force, PMF) or Na^+ gradient (sodium motive force, SMF) across the lipid bilayer. Very frequently, several redox reactions are performed in a consecutive manner, with the first reaction delivering a product which is used as substrate for the second redox reaction, resulting in a respiratory chain. From today's perspective, the (mostly) unicellular bacteria and archaea seem to be much simpler and less evolved when compared to multicellular eukaryotes. However, they are overwhelmingly complex with regard to the various respiratory chains which permit survival in very different habitats of our planet, utilizing a plethora of substances to drive metabolism. This includes nitrogen, sulfur and carbon compounds which are oxidized or reduced by specialized, respiratory enzymes of bacteria and archaea which lie at the heart of the geochemical N, S and C-cycles. This chapter gives an overview of general principles of microbial respiration considering thermodynamic aspects, chemical reactions and kinetic restraints. The respiratory chains of *Escherichia coli* and *Vibrio cholerae* are discussed as models for PMF-versus SMF-generating processes, respectively. We introduce main redox cofactors of microbial respiratory enzymes, and the concept of intra-and interelectron transfer. Since oxygen is an electron acceptor used by many respiratory chains, the formation and removal of toxic oxygen radicals is described. Promising directions of future research are respiratory enzymes as novel bacterial targets, and biotechnological applications relying on respiratory complexes.

Keywords Respiration · Electrochemical gradient · Electron transfer · Redox pump · Geochemical cycles · *Escherichia coli* · *Vibrio cholerae*

V. Muras · C. Toulouse · G. Fritz · J. Steuber (✉)
Institute of Microbiology, University of Hohenheim, Garbenstr. 30, 70599 Stuttgart, Germany
e-mail: julia.steuber@uni-hohenheim.de

A. Kuhn (ed.), *Bacterial Cell Walls and Membranes*, Subcellular Biochemistry 92, https://doi.org/10.1007/978-3-030-18768-2_10

Respiration: The Thermodynamic Perspective

Respiration is a fundamental process for energy conservation in living organisms of the three domains of life. The mechanisms elucidated with bacterial and archaeal systems can easily be transferred to eukaryotes as the basic processes are very similar, with the ultimate aim to form high-energy compounds like ATP. In a model bacterium such as *E. coli*, ATP is synthesized mainly by two routes. First, during substrate-level phosphorylation in glycolysis and the citric acid cycle and second during electron transport phosphorylation. The first pathway has the advantage of being independent of oxygen (or other exogenous electron acceptors, as described below). However, only limited amounts of ATP are produced. For example, in glycolysis two molecules of ATP are synthesized per mol glucose. During electron transport phosphorylation on the other hand, the yield is drastically increased, and up to 36 mol ATP are generated per mol glucose.

This is realized by the combination of electron donors with negative redox potentials, e.g. NADH, with electron acceptors with positive redox potentials, such as molecular oxygen. The electron transfer to the terminal electron acceptor does not occur in a single step reaction, but in several redox-reactions. In the respiratory chain, which involves a variety of proteins and a diverse set of soluble and protein bound cofactors, electrons are stepwise transferred in a series of redox-reactions from compounds with more negative redox potentials (electron donors) to compounds with more positive redox potentials (electron acceptors). The energy that is released during these redox-reactions is then coupled to the transport of ions (protons or sodium ions) across the membrane dielectric. Since the membrane is not permeable for the charged species and ions cannot cross the membrane on their own, this uphill transport of cations leads to the formation of an electric potential gradient, or membrane potential ($\Delta\Psi$), which in turn drives the ATP synthase, generating ATP (Fig. 10.1).

Clearly, the most intriguing question is how respiratory complexes located in microbial membranes are able to establish and maintain a form of energy, which is storable and which can be consumed as universal "fuel" at any time. This way it can be used in order to power cellular processes like flagellar rotation, to power substrate transporters for the efflux of cytotoxins or the influx of solutes, or for the production of high-energy compounds like ATP (Fig. 10.2). The solution to this problem is the semipermeable membrane, which acts as chemical and physical barrier separating the cytoplasm from the periplasm or the extracellular space, and which is virtually impermeable for ions on their own. Thus, once the ions are separated against their chemiosmotic gradient, the downhill transfer across the membrane can be used to perform work, e.g. for the synthesis of ATP (Fig. 10.1). This concept was first described as chemiosmotic theory by Peter Mitchell in 1961 and provides the basis for the understanding of energy conservation in all living organisms (Mitchell 1961; Mitchell and Moyle 1969; Mulkidjanian et al. 2008). According to Mitchell, the activity of the electron transport chain results in the uphill transport of protons across the membrane, which occurs when electrons flow from redox centers with more negative to redox centers with more positive midpoint potentials. The energy,

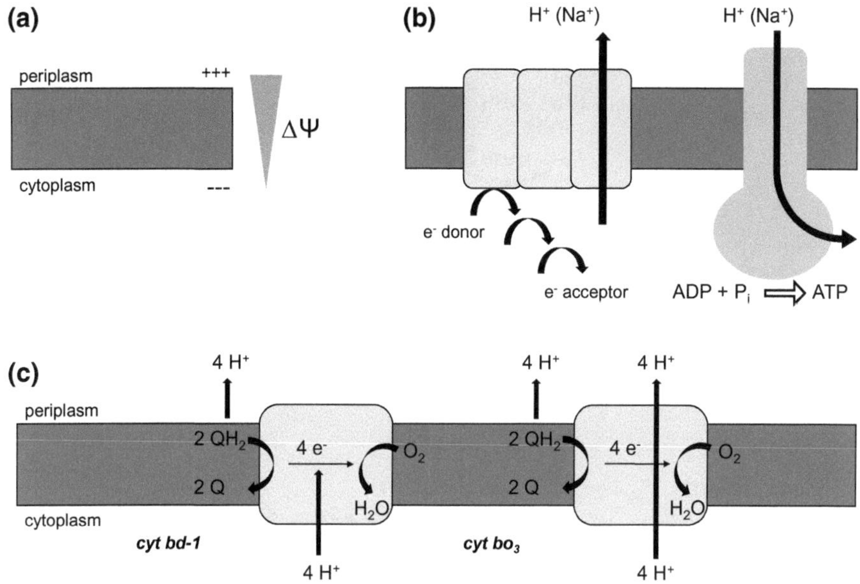

Fig. 10.1 Principles of respiratory energy conservation. **a** Energy is stored and provided as an electrochemical gradient across the membrane dielectric. **b** This energy is used to drive the formation of ATP by the F_1F_O ATP synthase. **c** The electrochemical gradient is formed during respiration with the help of different respiratory chain complexes, which determine overall efficiency of respiration, i.e. the cation/e^- ratio. Besides H^+, also Na^+ may be used as coupling cation in respiration. Some complexes, such as the cytochrome *bd* oxidase, do not act as redox pumps but contribute to the build-up of the electrochemical proton gradient by the uptake and release of protons and electrons at different sides of the membrane. In contrast, the cytochrome bo_3 complex represents a bona fide proton pump, thus increasing the overall H^+/e^- ratio

which is released as the electrons are transferred through the respiratory chain is coupled to the transfer of protons out of the bacterial cell (across the membrane), resulting in an electrochemical proton gradient, $\Delta\mu H^+$ (Eq. 10.1):

$$\Delta\mu H^+ = F\Delta + 2.3RT\ \Delta pH \tag{10.1}$$

The energy stored in the membrane thus consists of an electrical term, $\Delta\psi$, and the chemical gradient of protons, ΔpH. The unit is kJ per mol of translocated protons. To strengthen the fact that the electrical term, also called transmembrane voltage, represents the major component of the total electrochemical proton potential, Mitchell introduced the term "proton motive force" (PMF), which is defined as follows (Eq. 10.2):

$$PMF = \frac{\Delta\mu H^+}{F} - -\Delta\psi + \frac{2.3RT}{F}\Delta pH \tag{10.2}$$

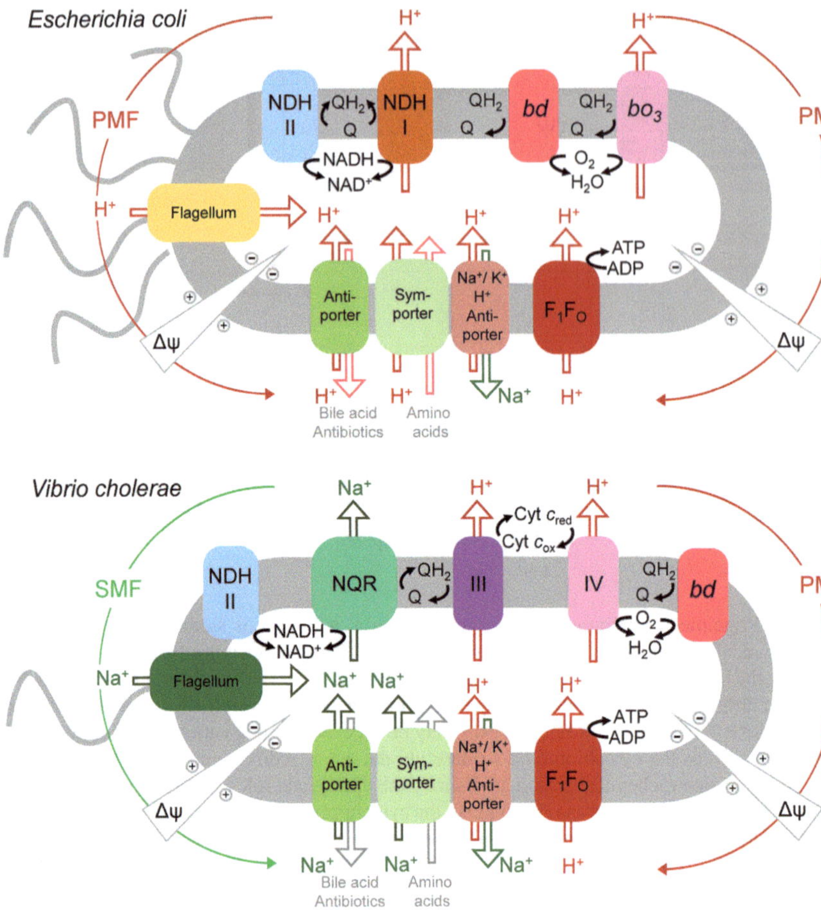

Fig. 10.2 The generation of PMF or SMF by aerobic respiration in *Escherichia coli* or *Vibrio cholerae* NADH serves as electron donor for the NADH-dehydrogenases NDH-II, NDH-I (or complex I), and NQR, which are either, non-electrogenic or electrogenic proton or sodium pumps. They promote the electron transfer to ubiquinone (UQ) which is reduced to ubiquinol (UQH_2). UQH_2 is a membrane associated electron carrier and transfers electrons to the ubiquinol oxidases. *E. coli bd*-type and bo_3-type ubiquinol oxidases subsequently transfer electrons from ubiquinol to the terminal electron acceptor O_2. In *V. cholerae* the *bc*1 complex (or complex III) donates electrons derived from ubiquinol to the electron carrier cytochrome *c* (Cyt *c*). Reduced Cyt *c* carries the electrons to the terminal oxidase (complex IV), where Cyt *c* is then reoxidized. In turn complex IV catalyzes the reduction of dioxygen to water. Alternatively, *V. cholerae* possesses also a *bd*-type ubiquinol oxidase, which accepts electrons from ubiquinol and shuttles them to oxygen. In *E. coli* NDH-I and bo_3, and in *V. cholerae* Complex III and IV, are linked to proton extrusion, generating a proton motive force (PMF). Beside the PMF, *V. cholerae* is able to utilize a sodium motive force (SMF), generated by sodium extrusion through the NQR. Consumers of the SMF and PMF facilitate transport of molecules (e.g. bile acids or antibiotics Steuber et al. 2015; Sulavik et al. 2001; Thanassi et al. 1997) and rotation of the flagellum. Both proton and sodium pumping respiratory complexes contribute to the build-up of the transmembrane voltage, which is the main driving force for ATP synthesis by the F_1F_O ATP synthetase

The unit of PMF is voltage. In respiring membranes from bacteria and archaea, the PMF amounts to approximately 180–220 mV, of which 150–180 mV are provided by the membrane potential, $\Delta\Psi$ with negative algebraic sign. F is the Faraday constant, referring to the charge of an electron.

Aerobic Versus Anaerobic Respiration

Let us first consider respiratory chains in the facultative aerobic bacterium *Escherichia coli*, an intensively studied model organism. Under aerobic conditions, the respiratory chain of *E. coli* is comprised of the two NADH dehydrogenase NDH-I and NDH-II, and two terminal quinol oxidases: quinol oxidase bo_3 and quinol oxidase *bd* (Anraku and Gennis 1987; Bogachev et al. 1996; Puustinen et al. 1989). The NADH dehydrogenases serve as electron input modules for electrons into the respiratory chain. The electron-accepting substrate of both NADH dehydrogenases is ubiquinone, mediating electron transfer to the terminal oxidases, which catalyze the electron transfer to exogenous acceptors like oxygen. During these steps, protons can be transported across the membrane, and in this way contribute to the electrochemical gradient. The exact number of H^+ transferred from the cytoplasmic to the periplasmic site per electron by the described pathway strongly depends on the exact route of the electrons and the activity of the individual complexes (Unden and Bongaerts 1997).

The first site, which is directly involved in the formation of $\Delta\Psi$ is the electrogenic complex I (or NDH-I), which couples the oxidation of NADH with ubiquinone to the transport of at least $3H^+$ per NADH oxidized (Bogachev et al. 1996). Like NDH-I, the terminal quinol oxidase bo_3 couples the reduction of O_2 to a directed transport of protons across the membrane. NDH-II on the other hand is non-electrogenic and does not exert proton pumping activity (Kerscher et al. 1999).

Although quinol oxidase *bd* and NADH dehydrogenase NDH-II are not able to pump protons, they still participate in the formation of an electrochemical gradient by charge separation, which is a result from two different substrate binding sites located at either side of the membrane. This can be readily visualized using the recently published structure of *G. thermodenitrificans bd* oxidase (Safarian et al. 2016). The three-subunit complex (CydA, CydB, CydS) contains three heme cofactors, and catalyzes the reduction of dioxygen to water without releasing reactive oxygen species. The electrons for this reaction are derived from ubiquinol, which binds to a conserved extracellular loop (Q-loop) in subunit CydA. Electrons from ubiquinol, are then transferred to the prosthetic heme groups, followed by the rapid reduction of dioxygen. When ubiquinol is oxidized at its designated binding site in CydA, two protons are released at the extracellular side of the membrane. Although protons are not pumped through the membrane, there are two postulated pathways, which allow access of protons from the cytoplasm to the oxygen binding site, where they are utilized for the formation of water. This reaction is equivalent to the transfer of one H^+ per electron (Fig. 10.1c).

Besides oxygen, other electron carriers can be employed as electron acceptors (Table 10.1). In the bacterial and archaeal kingdoms of life, there is a plethora of respiratory pathways utilizing many more electron donors and electron acceptors than NADH and O_2, the prominent substrates of respiration in the eukaryotic kingdom of life. As long as the free energy liberated by the overall redox reaction is sufficient to energize the respiratory membrane, chemiosmotic ATP synthesis by bacteria or archaea may utilize an impressive variety of organic or inorganic electron donor and acceptor pairs (Thauer et al. 1977). For example, under anoxic conditions, some bacteria can use nitrate, fumarate or DMSO and TMAO as terminal electron acceptor. This adaptation to the environment requires tightly regulated gene expression to optimize respiration under the given conditions. Thus, oxygen, which is the preferred electron acceptor due to its high (positive) midpoint redox potential, acts as repressor for the terminal reductases of anaerobic respiration (Unden and Bongaerts 1997).

Table 10.1 Midpoint standard redox potentials of intrinsic redox carriers or external electron acceptors utilized in various respiratory microorganisms. Typical electron donors and electron acceptors are listed in order of their midpoint redox potential. In aerobic respiration, oxygen can be used as final acceptor which provides the most effective electron acceptor in nature, due to the very positive midpoint potential. Bacteria and archaea may use other molecules as final acceptors to grow under exclusion of oxygen. If no specific organism is indicated in this Table, these electron carriers are widely distributed in plants, animals and bacteria

Redox couple (ox./red.)	$E^{0\prime}$ [V]	Microorganism
Ferredoxin (Fe^{3+}/Fe^{2+})	−0.43	
NAD^+/NADH	−0.32	
CO_2/acetate	−0.30	*Acetobacterium woodii*
S^0/HS^-	−0.27	*Desulfuromonadales*
CO_2/CH_4	−0.25	*Euryarchaeota* (e.g. *Methanobacteria*)
SO_4^{2-}/HS^-	−0.22	*Desulfovibrio*
Flavin—nucleotides	−0.19[a]	
Fumarate/succinate	+0.03	*Escherichia coli*, *Vibrio cholerae*
Ubiquinone/ubiquinol	+0.10	
Cytochrome *c* (Fe^{3+}/Fe^{2+})	+0.25	
Mn(IV)/Mn(II)	+0.41	*Desulfovibrio*
NO_2^-/NO_3^-	+0.43	*Paracoccus denitrificans*
½O_2, $2H^+/H_2O$	+0.82	Aerobic bacteria and archaea

[a]The redox potential of protein associated flavins strongly depends on the protein environment. Therefore this redox potential is given as representative (Franz Müller 1991). Table is adapted from (Oren 2005; Vella 1993)

Na^+ as Alternative Coupling Cation in Respiration

Bacteria and archaea do not necessarily rely solely on proton pumping respiratory enzymes, but may also rely on an electrochemical Na^+ gradient established by electrogenic Na^+ pumps (Steuber et al. 2015). This Na^+ gradient does not depend on secondary Na^+/H^+ antiporters, which may generate an electrochemical Na^+ gradient at the expense of the PMF (Steuber et al. 2015). In 1977, Unemoto and colleagues showed that oxidation of NADH with O_2 by the marine bacterium *Vibrio alginolyticus* was specifically stimulated by sodium ions (Unemoto and Hayashi 1977), and concluded that "*Na^+ probably acts as an essential cofactor both for the active transport of metabolites and for respiratory activity in marine bacteria*". In 1979, the same group reported that the Na^+-stimulated respiratory complex represented a NADH:quinone oxidoreductase (later identified as the Na^+-translocating NADH:quinone oxidoreductase, or NQR, described below), but the build-up of an electrochemical Na^+ gradient by this enzyme was not demonstrated at that time (Unemoto and Hayashi 1979). These findings were fundamental since they showed that an extension of Mitchell's chemiosmotic theory was necessary, now including Na^+ as a possible coupling ion in energized membranes capable of ATP synthesis. Indeed, Tokuda and Unemoto later showed that the Na^+-stimulated NADH:quinone oxidoreductase from *V. alginolyticus* actually translocates Na^+ ions (Tokuda and Unemoto 1981). This explained why the enzyme, strictly coupling the electron transfer reaction to the transport of Na^+ across the membrane, was stimulated by Na^+ in its NADH:quinone oxidoreduction activity. NADH:quinone oxidoreduction represents the "chemical" reaction of the pump, whereas Na^+ transport represents the "vectorial" reaction of the pump. However, the stoichiometry of the Na^+-NQR (2Na^+ transported/2 electrons transferred) is less efficient compared to complex I (or NDH-I), the proton pumping NADH dehydrogenase found in many bacteria. So why do bacteria operate NQRs with less efficiency, to build up a sodium motive force, and why do some bacteria, e.g. *Yersinia pestis* contain both respiratory enzymes, the sodium ion pumping Na^+-NQR and the proton pumping NDH-I (the bacterial complex I)?

A closer look at the natural habitats of these bacteria might, at least partially, answer this question (Getz et al. 2018). Primary sodium ion pumps (e.g. NQR from *Vibrio* species) are widespread in halophilic bacteria and bacteria living under alkaline conditions. These bacteria rely on a $\Delta\mu Na^+$ rather than a $\Delta\mu H^+$ for energy conservation. The use of a $\Delta\mu Na^+$ might confer several advantages in a given habitat. At alkaline conditions, i.e. at high external pH, the pH gradient is inverted with $pH_{in} < pH_{out}$. Under these conditions, it is advantageous to use other coupling ions than H^+. Thus, the use of Na^+ as coupling ion provides a reliable energy source in alkaline and saline environments. Since the 1980s, a correlation between alkaliphily of bacteria and the presence of Na^+ extrusion and Na^+ uptake systems (Na^+ cycling) has been reported, implementing a crucial role for Na^+ in pH homeostasis (Krulwich et al. 2001). It was observed that the regulation of the internal pH strongly depends on a set of Na^+/H^+ antiporters, which couple the extrusion of Na^+ to the influx of H^+.

This way, the sodium motive force allows for an optimized pH homeostasis (Hirst 2003).

Another finding by van de Vossenberg et al. indicates that especially thermophilic bacteria benefit from a sodium motive force. They reported, that liposomes prepared from lipids isolated from several bacterial species show increased leakage of protons but not sodium ions at elevated temperature (van de Vossenberg et al. 1995). Thus, using sodium as primary coupling ion in respiration increases the overall robustness, which allows rapid adaption to a wide range of pH. This could also explain the wide distribution of sodium ion pumps in a variety of pathogenic bacteria (Hayashi et al. 2001) which are frequently confronted with harsh environmental conditions in the host.

The unambiguous identification of the coupling cation of respiratory complexes (H^+ and/or Na^+) is difficult. For example, in complex I (NDH-I), the oxidation of 1 molecule NADH is coupled to the transport of 4 H^+ or $2H^+/e^-$ (Galkin et al. 2006). Interestingly, several studies suggest that complex I might also act as transporter of Na^+ under certain conditions, reviewed in (Castro et al. 2016; Steffen and Steuber 2013). Another example is the RNF (rhodobacter nitrogen fixation) complex which was first described as a Na^+ pump, yet recently, a member of this class of respiratory enzymes was found to act as a proton pump (Tremblay et al. 2013).

No Respiration Without Electron Transfer Reactions: Reactants and Principles

Non-metal Electron Carriers and Redox Cofactors

An important redox carrier that is found in all living cells is the reducing agent nicotinamide adenine dinucleotide (NADH). NADH constitutes an important electron donor fueling electrons into the respiratory chain to generate a proton- or sodium-motive force. It is a universal electron donor in respiratory chains of animals, plants and bacteria. NADH is converted to its oxidised form by different catabolic pathways in the cell. The enzymes that catalyze the exergonic oxidation of NADH with ubiquinone are the NADH:ubiquinone oxidoreductases (Fig. 10.2). The electrogenic NADH dehydrogenases couple this reaction to the translocation of protons or sodium ions. The products of this reaction are oxidized NAD^+ and ubiquinol (UQH_2). This regeneration of NAD^+ by respiration is important since only the constant supply of NAD^+ to metabolic hubs such as the TCA cycle guarantees survival and growth of the cell. Respiratory activities of microbial membranes catalyzing the conversion of NADH into NAD^+ are easily followed spectrophotometrically in a quantitative manner. In the UV/VIS range, the spectrum of NADH gives rise to two peaks at 340 nm and 260 nm, respectively. The adenine moiety absorbs at 260 nm, the intensity being insensitive to the oxidation state. However, the nicotinamide moiety has a strong absorbance maximum at 340 nm which decreases upon its oxidation. The concentration of NADH is determined with the law of Lambert-Beer, using the molar extinction coefficient of $\varepsilon_{340} = 6.22\ \text{mM}^{-1}\ \text{cm}^{-1}$. All NADH dehydrogenases, soluble and membrane-bound alike, rely on flavin redox cofactors as primary acceptor for the hydride (H^-) which is transferred to the isoalloxazine moiety of the flavin (Leys and Scrutton 2016). This two-reduction of the flavin cofactor is then followed by one-electron transfer steps into other cofactors such as FeS centers located in electron transfer distance to the flavin. Flavins are yellow-colored organic compounds with a basic structure of a 7,8-dimethyl-10-alkylisoalloxazine. The isoalloxazine moiety found in the flavins can readily undergo redox transitions, acting as one- or two-electron acceptor and participates as redox carrier in nearly all metabolic pathways. The full reduction of the flavin group requires two protons and two electrons and can occur in a two-step reaction or at once. The precursor of all biological relevant flavins is riboflavin (RF) a name derived from the ribityl side chain and the isoalloxazine ring giving rise to the yellow color. The molecule is also known as vitamin B_2. Phosphorylation of the sugar moiety yields riboflavin-5′phosphate, also called flavin mononucleotide (FMN). Flavin adenine dinucleotide (FAD) is a riboflavin with bound adenosine diphosphate (ADP).

Flavins are widely found in many respiratory complexes where they act as converter between two-electron and one-electron transfer reaction steps. Besides NADH dehydrogenases such as complex I (or NDH-I) (Parey et al. 2018), the NQR (Steuber et al. 2015) or the non-electrogenic NDH-II (Salewski et al. 2016; Marreiros et al. 2016b) utilize flavins as primary electron acceptor. Another prominent example for a flavin-containing respiratory complex from bacteria is the succinate dehydrogenase/fumarate reductase (Maklashina et al. 2013) related to complex II of the mitochondrial respiratory chain.

In respiratory complexes, flavins can readily be identified by UV/VIS spectroscopy. Flavins show strong absorbance bands around 220, 266, 375 and 447 nm. Both the redox potential of flavins and the optical spectra vary in response to the protein environment. To distinguish oxidized from reduced flavin, the decrease in absorbance around 447 nm can be monitored. The extinction coefficient ($\varepsilon_{447\ \text{nm}}$) in RF, FMN and FAD decreases from 12.2, 12.2, 11.3 mM^{-1} in the oxidized forms, to 0.78, 0.87, 0.98 mM^{-1} in the reduced forms (Weimar and Neims 1975). One-electron reduced flavin represents an organic radical and is therefore readily detected using electron paramagnetic resonance spectroscopy (EPR) (Barquera et al. 2003; Brosi et al. 2014). EPR spectroscopy is a useful tool to follow inter- and intra-electron transfer reactions in respiratory complexes. While flavins operate at the electron input site in respiratory NADH dehydrogenases, quinones act as electron acceptors, both in a protein-bound state, or as soluble carrier in the lipid bilayer (Galassi and Arantes 2015). They are comprised of a lipophilic quinone head group and a hydrophobic tail consisting of many isoprenoid units. Various quinones are used in microbial respiration which differ in the nature of their head group, such as ubiquinones or menaquinones. Ubiquinones (UQ) are a group of electron carriers found in eukaryotic and prokaryotic cells, which participate in electron transfer during cellular respiration. The ubiquinone molecule is a 5,6-dimethoxy-benzoquinone ring containing an additional methyl group at the C3 and a conjugated isoprenoid side chain at the C2. The basic structure of ubiquinone is a 1,4-benzoquinone which defines the quinone group of electron carriers. The hydrophobic side chain is composed of 6–10 isoprenoid moieties with the length varying in different species. The predominant ubiquinone in the yeast *Saccharomyces cerevisiae* is UQ6 (Gonzalez-Mariscal et al. 2014). In *E. coli* and *V. cholerae* it is UQ8 (Crane 2001; Denis et al. 1975; Xu et al. 2014). The benzoquinone head group of quinones can undergo two different redox transitions. It can either take up two electrons and two protons at once to form ubiquinol (UQH_2) or undergo consecutive one-electron reduction with the ubisemiquinone radical as intermediate (Deller et al. 2008). The midpoint potential of the Q/QH_2 redox couple is +0.1 V (Bogachev et al. 2009b). The midpoint potentials of the $Q/Q^{\bullet-}$ and $Q^{\bullet-}/QH_2$ pairs are around +0.04 V and +0.128 V, respectively (Miki et al. 1992). However, the midpoint potentials vary depending on the protein environment (Miki et al. 1992). As a result, quinones may act in different redox ranges in different respiratory complexes, which makes them versatile tools for promoting intra- and inter electron transfer reaction steps in respiratory chains. A prominent example is complex I: Using the enzyme from the yeast *Yarrowia lipolytica*, the position of bound ubiquinone within the complex was recently determined by cryo electron microscopy (Parey et al. 2018). The isoprenoid side chain determines the hydrophobicity of ubiquinones, which increases with an increasing chain length, anchoring the molecule to the cellular membrane. Ubiquinone plays an important role in the bacterial respiratory chain. The reduction by complex I (NDH-I) and subsequent oxidation goes along with the association and dissociation of two protons. Thus, redox-cycling of ubiquinone is a proposed mechanism for energy conversion (Fig. 10.1). Other proteins using ubiquinone as prosthetic groups are the succinate:quinone oxidoreductases, comprising succinate:quinone dehydrogenase or the highly related quinol:fumarate reductase, an enzyme which predominantly operates in the reoxidation of quinol (Maklashina et al. 2013; Tomitsuka et al. 2009) and ubiquinol:cytochrome *c* oxidoreductase (also called complex III, or cytochrome *bc*1 complex) (Crofts 2004). Beyond its role as electron carrier in the respiratory chain, UQH_2 and UQ are also involved in formation of free reactive oxygen species in the cell (Muras et al. 2016). It should be stressed that besides ubiquinone, other quinones are operative in the respiratory chains of bacteria or archaea, such as menaquinone and its methylated or demethylated forms (Hein et al. 2017). Following respiratory quinol formation activities of microbial membranes by spectrophotometry is not straightforward, since all regions of the spectrum contribute to absorption in both the oxidized and the reduced form of the redox carrier. Moreover, natural quinones exhibit very poor solubility in water, so very often, short chain ubiquinones with only one or two isoprenoid side chains (ubiquinone-1, ubiquinone-2) are used in enzymatic assays. In QH_2, a red shift of the maximum is observed and the molar extinction coefficient is decreased compared to oxidized Q (Trumpower 1982). Juarez and coworkers report a molar absorptivity at 282 nm of 2.7 mM^{-1} for ubiquinol and 14.5 mM^{-1} for ubiquinone (Juárez et al. 2009). In enzymatic assays for NADH dehydrogenases where both NADH and UQ are added, quinone can be followed at 282 nm because of the low interference with NADH absorption (Juárez et al. 2011). Alternatively, formation of UQH_2 can be followed using the difference in absorbance between 248 and 267 nm (Tokuda and Unemoto 1984).

Metal-Containing Electron Carriers and Redox Cofactors

Iron-sulfur centers—Respiratory electron transfer chains need to cover a large range in midpoint potentials, starting from the NADH/NAD couple operating around −320 mV, to redox pairs with highly positive midpoint potentials like H_2O/O_2 ($E_m^{0'} = +0.82$ V). In the segment from −320 to 0 mV, the above mentioned non-metal cofactors (flavins, quinones) are ideally suited to undergo reduction and re-oxidation, coupled to the generation of an electrochemical proton (or sodium) gradient. Typically, this redox range is also covered by iron-sulfur centers. For example, *E. coli* NDH-I (complex I) contains 9 different FeS centers (Gnandt et al. 2016).

Many different Fe–S clusters are known and about 50 different unique protein folds for the ligation of iron-sulfur clusters are reported (Meyer 2008). Usually, the iron-sulfur clusters are coordinated by thiol groups from cysteine residues, but ligation by histidine or serine residues is also common (Beinert 2000). Additionally, most iron-sulfur clusters contain two or more inorganic sulfur atoms that bridge the iron atoms. While some iron-centers confer stability to certain protein domains, they are also prone to oxidative damage. There are four major classes of iron-sulfur clusters, which are characterized by the number of Fe atoms and the amount of inorganic sulfur: [1Fe], [2Fe–2S], [3Fe–4S] and [4Fe–4S] clusters. They cover a broad range of midpoint potentials, allowing them to participate in different physiological reactions and redox transitions, including respiratory reactions. [1Fe] clusters or mononuclear iron-sulfur clusters contain one iron atom that is usually coordinated by four cysteine ligands. Therefore, they are often referred to as [Fe–S4] clusters, although they do not contain inorganic sulfur. The [1Fe] clusters undergo a transition between the 3^+ and the 2^+ valence state. The redox potential of [1Fe] clusters ranges from −100 to +200 mV (Meyer 2008). [1Fe] clusters are prevalent in rubredoxins, which are important electron carrier proteins in respiratory chains of sulfate-reducing bacteria (Wenk et al. 2017). There are several subgroups of [2Fe–2S] clusters found in vertebrates, plants, bacteria and archaea. [2Fe–2S] clusters comprise two iron atoms, which are linked via two inorganic sulfur atoms, the so called acid-labile sulfur. The iron atoms are usually ligated by two cysteine residues each. They are most common in ferredoxins, which carry out electron transport in many processes, for example to hydroxylating enzymes, such as P450 cytochromes in bacteria (Grinberg et al. 2000). The redox potential is in the range of −305 and −455 mV (Cammack et al. 1977) or −150 to −450 mV (Meyer 2008). In respiratory chains, a very prominent group of [2Fe–2S] clusters are found in the so-called Rieske proteins, originally discovered as part of complex III (also called ubiquinol:cytochrome *c* oxidoreductase, or *bc*1 complex). They share a unique structural feature, as one of the iron atoms is ligated by two histidine residues instead of two cysteines. This ligand exchange can lead to an upshift of the potential in Rieske proteins as high as +100 to +400 mV (Meyer 2008). [4Fe–4S] clusters are the most abundant iron-sulfur clusters found in bacterial ferredoxins. They are low potential electron carriers (redox potential: −250 to −650 mV) which undergo a $[4Fe{-}4S]^{2+,+}$ redox transition (Yoch and Carithers 1979). Due to their low redox potential, such ferredoxins are mostly involved in anaerobic metabolic pathways. However, [4Fe–4S] centers are also found in many respiratory complexes. The *E. coli* NDH-I contains seven [4Fe–4S] clusters and two [2Fe–2S] centers in its so-called peripheral arm which transfers electrons from the primary hydride acceptor, the FMN, to the ubiquinone located in the membrane arm of this respiratory proton pump (Gnandt et al. 2016). [3Fe–4S] clusters are structurally reminiscent to the cubane-forming [4Fe–4S] clusters. They are important cofactors in microbial respiratory complexes such as the succinate dehydrogenase/fumarate reductase (Maklashina et al. 2013), or hydrogenases (Beinert 1997).

Heme cofactors—The name "cytochrome" implies that these heme-containing proteins of cells were among the first colored proteins discovered in the early days of biochemistry and bioenergetics. Their typical shift in absorbance upon reduction (appearance of the α- and β-absorption maxima in the wavelength range of 550–600 nm) was paramount to study and understand the flow of electrons from NADH to oxygen in respiratory chains in energy-converting membranes (Chance 1972). In the respiratory chain from *E. coli* illustrated in Fig. 10.2, heme cofactors are found, and give the name to, the two different terminal oxidases reducing O_2 to H_2O, the *bo* oxidase (or complex IV), and the *bd* oxidase. Depending on the chemical composition of the side chains of the porphyrins which ligate the central Fe atom, the hemes are designated a, b, c or d (Nič et al. 2009). Comparing the four types of heme, it is evident that cytochrome c is outstanding in its way of interaction with its apoprotein. While heme a, b and d are bound via coordinate bonds to special amino acids, heme c is tightly attached to the protein by thioetherbonds.

E. coli is a versatile, facultatively aerobic organism with bipartite respiratory chains consisting of input and output respiratory complexes. Input modules feed electrons to the quinone pool, for example, the NADH dehydrogenases, succinate dehydrogenase, formate dehydrogenase, hydrogenase. Output respiratory complexes oxidize quinols and deliver electrons to the terminal electron acceptor, for example, the O_2-reducing oxidases, Trimethylamine N-oxide (TMAO) reductase, dimethyl sulfoxide (DMSO) reductase (Cheng and Weiner 2007), fumarate reductase (Maklashina et al. 2013), nitrate reductase (Coelho and Romão 2015). In Fig. 10.2, the components of the aerobic respiratory chain of *E. coli* are shown. In *E. coli*, electrons from quinol are delivered directly to the terminal reductases; in *V. cholerae*, threepartite respiratory chains exists which are reminiscent to the situation in mitochondria. Here, the redox gap between the quinone/quinol pair and the O_2/H_2O pair allows driving an intermediate redox-driven proton pump, the ubiquinol:cytochrome *c* reductase (*bc*1 complex, or complex III) (Crofts 2004) (Fig. 10.2). The complex III reduces the heme cofactor of cytochrome *c*, the small periplasmic electron carrier, which in turn is re-oxidized by the terminal oxidase of *V. cholerae* (Fig. 10.2).

Other transition metals—Many electron-input and -output modules of the various respiratory chains from *E. coli* which were discussed above as examples for flavin, heme or FeS containing respiratory complexes critically rely on additional, redox-active transition metals, such as copper (Cu), molybdenum (Mo), nickel (Ni) or tungsten (W). On the electron input side, the Ni-containing hydrogenases (Sargent 2016), and the Mo-containing formate dehydrogenase (Niks and Hille 2018), utilize the high-energy electron donors H_2 (midpoint redox potential $2H^+/H_2 = -0.42$ V) or formate (midpoint redox potential CO_2/formate = −0.43 V) (Fig. 10.3b) which are common by-products of fermentative microorganisms living in the same intestinal habitat as *E. coli*. Please note that *E. coli* also is capable of hydrogen production during fermentative growth, operating H_2-oxidizing and H_2-evolving hydrogenases depending on the preferred catabolic route (Niks and Hille 2018; Pinske and Sawers 2016). Considering the output modules of the *E. coli* respiratory chain, some enzymes rely on Cu, Mo or W to fulfill their functions. Binding and activation of O_2 by the terminal bo_3 oxidase of *E. coli* (Fig. 10.3a) requires a heterobimetallic center consisting of heme iron in close proximity to Cu (Borisov and Verkhovsky 2015). An Fe–Cu bimetallic redox center also represents the active site of the recently described heme c-copper(I) sulfite reductase (Hermann et al. 2015), a terminal reductase in dissimilatory (e.g., respiratory) sulfite reduction by the microaerobic bacterium *Wolinella succinogenes*. The reduction of oxidized sulfur compounds such as sulfate completes the geochemical cycle of sulfur (Fig. 10.3c). Another important respiratory process in the geochemical nitrogen cycle is the dissimilatory reduction of nitrate (Fig. 10.3b). In the first step, nitrate is reduced to nitrite by the respiratory nitrate reductase (Nar) under anaerobic growth (Sparacino-Watkins et al. 2014). The Nar belongs to the DMSO (dimethylsulfoxide) reductase family of enzymes which contain the molybpterin cofactor, a metal-organic compound consisting of Mo bound to a pyranopterin organic cofactor (Coelho and Romão 2015). The formate dehydrogenase introduced above as an example for an input module of anaerobic respiratory chains belongs to another family of molybdenum (Mo) or tungsten (W) containing respiratory enzymes which includes formate dehydrogenases and aldehyde:ferredoxin reductases. In fact, many microorganisms encode separate but very similar formate dehydrogenases containing Mo or W, respectively, with the central metal atom being coordinated by two pyranopterins (Niks and Hille 2018). While most bacterial formate dehydrogenases function in the direction of formate oxidation to CO_2, CO_2 reduction by methanogenic archaea critically depends on Mo- and/or W-containing formylmethanofuran dehydrogenases. These enzymes are similar to formate dehydrogenases with respect to structure and function, although the formate is not liberated but condenses with methanofuran (Niks and Hille 2018).

Principles of Electron Transfer Between Redox Centers of Respiratory Enzymes

The respiratory chains depicted in Figs. 10.2 and 10.3 are composed of redox-driven, electrogenic, H^+ (or Na^+)-translocating protein complexes (such as NDH-I or NQR),

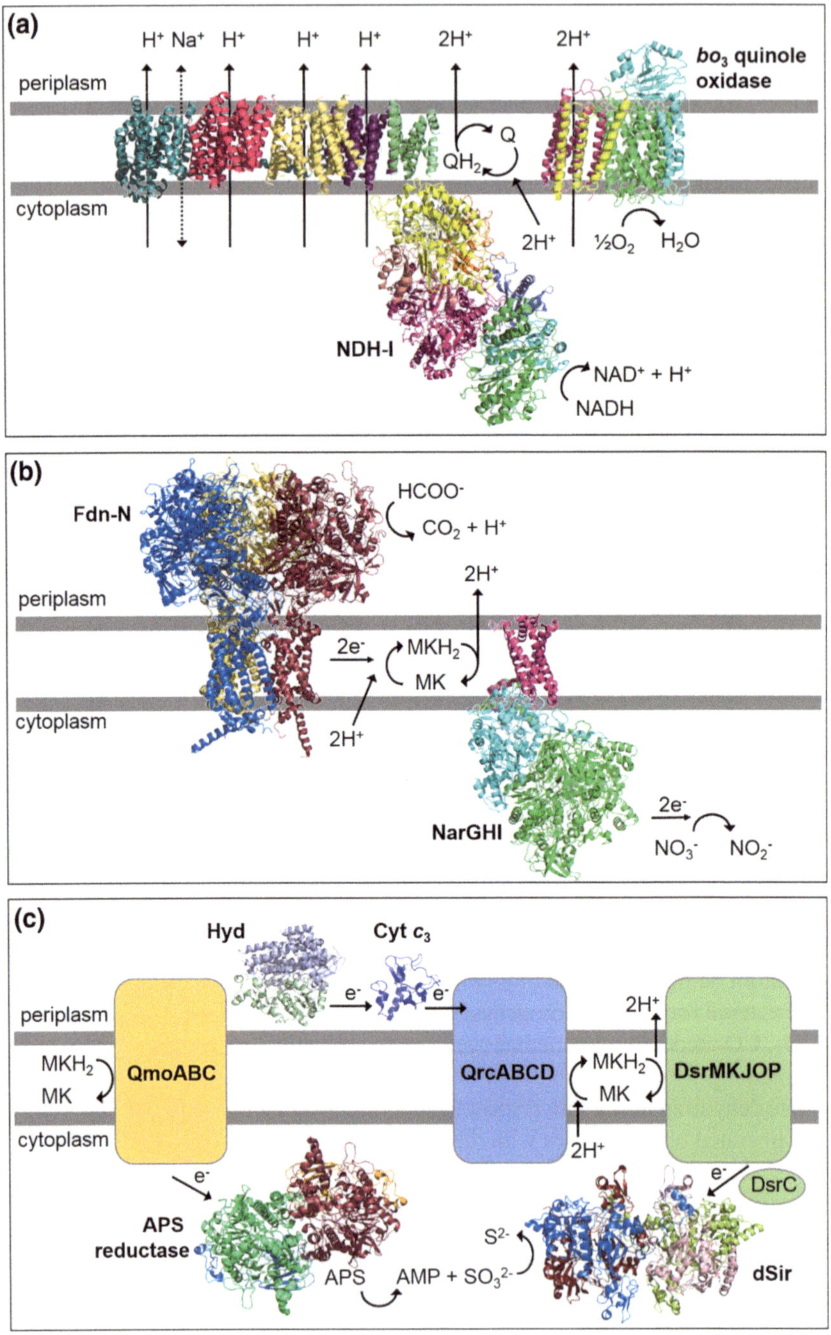
(a)
H+ Na+ H+ H+ H+ 2H+ 2H+
bo3 quinole oxidase
periplasm
Q
QH2
cytoplasm
2H+
½O2 H2O
NDH-I
NAD+ + H+
NADH
(b)
Fdn-N
HCOO-
CO2 + H+
2H+
periplasm
2e-
MKH2
MK
cytoplasm
2H+
NarGHI
2e-
NO3- NO2-
(c)
Hyd
Cyt c3
e-
e-
periplasm
2H+
MKH2
MK
QmoABC
QrcABCD
MKH2
MK
DsrMKJOP
cytoplasm
2H+
e-
e-
DsrC
APS reductase
S2-
APS
AMP + SO32-
dSir

◂**Fig. 10.3** A structural view of respiration. **a** Aerobic respiration by *E. coli*. The electrogenic NADH:quinone oxidoreductase (NDH-I, or complex I) oxidizes NADH and reduces ubiquinone (UQ) to ubiquinol (UQH_2) coupled to the translocation of up to $4H^+$ by its redox pump module, concomitant with the exchange of Na^+ by its antiporter module (Castro et al. 2016). The 3D structure shows the holo-NDH-I from *Thermus thermophilus* (PDB 3M9S, Efremov et al. 2010), which is highly similar to *E. coli* NDH-I (PDB 3M9C). UQH_2 is oxidized by the *E. coli* bo_3 oxidase (PDB 1FFT), an electrogenic proton pump (Abramson et al. 2000). **b** Anaerobic nitrate respiration with formate in *E. coli*. Nitrate is reduced to nitrite by NarGHI, a molybdenum- and FeS-containing respiratory complex (PDB 1Q16, Bertero et al. 2003), concomitant with the oxidation of menaquinol (MKH_2) to menaquinone (MK). The MKH_2 is formed upon formate oxidation by formate dehydrogenase N (FdnGHI) which also contains molybdenum- and FeS-cofactors (PDB 1KQF, Jormakka et al. 2002). The combined action of formate dehydrogenase and nitrate reductase, together with the uptake and release of protons during MKH_2 and MK formation, respectively, results in the generation of PMF by a redox-loop mechanism. **c** Respiratory complexes and electron carriers participating in dissimilatory sulfate reduction. Note that the depicted 3D structures do not originate from one single microorganism. Hyd, hydrogenase from *Desulfovibrio desulfuricans* (PDB 1E3D, Matias et al. 2001); Cyt *c*3, tetraheme type I cytochrome *c*3 from *D. desulfuricans* (PDB 1I77, Einsle et al. 2000); QrcABCD, quinone reductase complex (Venceslau et al. 2011). Hyd, Cyt *c*3 and Qrc form a supercomplex (Venceslau et al. 2011). DsrMKJOP, a MKH_2—oxidizing complex from *D. desulfuricans* (Pires et al. 2006); dSiR, dissimilatory sulfite reductase from *Archaeoglobus fulgidus* (PDB 3MMB, Parey et al. 2010). The menaquinol pool can either be oxidized by DsrMKJOP or QmoABC (Duarte et al. 2018). Electrons from DsrMKJOP are transferred to dSiR via the thiol-redox active DsrC protein component of dSiR (Santos et al. 2015). Sulfite ($SO_3{}^{2-}$) is formed by reduction of adenosine 5′phosphosulfate (APS) under release of adenosine monophosphate (AMP). The 3D structure of APS reductase from *A. fulgidus* (PDB 1JNR, Fritz et al. 2002) is shown. The QmoABC complex transfers electrons from MKH_2 to the APS reductase (Ramos et al. 2012)

of electron-transferring proteins which do not generate a transmembrane potential (such as cytochrome *c*) , and of small, redox-active organic molecules (such as ubiquinone-8, the substrate of NDH-I and NQR). A functional respiratory chain must catalyse substrate oxidation-reduction including bond breaking or bond formation at catalytic sites, inter- and/or intramolecular electron transfer, and primary energy conversion (that is, charge separation across the membrane dielectric). These different functions are performed by specialized components of the respiratory chains (Fig. 10.3). Unlike higher organisms, bacteria and archaea may combine different components of their respiratory chains to utilize various substrates for respiration, which is the basis for their exceptional metabolic diversity (Marreiros et al. 2016a). But despite their large differences, all these different respiratory chain components share a common function in catalysing electron transfer reactions. This is achieved with the help of protein-bound redox centers arranged in linear chains, as exemplified for the NQR sodium pump with its flavins and FeS centers (Fig. 10.4). The question if two redox centers located at a given distance within a protein complex may exchange electrons can be answered by applying the Marcus theory of electron transfer (Marcus and Sutin 1985). The Marcus theory states that electron transfer rates between the donor (for example, the FAD in subunit NqrF of the NQR) and the acceptor (for example, the [2Fe-2S] cluster in subunit NqrF of the NQR) (Fig. 10.4) depend on the difference in midpoint redox potentials of the two, the distance between them, and the input of energy which is required to change the geometry of the pre- to the post-electron transfer product(s). In subunit NqrF which catalyses the initial step of NADH oxidation in the NQR complex, the distance between the hydride-accepting FAD and the neighbouring [2Fe-2S] center is 9.8 Å (Steuber et al. 2014), and the midpoint redox potentials are $FADH_2/FAD = -200$ mV, and [2Fe-S] red/ox $= -270$ mV (Bogachev et al. 2009a). This example illustrates that midpoint redox potentials of cofactors determined under equilibrium conditions do not solely determine if electron transfer is possible, since electron transfer from the more positive donor (reduced FAD) to the more negative acceptor (oxidized [2Fe-2S] cluster) is observed experimentally upon reduction of the NQR by NADH (Bogachev et al. 2009a). As shown with many biological or synthetic redox-active proteins or compounds, productive electron transfer between two redox cofactors occurs from near van der Waals contact to about 13–14 Å (Page 2003). Applying this rule of thumb to the NQR, the distances between some cofactors which exchange electrons as confirmed in kinetic experiments (Bogachev et al. 2009a) are far too large for productive electron transfer when considering their observed position in the 3D structure of the NQR. This strongly suggests that the NQR undergoes conformational changes during turnover, and that the observed 3D structure represents one among different structural states of this respiratory machine (Steuber et al. 2014). It is concluded that the NQR is a conformational pump: redox reactions elicit movements of domains of the proteins to promote uphill Na^+ transport across the membrane.

Coupling of Electron Transfer to Cation (H^+ or Na^+) Translocation

The central question in respiration is: How does exergonic electron transfer drive the endergonic, i.e. against a prevailing electrochemical potential, transport of the coupling cation (H^+ or Na^+)?

Although the study of sodium-ion translocating pumps offers some benefits over the study of proton translocating pumps (see below), most of our recent understanding of the coupling process in redox driven ion translocation comes from the studies of the bacterial and mitochondrial complex I. A breakthrough clearly was the identification of the NDH-I structure when the redox steps could be linked to distinct structural features of the protein (Berrisford and Sazanov 2009; Efremov and Sazanov 2011; Sazanov and Hinchliffe 2006). The bacterial complex I usually consists of 14 subunits (or 13 for *E. coli*) and resembles an L-shape. Although bacterial complex I is very similar to the mitochondrial counterpart in terms of function, the latter is quite larger and comprises up to 46 subunits (>900 kDa) (Carroll et al. 2002). However, the 13 subunits, which are found in NDH-1 show a high homology to the according subunits found in mitochondrial complex I, and therefore seem to core module of this proton pump. The additional subunits found in the eukaryotic enzyme seem to exert mainly stabilizing effect. The structure as described by Efremov et al. (2010), shows that the protein complex can be divided in a hydrophilic and a hydrophobic part. The

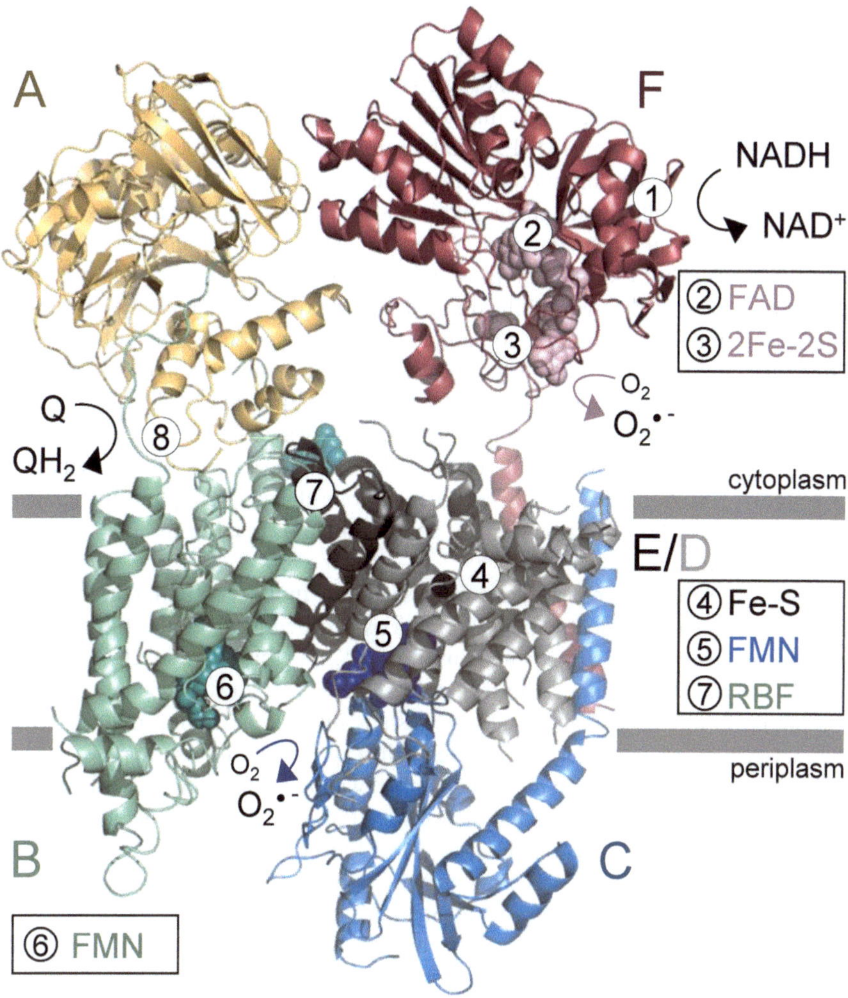

Fig. 10.4 Crystal structure of the NQR and intraelectron transfer from NADH to quinone. The electron donor NADH is oxidized to NAD at the cytoplasmic NqrF subunit (step 1) and transfers the electrons via the cofactors of subunits indicated with letters in lower case (steps 2–6 via FAD_F, $2Fe\text{-}2S_F$, $Fe\text{-}S_{D/E}$ FMN_C, FMN_B and RBF) to the electron acceptor ubiquinone (UQ8,), which is reduced to ubquinol (UQH_2) (step 8). Dioxygen is an alternative electron acceptor, which can be reduced to superoxide ($O_2^{\bullet -}$) by electrons deriving from FAD_F and FMN_B at the cytoplasm or periplasm, respectively. Subunits A, F show hydrophilic domains located towards the cytoplasm whereas subunit C is located towards the periplasm. NqrAFC have membrane associated anchor domains, whereas subunits B, D, E are rather hydrophobic and membrane associated. The transport of the sodium ion is assumed to take place at subunit B. Figure adapted from PDB file 4UAJ (Steuber et al. 2014)

subunits that make up the membrane domain are NuoA, NuoM, NuoN, NuoJ, NuoK, and NuoL, with the latter comprising a large amphipathic a-helix, spanning nearly the whole membrane domain (Efremov et al. 2010). The three subunits NuoM, NuoN and NuoL contain several conserved amino acids are highly homologous to each other and to cation/proton antiporters (MRP) (Mathiesen and Hägerhäll 2002). They are generally regarded the pumping module of complex I, with each subunit exhibiting two half channels through the membrane, which are thought to gate the protons from the one side of the membrane to the other. An additional channel is found at the interface between NuoN, NuoK, NuoJ (Efremov and Sazanov 2011). The electron input module is represented by the hydrophilic domain, which protrudes into the cytoplasm and which accommodates the redox cofactors—one FMN and nine iron sulfur clusters (Hinchliffe and Sazanov 2005). Electron transfer then terminates in ubiquinone and most likely occurs at the interface between the hydrophilic and hydrophobic domains involving subunits NuoA, NuoJ and NuoH (Efremov et al. 2010). This way the proton translocating module in the membrane and electron transfer are spatially separated by roughly 100 Å, addressing the need for a long-distance coupling mechanism. The discussion of how electron transfer and proton translocation are interconnected is still highly discussed.

The most plausible scenario at the time is a purely indirect coupling mode involving energy transfer of electrostatic interactions proceeding from the fully reduced ubiquinone headgroup (Efremov and Sazanov 2012). In this scenario, electrons from NADH are transferred along the iron-sulfur clusters to cluster N2 and finally to ubiquinone in the UQ chamber. The electrostatic interaction is relayed to the proton channels via a series of charged residues half way through the membrane, starting with an Asp/Glu quartet from the UQ chamber, which is referred to as the E-channel (Berrisford et al. 2016). These interactions lead to differentiated solvent exposure and protonation of key residues in each of the half-channels, which results in proton translocation. This way three protons are translocated through subunits NuoL, NuoM, NuoN, and the fourth proton beeing translocated through the E-channel at the interface between subunits NuoN, NuoK and NuoJ (Berrisford et al. 2016; Efremov and Sazanov 2011).

Other models for proton translocation by complex I combine direct and indirect mechanisms. Thus, ubiquinone besides its role as mediator for conformational change can act as "chemical gate" (Verkhovskaya and Bloch 2013) capturing and releasing (1 or 2) protons at opposite sides of the membrane (Treberg and Brand 2011; Nakamaru-Ogiso et al. 2010; Ohnishi et al. 2010b). In these models, the role of the membrane subunits NuoL, NuoN, and NuoM as proton translocating modules is not challenged, still being crucial for the translocation of the remaining protons.

Although progress is being made, the complete understanding of the mechanics, which links electron transfer to the translocation of protons requires intensive studies. In this context, it is worthwhile to study sodium ion -translocating rather than proton -translocating pumps, since the proton concentration is usually restricted to a very narrow pH range of 6–8, depending on the proton pump being studied. In contrast, variation of the coupling ion concentration is very easy in the case of Na^+ pumps. Therefore, Na^+ pumps offer the possibility to identify Na^+-dependent step(s) during the catalytic cycle. The human pathogen *Vibrio cholerae* is the causative agent of Cholera, a severe diarrheal disease (Clemens et al. 2017). *V. cholerae* does not possess an electrogenic NADH:quinone oxidoreductase (NDH-I) belonging to the complex I family of NADH dehydrogenases. Instead, it relies on a Na^+-translocating NADH:quinone oxidoreductase (NQR), the main producer of transmembrane voltage ($\Delta\Psi$) in *V. cholerae* (Vorburger et al. 2016) (Fig. 10.2). The NQR is a model system to study the coupling of redox reactions to Na^+ pumping, and elucidate general principles of respiratory pumps. It allows studying the Na^+ dependence of intraelectron transfer reactions, which led to the proposal that Na^+ uptake is preceded by electron transfer from the [2Fe–2S] cluster on subunit NqrF to the covalently attached flavin mononucleotide (FMN) on subunit NqrC (Juárez et al. 2010).

Per NADH oxidized and Q reduced to QH_2, NQR transports $2Na^+$ from the cytoplasm to the periplasm, as determined in vivo (Bogachev et al. 1997). Li^+ is also transported by NQR, but results in diminished $\Delta\Psi$ (Juárez et al. 2009), which could be due to some back-flow of the smaller Li^+ through a cation channel in NQR (Toulouse et al. 2017). The NQR is composed of six different subunits NqrABCDEF, which harbor FAD, FMN, riboflavin, ubiquinone-8 (Q8) and two FeS centers as redox cofactors (Barquera 2014; Belevich et al. 2016; Fritz and Steuber 2016; Reyes-Prieto et al. 2014; Steuber et al. 2014, 2015; Verkhovsky and Bogachev 2010). The current opinion on the electron transfer pathway in NQR is summarized in the following scheme (see also Fig. 10.4):

$$\text{NADH} \rightarrow \text{FAD}_{\text{NqrF}} \rightarrow [\text{2Fe–2S}]_{\text{NqrF}} \rightarrow \text{Fe}_{\text{NqrD/NqrE}} \rightarrow \text{FMN}_{\text{NqrC}} \rightarrow \text{FMN}_{\text{NqrB}} \rightarrow \text{Riboflavin}_{\text{NqrB}} \rightarrow \text{Q8}_{\text{NqrA}}$$

The X-ray structure of the entire Na^+-NQR complex at 3.5 Å resolution was determined and complemented with high resolution structures of NqrA, NqrC and NqrF (Steuber et al. 2014) (Fig. 10.4). With the exception of the bound UQ8 molecule, all redox cofactors were localized in the 3D structure of NQR. The number of quinone interaction sites in NQR is discussed controversially, but it seems likely that the cytoplasmic NqrA subunit harbors at least one molecule of UQ8 (Ito et al. 2017; Nedielkov et al. 2013; Strickland et al. 2014; Tuz et al. 2017). Very unexpectedly, the 3D structure of the NQR revealed that the largest membrane-bound subunit, NqrB, exhibits high structural similarity with the ammonium transporters Amt-1 and AmtB (Andrade et al. 2005; Khademi et al. 2004) and urea transporter (UT) (Levin et al. 2009), despite the lack of sequence similarity between NqrB and these secondary transporters. NqrB, Amt and UT comprise the so-called Amt/MEP/Rh ammonium transporter domain (interpro entry code IPR024041) which consists of two structural repeats, each with five transmembrane (TM) helices, yielding a 10 TM transporter module with one inner, central pore (Minocha et al. 2003). Amt is a homotrimer, providing a total of three pores for substrate transport, whereas the single NqrB subunit in NQR offers one channel through the complex, as predicted from the structural analysis of the NQR complex (Steuber et al. 2014). Amt is a secondary system which promotes cation flux along an existing electrochemical gradient, while NQR is a primary pump catalysing the uphill transport of Na^+ (or Li^+), most likely through subunit NqrB. To this end, NqrB acquired additional redox domains harbouring the covalently linked FMN at the periplasmic side, and the non-covalently bound riboflavin at the cytoplasmic side of NqrB (Steuber et al. 2014). The channel in NqrB comprises cytoplasmic and periplasmic vestibules which are connected by a narrow passage. This constriction in *V. cholerae* NqrB is formed by two conserved phenylalanine residues (F342 and F338) also found in the related NAD:ferredoxin oxidoreductase complex (Reyes-Prieto et al. 2014). F342 and F338 are located on TM VIII of NqrB which narrows the pore when compared with the channel in Amt. As a consequence, the aromatic side chains from F342 and F338 close the passage connecting the two vestibules of NqrB, as observed in the 3D structure of oxidized NQR (Steuber et al. 2014). To promote the uptake of Na^+, one expects negatively charged or polar residues in the entrance of the channel. Most likely, the highly conserved residue D346 located at the bottom of the cytoplasmic vestibule, above the presumed gate formed by the aromatic moiety of F342, is part of the Na^+ recruitment site in NqrB. Movement of the "gating residues" F338 and F342 in helix VIII out of the cation pathway to open the pore could be triggered by a redox-induced change in the position of transmembrane helix X, a key event in the pumping cycle of NQR (Steuber et al. 2014, 2015).

This proposed mechanism of redox-driven Na^+ transport by NQR needs to be further substantiated. Crystallographic and electron microscopic approaches to obtain an improved 3D structure of NQR at higher resolution, together with functional studies of variants of the NQR obtained by mutagenesis of predicted Na^+ ligands or cofactors, will reveal general insights into the mechanism of electron transfer-driven cation transport by respiratory complexes.

NqrB, NqrC, NqrD, and NqrE are homologous to subunits RnfD, RnfG, RnfE, and RnfA of the RNF (*rhodobacter* nitrogen fixation) complex. The genes encoding for RNF were first discovered by studying *Rhodobacter capsulatus* mutants deficient in nitrogen fixation, suggesting a role of the complex in electron transfer to nitrogenase (Schmehl et al. 1993). The RNF complex from *Acetobacterium woodii* was shown to catalyze redox-driven Na^+ transport (Hess et al. 2013; Westphal et al. 2018). The RNF complex uses ferredoxin (ox/red) and NAD(H) as substrates, and therefore is also termed NFO (NAD^+:ferredoxin oxidoreductase) (Boiangiu et al. 2005). Both Na^+-NQR and the RNF/NFO complexes are composed of six subunits, respectively, but only the central core of the complexes is conserved with regard to primary sequences and cofactor composition. The core of NQR and RNF consist of the three integral membrane subunits: RnfD (NqrB), RnfE (NqrD), and RnfA (NqrE), and the membrane-anchored RnfG (NqrC) (homologous Nqr subunits in brackets). This central part comprising four conserved subunits is considered to represent the minimal, functional unit of the NQR/RNF family of redox-driven cation pumps. NqrC is anchored via a single transmembrane helix to the membrane core of the NQR complex (Steuber et al. 2014). One can expect a similar structural orientation of RnfG in the RNF complex. NqrC/RnfG, like NqrB/RnfD, carries a covalently attached FMN cofactor (Backiel et al. 2008). According to the proposed mechanism of redox-driven Na^+ transport by NQR (Steuber, Vohl et al. 2014), the covalently attached FMN of NqrB might control the passage of Na^+ through the presumed channel in NqrB in a redox-dependent manner.

As the related RnfD subunit also carries a covalently attached FMN, it is tempting to speculate that both NQR and RNF complexes exhibit the very same mechanism of redox-driven Na^+ (or H^+) transport, despite their large differences with regard to their electron-accepting and electron-donating modules. Notably, subunits NqrA and RnfC are only weakly related. In case of RnfB, there is no homolog in the NQR complex (Reyes-Prieto et al. 2014). RnfB is proposed to represent the subunit which interacts with exogenous ferredoxin: depending on the direction of electron transfer and cation transport through the NFO/RNF complex, ferredoxin acts as electron donor or acceptor, respectively (Buckel and Thauer 2013). In the NQR complex, subunit NqrA participates in binding of the substrate quinone, thus comprises a quinone binding site (Nedielkov et al. 2013). In the RNF complex, RnfC shows some weak relation to NqrA and might represent the subunit for binding of NADH or NAD, either accepting or donating a hydride (Buckel and Thauer 2013). Other redox centers in the RNF complex which are not found in the NQR are two [4Fe–4S] centers and one FMN on subunit RnfC which interacts with NAD(H). RnfB which interacts with ferredoxin contains motifs for four [4Fe–4S] clusters and one [3Fe–4S] cluster (Buckel and Thauer 2013). It was also suggested that RnfB may ligate six [4Fe–4S] clusters (Reyes-Prieto et al. 2014). These assignments were based on sequence analyses and await biochemical verification.

Although RNF and NQR complexes differ in cofactor composition in their electron input and output modules (NqrF and NqrA, versus NfoC and NfoB), they utilize the same set of cofactors located in their highly conserved, membrane-embedded RnfAEGD or NqrEDCB subunits which are the key elements responsible for redox-dependent cation transport. There is recent evidence for proton transport by the RNF complex from *Clostridium ljungdahlii* (Tremblay et al. 2013), while all NQR complexes studied so far exclusively acted as Na^+ pumps. Comparing the molecular properties of H^+- with Na^+-translocating RNF complexes will give insight into the principles of cation selectivity in the NQR/RNF family of redox pumps.

Respiration and Oxidative Stress

The last step in aerobic respiration is the conversion of O_2 into H_2O. In aerobic and microaerophilic bacteria or archaea, this step is catalyzed by terminal oxidases. This enzyme catalyzes the reduction of O_2 into H_2O concomitant with the generation of a proton (or sodium) motive force. The full reduction of O_2 into two molecules of H_2O requires $4e^-$ and $4H^+$. For example, the reduction of O_2 by cytochrome *c* oxidase from *Paracoccus denitrificans* requires reduced cytochrome *c* as a one-electron carrier. Oxidation of cytochrome c goes along with the translocation of $1H^+/e^-$ across the membrane, since the *P. denitrificans* oxidase represents a true proton pump (Eq. 10.3):

$$4\text{Cyt}\ c_{\text{red}} + O_2 + 8H^+_{\text{in}} \rightarrow 4\text{Cyt}\ c_{\text{ox}} 2H_2O\ + 4H^+_{\text{out}} \tag{10.3}$$

During Earth's history rising oxygen content in the atmosphere introduced dioxygen (O_2) as a terminal electron acceptor for respiration. The advantages for aerobic respiration compared to anaerobic organisms are immense due to the high oxidizing power of dioxygen ($E_m^{0'} = + 0.82$ V) as compared to other electron acceptors, like S_0 ($E_m^{0'} = -0.27$ V) or CO_2 ($E_m^{0'} = -0.25$ V). On the downside, dioxygen can also undergo transformation into partially reduced forms, and form so called reactive oxygen species (ROS), which are capable to carry out deleterious reactions inside the cell.

Superoxide ($O_2^{\bullet -}$) and hydrogen peroxide (H_2O_2) are the predominant ROS molecules formed during stepwise electron reduction of dioxygen into superoxide, hydrogen peroxide, hydroxyl radical and water (Eq. 10.4):

$$O_2 \overset{e^-}{\rightarrow} O_2^{\bullet -} \overset{2H^+ + e^-}{\rightarrow} H_2O_2 \overset{H^+ + e^-}{\rightarrow} \bullet OH + H_2O \overset{H^+ + e^-}{\rightarrow} 2H_2O \qquad (10.4)$$

During aerobic respiration superoxide and hydrogen peroxide are formed when dioxygen receives electrons from reduced cofactors within proteins. This phenomenon is known for numerous electrogenic enzymes linked to respiration, e.g. complex I (Turrens and Boveris 1980), complex II (Zhang et al. 1998) and complex III (Boveris et al. 1976) of the respiratory chain. In complex I (or NDH-I), there are two putative sites, which are discussed to participate in the formation of superoxide. (I) The first site is the iron-sulfur cluster N2, which is the terminal FeS center in a chain comprising a total of 8 FeS centers, and which acts as electron donor for ubiquinone. Thus, ROS are produced by misguided electron transfer to protein bound ubiquinone or ubisemiquinone (Genova et al. 2001; Ohnishi et al. 2005). (II) The second site is the electron input module, i.e. the NADH binding pocket. The NADH binding pocket in complex I consists of a non-covalently bound FMN and a binuclear iron-sulfur cluster in close proximity. FMN acts as primary electron acceptor, receiving a hydride from NADH and transferring electrons to the iron-sulfur cluster. The reduced FMN cofactor, which is exposed to the surrounding medium is thought to accidently reduce dioxygen (Kussmaul and Hirst 2006; Ohnishi et al. 2010a; Pryde and Hirst 2011). It is still under debate whether the one- or the two- electron reduced FMN catalyzes this reaction. The NQR from *V. cholerae*, contains two cofactors, FMN in NqrB and FAD in NqrF with are believed to deliver electrons to the alternative electron acceptor dioxygen during aerobic respiration (Fig. 10.4) and thus produce ROS (Lin et al. 2007; Muras et al. 2016).

ROS radicals, especially hydroxyl radicals are known to induce DNA double strand breaks, protein damage and paves the way for lipid peroxidation (Halliwell 2006; Mailloux 2015). Lipid peroxidation is very likely to occur in eukaryotic cells, but less likely to happen in bacterial cells. This is explained by to the lipid composition in bacterial membranes. Bacterial lipids are mostly saturated or monounsaturated and not prone to peroxidation. Whereas polyunsaturated fatty acids provide with their methylene moieties surrounded by double bonds a target for lipid radical formation, resulting in a radical chain reaction and membrane decomposition. However, bacteria and archaea suffer from DNA damage by ROS. Neither $O_2^{\bullet -}$, nor H_2O_2 directly damage DNA, but hydroxyl radicals introduce DNA mutations by oxidizing ribose and bases (Imlay 2013). Protein damage by ROS is a result of the oxidation of cysteine residues and decomposition of protein iron cofactors. Oxidation of mononuclear iron or iron-sulfur centers results in loss of protein structure and function and may lead to mismetalation of those proteins as well as an increase in free iron (Imlay 2013). In aerobically respiring *E. coli*, increase in oxidative stress by exposure to redox-active chemicals such as paraquat results in growth retardation and auxotrophy for amino acids due to the inactivation of a plethora of proteins (Benov and Fridovich 1999; Boehm et al. 1976).

In primordial anaerobic life iron was easily accessible and was integrated in many redox proteins, because of its probability to switch between multiple oxidative stages. Thus, iron is still a common cofactor in respiratory enzymes such as NDH-I (complex I), NQR, or succinate dehydrogenase/fumarate reductase. In aerobic life the capability of free iron to react with oxygen species makes it also toxic (Valko et al. 2005). Fe^{2+} can react with hydrogen peroxide and forms reactive hydroxyl radicals in the so-called Fenton chemistry (Eq. 10.5):

$$Fe^{2+} + H_2O_2 \rightarrow Fe^{3+} + \bullet OH + OH^- \tag{10.5}$$

To protect from ROS-derived DNA or protein damage, mechanisms evolved in bacteria and archaea to overcome oxidative stress. These are (I) anaerobic or microaerophilic lifestyle, (II) conversion of ROS to harmless products, (III) repair mechanisms for DNA and proteins.

$O_2^{\bullet -}$ can be converted to hydrogen peroxide (H_2O_2) via the superoxide dismutase (SOD) (Fridovich 1978).

Superoxide Dismutase

Net reaction:	$2O_2^{\bullet -} + 2\,H^+ \rightarrow O_2 + H_2O_2$
	$O_2^{\bullet -} + SOD_{ox} \rightarrow O_2 + SOD_{red}$
	$O_2^{\bullet -} + SOD_{red} + 2\,H^+ \rightarrow H_2O_2 + SOD_{ox}$
$SOD_{red/ox}$: $Cu^{1+/2+}$, $Mn^{2+/3+}$, $Fe^{2+/3+}$, $Ni^{2+/3+}$	

H_2O_2 is converted to dioxygen and water via the catalase/peroxidase (Lushchak 2014; May 1901):

Catalase

Net reaction:	$2H_2O_2 \rightarrow O_2 + 2H_2O$
	H_2O_2 + Catalase-$Fe^{3+} \rightarrow H_2O + O =$ Catalase-Fe^{4+}
	$H_2O_2 + O =$ Catalase-$Fe^{4+} \rightarrow O_2 + H_2O +$ Catalase-Fe^{3+}

It should be stressed that ROS are a common side product of aerobic respiration. It is therefore not surprising that the reactivity of ROS towards FeS centers is utilized for O_2 signaling in many bacteria and archaea, allowing regulatory switches for the shift of aerobic to anaerobic respiratory pathways, and vice versa (Mettert and Kiley 2017).

Recently, it was shown that H_2O_2 acts as terminal electron acceptor in *E. coli* living in an anoxic environment. Under these conditions, and upon sudden exposure to low O_2 or H_2O_2, the periplasmic cytochrome *c* peroxidase acts as respiratory oxidase, allowing electron transfer from ubiquinol to H_2O_2 at rates comparable to anaerobic respiration rates (Khademian and Imlay 2017).

Bioenergetics—Lessons Learned and Future Perspectives

Respiratory Enzymes as Novel Targets for Antibiotics

Mankind faces a global healthcare crisis caused by antimicrobial resistance. There is desperate need to find novel drugs that effectively eradicate pathogens. One strategy to overcome resistances is to modify known drugs acting against one of the three target systems in bacteria (drug classes in brackets): (I) DNA structure/transcription (quinolones), (II) ribosome/translation (macrolides, tetracyclines, aminoglycosides), or (III) cell wall/peptidoglycan synthesis (β-lactams). Another strategy is to consider new, so far unexploited targets, e.g. the energy-producing, respiratory complexes of bacteria.

An example for a new drug, which targets the bacterial energy metabolism is bedaquiline. Bedaquiline showes antimicrobial activity by inhibiting the membrane-bound ATP synthase of *Mycobacterium tuberculosis* (Andries et al. 2005).

The Na^+-translocating NADH:quinone oxidoreductase (NQR) represents a novel, but well-characterized drug target that occurs only in Gram-negative bacteria. The NQR is widespread among many pathogens including *Pseudomonas aeruginosa* and *Klebsiella pneumoniae*, which top the list of dangerous pathogens published by the WHO. It has been shown that the NQR is essential for virulence of *Vibrio cholerae* (Merrell et al. 2002; Minato et al. 2014) and *Chlamydia trachomatis* (Dibrov et al. 2017). NQR catalysis an essential reaction that is analogous to mitochondrial complex I, but NQR exhibits a completely different architecture, different subunits and mechanism. In *P. aeruginosa*, subunit NqrA was identified as intracellular target for human antimicrobial peptides (Lee et al. 2016). Here, electron transfer is inhibited by 2-heptyl-4-hydroxyquinoline N-oxide (HQNO), a major secondary metabolite produced by *P. aeruginosa*. HQNO represents a quinone-type inhibitor which interferes with binding of the quinone substrate to NqrA, as shown by us using NMR spectroscopy (Nedielkov et al. 2013).

Another promising natural inhibitor against NQR-containing bacteria is Korormicin, which originally was obtained from *Pseudoalteromononas sp.* F-420 (Yoshikawa et al. 1997). Inhibitory action of Korormicin against the NQR complex was demonstrated in *Vibrio sp.* (Yoshikawa et al. 1999), but its action against NQR from *K. pneumoniae* and *P. aeruginosa* has not yet been tested. A synthetic compound derived thereof was developed and showed inhibitory properties on *Chlamydia trachomatis* (Dibrov et al. 2017). Another inhibitor of NQR is Ag^+ which binds at the active site of subunit NqrF. This subunit catalyzes the oxidation of NADH (Fadeeva et al. 2011).

Besides direct inactivation of the primary Na^+ pump, an indirect effect on energy-dependent efflux pumps is expected since the Na^+ pump is the major generator of the electrochemical gradient required for the expulsion of antibiotics (Fig. 10.2). This paves the way for a combinatorial therapeutic approach where known antibiotics are administered together with a second drug, acting against the primary Na^+ pump. Pathogens which are resistant due to the action of an efflux pump again will regain sensitivity towards an antibiotic because the electrochemical Na^+ gradient (outside positive) is no longer maintained by the NADH-oxidizing Na^+ pump. A successful example of a combinatorial therapy was reported for treatment of an infection with *K. pneumoniae* (extended spectrum β-lactamase producing strain) after co-administration of carbapenem (blocks cell wall synthesis) with an amino glycoside (blocks translation, Tamma et al. 2012). This therapeutic option is at risk in view of the rise in carbapenem-resistant, Gram-negative pathogens. Targeting respiratory enzymes from bacteria which differ sufficiently from the eukaryotic counterparts with novel inhibitors specific against bacterial respiration, but not harmful with respect to the host, is an opportunity for the development of antibiotics in the future.

The Past and the Future of Respiration

The cradle of life, or more precisely, the origin of the last universal common ancestor (LUCA), is a problem constantly debated, but all theories must deal with the problem of an energy-conserving, biological membrane: How could it evolve in such a manner as to protect the organism from the outside world, but at the same time allow passage of selected cations and substrates against, or along, electrochemical potentials by sophisticated proteins? One theory is highlighted here since it relies on the action of two very ancient respiratory complexes: the ancestors of today's ECH hydrogenase and F_1F_O ATP synthase. According to Nick Lane and coworkers, these two ancient respiratory complexes acted together in membranes still leaky for protons, relying on chemical proton gradients offered "for free" by geothermal activity. Their combined action provided energy in form of ATP to LUCA in a primordial world with hydrothermal vents where alkaline fluids were exposed to acidic ocean water (Sojo et al. 2016).

Player 1 is the ECH hydrogenase, a Ni- and FeS centers containing, membrane-bound enzyme catalyzing the endergonic oxidation of H_2 with ferredoxin, an FeS center-containing redox carrier, powered by the influx of protons into the primordial microorganism along the chemical gradient. Player 2 is the ATP synthase, catalyzing the endergonic synthesis of ATP from ADP and phosphate, also powered by the influx of protons from the acidic (external) to the alkaline (inner, or "cytoplasmic") milieu. Together, these two players provide the essentials of an early pathway of CO_2 fixation operative before true proton (or sodium) pumping across tight biological membranes evolved. Only later, these two complexes gained the capacity for the reverse reaction, e.g. proton (or Na^+) pumping across tight membranes, making LUCA independent from its first habitat in alkaline, hydrothermal vents. Why is this theory attractive from a bioenergetics perspective? ECH hydrogenase and ATP synthase are the respiratory complexes still operative in both prokaryotes and archaea today, and the core membrane module of ECH is evolutionary related to the central, redox-energy converting module of the electrogenic NDH-I (NADH dehydrogenase, or complex I), the first electron input machine in many respiratory chains (Yu et al. 2018).

Since chemiosmotic coupling is of central importance for every living cell, it is the biggest hurdle in ongoing attempts to generate a minimalistic, synthetic (micro)-organism. Once high energy electrons are provided (either from external electron donors such as H_2, or from a light quantum), their flow into membrane-embedded redox proteins must be coupled to gradient formation which in turn drives ATP synthesis. Synthetic biology and bio design rely on the knowledge derived from studying simpler respiratory enzymes and processes in the prokaryotic world (Anderson and Koder 2016). The design of a respiring, energy-converting microorganisms seems a far-fetched goal, but the application of bioenergetics principles derived from microbial respiratory enzymes led to the invention of simple, man-made cofactor-binding proteins ("maquettes") (Fry et al. 2016). These maquettes are promising, biotechnological tools for improved catalysis, energy conversion, fuel production or waste removal. They will help to optimize the bioenergetics conversion of chemical into electrical energy (Minteer 2016), hopefully providing solutions for sustainable production and usage of energy in the future.

Acknowledgements This work was supported by grant FR1488/8-1 (to GF) and grant FR 1321/6-1 (to JS) from the Deutsche Forschungsgemeinschaft.

References

Abramson J, Riistama S, Larsson G, Jasaitis A, Svensson-Ek M, Laakkonen L, Puustinen A, Iwata S, Wikström M (2000) The structure of the ubiquinol oxidase from *Escherichia coli* and its ubiquinone binding site. Nat Struct Biol 7:910–917. https://doi.org/10.1038/82824

Anderson R, Koder RL (2016) Biodesign for bioenergetics-the design and engineering of electron transfer cofactors, proteins and protein networks. Biochim Biophys Acta 1857:483–484. https://doi.org/10.1016/j.bbabio.2016.02.017

Andrade SLA, Dickmanns A, Ficner R, Einsle O (2005) Crystal structure of the archaeal ammonium transporter Amt-1 from *Archaeoglobus fulgidus*. Proc Natl Acad Sci USA 102:14994–14999. https://doi.org/10.1073/pnas.0506254102

Andries K, Verhasselt P, Guillemont J, Gohlmann HWH, Neefs J-M, Winkler H, van Gestel J, Timmerman P, Zhu M, Lee E, Williams P, de Chaffoy D, Huitric E, Hoffner S, Cambau E, Truffot-Pernot C, Lounis N, Jarlier V (2005) A diarylquinoline drug active on the ATP synthase of *Mycobacterium tuberculosis*. Science 307:223–227. https://doi.org/10.1126/science.1106753

Anraku Y, Gennis RB (1987) The aerobic respiratory chain of *Escherichia coli*. Trends Biochem Sci 12:262–266. https://doi.org/10.1016/0968-0004(87)90131-9

Backiel J, Juárez O, Zagorevski DV, Wang Z, Nilges MJ, Barquera B (2008) Covalent binding of flavins to RnfG and RnfD in the Rnf complex from *Vibrio cholerae*. Biochemistry 47:11273–11284. https://doi.org/10.1021/bi800920j

Barquera B (2014) The sodium pumping NADH:quinone oxidoreductase (Na^+-NQR), a unique redox-driven ion pump. J Bioenerg Biomembr 46:289–298. https://doi.org/10.1007/s10863-014-9565-9

Barquera B, Morgan JE, Lukoyanov D, Scholes CP, Gennis RB, Nilges MJ (2003) X- and W-band EPR and Q-band ENDOR studies of the flavin radical in the Na^+-translocating NADH:quinone oxidoreductase from *Vibrio cholerae*. J Am Chem Soc 125:265–275. https://doi.org/10.1021/ja0207201

Beinert H (1997) Iron-sulfur clusters: nature's modular, multipurpose structures. Science 277:653–659. https://doi.org/10.1126/science.277.5326.653

Beinert H (2000) Iron-sulfur proteins: ancient structures, still full of surprises. J Biol Inorg Chem 5:2–15

Belevich NP, Bertsova YV, Verkhovskaya ML, Baykov AA, Bogachev AV (2016) Identification of the coupling step in Na^+-translocating NADH:quinone oxidoreductase from real-time kinetics of electron transfer. Biochim Biophys Acta 1857:141–149. https://doi.org/10.1016/j.bbabio.2015.12.001

Benov L, Fridovich I (1999) Why superoxide imposes an aromatic amino acid auxotrophy on *Escherichia coli*. The transketolase connection. J Biol Chem 274:4202–4206. https://doi.org/10.1074/jbc.274.7.4202

Berrisford JM, Sazanov LA (2009) Structural basis for the mechanism of respiratory complex I. J Biol Chem 284:29773–29783. https://doi.org/10.1074/jbc.M109.032144

Berrisford JM, Baradaran R, Sazanov LA (2016) Structure of bacterial respiratory complex I. Biochim Biophys Acta 1857:892–901. https://doi.org/10.1016/j.bbabio.2016.01.012

Bertero MG, Rothery RA, Palak M, Hou C, Lim D, Blasco F, Weiner JH, Strynadka NCJ (2003) Insights into the respiratory electron transfer pathway from the structure of nitrate reductase A. Nat Struct Biol 10:681–687. https://doi.org/10.1038/nsb969

Boehm DE, Vincent K, Brown OR (1976) Oxygen and toxicity inhibition of amino acid biosynthesis. Nature 262:418–420. https://doi.org/10.1038/262418a0

Bogachev AV, Murtazina RA, Skulachev VP (1996) H^+/e^- stoichiometry for NADH dehydrogenase I and dimethyl sulfoxide reductase in anaerobically grown *Escherichia coli* cells. J Bacteriol 178:6233–6237

Bogachev AV, Murtazina RA, Skulachev VP (1997) The Na^+/e^- stoichiometry of the Na^+-motive NADH:quinone oxidoreductase in *Vibrio alginolyticus*. FEBS Lett 409:475–477. https://doi.org/10.1016/S0014-5793(97)00536-X

Bogachev AV, Belevich NP, Bertsova YV, Verkhovsky MI (2009a) Primary steps of the Na^+-translocating NADH:ubiquinone oxidoreductase catalytic cycle resolved by the ultrafast freeze-quench approach. J Biol Chem 284:5533–5538. https://doi.org/10.1074/jbc.M808984200

Bogachev AV, Bloch DA, Bertsova YV, Verkhovsky MI (2009b) Redox properties of the prosthetic groups of Na^+-translocating NADH:quinone oxidoreductase. 2. Study of the enzyme by optical spectroscopy. Biochemistry 48:6299–6304. https://doi.org/10.1021/bi900525v

Boiangiu CD, Jayamani E, Brügel D, Herrmann G, Kim J, Forzi L, Hedderich R, Vgenopoulou I, Pierik AJ, Steuber J (2005) Sodium ion pumps and hydrogen production in glutamate fermenting anaerobic bacteria. J Mol Microbiol Biotechnol 10:105–119. https://doi.org/10.1159/000091558

Borisov VB, Verkhovsky MI (2015) Oxygen as acceptor. EcoSal Plus 6. https://doi.org/10.1128/ecosalplus.esp-0012-2015

Boveris A, Cadenas E, Stoppani AO (1976) Role of ubiquinone in the mitochondrial generation of hydrogen peroxide. Biochem J 156:435–444. https://doi.org/10.1042/bj1560435

Brosi R, Bittl R, Engelhard C (2014) EPR on Flavoproteins. Methods Mol Biol 1146:341–360. https://doi.org/10.1007/978-1-4939-0452-5_13

Buckel W, Thauer RK (2013) Energy conservation via electron bifurcating ferredoxin reduction and proton/Na^+ translocating ferredoxin oxidation. Biochim et Biophys Acta (BBA)-Bioenerg 1827:94–113. https://doi.org/10.1016/j.bbabio.2012.07.002

Cammack R, Rao KK, Bargeron CP, Hutson KG, Andrew PW, Rogers LJ (1977) Midpoint redox potentials of plant and algal ferredoxins. Biochem J 168:205–209. https://doi.org/10.1042/bj1680205

Carroll J, Shannon RJ, Fearnley IM, Walker JE, Hirst J (2002) Definition of the nuclear encoded protein composition of bovine heart mitochondrial complex I. Identification of two new subunits. J Biol Chem 277:50311–50317. https://doi.org/10.1074/jbc.M209166200

Castro PJ, Silva AF, Marreiros BC, Batista AP, Pereira MM (2016) Respiratory complex I: a dual relation with H^+ and Na^+? Biochim Biophys Acta 1857:928–937. https://doi.org/10.1016/j.bbabio.2015.12.008

Chance B (1972) The nature of electron transfer and energy coupling reactions. FEBS Lett 23:3–20

Cheng VWT, Weiner JH (2007) S- and N-Oxide Reductases. EcoSal Plus 2. https://doi.org/10.1128/ecosalplus.3.2.8

Clemens JD, Nair GB, Ahmed T, Qadri F, Holmgren J (2017) Cholera. Lancet 390:1539–1549. https://doi.org/10.1016/S0140-6736(17)30559-7

Coelho C, Romão MJ (2015) Structural and mechanistic insights on nitrate reductases. Protein Sci 24:1901–1911. https://doi.org/10.1002/pro.2801

Crane FL (2001) Biochemical functions of coenzyme Q10. J Am Coll Nutr 20:591–598

Crofts AR (2004) The cytochrome *bc*1 complex: function in the context of structure. Annu Rev Physiol 66:689–733. https://doi.org/10.1146/annurev.physiol.66.032102.150251

Deller S, Macheroux P, Sollner S (2008) Flavin-dependent quinone reductases. Cell Mol Life Sci 65:141–160. https://doi.org/10.1007/s00018-007-7300-y

Denis FA, D'Oultremont PA, Debacq JJ, Cherel JM, Brisou J (1975) Distribution of ubiquinones (coenzyme Q) in Gram negative bacillae (Distribution des ubiquinones (coenzyme Q) chez les bacilles a Gram negatif). C R Seances Soc Biol Fil 169:380–383

Dibrov P, Dibrov E, Maddaford TG, Kenneth M, Nelson J, Resch C, Pierce GN (2017) Development of a novel rationally designed antibiotic to inhibit a nontraditional bacterial target. Can J Physiol Pharmacol 95:595–603. https://doi.org/10.1139/cjpp-2016-0505

Duarte AG, Catarino T, White GF, Lousa D, Neukirchen S, Soares CM, Sousa FL, Clarke TA, Pereira IAC (2018) An electrogenic redox loop in sulfate reduction reveals a likely widespread mechanism of energy conservation. Nat Commun 9:5448. https://doi.org/10.1038/s41467-018-07839-x

Efremov RG, Sazanov LA (2011) Structure of the membrane domain of respiratory complex I. Nature 476:414–420. https://doi.org/10.1038/nature10330

Efremov RG, Sazanov LA (2012) The coupling mechanism of respiratory complex I - a structural and evolutionary perspective. Biochim Biophys Acta 1817:1785–1795. https://doi.org/10.1016/j.bbabio.2012.02.015

Efremov RG, Baradaran R, Sazanov LA (2010) The architecture of respiratory complex I. Nature 465:441 EP. https://doi.org/10.1038/nature09066

Einsle O, Stach P, Messerschmidt A, Simon J, Kröger A, Huber R, Kroneck PM (2000) Cytochrome *c* nitrite reductase from *Wolinella succinogenes*. Structure at 1.6 Angström Å resolution, inhibitor binding, and heme-packing motifs. J Biol Chem 275:39608–39616. https://doi.org/10.1074/jbc.M006188200

Fadeeva MS, Bertsova YV, Euro L, Bogachev AV (2011) Cys377 residue in NqrF subunit confers Ag^+ sensitivity of Na^+-translocating NADH:quinone oxidoreductase from *Vibrio harveyi*. Biochemistry (Mosc) 76:186–195. https://doi.org/10.1134/S0006297911020040

Fridovich I (1978) The biology of oxygen radicals. Science 201:875–880. https://doi.org/10.1126/science.210504

Fritz G, Roth A, Schiffer A, Büchert T, Bourenkov G, Bartunik HD, Huber H, Stetter KO, Kroneck PMH, Ermler U (2002) Structure of adenylylsulfate reductase from the hyperthermophilic *Archaeoglobus fulgidus* at 1.6-A resolution. Proc Natl Acad Sci USA 99:1836–1841. https://doi.org/10.1073/pnas.042664399

Fritz G, Steuber J (2016) Sodium as coupling cation in respiratory energy conversion. Met Ions Life Sci 16:349–390. https://doi.org/10.1007/978-3-319-21756-7_11

Fry BA, Solomon LA, Leslie Dutton P, Moser CC (2016) Design and engineering of a man-made diffusive electron-transport protein. Biochim Biophys Acta 1857:513–521. https://doi.org/10.1016/j.bbabio.2015.09.008

Galassi VV, Arantes GM (2015) Partition, orientation and mobility of ubiquinones in a lipid bilayer. Biochim Biophys Acta 1847:1560–1573. https://doi.org/10.1016/j.bbabio.2015.08.001

Galkin A, Dröse S, Brandt U (2006) The proton pumping stoichiometry of purified mitochondrial complex I reconstituted into proteoliposomes. Biochim Biophys Acta 1757:1575–1581. https://doi.org/10.1016/j.bbabio.2006.10.001

Genova ML, Ventura B, Giuliano G, Bovina C, Formiggini G, Parenti Castelli G, Lenaz G (2001) The site of production of superoxide radical in mitochondrial complex I is not a bound ubisemiquinone but presumably iron-sulfur cluster N2. FEBS Lett 505:364–368. https://doi.org/10.1016/S0014-5793(01)02850-2

Getz EW, Tithi SS, Zhang L, Aylward FO (2018) Parallel evolution of genome streamlining and cellular bioenergetics across the marine radiation of a bacterial phylum. MBio 9. https://doi.org/10.1128/mbio.01089-18

Gnandt E, Dörner K, Strampraad MFJ, de Vries S, Friedrich T (2016) The multitude of iron-sulfur clusters in respiratory complex I. Biochim Biophys Acta 1857:1068–1072. https://doi.org/10.1016/j.bbabio.2016.02.018

Gonzalez-Mariscal I, Garcia-Teston E, Padilla S, Martin-Montalvo A, Pomares-Viciana T, Vazquez-Fonseca L, Gandolfo-Dominguez P, Santos-Ocana C (2014) Regulation of coenzyme Q biosynthesis in yeast: a new complex in the block. IUBMB Life 66:63–70. https://doi.org/10.1002/iub.1243

Grinberg AV, Hannemann F, Schiffler B, Müller J, Heinemann U, Bernhardt R (2000) Adrenodoxin: structure, stability, and electron transfer properties. Proteins 40:590–612

Halliwell B (2006) Reactive species and antioxidants. Redox biology is a fundamental theme of aerobic life. Plant Physiol 141:312–322. https://doi.org/10.1104/pp.106.077073

Hayashi M, Nakayama Y, Unemoto T (2001) Recent progress in the Na^+-translocating NADH-quinone reductase from the marine *Vibrio alginolyticus*. Biochim Biophys Acta 1505:37–44. https://doi.org/10.1016/S0005-2728(00)00275-9

Hein S, Klimmek O, Polly M, Kern M, Simon J (2017) A class C radical S-adenosylmethionine methyltransferase synthesizes 8-methylmenaquinone. Mol Microbiol 104:449–462. https://doi.org/10.1111/mmi.13638

Hermann B, Kern M, La Pietra L, Simon J, Einsle O (2015) The octahaem MccA is a haem c-copper sulfite reductase. Nature 520:706–709. https://doi.org/10.1038/nature14109

Hess V, Schuchmann K, Müller V (2013) The ferredoxin: NAD^+ oxidoreductase (Rnf) from the acetogen *Acetobacterium woodii* requires Na^+ and is reversibly coupled to the membrane potential. J Biol Chem 288:31496–31502. https://doi.org/10.1074/jbc.M113.510255

Hinchliffe P, Sazanov LA (2005) Organization of iron-sulfur clusters in respiratory complex I. Science 309:771–774. https://doi.org/10.1126/science.1113988

Hirst J (2003) The dichotomy of complex I: a sodium ion pump or a proton pump. Proc Natl Acad Sci 100:773. https://doi.org/10.1073/pnas.0330050100

Imlay JA (2013) The molecular mechanisms and physiological consequences of oxidative stress: lessons from a model bacterium. Nat Rev Microbiol 11:443–454. https://doi.org/10.1038/nrmicro3032

Ito T, Murai M, Ninokura S, Kitazumi Y, Mezic KG, Cress BF, Koffas MAG, Morgan JE, Barquera B, Miyoshi H (2017) Identification of the binding sites for ubiquinone and inhibitors in the Na^+-pumping NADH-ubiquinone oxidoreductase from *Vibrio cholerae* by photoaffinity labeling. J Biol Chem 292:7727–7742. https://doi.org/10.1074/jbc.M117.781393

Jormakka M, Törnroth S, Byrne B, Iwata S (2002) Molecular basis of proton motive force generation: structure of formate dehydrogenase-N. Science 295:1863–1868. https://doi.org/10.1126/science.1068186

Juárez O, Athearn K, Gillespie P, Barquera B (2009) Acid residues in the transmembrane helices of the Na^+-pumping NADH:quinone oxidoreductase from *Vibrio cholerae* involved in sodium translocation. Biochemistry 48:9516–9524. https://doi.org/10.1021/bi900845y

Juárez O, Morgan JE, Nilges MJ, Barquera B (2010) Energy transducing redox steps of the Na^+-pumping NADH:quinone oxidoreductase from *Vibrio cholerae*. Proc Natl Acad Sci USA 107:12505–12510. https://doi.org/10.1073/pnas.1002866107

Juárez O, Shea ME, Makhatadze GI, Barquera B (2011) The role and specificity of the catalytic and regulatory cation-binding sites of the Na^+-pumping NADH:quinone oxidoreductase from *Vibrio cholerae*. J Biol Chem 286:26383–26390. https://doi.org/10.1074/jbc.M111.257873

Kerscher SJ, Okun JG, Brandt U (1999) A single external enzyme confers alternative NADH:ubiquinone oxidoreductase activity in *Yarrowia lipolytica*. J Cell Sci 112(Pt 14):2347–2354

Khademi S, O'Connell J, Remis J, Robles-Colmenares Y, Miercke LJW, Stroud RM (2004) Mechanism of ammonia transport by Amt/MEP/Rh: structure of AmtB at 1.35 A. Science 305:1587–1594. https://doi.org/10.1126/science.1101952

Khademian M, Imlay JA (2017) *Escherichia coli* cytochrome *c* peroxidase is a respiratory oxidase that enables the use of hydrogen peroxide as a terminal electron acceptor. Proc Natl Acad Sci USA 114:E6922–E6931. https://doi.org/10.1073/pnas.1701587114

Krulwich TA, Ito M, Guffanti AA (2001) The Na^+-dependence of alkaliphily in *Bacillus*. Biochim Biophys Acta 1505:158–168

Kussmaul L, Hirst J (2006) The mechanism of superoxide production by NADH:ubiquinone oxidoreductase (complex I) from bovine heart mitochondria. Proc Natl Acad Sci USA 103:7607–7612. https://doi.org/10.1073/pnas.0510977103

Lee JTY, Wang G, Tam YT, Tam C (2016) Membrane-active epithelial keratin 6A fragments (KAMPs) are unique human antimicrobial peptides with a non-alphabeta structure. Front Microbiol 7:1799. https://doi.org/10.3389/fmicb.2016.01799

Levin EJ, Quick M, Zhou M (2009) Crystal structure of a bacterial homologue of the kidney urea transporter. Nature 462:757–761. https://doi.org/10.1038/nature08558

Leys D, Scrutton NS (2016) Sweating the assets of flavin cofactors: new insight of chemical versatility from knowledge of structure and mechanism. Curr Opin Struct Biol 41:19–26. https://doi.org/10.1016/j.sbi.2016.05.014

Lin P-C, Türk K, Häse CC, Fritz G, Steuber J (2007) Quinone reduction by the Na^+-translocating NADH dehydrogenase promotes extracellular superoxide production in *Vibrio cholerae*. J Bacteriol 189:3902–3908. https://doi.org/10.1128/JB.01651-06

Lushchak VI (2014) Classification of oxidative stress based on its intensity. EXCLI J 13:922–937

Mailloux RJ (2015) Teaching the fundamentals of electron transfer reactions in mitochondria and the production and detection of reactive oxygen species. Redox Biology 4:381–398. https://doi.org/10.1016/j.redox.2015.02.001

Maklashina E, Cecchini G, Dikanov SA (2013) Defining a direction: electron transfer and catalysis in *Escherichia coli* complex II enzymes. Biochim Biophys Acta 1827:668–678. https://doi.org/10.1016/j.bbabio.2013.01.010

Marcus RA, Sutin N (1985) Electron transfers in chemistry and biology. Biochim et Biophys Acta (BBA) Rev Bioenerg 811:265–322. https://doi.org/10.1016/0304-4173(85)90014-x

Marreiros BC, Calisto F, Castro PJ, Duarte AM, Sena FV, Silva AF, Sousa FM, Teixeira M, Refojo PN, Pereira MM (2016a) Exploring membrane respiratory chains. Biochim Biophys Acta 1857:1039–1067. https://doi.org/10.1016/j.bbabio.2016.03.028

Marreiros BC, Sena FV, Sousa FM, Batista AP, Pereira MM (2016b) Type II NADH:quinone oxidoreductase family: phylogenetic distribution, structural diversity and evolutionary divergences. Environ Microbiol 18:4697–4709. https://doi.org/10.1111/1462-2920.13352

Mathiesen C, Hägerhäll C (2002) Transmembrane topology of the NuoL, M and N subunits of NADH:quinone oxidoreductase and their homologues among membrane-bound hydrogenases and bona fide antiporters. Biochim Biophys Acta 1556:121–132. https://doi.org/10.1016/S0005-2728(02)00343-2

Matias PM, Soares CM, Saraiva LM, Coelho R, Morais J, Le Gall J, Carrondo MA (2001) NiFe hydrogenase from *Desulfovibrio desulfuricans* ATCC 27774: gene sequencing, three-dimensional structure determination and refinement at 1.8 A and modelling studies of its interaction with the tetrahaem cytochrome *c3*. J Biol Inorg Chem 6:63–81

May DW (1901) Catalase, a new enzym of general occurrence. Science 14:815–816

Merrell DS, Hava DL, Camilli A (2002) Identification of novel factors involved in colonization and acid tolerance of *Vibrio cholerae*. Mol Microbiol 43:1471–1491. https://doi.org/10.1046/j.1365-2958.2002.02857.x

Mettert EL, Kiley PJ (2017) Reassessing the structure and function relationship of the O_2 sensing transcription factor FNR. Antioxid Redox Signal. https://doi.org/10.1089/ars.2017.7365

Meyer J (2008) Iron-sulfur protein folds, iron-sulfur chemistry, and evolution. J Biol Inorg Chem 13:157–170. https://doi.org/10.1007/s00775-007-0318-7

Miki T, Yu L, Yu CA (1992) Characterization of ubisemiquinone radicals in succinate-ubiquinone reductase. Arch Biochem Biophys 293:61–66. https://doi.org/10.1016/0003-9861(92)90365-4

Minato Y, Fassio SR, Reddekopp RL, Häse CC (2014) Inhibition of the sodium-translocating NADH-ubiquinone oxidoreductase Na^+-NQR decreases cholera toxin production in *Vibrio cholerae* O1 at the late exponential growth phase. Microb Pathog 66:36–39. https://doi.org/10.1016/j.micpath.2013.12.002

Minocha R, Studley K, Saier MH Jr (2003) The urea transporter (UT) family: bioinformatic analyses leading to structural, functional, and evolutionary predictions. Recept Channels 9:345–352

Minteer SD (2016) Oxidative bioelectrocatalysis: from natural metabolic pathways to synthetic metabolons and minimal enzyme cascades. Biochim Biophys Acta 1857:621–624. https://doi.org/10.1016/j.bbabio.2015.08.008

Mitchell P (1961) Coupling of phosphorylation to electron and hydrogen transfer by a chemi-osmotic type of mechanism. Nature 191:144–148

Mitchell P, Moyle J (1969) Estimation of membrane potential and pH difference across the cristae membrane of rat liver mitochondria. Eur J Biochem 7:471–484. https://doi.org/10.1111/j.1432-1033.1969.tb19633.x

Müller F (1991) Chemistry and biochemistry of flavoenzymes, vol I. CRC Press, Boca Raton

Mulkidjanian AY, Dibrov P, Galperin MY (2008) The past and present of sodium energetics: may the sodium-motive force be with you. Biochim Biophys Acta 1777:985–992. https://doi.org/10.1016/j.bbabio.2008.04.028

Muras V, Dogaru-Kinn P, Minato Y, Häse CC, Steuber J (2016) The Na^+-translocating NADH:quinone oxidoreductase enhances oxidative stress in the cytoplasm of *Vibrio cholerae*. J Bacteriol 198:2307–2317. https://doi.org/10.1128/JB.00342-16

Nakamaru-Ogiso E, Kao M-C, Chen H, Sinha SC, Yagi T, Ohnishi T (2010) The membrane subunit NuoL(ND5) is involved in the indirect proton pumping mechanism of *Escherichia coli* complex I. J Biol Chem 285:39070–39078. https://doi.org/10.1074/jbc.M110.157826

Nedielkov R, Steffen W, Steuber J, Möller HM (2013) NMR reveals double occupancy of quinone-type ligands in the catalytic quinone binding site of the Na^+-translocating NADH:quinone oxidoreductase from *Vibrio cholerae*. J Biol Chem:jbc-M112. https://doi.org/10.1074/jbc.M112.435750

Nič M, Jirát J, Košata B, Jenkins A, McNaught A (2009) IUPAC compendium of chemical terminology. IUPAC, Research Triagle Park, NC

Niks D, Hille R (2018) Molybdenum- and tungsten-containing formate dehydrogenases and formylmethanofuran dehydrogenases: structure, mechanism and cofactor insertion. Protein Sci. https://doi.org/10.1002/pro.3498

Ohnishi ST, Ohnishi T, Muranaka S, Fujita H, Kimura H, Uemura K, K-i Y, Utsumi K (2005) A possible site of superoxide generation in the complex I segment of rat heart mitochondria. J Bioenerg Biomembr 37:1–15. https://doi.org/10.1007/s10863-005-4117-y

Ohnishi ST, Shinzawa-Itoh K, Ohta K, Yoshikawa S, Ohnishi T (2010a) New insights into the superoxide generation sites in bovine heart NADH-ubiquinone oxidoreductase (complex I): the significance of protein-associated ubiquinone and the dynamic shifting of generation sites between semiflavin and semiquinone radicals. Biochim Biophys Acta 1797:1901–1909. https://doi.org/10.1016/j.bbabio.2010.05.012

Ohnishi ST, Salerno JC, Ohnishi T (2010b) Possible roles of two quinone molecules in direct and indirect proton pumps of bovine heart NADH-quinone oxidoreductase (complex I). Biochim Biophys Acta 1797:1891–1893. https://doi.org/10.1016/j.bbabio.2010.06.010

Oren A (2005) Anaerobes. In: Encyclopedia of life sciences, vol 407. Wiley, Chichester, p 623

Page C (2003) Mechanism for electron transfer within and between proteins. Curr Opin Chem Biol 7:551–556. https://doi.org/10.1016/j.cbpa.2003.08.005

Parey K, Warkentin E, Kroneck PMH, Ermler U (2010) Reaction cycle of the dissimilatory sulfite reductase from *Archaeoglobus fulgidus*. Biochemistry 49:8912–8921. https://doi.org/10.1021/bi100781f

Parey K, Brandt U, Xie H, Mills DJ, Siegmund K, Vonck J, Kuhlbrandt W, Zickermann V (2018) Cryo-EM structure of respiratory complex I at work. Elife 7. https://doi.org/10.7554/elife.39213

Pinske C, Sawers RG (2016) Anaerobic formate and hydrogen metabolism. EcoSal Plus 7. https://doi.org/10.1128/ecosalplus.esp-0011-2016

Pires RH, Venceslau SS, Morais F, Teixeira M, Xavier AV, Pereira IAC (2006) Characterization of the *Desulfovibrio desulfuricans* ATCC 27774 DsrMKJOP complex-a membrane-bound redox complex involved in the sulfate respiratory pathway. Biochemistry 45:249–262. https://doi.org/10.1021/bi0515265

Pryde KR, Hirst J (2011) Superoxide is produced by the reduced flavin in mitochondrial complex I: a single, unified mechanism that applies during both forward and reverse electron transfer. J Biol Chem 286:18056–18065. https://doi.org/10.1074/jbc.M110.186841

Puustinen A, Finel M, Virkki M, Wikström M (1989) Cytochrome *o* (*bo*) is a proton pump in *Paracoccus denitrificans* and *Escherichia coli*. FEBS Lett 249:163–167. https://doi.org/10.1016/0014-5793(89)80616-7

Ramos AR, Keller KL, Wall JD, Pereira IAC (2012) The membrane QmoABC complex interacts directly with the dissimilatory adenosine 5′-phosphosulfate reductase in sulfate reducing bacteria. Front Microbiol 3:137. https://doi.org/10.3389/fmicb.2012.00137

Reyes-Prieto A, Barquera B, Juárez O (2014) Origin and evolution of the sodium-pumping NADH:ubiquinone oxidoreductase. PLoS ONE 9:e96696

Safarian S, Rajendran C, Müller H, Preu J, Langer JD, Ovchinnikov S, Hirose T, Kusumoto T, Sakamoto J, Michel H (2016) Structure of a *bd* oxidase indicates similar mechanisms for membrane-integrated oxygen reductases. Science 352:583–586. https://doi.org/10.1126/science.aaf2477

Salewski J, Batista AP, Sena FV, Millo D, Zebger I, Pereira MM, Hildebrandt P (2016) Substrate-protein interactions of type II NADH:quinone oxidoreductase from *Escherichia coli*. Biochemistry 55:2722–2734. https://doi.org/10.1021/acs.biochem.6b00070

Santos AA, Venceslau SS, Grein F, Leavitt WD, Dahl C, Johnston DT, Pereira IAC (2015) A protein trisulfide couples dissimilatory sulfate reduction to energy conservation. Science 350:1541–1545. https://doi.org/10.1126/science.aad3558

Sargent F (2016) The model NiFe-hydrogenases of *Escherichia coli*. Adv Microb Physiol 68:433–507. https://doi.org/10.1016/bs.ampbs.2016.02.008

Sazanov LA, Hinchliffe P (2006) Structure of the hydrophilic domain of respiratory complex I from *Thermus thermophilus*. Science 311:1430–1436. https://doi.org/10.1126/science.1123809

Schmehl M, Jahn A, Meyer zu Vilsendorf A, Hennecke S, Masepohl B, Schuppler M, Marxer M, Oelze J, Klipp W (1993) Identification of a new class of nitrogen fixation genes in *Rhodobacter capsulatus*: a putative membrane complex involved in electron transport to nitrogenase. Mol Gen Genet 241:602–615

Sojo V, Herschy B, Whicher A, Camprubi E, Lane N (2016) The origin of life in alkaline hydrothermal vents. Astrobiology 16:181–197. https://doi.org/10.1089/ast.2015.1406

Sparacino-Watkins C, Stolz JF, Basu P (2014) Nitrate and periplasmic nitrate reductases. Chem Soc Rev 43:676–706. https://doi.org/10.1039/c3cs60249d

Steffen W, Steuber J (2013) Cation transport by the respiratory NADH:quinone oxidoreductase (complex I): facts and hypotheses. Biochem Soc Trans 41:1280–1287. https://doi.org/10.1042/BST20130024

Steuber J, Vohl G, Casutt MS, Vorburger T, Diederichs K, Fritz G (2014) Structure of the *V. cholerae* Na^+-pumping NADH:quinone oxidoreductase. Nature 516:62–67. https://doi.org/10.1038/nature14003

Steuber J, Vohl G, Muras V, Toulouse C, Claussen B, Vorburger T, Fritz G (2015) The structure of Na^+-translocating of NADH:ubiquinone oxidoreductase of *Vibrio cholerae*: implications on coupling between electron transfer and Na^+ transport. Biol Chem 396:1015–1030. https://doi.org/10.1515/hsz-2015-0128

Strickland M, Juárez O, Neehaul Y, Cook DA, Barquera B, Hellwig P (2014) The conformational changes induced by ubiquinone binding in the Na^+-pumping NADH:ubiquinone oxidoreductase (Na^+-NQR) are kinetically controlled by conserved glycines 140 and 141 of the NqrB subunit. J Biol Chem 289:23723–23733. https://doi.org/10.1074/jbc.M114.574640

Sulavik MC, Houseweart C, Cramer C, Jiwani N, Murgolo N, Greene J, DiDomenico B, Shaw KJ, Miller GH, Hare R, Shimer G (2001) Antibiotic susceptibility profiles of *Escherichia coli* strains lacking multidrug efflux pump genes. Antimicrob Agents Chemother 45:1126–1136. https://doi.org/10.1128/AAC.45.4.1126-1136.2001

Tamma PD, Cosgrove SE, Maragakis LL (2012) Combination therapy for treatment of infections with Gram-negative bacteria. Clin Microbiol Rev 25:450–470. https://doi.org/10.1128/CMR.05041-11

Thanassi DG, Cheng LW, Nikaido H (1997) Active efflux of bile salts by *Escherichia coli*. J Bacteriol 179:2512–2518. https://doi.org/10.1128/jb.179.8.2512-2518.1997

Thauer RK, Jungermann K, Decker K (1977) Energy conservation in chemotrophic anaerobic bacteria. Bacteriol Rev 41:100–180

Tokuda H, Unemoto T (1981) A respiration-dependent primary sodium extrusion system functioning at alkaline pH in the marine bacterium *Vibrio alginolyticus*. Biochem Biophys Res Commun 102:265–271. https://doi.org/10.1016/0006-291X(81)91516-3

Tokuda H, Unemoto T (1984) Na^+ is translocated at NADH:quinone oxidoreductase segment in the respiratory chain of *Vibrio alginolyticus*. J Biol Chem 259:7785–7790

Tomitsuka E, Kita K, Esumi H (2009) Regulation of succinate-ubiquinone reductase and fumarate reductase activities in human complex II by phosphorylation of its flavoprotein subunit. Proc Jpn Acad Ser B Phys Biol Sci 85:258–265. https://doi.org/10.2183/pjab.85.258

Toulouse C, Claussen B, Muras V, Fritz G, Steuber J (2017) Strong pH dependence of coupling efficiency of the Na^+-translocating NADH:quinone oxidoreductase (Na^+-NQR) of *Vibrio cholerae*. Biol Chem 398:251–260. https://doi.org/10.1515/hsz-2016-0238

Treberg JR, Brand MD (2011) A model of the proton translocation mechanism of complex I. J Biol Chem 286:17579–17584. https://doi.org/10.1074/jbc.M111.227751

Tremblay P-L, Zhang T, Dar SA, Leang C, Lovley DR (2013) The Rnf complex of *Clostridium ljungdahlii* is a proton-translocating ferredoxin: NAD^+ oxidoreductase essential for autotrophic growth. MBio 4:e00406–e00412

Trumpower BL (ed) (1982) Function of quinones in energy conserving systems. Academic Press

Turrens JF, Boveris A (1980) Generation of superoxide anion by the NADH dehydrogenase of bovine heart mitochondria. Biochem J 191:421–427. https://doi.org/10.1042/bj1910421

Tuz K, Li C, Fang X, Raba DA, Liang P, Minh DDL, Juárez O (2017) Identification of the catalytic ubiquinone-binding Site of *Vibrio cholerae* sodium-dependent NADH dehydrogenase: a novel ubiquinone-binding motif. J Biol Chem 292:3039–3048. https://doi.org/10.1074/jbc.M116.770982

Unden G, Bongaerts J (1997) Alternative respiratory pathways of *Escherichia coli*: energetics and transcriptional regulation in response to electron acceptors. Biochim Biophys Acta 1320:217–234. https://doi.org/10.1016/s0005-2728(97)00034-0

Unemoto T, Hayashi M (1977) Na^+-dependent activation of NADH oxidase in membrane fractions from halophilic *Vibrio alginolyticus* and *V. costicolus*. J Biochem 82:1389–1395

Unemoto T, Hayashi M (1979) NADH:quinone oxidoreductase as a site of Na^+-dependent activation in the respiratory chain of marine *Vibrio alginolyticus*. J Biochem 85:1461–1467

Valko M, Morris H, Cronin MT (2005) Metals, toxicity and oxidative stress. Curr Med Chem 12:1161–1208. https://doi.org/10.2174/0929867053764635

van de Vossenberg JL, Ubbink-Kok T, Elferink MG, Driessen AJ, Konings WN (1995) Ion permeability of the cytoplasmic membrane limits the maximum growth temperature of bacteria and archaea. Mol Microbiol 18:925–932. https://doi.org/10.1111/j.1365-2958.1995.18050925.x

Vella F (1993) Principles of biochemistry by HR Horton, LA Moran, RS Ochs, JD Rawn and KG Scrimgeour. Neil Patterson Publishers, Prentice Hall Inc. Biochem. Educ. 21:165. https://doi.org/10.1016/0307-4412(93)90158-v

Venceslau SS, Matos D, Pereira IAC (2011) EPR characterization of the new Qrc complex from sulfate reducing bacteria and its ability to form a supercomplex with hydrogenase and TpI*c*3. FEBS Lett 585:2177–2181. https://doi.org/10.1016/j.febslet.2011.05.054

Verkhovskaya M, Bloch DA (2013) Energy-converting respiratory complex I: on the way to the molecular mechanism of the proton pump. Int J Biochem Cell Biol 45:491–511. https://doi.org/10.1016/j.biocel.2012.08.024

Verkhovsky MI, Bogachev AV (2010) Sodium-translocating NADH:quinone oxidoreductase as a redox-driven ion pump. Biochim Biophys Acta 1797:738–746. https://doi.org/10.1016/j.bbabio.2009.12.020

Vorburger T, Nedielkov R, Brosig A, Bok E, Schunke E, Steffen W, Mayer S, Götz F, Möller HM, Steuber J (2016) Role of the Na^+-translocating NADH:quinone oxidoreductase in voltage generation and Na^+ extrusion in *Vibrio cholerae*. Biochim Biophys Acta 1857:473–482. https://doi.org/10.1016/j.bbabio.2015.12.010

Weimar WR, Neims AH (1975) Physical and chemical properties of flavins; binding of flavins to protein and conformational effects; biosynthesis of riboflavin. In: Rivlin RS (ed) Riboflavin. Springer, US, Boston, MA, pp 1–47

Wenk CB, Wing BA, Halevy I (2017) Electron carriers in microbial sulfate reduction inferred from experimental and environmental sulfur isotope fractionations. ISME J. https://doi.org/10.1038/ismej.2017.185

Westphal L, Wiechmann A, Baker J, Minton NP, Müller V (2018) The Rnf complex is an energy-coupled transhydrogenase essential to reversibly link cellular NADH and ferredoxin pools in the acetogen *Acetobacterium woodii*. J Bacteriol 200:e00357–18. https://doi.org/10.1128/JB.00357-18

Xu W, Yang S, Zhao J, Su T, Zhao L, Liu J (2014) Improving coenzyme Q_8 production in *Escherichia coli* employing multiple strategies. J Ind Microbiol Biotechnol 41:1297–1303. https://doi.org/10.1007/s10295-014-1458-8

Yoch DC, Carithers RP (1979) Bacterial iron-sulfur proteins. Microbiol Rev 43:384–421

Yoshikawa K, Takadera T, Adachi K, Nishijima M, Sano H (1997) Korormicin, a novel antibiotic specifically active against marine Gram-negative bacteria, produced by a marine bacterium. J Antibiot (Tokyo) 50:949–953. https://doi.org/10.7164/antibiotics.50.949

Yoshikawa K, Nakayama Y, Hayashi M, Unemoto T, Mochida K (1999) Korormicin, an antibiotic specific for Gram-negative marine bacteria, strongly inhibits the respiratory chain-linked Na^+-translocating NADH:quinone reductase from the marine *Vibrio alginolyticus*. J Antibiot (Tokyo) 52:182–185. https://doi.org/10.7164/antibiotics.52.182

Yu H, Wu C-H, Schut GJ, Haja DK, Zhao G, Peters JW, Adams MWW, Li H (2018) Structure of an ancient respiratory system. Cell 173:1636–1649.e16. https://doi.org/10.1016/j.cell.2018.03.071
Zhang L, Yu L, Yu CA (1998) Generation of superoxide anion by succinate-cytochrome *c* reductase from bovine heart mitochondria. J Biol Chem 273:33972–33976

Chapter 11
Inner Membrane Translocases and Insertases

Jozefien De Geyter, Dries Smets, Spyridoula Karamanou and Anastassios Economou

Abstract The inner membrane of Gram-negative bacteria is a ~6 nm thick phospholipid bilayer. It forms a semi-permeable barrier between the cytoplasm and periplasm allowing only regulated export and import of ions, sugar polymers, DNA and proteins. Inner membrane proteins, embedded via hydrophobic transmembrane α-helices, play an essential role in this regulated trafficking: they mediate insertion into the membrane (insertases) or complete crossing of the membrane (translocases) or both. The Gram-negative inner membrane is equipped with a variety of different insertases and translocases. Many of them are specialized, taking care of the export of only a few protein substrates, while others have more general roles. Here, we focus on the three general export/insertion pathways, the secretory (Sec) pathway, YidC and the twin-arginine translocation (TAT) pathway, focusing closely on the *Escherichia coli* (*E. coli*) paradigm. We only briefly mention dedicated export pathways found in different Gram-negative bacteria. The Sec system deals with the majority of exported proteins and functions both as a translocase for secretory proteins and an insertase for membrane proteins. The insertase YidC assists the Sec system or operates independently on membrane protein clients. Sec and YidC, in common with most export pathways, require their protein clients to be in soluble non-folded states to fit through the translocation channels and grooves. The TAT pathway is an exception, as it translocates folded proteins, some loaded with prosthetic groups.

Keywords Protein secretion · Membrane insertion · Inner membrane · Sec · YidC · TAT · Signal peptide · Translocation · Protein folding · Export pathways · Co-translational · Post-translational · Chaperones

---

J. De Geyter · D. Smets · S. Karamanou · A. Economou (✉)
Laboratory of Molecular Bacteriology, Department of Microbiology and Immunology, Rega Institute for Medical Research, KU Leuven – University of Leuven, Herestraat 49, 3000 Leuven, Belgium
e-mail: tassos.economou@kuleuven.be

A. Kuhn (ed.), *Bacterial Cell Walls and Membranes*, Subcellular Biochemistry 92, https://doi.org/10.1007/978-3-030-18768-2_11

List of Abbreviations

APH	Amphipathic helix
BAM	B-barrel assembly machinery
CU	Chaperone-Usher
E. coli	*Escherichia coli*
IRA	Intramolecular regulator of ATPase
MD	Mature domain
MTS	Mature domain targeting signal
NBD	Nucleotide binding domain
OMVs	Outer membrane vesicles
PBD	Preprotein binding domain
PMF	Proton motive force
prl	Protein localization
REMPs	Redox enzyme maturation proteins
RNC	Ribosome-nascent chain complex
SD	Scaffold domain
Sec pathway	Secretory pathway
SP	Signal peptide
SRP	Signal recognition particle
TAM	Translocation and assembly module
TAT	Twin-arginine translocation
TF	Trigger factor
TMH	Transmembrane helix
TXSS	Type X secretion system

Introduction

All proteins are synthesized at cytoplasmic ribosomes. These unfolded amino acyl polymers need to find their correct location in the compartmentalized cell and acquire their native functional state (De Geyter et al. 2016). These processes are very important considering that both protein misfolding (Hartl 2017) and mislocalization (Hung and Link 2011) are deleterious and lead to disease.

In Gram-negative bacteria the inner membrane forms a semi-permeable barrier between the cytoplasm and the periplasm with tight control of the energy-dependent transport of ions and polymers. The outer membrane separates the periplasm from the extracellular environment and is also semi-permeable but with narrow porin channels and larger energized-channels that contact the inner membrane. In *Escherichia coli* (*E. coli*), the premier Gram-negative bacterial model system, ~35% of all proteins are located and function in extra-cytoplasmic locations (De Geyter et al. 2016; Orfanoudaki and Economou 2014). These have one thing in common: they must insert into or cross the inner membrane. Proteins that get inserted are membrane proteins and are collectively called the membranome. Proteins that fully cross the membrane are secretory proteins and collectively called the secretome. To accommodate the structural particularities of the many different proteins and subcellular locations, cells are equipped with several translocation systems and accompanying specialized pathways (De Geyter et al. 2016; Cline 2015; Tsirigotaki et al. 2017). The main translocation pathway in the bacterial inner membrane is the secretory (Sec) pathway. A smaller number of proteins use the YidC or the twin-arginine translocation (TAT) pathways (Fig. 11.1a, left).

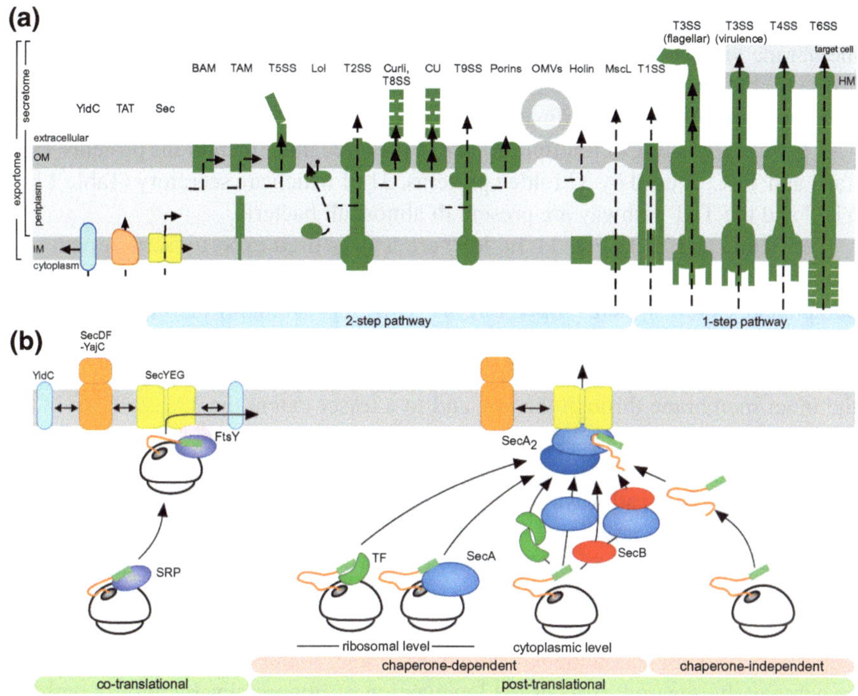

Fig. 11.1 Protein export systems in Gram-negative bacteria and co- and post-translational targeting to the Sec system. **a** Protein export systems in IM and OM for targeting, sorting and translocating the 'exportome'. YidC (light blue); TAT (orange) and general Sec (yellow) pathway form the core export systems in the IM (remaining export systems in dark green). **b** Targeting and sorting to the Sec system for 2-step pathways. Preproteins (orange) with SPs (green) synthesized at cytoplasmic ribosomes are sorted and targeted to Sec complex, composed of SecYEG (yellow) and SecDF-YajC (brown). Co-translational translocation involves YidC (light blue) and is directed by SRP (purple) to its receptor FtsY (pink). Post-translational targeting to SecYEG involves ATPase motor SecA (blue) and chaperones TF (dark green) and SecB (red). Abbreviations: β-barrel assembly machine (BAM); Chaperone-Usher (CU); Host Membrane (HM); Inner Membrane (IM); Localization of lipoprotein (Lol); mechano-sensitive channel (MscL); Outer Membrane (OM); Outer-Membrane Vesicles (OMVs); Signal Peptide (SP); Signal Recognition Particle (SRP); Translocation and Assembly module (TAM); Twin-Arginine Translocation (TAT); Type X Secretion System (TXSS); Trigger Factor (TF). *Figure 11.1 is based on an original from (Tsirigotaki* et al. 2017*)* and updated

In *E. coli* K12 1467 out of a total of 1520 non-cytoplasmic proteins that constitute the exportome (i.e. 97%) use the essential and ubiquitous Sec pathway (Orfanoudaki and Economou 2014; Tsirigotaki et al. 2017) (Table 11.1; http://stepdb.eu). At its core the Sec system comprises a heterotrimeric membrane protein—SecY, SecE and SecG—that forms a hydrophilic pore with a lateral gate in the hydrophobic membrane, allowing passage of proteins across or into the membrane. SecYEG also has a periplasmic plug that controls vectorial crossing. Practically all inner membrane proteins are inserted via the Sec pathway in a co-translational fashion with folding and membrane insertion often going hand in hand. Most secretory proteins use the Sec pathway for crossing the inner membrane completely in a non-folded conformation, typically post-translationally. Up to 70 proteins may be potentially inserted into the membrane via Sec-YidC in a co-translational fashion (Orfanoudaki and Economou 2014) (Table 11.1). A foldase-like activity of YidC assures that protein folding and membrane insertion are coordinated (Dalbey et al. 2014; Kiefer and Kuhn 2018). The TAT pathway, existing of multiple copies of the three inner membrane proteins TatA, TatB and TatC, is used by 32 folded proteins, 31 of which are secretory (Table 11.1). YidC and the TAT pathway are present in almost all bacteria.

Sec, YidC and TAT (Fig. 11.1a, left) are widely used export/insertion pathways found in the inner membrane. Gram-negative bacteria also have specialized, multi-component export systems, operating as 'two-step' (middle) or 'one-step' (right) mechanisms (De Geyter et al. 2016; Tsirigotaki et al. 2017). Two-step pathways take care of proteins that have not reached their final destination yet after crossing the inner membrane through the Sec and to a lesser extent the TAT pathway. One-step pathways take exported proteins directly from the cytoplasm to the extracellular milieu or even to a host cell (De Geyter et al. 2016; Tsirigotaki et al. 2017).

Twelve two-step pathways take periplasmic proteins to the outer membrane and beyond (Fig. 11.1a, middle): (i) the β-barrel assembly machinery (BAM) inserts and folds outer membrane proteins; (ii) BAM together with the translocation and assembly module (TAM) assist Type 5 Secretion System (T5SS) biogenesis; (iii) the T5SS comprises an outer membrane protein with a C-terminal β-barrel membrane pore through which the N-terminal passenger domain can be surface-exposed or completely released; (iv) correct localization of lipoproteins by the Lol pathway in the inner leaflet of the outer membrane or the outer leaflet of the inner membrane; (v) the T2SS takes Sec-translocated, periplasmically folded proteins across the outer membrane; (vi) the Curli pathway or T8SS exports structural components from the periplasm to the extracellular space and mediates their polymerization into a filamentous, outwardly facing amyloid proteins attached to the outer membrane surface. Curli function in cell adhesion, biofilm formation and host invasion; (vii) the Chaperone-Usher (CU) pathway exports four proteins from the periplasm to the extracellular space and mediates their polymerization into pili or fimbriae at the outer membrane, similarly to the curli pathway. *E. coli* K12 harbours 12 putative, cryptic fimbrial CU operons; (viii) the T9SS takes proteins from the periplasm across the outer membrane, either fully releasing them or attaching them to the bacterial surface and has a role in motility and virulence (Lasica et al. 2017); (ix) an outer membrane protein X, F and C porin complex exports YebF, a colicin M immunity

Table 11.1 Subcellular distribution of the *E. coli* K-12 proteome and number of exported proteins using specific export systems. Overview of the number of proteins secreted by the different export systems of the K-12 proteome, including the flagellar type 3 secretion system (fT3SS) that is not discussed here. See http://stepdb.eu for a comprehensive manually curated annotation of K-12 protein topologies, biophysical properties, structures and break down in additional sub-cellular locations and references (Orfanoudaki and Economou 2014; Papanastasiou et al. 2013, 2016) for more details on the subcellular topologies of *E. coli* proteins

(Sub)cellular location	Secretion system/*targeting signal*	No of proteins
Total proteome		4313
Exportome	Multiple	1520
Secretome[a]		548
	Sec Sec SP	498
	TAT TAT SP piggy-back	31 26 5
	fT3SS	13
	Non-classical No SP	6
IM Membranome		972
	Sec *SP or TMH1*	901
	YidC *SP or TMH1*	2[b]
	Sec-YidC	68[c]
	TAT	1

[a]includes proteins destined for the outer membrane of K-12
[b]Only 2 YidC-only membrane proteins have been experimentally confirmed so far
[c]This number has been implied from negative experimental data and should be considered tenuous: it represents "potential YidC substrates" on the basis of aberrant localization (GFP-tagged membrane proteins followed using fluorescence microscopy) of polypeptides included in the group of 70, when YidC is depleted (by placing yidC behind an arabinose-inducible promoter and growing cells in the absence of arabinose, resulting in non-detectable YidC amounts) (Gray et al. 2011). More integral membrane proteins might be YidC-dependent or even YidC-only substrates
Sec SP: cleavable Sec pathway signal peptide; TAT SP: cleavable TAT-pathway signal peptide; piggy-back: polypeptides that do not harbour a TAT-specific SP but that cross the inner membrane bound on another protein that is a bona fide TAT pathway substrate and carries the necessary TAT signal peptide; fT3SS: flagellar type 3 secretion system, a fourth export system in K-12 next to Sec, YidC and TAT that exports 14 structural or regulatory components of the flagellum, a motility organelle (De Geyter et al. 2016); Non-classical: generic terminology used to refer to polypeptides that have been experimentally identified to be secreted but have no known SP or other identifiable export signals to explain their secretion. IM: Inner membrane. TMH1: the first transmembrane helix of a typical inner membrane protein

family protein, from the periplasm across the outer membrane; (x) some proteins leave the periplasm through outer membrane vesicles (OMVs) and can enter target cells upon vesicle fusion; (xi) holins are also implied to be involved in protein secretion (Desvaux 2012). Chitinases are secreted using a holin-like protein ChiW, a peptidoglycan remodeling transpeptidase ChiX and probably other unidentified proteins (Hamilton et al. 2014); and (xii) under stress conditions some cytoplasmic proteins can be taken across the inner (and outer) membrane to the extracellular environment via the mechanosensitive MscL channel (Morra et al. 2018).

Five one-step pathways have been identified (Fig. 11.1a, right): (i) the T1SS takes pathogenesis-related (such as haemolysin α) and nutrient scavenging proteins from the cytoplasm to the extracellular space; (ii) the flagellar T3SS exports 14 structural or regulatory components of the flagellum, a motility organelle; (iii) the homologous virulence T3SS releases toxins into the extracellular environment; (iv) the T4SS injects proteins and nucleic acids into other cells as virulence factors or to promote genetic exchange; and (v) the T6SS injects pathogenic effectors into target cells using a contractile phage-like tail (Leiman et al. 2009).

The Sec Translocase

Inner Membrane Insertion or Crossing via the Sec Pathway

The essential Sec system, ubiquitous in all domains of life, comprises three inner membrane proteins SecY, SecE and SecG that assemble to form a narrow (16–22 Å) (Cranford-Smith and Huber 2018) translocase pore in the membrane allowing unfolded proteins to enter into the membrane co-translationally or to cross it completely post-translationally (Tsirigotaki et al. 2017). In prokaryotes and eukaryotes, the signal recognition particle (SRP) pathway cooperates with SecYEG to accomplish co-translational translocation (Steinberg et al. 2018) (Fig. 11.1b, left). In bacteria, the cytoplasmic ATPase motor SecA also associates with SecYEG at the inner membrane and recognizes substrates with high affinity (Tsirigotaki et al. 2017; Steinberg et al. 2018). SecYEG and A form the translocase holoenzyme of post-translational translocation (Tsirigotaki et al. 2017; Cranford-Smith and Huber 2018; Steinberg et al. 2018; Crane and Randall 2017) (Fig. 11.1b, right).

SecYEG can be assisted by auxiliary components like SecDFYajC and YidC to achieve a higher translocation efficiency (Fig. 11.1b). SecDF is a two-subunit inner membrane protein with 12 transmembrane helices (TMHs) and a large periplasmic domain that enhances and completes post-translational protein translocation through SecYEG (Tsukazaki 2018). SecDF can also have a role in releasing protein substrates after translocation is completed (Tsirigotaki et al. 2017). In *E. coli,* the small protein YajC of unknown function copurifies with SecDF (Crane and Randall 2017). YidC can cooperate with the Sec system to insert proteins into the IM (Tsirigotaki et al. 2017; Steinberg et al. 2018).

All of the Sec-dependent secretome and some members of the membranome are synthesized as preproteins (Tsirigotaki et al. 2017). Preproteins have cleavable N-terminal signal peptides (SP) fused to their mature domains (MD) (Chatzi et al. 2013; von Heijne 1990). The SP has three regions: an N-terminal positively charged region, a central hydrophobic region with helix-forming propensity and a C-terminal hydrophilic region that contains the signal peptidase recognition site (Cranford-Smith and Huber 2018). The MD will form the natively folded functional protein (Tsirigotaki et al. 2017). The SP functions as a sorting and targeting signal in both the co- and post-translational pathways (Tsirigotaki et al. 2017; Cranford-Smith and Huber 2018; Chatzi et al. 2013). In the post-translational pathway, it also serves as an allosteric activator of the Sec translocase (Gouridis et al. 2009). Hydrophobic regions in the MD that are recognized by SecA serve as additional post-translational mature domain targeting signals (MTSs) (Chatzi et al. 2017).

Inner membrane proteins are targeted to the Sec translocase as they are still being translated by the ribosome (Tsirigotaki et al. 2017; Steinberg et al. 2018; Saraogi and Shan 2014). This co-translational targeting is mediated by the SRP that recognizes, on one hand, the SP emerging from the translating ribosome and, on the other, its receptor at the membrane, FtsY (Tsirigotaki et al. 2017; Steinberg et al. 2018; Saraogi and Shan 2014; Kuhn et al. 2017; Akopian et al. 2013) (Fig. 11.1b, left). SRP-FtsY binding brings the ribosome-nascent chain complex (RNC) close to SecYEG (Saraogi et al. 2014; Shen et al. 2012) and consequently co-translational insertion is limited by the translation rate (Steinberg et al. 2018; Saraogi and Shan 2014). Co-translational translocation is energized by the translation process from the ribosome (Tsirigotaki et al. 2017).

Secretory proteins are targeted post-translationally: most of their chain is already synthesized when targeting starts (Tsirigotaki et al. 2017). SecA (cytoplasmic or ribosome-bound), the cytoplasmic export-specific chaperone SecB or the general chaperone trigger factor (TF; cytoplasmic or ribosome-bound) may assist secretory preproteins to become targeted to the Sec translocase or secretory preproteins may find their way from the ribosome to the SecYEG-SecA holoenzyme independently of chaperones (Tsirigotaki et al. 2017) (Fig. 11.1b, right). Post-translational translocation is driven by ATP hydrolysis by SecA and by the proton motive force (PMF) (Tsirigotaki et al. 2017).

After membrane insertion or crossing, the generic signal peptidase or the one specialized for lipoproteins, proteolytically cleave the SP, releasing the MD in the membrane or in the periplasm (Tsirigotaki et al. 2017). Membrane proteins typically acquire their native folded state as they are being inserted in the membrane (Dalbey et al. 2011). The MD of secretory proteins only folds completely after being released into the periplasm, becoming a functional periplasmic resident, soluble or peripherally associated with the inner or outer membrane, or continuing its trafficking through the outer membrane component of the two-step secretion systems (Fig. 11.1a, middle). Secretory proteins that continue their trafficking are assisted by periplasmic chaperones and targeting factors to remain soluble, unfolded and to reach their second export system and final localization (De Geyter et al. 2016; Tsirigotaki et al. 2017).

Many years of research and many structural, biochemical and biophysical studies have given us significant insight into the intermolecular interactions between and conformational changes of Sec pathway components. Many steps of the Sec-mediated translocation mechanism are being gradually better understood and also proofreading mechanisms are being unraveled: only proteins with both a SP and hydrophobic MTSs will be allowed to proceed to secretion (Gouridis et al. 2009; Chatzi et al. 2017; Bieker et al. 1990). Yet, many questions still remain to be elucidated and a deeper understanding of the Sec system will contribute to the production of biopharmaceuticals and the development of antibiotics and vaccines.

Structure of the Sec Translocase

The SecY channel pore covers the complete width of the inner membrane, allowing secretory proteins to cross the hydrophobic lipid bilayer in a protected, hydrophilic environment provided by the pore (Tsirigotaki et al. 2017) (Fig. 11.2b, c). SecY also contains a lateral gate (Fig. 11.2b) that allows inner membrane protein TMHs and SPs that enter from the cytoplasm to slip laterally towards the hydrophobic core of the membrane (Tsirigotaki et al. 2017). In the resting channel, a helical plug seals the SecYEG pore at the periplasmic side (Fig. 11.2b, c), maintaining the membrane permeability barrier (Tsirigotaki et al. 2017).

SecY has 10 TMHs, in two structurally similar quintets, each providing one half of the channel, and allowing a hydrophilic pore between them (Fig. 11.2b, c). The pore is centrally restricted via a ring formed by hydrophobic residues resulting in an hourglass-shape (Tsirigotaki et al. 2017; Crane and Randall 2017). The lipid-facing lateral gate is formed by TMH2,3 and TMH7,8 (Fig. 11.2b) (Tsirigotaki et al. 2017). SecE and SecG associate with SecY on opposite sides (Fig. 11.2b) (Tsirigotaki et al. 2017) and are believed to stabilize SecY in resting conditions and increase the translocation rate (Cranford-Smith and Huber 2018). SecE has two TMHs and one amphipathic helix (APH), SecG consists of two TMHs. A cytoplasmic loop of SecG seals the SecY pore but SecG is non-essential under most conditions (Tsirigotaki et al. 2017). Structural and crosslinking studies have shown both monomeric and oligomeric SecYEGs but a single SecYEG heterotrimer is functional (Kedrov et al. 2011). Ribosomes and SecA (Fig. 11.2c) interact with the cytoplasmic protrusions of SecY and induce long-range conformational changes to the plug, ring and lateral gate resulting in a stabilized open state (Tsirigotaki et al. 2017).

The bacterial SRP is a ribonucleoprotein complex composed of the Ffh GTPase and a 4.5S RNA (Steinberg et al. 2018; Saraogi and Shan 2014; Kuhn et al. 2017) (Fig. 11.2a). The SRP receptor FtsY is also a GTPase and associates with the membrane through interactions with lipids or SecY (Steinberg et al. 2018; Kuhn et al. 2017) (Fig. 11.2a). Ffh and FtsY both have mainly α-helical, three-domain structures (Steinberg et al. 2018; Kuhn et al. 2017). Their N- and G-domains are homologous (Steinberg et al. 2018; Saraogi and Shan 2014; Kuhn et al. 2017) (Fig. 11.2a). Ffh-FtsY complex formation via their NG-domains stimulates GTPase activity (Steinberg

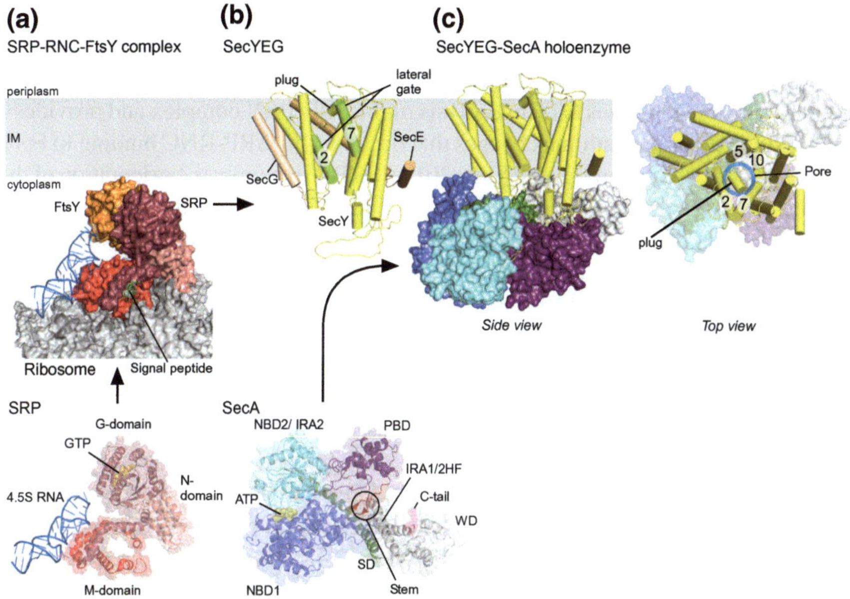

Fig. 11.2 Architecture of the Sec pathway components. **a** Architecture of co-translational targeting components. Top, surface model of SRP (purple)-RNC-FtsY (orange) complex (PDB entry 5GAD). Bottom, SRP (ribbons) composed of the helical N-domain (pink), G-domain (light purple) that binds GTP (yellow spheres); M-domain (red) and hairpin-structured 4.5S RNA (blue) (Tsirigotaki et al. 2017). **b** Architecture of the SecYEG and SecA. Top, side view of SecYEG (PDB entry 3DIN, helices depicted as cylinders) composed of SecY (yellow) and surrounded by SecE (brown) and SecG (light brown). SecY contains a periplasmic plug and lateral gate facing the lipid bilayer (light green) (Tsirigotaki et al. 2017). TMH1 of *E. coli* K12 SecE and SecG proteins are not present in the proteins crystallized here. Bottom, the ATPase motor SecA has a PBD (purple); WD (grey); SD (green); stem (red); C-tail (dark purple); NBD1 (dark blue); IRA1/2HF (grey helix) and ATPase2 motor (IRA2/NBD2, light purple) domain of which NBD1 and IRA2 sandwich a single nucleotide (ATP or ADP, yellow spheres). **c** Architecture of the SecYEG-SecA holoenzyme. Left, side view of the holoenzyme with SecYEG (helices depicted as cylinders) bound to SecA (surface model). Right, top view (from periplasm) of holoenzyme visualizing the pore, formed by hydrophobic residues on TMH 2, 5, 7 and 10. The helical 'plug' seals the pore (blue) (Tsirigotaki et al. 2017). Abbreviations: Adenosine triphosphate (ATP), Adenosine diphosphate (ADP), GTPase domain (G-domain); Guanosine triphosphate (GTP); Intra-molecular regulator of ATPase2 (IRA2); Nucleotide-binding domain2 (NBD2); Intra-molecular regulator of ATPase1 (IRA1); 2-helix finger (2HF); Methionine-rich domain (M-domain); Nucleotide-binding domain1 (NBD1); Ribosome Nascent Chain (RNC); Signal Recognition Particle (SRP); Scaffold domain (SD);; wing domain (WD); the preprotein binding domain (PBD). *Figure 11.2 is based on an original from (Tsirigotaki* et al. 2017*) and adapted*

et al. 2018; Kuhn et al. 2017). The NG-domains of FtsY contain SecY binding sites (Steinberg et al. 2018). The M-domain of Ffh binds the SP and SRP RNA (Steinberg et al. 2018; Saraogi and Shan 2014; Kuhn et al. 2017). The SRP RNA has a hairpin-like structure and regulates GTP hydrolysis of the SRP-FtsY complex and provides a scaffold for the conformational changes that occur during SRP-RNC binding to FtsY (Steinberg et al. 2018; Saraogi and Shan 2014). SRP is dynamic: the orientation of the Ffh NG-domain can change relative to that of the RNA/M-domain, possibly due to the flexible linker connecting the Ffh M- and NG-domains (Saraogi and Shan 2014). The A-domain of FtsY is not essential but contains lipid and SecY interaction sites and prevents fruitless SRP-FtsY interactions in the absence of SecYEG (Steinberg et al. 2018; Kuhn et al. 2017).

The ATPase motor SecA has four domains (Tsirigotaki et al. 2017) (Fig. 11.2b). It contains a nucleotide binding domain (NBD or NBD1) and an intramolecular regulator of ATPase 2 (IRA2 or NBD2) that sandwich ATP. The preprotein binding domain (PBD) contributes to recognition of both SPs and MDs and the C-terminal domain is a core structure with multiple roles: it controls ATPase motor dynamics and hydrolysis, provides surfaces for MD recognition and docks SecA onto SecY via its scaffold helix and its helix-turn-helix intramolecular switch IRA1 (or 2-helix finger). A narrow stem connects the PBD to NBD. The helical C-domain comprises the scaffold domain (SD), IRA1, the wing domain (WD) and a C-terminal tail that folds onto the stem in the form of a pseudo-MD that auto-inhibits preprotein access (Gelis et al. 2007). IRA1 may facilitate nascent chain threading through the SecY pore (Cranford-Smith and Huber 2018). Cytoplasmic SecA is dimeric; when ribosome-bound it is either monomeric or an elongated dimer (Tsirigotaki et al. 2017). SecA binds as an asymmetric dimer with one protomer to SecYEG (Fig. 11.2c) and monomerizes during the translocation process (Gouridis et al. 2009). Interaction of SecA with SecYEG or a secretory preprotein causes conformational changes in SecA and its intrinsically dynamic PBD can sample wide open, open and closed states (Vandenberk 2018, in press). In SecYEG-bound SecA, during nucleotide-driven strokes, large parts of SecA become stabilized and proteolytically protected (Economou and Wickner 1994).

Sorting and Targeting of Preproteins to the Sec Translocase

Sec-routed preproteins are sorted from cytoplasmic resident proteins and targeted to the Sec translocase at the membrane via distinct proteinaceous factors that exercise 'sorting control' on the basis of physicochemical and sequence features at the ribosomal channel exit and via inherent structural features of the SP and MD (Tsirigotaki et al. 2017).

In co-translational insertion, the hydrophobic SP and/or the N-terminal-most highly hydrophobic TMH of inner membrane proteins are recognized by ribosome-bound SRP as they emerge from the ribosome (Tsirigotaki et al. 2017; Steinberg et al. 2018; Saraogi and Shan 2014; Kuhn et al. 2017), thus segregating away these molecules (Fig. 11.3a, step 1). SRP needs to be ribosome-bound to achieve a high affinity interaction with the SP/nascent chain (Saraogi and Shan 2014). Binding of the SP to SRP induces a first conformational change in SRP, facilitating the interaction with FtsY (Saraogi and Shan 2014; Kuhn et al. 2017). FtsY exchanges between cytoplasm and the membrane where it can associate with SecYEG (Tsirigotaki et al. 2017; Steinberg et al. 2018). Only in SecY-associated FtsY the FtsY A-domain moves away and frees the SRP binding site located on its NG-domains (Steinberg et al. 2018; Kuhn et al. 2017). Both SRP and FtsY need to be GTP-bound to achieve RNC targeting. SRP binding to FtsY brings the RNC in close proximity to SecYEG and leads to the formation of an unstable early SRP-RNC-FtsY intermediate complex (Tsirigotaki et al. 2017; Kuhn et al. 2017) (Fig. 11.3a, step 2). SRP-FtsY binding induces GTP hydrolysis and a second set of conformational changes so that a stable closed SRP-RNC-FtsY complex is formed (Steinberg et al. 2018; Saraogi and Shan 2014; Kuhn et al. 2017). In the latter, SRP and FtsY domains get rearranged and hence SRP is displaced from the ribosome and FtsY is displaced from SecY, allowing the translating ribosome to bind the SecY directly (Steinberg et al. 2018; Saraogi and Shan 2014; Kuhn et al. 2017) (Fig. 11.3a, step 3). These events induce SP release from SRP and promote nascent chain entrance into the SecYEG pore (Tsirigotaki et al. 2017; Steinberg et al. 2018; Kuhn et al. 2017) (Fig. 11.3a, step 3). Continuing GTP hydrolysis destabilizes and disassembles SRP-FtsY complexes (Kuhn et al. 2017) (Fig. 11.3a, step 4). The small cytoplasmic pool of FtsY, might interact with the SRP-RNC complexes in the cytoplasm and then bias their association with SecYEG (Steinberg et al. 2018; Saraogi and Shan 2014) (Fig. 11.3a, step 1). A co-translational SRP-independent targeting route of single-spanning membrane proteins involving SecA has also been proposed (Steinberg et al. 2018). Nascent chain capture by SRP at an early stage of translation avoids premature folding or aggregation and assures the delivery of an unfolded protein substrate to the Sec translocase (Tsirigotaki et al. 2017; Steinberg et al. 2018). The SRP pathway is also used to target some inner membrane proteins to the YidC insertase. (Steinberg et al. 2018; Kuhn et al. 2017)

In post-translational translocation, nascent secretory preproteins are already largely synthesized when targeting starts and hence it is more challenging to secure soluble non-folded states that are translocation-competent (Tsirigotaki et al. 2017). Some SPs delay folding of fast folding preproteins and together with chaperone interactions maintain preproteins in soluble non-folded states (Tsirigotaki et al. 2017; Cranford-Smith and Huber 2018). Less hydrophobic SPs that escape SRP surveillance and/or the MD signals of non-folded secretory proteins can be recognized by the ribosome-bound chaperones SecA and TF (Tsirigotaki et al. 2017; Crane and Randall 2017). Ribosome-bound SecA (Huber et al. 2017; Collinson 2017) or TF (Oh et al. 2011) recognize nascent chains co-translationally after ~110 amino acyl residues have been synthesized. However, many secretory preproteins do not fold fast and do not aggregate but maintain a loose non-folded 3D structure (Tsirigotaki et al. 2018). These nascent secretory chains can leave the ribosome unaccompanied and interact in the cytoplasm with SecA, TF, SecB or SecA-SecB complexes prior to targeting to SecYEG or SecYEG-bound SecA (Tsirigotaki et al. 2017; Crane and Randall 2017). The precise role of TF and SecB, both non-essential proteins

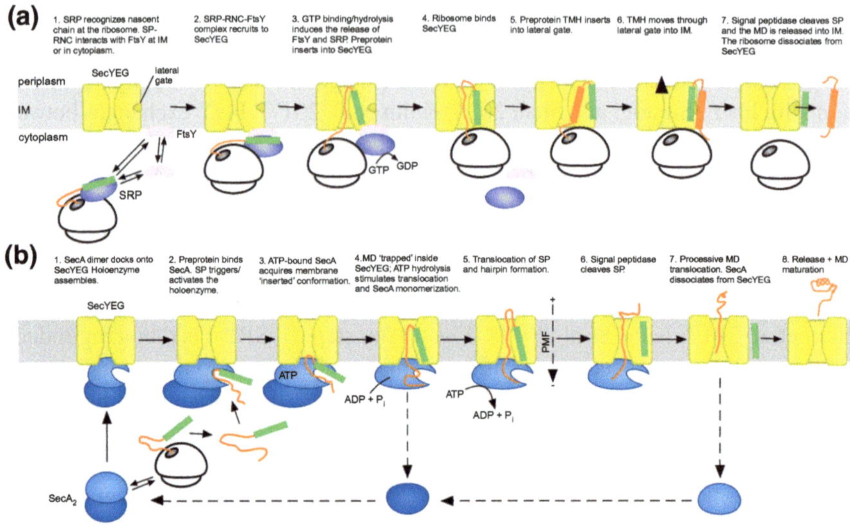

Fig. 11.3 Model of the Sec-dependent co- and post-translational translocation and release. **a** Co-translational targeting, translocation and insertion into the IM of a nascent chain (orange) with SP (green) (Tsirigotaki et al. 2017; Steinberg et al. 2018; Kuhn et al. 2017). The complex is composed of SRP (purple), FtsY (pink) and SecYEG (yellow). **b** Post-translational targeting, translocation and release into the periplasm (Tsirigotaki et al. 2017; Crane and Randall 2017). The complex now requires the ATPase motor SecA (blue–dark blue). The steps of post- and co-translational translocation are explained in more depth in the main text. *Abbreviations* Adenosine triphosphate (ATP); Adenosine diphosphate (ADP); Inner Membrane (IM); Mature Domain (MD); Proton Motive Force (PMF); Signal Peptide (SP); Signal Recognition Particle (SRP); Transmembrane Helix (TMH). *Figure 11.3b is reproduced from (Tsirigotaki* et al. 2017*)*

(Guthrie and Wickner 1990; Sala et al. 2014), remains to be determined since in the absence of the two chaperones, secretory preproteins remain in soluble non-folded translocation-competent states in vitro (Tsirigotaki et al. 2018) and are targeted to SecYEG-SecA via their SP and inherent MD features—MTSs—in vitro and in vivo (Tsirigotaki et al. 2017; Crane and Randall 2017). Even in the absence of the SP, secretory MDs can remain soluble and unfolded and be targeted to SecYEG-SecA, and in the case of protein localization (*prl*) translocases, even become secreted, indicating that the MD has inherent features that keep it in soluble non-folded states and has targeting signals for SecYEG-bound SecA (Tsirigotaki et al. 2017). These MTSs are hydrophobic patches dispersed over the MD that bind to SecA on a groove lying adjacent to the SP binding cleft (Chatzi et al. 2017).

Co-translational Insertion of Inner Membrane Proteins Through the Sec Pathway

Upon proper SRP-mediated sorting and targeting of inner membrane proteins to SecYEG (Fig. 11.3a, steps 1–2), the SP of the delivered RNC binds the lateral gate of SecY (Kuhn et al. 2017) (Fig. 11.3a, steps 3–4). The nascent chain inserts into SecY as a hairpin loop (Kuhn et al. 2017). Periplasmic hydrophilic regions remain unfolded in the hydrophilic pore and get translocated across the membrane while TMHs occupy the lateral gate (Tsirigotaki et al. 2017; Kuhn et al. 2017) (Fig. 11.3a, step 5). Cytoplasmic regions get released from the Sec translocon (Kuhn et al. 2017). TMHs either leave the lateral gate and go into the membrane one by one or they accumulate in the lateral gate and slip bundled to the lipid phase (Tsirigotaki et al. 2017; Kuhn et al. 2017) (Fig. 11.3a, step 6). The SP gets cleaved by the signal peptidase (Fig. 11.3a, step 7). Some single-spanning membrane proteins and membrane proteins with large periplasmic regions are co-translationally inserted by intervening SecAs (Kuhn et al. 2017)

Post-translational Translocation of Secretory Proteins Through the Sec Pathway

The Sec translocase exists of one or two copies of SecYEG with a SecA dimer bound to one SecYEG (Tsirigotaki et al. 2017; Cranford-Smith and Huber 2018) (Fig. 11.3b, step 1). SecA binds SecYEG with only one of its protomers and becomes asymmetric (Gouridis et al. 2009, 2013). Only one SecYEG copy is active and will be used as a pore (Hizlan et al. 2012; Osborne and Rapoport 2007). The SecYEG-bound SecA protomer binds a preprotein with high affinity and has distinct binding sites for the SP and MD (Gouridis et al. 2009, 2013; Chatzi et al. 2017; Sardis et al. 2017) (Fig. 11.3b, step 2). The preprotein reaches SecYEG-SecA independently or a SecA-preprotein complex formed at the ribosome or in the cytoplasm finds its way to SecYEG at the membrane or SecB or TF or other chaperones target the preprotein to SecYEG-bound SecA (Tsirigotaki et al. 2017) (Fig. 11.1b, right). SecB delivers the preprotein to SecA by binding the C-terminal tail of SecA (Tsirigotaki et al. 2017; Crane and Randall 2017). There is no known interaction between TF and SecA and hence it is not known if and how TF delivers preproteins to SecA (Tsirigotaki et al. 2017). Multiple chaperone relay mechanisms may also be active (Tsirigotaki et al. 2017; Cranford-Smith and Huber 2018). Preproteins bind to SecA as bivalent ligands: both the SP and MTSs bind synergistically to different binding sites on SecA (Chatzi et al. 2017; Sardis et al. 2017). SP binding changes the conformation of SecA: the protein gets elongated and looser, transmits these changes to SecYEG and the activation energy of the holoenzyme is decreased (Gouridis et al. 2009, 2013; Taufik et al. 2013; Lill et al. 1990; Karamanou et al. 2007). This important 'triggering' effect on the holoenzyme can be mimicked by *prl* mutations that hence by-pass the SP requirement (Bieker et al. 1990; Gouridis et al. 2013; Smith et al. 2005; Derman et al. 1993). ATP binding tightens the association between SecA C-domain parts and the channel and short preprotein segments insert in the pore (Economou and Wickner 1994; Schiebel et al. 1991) (Fig. 11.3b, step 3). The triggered SecA protomer cannot undergo multiple rounds of ATP hydrolysis. For this the second regulatory protomer, that is not bound to SecYEG, must be present (Gouridis et al. 2013). The preprotein MD becomes 'trapped' in the SecYEG-SecA translocase (Schiebel et al. 1991) and SecA monomerizes (Gouridis et al. 2013; Alami et al. 2007) (Fig. 11.3b, step 4). The SP reaches the lateral gate, with its C-terminus pointing towards the periplasm, and the MD gets threaded along the SecY pore (Hizlan et al. 2012; Li et al. 2016) (Fig. 11.3b, step 5). Shifting of the plug outwards and pore ring dilation facilitate passage of the translocating polypeptide in a loop-like configuration (Li et al. 2016). The loop configuration with the C-terminal SP segment sealing the inside of the channel from the hydrophobic membrane interior may ensure membrane integrity (Li et al. 2016). Cycles of ATP hydrolysis by SecA promote the forward segmental translocation of the MD (Economou and Wickner 1994; Schiebel et al. 1991). The PMF ensures translocation of MD segments whenever they are unbound to SecA (Schiebel et al. 1991). With every round of ATP binding and hydrolysis ~30 amino acids are translocated (Cranford-Smith and Huber 2018). As the MD is completely threaded through the Sec translocase, the SP gets cleaved by the signal peptidase (Auclair et al. 2012) (Fig. 11.3b, step 6) and the SecA-ADP protomer dissociates from SecYEG and dimerizes again in the cytoplasm (Economou and Wickner 1994; Gouridis et al. 2013) (Fig. 11.3b, step 7). The MD is released completely (Fig. 11.3b, step 8) and can adopt its final cell envelope localization and structure via other secretion systems and periplasmic folding and targeting factors (De Geyter et al. 2016; Tsirigotaki et al. 2017).

The YidC Insertase

Membrane Protein Insertion and Assembly by YidC

The insertase YidC is present in all three domains of life and operates next to and—when the Sec system is present—together with the Sec translocase to mediate co-translational insertion and folding of inner membrane proteins (Dalbey et al. 2014). YidC, itself an inner membrane protein, also assists in membrane protein oligomerization (van der Laan et al. 2003; Wickstrom et al. 2011), like for the F_0 domain of the ATP synthase and the cytochrome bo oxidase complex (van der Laan et al. 2003), and has a role in the quality control of membrane proteins (Dalbey et al. 2014; Kiefer and Kuhn 2018; van Bloois et al. 2008). YidC can insert 500 molecules per second (Winterfeld et al. 2009) and for this does not use, unlike the Sec translocase, any energy source like ATP or the PMF (Dalbey et al. 2014; Kuhn and Kiefer 2017). YidC depletion induces the phage shock (van der Laan et al. 2003;

Wang et al. 2010) and Cpx response (Shimohata et al. 2007) pathways, both involved in detecting and coping with membrane-defects, and many chaperones (Wickstrom et al. 2011; Wang et al. 2010; Price et al. 2010; Kihara et al. 1996). YidC plays a role in cell division and cell motility (Wang et al. 2010) as its depletion resulted in cell death after several hours due to defective assembly of energy-transducing membrane complexes (Kuhn and Kiefer 2017). This renders YidC essential for viability and a potential antibiotic target.

Cooperation with the Sec Translocase

In *E. coli* K12 only 2 of the 70 YidC-dependent substrates are identified so far to be inserted in the membrane via a YidC-only mechanism, independent of the Sec translocase (Dalbey et al. 2014; Kiefer and Kuhn 2018) (Table 11.1). It is indeed possible that many more Sec substrates also contact/use YidC during membrane integration but there is currently no hard experimental evidence for this. YidC-only substrates typically have one or two very short transmembrane regions, such that YidC may handle them alone (Dalbey et al. 2014; Kiefer and Kuhn 2018). The energetic gain coming from inserting a hydrophobic TMH into the membrane may be used to overcome the energetic cost of transferring a hydrophilic region across the membrane (Hennon et al. 2015). When the energetic cost is too high, YidC cannot insert the protein substrate (Kuhn and Kiefer 2017; Hennon et al. 2015; Soman et al. 2014). Membrane integration of these more complex proteins is cooperatively catalysed by YidC and the Sec translocase (Kiefer and Kuhn 2018; Soman et al. 2014). In its Sec-dependent mode YidC may facilitate membrane protein integration, help with the assembly and packing of TMHs of multi-spanning membrane proteins to generate their helical bundles and promote the removal of TMHs of inserting membrane proteins from the Sec channel (Dalbey et al. 2014; Hennon et al. 2015). Cooperation with the Sec translocase requires YidC-SecYEG complexes: YidC either directly contacts the Sec translocase through interactions with SecY or SecDFYajC, used as a bridge between SecYEG and YidC (Dalbey et al. 2014). YidC interactions with Sec components are weak and transient (Kiefer and Kuhn 2018). YidC was suggested to contact the lateral gate of SecY thus facilitating substrate exchange/sharing (Kiefer and Kuhn 2018; Hennon et al. 2015; Petriman et al. 2018). Interestingly, YidC-only substrates can also be crosslinked to the Sec translocase, suggesting that even during YidC-only insertion, SecYEG is proximal and may have a 'surveying' role (Dalbey et al. 2014).

YidC Structure

In Gram-negative bacteria, YidC has six TMHs, a large periplasmic domain between the first two and a flexible coiled coil cytoplasmic region between the second and third

TMH (Dalbey et al. 2014; Kiefer and Kuhn 2018; Hennon et al. 2015) (Fig. 11.4a, left). The TMHs form a helical bundle (Kiefer and Kuhn 2018). The five C-terminal TMHs seem important for the insertase function, while the periplasmic domain, having a super-β-sandwich fold, is not required for function (Dalbey et al. 2014; Hennon et al. 2015). The periplasmic domain possibly assists the folding of inserted substrates (Dalbey et al. 2014; Hennon et al. 2015) and interacts with SecF (Xie et al. 2006). The cytoplasmic coiled coil domain contacts SRP, FtsY, SecY and SecE and SecG weakly (Petriman et al. 2018). Gram-positive YidC, like mitochondrial and chloroplast homologues, have five TMHs (Dalbey et al. 2014; Hennon et al. 2015). Low-resolution YidC structures, with and without ribosomes/RNCs, harbour either dimeric or monomer YidC with a hydrophobic slide through which the substrate may enter and exit but the functional unit of YidC seems to be a monomer (Dalbey et al. 2014; Hennon et al. 2015). The high-resolution structure of YidC2 from the Gram-positive *Bacillus halodurans* (Kumazaki et al. 2014a), revealed that YidC does not form a pore like the Sec translocase does but has a hydrophilic groove spanning the inner leaflet of the inner membrane with an opening to the cytoplasm and to the membrane interior (Kumazaki et al. 2014a; Dalbey and Kuhn 2014) (Fig. 11.4a, middle, right and Fig. 11.4b). At the entrance of the hydrophilic groove there is a typical helical hairpin structure (Kumazaki et al. 2014a) (Fig. 11.4a, middle). Inside the hydrophilic groove there is a conserved positively-charged arginine residue (Fig. 11.4b). Hydrophobic residues on TMH3 and TMH5 form a hydrophobic slide (Fig. 11.4a, right and Fig. 11.4b).

Mechanism of YidC-Mediated Membrane Protein Insertion

YidC substrates. YidC substrates are typically small with only one or two TMHs (Kuhn and Kiefer 2017). Studies have shown that YidC assists substrates with rather low hydrophobicity TMHs and increasing the hydrophobicity of the TMH of a YidC substrate allows YidC-independent insertion (Dalbey et al. 2014; Hennon et al. 2015; Ernst et al. 2011). Another YidC determinant is the presence/distribution of charged residues: negatively charged residues either in their TMH or the periplasmic regions (Price and Driessen 2010) or in some cases and unbalanced charge distribution around one or more TMHs (Dalbey et al. 2014; Hennon et al. 2015; Dalbey and Kuhn 2014; Gray et al. 2011). Substrates with positively charged residues are more likely to be inserted by a YidC-Sec mechanism (Zhu et al. 2013) and balanced charge distributions around TMHs result in YidC-independent membrane insertion (Dalbey et al. 2014; Hennon et al. 2015). YidC-only membrane proteins commonly have small (<100 amino acids) periplasmic domains (Wickstrom et al. 2011; Kuhn 1988). When periplasmic domains are larger, substrates become strictly Sec-dependent (Kuhn and Kiefer 2017; Kuhn 1988). The cytoplasmic side of the YidC TMHs is very flexible, probably allowing it to accommodate different-sized substrates (Hennon and Dalbey 2014).

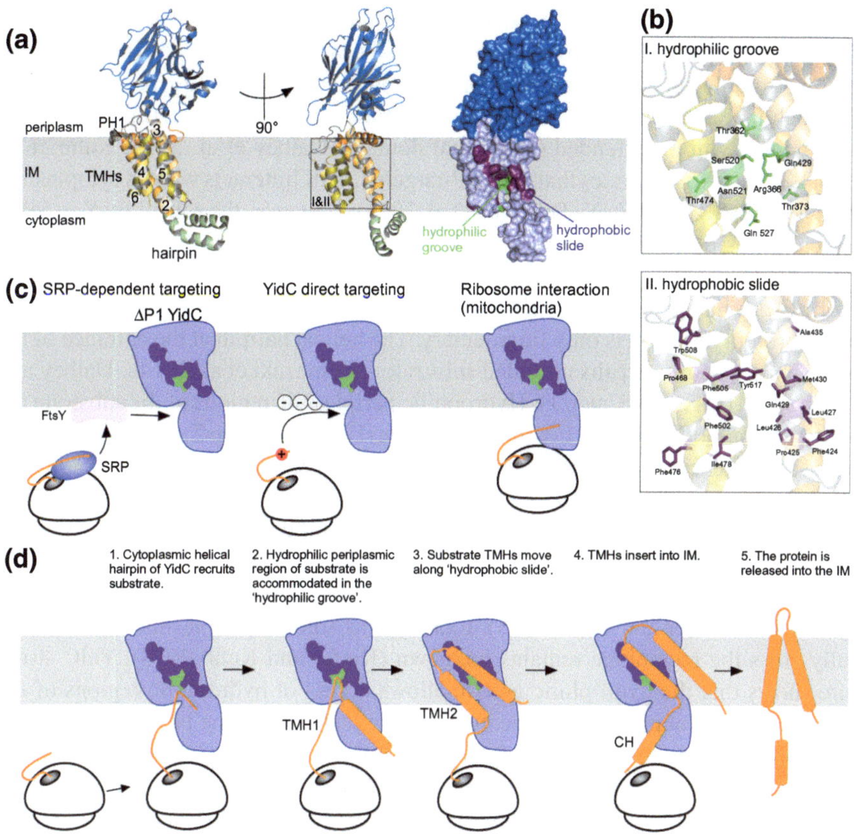

Fig. 11.4 Architecture, targeting and protein export mechanism of the membrane insertase YidC. **a** YidC (PDB entry 3WVF) is composed of a periplasmic domain (P1, blue) and a membrane segment of 1 periplasmic helix (PH1) and 5 TMHs (2–6, shades of orange) (Dalbey et al. 2014; Kuhn et al. 2017; Hennon et al. 2015). Left and middle, side view of YidC in ribbons and rotation of 90 °C is displayed, visualizing the cleft. Right, surface model of the membrane segment with important residues within the hydrophilic groove and hydrophobic slide (Hennon et al. 2015; Kumazaki et al. 2014b). **b** The cleft contains two domains (marked in **a**) (I) the 'hydrophilic groove', dominated by Arg366 (residues in green (Hennon et al. 2015; Kumazaki et al. 2014b)) and (II) the 'hydrophobic slide' (residues in purple (Kumazaki et al. 2014b)). **c** Models of YidC targeting (YidC without its periplasmic domain P1 is displayed). Left, co-translational targeting with SRP-RNC-FtsY complex; middle, nascent chains are targeted directly to the membrane via electrostatic interactions between positively charged groups and negatively charged lipid head groups (Kuhn and Kiefer 2017; Dalbey and Kuhn 2014). Right, direct interaction between the RNC and YidC (Dalbey et al. 2014; Kuhn et al. 2017). **d** Model of YidC-mediated membrane insertion of an exported substrate containing two TMHs and a C-terminal cytoplasmic helix (orange, helices depicted as cylinders (Kiefer and Kuhn 2018; Kuhn et al. 2017; Dalbey and Kuhn 2014)). The substrate shown translocating is modeled after MscL of *E. coli*. MscL is a SP-less, YidC-only substrate and its first hydrophobic TMH is recognized by the SRP. MscL acquires a N_{in}-C_{in} orientation (Facey et al. 2007). Other YidC-substrates may adopt other orientations and can have a different number of TMHs (Dalbey et al. 2014). The steps are explained in more depth in the main text. Abbreviations: Cytoplasmic Helix (CH); Inner Membrane (IM); Ribosome Nascent Chain (RNC); Signal Peptide (SP); Signal Recognition Particle (SRP); Transmembrane Helix (TMH)

Targeting to YidC. YidC-dependent proteins can be SRP-targeted to the membrane (Fig. 11.4c, left), via electrostatic interactions between the substrate and lipid head groups (middle) or via interactions of YidC with the translating ribosome (right), although the latter option is mostly seen in mitochondria and bacteria with a YidC homologue with an extended C-terminal domain (Dalbey et al. 2014; Kuhn et al. 2017). To receive substrates that are SRP-targeted, YidC interacts with its cytoplasmic domain with SRP and FtsY (Petriman et al. 2018). However, the role of FtsY in SRP-targeting to YidC might be more limited than that in SRP-targeting to SecYEG (Kuhn et al. 2017).

YidC-dependent insertion and folding. The current model of YidC-mediated membrane protein insertion is only fragmentary. The helical hairpin at the entrance of the hydrophilic groove recruits unfolded substrates (Kumazaki et al. 2014a; Dalbey and Kuhn 2014) (Fig. 11.4d, step 1). Hydrophilic periplasmic regions of the substrate are at first accommodated in the hydrophilic groove (Fig. 11.4d, step 2) and are eventually released to the trans side of the membrane (Kuhn and Kiefer 2017; Dalbey and Kuhn 2014; Kumazaki et al. 2014b). Negative charges on YidC substrates may be recognized by a conserved arginine residue in the hydrophilic groove of YidC (Kumazaki et al. 2014a; Dalbey and Kuhn 2014) (Fig. 11.4a, b). The size of the hydrophilic groove was suggested to exert a limit on the size of periplasmic domains that YidC can handle (Kiefer and Kuhn 2018). Exactly how soluble periplasmic domains can fully cross the membrane remains unknown (Kiefer and Kuhn 2018). YidC structure shows that the hydrophilic groove allows access of hydrophilic regions of the substrate in the membrane—in the cytoplasmic leaflet—but the periplasmic leaflet, and hence the way to the *trans* side of the membrane, appears 'closed' (Kiefer and Kuhn 2018). Conformational changes upon substrate binding and membrane thinning around YidC, might make the hydrophobic region between the hydrophilic groove and the periplasm smaller, and promote transition of the soluble periplasmic domains to the *trans* side (Dalbey et al. 2014; Kiefer and Kuhn 2018; Hennon et al. 2015). YidC also makes hydrophobic contacts with substrates using its TMHs (Dalbey et al. 2014) (Fig. 11.4d, step 3). The TMHs of substrates are inserted into the membrane via a hydrophobic sliding mechanism (Dalbey et al. 2014; Kiefer and Kuhn 2018; Kumazaki et al. 2014a, b; Dalbey and Kuhn 2014) (Fig. 11.4b, d, step 3). TMH3 and TMH5 of YidC form a substrate-capturing clamp in the membrane. Hydrophobic contacts between TMH3 and 5 of YidC and the TMH of the substrate allow the substrate to move, 'slide', along YidC and eventually be inserted in the membrane (Fig. 11.4b, d, step 4). The final maturation step is the cleavage of the SP by the signal peptidase (Fig. 11.4d, step 5).

The TAT System

Translocation of Folded Proteins

Whereas the Sec system translocates unfolded protein substrates across the inner membrane, other, fully folded proteins—often multimeric or containing co-factors—are translocated across the inner membrane via the TAT pathway (Cline 2015; Berks 2015). Similar to the Sec system, TAT-routed translocation is an active, PMF-requiring process (Berks 2015). The TAT pathway is present in the majority of prokaryotes, all chloroplasts and some mitochondria and plays a crucial role in the biosynthesis of photosynthetic and respiratory energy metabolism components and virulence factors (Berks 2015).

After TAT-mediated translocation, its substrates become soluble periplasmic residents or periplasmic proteins peripherally associated with the inner membrane or are lipoproteins. Four fast folding and co-factor containing lipoproteins, having a TAT-like SP with a lipobox, are exported through TAT in *E. coli* (Shruthi et al. 2010a, b). These are often substrate binding proteins, transporters and enzymes (Shruthi et al. 2010a, b).

The need of some proteins to fold in the cytoplasm comes from the absence of folding and maturation machinery at the *trans* side of the membrane or/and efficient selection of the desired metal cofactor (Cline 2015). Folded proteins are more challenging to translocate due to their larger diameter: this requires a large pore while preventing membrane leakage of smaller molecules and ions (Berks 2015). TAT substrates range in size from 20 to 70 Å and hence the pore diameter also needs to be adjustable. Substrates are recognized with affinities in the low nM to low μM range (Cline 2015; Wojnowska et al. 2018).

When a protein substrate has not acquired its proper final folded state, TAT-mediated transport will not occur (Berks 2015): TAT has a proofreading mechanism assessing the conformational flexibility of substrates and will only export the more rigid, less dynamic ones (Jones et al. 2016). Even adding a short unstructured region to a folded protein blocks TAT export (Jones et al. 2016). TAT senses the degree of substrate stability and folding: export efficiency is anti-correlated to conformational flexibility (Sutherland et al. 2018). However, the multicopper oxidase CueO seems to be an exception (Stolle et al. 2016). CueO is translocated through the TAT pathway in a folded, with secondary structure, yet flexible state. Only in the periplasm copper ions—toxic in the cytoplasm—bind, inducing the final stable structure (Stolle et al. 2016). This is an additional protective mechanism, next to the main function of CueO: oxidation of toxic Cu(I) in the periplasm, before it can enter the cytoplasm (Stolle et al. 2016).

The structure of and interactions between the TAT components are being increasingly better understood thanks to high-resolution structures and crosslinking experiments. The first step of TAT-routing, SP binding, has also been dissected in some detail concerning binding sites, interacting residues and (non-)essential features of both the SP and the receptor (Huang et al. 2017; Huang and Palmer 2017; Lausberg et al. 2012; Ulfig and Freudl 2018; Ulfig et al. 2017). Subsequent steps are more difficult to investigate: the TAT translocase assembles on demand only after substrate binding and then disassembles again after translocation work is done (Cline 2015; Berks 2015) (Fig. 11.5d). Inhibitors could be employed to freeze the translocation process, providing more opportunities for biophysical and structural dissection, but no such inhibitors are known (Bageshwar et al. 2016; Ghosh et al. 2017). TAT is important in the virulence of pathogenic bacteria and might be targeted with new antibacterials (Bageshwar et al. 2016; Ghosh et al. 2017; Ball et al. 2016; Bhuwan et al. 2016). The TAT pathway may also find applications in biotechnology in producing complex, natively folded therapeutics (Ghosh et al. 2017; Branston et al. 2012). TAT SP-driven export of heterologous proteins (Bagherinejad et al. 2016; Matos et al. 2012) and TAT overexpression giving higher export yields (Branston et al. 2012; Bagherinejad et al. 2016) are steps in this direction.

The TAT System Components

Five TAT pathway components have been identified: TatA, TatB, TatC, TatD and TatE (Palmer and Berks 2010). Not all five always co-exist in one particular organism (Palmer and Berks 2012). The *E. coli* TAT pathway is, like that of many Gram-negatives, built of multiple copies of the inner membrane proteins TatA, TatB—two homologous proteins—and TatC, (Hamsanathan and Musser 2018) (Fig. 11.5a). TatA has an L-shape with one N-terminal TMH and one C-terminal APH (Cline 2015; Zhang et al. 2014a) (Fig. 11.5a, left). TatB also has an L-shaped structure consisting of four α-helices connected to each other with flexible linkers (Cline 2015; Zhang et al. 2014b) (Fig. 11.5a, middle). The first N-terminal helix is a TMH forming a rigid structure with the second helix, an APH lying on the lipid bilayer-cytoplasm border (Zhang et al. 2014b) (Fig. 11.5a, middle). Cytoplasmic helices 3 and 4 are more flexible and are suggested to acquire different conformational states (Zhang et al. 2014b) (Fig. 11.5a, middle). TatC, itself inserted into the membrane by SecYEG and YidC (Zhu et al. 2012), has six TMHs that form a concave cavity, resembling a cupped hand (Berks 2015) (Fig. 11.5a, right). Phylogenetic analysis and approaches combining synonymous/non-synonymous distributions and amino acid conservation from alignments revealed 11 conserved residues that have not been studied before and also showed that TatC diversified among different bacterial classes, allowing the TAT system to transport specialized substrates or adapting to specific environmental niches (Simone et al. 2013). TatE, an inner membrane protein belonging to the TatA/TatB family, could be a functional partner of the *E. coli* TAT pathway: it functionally overlaps with TatA and TatB and physically interacts with TatA, TatB and TatC in resting (non-translocating) and active (translocating) conditions (Eimer et al. 2018). TatD, the only water-soluble cytoplasmic protein of the TAT system, is seemingly without any function in TAT-mediated protein translocation (Palmer and Berks 2010). Most Gram-positives have a minimal TAT translocase, comprising TatA—taking over the function of TatB—and TatC (van der Ploeg et al. 2011).

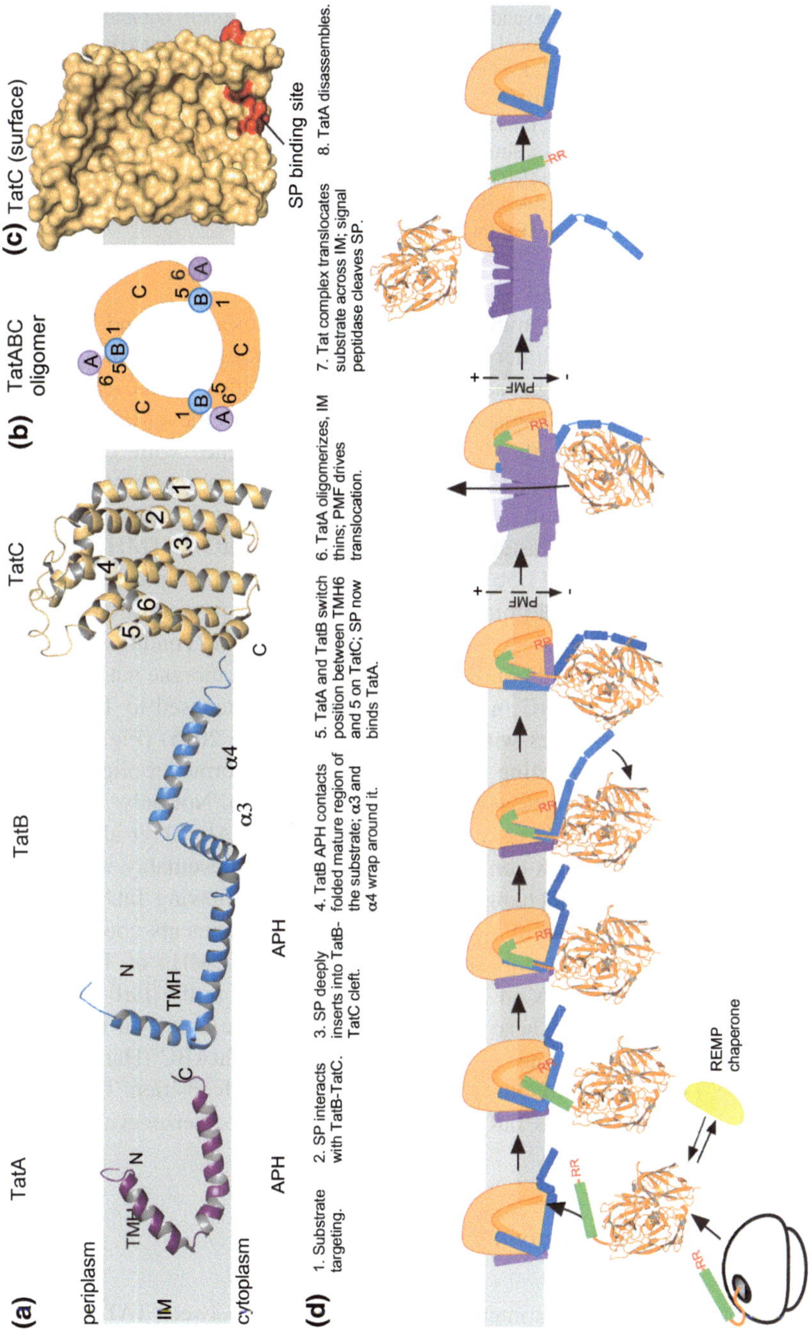
(a)
periplasm
IM
cytoplasm
TatA
TMH
N
C
APH
TatB
N
TMH
APH
α3
α4
TatC
1
2
3
4
5
6
C
(b) TatABC oligomer
(c) TatC (surface)
SP binding site
(d)
1. Substrate targeting.
2. SP interacts with TatB-TatC.
3. SP deeply inserts into TatB-TatC cleft.
4. TatB APH contacts folded mature region of the substrate; α3 and α4 wrap around it.
5. TatA and TatB switch position between TMH6 and 5 on TatC; SP now binds TatA.
6. TatA oligomerizes, IM thins; PMF drives translocation.
7. Tat complex translocates substrate across IM; signal peptidase cleaves SP.
8. TatA disassembles.
RR
REMP chaperone
PMF

◂**Fig. 11.5** Architecture of the TAT export system, TatABC oligomerization and translocation mechanism. **a** Structure of the TAT translocase components: TatA (PDB entry 2MN7, purple), TatA homologue TatB (PDB entry 2MI2, blue) and TatC (PDB entry 4B4A, *A. aeolicus* homologue, light orange) with its 6 TMHs indicated. **b** Model of the TatABC complex oligomerization (Habersetzer et al. 2017). **c** Surface model of TatC displaying the cleft and SP binding site in red (Rollauer et al. 2012). **d** Translocation mechanism of a folded preprotein (Cline 2015; Ulfig and Freudl 2018; Alcock et al. 2016; Frobel et al. 2011) (Tat substrate FtsP; PDB entry 2UXT) through the TAT system (only one copy of TatB and TatC are shown for simplicity). TAT substrates contain a SP (light green) with a N-terminal twin-arginine (RR) motif (red). See text for more details. Abbreviations: Amphipathic Helix (APH); Signal Peptide (SP); Twin-Arginine Transport (TAT); TransMembrane Helix (TMH); Proton Motive Force (PMF); Inner Membrane (IM)

The TAT translocase contains multiple copies (2–8 (Hamsanathan and Musser 2018)) of the TatB-TatC heterodimer (Cline 2015) (Fig. 11.5b). Crosslinking data suggest that TatCs assemble face-to-face, with the concave cavities juxtaposed (Cline 2015) (Fig. 11.5b). TatBs may connect TMH5 of one TatC to the TMH2-TMH4 cleft of the adjacent TatC with the TMH of TatB being aligned with TMH5 of TatC (Cline 2015) (Fig. 11.5b). A similar model of TatBC, was presented based on evolutionary and experimental data, in which TatB is sandwiched between two TatCs and connects the polar cluster TMH5/6 site of one TatC to TMH1 of the other (Alcock et al. 2016) (Fig. 11.5b). Despite the presence of multiple TatB-TatC heterodimers, and hence multiple binding sites, only one or two substrates bind per translocase (Wojnowska et al. 2018). Under non-transporting conditions, TatA often assembles in small homo-oligomers dispersed over the membrane but can also be associated with TatB and TatC in which case TatA is juxtaposed to TatB and the TMH of TatA also associates with TMH5 of TatC (Cline 2015) (Fig. 11.5b). In a substrate-activated, transporting TAT translocase, TatA forms an oligomer at the platform formed by TatB-TatC heterodimers (Cline 2015). Not only TatB but also TatA can bind the TMH5/6 polar cluster site on TatC (Alcock et al. 2016). This suggests that TatB and TatA compete for TatC binding: presumably, substrate binding induces conformational changes displacing TatB and allowing TatA to bind the polar cluster site (Alcock et al. 2016). Crosslinking experiments confirm the binding of the TatB TMH to a polar cluster on one face of TMH5 of TatC and identified a binding site for TatA on TMH6 of TatC, adjacent to the TatB binding site (Habersetzer et al. 2017). These observations updated the model of the non-translocating TAT translocase containing multiple copies of TatABC (Habersetzer et al. 2017) (Fig. 11.5b). Upon SP binding, TatA and TatB switch position: TatB now binds TatC TMH6 and TatA binds TatC TMH5 (see below) (Habersetzer et al. 2017).

The TAT Pathway Mechanism

Despite increasing knowledge on intermolecular interactions between TAT components and with the protein substrate, the mechanism of TAT-mediated protein translocation remains poorly understood. Three models have been proposed but none satisfies all experimental observations (Hamsanathan and Musser 2018). In this section, we divide the TAT pathway in three steps: SP binding, TatA oligomerization and the actual translocation and we describe current knowledge in a model (Fig. 11.5d).

SP binding. TAT substrates have N-terminal SPs that, like the Sec ones, consist of three regions (N-, H- and C-) (Cline 2015; Berks 2015). Some TAT substrates have SPs with extended N regions that may function as timers to tune translocation and folding (Cline 2015). At the N-/H-junction lie the twin-arginines, after which the pathway was named (Cline 2015). Substrates are targeted to the TatB-TatC receptor. TatC binds the twin-arginine motif with its cytosolic N-terminal loop and the loop between TMH2 and 3 (Zoufaly et al. 2012) (Fig. 11.5c) and TatB recognizes the H-region (Cline 2015; Huang et al. 2017) (Fig. 11.5d, step 1–2).

The twin-arginine motif is not essential for TAT translocase recognition: it can be bypassed by increasing H-region hydrophobicity (Huang and Palmer 2017; Ulfig and Freudl 2018). Moreover, replacing the TAT SP by a more hydrophobic Sec SP also results in TAT-routing (Huang and Palmer 2017). Mutations in the twin-arginine binding site of TatC also abolish protein export but this can, again, be alleviated by hydrophobicity-enhancing mutations (Huang and Palmer 2017). TatB mutations can also restore TAT export in the absence of (a) proper SP (binding) (Huang et al. 2017). These mutations suppress the export defect when the twin-arginine motif or the TatC SP binding site is mutated by inducing a conformational change (Huang et al. 2017). They are analogous to *prl* mutations (Bieker et al. 1990) that bring about the same allosteric change to the Sec translocase as the SPs do (Gouridis et al. 2009) and render the Sec translocase SP-independent. In a similar vein, a TatB mutation allowed export of signal-less substrates and substrate-independent TatA oligomerization and hence TAT activation (Huang et al. 2017). These observations indicate two things: (i) TAT SPs—like Sec ones—have two functions: they target substrates to the TatB-TatC receptor and allosterically prime the respective translocase (Gouridis et al. 2013; Huang et al. 2017) and (ii) TAT substrate mature regions must contain some targeting information (Ulfig and Freudl 2018), as determined for the Sec system (Chatzi et al. 2017). TatC mutations that restore export via TAT of defective SPs have also been identified (Lausberg et al. 2012). TatB and TatC mutations even act synergistically, further indicating that TatB and TatC form an allosteric ensemble and cooperate in SP recognition and binding (Lausberg et al. 2012).

The H-region is more important in receptor binding than initially thought (Ulfig et al. 2017). Introducing negatively-charged residues or truncations in the H-region abolishes TAT-routed export (Ulfig et al. 2017). Moreover, increasing its hydrophobicity increases translocation efficiency (Ulfig et al. 2017). Crosslinking experiments provide evidence that the SP contacts the periplasmic N-terminus of TatB and the distal part of the TatC TMH5 (Ulfig et al. 2017) (Fig. 11.5d, step 3). Taken together, these data suggest that the SP inserts deeply into the hydrophobic intramembrane TatB-TatC binding cavity, as seen in the thylakoid TAT system, with the recognition of the twin-arginine motif by the cytoplasmic parts of TatC being a first step in the binding process to prime the system for translocation (Berks 2015; Ulfig et al. 2017) (Fig. 11.5d, step 3). Also, the twin-arginine motif optimizes—together with positively charged amino acids in the C-region—recognition of the less hydrophobic SP by the TAT translocase and avoids TAT substrates being mistakenly targeted to the Sec system (Huang and Palmer 2017).

The early mature region after the SP has also been shown to interact with TatB-TatC (Ulfig and Freudl 2018). More specifically, the cytoplasmic helices of TatB would bind the early mature region and that way capture TAT substrates (Fig. 11.5d, step 4) (Ulfig and Freudl 2018). This is in line with the previous observation of the TatB APH forming a platform for the folded domain of the substrate (Cline 2015) and with the fact that SP-less substrates can be exported through the TAT system (Huang et al. 2017). The highly flexible helix 3 and 4 of TatB could wrap around the folded mature domains of substrates, changing their conformation according to substrate size (Zhang et al. 2014b) (Fig. 11.5d, step 4). That way they can promote binding to the TAT system or they keep the substrate correctly positioned for membrane crossing (Zhang et al. 2014b).

TAT substrates bind reversibly to TatB-TatC in seconds (Whitaker et al. 2012). Certain phospholipids could also facilitate TAT substrate targeting by recognizing them (Sikdar and Doerrler 2010). TatB and TatA switch position, probably induced by SP binding to TatB (Fig. 11.5d, step 5), and this induces a conformational change that turns TatB-TatC heterodimers into TatB and TatC homodimers and brings TatA closer to the TatC cavity where it can polymerize to form the active, translocating TAT complex (Habersetzer et al. 2017) (Fig. 11.5d, step 6). Only after binding to TatB-TatC, SPs contact a TatA monomer (Cline 2015; Frobel et al. 2011) (Fig. 11.5d, step 5). TatA monomers could already be present in the TAT translocase or SP binding helps recruit the first copy to TatB-TatC, serving to nucleate TatA oligomerization (Frobel et al. 2011). TatA-SP binding only happens in the presence of the PMF (Cline 2015; Frobel et al. 2011). Binding to TatB and TatC positions the SP so that it can contact the N-proximal part of the TMH of TatA (Frobel et al. 2011) (Fig. 11.5d, step 5).

Export of many TAT substrates is assisted by chaperones, referred to as redox enzyme maturation proteins (REMPs). Chaperones bind the SP and protect it from degradation, prevent premature (before substrate folding/multimerization/co-factor insertion) translocation and assist targeting to TatB-TatC (Cherak and Turner 2017; Kuzniatsova et al. 2016; Stevens and Paetzel 2012). Some substrates use a 'piggy-back' mechanism in which only one protein has a TAT SP and the second partner protein hitchhikes along (Orfanoudaki and Economou 2014; Rodrigue et al. 1999).

TatA oligomerization. SP binding, the first step in the TAT pathway, triggers TatA recruitment and oligomerization (Fig. 11.5d, step 6). For this, a functional TatB-TatC receptor, bound SP and the PMF are essential (Cline 2015; Alcock et al. 2013; Rose et al. 2013). The TatA TMH forms a new hydrogen bond with TMH4 of TatC, initiating TatA oligomerization and formation of the transport-active translocase (Cline 2015) (Fig. 11.5d, step 6). TatA oligomerization is not precisely understood but the monomers seem to contact each other through the TMH and the APH (Berks 2015; Zhang et al. 2014a). A 'charge zipper' model for TatA oligomerization was proposed: complementary charge patterns on the TatA APH and the unstructured C-tail of TatA would bring the two domains together, resulting in TatA oligomers around a sort of pore (Walther et al. 2013). The physiological relevance of this model has been challenged and revisited based on in vivo assays (Alcock et al. 2017). Those data showed that complementary charges on the TatA APH and C-tail are not necessary for TatA oligomerization, but yet they are important for TAT transport: mutating the charged residues results in stable TatA oligomers that cannot disassemble after protein substrate translocation (Alcock et al. 2017). TatA oligomers containing 8–25 copies (Cline 2015; Alcock et al. 2013) have been visualized using live cell imaging (Alcock et al. 2013; Rose et al. 2013) preferentially forming at cell poles (Rose et al. 2013). This substrate-driven 'on demand' formation of functional TAT translocases in the presence of substrates is supported by the observation that more substrates result in the formation of more TatA complexes (Alcock et al. 2013; Rose et al. 2013). Substrate-dependent accumulation of TatC, independently of the presence of other TAT components and the PMF, was also observed (Rose et al. 2013). Assembling active and functional translocases by TatA oligomerization on demand may serve two purposes: (i) it guarantees inner membrane integrity when no translocation is needed or/and (ii) the TatA oligomerization might be necessary to convert PMF energy into translocation work (Berks 2015).

Actual translocation. The newly assembled translocase translocates substrates to the *trans* side of the membrane slowly (minute range) (Cline 2015; Berks 2015; Whitaker et al. 2012) (Fig. 11.5d, step 7). The SP is cleaved, releasing the substrate, and TatA oligomers dissociate (Fig. 11.5d, step 8). Translocases that export non-folded substrates consist of channels with sealable hydrophilic membrane pores. In contrast, the TAT translocase does not contain a pore, probably because such a pore would have to be too large to allow passage of folded proteins and would not be easy to seal when resting (Berks 2015). TatA oligomers may weaken and thin the lipid bilayer, to facilitate translocation (Cline 2015) (Fig. 11.5d, step 7). How TatA might cause this is unclear but two models have been suggested: (i) the short TatA TMH—placed asymmetrically in the membrane due to association with TatC—does not fully span the width of the membrane causing a hydrophobic-hydrophilic mismatch; and (ii) interactions of TatA APH with the cytoplasmic leaflet of the inner membrane (Cline 2015; Berks 2015). The translocation efficiency of the TAT system is also dependent on the phospholipid composition of the inner membrane, suggesting that phospholipid-protein interactions could contribute to the translocation process (Berks 2015; Sikdar and Doerrler 2010).

Acknowledgements Our research is funded through the: Research Foundation Flanders (FWO) [grant #G.0B49.15 (to SK); grant #G0C6814N RiMembR (to AE); FWO/F.R.S.-FNRS "Excellence of Science-EOS" programme grant #30550343 (to AE)], EU (FP7 KBBE.2013.3.6-02: Synthetic Biology towards applications; #613877 StrepSynth; to AE), RUN (#RUN/16/001 KU Leuven; to AE) and C1 (ZKD4582—C16/18/008 KU Leuven; to SK and AE). JDG is an FWO doctoral fellow.

References

Akopian D et al (2013) Signal recognition particle: an essential protein-targeting machine. Annu Rev Biochem 82:693–721

Alami M et al (2007) Nanodiscs unravel the interaction between the SecYEG channel and its cytosolic partner SecA. EMBO J 26(8):1995–2004

Alcock F et al (2016) Assembling the Tat protein translocase. Elife 5

Alcock F et al (2017) In vivo experiments do not support the charge zipper model for Tat translocase assembly. Elife 6

Alcock F et al (2013) Live cell imaging shows reversible assembly of the TatA component of the twin-arginine protein transport system. Proc Natl Acad Sci U S A 110(38):E3650–E3659

Auclair SM, Bhanu MK, Kendall DA (2012) Signal peptidase I: cleaving the way to mature proteins. Protein Sci 21(1):13–25

Bageshwar UK et al (2016) High throughput screen for escherichia coli twin Arginine translocation (Tat) inhibitors. PLoS ONE 11(2):e0149659

Bagherinejad MR et al (2016) Twin arginine translocation system in secretory expression of recombinant human growth hormone. Res Pharm Sci 11(6):461–469

Ball G et al (2016) Contribution of the twin Arginine translocation system to the exoproteome of Pseudomonas aeruginosa. Sci Rep 6:27675

Berks BC (2015) The twin-arginine protein translocation pathway. Annu Rev Biochem 84:843–864

Bhuwan M et al (2016) Interaction of mycobacterium tuberculosis virulence factor RipA with Chaperone MoxR1 is required for transport through the TAT secretion system. MBio 7(2):e02259

Bieker KL, Phillips GJ, Silhavy TJ (1990) The sec and prl genes of Escherichia coli. J Bioenerg Biomembr 22(3):291–310

Branston SD et al (2012) Investigation of the impact of Tat export pathway enhancement on E. coli culture, protein production and early stage recovery. Biotechnol Bioeng 109(4):983–91

Chatzi KE et al (2013) Breaking on through to the other side: protein export through the bacterial Sec system. Biochem J 449(1):25–37

Chatzi KE et al (2017) Preprotein mature domains contain translocase targeting signals that are essential for secretion. J Cell Biol 216(5):1357–1369

Cherak SJ, Turner RJ (2017) Assembly pathway of a bacterial complex iron sulfur molybdoenzyme. Biomol Concepts 8(3–4):155–167

Cline K (2015) Mechanistic aspects of folded protein transport by the twin arginine translocase (Tat). J Biol Chem 290(27):16530–16538

Collinson I (2017) SecA-a new twist in the tale. J Bacteriol 199(2)

Crane JM, Randall LL (2017) The Sec system: protein export in Escherichia coli. EcoSal Plus 7(2)

Cranford-Smith T, Huber D (2018) The way is the goal: how SecA transports proteins across the cytoplasmic membrane in bacteria. FEMS Microbiol Lett 365(11)

Dalbey RE, Kuhn A (2014) How YidC inserts and folds proteins across a membrane. Nat Struct Mol Biol 21(5):435–436

Dalbey RE, Wang P, Kuhn A (2011) Assembly of bacterial inner membrane proteins. Annu Rev Biochem 80:161–187

Dalbey RE et al (2014) The membrane insertase YidC. Biochim Biophys Acta 1843(8):1489–1496

De Geyter J et al (2016) Protein folding in the cell envelope of Escherichia coli. Nat Microbiol 1(8):16107

Derman AI et al (1993) A signal sequence is not required for protein export in prlA mutants of Escherichia coli. EMBO J 12(3):879–888

Desvaux M (2012) Contribution of holins to protein trafficking: secretion, leakage or lysis? Trends Microbiol 20(6):259–261

Economou A, Wickner W (1994) SecA promotes preprotein translocation by undergoing ATP-driven cycles of membrane insertion and deinsertion. Cell 78(5):835–843

Eimer E et al (2018) Unanticipated functional diversity among the TatA-type components of the Tat protein translocase. Sci Rep 8(1):1326

Ernst S et al (2011) YidC-driven membrane insertion of single fluorescent Pf3 coat proteins. J Mol Biol 412(2):165–175
Facey SJ et al (2007) The mechanosensitive channel protein MscL is targeted by the SRP to the novel YidC membrane insertion pathway of Escherichia coli. J Mol Biol 365(4):995–1004
Frobel J, Rose P, Muller M (2011) Early contacts between substrate proteins and TatA translocase component in twin-arginine translocation. J Biol Chem 286(51):43679–43689
Gelis I et al (2007) Structural basis for signal-sequence recognition by the translocase motor SecA as determined by NMR. Cell 131(4):756–769
Ghosh D et al (2017) Evaluating a new high-throughput twin-Arginine translocase assay in bacteria for therapeutic applications. Curr Microbiol 74(11):1332–1336
Gouridis G et al (2009) Signal peptides are allosteric activators of the protein translocase. Nature 462(7271):363–367
Gouridis G et al (2013) Quaternary dynamics of the SecA motor drive translocase catalysis. Mol Cell 52(5):655–666
Gray AN et al (2011) Unbalanced charge distribution as a determinant for dependence of a subset of Escherichia coli membrane proteins on the membrane insertase YidC. MBio 2(6)
Guthrie B, Wickner W (1990) Trigger factor depletion or overproduction causes defective cell division but does not block protein export. J Bacteriol 172(10):5555–5562
Habersetzer J et al (2017) Substrate-triggered position switching of TatA and TatB during Tat transport in Escherichia coli. Open Biol 7(8)
Hamilton JJ et al (2014) A holin and an endopeptidase are essential for chitinolytic protein secretion in Serratia marcescens. J Cell Biol 207(5):615–626
Hamsanathan S, Musser SM (2018) The Tat protein transport system: intriguing questions and conundrums. FEMS Microbiol Lett 365(12)
Hartl FU (2017) Protein misfolding diseases. Annu Rev Biochem 86:21–26
Hennon SW, Dalbey RE (2014) Cross-linking-based flexibility and proximity relationships between the TM segments of the Escherichia coli YidC. Biochemistry 53(20):3278–3286
Hennon SW et al (2015) YidC/Alb3/Oxa1 Family of Insertases. J Biol Chem 290(24):14866–14874
Hizlan D et al (2012) Structure of the SecY complex unlocked by a preprotein mimic. Cell Rep 1(1):21–28
Huang Q, Palmer T (2017) Signal peptide hydrophobicity modulates interaction with the twin-Arginine translocase. MBio 8(4)
Huang Q et al (2017) A signal sequence suppressor mutant that stabilizes an assembled state of the twin arginine translocase. Proc Natl Acad Sci USA 114(10):E1958–E1967
Huber D et al (2017) SecA cotranslationally interacts with nascent substrate proteins in vivo. J Bacteriol 199(2)
Hung MC, Link W (2011) Protein localization in disease and therapy. J Cell Sci 124(Pt 20):3381–3392
Jones AS et al (2016) Proofreading of substrate structure by the Twin-Arginine translocase is highly dependent on substrate conformational flexibility but surprisingly tolerant of surface charge and hydrophobicity changes. Biochim Biophys Acta 1863(12):3116–3124
Karamanou S et al (2007) Preprotein-controlled catalysis in the helicase motor of SecA. EMBO J 26(12):2904–2914
Kedrov A et al (2011) A single copy of SecYEG is sufficient for preprotein translocation. EMBO J 30(21):4387–4397
Kiefer D, Kuhn A (2018) YidC-mediated membrane insertion. FEMS Microbiol Lett 365(12)
Kihara A, Akiyama Y, Ito K (1996) A protease complex in the Escherichia coli plasma membrane: HflKC (HflA) forms a complex with FtsH (HflB), regulating its proteolytic activity against SecY. EMBO J 15(22):6122–6131
Kuhn A (1988) Alterations in the extracellular domain of M13 procoat protein make its membrane insertion dependent on secA and secY. Eur J Biochem 177(2):267–271
Kuhn A, Koch HG, Dalbey RE (2017) Targeting and insertion of membrane proteins. EcoSal Plus 7(2)

Kuhn A, Kiefer D (2017) Membrane protein insertase YidC in bacteria and archaea. Mol Microbiol 103(4):590–594
Kumazaki K et al (2014a) Structural basis of Sec-independent membrane protein insertion by YidC. Nature 509(7501):516–520
Kumazaki K et al (2014b) Crystal structure of Escherichia coli YidC, a membrane protein chaperone and insertase. Sci Rep 4:7299
Kuzniatsova L, Winstone TM, Turner RJ (2016) Identification of protein-protein interactions between the TatB and TatC subunits of the twin-arginine translocase system and respiratory enzyme specific chaperones. Biochim Biophys Acta 1858(4):767–775
Lasica AM et al (2017) The type IX secretion system (T9SS): highlights and recent insights into its structure and function. Front Cell Infect Microbiol 7:215
Lausberg F et al (2012) Genetic evidence for a tight cooperation of TatB and TatC during productive recognition of twin-arginine (Tat) signal peptides in Escherichia coli. PLoS ONE 7(6):e39867
Leiman PG et al (2009) Type VI secretion apparatus and phage tail-associated protein complexes share a common evolutionary origin. Proc Natl Acad Sci U S A 106(11):4154–4159
Li L et al (2016) Crystal structure of a substrate-engaged SecY protein-translocation channel. Nature 531(7594):395–399
Lill R, Dowhan W, Wickner W (1990) The ATPase activity of SecA is regulated by acidic phospholipids, SecY, and the leader and mature domains of precursor proteins. Cell 60(2):271–280
Matos CF et al (2012) High-yield export of a native heterologous protein to the periplasm by the tat translocation pathway in Escherichia coli. Biotechnol Bioeng 109(10):2533–2542
Morra R et al (2018) Translation stress positively regulates MscL-dependent excretion of cytoplasmic proteins. MBio 9(1)
Oh E et al (2011) Selective ribosome profiling reveals the cotranslational chaperone action of trigger factor in vivo. Cell 147(6):1295–1308
Orfanoudaki G, Economou A (2014) Proteome-wide subcellular topologies of E. coli polypeptides database (STEPdb). Mol Cell Proteomics 13(12):3674–3687
Osborne AR, Rapoport TA (2007) Protein translocation is mediated by oligomers of the SecY complex with one SecY copy forming the channel. Cell 129(1):97–110
Palmer T, Berks BC (2012) The twin-arginine translocation (Tat) protein export pathway. Nat Rev Microbiol 10(7):483–496
PalmerT, Sargent F, Berks BC (2010) The Tat protein export pathway. EcoSal Plus 4(1)
Papanastasiou M et al (2016) Rapid label-free quantitative analysis of the E. coli BL21(DE3) inner membrane proteome. Proteomics 16(1):85–97
Papanastasiou M et al (2013) The Escherichia coli peripheral inner membrane proteome. Mol Cell Proteomics 12(3):599–610
Petriman NA et al (2018) The interaction network of the YidC insertase with the SecYEG translocon, SRP and the SRP receptor FtsY. Sci Rep 8(1):578
Price CE, Driessen AJ (2010) Conserved negative charges in the transmembrane segments of subunit K of the NADH: ubiquinone oxidoreductase determine its dependence on YidC for membrane insertion. J Biol Chem 285(6):3575–3581
Price CE et al (2010) Differential effect of YidC depletion on the membrane proteome of Escherichia coli under aerobic and anaerobic growth conditions. Proteomics 10(18):3235–3247
Rodrigue A et al (1999) Co-translocation of a periplasmic enzyme complex by a hitchhiker mechanism through the bacterial tat pathway. J Biol Chem 274(19):13223–13228
Rollauer SE et al (2012) Structure of the TatC core of the twin-arginine protein transport system. Nature 492(7428):210–214
Rose P et al (2013) Substrate-dependent assembly of the Tat translocase as observed in live Escherichia coli cells. PLoS ONE 8(8):e69488
Sala A, Bordes P, Genevaux P (2014) Multitasking SecB chaperones in bacteria. Front Microbiol 5:666
Saraogi I, Shan SO (2014) Co-translational protein targeting to the bacterial membrane. Biochim Biophys Acta 1843(8):1433–1441

Saraogi I, Akopian D, Shan SO (2014) Regulation of cargo recognition, commitment, and unloading drives cotranslational protein targeting. J Cell Biol 205(5):693–706
Sardis MF et al (2017) Preprotein conformational dynamics drive bivalent translocase docking and secretion. Structure 25(7):1056–1067 e6
Schiebel E et al (1991) Delta mu H+ and ATP function at different steps of the catalytic cycle of preprotein translocase. Cell 64(5):927–939
Shen K et al (2012) Activated GTPase movement on an RNA scaffold drives co-translational protein targeting. Nature 492(7428):271–275
Shimohata N et al (2007) SecY alterations that impair membrane protein folding and generate a membrane stress. J Cell Biol 176(3):307–317
Shruthi H et al (2010a) Twin arginine translocase pathway and fast-folding lipoprotein biosynthesis in E. coli: interesting implications and applications. Mol Biosyst 6(6):999–1007
Shruthi H, Babu MM, Sankaran K (2010b) TAT-pathway-dependent lipoproteins as a niche-based adaptation in prokaryotes. J Mol Evol 70(4):359–370
Sikdar R, Doerrler WT (2010) Inefficient Tat-dependent export of periplasmic amidases in an Escherichia coli strain with mutations in two DedA family genes. J Bacteriol 192(3):807–818
Simone D et al (2013) Diversity and evolution of bacterial twin arginine translocase protein, TatC, reveals a protein secretion system that is evolving to fit its environmental niche. PLoS ONE 8(11):e78742
Smith MA et al (2005) Modeling the effects of prl mutations on the Escherichia coli SecY complex. J Bacteriol 187(18):6454–6465
Soman R et al (2014) Polarity and charge of the periplasmic loop determine the YidC and sec translocase requirement for the M13 procoat lep protein. J Biol Chem 289(2):1023–1032
Steinberg R et al (2018) Co-translational protein targeting in bacteria. FEMS Microbiol Lett 365(11)
Stevens CM, Paetzel M (2012) Purification of a Tat leader peptide by co-expression with its chaperone. Protein Expr Purif 84(1):167–172
Stolle P, Hou B, Bruser T (2016) The Tat substrate CueO is transported in an incomplete folding state. J Biol Chem 291(26):13520–13528
Sutherland GA et al (2018) Probing the quality control mechanism of the Escherichia coli twin-arginine translocase with folding variants of a de novo-designed heme protein. J Biol Chem 293(18):6672–6681
Taufik I et al (2013) Monitoring the activity of single translocons. J Mol Biol 425(22):4145–4153
Tsirigotaki A et al (2018) Long-lived folding intermediates predominate the targeting-competent secretome. Structure 26(5):695–707 e5
Tsirigotaki A et al (2017) Protein export through the bacterial Sec pathway. Nat Rev Microbiol 15(1):21–36
Tsukazaki T (2018) Structure-based working model of SecDF, a proton-driven bacterial protein translocation factor. FEMS Microbiol Lett 365(12)
Ulfig A, Freudl R (2018) The early mature part of bacterial twin-arginine translocation (Tat) precursor proteins contributes to TatBC receptor binding. J Biol Chem 293(19):7281–7299
Ulfig A et al (2017) The h-region of twin-arginine signal peptides supports productive binding of bacterial Tat precursor proteins to the TatBC receptor complex. J Biol Chem 292(26):10865–10882
van Bloois E et al (2008) Detection of cross-links between FtsH, YidC, HflK/C suggests a linked role for these proteins in quality control upon insertion of bacterial inner membrane proteins. FEBS Lett 582(10):1419–1424
van der Laan M et al (2003) A conserved function of YidC in the biogenesis of respiratory chain complexes. Proc Natl Acad Sci USA 100(10):5801–5806
van der Ploeg R et al (2011) Salt sensitivity of minimal twin arginine translocases. J Biol Chem 286(51):43759–43770
von Heijne G (1990) The signal peptide. J Membr Biol 115(3):195–201
Walther TH et al (2013) Folding and self-assembly of the TatA translocation pore based on a charge zipper mechanism. Cell 152(1–2):316–326

Wang P, Kuhn A, Dalbey RE (2010) Global change of gene expression and cell physiology in YidC-depleted Escherichia coli. J Bacteriol 192(8):2193–2209

Whitaker N, Bageshwar UK, Musser SM (2012) Kinetics of precursor interactions with the bacterial Tat translocase detected by real-time FRET. J Biol Chem 287(14):11252–11260

Wickstrom D et al (2011) Characterization of the consequences of YidC depletion on the inner membrane proteome of E. coli using 2D blue native/SDS-PAGE. J Mol Biol 409(2):124–35

Winterfeld S et al (2009) Substrate-induced conformational change of the Escherichia coli membrane insertase YidC. Biochemistry 48(28):6684–6691

Wojnowska M et al (2018) Precursor-Receptor Interactions in the Twin Arginine Protein Transport Pathway Probed with a New Receptor Complex Preparation. Biochemistry 57(10):1663–1671

Xie K et al (2006) Different regions of the nonconserved large periplasmic domain of Escherichia coli YidC are involved in the SecF interaction and membrane insertase activity. Biochemistry 45(44):13401–13408

Zhang Y et al (2014a) Structural basis for TatA oligomerization: an NMR study of Escherichia coli TatA dimeric structure. PLoS ONE 9(8):e103157

Zhang Y et al (2014b) Solution structure of the TatB component of the twin-arginine translocation system. Biochim Biophys Acta 1838(7):1881–1888

Zhu L et al (2012) Both YidC and SecYEG are required for translocation of the periplasmic loops 1 and 2 of the multispanning membrane protein TatC. J Mol Biol 424(5):354–367

Zhu L et al (2013) Charge composition features of model single-span membrane proteins that determine selection of YidC and SecYEG translocase pathways in Escherichia coli. J Biol Chem 288(11):7704–7716

Zoufaly S et al (2012) Mapping precursor-binding site on TatC subunit of twin arginine-specific protein translocase by site-specific photo cross-linking. J Biol Chem 287(16):13430–13441

Part IV
Pili

Chapter 12
The Biosynthesis and Structures of Bacterial Pili

Magdalena Lukaszczyk, Brajabandhu Pradhan and Han Remaut

Abstract To interact with the external environments, bacteria often display long proteinaceous appendages on their cell surface, called pili or fimbriae. These non-flagellar thread-like structures are polymers composed of covalently or non-covalently interacting repeated pilin subunits. Distinct pilus classes can be identified on basis of their assembly pathways, including chaperone-usher pili, type V pili, type IV pili, curli and fap fibers, conjugative and type IV secretion pili, as well as sortase-mediated pili. Pili play versatile roles in bacterial physiology, and can be involved in adhesion and host cell invasion, DNA and protein secretion and uptake, biofilm formation, cell motility and more. Recent advances in structure determination of components involved in the various pilus systems has enabled a better molecular understanding of their mechanisms of assembly and function. In this chapter we describe the diversity in structure, biogenesis and function of the different pilus systems found in Gram-positive and Gram-negative bacteria, and review their potential as anti-microbial targets.

Keywords Pili · Fimbriae · Chaperone-usher · Curli · Sortase · Secretion · Adhesion · Biofilm · Conjugation

Introduction

Bacterial cells are frequently decorated with non-flagellar proteinaceous cell surface appendages, referred to as pili or fimbriae. The appendages usually have low nanometer scale width, but can be multiple microns in length, often exceeding the diameter of the producing bacterium. The structures are made up of many hundreds or thousands

---

Magdalena Lukaszczyk, Brajabandhu Pradhan—Authors contributed equally.

M. Lukaszczyk · B. Pradhan · H. Remaut (✉)
Structural Biology Brussels, Vrije Universiteit Brussel, Pleinlaan 2, 1050 Brussels, Belgium
e-mail: han.remaut@vub.be

Structural and Molecular Microbiology, Structural Biology Research Center, VIB, Pleinlaan 2, 1050 Brussels, Belgium

A. Kuhn (ed.), *Bacterial Cell Walls and Membranes*, Subcellular Biochemistry 92, https://doi.org/10.1007/978-3-030-18768-2_12

of pilus subunits, which are covalently or non-covalently associated depending on the pilus system. The function of bacterial pili is varied. The vast majority is implicated in adherence and/or multicellular behaviour. Pili frequently mediate adherence and/or invasion (in)to eukaryotic host cells, but can also be implication in biofilm formation through pilus self-association, binding of neighbouring cells or giving shape to the extracellular matrix. In other systems, though, the pili serve a role as a hollow conduit or a scaffolding structure for the secretion or uptake of proteins and nucleic acids, or in rare instances extracellular electron transport. In this review we summarize the current molecular and structural understanding of the function, build-up and assembly of the major pilus systems found in Gram-positive and Gram-negative bacteria. For each system, we provide a review of the main architecture of the pili and its constituent components, as well as the prevailing mechanistic understanding of the assembly pathways. Since pili are frequently first line virulence factors, a great interest has been gathered for their chemical inhibition as a means towards the development of non-antibiotic, virulence targeted antibacterials (Steadman et al. 2014; Ruer et al. 2015). For some systems, like the type 1 fimbriae implicated in urinary tract infections (UTIs), significant progress has been made, to the point that anti-adhesive compounds are undergoing clinical trials and may reach the market in coming years. For many more systems, the molecular understanding is now such that selective inhibitory compounds can be sought and are being identified. Coming years will show if more of these pathways can be targeted in a clinical setting.

The Chaperone-Usher Pilus System

Arguably the most abundant pilus assembly system in Gram-negative bacteria is the chaperone-usher (CU) pathway. It is a conserved protein secretion-assembly system found in the Gram-Negative classes α-, β-, γ- and δ-proteobacteria, where it is primarily associated with human and animal pathogenic genera including *Escherichia*, *Shigella*, *Proteus*, *Klebsiella*, *Salmonella*, *Pseudomonas, Yersinia* and many others (Sauer et al. 2004). CU fimbriae are linear non-covalent multi-subunit polymers that require two accessory proteins for their assembly and translocation to the cell surface: a periplasmic chaperone and an outer membrane usher. The chaperone stabilizes fimbrial subunits in the periplasm and targets them to the usher, a pilus assembly platform in the outer membrane (OM) that facilitates subunit polymerization and transport to the cell surface (Thanassi et al. 1998).

Pili assembled by the chaperone-usher pathway are prime virulence factors of proteobacteria. They contribute to the establishment and persistence of the infection by mediating host- and tissue-specific adherence, and can play a role in the evasion of host defence mechanisms by contributing to biofilm formation or inducing host cell invasion (Sauer et al. 2004). Chaperone-usher pili are associated with a wide range of diseases, such as urinary and gastrointestinal tract infections, meningitis and sepsis (Proft and Baker 2009). Amongst the earliest and most comprehensively studied chaperone-usher systems are type 1 and P fimbriae produced by uropathogenic *E.*

coli (UPEC) (Hultgren et al. 1993). The type 1 fimbrial adhesin FimH binds to the D-mannosylated receptors of the human and animal bladder, whereas the P pilus adhesin PapG binds galabiose-containing glycosphingolipids in the kidney epithelium. These two systems formed the basis of extensive structural and molecular studies, which together with those on the *Yersinia pestis* capsular antigen Caf, have revealed the canonical principles of CU biogenesis (Sauer et al. 2004).

Morphology and Structure of CU Pili

CU pili are encoded in gene clusters comprising a cognate chaperone and usher, and up to seven different pilus subunits of approximately 10–30 kDa (Sauer et al. 2004; Nuccio and Bäumler 2007). The assembled pili are homo- or heteropolymers of hundreds to thousands of pilus subunits organized into linear single-start filaments. These filaments can undergo additional quaternary condensation to form rigid, helically wound rods or are found as long flexible filaments that often collapse into a dense capsular mass on the bacterial cell surface. Archetypal examples of rod-forming pili are type 1 and P pili of *E. coli*. These are monoadhesive structures, capped by a single copy of an adhesive subunit. In this review, the general description of pilus assembly by chaperone-usher pathway will be based primarily on the type 1 pilus, encoded by the *fim* operon. The type 1 pilus consists of a long, rigid and helical pilus rod composed of several thousand copies of the major pilus subunit FimA (Fig. 12.1) and is terminated with a short flexible tip fibrillum built of the two adaptor subunits FimG and FimF and the adhesin FimH, which is located in a single copy at the distal tip of the pilus. Consecutive subunits interact through non-covalent contacts. Each pilin subunit is characterized by an incomplete immunoglobulin-like fold, lacking its C-terminal β-strand, and by the presence of an unstructured N-terminal extension (Nte) of 10–20 amino acids (Fig. 12.1). In the mature pilus, the Nte of one subunit complements the incomplete Ig-fold of the consecutive subunit in a mechanism called donor strand exchange (DSE) (Choudhury et al. 1999; Sauer et al. 1999) (Fig. 12.1d). Although non-covalent, these fold complementation interactions between pilus subunits and their complementary Nte peptides have extremely high activation barriers for dissociation, displaying extrapolated dissociation half-lives of 10^9–10^{11} years (Puorger et al. 2008, 2011). The long-lived fold complementation interactions effectively protect the pili from loss-of-function by breakage at one of the several hundred DSE contacts. In addition, the pilus rods are dynamic structures characterized by remarkable spring-like properties (Fällman et al. 2005). Force spectroscopy and EM imaging have shown reversible uncoiling of the helical structure in the pilus rod. The recent atomic cryoEM models of both type 1 and P pilus provide the details of the extensive subunit-subunit interaction network within the rod (Hospenthal et al. 2016, 2017; Spaulding et al. 2018). In their coiled state, both pili are right-handed superhelical structures, where the Nte of one pilus subunit complements the hydrophobic groove of the adjacent subunit. The type 1 pilus rod has a diameter of ~70 Å with a ~14 Å wide central hollow lumen, whereas

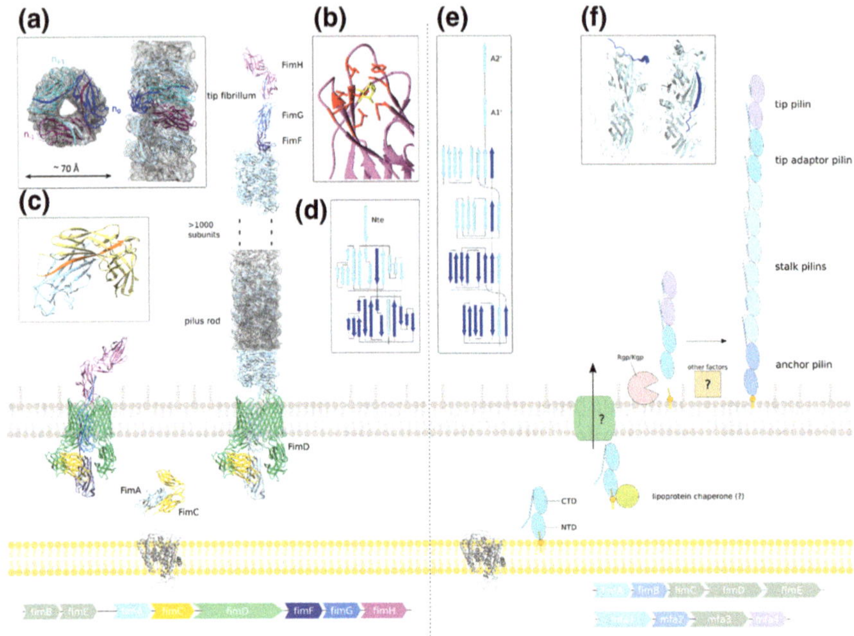

Fig. 12.1 The pilus architecture and schematic assembly pathways for chaperone-usher pilus (left) and type V pilus systems (right). Fimbrial subunits are shown in blue, tip adhesins/pilins in pink, transmembrane pores in green, chaperones in yellow. At the bottom representative operons coding for both systems are shown. **Left**: Type 1 pilus biogenesis as model chaperone-usher assembly pathway. The chaperone FimC (yellow) binds the fimbrial subunit FimA (light blue) (PDB entry: 4DWH) in the periplasm. The pilus tip fibrillum complex composed of FimC:FimF:FimG:FimH traverses the outer membrane usher FimD (PDB entry: 4j3O). Type 1 pilus is composed of a pilus rod (EMDB entry: EMD-7342, PDB entry: 6c53) built of over 1000 FimA subunits and a tip fibrillum (FimF:FimG:FimH) (PDB entry: 3JWN). **a** Cryo-EM map of type 1 pilus rod (EMDB entry: EMD-7342, PDB entry: 6c53) with the neighbouring subunits n_{-1}, n_0, n_{+1}. **b** Structure of FimH binding pocket interacting with α-D-mannose (PDB entry: 1KLF). The D-mannose (yellow) and the mannose-interacting residues (Phe1, Asn46, Asp47, Asp54, Gln133, Asn135, Asp140 and Phe142) (red) are shown in ball and stick model. **c** Donor strand complementation (DSC) of the FimA (blue) pilus subunit by the G1 strand (in orange) of the chaperone FimC (yellow) (PDB entry: 4DWH). G1 strand of the FimC is located parallel to the strand F of FimA during DSC. **d** A topological diagram of donor strand exchange (DSE) of adjacent subunits of type 1 pilus. The incomplete Ig-fold of one subunit (light blue) is complemented by an Nte of the previous subunit (dark blue). **Right**: The proposed pathway scheme of type V pilus biogenesis. **e** A topological diagram of proposed strand-exchange mechanism of adjacent subunits of type V pilus. The hydrophobic groove exposed along both NTD and CTD of one subunit (light blue) is complemented by the A1′ and A2′ strands of the consecutive subunit. **f** Left: "open" conformation of anchor pilin BovFim4b (PDB entry: 5CAG). Right: "closed" conformation of FimA4 (PDB entry: 4Q98). The conserved C-terminal appendage composed of A1′ and additional disordered region is shown in dark blue. In the "open" conformation this appendage is extended, while in the "closed" conformation it folds back to the C-terminal domain

the P pilus has a diameter of ~81 Å with a ~21 Å lumen. In both structures the pilin subunits form a continuous ascending path, with the N-terminal portion of Ntes facing towards the pilus exterior and the C-terminal part towards the lumen. Both pili have a similar helical pitch of ~25 Å, though the type 1 structure comprises 3.13 FimA pilin subunits per turn, compared to 3.28 PapA subunits in the more tightly packed P pilus. Subsequently, the axial rise of type 1 pilus is slightly higher (8 Å for type 1 and 7.7 Å for P pilus). In the helical rod, each PapA subunit interacts with five preceding and five succeeding subunits (Hospenthal et al. 2016), whereas each FimA subunit interacts with four preceding and four succeeding subunits (Spaulding et al. 2018); (Hospenthal et al. 2017). The overall interaction network responsible for maintaining the helical quaternary structure involves mostly weak hydrophilic contacts (Hospenthal et al. 2016). The combination of both strong DSE interaction and weak hydrophilic forces explains the ability of pilus rod to uncoil elastically when the shear force is applied, without breaking apart. At the distal tip, FimF and FimG provide a short flexible linker between the rigid FimA pilus rod and the FimH adhesin. FimH lacks an N-terminal extension. Instead, it has a two-domain structure with a C-terminal pilin domain and a full N-terminal β-sandwich domain that carries the adhesive function (Choudhury et al. 1999). The pilin domain interacts with the penultimate pilus subunit (FimG) by DSE, whereas the lectin domain mediates binding to the host receptors (Fig. 12.1). In the P pilus, the PapA pilus rod is separated from the PapG tip adhesin by a longer flexible fibrillum made up of the linker subunits PapK, PapE and PapF (Rose et al. 2008). Most chaperone-usher pili are monoadhesive, displaying rigid pili with tip-localized two-domain adhesins similar to the type 1 pilus (Nuccio and Bäumler 2007). The adhesins' lectin domains share an elongated β-jelly-roll topology. However, the structures of their ligand binding pocket may differ remarkably (Moonens and Remaut 2017). For example, FimH adhesin has a deep, negatively charged binding pocket which is located at the tip of the lectin domain (Hung et al. 2002) whereas the PapG adhesin has a shallow cavity localized on the side of the lectin domain (Dodson et al. 2001).

In polyadhesive CU pili, subunits polymerize through DSE-mediated subunit contacts, but lack the coil-forming quaternary interaction seen in the pilus rods of monoadhesive pili. They form flexible filaments of one or more types of a major structural subunit, which each contain one to two independent receptor binding sites (Zav'yalov 2013). In such a way, the entire pilus structure is directly involved in the adhesion. Examples include *E. coli* F4 (Fae) and Afa/Dr fimbriae (Moonens et al. 2015; Keller et al. 2002; Anderson et al. 2004), or the *Y. pestis* F1 (Caf) or pH6 (Psa) antigens (Zavialov et al. 2003; Bao et al. 2013). Although the individual subunits have low micromolar to millimolar affinity for their glycan or protein ligands, their high valency results in high binding avidity (Moonens and Remaut 2017).

Biogenesis of CU Pili

Chaperone-usher pili are encoded in gene clusters that hold both the structural subunits as well as the assembly machinery, consisting of a pilus-specific chaperone and usher protein. The individual pilin subunits are translocated across the inner membrane via the SecYEG translocon, after which chaperone and usher are sufficient for the folding and stabilization of pilus subunits, their ordered assembly into pili, and translocation to the cell surface (Sauer et al. 2004). The secretion-assembly pathway is here described based on the *fim* operon. CU pilus subunits do not stably fold in the periplasm as a result of an incomplete immunoglobulin (Ig)-like fold that lacks seventh, C-terminal β-strand. In the periplasm, initial subunit folding is catalysed by the chaperone (FimC) (Vetsch et al. 2004; Bann et al. 2004). Prior to incorporation into the pilus, the missing C-terminal β-strand in the incomplete Ig-like fold of pilus subunits is complemented by an extended β strand (G1) of chaperone (Fig. 12.1c), an interaction referred to as "donor strand complementation" (DSC) (Choudhury et al. 1999; Sauer et al. 1999). However, the G1 strand of the chaperone lies parallel to the subunit strand F during DSC, rather than the antiparallel pairing in a canonical Ig-fold and seen for the Nte during the DSE interaction. As a result, the chaperone traps pilus subunits into a more loosely packed, higher energy folding intermediate compared to the subunit–Nte interaction (Sauer et al. 2002; Zavialov et al. 2005). The chaperone:subunit complexes are recruited to the outer membrane usher (FimD), which acts as a platform for pilus polymerization and OM translocation. The usher is a five domain protein composed of a 24-stranded β-barrel that forms the pilus translocation channel, a plug domain that occludes the pore in the resting state, two C-terminal (CTD1 and CTD2) and an N-terminal domain (NTD) that reside in the periplasm and forms the chaperone:subunit binding sites (Remaut et al. 2008; Phan et al. 2011). The usher NTD domain recruits the chaperone:adhesin complex and this interaction is mostly mediated by the chaperone (Nishiyama et al. 2005). The subunits are sequentially incorporated at the base of the growing pilus, starting from the tip adhesin. In type 1 and P pilus assembly, the affinity of the usher (FimD/PapC) NTD is highest for the chaperone:adhesin complex (FimC:FimH/PapD:PapG) and decreases for later subunits (Dodson et al. 1993). Together with the differential affinities of the usher for different subunit:chaperone complexes, favourable DSE kinetics of cognate versus non-cognate subunit–Nte interactions direct the order of pilus assembly (Nishiyama et al. 2003; Rose et al. 2008). Binding of the chaperone:adhesin complex primes the usher for pilus assembly (Nishiyama et al. 2008). The pre-initiation step was recently captured for in a crystal structure of the P pilus usher PapC bound to the PapD:PapG chaperone:adhesin complex (Omattage et al. 2018). In this pre-initiation complex, the chaperone:subunit is bound to the usher NTD, whilst the usher is still in a locked conformation with the plug domain residing inside the lumen of the β-barrel. A previous crystal structure of the FimD usher bound to the FimC:FimH chaperone:adhesin complex capture the usher in its primed conformation and gave exquisite insight into the mechanism of pilus assembly at the usher (Phan et al. 2011) (Fig. 12.1). In this structure, the lectin domain of FimH displaces the usher plug domain and traverses the lumen of the usher β-barrel, whilst its C-terminal pilin domain remains bound to the FimC chaperone, itself bound to the usher by the CTD1 and CTD2 domains. This primed usher configuration leaves the NTD, the primary chaperone:subunit recruitment platform of the usher, accessible to bind the next chaperone:subunit (FimC:FimG). When the new chaperone:subunit complex binds the NTD, the subunit is oriented such that its N-terminal extension is ideally positioned to undergo DSE with the subunit at the base of the growing fiber and bound at the CTDs. During this DSE reaction, the Nte of the incoming subunit competes out the donor β-strand of the chaperone in the chaperone:subunit at the CTDs (Remaut et al. 2006; Phan et al. 2011). DSE is a concerted zip-in-zip-out process that initiates at the so-called P5 pocket, a hydrophobic pocket accepting the incoming Nte and not occupied by the chaperone donor strand (Remaut et al. 2006). At the end of DSE process the chaperone dissociates from the now penultimate subunit and the usher CTDs, after which the newly added chaperone:subunit complex shifts from the NTD to the now liberated binding site at the CTD. This makes the NTD accessible for a new recruitment round and translocates the pilus outward with one subunit step. Consecutive iterations of the recruitment, DSE and translocation steps result in the stepwise build-up of the pili. At all times, the pilus is tethered to the usher via the last incorporated chaperone:subunit complex. In type 1 and P pili, the length of individual pili is rather uniform and depends on the incorporation of a dedicated terminator subunit (FimI and PapH, respectively) (Verger et al. 2006; Bečárová 2015). These terminator subunits show occluded P5 pockets, so that they cannot undergo DSE and remain anchored in the usher. In many operons, however, there is no indication for a terminator subunit and pilus growth is likely controlled by protein expression and the availability of free subunits.

CU Pili as Antibacterial Targets

Because of their prime importance in bacterial virulence, chaperone-usher pili have raised considerable interest as novel target for vaccine or antibacterial drug development (Steadman et al. 2014). Let us consider the example of uropathogenic *E. coli*, where type 1 pili are strongly associated with symptomatic bladder infections (cystitis) by mediating strong attachment and invasion of the superficial umbrella cells in the urothelium, resulting in tissue damage and inflammation (Mulvey et al. 1998). S and P pili are associated with ascending UTIs by the binding of sialosyl oligosaccharides and globoside receptors in the kidney epithelium, respectively, leading to pyelonephritis (Roberts et al. 1994; Korhonen et al. 1986). More recently F9 fimbriae were shown important for inflammation-associated adherence in ongoing and chronic bladder infections (Conover et al. 2016), and the type 1 and F17-like pili were associated with the establishment of a gut reservoir of uropathogenic *E. coli* (Spaulding et al. 2017). Strains lacking these pili are severely compromised in the initiation, persistence and recurrence of UTIs. Indeed, a promising approach has been proposed to use anti-adhesive agents that interfere with the bacterial adherence to the host tissue (Ruer et al. 2015). As such drugs would target bacterial virulence factor rather than being bactericidal, the spread of antibiotic resistance mechanisms is thought less likely to occur. Competitive inhibitors of FimH adhesin in the form of α-D-mannose derivatives called mannosides, are highly efficacious in murine models of UTI (Cusumano et al. 2011; Spaulding et al. 2017). Mannosides reach low nanomolar affinities and can be formulated as orally available anti-adhesive compounds that successfully treat UPEC UTI. Anti-adhesive receptor-analogues have also been reported for the P and F9 pili (Ohlsson et al. 2002; Kalas et al. 2018). An alternative to competitive inhibition of the adhesin is the chemical attenuation of pilus biogenesis. This has been achieved for type 1, P and S pili by inhibiting chaperone docking to the usher using a family of bicyclic 2-pyridones, called pilicides (Pinkner et al. 2006), or by competitive inhibition of the DSE reaction with organic compounds (AL1) binding the subunit P5 pocket of FimH (Lo et al. 2014). Chaperone-usher pili also make promising vaccine candidates. The immunization of mice and primates with type 1 pilus adhesin FimH reduced the in vivo colonization of bladder as well as recurrent UTIs (Langermann et al. 1997, 2000). In animal husbandry, immunization of sows with cocktails including F4, F5, F6 and F41 fimbrial subunits from enterotoxigenic *E. coli* protects the litter from neonatal diarrhea (Matias et al. 2017), and passive immunization by addition of neutralizing antibodies of the F4 adhesin FaeG to pig feed protects young animals from post-weaning diarrhea (Virdi et al. 2013).

Type V Pilus

In 2016 Xu et al. characterized a novel type of pili with a biogenesis mechanism distinct from all known pilus systems and described as proteinase-mediated donor-strand exchange (Xu et al. 2016). These unique pilus system, termed the Type V or Bacteroidia pilus, is found exclusively in the Bacteroidia class. They were first coined in *Porphyromonas gingivalis*, a human oral pathogen associated with severe adult periodontitis and gingivitis (Xu et al. 2016). Bacteroidia pili play a role in bacterial adhesion, co-aggregation and biofilm formation. Two morphologically different pili of *P. gingivalis* have been described: major or long (0.3–1.6 μm), with a major pilin subunit FimA (not to be confused with FimA major subunit of CU type 1 pilus), and minor or short (80–120 nm), with a major pilus subunit Mfa1 (Hamada et al. 1996; Yoshimura et al. 1984). Both major and minor pili are encoded by similar operons, containing the genes of structural pilins forming the pilus stalk (FimA or Mfa1, respectively), anchoring pilins (FimB and Mfa2, respectively), tip pilins (Mfa4), as well as other accessory subunits and regulatory elements.

Type V Pilus Structure and Biogenesis

Type V pilins are composed of two domains: an N-terminal domain (NTD) and a slightly larger C-terminal domain (CTD). Both domains have a transthyretin-like fold that is composed of seven core β-strands organized into two β-sheets. The NTD has an archetypal fold with 7 β-strands (A1–G1), whereas the CTD (except for Mfa4) possess an C-terminal extension of two highly conserved amphipathic β-strands (A1′ and A2′). Crystal structures of the *P. gingivalis* FimA superfamily stalk subunits showed that this appendage may be present in two conformations: "open"—exposed along the CTD, and "closed"—folded back to the CTD, with a flexible loop between A1′ and A2′ strands (Xu et al. 2016) (Fig. 12.1f). The structural characterization of type V tip pilins show that these subunits terminate the pilus structure, as they either lack the A1′–A2′ appendage (e.g. Mfa4), or the appendage interacts with a fused C-terminal lectin domain (e.g. BovFim1C) (Kloppsteck et al. 2016; Xu et al. 2016).

The biogenesis of type V pilus superficially resembles the donor strand-mediated fold complementation mechanism seen in CU pilus, however, it additionally requires the lipoprotein precursors of pilin subunits and outer membrane proteinase. It is thought that the mechanism of type V pilus assembly, both in *P. gingivalis* and *B. fragilis*, is based on the lipoprotein sorting pathway, although this has not been unambiguously demonstrated (Shoji et al. 2004). Type V prepilins are produced as lipoprotein precursors. Importantly, prepilins possess exceptionally long signal peptides, as compared to other bacterial pilins. The C-terminal part of the signal peptide is called a lipobox and contains the lipidated cysteine residue. Pilin subunits are transported by SecYEG machinery to the periplasmic side of inner membrane, where they are folded, and their signal peptide is cleaved off by a type II signal proteinase. Subsequently, the C-terminal cysteine residue is lipidated and the modified pilin subunits are secreted to the extracellular environment. How the periplasmic transport and outer membrane translocation step occur is presently unknown. Whether this requires the help of a periplasmic chaperone and outer membrane usher-like protein remain to be identified. The folded pilins occur in the periplasm in the "closed" state, therefore they are predicted to be stable and the presumed chaperone most likely binds the lipid moiety present on the pilin N-terminus. The mechanism of type V pilus assembly at the outer membrane has not been described yet, but it is hypothesized to be based on the lipoprotein-sorting machinery. The first pilin subunit bound to the outer membrane is an anchor pilin, which is docked in the membrane via the N-terminal lipid moiety. Next pilin subunits, stalk and tip pilins, undergo the cleavage of the N-terminal short peptide, by trypsin-like arginine and lysine specific proteinase located in the outer membrane: gingipain R or gingipain K (Rgp or Kgp). The proteolytic step yields the mature, assembly-prone form of pilin (Nakayama et al. 1996). During proteolytic maturation, the A1 β-strand is removed, and an extended hydrophobic groove is generated in the NTD. This groove can now be occupied by the β-strand from the neighbouring subunit, similarly as in the chaperone-usher pilus. Although both N-terminal and C-terminal regions of the adjacent pilin have been suggested to act as donors of complementing strand (Kloppsteck et al. 2016; Xu et al. 2016), the involvement of the flexible C-terminal appendage of A1′ and A2′ was validated by the cross-linking experiments (Xu et al. 2016). Accordingly, the hydrophobic groove of NTD can be occupied by the extended C-terminal appendage of the neighbouring subunit, with the A1′ strand filling the NTD groove to restore seven β-stranded fold, and the A2′ strand complementing the CTD groove, parallel to G2 strand, which in "closed" conformation was occupied by its own A1′ and A2′ strands (Xu et al. 2016) (Fig. 12.1e). A recent study confirmed the crucial role of the conserved sequences in both N- and C-termini of the pilus subunit Mfa1 in subunit polymerization of Mfa fimbriae. Treatment of *P. gingivalis* with peptides analogous to these sequences inhibits the Mfa fimbriae assembly and impedes biofilm formation (Alaei et al. 2019).

Type IV Pili

Type IV pili (T4P) are several micrometres long, flexible surface appendages that are bacterial virulence factors, widely distributed in many Gram-negative bacteria, including *Pseudomonas aeruginosa, Neisseria gonorrhoeae, N. meningitidis, Myxococcus xanthus*. Some of the bacterial species, like *Vibrio cholerae* and enteropathogenic *E. coli* (EPEC), produce bundle-forming pili, required for adherence to epithelial cells and auto-aggregation (Ramboarina et al. 2005). Apart from Gram-negative bacteria, type IV-related pili have also been identified in Gram-positive genera *Clostridia* and *Ruminococcus*, in *Cyanobacteria* and in archaea, where they form the archellum, suggesting an early evolutionary origin of this system (Imam et al. 2011; Proft and Baker 2009; Szabó et al. 2007). Type IV pili are multifunctional organelles with a distinguishing ability to extend and retract by reversible polymerization and depolymerization. They play a role in adhesion to host cells and solid substrates, in biofilm formation, DNA and phage uptake, cell motility, cellular invasion as well as microcolony formation. Apart from their function in virulence, type IV pili machinery powers a flagella-independent type of bacterial movement known as twitching motility (Mattick 2002). This is possible thanks to the repeated cycles of extension, adhesion and retraction powered by cytoplasmic adenosine triphosphatases (ATPases). Although T4P are very thin (6–9 nm) structures, they are remarkably strong molecular machines that can endure extensive forces of over 100 pN (Maier et al. 2002).

Type IV pilus subunits share a distinguishing *N*-methylated N-terminus, and a conserved hydrophobic N-terminal 25-residue α-helical domain and a C-terminal disulphide bonded β-domain (Craig et al. 2004). Based on the sequence similarity and length, type IV pilins are divided into two subclasses: type IVa and type IVb. The type IVa pilins are shorter (average length of mature protein: 150 amino acids) than the type IVb pilins (190 amino acids), and hold a shorter signal peptide (5–6 amino acids for type IVa and 15–30 amino acids for type IVb) (Craig and Li 2008). The *N*-methylated N-terminal residue of type IVa pilins is always phenylalanine, whereas for type IVb it may be methionine, leucine or valine. Comparison of the available pilin structures shows notably different protein topologies in the β-domain of type IVa and type IVb pilins. The conserved N-terminal hydrophobic α-helix of Type IV pilus subunits acts both as a transmembrane and coiled-coil protein interaction domain in the structural core of the pili (Giltner et al. 2012) (Fig. 12.2). Although the general build-up is equivalent, Type IVa and type IVb pilins form pili that differ in diameter and helical structure. The occurrence of these two pilus types is also different: type IVa are found in many Gram-negative bacteria, including pathogens like *Neisseria* spp. or *P. aeruginosa*, whereas type IVb pili have only been found in human enteric bacteria, like *V. cholerae*, *Salmonella enterica* serovar Typhi, enteropathogenic *E. coli* (EPEC) and enterotoxigenic *E. coli* (ETEC). The host range of type IVa pili expressing pathogens is much broader and includes humans and other mammals, plants and possibly other bacteria (Craig et al. 2004). Interestingly, the type IVa pili assembly requires a complex machinery built of many components, which genes are scattered around the whole bacterial genome. In contrast, type IVb pili assembly systems are composed of smaller number of proteins usually encoded by gene clusters, sometimes present on the plasmids (Pelicic 2008).

Overall, the architecture of both type IVa and type IVb pilins is similar. They are small (15–20 kDa) proteins having a conserved lollipop-like fold with an extended hydrophobic N-terminal α-helical spine (α1) and a globular C-terminal head domain, typically composed of four to seven-stranded antiparallel β-sheet. The hydrophobic N-terminal part of α1 helix, α1-N, protrudes from the protein core, while the amphipathic C-terminal part, α1-C, is embedded in the globular domain and packs against it. The head domains hold two regions involved in T4P interactions: a D-region, containing the conserved cysteines, and exposed αβ-loop that can undergo post-translational modifications (Fig. 12.2b). For example, the *N. gonorrhoeae* gonococcal (GC) pilin D-region forms a ridge that displays hypervariable surface regions, and the second ridge is formed by αβ-loop containing two post-translational modifications (Craig and Li 2008). The structural diversity of type IV pilins results from the loop sequence differences as well as the topology of the secondary structure elements in the globular domain (Craig et al. 2004; Giltner et al. 2012). Type IVa pilins are characterized by four contiguous β-strands as in the *N. gonorrhoeae* GC pilin (Parge et al. 1995), whereas the type IVb pilins display more variable β-sheet topology, with five to seven β strands, as in the case of EPEC BfpA pilin or *V. cholerae* TcpA pilin (Craig and Li 2008; Craig et al. 2003).

Type IV Pilus Assembly Components and Biogenesis

Type IV pilus (T4P) machinery is a multimeric protein assembly that spans both inner and outer membrane in Gram-negative bacteria. It is homologous to the type 2 secretion system (T2SS), which is involved in the transport of folded proteins from the periplasm to the extracellular environment across the cell membrane (Berry and Pelicic 2015). T2SS assembles a periplasmic pseudopilus that is implicated in the secretion of toxins and hydrolytic enzymes, whereas type IV pilus machinery is responsible for the assembly and disassembly of pilin subunits in the core of the pilus.

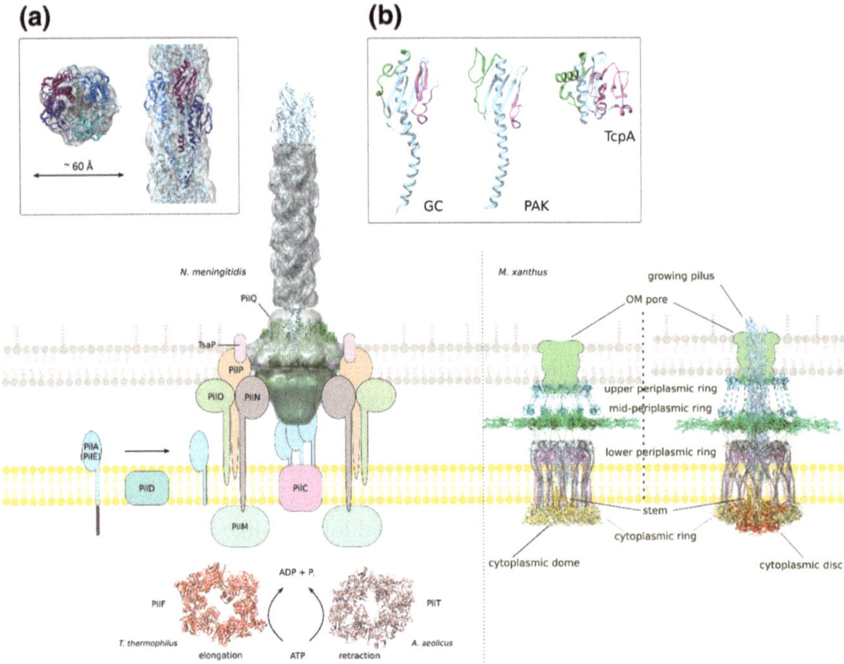

Fig. 12.2 The architecture of type IV pilus systems. **Left**: Schematic representation of the archetypal type IV pilus system. The structures presented in the figure come from different organisms. The pilus rod is composed of thousands of helically arranged pilin subunits (represented by PilE from *N. meningitidis*, EMDB entry: EMD-8287, PDB entry: 5KUA). The pilus traverses the outer membrane pore (PilQ from *N. meningitidis*, EMDB entry: EMD-2105, PDB entry: 4AV2). The pilus elongation and retraction is catalysed by cytoplasmic ATPases (PilF from *T. thermophilus*, PDB entry: 5OIU and PilT from *A. aeolicus*, PDB entry: 2GSZ). **a** Cryo-EM map of type IV pilus rod of *N. meningitidis* (EMDB entry: EMD-8287, PDB entry: 5KUA) with the neighbouring PilE subunits shown. **b** Structures of type IVa pilins: pilin GC from *N. gonorrhoeae* (PDB entry: 2PIL) and PAK pilin from *P. aeruginosa* (PDB entry: 1OQW) and type IVb pilin: truncated TcpA pilin from *Vibrio cholerae* (PDB entry: 1OQV). The αβ-loops are shown in green and the D-regions are shown in magenta. **Right**: The architectural model of the type IVa pilus machinery from *M. xanthus* in a non-piliated and piliated state (PDB entries: 3JC9 and 3JC8, respectively). PilA pilus subunits are shown in light blue. The components of secretin subcomplex, N-terminal motif of TsaP, LysM (turquoise), together with PilQ AMIN domains (green) anchor the entire basal body to the peptidoglycan layer. The alignment subcomplex components, PilN (grey) and PilO (pink) form coiled coils that is a link between the secretin and motor complexes. Lipoprotein PilP (khaki) is anchored in the inner membrane and PilM (beige) is located in the cytoplasm. The motor complex components, hexameric ATPase PilB (red) and inner membrane PilC (orange) are present at the base of the system. Strikingly, the distance between inner and outer membrane has been found to be longer in the piliated state of the system

Moreover, in contrast to CU pili and curli, the biogenesis of type IV pili requires ATP as an energy source (Turner et al. 1993). The same holds true for pilus disassembly (Merz et al. 2000). The nomenclature of T4P components varies drastically amongst different organisms, in this chapter the *P. aeruginosa* nomenclature will be used.

The biogenesis of type IV pilus requires the action of a specialized machinery, including an outer membrane secretin subcomplex (PilQ, PilF, TsaP), an inner membrane motor subcomplex (PilC, PilB, PilT) and an alignment subcomplex (PilM, PilN, PilO, PilP) (Leighton et al. 2015a). The outer membrane secretin subcomplex is composed of two elements: PilQ, a large gated multimeric outer membrane pore and a pilotin PilF, which function is to ensure correct localization, assembly and outer membrane insertion of the secretin subcomplex. In *N. gonorrhoeae* and *M. xanthus* there is another component of secretin subcomplex present, the T4P secretin-associated protein, TsaP. This component was suggested to anchor the outer membrane secretin subcomplex to the peptidoglycan cell wall by its LysM motif (Siewering et al. 2014), but its role requires further investigation. The motor subcomplex is made up of the inner membrane protein PilC and the cytoplasmic ATPases PilB and PilT, which drive the processes of pilus elongation and retraction, respectively (Whitchurch et al. 1991; Chiang et al. 2005). The alignment subcomplex, composed of highly conserved PilM, PilN, PilO and PilP proteins, functions as a platform connecting the secretin and the motor complexes (Thanassi et al. 2012; Takhar et al. 2013). The function of particular subunits has not been described, however recent study show that PilNO complex is a dynamic link between the secretin and motor complexes, involved in both elongation and retraction of type IV pilus (Leighton et al. 2015b). PilP subunit is bound to the PilQ pore, which stabilizes its assembly during pilus secretion as the passage of the pilus through the pore requires the disassembly of membrane-spanning domains of PilQ (Berry et al. 2012). PilN and PilO subunits form a dimer via their periplasmic domains. This dimer serves as an anchor of the whole alignment subcomplex in the inner membrane and binds the periplasmic lipoprotein PilP, which interacts with the PilQ pore. Finally, PilN binds the actin-like protein PilM, so that all the cellular compartments are connected via the PilMNOPQ protein interaction network. This trans-envelope complex interacts also with the PilA major pilus subunit and facilitates the passage of the pilus through the periplasm (Tammam et al. 2013). The final pilus rod is made of major subunit PilA (PilE in *N. meningitidis*) and minor pilus subunits, FimU, PilV, PilW, PilX and PilE, which together form a pilus rod, organized in a helical manner.

The pilus formation starts from the insertion of the prepilin subunits into the inner membrane by the SecYEG translocon. Prepilin maturation requires the cleavage of the signal peptide and methylation of the mature pilin N-terminus by the action of prepilin peptidase PilD (Zhang et al. 1994; Lory and Strom 1997). Prior to incorporation into the pilus, the conserved hydrophobic N-terminal α-helix in the mature pilins anchors them in the inner membrane, with the globular domain exposed to the periplasm. Mature pilin subunits are then extracted from the inner membrane and incorporated into the base of a growing pilus by the assembly machinery, in a process that involves coiled-coil interactions between the N-terminal α-helices in the central hydrophobic core of the pili (Craig et al. 2006). The assembled T4P core passes through the outer membrane secretin PilQ, which is a multimeric gated channel, homologs of which are also present in type II and type III secretion systems (Bitter et al. 1998). The assembly-disassembly of the pilus core requires ATP hydrolysis, by the action of the cytoplasmic ATPases, PilB (PilF in *T. thermophilus*) and PilT, respectively (Misic et al. 2010; Collins et al. 2018; Turner et al. 1993). Finally, the T4P assembly requires an integral inner membrane protein PilG, as well as several minor pilins (PilH-K) (Tonjum et al. 1995; Winther-Larsen et al. 2005).

Structure of Type IV Pili and Their Assembly Machinery

Since the T4P machinery is a large multiprotein assembly, the reconstruction of its overall structure in an intact state is a challenging task, despite the many structures of individual T4P components being available (Misic et al. 2010; Karuppiah and Derrick 2011; Kim et al. 2006; Berry et al. 2012; Craig et al. 2003). Cryo-electron tomography (cryo-ET) enabled reconstruction of type IV pilus machinery from *M. xanthus* and *T. thermophilus* intact cells, at resolutions 30–40 Å and 32–45 Å, respectively (Chang et al. 2016; Gold et al. 2015). The structures from both organisms showed the T4P machineries to span the entire cell envelope and to be present in a closed state in the absence of pilus and an open state when the pilus is present. Models with the available structure of individual components docked into the cryoET volumes indicate that in general the same core components form the overall T4P machinery in both organisms (Fig. 12.2). The differences include longer PilQ secretin in *T. thermophilus*, which is characterized by larger distance between inner and outer membrane. Moreover, *Thermus* lacks the PilP homologue, which in *M. xanthus* is a part of mid-periplasmic ring. The T4P secretion apparatus is made of four interconnected rings, three in the periplasm and one in the cytoplasm, a cytoplasmic disc and dome, a periplasmic stem and an outer membrane pore (Fig. 12.2). In the resting, non-piliated state the PilA major pilus subunits form the short stem, whereas in the piliated state this structure transverse through the periplasmic rings and the PilQ pore to the external environment. The pilus arises at the cytoplasmic dome formed by cytoplasmic domains of IM PilC, surrounded by PilM cytoplasmic ring. The coiled coils across the inner membrane connect the cytoplasmic ring with the lower periplasmic ring, which is composed of the globular domains of PilO and PilN. Periplasmic domains of PilQ with PilP form the mid-periplasmic ring and the upper periplasmic ring is composed of TsaP around PilQ. The structure docking suggests a 1:1 stoichiometry of ring components PilP, PilN, PilO, PilM and TsaP (Fig. 12.2). Although the modelling suggests 12 copies of each subunit, the exact number of particular subunits is still in question (Leighton et al. 2015a; Tammam et al. 2013). The hexameric ATPases PilB and PilT bind interchangeably to the base of the T4P machinery and occur as a cytoplasmic disc in the piliated structure during elongation and retraction, respectively. The other differences between piliated and non-piliated state include the opening of the PilQ gate (PilQ of *T. thermophilus* has 2 gates) and the pilus traversing the periplasm and PilQ pore into the extracellular environment.

Interestingly, the distance between inner and outer membrane is longer in the piliated state, possibly due to the presence of the pilus. The model of pilus assembly was proposed, where ATP hydrolysis by PilB induces PilC rotation and relocation to facilitate incorporation of PilA major pilus subunits, one at a time, from the IM to the base of nascent pilus. In contrast, PilT bound to the basal body induces pilus retraction, by rotating PilC into location that induces expulsion of PilA subunits (Chang et al. 2016). Furthermore, PilNO heterodimer was found to participate in both pilus retraction and elongation (Leighton et al. 2015a). The structure of type IV pilus rod from *N. meningitidis* was modelled by fitting the 1.44 Å crystal structure of major pilin PilE subunit into the ~6 Å cryo-EM density map (Fig. 12.2). The fully assembled T4P core is composed of the N-terminal α-helices packed together in a coiled arrangement in the centre of the structure, with the globular domains facing to the outside. Notably, the α-helical order is lost in the central part of α1 in the intact pilus (between residues Gly14 and Pro22), making this region more flexible. The extension of this region may be responsible for the spring-like properties of type IV pilus under shear forces (Kolappan et al. 2016). Importantly, the glutamate residue Glu5 of each subunit was confirmed to be essential for the assembly of T4P, as it forms a salt bridge with the positively charged N-terminal amine of the adjacent subunit. This interaction is believed to be involved in the incorporation of subsequent pilins from the inner membrane into the nascent pilus (Craig et al. 2006; Kolappan et al. 2016).

Mycobacterial Pili

For a long time, there was no evidence of pili produced by mycobacteria. However, in 2005 Alteri observed by negative staining and transmission electron microscopy (TEM) that mycobacteria produce two morphologically distinct types of pili: type IV- and curli-like pili (MTP) (Alteri 2005). Mycobacterium type IV pili are flexible appendages forming rope-like bundles. The analysis of *M. tuberculosis* genome showed that this organism contains the type IVb pili gene cluster, encoding small type IVb prepilins of the Flp pili family (**f**imbrial **l**ow-molecular weight **p**rotein). The Flp protein expression and secretion by *M.* tuberculosis was confirmed by gene expression analysis and immunofluorescent microscopy. *M. tuberculosis* type IVb pili are encoded by a 5-kb genomic island, which contains seven genes, including type IVb flp prepilin, a transmembrane protein and secreted proteins (Danelishvili et al. 2010). In addition, two Flp family prepilin peptidases were found distant from the *flp* locus. These two proteins encoded by Rv0990c and Rv2551c ORFs are thought to be involved in the secretion or cleavage of the Flp prepilin substrate. The flp locus of *M. tuberculosis* is homologous to flp-tad locus of *Aggregatibacter actinomycetemcomitans* and was presumably acquired by horizontal gene transfer. In many pathogens Flp/Tad pili serve as colonization factors and promote biofilm formation (Ramsugit and Pillay 2015). As *M. tuberculosis* is a non-motile organisms, its type IV pilus potentially functions as adherent factor, which was also initially proved by Alteri (2005). Its function, however, needs to be further elucidated.

Bacterial Amyloid Fibers

Curli and Fap are functional bacterial amyloid fibers secreted by many Gram-negative bacteria as part of an extracellular matrix that binds cells together to form bacterial communities known as biofilms. Biofilms are a commonly found sedimentary life-form for bacteria. The extracellular matrix helps adherence to the substrate and provides a protective shield that helps cells cope with various environmental stress such as oxidative damage, desiccation, antibiotics as well as host immune responses (Branda et al. 2005; Depas et al. 2015; Hall-Stoodley and Stoodley 2009). Biofilms can be found on both biotic and abiotic surfaces in almost all kinds of environments (Chai et al. 2013; Jeter and Matthysse 2005; Otter et al. 2015; Ryu and Beuchat 2005; Uhlich et al. 2006). The secreted extracellular matrix is composed of polysaccharides, nucleic acids and protein fibers. Functional amyloids can constitute a major fraction (up to 85%) of the biofilm matrix (Larsen et al. 2008; Reichhardt and Cegelski 2014). Historically, amyloids are best known for their association with neurodegenerative diseases such as Alzheimer's, Parkinson's and Huntington's disease and are considered off-pathway protein folding products that are associated with cytotoxicity (Chiti and Dobson 2006). However, over the last two decades an increasing number of examples emerged of so-called functional amyloids, where the amyloid state comprises the native state of the protein and fulfils an adapted biological function (Blanco et al. 2012). In bacteria, several such functional amyloids are involved in biofilm formation, adherence, persistence and pathogenesis (Fowler et al. 2007; Van Gerven et al. 2018). Curli is one of the first and most extensively studied bacterial amyloids. Curli were first identified in *Salmonella* (Grund and Weber 1988) and *E. coli* (Olsén et al. 1989) biofilms and later found to be broadly distributed amongst Proteobacteria and Bacteroidetes (Dueholm et al. 2012). A recently discovered bacterial amyloid system is Fap, found in *Pseudomonas* (Dueholm et al. 2010). Curli and Fap both form surface-localized linear fibers implicated in biofilm formation. Although non-homologous, both systems share a number of common characteristics, such as an amyloid-like cross β architecture of the fibers, the presence of a minor and major subunit that act as a specific nucleator and the main polymerizing subunit, respectively, as well as diffusion-driven secretion channels in the outer membrane. This section will discuss in detail, the structure and function of components of Curli and Fap systems.

Curli

The curli pathway is one of the most extensively studied functional amyloid system. Curli subunits are secreted through the type VIII secretion system, also known as the nucleation-precipitation pathway (Chapman et al. 2002; Olsén et al. 1989). In *E. coli*, curli subunits and assembly machinery are encoded in two *csg* (curli structural gene) operons: *csgBAC* and *csgDEFG* (Chapman et al. 2002; Hammar et al. 1995). CsgD is the master regulator of the curli biogenesis as it activates the transcription of *csgBAC* operon through a coordinated complex signalling network (Brombacher et al. 2003; Chirwa and Herrington 2003; Gerstel and Römling 2003). CsgA is the major structural component of the curli fibers while CsgB is the nucleator of amyloid self-assembly (Olsén et al. 1989; Chapman et al. 2002). CsgC is a periplasmic protein responsible for inhibition of premature polymerization of CsgA and therefore possibly helps the bacteria in keeping a check on the cytotoxic aspects of amyloids (Evans et al. 2015). CsgG is an outer membrane protein forming a pore responsible for transport of the curli subunits from the periplasm to the extracellular space. It forms a pore complex by interacting with CsgE in the periplasmic side and CsgF in the extracellular side of the outer membrane. CsgE helps in fine-tuning the regulation of transport of CsgA and CsgB through the pore complex (Goyal et al. 2014; Nenninger et al. 2011) while CsgF interacts with CsgG and CsgB to possibly act as the interface between the assembled fibers and the outer membrane (Nenninger et al. 2009; Schubeis et al. 2018).

Architecture of Curli Fibers

Curli fibers are composed of two components, the major structural component CsgA, and minor nucleator component CsgB. In *E. coli*, both subunits are approximately 13 kDa polypeptides. A 20 residue long N-terminal signal peptide targets the subunits for export to the periplasm by the SecYEG translocon, after which it is cleaved off (Chapman et al. 2002; Olsén et al. 1989). The first 22 residues of the mature subunits (after proteolytic removal of the signal peptide), termed N22, are believed to form a pathway-specific targeting signal. N22 is sufficient to direct native and non-native proteins to the outer surface through the curli secretion machinery (Robinson et al. 2006). N22 is followed by the amyloidogenic core domain of the subunits, formed by five pseudo repeat sequences of ~22 amino acids, termed R1 to R5. In other species, CsgA can be considerably longer, encompassing over twenty pseudo-repeats. The pseudo-repeats hold an average 30 percent pairwise sequence identity. Isolated pseudo-repeat peptides are in themselves amyloidogenic, though can differ strongly in their assembly propensity. In *E. coli* CsgA for instance R1, R3 and R5 will readily self-assemble, whilst R2 and R4 only poorly do so (Hammer et al. 2007; Wang et al. 2007, #128). Although in vitro CsgA will self-assemble into curli-like fibers, the minor subunit CsgB acts as a potent nucleator for CsgA fibrillation. In vivo, the formation of cell surface associated curli requires the nucleator CsgB, in a process that is dependent on the curli accessory protein CsgF (Nenninger et al. 2009). R4 and R5 of CsgB (with its markedly different primary sequence containing four positively charged residues Lys133, Arg140, Arg147, and Arg151) has been implicated in interaction of CsgB with CsgF (Hammer et al. 2012). Deletion of R4 or R5 in CsgB results in secretion of CsgA and CsgB to the exterior without any association to the cell surface (Hammer et al. 2012). Although CsgB is required for nucleation of CsgA in vivo, both CsgA and CsgB can form fibers independently in vitro (Shu et al. 2012).

Experimental structures of the CsgA or CsgB subunits and the assembled curli are currently lacking. Staining with amyloid-responsive dyes (Congo Red and thioflavin T), circular dichroism spectra and X-ray diffraction experiments on CsgA fibers point to the presence of a cross-β structure with nucleation-dependent growth characteristics, typical of amyloids (Chapman et al. 2002; Shewmaker et al. 2009). Solid state NMR studies, however, indicate that unlike classical amyloids, CsgA does not have an in-register parallel β sheet architecture (Shewmaker et al. 2009). Using calculations of amino acid contacts based on covariation of amino acids in various CsgA homologues, a model has been proposed. According to this model CsgA is expected to have a β helical architecture, where each pseudo-repeat forms one helical turn (Tian et al. 2015). This model is consistent with results of simulations designed to find the lowest energy conformation of a single molecule. Another model proposed by Louros et al. (2016) based on the Salmonella CsgA homology model depicts the curli fibers as two-start filaments made of stacked β-helical solenoids (Collinson et al. 1993). However, there is ambiguity in the handedness of the helix. Experimental evidence supporting these models is still missing. In vitro studies of CsgA self-assembly using high speed atomic force microscopy has revealed polarity in growth of the fiber, with one end growing visibly faster than the other. Also, they exhibit stop-and-go dynamics in growth where periods of stagnations are interrupted by steady bursts of elongation (Sleutel et al. 2017).

The Curli Assembly Machinery

Secretion and assembly of curli requires a dedicated transport channel in the outer membrane, CsgG, and two accessory proteins CsgE and CsgF (Loferer et al. 1997; Nenninger et al. 2009; Robinson et al. 2006). In *E. coli* CsgG is a 262-residue long lipoprotein that forms a nonameric transport complex that traverses the OM through a 36-stranded β-barrel (Cao et al. 2014; Goyal et al. 2014) (Fig. 12.3). Prior to insertion in the OM, CsgG is found as a soluble monomeric protein. The transmembrane β-barrel is formed upon circular oligomerization of nine subunits, each contributing two β-hairpins to the channel (Goyal et al. 2014). The pore volume is separated into two 3–4 nm wide cavities on the periplasmic and extracellular side of the outer membrane, separated by a diaphragm-like 1 nm wide constriction formed by the lateral packing of a conserved 12-residue 'constriction loop' (CL) in the nine subunits. The luminal lining of the constriction is composed of three stacked concentric rings formed by the side chains of residues Y51, N55, and F56 (Goyal et al.

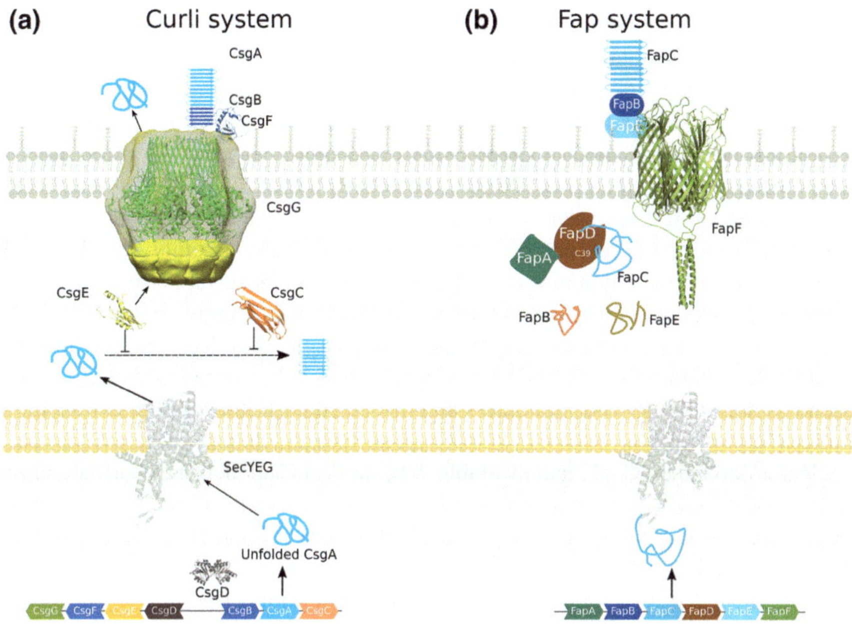

Fig. 12.3 The biogenesis pathways of the bacterial amyloids curli and Fap. **a** Schematic representation of the curli assembly pathway and the *E. coli csgBAC* and *csgDEFG* gene clusters. Blue and dark blue: major curli subunit CsgA and nucleator subunit CsgB, respectively. CsgG forms the OM curli translocation channel (green; PDB entry: 4UV3), bound on the outer side by the accessory protein CsgF (dark green; PDB entry: 5M1U), and the on the periplasmic side by the secretion factor CsgE (yellow; PDB entry: 2NA4). In the periplasm, the amyloid chaperone CsgC (orange, PDB entry: 2Y2Y) prevents premature curli fiber formation. **b** Schematic representation and operon structure of the Fap assembly pathway. The major Fap subunit FapC and presumed Fap nucleator subunit FapB are shown in blue and dark blue, respectively. FapF (dark green, PDB entry: 5O65)

2014). The latter is strictly conserved and was found important for curli secretion. A structurally equivalent phenylalanine "clamp" is observed in the Anthrax protective antigen, where it facilitates recruitment and translocation of unfolded protein substrate to the secretion channel (Krantz et al. 2005). Single channel conductance experiments and structural data suggest CsgG operates as an ungated peptide diffusion channel (Goyal et al. 2014). The driving force for protein translocation in the curli pathway is currently unknown. Under physiological concentrations, secretion of curli subunits requires the 12 kDa periplasmic accessory protein CsgE (Nenninger et al. 2011). CsgE provides specificity to the CsgG channel and has been found to form a periplasmic plug to the channel (Nenninger et al. 2011; Goyal et al. 2014, #122). In vitro, CsgE maintains equilibrium between a monomeric and nonameric species. The latter forms a dynamic complex with CsgG, forming a cap-like structure that closes off the periplasmic vestibule of the channel (Goyal et al. 2014). CsgE inhibits CsgA polymerization when added in stoichiometric amounts, indicating it can directly interact with the secretion substrate (Nenninger et al. 2011). The solution NMR structures of CsgE W48A/F79A, a mutant that stabilizes the CsgE monomer, reveal a globular mixed α/β fold with three distinct electrostatic surfaces, (i) a positively charged "head," (ii) a negatively charged "stem" and (iii) a negatively charged "tail" region (Klein et al. 2018; Shu et al. 2016). The positively charged head region has been implicated in CsgE-CsgA interaction. However, how and at what time during the secretion process CsgE interacts with the secretion substrates is currently unknown. It is speculated that CsgE recruits CsgA and CsgB to the secretion complex. A second accessory factor to the curli secretion-assembly machinery is CsgF, a 13 kDa protein that is localized on the extracellular surface (Nenninger et al. 2009). CsgF acts a coupling factor between secretion and assembly of curli subunits. Although *csgF* null mutants secrete CsgA and CsgB, they lack surface-attached curli fibers (Chapman et al. 2002; Nenninger et al. 2009). Instead, secreted CsgA is found as non-polymerized monomers and dispersed curli fibers in the extracellular milieu (Nenninger et al. 2009). In vivo, CsgB surface attachment and activity as curli nucleator depend on CsgF, and available data suggest CsgF acts as a curli assembly chaperone through direct interaction with the curli nucleator CsgB, an interaction that is found to depend on CsgB's C-terminal repeats R4 and R5 (Nenninger et al. 2009). A recently determined solution NMR structure of CsgF shows a small α/β domain, preceded by a long, highly disordered N-terminus (Schubeis et al. 2018). How the protein interacts with the secretion channel and the secretion substrates is unknown.

A last component in the curli assembly-secretion pathway is the CsgC (also referred to as CsgH outside γ-proteobacteria), an 11 kDa periplasmic protein (Hammar et al. 1995; Dueholm et al. 2012; Evans et al. 2015; Taylor et al. 2016). CsgC is dispensable for curli assembly, and for long, the physiological role of the protein remained obscure (Gibson et al. 2007; Taylor et al. 2011). A recent study showed the protein protects bacterial cells from toxicity associated with an accidental periplasmic accumulation of curli subunits (Evans et al. 2015). It does so by inhibiting CsgA and CsgB fiber formation at sub-stoichiometric concentrations. The crystal structure of CsgC reveals a 7-stranded β-sandwich protein with a striking CxC motif similar to its structural homolog DsbD (Taylor et al. 2011). However, this CXC motif does not appear to be involved in CsgC's amyloid inhibitory activity, which relies on electrostatic interaction with the curli subunits (Taylor et al. 2016). The exact mode of action of the amyloid inhibitor remains unclear. Whilst Taylor and co-workers suggested CsgC to act on an amyloid precursor (Taylor et al. 2016), a study that followed CsgA fibrillation in presence of CsgC using atomic force microscopy indicated the protein acts primarily by inhibiting fiber elongation, presumably by capping the fiber growth poles (Sleutel et al. 2017). Strikingly, the inhibitor is also able to attenuate fibrillation in other amyloids such as α-synuclein, albeit at higher concentrations (Evans et al. 2015).

Fap Fibers

Fap is another functional bacterial amyloid system found in proteobacteria (β-, δ-, and γ-proteobacteria). Although evolutionarily unrelated to curli, it too uses a dedicated pathway for secretion and assembly of cell-surface localized amyloid fibers. Similar to curli, Fap are part of the extracellular matrix of biofilm-associated cells, where they increase colony hydrophobicity and stiffness to enhance survival under various harsh environments (Zeng et al. 2015). Fap was initially identified in pathogenic and non-pathogenic *Pseudomonas* species, and subsequently found in the genomes of at least 39 additional genera of proteobacteria (Dueholm et al. 2010; Rouse et al. 2018a). The Fap pathway is encoded by the *fapABCDEF* operon, which includes the structural fiber proteins (FapB, FapC and FapE), as well as the proteins required for safe guidance across the periplasm (FapD, FapA) and secretion across the outer membrane (FapF) (Dueholm et al. 2010). Fap components are exported to periplasm via the SecYEG translocon where the signal sequence is cleaved off. Mature FapB, FapC and FapE are then secreted through FapF, a trimeric outer membrane pore composed of three 12-stranded β-barrels, each plugged by a 13-residue α-helix that join into a periplasmic asymmetric triple coiled-coil via a 40–50-residue long flexible connector (Rouse et al. 2017, 2018b). The N terminal coiled coil domain and helical plug are thought to present a mechanism to regulate substrate transport, although if and how the plugged channels are gated remains unknown. FapC is the major structural component of the amyloid fiber, analogous to CsgA in curli. The protein is composed of three amyloid-prone pseudo-repeat regions of approximately 30 residues, interrupted by two linker regions (Bleem et al. 2018; Dueholm et al. 2010, 2013). FapB is a minor structural component of Fap, has shorter repeats compared to FapC, and is proposed to act as nucleator, analogous to CsgB (Dueholm et al. 2010). FapD and FapF are two periplasmic accessory proteins (Rouse et al. 2017, 2018b). FapD is a C39-family protease that is proposed to play a key role in selectivity of the FapF pore (Rouse et al. 2017). Deletion of FapD or mutation in its cysteine active site resulted in the loss of FapC secretion, suggesting FapD is indeed a peptidase and has a regulatory role (Rouse et al. 2017). FapA is speculated to be a periplasmic inhibitor of Fap assembly, countering cytotoxic effects of premature amyloid polymerization of FapC amyloid, similar to the role of CsgC in the curli system (Rouse et al. 2018a).

Type IV Secretion Pilus

Type IV secretion(T4SS) pili are the extracellular, tubular filaments of Type IV secretion machineries, capable of transporting DNA, proteins and nucleoprotein complexes in and out bacteria and archaea. These systems serve three major purposes in the bacteria, (i) conjugation, (ii) DNA uptake (competence) and release, and (iii) delivery of effector proteins (Grohmann et al. 2018). Conjugation is the process by which a donor cell delivers genetic material (plasmids, Integrative and

Conjugative Elements (ICEs)) into a recipient cell in a contact dependent manner (Waksman 2019). Two well-known examples are the F plasmid transfer system (Tra operon) in *E. coli*, mediating horizontal gene transfer and propagating antibiotic resistance genes and the Ti plasmid injection system in *Agrobacterium tumefaciens*, associated with crown gall disease in plant hosts. The second function of the type IV secretion systems is DNA uptake, also known as bacterial competence (ComB system in *Helicobacter pylori*) and release (Tra system in *N. gonorrhoeae*), for fitness, pathogenicity and virulence of the species. A third group of T4SSs found in bacterial pathogens functions to deliver protein effectors into eukaryotic host cells. Examples include the VirB/VirD4 and Trw T4SSs in *Bartonella* species, and the Cag secretion system in the gastrointestinal pathogen *H. pylori*, which mediates the injection of the oncogenic effector protein CagA (Backert et al. 2017; Wagner and Dehio 2019).

Type IV secretion systems are broadly distributed and diverse in composition and function. In this review, we will focus on the VirB/VirD4 T4SS of *A. tumefaciens* and the F plasmid conjugative system of *E. coli*.

Architecture of Type IV Secretion Pili

Recent cryo-EM structures of F-family pili encoded by the pOX38 and pED208 conjugative plasmids have greatly enhanced our understanding of the build-up of these surface appendages (Costa et al. 2016). In this chapter, we will use these conjugative systems as the point of reference to describe the architecture of Type IV secretion pili. However, T4SSs differ strongly in the composition and complexity of their associated pili, as well as in the nature of the translocation substrates. Additional structures will thus be needed to determine how representative the conjugative pilus structures are for the architecture of Type IV secretion pili in general. F pili are tubular appendages on the outer membrane responsible for conjugation. The F pilus establishes contact between the donor and recipient cell and acts as a conduit for the exchange of genetic material between the cells, in the form of ssDNA (Waksman 2019). The pili are composed of thousands of TraA subunits, which can assemble and disassemble to extend or retract the appendage (Clarke et al. 2008a). Both the pOX38 and pED208 conjugative pili are hollow cylinders of comparable dimension. The pED208 encoded pilus is 87 Å thick with a lumenal diameter of 28 Å. Both conjugative pili show as five-start helical filaments (Fig. 12.4). The pED208 pilus shows twist angle of 28.2° and a rise of 12.1 Å, while the F pilus (encoded by pOX38) is found two distinctive conformational populations, one with twist of 27.9° and a rise of 13.2 Å, and a second with a twist of 28.1° and rise of 12.5 Å. The conjugative pili are homopolymers of the TraA pilin, which is a ~64 AA long peptide that folds into an all helical structure comprising three helices ($\alpha 1$, $\alpha 2$, and $\alpha 3$). TraA is synthesized as a pro-pilin with an unusual leader peptide of ~50 AA (Majdalani et al. 1996). The pro-pilin is inserted into the inner membrane with the help of TraQ, a transmembrane protein encoded by the F plasmid, in a process that requires ATP and proton motive force. After its insertion, the 5.5 kDa leader peptide is cleaved off by the periplasmic

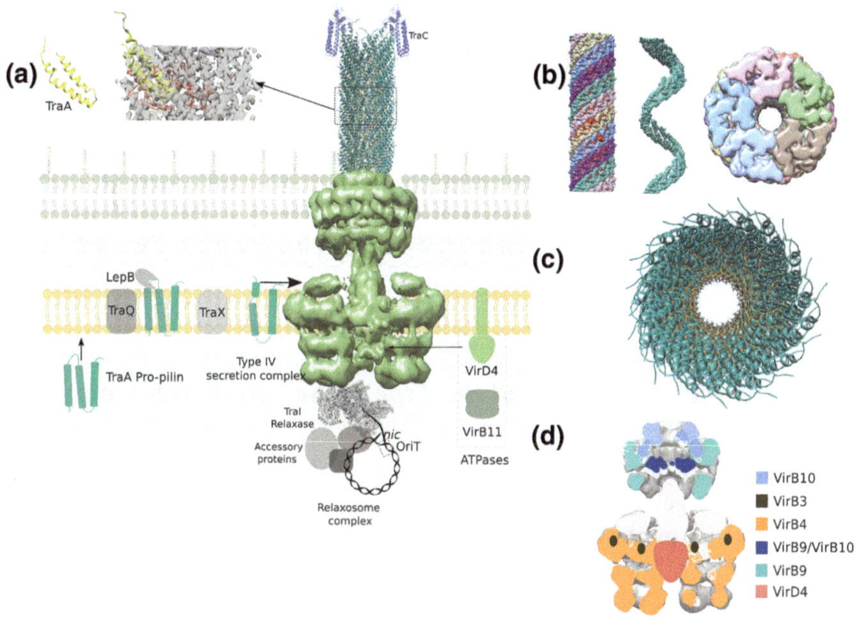

Fig. 12.4 Structure and assembly pathway of conjugative Type IV secretion pili. The shown structures are a collage of different studies and systems. **a** Ribbon representation of the pED208 TraA monomer (yellow) shown in isolation and docked into the cyoEM density of the pED208 F-pilus (PDB entry: 5LEG; EMDB entry: EMD-4042; the TraA bound lipid is shown in red). CryoEM structure of the R388 conjugative T4SS complex (green, EMDB entry: EMD-2567), and ribbon representation of the TraI relaxase of the F/R1 plasmid (grey, PDB entry: 5N8O). **b** Segmented cryoEM volume for the F conjugative pilus in side and top cross-sectional view (EMDB entry EMD-4042). The individual filaments of the five-start helices are differentiated in colour. **c** Ribbon representation of the pED208 F-pilus shown in top view (PDB entry: 5LEG). **d** Segmented cryoEM volume of the R388 conjugative T4SS complex (EMDB entry: EMD-2567) with a colour assignment of the different components, making use of the Vir nomenclature

peptidase LepB, resulting in arrangement where the $\alpha2$–$\alpha3$ loop is pointing towards the cytoplasm while N- and C-termini of $\alpha1$ and $\alpha3$ are in the periplasm (Majdalani and Ippen-Ihler 1996). The N-terminus of mature TraA is then acetylated by TraX, another protein encoded by the F plasmid (Maneewannakul et al. 1995).

A remarkable feature that has been discovered during helical reconstruction of the conjugative pili is the a stoichiometric binding of a phospholipids by the TraA subunits. Mass spectrometry analysis confirmed these to be phosphatidylglycerol (PG) species, primarily PG 32:1, PG33:1 and PG 34:1. Interestingly the cell membrane also contains these two phospholipids. However, other constituents of the membrane such as phosphatidylethanolamine (PE) and cardiolipin were absent, indicating TraA pilin subunits might have special preference for PGs. The head groups of the lipids are exposed to the lumen, whilst the acyl chains are completely buried between the helical segments of the TraA subunits. This coating of the pilus lumen with phospholipid head groups makes it electronegative, which thought to help in transport of the ssDNA. Mutation in the phospholipid binding sites (i.e. Y37V and A28F in pED208) switched off pilus assembly at the membrane surface, suggesting that the lipids have an additional role of maintaining integrity of the pilus assembly (Costa et al. 2016). The bound lipids may also assist in insertion of the appendage into the host membrane during establishment of initial contact, and may lower the energy barrier for subunit extraction during retraction of conjugative pili (Clarke et al. 2008b) (Hospenthal et al. 2017). Further studies are required to prove or disprove these speculative roles of lipids in the pilus.

Structure and Mechanism of the Type IV Secretion Machinery

The assembly and disassembly of conjugative pili and other Type IV secretion pili, as well as the passage of the secreted substrate depends on the trans-envelope spanning TypeIV secretion machinery. One of the canonical T4SS is the tDNA injecting system in *A. tumefaciens*. The components of this system are arranged in two operons, the VirB operon encoding 11 structural genes (VirB1–VirB11) and the VirD operon encoding 4 (VirD1–VirD4) genes with regulatory and accessory functions. Although different amongst species and from the type of system, a typical T4SS generally has the following building blocks (i) the transmembrane secretion complex composed of energy components (VirB4 and VirB11) and structural components (VirB3, VirB6, and VirB8 in the IM, and VirB7, VirB9, and VirB10 in the outer membrane (OM)), Hydrolase VirB1, (ii) the type 4 coupling protein(T4CP), (iii) the pilus (Waksman 2019).

X-ray structures of T4SS subcomplexes and the recent cryo-EM structure of the nearly complete T4SS secretion complex of conjugative plasmid R388 has vastly enhanced our understanding of the arrangement of various components in the 3 Mda complex (Low et al. 2014). It consists of an outer membrane embedded core complex (OMC), an inner membrane core complex (IMC) with 2 barrel-like legs inserting into the cytoplasm (Fig. 12.4). These components span the full Gram-negative cell envelope, going from the cytoplasmic side of the inner membrane to the extracellular side outer membrane, together forming the T4SS complex. The 1 Mda OMC is connected to the 2.6 Mda IMC through a connecting stalk. The OMC is composed of 14 copies each of VirB7, VirB9 and VirB10 while the IMC is made of 12 copies each of VirB3, VirB4, VirB5, VirB8 and 24 copies of VirB6. A segment of VirB10 is inserted into the inner membrane likely to regulate the conformational changes induced by various ATPases present in the inner membrane. The OMC can be further divided into two layers, an outer membrane O layer and an Inner membrane I layer (Marlovits et al. 2004; Rivera-Calzada et al. 2013). The O layer forms a channel in the outer membrane that is composed of an α helical barrel, where each α hairpin motif is contributed by 14 VirB10 units (Chandran et al. 2009). The IMC in the periplasm is composed of two arches connected to two barrel-like leg structures span the inner membrane and project into the cytoplasm. The leg like structures are made of 6 copies each of VirB4, an ATPase, constituting trimer of dimers (12 VirB4/TrwK subunits in total) (Low et al. 2014). VirB11, another ATPase, belonging to the large family of AAA+ hexameric traffic ATPases is responsible for energising the secretion process (Rashkova et al. 2000) along with VirB4. It is found embedded in the cytoplasmic membrane and has nucleoside triphosphatase activity that gets enhanced in the presence of lipids (Planet et al. 2001; Rivas et al. 1997). VirB11 of *H pylori* and *Brucella suis* is reported to form double hexameric rings (Hare et al. 2006; Savvides et al. 2003; Yeo et al. 2000). Although it has been structurally well characterized, its exact location in the assembled complex and its structure-function relationship in it is not clear. VirD4 ATPase is the type IV coupling proteins (T4CP) located between the two barrel like legs of the complex (Redzej et al. 2017). It recruits the relaxosome processed substrates(DNA/protein) that is to be transported across the secretion machinery (Cabeźn et al. 1997; Lang et al. 2010; Vergunst et al. 2000). Two copies of VirD4 dimers are located opposite to each other in the cleft formed by VirB4 hexamers in the cytoplasmic side of the inner membrane. As known from its crystal structure (Gomis-Rüth et al. 2001). It has a N-terminal transmembrane domain anchored to the cell membrane that provides stability to the cytoplasmic C terminal domain. The C terminal domain comprises of a catalytic DNA binding domain(NDB) and all helical α domain(AAD) (Chandran Darbari and Waksman 2015).

VirD4 can sense both intracellular and extracellular signal to regulate the opening/closing of the T4SS machinery to allow exchange of substrates (Berry and Christie 2011). It plays the crucial role of recruiting substrates for secretion through the machinery. For translocation of DNA substrate it first undergoes pre-processing by a relaxosome complex composed of 3–4 cytoplasmic proteins (Ilangovan et al. 2017) one of them being the relaxase. Relaxase is an enzyme that can execute two activities namely trans-esterase and helicase. Two relaxase molecules attaches to the origin of Transfer region (*oriT*) of the plasmid DNA with the help of other proteins of the relaxase complex. A first relaxase molecule makes a single strand nick in the transfer strand of the plasmid DNA (T strand) at the *nic* site of the *oriT* with its trans-esterase activity, and gets covalently linked to the free 5′ phosphate. A second relaxase molecule then attaches to the 3′ end of the other strand and unwinds the DNA by its helicase activity (Ilangovan et al. 2017). VirD4 subsequently captures the transfer complex through the relaxase attached to the *nic* site (Atmakuri et al. 2004). Notably, several T4SSs such as that of *Bartonella tribocorum*, *Brucella sp.*, and *Bordetella pertussis* lack ViD4, suggesting VirD4 might have a redundant role in pilus biogenesis. The T4SS is thought be responsible for both pilus biogenesis and conjugation, however mechanistic details of pilus biogenesis are still not clear. Likely, after assembling the pilus it switches to DNA transfer mode (Ilangovan et al. 2017; Zechner et al. 2012). VirD4 and the relaxase complex remain in a dormant state until some contact has been established with the recipient cell through the conjugation pilus. Upon contact the ATPases and the relaxases are activated which creates a ssDNA bubble on the 3′ side exposing the *nic* site that is further processed by the relaxase. The pathway of the transfer DNA through the secretion machinery, and the mechanism of transfer into the recipient cell remain unknown in molecular detail.

Sortase Mediated Pili of Gram-Positive Bacteria

Gram-positive bacteria lack the diderm cell envelope of Gram-negative bacteria, so that secretion and attachment of pili to the cell surface can be expected to be fundamentally different. In these monoderm systems, pilus subunits are secreted through the general secretory pathway, but require an alternative mechanism for polymerization and attachment to the cell surface. Both these steps rely on the transpeptidation activity a family of enzymes called sortases. Sortase mediated pili were first identified in Gram-positive bacteria in 1968 (Yanagawa et al. 1968), however, their detailed molecular characterization started relatively recently (Ton-That et al. 2004). They lack homology to pili of Gram negative bacteria and are usually much thinner and flexible (Hae et al. 2007). Sortase mediated pili are linear polymers of covalently associated subunits, and are attached to the cell surface by covalent association to the peptidoglycan layer. The intersubunit contacts and cell wall association are formed by a transpeptidase reaction catalysed by surface-localized sortases. Typically, the pili contain two to three subunit types, referred to as basal, backbone and auxiliary pilins, each with their own specific function (Pansegrau and Bagnoli 2017). The structural data available for sortase mediated pili from various bacterial species have shown the presence of intramolecular isopeptide bonds in the pilins (Hae et al. 2007). These isopeptide bonds are formed autocatalytically and are thought to be the functional equivalent of disulphide bonds in other systems, i.e. providing extra stability to the proteins by crosslinking (Kang and Baker 2009). Another interesting feature of these systems is the formation of thioester bonds between the tip pilin and host receptor proteins (Linke-Winnebeck et al. 2014; Pointon et al. 2010).

Pilus Architectures

The pili of Gram-positive bacteria are composed of two or three types of multidomain pilins, called backbone pilin, basal pilin, and tip pilin. *Corynebacterium diphtheria*, for example, has three distinct pilus systems, SpaABC, SpaDEF, and SpaHIG (Spa for sortase-mediated pilus assembly), each with three distinct subunits. In these trimeric systems SpaA, SpaD and SpaH are the major or the backbone pilins, while SpaB, SpaE, SpaI, and SpaC, SpaF, SpaG are basal and tip pilins, respectively (Ton-That and Schneewind 2003). The Spa operon model has been found conserved across many other Gram-positive bacteria such as *Streptococcus agalactiae*, *S. pyogenes*, *S. pneumonia*, *Enterococcus faecalis*, *E. faecium*, *Bacillus cereus*, and *Actinomyces naeslundii* (Mandlik et al. 2008b). The structures of a large number of pilins have now been studied, a detailed review of which can be found in (Krishnan 2015). Despite having limited primary sequence homology most pilins have a similar core tertiary structure. The pilins are modular multidomain proteins that are composed of multiple immunoglobulin like β-sandwich domains, which can be grouped into two main types: CnaA and CnaB domains. The CnaA domain was first identified

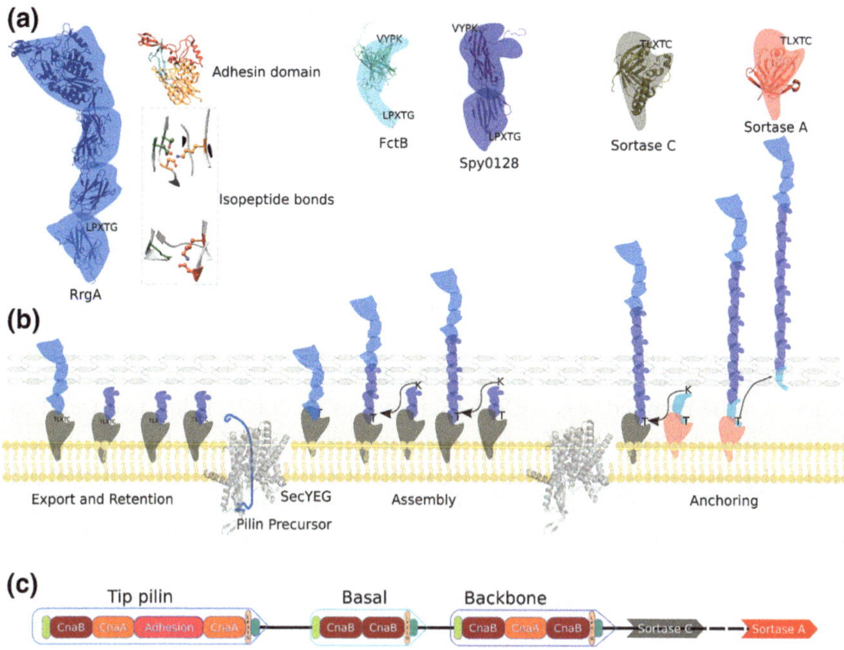

Fig. 12.5 Schematic representation of sortase-mediated pilus assembly. **a** Representative structures of adhesins and pilus subunits. From left to right: the tip pilin or pilus adhesin RrgA adhesin from *S. pneumoniae* (blue, PDB entry: 2WW8), the basal pilin FctB (cyan, PDB entry: 3KLQ) and backbone pilin Spy0128 (magenta, PDB entry: 3B2M) from *S. pyogenes*, and the sortase C of *S. pneumoniae* (dark grey, PDB entry: 2W1J) and sortase A (red, PDB entry:1T2P) of *S. aureus*. **b** Schematic diagram of sortase-mediated crosslinking and extension of Gram-positive pili (tip, backbone and basal pilins are coloured as in (**a**). The SecYEG translocase is shown in grey (PDB entry 5AWW). **c** Archetypical domain organization in tip, backbone and basal pilins

and coined in the ligand-binding region A of the collagen adhesin CNA of *S. aureus* (Symersky et al. 1997). In addition to the CnaA and CnaB domains, the pilins can also include a von Willebrand A (VWA)-like domain(vAPD) or thioester containing adhesin pilin domain(TAPD) (Fig. 12.5).

Backbone pilins. The backbone pilins are the major pilin subunit forming the Gram-positive pili, present in multiple tens to hundreds of copies. These subunits generally consist of two to four CnaA or CnaB domains that form intramolecular isopeptide bonds between the ε-amino group of a lysine and the carbonyl of a glutamic or aspartic acid, or the amide group of an asparagine residue (Kang et al. 2007; Kang and Baker 2011). These bonds are often made such that they link the first and last β strand of the domains, thereby resulting in a tightly packed β sheet structure with high thermal stability and protease resistance (Hae et al. 2007). The N terminal domain of backbone pilins, called D1, often holds a YPKN motif, of which the lysine is involved in the intermolecular transpeptide reaction during polymerization. Backbone pilins of several Gram-positive bacteria have been structurally characterized (Krishnan 2015), showing different architectures comprising pilins with two domains like Spy012 of *Streptococcus pyogenes* (Kang et al. 2007), three domains like SpaA from *Corynebacterium diphtheriae* (Kang et al. 2009), or 4 domains such as in BcpA in *Bacillus cereus* (Budzik et al. 2009).

Basal pilins. The basal pilins are responsible for anchoring of the pilus to the peptidoglycan layer. They are the last to be incorporated into the pilus and along with the housekeeping sortase (SrtF) act as the termination signal (Mandlik et al. 2008a). Basal pilins are often made of one to three CnaB type domains with or without isopeptide bonds. Basal pilins have a conserved proline rich C terminal tail that anchors to the cell wall (Fig. 12.5). The C terminal sorting motif of basal pilins have a triple glycine (TG) motif which possibly is required for sortase specificity, as demonstrated in *L. rhamnosus* GG (Douillard et al. 2014). The first basal pilin structure to be reported was GBS52 (Krishnan et al. 2007). It has two CnaB like domains linked head to tail through a hydrophobic interface. It has a IYPK pilin motif, of which lysine is thought to be responsible for incorporation into the pilus shaft by forming transpeptide bond with a corresponding backbone pilus C terminal. The crystal structure of another basal pilin FctB reveals the C terminal proline rich tail to fold into a polyproline-II helix (Linke et al. 2010). And again, FctB also has a lysine in the last strand that is responsible for the transpeptide bond formation. Interestingly, RrgC, a basal pilin of *Streptococcus pneumoniae* has three CnaB type domains instead of two (Shaik et al. 2014). It has two isopeptide bonds in domain 2 and domain 3 and a short polyproline C terminal tail. In all the reported basal pilin structures there is no isopeptide bond in the domain 1, presumably making it more flexible, which might be a necessity for cell wall anchoring of the pilus.

Tip pilins. Tip pilins are responsible for adhesion of the pilin to various host surfaces. These pilins are usually the largest among the pilus subunits, with 4 domains, which can include both CnaA and CnaB domains, as well as a specialized domain for adhesion: vAPD or TAPD, present at the pilus tip. CnaA and CnaB domains provide support to the adhesive domain and connect it to the backbone pilin (Fig. 12.5). The first tip pilin to be structurally characterized was RrgA (Izoré et al. 2010). It has 4 domains with D3 being a vAPD domain that is located at the tip and has adhesive properties. The D4 domain is responsible for connecting to the pilus shaft. The typical vAPD domain consists of a six stranded β sheets surrounded by α helices and resembles collagen-binding eukaryotic integrin I-domains and the MIDAS-containing vWFA-like subdomains of complement convertases. The MIDAS motif in RrgA is surrounded by two arm-like structures made of β hairpins and long loops that are implicated in collagen and metal ion binding, and can provide a binding site for anionic extracellular matrix molecules. Similar to RrgA two other tip pilins GBS104 (Krishnan et al. 2013) and Cpa (Linke-Winnebeck et al. 2014) have been partially characterized. Both have 4 domains, though while Cpa differs markedly in term of supertertiary organisation of its domains and its adhesive domain into Y shaped molecule with 2 thioester containing adhesion domains. The Y shape is formed by the centrally located D2 connecting D1 and D3 that form the binding arms and the D4 acts as the stalk. The adhesive domains, D1 and D3, share a common fold comprising of a β sandwich in a helical bundle where the β sandwich contains the thioester binding domain. Apart from that, thioester bonds can be found in adhesion domains of some tip pilins, e.g. Cpa from *S. pyogenes*. These bonds are formed between the side chains of cysteine and glutamine and are believed to mediate interaction with ligands present on the host cells.

Biogenesis and Secretion of Sortase Mediate Pili

Pilus biogenesis in Gram-positive bacteria is a biphasic process. A class of transpeptidases called sortases are responsible for biogenesis and housekeeping of the pili. Gram-positive bacteria encode several sortase types, each with diverse roles. Based on sequence homology, sortases have been classified into 6 main classes, sortase A–F (Spirig et al. 2011). Among these, class A and class C sortases are crucial from the perspective of pilus biogenesis and maintenance. Class C sortases catalyse the covalent linking reactions during initiation and elongation of pilus assembly (Budzik et al. 2007; Marraffini et al. 2006). Class A sortases are housekeeping enzymes that are responsible for covalent linking of the pili to the peptidoglycan layer (Bradshaw et al. 2015). Apart from the above-mentioned classes, in some species (e.g. *Corynebacterium diphtheriae*) Class E sortases act as the housekeeping enzyme instead of Class A (Comfort and Clubb 2004; Ton-That and Schneewind 2003). In general, substrates of sortases are transported to the cell surface via the Sec translocon to eventually get anchored in the cell membrane through its C-terminal part. The C-terminal part contains a cell wall targeting signal sequence (CWSS), which consists of a pentapeptide sorting motif followed by a hydrophobic domain which precedes a positively charged short chain (Schneewind and Missiakas 2014). The sorting motif is generally LPXTG that gets cleaved between threonine and glycine by a suitable sortase and produces an acyl intermediate which later gets incorporated in the growing pili.

Proteins that are targets of sortases often have an N-terminal signal peptide and C-terminal cell wall targeting signal sequence (CWSS) with LPXTG motif. In pilins, there is an additional conserved motif in the N-terminal region with sequence YPKN, also known as the pilin motif. After transport through the SecYEG translocon, the N-terminal signal sequence gets cleaved off and the pre-pilin subunit gets inserted into the cell membrane through the hydrophobic domain of the CWSS while the LPXTG motif is exposed to the exterior. A sortase enzyme present on the cell membrane recognizes the anchored pre-pilins through this LPXTG motif. Upon recognition, the cysteine of the TLXTC carries out a nucleophilic attack on the peptide bond between threonine and glycine in the LPXTG motif to break the pre-pilin into two parts (i) a thioester acyl intermediate containing the long N-terminal fragment and (ii) a short C-terminal cleavage product that is released (Marraffini et al. 2006). The resultant acyl intermediate forms a stable product by a nucleophilic substitution reaction, the nucleophile for which is variable depending upon the respective sortase in action. This nucleophile can either be an amine-containing residue in the peptide moiety of a lipid II molecule (class A sortase), meso-2,6-diaminopimelic acid (mDap) in the peptidoglycan cross-bridge (class A, D and E sortases), a lysine residue in a peptidoglycan cross-bridge (class B sortases), or a lysine residue that is part of the "YPKN" motif of a pilin precursor (class C sortases and certain class B sortases) (Frankel et al. 2005; Ton-That et al. 2004). The nucleophilic substitution marks the completion of the transpeptidation reaction, which either results in elongation of the polymerising pili or its covalent attachment to the peptidoglycan layer. These two processes are catalysed by two classes of sortases, the Class C type sortase for initiating and elongating the pilus and the housekeeping class A type sortase for terminating and anchoring of pili to the cell wall. These reactions require that both sortases and pilins are inserted in the cytoplasmic membrane. For this purpose, the pilins have a C-terminal CWSS, whilst sortases have a transmembrane domain in their N terminal regions (Schneewind and Missiakas 2014). The pilus assembly reaction starts when the LPXTG motif of a tip pilin is recognized and cleaved by a class C sortase present in the membrane. The resulting acyl intermediate then undergoes nucleophilic attack by the ε amine group of a specific lysine residue in a second pilus subunit, most often from the conserved YPKN motif in the N-terminus of a backbone pilin, resulting in the formation of an isopeptide bond between the tip pilin and the backbone pilin (Ton-That and Schneewind 2004). The backbone pilin is recognized by a separate C type sortase, which catalyses the transpeptidation reaction with subsequent backbone pilins until it reaches a termination signal by transpeptidation to a basal pilin. The latter is recognized and coupled to a class A sortase, which catalyses transpeptidation to the peptidoglycan precursor lipid II. In a second group of organisms such as in *Bacillus cereus* there is no distinct basal pilin (Budzik et al. 2007). In such cases the termination occurs when sortase A recognizes a backbone pilin and gets coupled to it by cleaving the LPXTG motif. This happens stochastically after a few rounds of polymerization. The acyl intermediate formed in this transpeptidation reaction can only be resolved by a nucleophilic reaction with the N-terminal glycine of peptide side chains in lipid II molecules. Therefore, addition of a sortase A coupled backbone pilin acts as the termination signal. As a final step, the pilus filaments coupled to lipid II are then incorporated into the peptidoglycan layer, resulting in their covalent attachment to the cell wall.

Acknowledgements This work was supported by VIB and the ERC under consolidator grant BAS-SBBT.

Addendum on e-pili

Geobacter species such as *Geobacter sulfurreducens* possess micrometer long conductive filaments or "nanowires" that mediate extracellular electron transport (Reguera et al. 2005), and play a role in respiration (Malvankar et al. 2011) and interspecific electron exchange (Summers et al. 2010). Nanowire-expressing bacteria are associated with many important redox processes such as carbon and mineral recycling in soil, metal corrosion, conversion of organic waste to methane and electricity (Malvankar and Lovley 2014). The molecular nature of these nanowires has been obscure, however, and for a long time they were considered to be composed of Type IV pilus-like PilA subunits (Childers et al. 2002), with the conductive properties proposed to be coming from stacked aromatic residues in the pilus subunits, or through association with the C-type cytochrome OmcS (Leang et al. 2013).

The recent cryoEM structures of isolated conductive pili or "e-pili" from *G. sulfurreducens* have now revealed their unique structure, composition and electron transport mechanism (Filman et al. 2018; Wang et al. 2019). This study shows *Geobacter* e-pili are composed of a unique noncovalent linear polymer of the cytochrome OmcS (Fig. 12.6). OmcS was previously known to be essential for bacterial growth on insoluble electron acceptors such as Fe (III) oxide and electrodes, but had not been directly implicated as the main nanowire component (Holmes et al. 2006; Mehta et al. 2005). The structures show each OmcS subunit contains six stacked hemes that are placed within 3.5–6 Å to each other in a trajectory that spans the length of the nanowire and results in electric coupling within and across subunits. The hemes are found in parallel pairs, each pair oriented perpendicular to the next. In the parallel arrangement, the hemes are present within a range of 3.4–4.1 Å while in the perpendicular arrangement they are within a range of 5.4–6.1 Å (Fig. 12.6). The parallel arrangement is expected to maximize electron coupling while the perpendicular arrangement is thought to enhance the structural stability of the subunit. For each heme group, two histidines situated axially form coordination bonds with the iron atom present at the center, and the vinyl groups form covalent thioester bonds with cysteine. In the filament, subunits associate in a head-to-tail arrangement, with each subunit in contact with only one preceding and succeeding subunit, through an inter-subunit contact comprising ~2,600 $Å^2$ surface area. Histidine 16 of each subunit coordinates an adjacent heme group present in the succeeding subunit. These inter-subunit coordination bonds are proposed to maximize the structural stability of the nanowire. This cross-interface heme coordination is speculated to result from a domain swapping of the C-terminal helix, a process that has been observed in induced cytochrome polymerization (Hirota et al. 2010; Wang et al. 2019). Little is known about the secretion and in vivo assembly pathway of OmcS e-pili. Although it is now evident that PilA is not the main structural subunit of the nanowires, previous genetic studies clearly associate it with the presence of *Geobacter*'s e-pili and the secretion of OmcS to the extracellular space (Liu et al. 2018; Richter et al. 2012). Deletion or mutation of PilA leads to the absence of OmcS on the surface (Reguera et al. 2005) while overexpression of PilA leads to overproduction of nanowires (Leang et al. 2013; Summers et al. 2010). The cryoEM structures now clearly demonstrate PilA is not found as a component of e-pili (Filman et al. 2018; Wang et al. 2019). Whether PilA is found as an additional extracellular fiber or acts as a pseudopilus of a T2SS, and where and how OmcS subunits polymerize and associate with the cell surface will require further study.

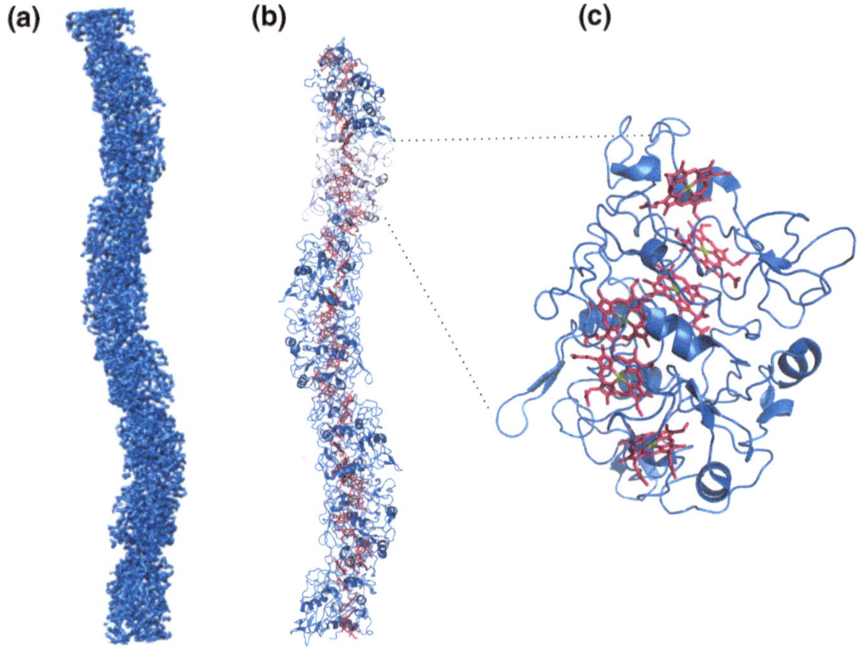

Fig. 12.6 **a** 3.7 Å cryo-EM map of *Geobacter sulfurreducens* e-pili or nanowires (EMD_9046). **b** atomic model of OmcS subunits arranged in the e-pili (width 55 Å, PDB id: 6ef8), and **c** atomic model of a single OmcS subunit, with its chain of six stacked heme groups shown in pink and the iron atom in green

References

Alaei SR, Park JH, Walker SG, Thanassi DG (2019) Peptide-based inhibitors of fimbrial biogenesis in Porphyromonas gingivalis. Infect Immun. https://doi.org/10.1128/iai.00750-18

Alteri CJ (2005) Novel pili of Mycobacterium tuberculosis. PhD thesis, The University of Arizona

Anderson KL, Billington J, Pettigrew D, Cota E, Simpson P, Roversi P, Chen HA, Urvil P, du Merle L, Barlow PN, Medof ME, Smith RAG, Nowicki B, Le Bouguénec C, Lea SM, Matthews S (2004) An atomic resolution model for assembly, architecture, and function of the Dr. Adhesins. Mol Cell 15(4):647–657. https://doi.org/10.1016/j.molcel.2004.08.003

Atmakuri K, Cascales E, Christie PJ (2004) Energetic components VirD4, VirB11 and VirB4 mediate early DNA transfer reactions required for bacterial type IV secretion. Mol Microbiol 54(5):1199–1211. https://doi.org/10.1111/j.1365-2958.2004.04345.x

Backert S, Haas R, Gerhard M, Naumann M (2017) The helicobacter pylori type IV secretion system encoded by the cag Pathogenicity Island: architecture, function, and signaling, vol 413, pp 187–220. https://doi.org/10.1007/978-3-319-75241-9_8

Bann JG, Pinkner JS, Frieden C, Hultgren SJ (2004) Catalysis of protein folding by chaperones in pathogenic bacteria. PNAS 101(50):17389–17393. https://doi.org/10.1073/pnas.0408072101

Bao R, Nair MKM, W-k Tang, Esser L, Sadhukhan A, Holland RL, Xia D, Schifferli DM (2013) Structural basis for the specific recognition of dual receptors by the homopolymeric pH 6 antigen (Psa) fimbriae of Yersinia pestis. Proc Natl Acad Sci USA 110(3):1065–1070. https://doi.org/10.1073/pnas.1212431110

Bečárová Z (2015) Mechanism of FimI, the assembly termination subunit of the type 1 pili from uropathogenic Escherichia coli. Doctoral thesis, ETH Zurich,

Berry J-L, Pelicic V (2015) Exceptionally widespread nanomachines composed of type IV pilins: the prokaryotic Swiss Army knives. FEMS Microbiol Rev 39(1):134–154. https://doi.org/10.1093/femsre/fuu001

Berry J-L, Phelan MM, Collins RF, Adomavicius T, Tønjum T, Frye SA, Bird L, Owens R, Ford RC, Lian L-Y, Derrick JP (2012) Structure and assembly of a trans-periplasmic channel for type IV Pili in Neisseria meningitidis. PLoS Pathog 8(9):e1002923. https://doi.org/10.1371/journal.ppat.1002923

Berry TM, Christie PJ (2011) Caught in the act: the dialogue between bacteriophage R17 and the type IV secretion machine of plasmid R1. Mol Microbiol 82(5):1039–1043. https://doi.org/10.1111/j.1365-2958.2011.07870.x

Bitter W, Koster M, Latijnhouwers M, de Cock H, Tommassen J (1998) Formation of oligomeric rings by XcpQ and PilQ, which are involved in protein transport across the outer membrane of Pseudomonas aeruginosa. Mol Microbiol 27(1):209–219

Blanco LP, Evans ML, Smith DR, Badtke MP, Chapman MR (2012) Diversity, biogenesis and function of microbial amyloids. Trends Microbiol 20(2):66–73. https://doi.org/10.1016/j.tim.2011.11.005

Bleem A, Christiansen G, Madsen DJ, Maric H, Strømgaard K, Bryers JD, Daggett V, Meyer RL, Otzen DE (2018) Protein engineering reveals mechanisms of functional amyloid formation in Pseudomonas aeruginosa biofilms. J Mol Biol 430(20):3751–3763. https://doi.org/10.1016/j.jmb.2018.06.043

Bradshaw WJ, Davies AH, Chambers CJ, Roberts AK, Shone CC, Acharya KR (2015) Molecular features of the sortase enzyme family. FEBS J 282(11):2097–2114. https://doi.org/10.1111/febs.13288

Branda SS, Vik Å, Friedman L, Kolter R (2005) Biofilms: the matrix revisited. Trends Microbiol 13(1):20–26. https://doi.org/10.1016/j.tim.2004.11.006

Brombacher E, Dorel C, Zehnder AJB, Landini P (2003) The curli biosynthesis regulator CsgD co-ordinates the expression of both positive and negative determinants for biofilm formation in Eschericia coli. Microbiology 149(10):2847–2857. https://doi.org/10.1099/mic.0.26306-0

Budzik JM, Marraffini LA, Schneewind O (2007) Assembly of pili on the surface of Bacillus cereus vegetative cells. Mol Microbiol 66(2):495–510. https://doi.org/10.1111/j.1365-2958.2007.05939.x

Budzik JM, Poor CB, Faull KF, Whitelegge JP, He C, Schneewind O (2009) Intramolecular amide bonds stabilize pili on the surface of bacilli. Proc Natl Acad Sci USA 106(47):19992–19997. https://doi.org/10.1073/pnas.0910887106

Cabeźn E, Ignacio Sastre J, De La Cruz F (1997) Genetic evidence of a coupling role for the TraG protein family in bacterial conjugation. Mol Gen Genet 254(4):400–406. https://doi.org/10.1007/s004380050432

Cao B, Zhao Y, Kou Y, Ni D, Zhang XC, Huang Y (2014) Structure of the nonameric bacterial amyloid secretion channel. PNAS 111(50):E5439–E5444. https://doi.org/10.1073/pnas.1411942111

Chai L, Romero D, Kayatekin C, Akabayov B, Vlamakis H, Losick R, Kolter R (2013) Isolation, characterization, and aggregation of a structured bacterial matrix precursor. J Biol Chem 288(24):17559–17568. https://doi.org/10.1074/jbc.M113.453605

Chandran Darbari V, Waksman G (2015) Structural biology of bacterial type IV secretion systems. Annu Rev Biochem 84(1):603–629. https://doi.org/10.1146/annurev-biochem-062911-102821

Chandran V, Fronzes R, Duquerroy S, Cronin N, Navaza J, Waksman G (2009) Structure of the outer membrane complex of a type IV secretion system. Nature 462(7276):1011–1015. https://doi.org/10.1038/nature08588

Chang Y-W, Rettberg LA, Treuner-Lange A, Iwasa J, Søgaard-Andersen L, Jensen GJ (2016) Architecture of the type IVa pilus machine. Science 351(6278):aad2001. https://doi.org/10.1126/science.aad2001

Chapman MR, Robinson LS, Pinkner JS, Roth R, Heuser J, Hammar M, Normark S, Hultgren SJ (2002) Role of fiber Escherichia in coli curli operons directing formation amyloid. Science 295:851–855. https://doi.org/10.1126/science.1067484

Chiang P, Habash M, Burrows LL (2005) Disparate subcellular localization patterns of Pseudomonas aeruginosa Type IV pilus ATPases involved in twitching motility. J Bacteriol 187(3):829–839. https://doi.org/10.1128/JB.187.3.829-839.2005

Childers SE, Ciufo S, Lovley DR (2002) *Geobacter metallireducens* accesses insoluble Fe(III) oxide by chemotaxis. Nature 416:767–769

Chirwa NT, Herrington MB (2003) CsgD, a regulator of curli and cellulose synthesis, also regulates serine hydroxymethyltransferase synthesis in Escherichia coli K-12. Microbiology 149(2):525–535. https://doi.org/10.1099/mic.0.25841-0

Chiti F, Dobson CM (2006) Protein misfolding, functional amyloid, and human disease. Annu Rev Biochem 75(1):333–366. https://doi.org/10.1146/annurev.biochem.75.101304.123901

Choudhury D, Thompson A, Stojanoff V, Langermann S, Pinkner J, Hultgren SJ, Knight SD (1999) X-ray structure of the FimC-FimH chaperone-adhesin complex from uropathogenic Escherichia coli. Science 285(5430):1061–1066

Clarke M, Maddera L, Harris RL, Silverman PM (2008a) F-pili dynamics by live-cell imaging. Proc Natl Acad Sci USA 105(46):17978–17981. https://doi.org/10.1073/pnas.0806786105

Clarke M, Maddera L, Harris RL, Silverman PM (2008b) F-pili dynamics by live-cell imaging. PNAS 105(46):17978–17981. https://doi.org/10.1073/pnas.0806786105

Collins R, Karuppiah V, Siebert CA, Dajani R, Thistlethwaite A, Derrick JP (2018) Structural cycle of the Thermus thermophilus PilF ATPase: the powering of type IVa pilus assembly. Sci Rep 8(1):14022. https://doi.org/10.1038/s41598-018-32218-3

Collinson SK, Doig PC, Doran JL, Clouthier S, Trust TJ, Kay WW (1993) Thin, aggregative fimbriae mediate binding of Salmonella enteritidis to fibronectin. J Bacteriol 175(1):12–18. https://doi.org/10.1128/jb.175.1.12-18.1993

Comfort D, Clubb RT (2004) A comparative genome analysis identifies distinct sorting pathways in Gram-positive bacteria. Infect Immun 72(5):2710–2722. https://doi.org/10.1128/IAI.72.5.2710-2722.2004

Conover Matt S, Ruer S, Taganna J, Kalas V, De Greve H, Pinkner Jerome S, Dodson Karen W, Remaut H, Hultgren Scott J (2016) Inflammation-induced adhesin-receptor interaction provides a fitness advantage to uropathogenic E. coli during chronic infection. Cell Host Microbe 20(4):482–492. https://doi.org/10.1016/j.chom.2016.08.013

Costa TRD, Ilangovan A, Ukleja M, Redzej A, Santini JM, Smith TK, Egelman EH, Waksman G (2016) Structure of the bacterial sex F pilus reveals an assembly of a stoichiometric protein-phospholipid complex. Cell 166(6):1436–1444.e1410. https://doi.org/10.1016/j.cell.2016.08.025

Craig L, Li J (2008) Type IV pili: paradoxes in form and function. Curr Opin Struct Biol 18(2):267–277. https://doi.org/10.1016/j.sbi.2007.12.009

Craig L, Pique ME, Tainer JA (2004) Type IV pilus structure and bacterial pathogenicity. Nat Rev Microbiol 2(5):363–378. https://doi.org/10.1038/nrmicro885

Craig L, Taylor RK, Pique ME, Adair BD, Arvai AS, Singh M, Lloyd SJ, Shin DS, Getzoff ED, Yeager M, Forest KT, Tainer JA (2003) Type IV pilin structure and assembly: X-ray and EM analyses of Vibrio cholerae toxin-coregulated pilus and Pseudomonas aeruginosa PAK pilin. Mol Cell 11(5):1139–1150

Craig L, Volkmann N, Arvai AS, Pique ME, Yeager M, Egelman EH, Tainer JA (2006) Type IV pilus structure by cryo-electron microscopy and crystallography: implications for pilus assembly and functions. Mol Cell 23(5):651–662. https://doi.org/10.1016/j.molcel.2006.07.004

Cusumano CK, Pinkner JS, Han Z, Greene SE, Ford BA, Crowley JR, Henderson JP, Janetka JW, Hultgren SJ (2011) Treatment and prevention of urinary tract infection with orally active FimH inhibitors. Sci Transl Med 3(109):109ra115. https://doi.org/10.1126/scitranslmed.3003021

Danelishvili L, Yamazaki Y, Selker J, Bermudez LE (2010) Secreted Mycobacterium tuberculosis Rv3654c and Rv3655c proteins participate in the suppression of macrophage apoptosis. PLoS ONE 5(5). https://doi.org/10.1371/journal.pone.0010474

Depas WH, Chapman MR, Hufnagel DA (2015) The biology of the Escherichia coli extracellular matrix. In: Microbial biofilms, 2nd edn, pp 249–267. https://doi.org/10.1128/microbiolspec.mb-0014-2014
Dodson KW, Jacob-Dubuisson F, Striker RT, Hultgren SJ (1993) Outer-membrane PapC molecular usher discriminately recognizes periplasmic chaperone-pilus subunit complexes. Proc Natl Acad Sci USA 90(8):3670–3674
Dodson KW, Pinkner JS, Rose T, Magnusson G, Hultgren SJ, Waksman G (2001) Structural basis of the interaction of the Pyelonephritic E. coli adhesin to its human kidney receptor. Cell 105(6):733–743. https://doi.org/10.1016/s0092-8674(01)00388-9
Douillard FP, Rasinkangas P, Von Ossowski I, Reunanen J, Palva A, De Vos WM (2014) Functional identification of conserved residues involved in Lactobacillus rhamnosus strain GG sortase specificity and pilus biogenesis. J Biol Chem 289(22):15764–15775. https://doi.org/10.1074/jbc.M113.542332
Dueholm MS, Albertsen M, Otzen D, Nielsen PH (2012) Curli functional amyloid systems are phylogenetically widespread and display large diversity in operon and protein structure. PLoS ONE 7(12):1–10. https://doi.org/10.1371/journal.pone.0051274
Dueholm MS, Petersen SV, Sønderkær M, Larsen P, Christiansen G, Hein KL, Enghild JJ, Nielsen JL, Nielsen KL, Nielsen PH, Otzen DE (2010) Functional amyloid in pseudomonas. Mol Microbiol 77(4):1009–1020. https://doi.org/10.1111/j.1365-2958.2010.07269.x
Dueholm MS, Søndergaard MT, Nilsson M, Christiansen G, Stensballe A, Overgaard MT, Givskov M, Tolker-Nielsen T, Otzen DE, Nielsen PH (2013) Expression of Fap amyloids in Pseudomonas aeruginosa, P. fluorescens, and P. putida results in aggregation and increased biofilm formation. MicrobiologyOpen 2(3):365–382. https://doi.org/10.1002/mbo3.81
Evans ML, Chorell E, Taylor JD, Åden J, Götheson A, Li F, Koch M, Sefer L, Matthews SJ, Wittung-Stafshede P, Almqvist F, Chapman MR (2015) The bacterial curli system possesses a potent and selective inhibitor of amyloid formation. Mol Cell 57(3):445–456. https://doi.org/10.1016/j.molcel.2014.12.025
Fällman E, Schedin S, Jass J, Uhlin B-E, Axner O (2005) The unfolding of the P pili quaternary structure by stretching is reversible, not plastic. EMBO Rep 6(1):52–56. https://doi.org/10.1038/sj.embor.7400310
Filman DJ, Marino SF, Ward JE, Yang L, Mester Z, Bullitt E, Lovley DR, Strauss M (2018). Structure of a cytochrome-based bacterial nanowire. 492645
Fowler DM, Koulov AV, Balch WE, Kelly JW (2007) Functional amyloid—from bacteria to humans. Trends Biochem Sci 32(5):217–224. https://doi.org/10.1016/j.tibs.2007.03.003
Frankel BA, Kruger RG, Robinson DE, Kelleher NL, McCafferty DG (2005) Staphylococcus aureus sortase transpeptidase SrtA: insight into the kinetic mechanism and evidence for a reverse protonation catalytic mechanism. Biochemistry 44(33):11188–11200. https://doi.org/10.1021/bi050141j
Gerstel U, Römling U (2003) The csgD promoter, a control unit for biofilm formation in Salmonella typhimurium. Res Microbiol 154(10):659–667. https://doi.org/10.1016/j.resmic.2003.08.005
Gibson DL, White AP, Rajotte CM, Kay WW (2007) AgfC and AgfE facilitate extracellular thin aggregative fimbriae synthesis in Salmonella Enteritidis. Microbiology 153(4):1131–1140. https://doi.org/10.1099/mic.0.2006/000935-0
Giltner CL, Nguyen Y, Burrows LL (2012) Type IV pilin proteins: versatile molecular modules. Microbiol Mol Biol Rev 76(4):740–772. https://doi.org/10.1128/MMBR.00035-12
Gold VAM, Salzer R, Averhoff B, Kühlbrandt W (2015) Structure of a type IV pilus machinery in the open and closed state. eLife
Gomis-Rüth FX, Moncalián G, Pérez-Luque R, González A, Cabezón E, De La Cruz F, Coll M (2001) The bacterial conjugation protein TrwB resembles ring helicases and F1-atpase. Nature 409(6820):637–641. https://doi.org/10.1038/35054586
Goyal P, Krasteva PV, Van Gerven N, Gubellini F, Van Den Broeck I, Troupiotis-Tsaïlaki A, Jonckheere W, Péhau-Arnaudet G, Pinkner JS, Chapman MR, Hultgren SJ, Howorka S, Fronzes R, Remaut H (2014) Structural and mechanistic insights into the bacterial amyloid secretion channel CsgG. Nature 516(7530):250–253. https://doi.org/10.1038/nature13768

Grohmann E, Christie PJ, Waksman G, Backert S (2018) Type IV secretion in Gram-negative and Gram-positive bacteria. Mol Microbiol 107(4):455–471. https://doi.org/10.1111/mmi.13896

Grund S, Weber A (1988) A new type of fimbriae on Salmonella typhimurium. J Vet Med Ser B 35(1–10):779–782. https://doi.org/10.1111/j.1439-0450.1988.tb00560.x

Hae JK, Coulibaly F, Clow F, Proft T, Baker EN (2007) Stabilizing isopeptide bonds revealed in Gram-positive bacterial pilus structure. Science 318(5856):1625–1628. https://doi.org/10.1126/science.1145806

Hall-Stoodley L, Stoodley P (2009) Evolving concepts in biofilm infections. Cell Microbiol 11(7):1034–1043. https://doi.org/10.1111/j.1462-5822.2009.01323.x

Hamada N, Sojar HT, Cho MI, Genco RJ (1996) Isolation and characterization of a minor fimbria from Porphyromonas gingivalis. Infect Immun 64(11):4788–4794

Hammar M, Arnqvist A, Bian Z, Olsen A, Normark S (1995) Expression of two csg operons is required for production of fibronectin- and congo red-binding curli polymers in Escherichia coli K-12. Mol Microbiol 18(4):661–670

Hammer ND, McGuffie BA, Zhou Y, Badtke MP, Reinke AA, Brännström K, Gestwicki JE, Olofsson A, Almqvist F, Chapman MR (2012) The C-terminal repeating units of CsgB direct bacterial functional amyloid nucleation. J Mol Biol 422(3):376–389. https://doi.org/10.1016/j.jmb.2012.05.043

Hammer ND, Schmidt JC, Chapman MR (2007) The curli nucleator protein, CsgB, contains an amyloidogenic domain that directs CsgA polymerization. PNAS 104(30):12494–12499. https://doi.org/10.1073/pnas.0703310104

Hare S, Bayliss R, Baron C, Waksman G (2006) A large domain swap in the VirB11 ATPase of Brucella suis leaves the hexameric assembly intact. J Mol Biol 360(1):56–66. https://doi.org/10.1016/j.jmb.2006.04.060

Hirota S, Hattori Y, Nagao S, Taketa M, Komori H, Kamikubo H, Wang Z, Takahashi I, Negi S, Sugiura Y, et al (2010) Cytochrome *c* polymerization by successive domain swapping at the Cterminal helix. Proc Nat Acad Sci US 107:12854–12859

Holmes DE, Chaudhuri SK, Nevin KP, Mehta T, Methé BA, Liu A, Ward JE, Woodard TL, Webster J, Lovley DR (2006) Microarray and genetic analysis of electron transfer to electrodes in *Geobacter sulfurreducens*. Environ Microbiol 8:1805–1815

Hospenthal MK, Costa TRD, Waksman G (2017) A comprehensive guide to pilus biogenesis in Gram-negative bacteria. Nat Rev Microbiol 15(6):365–379. https://doi.org/10.1038/nrmicro.2017.40

Hospenthal Manuela K, Redzej A, Dodson K, Ukleja M, Frenz B, Rodrigues C, Hultgren Scott J, DiMaio F, Egelman Edward H, Waksman G (2016) Structure of a chaperone-usher pilus reveals the molecular basis of rod uncoiling. Cell 164(1–2):269–278. https://doi.org/10.1016/j.cell.2015.11.049

Hultgren SJ, Abraham S, Caparon M, Falk P, St Geme JW, Normark S (1993) Pilus and nonpilus bacterial adhesins: assembly and function in cell recognition. Cell 73(5):887–901

Hung C-S, Bouckaert J, Hung D, Pinkner J, Widberg C, DeFusco A, Auguste CG, Strouse R, Langermann S, Waksman G, Hultgren SJ (2002) Structural basis of tropism of Escherichia coli to the bladder during urinary tract infection. Mol Microbiol 44(4):903–915. https://doi.org/10.1046/j.1365-2958.2002.02915.x

Ilangovan A, Kay CWM, Roier S, El Mkami H, Salvadori E, Zechner EL, Zanetti G, Waksman G (2017) Cryo-EM structure of a relaxase reveals the molecular basis of DNA unwinding during bacterial conjugation. Cell 169(4):708–721.e712. https://doi.org/10.1016/j.cell.2017.04.010

Imam S, Chen Z, Roos DS, Pohlschröder M (2011) Identification of surprisingly diverse type IV Pili, across a broad range of Gram-positive bacteria. PLoS ONE 6(12). https://doi.org/10.1371/journal.pone.0028919

Izoré T, Contreras-Martel C, El Mortaji L, Manzano C, Terrasse R, Vernet T, Di Guilmi AM, Dessen A (2010) Structural basis of host cell recognition by the pilus adhesin from Streptococcus pneumoniae. Structure 18(1):106–115. https://doi.org/10.1016/j.str.2009.10.019

Jeter C, Matthysse AG (2005) Characterization of the binding of diarrheagenic strains of *E. coli* to plant surfaces and the role of curli in the interaction of the bacteria with alfalfa sprouts. Mol Plant-Microbe Interact 18(11):1235–1242. https://doi.org/10.1094/mpmi-18-1235

Kalas V, Hibbing ME, Maddirala AR, Chugani R, Pinkner JS, Mydock-McGrane LK, Conover MS, Janetka JW, Hultgren SJ (2018) Structure-based discovery of glycomimetic FmlH ligands as inhibitors of bacterial adhesion during urinary tract infection. Proc Natl Acad Sci USA 115(12):E2819–E2828. https://doi.org/10.1073/pnas.1720140115

Kang HJ, Baker EN (2009) Intramolecular isopeptide bonds give thermodynamic and proteolytic stability to the major Pilin protein of Streptococcus pyogenes. J Biol Chem 284(31):20729–20737. https://doi.org/10.1074/jbc.M109.014514

Kang HJ, Baker EN (2011) Intramolecular isopeptide bonds: Protein crosslinks built for stress? Trends Biochem Sci 36(4):229–237. https://doi.org/10.1016/j.tibs.2010.09.007

Kang HJ, Paterson NG, Gaspar AH, Ton-That H, Baker EN (2009) The Corynebacterium diphtheriae shaft pilin SpaA is built of tandem Ig-like modules with stabilizing isopeptide and disulfide bonds. PNAS 106(40):16967–16971. https://doi.org/10.1073/pnas.0906826106

Karuppiah V, Derrick JP (2011) Structure of the PilM-PilN inner membrane type IV pilus biogenesis complex from Thermus thermophilus. J Biol Chem 286(27):24434–24442. https://doi.org/10.1074/jbc.M111.243535

Keller R, Ordoñez JG, de Oliveira RR, Trabulsi LR, Baldwin TJ, Knutton S (2002) Afa, a diffuse adherence fibrillar adhesin associated with enteropathogenic Escherichia coli. Infect Immun 70(5):2681–2689. https://doi.org/10.1128/IAI.70.5.2681-2689.2002

Kim K, Oh J, Han D, Kim EE, Lee B, Kim Y (2006) Crystal structure of PilF: functional implication in the type 4 pilus biogenesis in Pseudomonas aeruginosa. Biochem Biophys Res Commun 340(4):1028–1038. https://doi.org/10.1016/j.bbrc.2005.12.108

Klein RD, Shu Q, Cusumano ZT, Nagamatsu K, Gualberto NC, Lynch AJL, Wu C, Wang W, Jain N, Pinkner JS, Amarasinghe GK, Hultgren SJ, Frieden C, Chapman MR (2018) Structure-function analysis of the curli accessory protein CsgE defines surfaces essential for coordinating amyloid fiber formation. mBio 9(4). https://doi.org/10.1128/mbio.01349-18

Kloppsteck P, Hall M, Hasegawa Y, Persson K (2016) Structure of the fimbrial protein Mfa4 from Porphyromonas gingivalis in its precursor form: implications for a donor-strand complementation mechanism. Sci Rep 6. https://doi.org/10.1038/srep22945

Kolappan S, Coureuil M, Yu X, Nassif X, Egelman EH, Craig L (2016) Structure of the Neisseria meningitidis Type IV pilus. Nat Commun 7:13015. https://doi.org/10.1038/ncomms13015

Korhonen TK, Parkkinen J, Hacker J, Finne J, Pere A, Rhen M, Holthofer H (1986) Binding of Escherichia coli S fimbriae to human kidney epithelium. Infect Immun 54(2):322–327

Krantz BA, Melnyk RA, Zhang S, Juris SJ, Lacy DB, Wu Z, Finkelstein A, Collier RJ (2005) A phenylalanine clamp catalyzes protein translocation through the anthrax toxin pore. Science 309(5735):777–781. https://doi.org/10.1126/science.1113380

Krishnan V (2015) Pilins in Gram-positive bacteria: a structural perspective. IUBMB Life 67(7):533–543. https://doi.org/10.1002/iub.1400

Krishnan V, Dwivedi P, Kim BJ, Samal A, MacOn K, Ma X, Mishra A, Doran KS, Ton-That H, Narayana SVL (2013) Structure of Streptococcus agalactiae tip pilin GBS104: a model for GBS pili assembly and host interactions. Acta Crystallogr D Biol Crystallogr 69(6):1073–1089. https://doi.org/10.1107/S0907444913004642

Krishnan V, Gaspar AH, Ye N, Mandlik A, Ton-That H, Narayana SVL (2007) An IgG-like domain in the minor pilin GBS52 of Streptococcus agalactiae mediates lung epithelial cell adhesion. Structure 15(8):893–903. https://doi.org/10.1016/j.str.2007.06.015

Lang S, Gruber K, Mihajlovic S, Arnold R, Gruber CJ, Steinlechner S, Jehl MA, Rattei T, Fröhlich KU, Zechner EL (2010) Molecular recognition determinants for type IV secretion of diverse families of conjugative relaxases. Mol Microbiol 78(6):1539–1555. https://doi.org/10.1111/j.1365-2958.2010.07423.x

Langermann S, Möllby R, Burlein JE, Palaszynski SR, Auguste CG, DeFusco A, Strouse R, Schenerman MA, Hultgren SJ, Pinkner JS, Winberg J, Guldevall L, Söderhäll M, Ishikawa K,

Normark S, Koenig S (2000) Vaccination with FimH adhesin protects cynomolgus monkeys from colonization and infection by uropathogenic Escherichia coli. J Infect Dis 181(2):774–778. https://doi.org/10.1086/315258
Langermann S, Palaszynski S, Barnhart M, Auguste G, Pinkner JS, Burlein J, Barren P, Koenig S, Leath S, Jones CH, Hultgren SJ (1997) Prevention of mucosal Escherichia coli infection by FimH-adhesin-based systemic vaccination. Science 276(5312):607–611
Larsen P, Nielsen JL, Otzen D, Nielsen PH (2008) Amyloid-like adhesins produced by floc-forming and filamentous bacteria in activated sludge. Appl Environ Microbiol 74(5):1517–1526. https://doi.org/10.1128/AEM.02274-07
Leang C, Malvankar NS, Franks AE, Nevin KP, Lovley DR (2013) Engineering *Geobacter sulfurreducens* to produce a highly cohesive conductive matrix with enhanced capacity for current production. Energy Environ Sci 6:1901–1908
Leighton TL, Buensuceso RNC, Howell PL, Burrows LL (2015a) Biogenesis of Pseudomonas aeruginosa type IV pili and regulation of their function. Environ Microbiol 17(11):4148–4163. https://doi.org/10.1111/1462-2920.12849
Leighton TL, Dayalani N, Sampaleanu LM, Howell PL, Burrows LL (2015b) Novel role for PilNO in type IV pilus retraction revealed by alignment subcomplex mutations. J Bacteriol 197(13):2229–2238. https://doi.org/10.1128/JB.00220-15
Linke C, Young P, Kang H, Bunker R, Middleditch M, Caradoc-Davies T, Proft T, Baker E (2010) Crystal structure of the minor pilin FctB reveals determinants of Group A streptococcal pilus anchoring. J Biol Chem 285(26):20381–20389. https://doi.org/10.1074/jbc.M109.089680
Linke-Winnebeck C, Paterson NG, Young PG, Middleditch MJ, Greenwood DR, Witte G, Baker EN (2014) Structural model for covalent adhesion of the Streptococcus pyogenes pilus through a thioester bond. J Biol Chem 289(1):177–189. https://doi.org/10.1074/jbc.M113.523761
Liu X, Zhuo S, Rensing C, Zhou S (2018) Syntrophic growth with direct interspecies electron transfer between pili-free *Geobacter* species. ISME J 12:2142–2151
Lo AWH, Van de Water K, Gane PJ, Chan AWE, Steadman D, Stevens K, Selwood DL, Waksman G, Remaut H (2014) Suppression of type 1 pilus assembly in uropathogenic Escherichia coli by chemical inhibition of subunit polymerization. J Antimicrob Chemother 69(4):1017–1026. https://doi.org/10.1093/jac/dkt467
Loferer H, Hammer M, Normark S (1997) Availability of the fibre subunit CsgA and the nucleator protein CsgB during assembly of fibronectin-binding curli is limited by the intracellular concentration of the novel lipoprotein CsgG. Mol Microbiol 26(1):11–23. https://doi.org/10.1046/j.1365-2958.1997.5231883.x
Lory S, Strom MS (1997) Structure-function relationship of type-IV prepilin peptidase of Pseudomonas aeruginosa—a review. Gene 192(1):117–121
Louros NN, Bolas GMP, Tsiolaki PL, Hamodrakas SJ, Iconomidou VA (2016) Intrinsic aggregation propensity of the CsgB nucleator protein is crucial for curli fiber formation. J Struct Biol 195(2):179–189. https://doi.org/10.1016/j.jsb.2016.05.012
Low HH, Gubellini F, Rivera-Calzada A, Braun N, Connery S, Dujeancourt A, Lu F, Redzej A, Fronzes R, Orlova EV, Waksman G (2014) Structure of a type IV secretion system. Nature 508(7497):550–553. https://doi.org/10.1038/nature13081
Maier B, Potter L, So M, Seifert HS, Sheetz MP (2002) Single pilus motor forces exceed 100 pN. PNAS 99(25):16012–16017. https://doi.org/10.1073/pnas.242523299
Majdalani N, Ippen-Ihler K (1996) Membrane insertion of the F-pilin subunit is Sec independent but requires leader peptidase B and the proton motive force. J Bacteriol 178(13):3742–3747. https://doi.org/10.1128/jb.178.13.3742-3747.1996
Majdalani N, Moore D, Maneewannakul S, Ippen-Ihler K (1996) Role of the propilin leader peptide in the maturation of F pilin. J Bacteriol 178(13):3748–3754. https://doi.org/10.1128/jb.178.13.3748-3754.1996
Malvankar NS, Lovley DR (2014) Microbial nanowires for bioenergy applications. Curr Opin Biotechnol 27:88–95

Malvankar NS, Vargas M, Nevin KP, Franks AE, Leang C, Kim B-C, Inoue K, Mester T, Covalla SF, Johnson JP, et al (2011) Tunable metallic-like conductivity in microbial nanowire networks. Nat Nanotechnol 6:573

Mandlik A, Das A, Ton-That H (2008a) The molecular switch that activates the cell wall anchoring step of pilus assembly in Gram-positive bacteria. Proc Natl Acad Sci USA 105(37):14147–14152. https://doi.org/10.1073/pnas.0806350105

Mandlik A, Swierczynski A, Das A, Ton-That H (2008b) Pili in Gram-positive bacteria: assembly, involvement in colonization and biofilm development. Trends Microbiol 16(1):33–40. https://doi.org/10.1016/j.tim.2007.10.010

Maneewannakul K, Maneewannakul S, Ippen-Ihler K (1995) Characterization of traX, the F plasmid locus required for acetylation of F-pilin subunits. J Bacteriol 177(11):2957–2964. https://doi.org/10.1128/jb.177.11.2957-2964.1995

Marlovits TC, Kubori T, Sukhan A, Thomas DR, Galán JE, Unger VM (2004) Structural insights into the assembly of the type III secretion needle complex. Science 306(5698):1040–1042. https://doi.org/10.1126/science.1102610

Marraffini LA, DeDent AC, Schneewind O (2006) Sortases and the art of anchoring proteins to the envelopes of Gram-positive bacteria. Microbiol Mol Biol Rev 70(1):192–221. https://doi.org/10.1128/MMBR.70.1.192-221.2006

Matias J, Berzosa M, Pastor Y, Irache JM, Gamazo C (2017) Maternal vaccination. Immunization of sows during pregnancy against ETEC Infections. Vaccines (Basel) 5(4). https://doi.org/10.3390/vaccines5040048

Mattick JS (2002) Type IV pili and twitching motility. Annu Rev Microbiol 56:289–314. https://doi.org/10.1146/annurev.micro.56.012302.160938. Epub 2002 Jan 30

Mehta T, Coppi MV, Childers SE, Lovley DR (2005) Outer membrane *c*-type cytochromes required for Fe(III) and Mn(IV) oxide reduction in *Geobacter sulfurreducens*. Appl Environ Microbiol 71:8634–8641

Merz AJ, So M, Sheetz MP (2000) Pilus retraction powers bacterial twitching motility. Nature 407(6800):98–102. https://doi.org/10.1038/35024105

Misic AM, Satyshur KA, Forest KT (2010) P. aeruginosa PilT structures with and without nucleotide reveal a dynamic type IV pilus retraction motor. J Mol Biol 400(5):1011–1021. https://doi.org/10.1016/j.jmb.2010.05.066. Epub 2010 June 1

Moonens K, Remaut H (2017) Evolution and structural dynamics of bacterial glycan binding adhesins. Curr Opin Struct Biol 44:48–58. https://doi.org/10.1016/j.sbi.2016.12.003

Moonens K, Van den Broeck I, De Kerpel M, Deboeck F, Raymaekers H, Remaut H, De Greve H (2015) Structural and functional insight into the carbohydrate receptor binding of F4 fimbriae-producing enterotoxigenic Escherichia coli. J Biol Chem 290(13):8409–8419. https://doi.org/10.1074/jbc.M114.618595

Mulvey MA, Lopez-Boado YS, Wilson CL, Roth R, Parks WC, Heuser J, Hultgren SJ (1998) Induction and evasion of host defenses by type 1-piliated uropathogenic Escherichia coli. Science 282(5393):1494–1497

Nakayama K, Yoshimura F, Kadowaki T, Yamamoto K (1996) Involvement of arginine-specific cysteine proteinase (Arg-gingipain) in fimbriation of Porphyromonas gingivalis. J Bacteriol 178(10):2818–2824

Nenninger AA, Robinson LS, Hammer ND, Epstein EA, Badtke MP, Hultgren SJ, Chapman MR (2011) CsgE is a curli secretion specificity factor that prevents amyloid fibre aggregation. Mol Microbiol 81(2):486–499. https://doi.org/10.1111/j.1365-2958.2011.07706.x

Nenninger AA, Robinson LS, Hultgren SJ (2009) Localized and efficient curli nucleation requires the chaperone-like amyloid assembly protein CsgF. PNAS 106(3):900–905. https://doi.org/10.1073/pnas.0812143106

Nishiyama M, Horst R, Eidam O, Herrmann T, Ignatov O, Vetsch M, Bettendorff P, Jelesarov I, Grütter MG, Wüthrich K, Glockshuber R, Capitani G (2005) Structural basis of chaperone-subunit complex recognition by the type 1 pilus assembly platform FimD. EMBO J 24(12):2075–2086. https://doi.org/10.1038/sj.emboj.7600693

Nishiyama M, Ishikawa T, Rechsteiner H, Glockshuber R (2008) Reconstitution of pilus assembly reveals a bacterial outer membrane catalyst. Science 320(5874):376–379. https://doi.org/10.1126/science.1154994

Nishiyama M, Vetsch M, Puorger C, Jelesarov I, Glockshuber R (2003) Identification and characterization of the chaperone-subunit complex-binding domain from the type 1 pilus assembly platform FimD. J Mol Biol 330(3):513–525. https://doi.org/10.1016/S0022-2836(03)00591-6

Nuccio S-P, Bäumler AJ (2007) Evolution of the chaperone/usher assembly pathway: fimbrial classification goes Greek. Microbiol Mol Biol Rev 71(4):551–575. https://doi.org/10.1128/MMBR.00014-07

Ohlsson J, Jass J, Uhlin BE, Kihlberg J, Nilsson UJ (2002) Discovery of potent inhibitors of PapG adhesins from uropathogenic Escherichia coli through synthesis and evaluation of galabiose derivatives. ChemBioChem 3(8):772–779. https://doi.org/10.1002/1439-7633(20020802)3:8%3c772:AID-CBIC772%3e3.0.CO;2-8

Olsén A, Jonsson A, Normark S (1989) Fibronectin binding mediated by a novel class of surface organelles on Escherichia coli. Nature 338(6217):652–655. https://doi.org/10.1038/338652a0

Omattage NS, Deng Z, Pinkner JS, Dodson KW, Almqvist F, Yuan P, Hultgren SJ (2018) Structural basis for usher activation and intramolecular subunit transfer in P pilus biogenesis in Escherichia coli. Nat Microbiol 3(12):1362. https://doi.org/10.1038/s41564-018-0255-y

Otter JA, Vickery K, Walker JT, deLancey Pulcini E, Stoodley P, Goldenberg SD, Salkeld JA, Chewins J, Yezli S, Edgeworth JD (2015) Surface-attached cells, biofilms and biocide susceptibility: implications for hospital cleaning and disinfection. J Hosp Infect 89(1):16–27. https://doi.org/10.1016/j.jhin.2014.09.008

Pansegrau W, Bagnoli F (2017) Pilus assembly in Gram-positive bacteria, vol 404, pp 203–233. https://doi.org/10.1007/82_2015_5016

Parge HE, Forest KT, Hickey MJ, Christensen DA, Getzoff ED, Tainer JA (1995) Structure of the fibre-forming protein pilin at 2.6 A resolution. Nature 378(6552):32–38. https://doi.org/10.1038/378032a0

Pelicic V (2008) Type IV pili: e pluribus unum? Mol Microbiol 68(4):827–837. https://doi.org/10.1111/j.1365-2958.2008.06197.x

Phan G, Remaut H, Wang T, Allen WJ, Pirker KF, Lebedev A, Henderson NS, Geibel S, Volkan E, Yan J, Kunze MBA, Pinkner JS, Ford B, Kay CWM, Li H, Hultgren S, Thanassi DG, Waksman G (2011) Crystal structure of the FimD usher bound to its cognate FimC:FimH substrate. Nature 474(7349):49–53. https://doi.org/10.1038/nature10109

Pinkner JS, Remaut H, Buelens F, Miller E, Aberg V, Pemberton N, Hedenström M, Larsson A, Seed P, Waksman G, Hultgren SJ, Almqvist F (2006) Rationally designed small compounds inhibit pilus biogenesis in uropathogenic bacteria. Proc Natl Acad Sci USA 103(47):17897–17902. https://doi.org/10.1073/pnas.0606795103

Planet PJ, Kachlany SC, DeSalle R, Figurski DH (2001) Phylogeny of genes for secretion NTPases: identification of the widespread tadA subfamily and development of a diagnostic key for gene classification. PNAS 98(5):2503–2508. https://doi.org/10.1016/j.jad.2013.02.029

Pointon JA, Smith WD, Saalbach G, Crow A, Kehoe MA, Banfield MJ (2010) A highly unusual thioester bond in a pilus adhesin is required for efficient host cell interaction. J Biol Chem 285(44):33858–33866. https://doi.org/10.1074/jbc.M110.149385

Proft T, Baker EN (2009) Pili in Gram-negative and Gram-positive bacteria—structure, assembly and their role in disease. Cell Mol Life Sci 66(4):613–635. https://doi.org/10.1007/s00018-008-8477-4

Puorger C, Eidam O, Capitani G, Erilov D, Grütter MG, Glockshuber R (2008) Infinite kinetic stability against dissociation of supramolecular protein complexes through donor strand complementation. Structure 16(4):631–642. https://doi.org/10.1016/j.str.2008.01.013

Puorger C, Vetsch M, Wider G, Glockshuber R (2011) Structure, folding and stability of FimA, the main structural subunit of type 1 pili from uropathogenic Escherichia coli strains. J Mol Biol 412(3):520–535. https://doi.org/10.1016/j.jmb.2011.07.044

Ramboarina S, Fernandes PJ, Daniell S, Islam S, Simpson P, Frankel G, Booy F, Donnenberg MS, Matthews S (2005) Structure of the bundle-forming pilus from enteropathogenic Escherichia coli. J Biol Chem 280(48):40252–40260. https://doi.org/10.1074/jbc.M508099200

Ramsugit S, Pillay M (2015) Pili of Mycobacterium tuberculosis: current knowledge and future prospects. Arch Microbiol 197(6):737–744. https://doi.org/10.1007/s00203-015-1117-0

Rashkova S, Zhou XR, Chen J, Christie PJ (2000) Self-assembly of the Agrobacterium tumefaciens VirB11 traffic ATPase. J Bacteriol 182(15):4137–4145. https://doi.org/10.1128/JB.182.15.4137-4145.2000

Redzej A, Ukleja M, Connery S, Trokter M, Felisberto-Rodrigues C, Cryar A, Thalassinos K, Hayward RD, Orlova EV, Waksman G (2017) Structure of a VirD4 coupling protein bound to a VirB type IV secretion machinery. EMBO J 36(20):e201796629–e201796629. https://doi.org/10.15252/embj.201796629

Reguera G, McCarthy KD, Mehta T, Nicoll JS, Tuominen MT, Lovley DR (2005) Extracellular electron transfer via microbial nanowires. Nature 435:1098–1101

Reichhardt C, Cegelski L (2014) Solid-state NMR for bacterial biofilms. Mol Phys 112(7):887–894. https://doi.org/10.1080/00268976.2013.837983

Remaut H, Rose RJ, Hannan TJ, Hultgren SJ, Radford SE, Ashcroft AE, Waksman G (2006) Donor-strand exchange in chaperone-assisted pilus assembly proceeds through a concerted beta strand displacement mechanism. Mol Cell 22(6):831–842. https://doi.org/10.1016/j.molcel.2006.05.033

Remaut H, Tang C, Henderson NS, Pinkner JS, Wang T, Hultgren SJ, Thanassi DG, Waksman G, Li H (2008) Fiber formation across the bacterial outer membrane by the chaperone/usher pathway. Cell 133(4):640–652. https://doi.org/10.1016/j.cell.2008.03.033

Richter LV, Sandler SJ, Weis RM (2012) Two isoforms of *Geobacter sulfurreducens* PilA have distinct roles in pilus biogenesis, cytochrome localization, extracellular electron transfer, and biofilm formation. J Bacteriol 194:2551–2563

Rivas S, Bolland S, Cabezón E, Goñi FM, De La Cruz F (1997) TrwD, a protein encoded by the IncW plasmid R388, displays an ATP hydrolase activity essential for bacterial conjugation. J Biol Chem 272(41):25583–25590. https://doi.org/10.1074/jbc.272.41.25583

Rivera-Calzada A, Fronzes R, Savva CG, Chandran V, Lian PW, Laeremans T, Pardon E, Steyaert J, Remaut H, Waksman G, Orlova EV (2013) Structure of a bacterial type IV secretion core complex at subnanometre resolution. EMBO J 32(8):1195–1204. https://doi.org/10.1038/emboj.2013.58

Roberts JA, Marklund BI, Ilver D, Haslam D, Kaack MB, Baskin G, Louis M, Mollby R, Winberg J, Normark S (1994) The Gal(alpha 1–4)Gal-specific tip adhesin of Escherichia coli P-fimbriae is needed for pyelonephritis to occur in the normal urinary tract. Proc Natl Acad Sci USA 91(25):11889–11893

Robinson LS, Ashman EM, Hultgren SJ, Chapman MR (2006) Secretion of curli fibre subunits is mediated by the outer membrane-localized CsgG protein. Mol Microbiol 59(3):870–881. https://doi.org/10.1111/j.1365-2958.2005.04997.x

Rose RJ, Verger D, Daviter T, Remaut H, Paci E, Waksman G, Ashcroft AE, Radford SE (2008) Unraveling the molecular basis of subunit specificity in P pilus assembly by mass spectrometry. PNAS 105(35):12873–12878. https://doi.org/10.1073/pnas.0802177105

Rouse SL, Hawthorne WJ, Berry JL, Chorev DS, Ionescu SA, Lambert S, Stylianou F, Ewert W, Mackie U, Morgan RML, Otzen D, Herbst FA, Nielsen PH, Dueholm M, Bayley H, Robinson CV, Hare S, Matthews S (2017) A new class of hybrid secretion system is employed in Pseudomonas amyloid biogenesis. Nat Commun 8(1):263–263. https://doi.org/10.1038/s41467-017-00361-6

Rouse SL, Matthews SJ, Dueholm MS (2018a) Ecology and biogenesis of functional amyloids in pseudomonas. J Mol Biol 430(20):3685–3695. https://doi.org/10.1016/j.jmb.2018.05.004

Rouse SL, Stylianou F, Wu HYG, Berry JL, Sewell L, Morgan RML, Sauerwein AC, Matthews S (2018b) The FapF amyloid secretion transporter possesses an atypical asymmetric coiled coil. J Mol Biol 430(20):3863–3871. https://doi.org/10.1016/j.jmb.2018.06.007

Ruer S, Pinotsis N, Steadman D, Waksman G, Remaut H (2015) Virulence-targeted antibacterials: concept, promise, and susceptibility to resistance mechanisms. Chem Biol Drug Des 86(4):379–399. https://doi.org/10.1111/cbdd.12517. Epub 2015 Feb 6

Ryu JH, Beuchat LR (2005) Biofilm formation by Escherichia coli O157:H7 on stainless steel: effect of exopolysaccharide and curli production on its resistance to chlorine. Appl Environ Microbiol 71(1):247–254. https://doi.org/10.1128/AEM.71.1.247-254.2005

Sauer FG, Fütterer K, Pinkner JS, Dodson KW, Hultgren SJ, Waksman G (1999) Structural basis of chaperone function and pilus biogenesis. Science 285(5430):1058–1061

Sauer FG, Pinkner JS, Waksman G, Hultgren SJ (2002) Chaperone priming of pilus subunits facilitates a topological transition that drives fiber formation. Cell 111(4):543–551

Sauer FG, Remaut H, Hultgren SJ, Waksman G (2004) Fiber assembly by the chaperone–usher pathway. Biochimica et Biophysica Acta (BBA)—Mol Cell Res 1694(1–3):259–267. https://doi.org/10.1016/j.bbamcr.2004.02.010

Savvides SN, Yeo HJ, Beck MR, Blaesing F, Lurz R, Lanka E, Buhrdorf R, Fischer W, Haas R, Waksman G (2003) VirB11 ATPases are dynamic hexameric assemblies: new insights into bacterial type IV secretion. EMBO J 22(9):1969–1980. https://doi.org/10.1093/emboj/cdg223

Schneewind O, Missiakas D (2014) Sec-secretion and sortase-mediated anchoring of proteins in Gram-positive bacteria. Biochimica et Biophysica Acta—Mol Cell Res 1843(8):1687–1697. https://doi.org/10.1016/j.bbamcr.2013.11.009

Schubeis T, Spehr J, Viereck J, Köpping L, Nagaraj M, Ahmed M, Ritter C (2018) Structural and functional characterization of the Curli adaptor protein CsgF. FEBS Lett 592(6):1020–1029. https://doi.org/10.1002/1873-3468.13002

Shaik MM, Maccagni A, Tourcier G, Di Guilmi AM, Dessen A (2014) Structural basis of pilus anchoring by the ancillary pilin RrgC of Streptococcus pneumoniae. J Biol Chem 289(24):16988–16997. https://doi.org/10.1074/jbc.M114.555854

Shewmaker F, McGlinchey RP, Thurber KR, McPhie P, Dyda F, Tycko R, Wickner RB (2009) The functional curli amyloid is not based on in-register parallel β-sheet structure. J Biol Chem 284(37):25065–25076. https://doi.org/10.1074/jbc.M109.007054

Shoji M, Naito M, Yukitake H, Sato K, Sakai E, Ohara N, Nakayama K (2004) The major structural components of two cell surface filaments of Porphyromonas gingivalis are matured through lipoprotein precursors. Mol Microbiol 52(5):1513–1525. https://doi.org/10.1111/j.1365-2958.2004.04105.x

Shu Q, Crick SL, Pinkner JS, Ford B, Hultgren SJ, Frieden C (2012) The E. coli CsgB nucleator of curli assembles to -sheet oligomers that alter the CsgA fibrillization mechanism. PNAS 109(17):6502–6507. https://doi.org/10.1073/pnas.1204161109

Shu Q, Krezel AM, Cusumano ZT, Pinkner JS, Klein R, Hultgren SJ, Frieden C (2016) Solution NMR structure of CsgE: structural insights into a chaperone and regulator protein important for functional amyloid formation. PNAS 113(26):7130–7135. https://doi.org/10.1073/pnas.1607222113

Siewering K, Jain S, Friedrich C, Webber-Birungi MT, Semchonok DA, Binzen I, Wagner A, Huntley S, Kahnt J, Klingl A, Boekema EJ, Søgaard-Andersen L, van der Does C (2014) Peptidoglycan-binding protein TsaP functions in surface assembly of type IV pili. Proc Natl Acad Sci USA 111(10):E953–E961. https://doi.org/10.1073/pnas.1322889111

Sleutel M, Van Den Broeck I, Van Gerven N, Feuillie C, Jonckheere W, Valotteau C, Dufrêne YF, Remaut H (2017) Nucleation and growth of a bacterial functional amyloid at single-fiber resolution. Nat Chem Biol 13(8):902–908. https://doi.org/10.1038/nchembio.2413

Spaulding CN, Iv HLS, Zheng W, Dodson KW, Hazen JE, Conover MS, Wang F, Svenmarker P, Luna-Rico A, Francetic O, Andersson M, Hultgren S, Egelman EH (2018) Functional role of the type 1 pilus rod structure in mediating host-pathogen interactions. eLife

Spaulding CN, Klein RD, Ruer S, Kau AL, Schreiber HL, Cusumano ZT, Dodson KW, Pinkner JS, Fremont DH, Janetka JW, Remaut H, Gordon JI, Hultgren SJ (2017) Selective depletion of uropathogenic E. coli from the gut by a FimH antagonist. Nature 546(7659):528–532. https://doi.org/10.1038/nature22972. Epub 2017 June 14

Spirig T, Weiner EM, Clubb RT (2011) Sortase enzymes in Gram-positive bacteria. Mol Microbiol 82(5):1044–1059. https://doi.org/10.1111/j.1365-2958.2011.07887.x

Steadman D, Lo A, Waksman G, Remaut H (2014) Bacterial surface appendages as targets for novel antibacterial therapeutics. Future Microbiol 9(7):887–900. https://doi.org/10.2217/fmb.14.46
Summers ZM, Fogarty HE, Leang C, Franks AE, Malvankar NS, Lovley DR (2010) Direct exchange of electrons within aggregates of an evolved syntrophic coculture of anaerobic bacteria. Science 330:1413–1415
Symersky J, Patti JM, Carson M, House-Pompeo K, Teale M, Moore D, Jin L, Schneider A, Delucas LJ, Hook M, Narayana SVL (1997) Structure of the collagen-binding domain from a Staphylococcus aureus adhesin. Nat Struct Biol 4(10):833–838. https://doi.org/10.1038/nsb1097-833
Szabó Z, Stahl AO, Albers S-V, Kissinger JC, Driessen AJM, Pohlschröder M (2007) Identification of diverse archaeal proteins with Class III signal peptides cleaved by distinct archaeal prepilin peptidases. J Bacteriol 189(3):772–778. https://doi.org/10.1128/JB.01547-06
Takhar HK, Kemp K, Kim M, Howell PL, Burrows LL (2013) The platform protein is essential for type IV pilus biogenesis. J Biol Chem 288(14):9721–9728. https://doi.org/10.1074/jbc.M113.453506. Epub 2013 Feb 14
Tammam S, Sampaleanu LM, Koo J, Manoharan K, Daubaras M, Burrows LL, Howell PL (2013) PilMNOPQ from the Pseudomonas aeruginosa Type IV pilus system form a transenvelope protein interaction network that interacts with PilA. J Bacteriol 195(10):2126–2135. https://doi.org/10.1128/JB.00032-13
Taylor JD, Hawthorne WJ, Lo J, Dear A, Jain N, Meisl G, Andreasen M, Fletcher C, Koch M, Darvill N, Scull N, Escalera-Maurer A, Sefer L, Wenman R, Lambert S, Jean J, Xu Y, Turner B, Kazarian SG, Chapman MR, Bubeck D, De Simone A, Knowles TPJ, Matthews SJ (2016) Electrostatically-guided inhibition of Curli amyloid nucleation by the CsgC-like family of chaperones. Sci Rep 6:1–11. https://doi.org/10.1038/srep24656
Taylor JD, Zhou Y, Salgado PS, Patwardhan A, McGuffie M, Pape T, Grabe G, Ashman E, Constable SC, Simpson PJ, Lee WC, Cota E, Chapman MR, Matthews SJ (2011) Atomic resolution insights into curli fiber biogenesis. Structure 19(9):1307–1316. https://doi.org/10.1016/j.str.2011.05.015
Thanassi DG, Bliska JB, Christie PJ (2012) Surface organelles assembled by secretion systems of Gram-negative bacteria: diversity in structure and function. FEMS Microbiol Rev 36(6):1046–1082. https://doi.org/10.1111/j.1574-6976.2012.00342.x
Thanassi DG, Saulino ET, Hultgren SJ (1998) The chaperone/usher pathway: a major terminal branch of the general secretory pathway. Curr Opin Microbiol 1(2):223–231
Tian P, Boomsma W, Wang Y, Otzen DE, Jensen MH, Lindorff-Larsen K (2015) Structure of a functional amyloid protein subunit computed using sequence variation. J Am Chem Soc 137(1):22–25. https://doi.org/10.1021/ja5093634
Ton-That H, Marraffini LA, Schneewind O (2004) Sortases and pilin elements involved in pilus assemly of Corynebacterium diphtheriae. Mol Microbiol 53(1):251–261. https://doi.org/10.1111/j.1365-2958.2004.04117.x
Ton-That H, Schneewind O (2003) Assembly of pili on the surface of Corynebacterium diphtheriae. Mol Microbiol 50(4):1429–1438. https://doi.org/10.1046/j.1365-2958.2003.03782.x
Ton-That H, Schneewind O (2004) Assembly of pili in Gram-positive bacteria. Trends Microbiol 12(5):228–234. https://doi.org/10.1016/j.tim.2004.03.004
Tonjum T, Freitag NE, Namork E, Koomey M (1995) Identification and characterization of pilG, a highly conserved pilus-assembly gene in pathogenic Neisseria. Mol Microbiol 16(3):451–464
Turner LR, Lara JC, Nunn DN, Lory S (1993) Mutations in the consensus ATP-binding sites of XcpR and PilB eliminate extracellular protein secretion and pilus biogenesis in Pseudomonas aeruginosa. J Bacteriol 175(16):4962–4969
Uhlich GA, Cooke PH, Solomon EB (2006) Analyses of the red-dry-rough phenotype of an Escherichia coli O157:H7 strain and its role in biofilm formation and resistance to antibacterial agents. Appl Environ Microbiol 72(4):2564–2572. https://doi.org/10.1128/AEM.72.4.2564
Van Gerven N, Van der Verren SE, Reiter DM, Remaut H (2018) The role of functional amyloids in bacterial virulence. J Mol Biol 430(20):3657–3684. https://doi.org/10.1016/j.jmb.2018.07.010

Verger D, Miller E, Remaut H, Waksman G, Hultgren S (2006) Molecular mechanism of P pilus termination in uropathogenic Escherichia coli. EMBO Rep 7(12):1228–1232. https://doi.org/10.1038/sj.embor.7400833

Vergunst AC, Schrammeijer B, Den Dulk-Ras A, De Vlaam CMT, Regensburg-Tuink TJG, Hooykaas PJJ (2000) VirB/D4-dependent protein translocation from Agrobacterium into plant cells. Science 290(5493):979–982. https://doi.org/10.1126/science.290.5493.979

Vetsch M, Puorger C, Spirig T, Grauschopf U, Weber-Ban EU, Glockshuber R (2004) Pilus chaperones represent a new type of protein-folding catalyst. Nature 431(7006):329–333. https://doi.org/10.1038/nature02891

Virdi V, Coddens A, De Buck S, Millet S, Goddeeris BM, Cox E, De Greve H, Depicker A (2013) Orally fed seeds producing designer IgAs protect weaned piglets against enterotoxigenic Escherichia coli infection. Proc Natl Acad Sci USA 110(29):11809–11814. https://doi.org/10.1073/pnas.1301975110. Epub 2013 June 25

Wagner A, Dehio C (2019) Role of distinct Type-IV-secretion systems and secreted effector sets in host adaptation by pathogenic *Bartonella* species. Cell Microbiol e13004–e13004. https://doi.org/10.1111/cmi.13004

Waksman G (2019) From conjugation to T4S systems in Gram-negative bacteria: a mechanistic biology perspective. EMBO Rep e47012–e47012. https://doi.org/10.15252/embr.201847012

Wang X, Smith DR, Jones JW, Chapman MR (2007) In vitro polymerization of a functional Escherichia coli amyloid protein. J Biol Chem 282(6):3713–3719. https://doi.org/10.1074/jbc.M609228200

Wang F, Gu Y, O'Brien JP, Yi SM, Yalcin SE, Srikanth V, Shen C, Vu D, Ing NL, Hochbaum AI, et al (2019) Structure of microbial nanowires reveals stacked hemes that transport electrons over micrometers. Cell 177:361–369.e310

Whitchurch CB, Hobbs M, Livingston SP, Krishnapillai V, Mattick JS (1991) Characterisation of a Pseudomonas aeruginosa twitching motility gene and evidence for a specialised protein export system widespread in eubacteria. Gene 101(1):33–44

Winther-Larsen HC, Wolfgang M, Dunham S, van Putten JP, Dorward D, Lovold C, Aas FE, Koomey M (2005) A conserved set of pilin-like molecules controls type IV pilus dynamics and organelle-associated functions in Neisseria gonorrhoeae. Mol Microbiol 56(4):903–917. https://doi.org/10.1111/j.1365-2958.2005.04591.x

Xu Q, Shoji M, Shibata S, Naito M, Sato K, Elsliger M-A, Grant JC, Axelrod HL, Chiu H-J, Farr CL, Jaroszewski L, Knuth MW, Deacon AM, Godzik A, Lesley SA, Curtis MA, Nakayama K, Wilson IA (2016) A distinct type of pilus from the human microbiome. Cell 165(3):690–703. https://doi.org/10.1016/j.cell.2016.03.016

Yanagawa R, Otsuki K, Tokui T (1968) Electron microscopy of fine structure of Corynebacterium renale with special reference to pili. Jpn J Vet Res 16(1):31–37

Yeo HJ, Savvides SN, Herr AB, Lanka E, Waksman G (2000) Crystal structure of the hexameric traffic ATPase of the Helicobacter pylori type IV secretion system. Mol Cell 6(6):1461–1472. https://doi.org/10.1016/S1097-2765(00)00142-8

Yoshimura F, Takahashi K, Nodasaka Y, Suzuki T (1984) Purification and characterization of a novel type of fimbriae from the oral anaerobe Bacteroides gingivalis. J Bacteriol 160(3):949–957

Zav'yalov VP (2013) Polyadhesins: an armory of Gram-negative pathogens for penetration through the immune shield

Zavialov AV, Berglund J, Pudney AF, Fooks LJ, Ibrahim TM, MacIntyre S, Knight SD (2003) Structure and biogenesis of the capsular F1 antigen from Yersinia pestis: preserved folding energy drives fiber formation. Cell 113(5):587–596

Zavialov AV, Tischenko VM, Fooks LJ, Brandsdal BO, Aqvist J, Zav'yalov VP, Macintyre S, Knight SD (2005) Resolving the energy paradox of chaperone/usher-mediated fibre assembly. Biochem J 389(Pt 3):685–694. https://doi.org/10.1042/BJ20050426

Zechner EL, Lang S, Schildbach JF (2012) Assembly and mechanisms of bacterial type IV secretion machines. Philos Trans R Soc B: Biol Sci 367(1592):1073–1087. https://doi.org/10.1098/rstb.2011.0207

Zeng G, Vad BS, Dueholm MS, Christiansen G, Nilsson M, Tolker-Nielsen T, Nielsen PH, Meyer RL, Otzen DE (2015) Functional bacterial amyloid increases Pseudomonas biofilm hydrophobicity and stiffness. Front Microbiol 6:1099–1099. https://doi.org/10.3389/fmicb.2015.01099
Zhang HZ, Lory S, Donnenberg MS (1994) A plasmid-encoded prepilin peptidase gene from enteropathogenic Escherichia coli. J Bacteriol 176(22):6885–6891

Part V
Cell Walls of Gram-Positive Bacteria and Archaea

Chapter 13
Cell Walls and Membranes of Actinobacteria

Kathryn C. Rahlwes, Ian L. Sparks and Yasu S. Morita

Abstract Actinobacteria is a group of diverse bacteria. Most species in this class of bacteria are filamentous aerobes found in soil, including the genus *Streptomyces* perhaps best known for their fascinating capabilities of producing antibiotics. These bacteria typically have a Gram-positive cell envelope, comprised of a plasma membrane and a thick peptidoglycan layer. However, there is a notable exception of the Corynebacteriales order, which has evolved a unique type of outer membrane likely as a consequence of convergent evolution. In this chapter, we will focus on the unique cell envelope of this order. This cell envelope features the peptidoglycan layer that is covalently modified by an additional layer of arabinogalactan. Furthermore, the arabinogalactan layer provides the platform for the covalent attachment of mycolic acids, some of the longest natural fatty acids that can contain ~100 carbon atoms per molecule. Mycolic acids are thought to be the main component of the outer membrane, which is composed of many additional lipids including trehalose dimycolate, also known as the cord factor. Importantly, a subset of bacteria in the Corynebacteriales order are pathogens of human and domestic animals, including *Mycobacterium tuberculosis*. The surface coat of these pathogens are the first point of contact with the host immune system, and we now know a number of host receptors specific to molecular patterns exposed on the pathogen's surface, highlighting the importance of understanding how the cell envelope of Actinobacteria is structured and constructed. This chapter describes the main structural and biosynthetic features of major components found in the actinobacterial cell envelopes and highlights the key differences between them.

Keywords Actinobacteria · Arabinogalactan · Cell envelope · Corynebacteria · Glycolipid · Membrane · Mycobacteria · Mycolic acid · Peptidoglycan · Phospholipid · Streptomyces

---

K. C. Rahlwes · I. L. Sparks · Y. S. Morita (✉)
Department of Microbiology, University of Massachusetts, 639 North Pleasant Street, Amherst, MA 01003, USA
e-mail: ymorita@microbio.umass.edu

A. Kuhn (ed.), *Bacterial Cell Walls and Membranes*, Subcellular Biochemistry 92, https://doi.org/10.1007/978-3-030-18768-2_13

Introduction

Actinobacteria is a vast and variable class of bacteria. One unifying feature of this class is the high GC content, generally ranging between 55 and 75%. A morphological feature traditionally used to classify Actinobacteria was filamentous growth, but a phylogenetic analysis using 16S rRNA gene has revealed that this group is much more morphologically diverse than it was previously thought. Their lifestyle is also immensely diverse. Many are environmental species that live in soil and aquatic environments where nutrient availability fluctuates. *Streptomyces* species, for example, have the robust ability to grow using a wide variety of nutrients, carrying numerous genes for metabolic regulation, polysaccharide degradation, and carbohydrate transport (Hodgson 2000; Bertram et al. 2004; Bentley et al. 2002). In contrast, there are symbionts, such as the plant symbiont *Frankia* or the human pathogen *Mycobacterium tuberculosis*, which have a limited number of membrane transporters, implying more restricted strategies to acquire nutrients from the host (Niederweis 2008; Normand et al. 2007). Another important feature of Actinobacteria is the Gram-positive cell wall. Most bacteria in the Actinobacteria class carry typical monoderm cell envelope with a relatively thick peptidoglycan layer. However, some members, such as the well-known *Mycobacterium* species, have evolved a diderm cell envelope (Fig. 13.1). Their unusual cell envelope structure makes Gram staining unreliable and is more readily distinguished from the other envelope types by acid-fast staining. This chapter discusses the cell envelope of Actinobacteria, one of six classes within the Actinobacteria phylum (Gao and Gupta 2012). We will primarily focus on the diderm cell envelope of the Corynebacteriales order and compare with those from other orders within the Actinobacteria class.

Of the Corynebacteriales order, *Mycobacterium* and *Corynebacterium* are the best studied genera mainly due to their medical and industrial importance. To name a few, *M. tuberculosis*, *Mycobacterium leprae*, *Mycobacterium bovis*, and *Corynebacterium diphtheriae* are the etiologic agents of tuberculosis (TB), leprosy, bovine TB, and diphtheria, respectively. *Corynebacterium glutamicum* is an industrial source of producing glutamic acid. Other genera such as *Nocardia*, *Rhodococcus*, *Gordonia*, and *Tsukamurella* also include some pathogenic species, which often infects immunocompromised individuals. However, most species in this order are environmental, with some species, such as *Rhodococcus olei*, showing potential industrial use in degrading petroleum oil in contaminated soil (Chaudhary and Kim 2018). Even in the genus of *Mycobacterium* where you find many pathogens, most species are nonpathogenic. For instance, *Mycobacterium smegmatis* is a nonpathogenic saprophyte. This species has become an established model for mycobacteria research because it is a fast grower, in contrast to the slow-growing pathogenic species, and many aspects of cellular physiology, including the cell envelope structures, are comparable to the pathogens.

Among infectious diseases caused by actinobacterial species, TB is the most devastating, currently being one of the top ten causes of death worldwide. In 2017, 10 million people worldwide fell ill with the disease and 1.6 million died (World

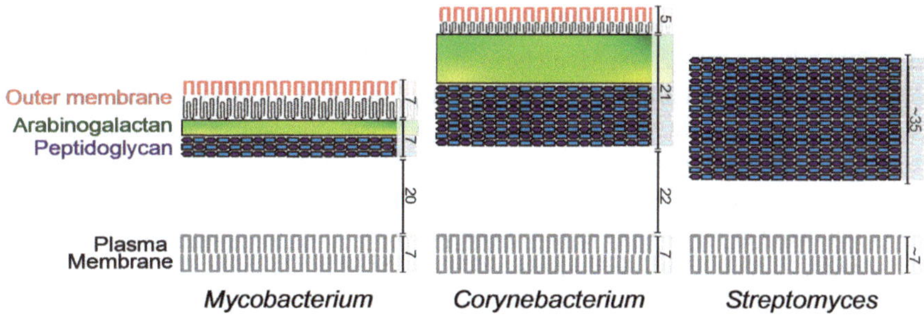

Fig. 13.1 Schematic of the cell envelopes of *Mycobacterium*, *Corynebacterium*, and *Streptomyces*. Bars indicate the height of each layer drawn to scale. In *M. smegmatis*, the plasma membrane, periplasmic space, peptidoglycan-arabinogalactan layer and outer membrane has a thickness of ~7, ~20, ~7 and ~7 nm (Zuber et al. 2008). For *C. glutamicum*, the peptidoglycan-arabinogalactan layer is ~3 times thicker than that of *M. smegmatis*, and outer membrane thinner than that of *M. smegmatis*, presumably due to short mycolic acids (Zuber et al. 2008). The thickness of each layer within the peptidoglycan-arabinogalactan layer is unknown. For *Streptomyces*, there are no precise estimates, but from published images, the peptidoglycan cell wall extends ~35 nm out from the plasma membrane, which appears comparable in thickness to *Mycobacterium* and *Corynebacterium* (Lerat et al. 2012; Yague et al. 2016; Celler et al. 2016)

Health Organization 2018). Additionally, estimated 1.7 billion people are latently infected with *M. tuberculosis* and therefore may develop the disease in their lifetime. Current regimens for the treatment of TB are combinations of the first-line drugs: isoniazid, rifampin, pyrazinamide, and ethambutol, for at least six months. Among them, isoniazid and ethambutol target the biosynthesis of cell envelope components, making cell envelope biosynthesis a proven target of TB chemotherapy. Similar to other microbial infections, the rise in multi-drug resistant *M. tuberculosis* is a global concern. Understanding the diderm cell envelope of *M. tuberculosis* and other Corynebacteriales is important not only for the sake of the unique biology that has evolved in this bacterial lineage, but also from the perspective of identifying novel drug targets to treat the devastating diseases they cause.

The biosynthesis of cell envelope is a tightly regulated process, requiring temporal and spatial controls. In this regard, certain similarities in the actinobacterial cell growth and division are noteworthy. Rapid mechanical separation of daughter cells is found in diverse lineages of Actinobacteria, including *Micrococcus luteus, Brachybacterium faecium, C. glutamicum, M. smegmatis,* and *Streptomyces venezuelae* (Zhou et al. 2016, 2019), suggesting that this mechanism of cell separation is widely conserved in Actinobacteria. From Streptomycetales to Corynebacteriales, polar growth is another well-conserved feature (Daniel and Errington 2003; Thanky et al. 2007; Ramos et al. 2003). Many proteins show specific subcellular localizations, which are likely critical for the function of these proteins and the spatially coordinated cell growth (Puffal et al. 2018). One prominent example is DivIVA, which localizes to the polar ends of *Mycobacterium*, *Corynebacterium*, *Brevibacterium*, and *Streptomyces* cells and helps to coordinate the polar cell envelope biosynthesis

(Ramos et al. 2003; Flärdh 2003; Letek et al. 2008; Hempel et al. 2008; Nguyen et al. 2007; Kang et al. 2008; Meniche et al. 2014; Donovan et al. 2012; Melzer et al. 2018). Spatial coordination is not only dictated by proteins. Plasma membrane has recently been shown to be segregated into functional domains in mycobacteria, and many cell envelope biosynthetic reactions are compartmentalized within the membrane (Hayashi et al. 2016, 2018), further highlighting the intricate spatial controls. More detailed review articles are available on spatial coordination and regulations of actinobacterial cell division and polar envelope growth (Puffal et al. 2018; Donovan and Bramkamp 2014; Logsdon and Aldridge 2018; Flärdh et al. 2012).

In this chapter, we focus primarily on the structure and biosynthesis of the actinobacterial cell envelope (Fig. 13.1), starting with the innermost layer, plasma membrane, followed by the peptidoglycan layer. In Corynebacteriales, the peptidoglycan layer is covalently linked to an arabinogalactan layer, which is covalently linked to a mycolic acid layer. Mycolic acids are long fatty acids and are a core component of the outer membrane. While much less is known, we will also compare and contrast the capsule layer of Actinobacteria.

Plasma Membrane

The plasma membrane is the fundamental innermost layer of the cell envelope. The cryo-electron micrograph of mycobacterial plasma membrane has indicated its thickness to be about 7 nm (Zuber et al. 2008; Hoffmann et al. 2008). The major structural components of the actinobacterial plasma membrane are glycerophospholipids. In Actinobacteria in general, the core plasma membrane is composed of cardiolipin (CL), phosphatidylglycerol (PG), phosphatidylethanolamine (PE), phosphatidylinositol (PI), phosphatidylinositol mannosides (PIMs), and other less abundant lipids such as ornithine lipids (OL), and menaquinones (Fig. 14.2a).

CLs and PGs

CL represents one of the most abundant plasma membrane components, constituting roughly 10–50% of the total phospholipids in Actinobacteria (Jackson et al. 2000; Yano et al. 1969; Nampoothiri et al. 2002; Kimura et al. 1967). It is generally considered a plasma membrane phospholipid, but a large amount of cell wall-associated CL has been found in *C. glutamicum* (Bansal-Mutalik and Nikaido 2011). In non-actinobacterial species, CL is typically synthesized by a CL synthase that facilitates transesterification between two PG molecules, producing one CL and one glycerol. In contrast, Actinobacteria use eukaryotic type CL synthase, which produces CL from PG and CDP-diacylglycerol (CDP-DAG), a high-energy molecule that can act as a donor of DAG (Sandoval-Calderon et al. 2009). The use of CDP-DAG makes this reaction energetically favorable, and being consistent with this reaction kinetics, PG does not significantly accumulate in either *Mycobacterium* or *Streptomyces* (Jackson et al. 2000; Sandoval-Calderon et al. 2009; Hoischen et al. 1997; Mathur et al. 1976; Zuneda et al. 1984; Lechevalier et al. 1977). This biosynthetic approach is energetically more demanding for the cell, likely suggesting evolutionary needs for these lineages of bacteria to have a high CL/PG ratio. While the physiological reasons for this biosynthetic mechanism are unknown, CL is essential for hyphal growth and spore formation in *Streptomyces* (Jyothikumar et al. 2012). Interestingly, *C. glutamicum* does not possess the eukaryotic type CL synthase (Nampoothiri et al. 2002). Being consistent with the lack of the eukaryotic type enzyme, this organism has a low CL/PG ratio (Bansal-Mutalik and Nikaido 2011).

Aminolipids

Aminolipids are widely found as a structural component of the plasma membranes in Actinobacteria, but their abundance can vary. For example, nitrogen-containing lipids are not found in the protoplast of *M. luteus* (Gilby et al. 1958). PE is often a dominant phospholipid in Actinobacteria, but corynebacteria cannot synthesize it (Brennan and Lehane 1971). Two sequential enzymatic reactions synthesize PE: phosphatidylserine synthase produces phosphatidylserine from serine and CDP-DAG, and then phosphatidylserine decarboxylase removes carbon dioxide from phosphatidylserine, producing PE. These two enzymatic activities are differentially enriched in specific membrane domains of *M. smegmatis* (Morita et al. 2005), although the significance of this spatial segregation is unknown (see below for more details on membrane compartmentalization).

Another class of aminolipids includes ornithine- and lysine-amide lipids, in which a long-chain β-hydroxy-fatty acid acylates the α-amino group of ornithine or lysine, and the β-hydroxy group of the long-chain fatty acid is further modified by another fatty acid. These phosphorus-free lipids have been detected in several species of *Mycobacterium* and *Streptomyces* (Kimura et al. 1967; Kawanami et al. 1968; Batrakov and Bergelson 1978; Laneelle et al. 1990). The precise function of these lipids remains unknown, but they are widespread in Gram-negative bacteria and Actinobacteria (Vences-Guzman et al. 2012). In *Streptomyces coelicolor*, ornithine-amide lipids accumulate during phosphorus starvation or sporulation stages (Sandoval-Calderon et al. 2015). Under these conditions where ornithine-amide lipids are abundant, the level of PE becomes negligible. It has been suggested that ornithine-amide lipids are present in both the inner and the outer membranes in Gram-negative bacteria. Therefore, it is possible that these lipids constitute the outer membrane of mycobacteria as well.

Aminolipids can also be produced by amino acid modifications of glycerolipids. *M. tuberculosis* produces lysinylated PG, and the deletion of the biosynthetic enzyme LysX results in defective cell envelope integrity and less effective establishment of infection in mice lungs (Maloney et al. 2009, 2011). A similar modification can occur in *Mycobacterium phlei*, where DAG rather than PG is lysinylated

(Lerouge et al. 1988). Similarly, *C. glutamicum* produces alanylated PG as well as alanylated DAG, and their synthesis is tightly regulated with the synthesis and trafficking of trehalose corynomycolates (Klatt et al. 2018).

Phosphatidylinositols and Mannolipids

PI is a rather unusual phospholipid species to be found in the *Bacteria* domain but constitutes a significant fraction of the total phospholipids in Actinobacteria (Morita et al. 2011). A fascinating feature of PI in eukaryotes is the versatile modifications of the functional head group, inositol. For example, proteins and glycans are anchored to the eukaryotic cell surface through glycosylphosphatidylinositols, playing numerous critical roles on the cell surface. Furthermore, intracellular signaling and membrane trafficking are mediated through phosphorylated PIs. Mycobacterial PI is also modified in similar ways, and the discovery of actinobacterial PIMs predates that of eukaryotic glycosylphosphatidylinositol anchors (Kimura et al. 1967; Lee and Ballou 1964; Ballou et al. 1963; Minnikin et al. 1977). The biosynthetic pathway starts with the production of inositol 3-phosphate from glucose (Glc) 6-phosphate, mediated by the inositol 3-phosphate synthase Ino1 (Haites et al. 2005; Movahedzadeh et al. 2004). Being the first key enzyme, the expression of *ino1* is tightly controlled by the transcription factor IpsA, which senses the decreased cellular levels of inositol and activates the *ino1* gene transcription (Baumgart et al. 2013). PI is produced from inositol and CDP-diacylglycerol, and this energetically favorable reaction is mediated by the PI synthase PgsA, which is an essential enzyme in *M. smegmatis* (Jackson et al. 2000). In mycobacteria, the major PIM species are PIM2 and PIM6, containing 2 and 6 mannose (Man) residues, respectively (Fig. 13.2a). These PIM species can be modified by up to four fatty acids: one at the 6-OH of the Man residue attached to the 6-OH of inositol ring, and another at the 3-OH of the inositol ring, in addition to the DAG moiety of PI. AcPIM2 and AcPIM6 are the most abundant PIM products in *M. smegmatis* grown under standard laboratory conditions, and they have an additional fatty acid attached to the Man residue. AcPIM2 is synthesized by sequential additions of Man residues mediated by PimA and PimB' (Guerin et al. 2009; Lea-Smith et al. 2008; Kordulakova et al. 2002), followed by the acyl addition to the Man residue mediated by PatA (Albesa-Jove et al. 2016; Kordulakova et al. 2003). There has been no report of AcPIM2 or AcPIM6 having the acyl modification on the inositol ring instead of the Man acyl chain, indicating that the inositol acylation occurs only after Man residue is acylated.

PimA and PimB' are GDP-Man-dependent enzymes, suggesting that the reactions take place on the cytoplasmic side of the plasma membrane. In contrast, the later mannosylation steps to produce AcPIM6, lipomannan (LM) and lipoarabinomannan (LAM) are dependent on a lipidic Man donor, which is known as polyprenol-phosphate-Man (PPM). PPM is produced from GDP-Man and polyprenol-phosphate by a membrane-bound PPM synthase, Ppm1, in *M. tuberculosis*. Interestingly, in other species of *Mycobacterium* such as *M. smegmatis*, *Mycobacterium avium*, and

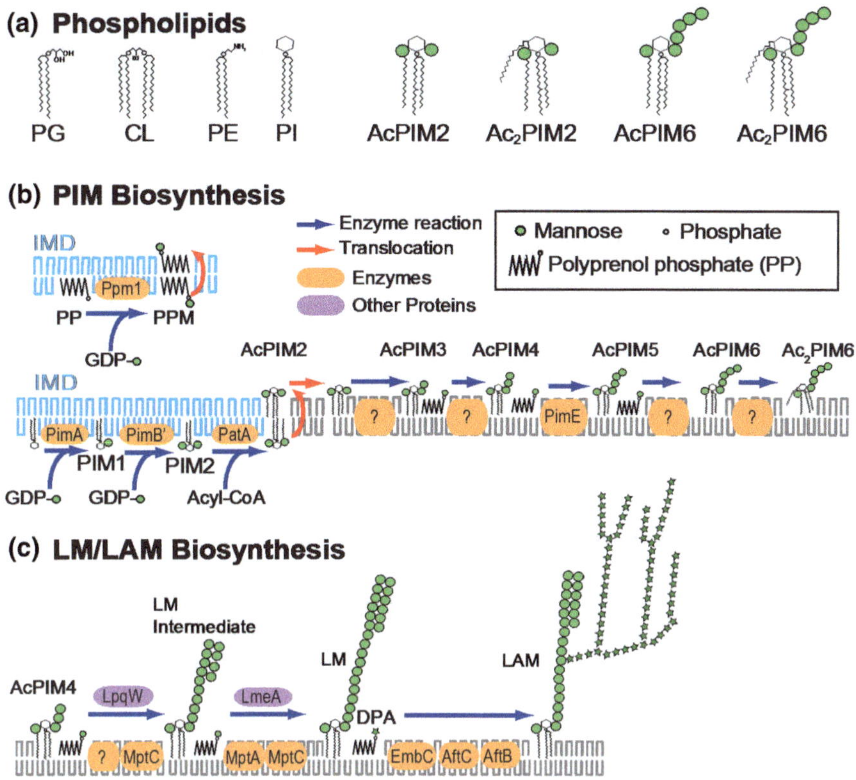

Fig. 13.2 Phospholipids of Actinobacteria. **a** Structures of major phospholipids. The acyl chain compositions can vary. PG, phosphatidylglycerol; CL, cardiolipin; PE, phosphatidylethanolamine; PI, phosphatidylinositol. AcPIM6 and Ac_2PIM6 are found in *Mycobacterium* and not in *Corynebacterium* (see text for details). **b** Biosynthesis of polyprenol phosphate Man (PPM) and PIMs. The question marks indicate currently undetermined or unconfirmed enzymes. The biosyntheses of PPM and the early steps of PIM are proposed to take place in the IMD (Hayashi et al. 2016; Morita et al. 2005). **c** Biosynthesis of lipomannan (LM) and lipoarabinomannan (LAM). Polyprenol phosphate Man (PPM) and decaprenol phosphate β-D-Ara*f* (DPA) are the Man and Ara donors for LM and LAM biosynthesis, respectively

M. leprae as well as in several *Corynebacterium* species, two genes encode separate domains of PPM synthase (Gurcha et al. 2002). Ppm synthase is essential in mycobacteria (Rana et al. 2012), and its deletion in *C. glutamicum* results in a reduced growth rate (Gibson et al. 2003), indicating the critical importance of surface mannosylation. GDP-Man and PPM biosynthetic pathways are also conserved in the genus *Streptomyces* and are important for protein *O*-mannosylation (Wehmeier et al. 2009; Howlett et al. 2018). The disruption of these pathways in *Streptomyces* results in increased antibiotic sensitivities, suggesting the roles of mannosylated proteins in cell envelope integrity.

The mannosyltransferase(s) that extends AcPIM2 to a more polar AcPIM4 is currently unknown in mycobacteria. In corynebacteria, AcPIM2 is produced by a similar pathway and is further elongated by MptB, a mannosyltransferase that is presumably a processive enzyme to create an α1,6 mannan chain (Mishra et al. 2008a). Corynebacteria do not produce PIM6 species but do produce LM and LAM (Crellin et al. 2013), and MptB is involved in the production of these more extensively mannosylated species (see below). The gene encoding the ortholog of corynebacterial MptB was deleted in *M. smegmatis*, and the gene deletion showed no defects in PIMs/LM/LAM synthesis (Mishra et al. 2008a), making it unclear if MptB is redundant with another enzyme in *M. smegmatis*, or if there is another enzyme that mediates the reaction in *M. smegmatis*.

A key mannosyltransferase that drives the synthesis of PIM6 species is PimE, which adds the fifth Man to PIM4 using PPM as a Man donor (Fig. 13.2b). As mentioned above, corynebacteria also produce PIM species, but lack PIM6. Being consistent with this observation, there is no apparent ortholog of PimE in corynebacteria. LpqW, a lipoprotein, is suggested to be involved in regulating the AcPIM4 biosynthetic branch point in *M. smegmatis* (Kovacevic et al. 2006). When the *lpqW* gene was deleted, the *M. smegmatis* mutant became defective in producing LM/LAM and growth was retarded. Suppressor mutants that restored rapid growth were isolated from Δ*lpqW*, and all carried mutations in the *pimE* gene (Crellin et al. 2008). The *pimE* mutation blocked the synthesis of AcPIM6 but allowed the increased production of LM/LAM in Δ*lpqW*. These observations suggested that LpqW is a key regulatory protein controlling the two alternative pathways, AcPIM6 or LM/LAM. This proposed role of LpqW in *M. smegmatis* somewhat contradicts the fact that LpqW is present in corynebacteria yet AcPIM6 is not produced in these bacteria. Instead, LpqW is proposed as a regulatory protein essential for the activity of MptB mannosyltransferase in *C. glutamicum* (Rainczuk et al. 2012).

Once the mannan chain of LM is extended to an intermediate length of 5–20 residues, another processive mannosyltransferase, MptA, elongates the α1,6 mannan chain to a mature size of 21–34 residues (Kaur et al. 2007; Mishra et al. 2007). The α1,6 mannan backbone is decorated by α1,2 mono-Man branches, and this reaction is mediated by the α1,2 mannosyltransferase MptC (Kaur et al. 2006, 2008; Sena et al. 2010). Two proteins are involved in regulating mannan elongation: one membrane protein encoded by the *C. glutamicum NCgl2760* gene is proposed to play a role in mannan elongation at, or immediately prior to the MptA-dependent elongation step (Cashmore et al. 2017). The deletion of the *M. smegmatis* ortholog, *MSMEG_0317*, was not possible, suggesting that it is an essential gene. This is somewhat surprising as *mptA* is not an essential gene in *M. smegmatis* (Kaur et al. 2007; Fukuda et al. 2013). Another protein, termed LM elongation factor A (LmeA), is an *M. smegmatis* periplasmic protein necessary for the α1,6 mannan elongation mediated by MptA (Rahlwes et al. 2017). LmeA is not essential for in vitro growth, and genetic studies suggested that MptA is epistatic to LmeA, but the precise function of this protein remains unknown.

A single arabinan made of ~70 arabinofuranose (Ara*f*) residues is attached to the mannan backbone of LAM (Kaur et al. 2014). It is composed of a linear α1,5 chain with α1,3 branches and terminated with a linear tetra-Ara*f* or branched hexa-Ara*f* motif. Both of these terminal motifs end with the non-reducing β1,2 Ara*f* residue. The first arabinosyltransferase that primes the mannan chain is unknown. EmbC is the processive α1,5 arabinosyltransferase that elongates the primed arabinose (Ara) (Zhang et al. 2003; Shi et al. 2006) and is an essential enzyme in *M. tuberculosis* (Goude et al. 2008) (Fig. 13.2c). AftC is an α1,3 arabinosyltransferase that creates branching in LAM biosynthesis (Birch et al. 2008, 2010). AftB is the β1,2 arabinosyltransferase, which forms the terminal motif (Jankute et al. 2017). Both AftC and AftB are also involved in arabinogalactan biosynthesis (see below).

The presence of PI and mannolipid species beyond mycobacteria and corynebacteria becomes less well described. PIM1 and PIM2 are present in *Streptomyces* and can comprise 2–21% of the main polar lipids in the plasma membrane (Kimura et al. 1967; Sandoval-Calderon et al. 2015; Nguyen and Kim 2015). While LAM is suggested to be present in the outer membrane in mycobacteria, LAM can be produced without the outer membrane: *Corynebacterium otitidis* (formerly *Turicella otitidis)*, which lack mycolic acids and therefore the mycolate-based outer membrane, has the ability to synthesize LAM (Gilleron et al. 2005). Furthermore, *Amycolatopsis sulphurea* and *Lechevalieria aerocolonigenes*, which belong to the order Pseudonocardiales (Gibson et al. 2005), do not possess the outer membrane but are known to synthesize LAM. These observations suggest that PIMs/LM/LAM are plasma membrane mannolipids, predating the evolution of the outer membrane.

What are the potential reasons for having these glycolipids in actinobacterial plasma membrane? The method of reverse micellar solution extraction indicated that PIM species are found in the plasma membrane of mycobacteria and corynebacteria (Bansal-Mutalik and Nikaido 2011, 2014). Our previous study in *M. smegmatis* is consistent with this notion because the deletion of *pimE*, the fifth mannosyltransferase of AcPIM6 biosynthesis, resulted in abnormal mesosome-like membrane accumulation in the cytoplasm (Morita et al. 2006). Simultaneously, the *pimE* deletion mutant becomes hypersensitive to various antibiotics as well as a low concentration of copper, which is often included in standard mycobacterial growth media (Eagen et al. 2018). These observations suggest the importance of PimE in the structural integrity of the plasma membrane. It remains unknown if the accumulation of AcPIM4 or the lack of AcPIM6 in the *pimE* deletion mutant is toxic to the cell. One possibility is that AcPIM6 plays a structural role in anchoring the plasma membrane to the peptidoglycan layer, and the lack of anchoring results in destabilization of the plasma membrane, leading to the invagination and mesosome formation. LAM is also suggested to be a plasma membrane lipid in mycobacteria (Hunter et al. 1986), and visualization of surface-exposed LAM using atomic force microscopy indicates that LAM is not exposed on the cell surface unless outer membrane integrity is perturbed by antibiotics (Alsteens et al. 2008). These observations are consistent with the idea that LAM is anchored to the plasma membrane and glycan moiety intercalates the cell wall peptidoglycan, similar to the functions of (lipo)teichoic acids in Gram-positive bacteria (Weidenmaier and Peschel 2008) (see below for the discussion on outer membrane LM/LAM).

Although not PI-anchored, notable DAG-anchored mannolipids are also found in *Corynebacterium* and *Micrococcus*. *C. glutamicum* produces two LM species, termed Cg-LM-A and Cg-LM-B, and DAG-anchored Cg-LM-B is more predominant than the PI-anchored Cg-LM-A (Lea-Smith et al. 2008). Cg-LM-B carries 8–22 mannosyl residues and is anchored to the plasma membrane by α-D-glucopyranosyluronic acid DAG (Lea-Smith et al. 2008; Mishra et al. 2008a, b; Tatituri et al. 2007). The biosynthetic pathway is distinct from that of PI-anchored Cg-LM-A, involving the first priming mannosyltransferase, MgtA, which transfers Man onto the α-D-glucopyranosyluronic acid residue of the lipid precursor (Tatituri et al. 2007). Once MgtA adds the first Man, the same PPM-dependent MptB and MptA that produce Cg-LM-A extend the mannan chain (Mishra et al. 2008a). *Micrococcus* also produces DAG-anchored mannosides: D-mannosyl- α1,3-DAG (Man1-DAG), D-mannosyl- α1,3- D-mannosyl- α1,3-DAG (Man2-DAG) as well as much larger DAG-anchored LM carrying ~50 Man residues (Scher and Lennarz 1969; Lennarz and Talamo 1966; Pakkiri et al. 2004; Powell et al. 1975). *Micrococcus* species lack lipoteichoic acids, and LM is suggested to play a structural role in the cell wall. The biosynthetic pathway of these *Micrococcus* LM is not fully understood, but the first two mannoses are added using GDP-Man in the cytoplasmic side, and the Man2-DAG is proposed to flip to the periplasmic side of the plasma membrane to serve as the lipid anchor for further Man extension using PPM as the Man donor (Pakkiri et al. 2004; Pakkiri and Waechter 2005).

Plasma Membrane Compartmentalization

In *M. smegmatis*, a lipid domain termed the intracellular membrane domain (IMD) has been recently reported (Hayashi et al. 2016) and is suggested to form areas within the plasma membrane that are spatially distinct from the conventional plasma membrane. The IMD is enriched in metabolic enzymes, and many of them are involved in cell envelope biosynthesis. Furthermore, the IMD localizes to the polar region where the active elongation of the cell envelope takes place, suggesting that it is a strategic positioning of membrane-associated enzymes to the locations where the biosynthetic products are needed. Notably, the IMD is a dynamic entity which responds to environmental stresses and repositions its subcellular localization from polar enrichment during active growth to more sidewall localizations under stress exposure (Hayashi et al. 2018). There are many remaining questions in this research area: (1) what are the molecular mechanisms of protein localization to specific plasma membrane regions? (2) how are lipid intermediates able to translocate from one membrane domain to another? (3) what is the molecular mechanism of lipid domain formation? (4) what is the signaling mechanism for the spatial repositioning and how does the IMD relocate its subcellular location? We have provided a more detailed overview of the spatial control of the mycobacterial cell envelope in a recent review (Puffal et al. 2018).

Peptidoglycan

Peptidoglycan is a mesh of carbohydrate polymers crosslinked by short peptide side chains. It acts as an exoskeleton of bacteria, giving cells their shape and strength. This is illustrated by the demonstration that the digestion of the peptidoglycan layer results in the formation of spheroplasts in many rod-shaped bacteria, including Actinobacteria such as corynebacteria and mycobacteria (Melzer et al. 2018; Verma et al. 1989; Udou et al. 1983). The peptidoglycan layer is composed of repeating units of β1,4-linked *N*-acetylglucosamine (GlcNAc) and *N*-acetylmuramic acid (MurNAc) and tetrapeptides extending from MurNAc residues. Since the original proposal of peptidoglycan types for taxonomic classification (Schleifer and Kandler 1972), the amino acid composition of peptide stems has been widely used as a critical phenotypic feature for species identification in Actinobacteria. In general, the actinobacterial peptide stem comprises L-alanine-D-isoglutamine-L-diamino acid-D-alanine, where the third position L-diamino acid varies in different lineages. For example, the third position is *meso*-2,6-diaminopimelic acid (DAP) in mycobacteria and *C. diphtheriae* (Petit et al. 1969; Kato et al. 1968), while other diamino acids, such as L-2,4-diaminobutyrate, L, L-diaminopimelic acid, L-lysine and L-ornithine, or monoamino acids, such as L-homoserine, are found in various species of *Corynebacterium*, *Streptomyces*, and *Bifidobacterium* (Perkins and Cummins 1964; Perkins 1971; Koch et al. 1970; Veerkamp 1971; Leyh-Bouille et al. 1970).

In contrast to the highly variable third position, the first position L-alanine is almost invariable. Nonetheless, L-serine is found in this first position in some species of the order Micrococcales (von Wintzingerode et al. 2001; Hamada et al. 2009), and glycine substitutes the first position L-alanine in *M. leprae* (Draper et al. 1987). Beyond this core structure, the glycan units and peptide stems are subject to additional modifications, creating further variations in the peptidoglycan structure.

In addition to compositional differences, different types of peptide cross-linking are found not only among various lineages of Actinobacteria but also within a single species. The most common linkage is between the ω-amino group of the position 3 diamino acid and the position 4 D-alanine, mediated by two conventional penicillin-binding proteins (PBP1 and PBP2, encoded by *ponA1* and *ponA2* genes), which catalyze the D,D-transpeptidase reaction. In addition to this 3-4 cross-linking, mycobacteria and some species of *Streptomyces* and *Corynebacterium* insert the 3-3 cross-linking between two residues of diamino acid, catalyzed by L,D-transpeptidases (Leyh-Bouille et al. 1970; Wietzerbin et al. 1974; Lavollay et al. 2009, 2011; Kumar et al. 2012). In fact, this unusual 3-3 cross-linking is the predominant linkage found in *M. tuberculosis* (Kumar et al. 2012; Lavollay et al. 2008). Consistent with the abundance of the 3-3 cross-linking, the genomes of *M. tuberculosis* and *M. smegmatis* respectively carry five and six homologs of L,D-transpeptidases that mediate this reaction. Several of these play critical non-redundant roles in maintaining cell wall integrity, antibiotic resistance, and the establishment of *M. tuberculosis* infection in animal models (Gupta et al. 2010; Schoonmaker et al. 2014; Brammer Basta et al. 2015; Kieser et al. 2015; Sanders et al. 2014). In particular, genome-wide TnSeq analyses demonstrated that one of the L,D-transpeptidases, LdtB, and two penicillin-binding proteins genetically interacts with distinct sets of genes, suggesting non-redundant functions of these two peptide bridges (Kieser et al. 2015). In addition to the genetic studies, recent cell biological analyses further revealed that cross-linking by L,D-transpeptidases are particularly necessary for creating 3-3 crosslinking in the aging cell wall along the sidewall while D,D-transpeptidases such as PBP1 and PBP2 are critical for polar elongation and repair of damaged sidewall peptidoglycan (Baranowski et al. 2018; Garcia-Heredia et al. 2018). These recent studies reinforce the concept that these enzymes play distinct roles. For more details of the peptidoglycan structure, comprehensive reviews are available (Schleifer and Kandler 1972; Vollmer et al. 2008; Mainardi et al. 2008).

The biosynthesis of peptidoglycan in Actinobacteria is generally similar to the evolutionarily conserved pathway found in other bacteria. It is separated into cytoplasmic and membrane steps: cytoplasmic enzymes make pentapeptidyl MurNAc, and membrane-bound enzymes produce the polyprenol-linked peptidoglycan precursor lipid II on the cytoplasmic side of the plasma membrane (Fig. 13.3a). MurF is the last enzyme of the cytoplasmic steps, mediating the UDP-MurNAc-tripeptide-D-alanyl-D-alanine ligase. One of the MurF substrates, D-alanyl-D-alanine, is synthesized by a D-alanine-D-alanine ligase. In *Mycobacterium* and *Streptomyces* species, this ligase belongs to the DdlA group (Noda et al. 2004). In contrast, *Amycolatopsis orientalis*, an actinobacterial species in the order Pseudonocardiales, uses VanA group D-alanine-D-lactate ligase, producing a lipid II capped with D-lactate instead of D-alanine (Marshall and Wright 1998). This bacterium is the producer of the antibiotic vancomycin, which binds the terminal D-alanine-D-alanine residue of lipid II to prevent its utilization for peptidoglycan biosynthesis. This VanA-mediated modification is a clever way for this bacterium to prevent its antibiotic product from inhibiting its lipid II. The first membrane step is mediated by MraY, the polyprenyl transferase, which conjugates pentapeptidyl MurNAc to a polyprenol phosphate. The resulting intermediate, termed lipid I, is then modified by GlcNAc to become a complete precursor, lipid II, mediated by the GlcNAc transferase, MurG. MurJ is the proposed flippase, which translocates lipid II from the cytoplasmic side to the periplasmic side of the plasma membrane. PBP1 and PBP2 function as the trans-glycosylases that transfer the de novo synthesized peptidoglycan subunit to the elongating chain, and also as the D,D-transpeptidases to introduce the classical 3-4 peptide bridges (Fig. 13.3a).

Given its structural importance, it is not surprising that many regulatory mechanisms control the peptidoglycan biosynthesis. De novo synthesis of peptidoglycan precursor starts in the cytoplasm, with the committing step mediated by MurA. This enzyme, UDP-*N*-GlcNAc enolpyruvyl transferase, is regulated in response to nutrient availability through physical interactions with a cytoplasmic regulator CwlM. The serine/threonine kinase PknB phosphorylates CwlM, and this phosphorylated form activates MurA (Boutte et al. 2016). Since PknB carries the extracellular peptidoglycan-binding domain known as the PASTA domain, it seems that intracytoplasmic MurA is regulated by sensing the periplasmic peptidoglycan biosynthetic activities. CwlM orthologs are widely found in Actinobacteria, suggesting

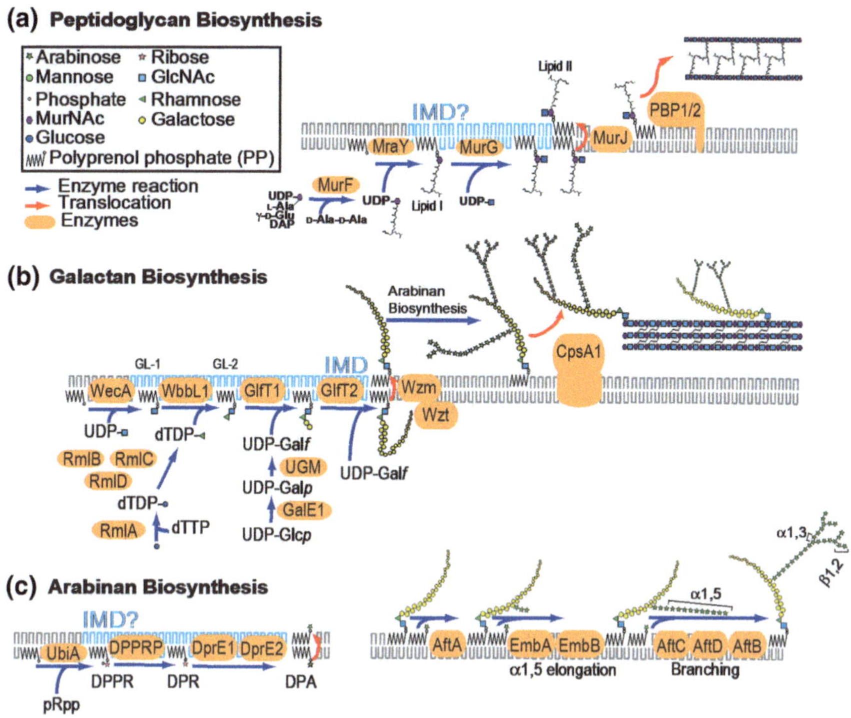

Fig. 13.3 Biosynthesis of peptidoglycan-arabinogalactan layer. **a** De novo synthesis of peptidoglycan. The pentapeptide compositions vary (see text for details). Remodeling of peptidoglycan layer by hydrolases and L,D-transpeptidases is not shown (see text for details). The IMD association of MurG is suggested from the proteomic analyses (Hayashi et al. 2016). **b** Biosynthesis of galactan. The biosyntheses of galactan and arabinan (see panel **c**) are found in species of the Corynebacteriales order. The representative pathway in *Mycobacterium* is shown. UGM, UDP-Gal*p* mutase. **c** Biosynthesis of decaprenol-phosphate-β-D-Ara*f* (DPA) and arabinan. pRpp, 5-phosphoribosyl-1-pyrophosphate; DPPR, decaprenol-phosphate-5-β-D-phosphoribofuranose; DPR, decaprenol-phosphate-β-D-ribofuranose. In *M. tuberculosis*, Rv3789 is a proposed DPA flippase (not shown). In *M. smegmatis*, a DPPR phosphatase (DPPRP) candidate (MSMEG_6402), DprE1 and DprE2 are suggested to associate with the IMD based on the proteomic analyses (Hayashi et al. 2016). The biosynthesis of arabinan follows that of galactan (see panel **b**)

that this is a highly conserved regulatory mechanism (Boutte et al. 2016). Furthermore, there are additional examples of potential regulations of peptidoglycan biosynthesis by phosphorylation. MurC, which mediates the addition of L-alanine onto UDP-MurNAc, is phosphorylated by the protein kinase PknA in *C. glutamicum*, and the phosphorylation results in decreased enzymatic activity (Fiuza et al. 2008). In *M. tuberculosis*, PknA phosphorylates the next enzyme in the pathway, MurD, which mediates the addition of D-glutamate onto UDP-*N*-acetylmuramoyl-L-alanine (Thakur and Chakraborti 2008). In addition to CwlM, PknB also phosphorylates PBP1 in *M. tuberculosis*, and the phosphorylation plays a critical role in polar cell envelope elongation (Kieser et al. 2015). These observations collectively suggest that phosphorylation by serine/threonine kinases is a critical mechanism of regulating peptidoglycan biosynthesis at multiple different steps. *S. coelicolor* possesses 34 putative serine/threonine kinases, and some of them carry the extracellular PASTA peptidoglycan-binding domain (Petrickova 2003), suggesting that similar regulation of peptidoglycan biosynthesis by phosphorylation of key enzymes may occur in *Streptomyces* species.

Mycobacteria grow from the polar ends, and polar peptidoglycan synthesis supports cell elongation. When dividing, peptidoglycan synthesis is needed for the septum formation. Furthermore, sidewall peptidoglycan synthesis occurs in response to cell wall damage (Garcia-Heredia et al. 2018). These observations indicated that the synthesis of peptidoglycan is a controlled process likely requiring the coordination of biosynthetic enzymes as well as hydrolyzing enzymes. RipA is a peptidoglycan endopeptidase, which forms a complex with the peptidoglycan hydrolases RpfB and RpfE (Hett et al. 2007, 2008), or with the biosynthetic enzyme PBP1 (Hett et al. 2010). When fluorescent protein-tagged *M. tuberculosis* RipA and RpfB are heterologously expressed in *M. smegmatis*, they localized to the septum (Hett et al. 2007), and these proteins are suggested in coordinating septum resolution and cell separation (Chao et al. 2013). Being consistent with a known interaction with PBP1, an *M. smegmatis* mutant lacking all four Rpf homologs showed a reduced level of 4-3 crosslinking (Ealand et al. 2018). Other peptidoglycan modifying enzymes also play important roles. The peptidoglycan hydrolase, ChiZ (Rv2719c), is a protein in the mycobacterial divisomal complex (Chauhan et al. 2006; Vadrevu et al. 2011). Four peptidoglycan degrading amidases (Ami1-4) are present in the genomes of *M. tuberculosis* and *M. smegmatis* (Machowski et al. 2014). One of them is CwlM (Ami2), the above mentioned cytoplasmic regulator for which the amidase activity is not essential (Boutte et al. 2016). Of the remaining three, Ami1 (MSMEG_6281) is required for cell division (Senzani et al. 2017). Finally, DacB2 is a mycobacterial enzyme that shows D,D-carboxypeptidase and D,D-endopeptidase activities, and is proposed to play a role in converting the peptide bridges from 4-3 to 3-3 crosslinking (Baranowski et al. 2018; Bansal et al. 2015). The recent discoveries of many peptidoglycan hydrolyzing enzymes indicate the presence of complex mechanisms to maintain the peptidoglycan integrity.

Do Actinobacteria produce teichoic acids, which are important structural components of Gram-positive peptidoglycan layer? It is well established that teichoic, teichuronic and teichulosonic acids are widespread in bacteria such as *Streptomyces*, *Micrococcus*, *Propionibacterium*, *Kribbella*, *Catellatospora*, and *Actinoplanes* (Tul'skaya et al. 2011; Naumova et al. 1980), which do not have mycolic acid-based outer membrane (see below). Furthermore, lipoteichoic acids have been reported in *Streptomyces*, *Agromyces*, and *Thermobifida* species (Rahman et al. 2009; Cot et al. 2011) (and references therein). These molecules appear to be absent in bacteria that produce mycolic acid-based outer membrane. These mycolic acid-producing bacteria are sometimes called "mycolata," and are found in all known families within the Corynebacteriales order: *Corynebacteriaceae*, *Dietziaceae*, *Gordoniaceae*, *Mycobacteriaceae*, *Nocardiaceae*, *Segniliparaceae*, *Tsukamurellaceae*, and *Williamsia*. With a recent demonstration of the Gram-negative outer membrane having a load-bearing function (Rojas et al. 2018), the evolution of the unique outer membrane in mycolata bacteria (see below) might have provided an alternative load-bearing function to the cell envelope and might have made teichoic acids and related molecules unnecessary.

Arabinogalactan

The arabinogalactan layer is composed primarily of a galactose (Gal) polymer of repeating β1,5- and β1,6-linked D-galactofuranose (Gal*f*) units, covalently modified by stretches of α1,5-linked Ara*f* residues which are branched by α1,3 branching sites (Daffe et al. 1990; Jankute et al. 2015; Angala et al. 2014). The linear galactan chain of the arabinogalactan is attached to the MurNAc residue of peptidoglycan through the α-L-Rha*p*-α1,3-D-GlcNAc-1-phosphate linker (McNeil et al. 1990). This layer is found in mycolata and is not known to be present in conventional Gram-positive Actinobacteria such as *Streptomyces* or *Micrococcus* species. It is best-studied in *Mycobacterium* and *Corynebacterium* species. In particular, *C. glutamicum* has been a useful model for delineating the structure and biosynthesis of this layer because arabinan biosynthesis is dispensable under laboratory growth conditions (Alderwick et al. 2005). The complete absence of arabinan biosynthesis in *Corynebacterium* results in a slower growing but viable mutant (Jankute et al. 2018). Even though the general structures are similar among mycolata, there are some notable differences. The galactan chain consists of ~30 residues in mycobacteria while it is much shorter in corynebacteria (Daffe et al. 1990; Alderwick et al. 2005). More drastic differences are found in *Nocardia* and *Rhodococcus*, in which galactan is primarily composed of linear β1,5-D-Gal*f* without the alternating β1,6 linkages (Daffe et al. 1993). Furthermore, the Gal*f* residues are partially modified by β1,6 mono-Glc side chains in *Nocardia* or by β1,2- and β1,3-linked Gal*f* residues in *Rhodococcus*. The arabinan portion of arabinogalactan is also differentially modified: galactosamine and succinate in *M. tuberculosis* and some slow-growing mycobacteria (Bhamidi et al. 2008; Draper et al. 1997; Lee et al. 2006; Peng et al. 2012) or rhamnose (Rha) in *C. glutamicum* (Alderwick et al. 2005). In some *Corynebacterium* species, arabinogalactan can be additionally modified by other monosaccharides such as Man and Glc (Abou-Zeid et al. 1982). The non-reducing termini of the arabinan moiety appear to be the most complex in mycobacteria, consisting of the branching motif: Ara*f*-β1,2-Ara*f*-α1,5-(Ara*f*-β1,2-Ara*f*-α1,3-)Ara*f*-α1,5-Ara*f*-α1-. Other mycolata bacteria tend to have simpler terminal ends. For example, *Nocardia* species cap the non-reducing ends of the arabinan residue by the linear motif: Ara*f*-β1,2-Ara*f*-α1,5-Ara*f*-α1-.

Galactan biosynthesis starts from building the linker moiety on a decaprenol-phosphate lipid. In mycobacteria, a homolog of WecA is the proposed first enzyme that transfers GlcNAc-phosphate from UDP-GlcNAc to the polyprenol lipid, forming an intermediate termed GL-1 (Jin et al. 2010). The deletion of the encoding gene, MSMEG_4947, in *M. smegmatis* resulting in severe morphological changes is consistent with its role in the cell wall galactan biosynthesis. The rhamnosyltransferase WbbL adds L-rhamnopyranose (Rha*p*) from dTDP-Rha*p* to the GlcNAc residue of GL-1, forming GL-2 (Mills et al. 2004). The Rha*p* donor, dTDP-Rha*p*, is synthesized by sequential actions of four enzymes, RmlA-D, starting from dTTP and D-Glc-α1-phosphate substrates (Ma et al. 2001; Li et al. 2006; Ma et al. 1997, 2002; Qu et al. 2007; Stern et al. 1999) (Fig. 13.3b).

The next step in the pathway is to prime the GL-2 with two Gal*f* residues. The donor of Gal*f* is UDP-Gal*f*, which is produced by the sequential actions of two enzymes. The first enzyme, GalE1, is the UDP-Glc 4-epimerase, which epimerizes UDP-glucopyranose (UDP-Glc*p*) to UDP-galactopyranose (UDP-Gal*p*) (Weston et al. 1997; Pardeshi et al. 2017). UDP-Gal*p* is then converted to UDP-Gal*f* by an essential enzyme, UDP-Gal*p* mutase (Weston et al. 1997; Pan et al. 2001), for which the atomic resolution structure was recently revealed (van Straaten et al. 2015). GlfT1 is the galactosyltransferase responsible for adding the first two Gal*f* residues to GL-2 forming Gal*f*-Gal*f*-Rha*p*-GlcNAc-phosphate-decaprenol (GL-4) (Mikusova et al. 2006; Alderwick et al. 2008; Belanova et al. 2008). GlfT2 is the processive galactosyltransferase, which extends the galactan polymer (Szczepina et al. 2009; Wheatley et al. 2012). A comparative study between *Mycobacterium* and *Corynebacterium* showed that GlfT2 can dictate the chain length of galactan (Wesener et al. 2017). We and others have demonstrated that GlfT2 is enriched in the polar region of mycobacterial cells and bound to the IMD (Meniche et al. 2014; Hayashi et al. 2016), where the enzyme perhaps coordinates and facilitates the spatially localized biosynthesis of the galactan layer. Once the decaprenol-linked galactan precursor is synthesized, it is flipped to the periplasmic side of the plasma membrane, and a putative ABC transporter, composed of two proteins, Wzm and Wzt, is implicated in this process (Dianiskova et al. 2011).

Once the decaprenol-linked galactan precursor is translocated to the periplasmic side, the galactan chain is modified by several arabinans. An earlier study suggested that there are three arabinans attached in one galactan polymer (Alderwick et al. 2005), but a more recent study suggests that only two arabinan chains are present per galactan (Bhamidi et al. 2011). The donor of Ara is decaprenol-phosphate-β-D-Ara*f* (DPA) (Alderwick et al. 2005; Lee et al. 1995, 1997; Wolucka and de Hoffmann 1995; Wolucka et al. 1994; Xin et al. 1997). DPA biosynthesis starts with UbiA, a 5-phospho-α-D-ribose-1-pyrophosphate:decaprenol phosphate 5-phosphoribosyltransferase, which produces decaprenol-phosphate-5-β-D-phosphoribofuranose (DPPR) using 5-phosphoribosyl-1-pyrophosphate (pRpp) and decaprenol phosphate as substrates (Scherman et al. 1995; Alderwick et al. 2011). The deletion of *ubiA* gene in *C. glutamicum* results in the complete abrogation of cell wall arabinan, suggesting that this enzyme is the sole enzyme that diverts pRpp into the biosynthesis of arabinan (Alderwick et al. 2005). In *M. tuberculosis*, the genome region spanning Rv3789-Rv3809c are dedicated to arabinan biosynthesis, and a putative phosphorylase in this region, Rv3807c, is proposed to act as the next enzyme, DPPR phosphatase, which forms decaprenol-phosphate-β-D-ribofuranose (DPR). However, the deletion of the ortholog in *M. smegmatis* showed only a mild impact on arabinan content of arabinogalactan (Jiang et al. 2011), suggesting that there are phosphatases that can surrogate the function of this enzyme. The third and fourth enzymes in the DPA biosynthesis pathway are oxidoreductases: DprE1 oxidizes DPR to decaprenol-phosphate-2-keto-β-D-*erythro*-pentofuranose (DPK), which is then reduced by DprE2 to DPA (Mikusova et al. 2005) (Fig. 13.3c). A homolog of *dprE2* is present in *C. glutamicum*, making it a redundant gene, and this second gene is also present in *M. tuberculosis* (Rv2073c) (Meniche et al. 2008). Inhibition of DprE1 results in the accumulation of DPR and kills *M. tuberculosis*, validating this enzyme as a potential target of TB chemotherapy (Grover et al. 2014; Makarov et al. 2009). Since DprE2 is an NADH-dependent oxidoreductase (Mikusova et al. 2005), it seems likely that biosynthetic enzymes upstream of DprE2 are active on the cytoplasmic side of the plasma membrane. DprE1 does not appear to carry trans-membrane domains while DprE2 does (data not shown). Nonetheless, both have been identified as IMD-associated proteins by proteomic analysis (Hayashi et al. 2016), likely suggesting a peripheral association of DprE1 with the membrane domain. Once DPA is produced, it is flipped to the periplasmic side of the plasma membrane, and Rv3789 has been proposed as the flippase candidate (Larrouy-Maumus et al. 2012).

Using DPA as the Ara donor, coordinated actions of multiple arabinosyltransferases, which belong to the GT-C glycosyltransferase superfamily, drive the biosynthesis of arabinan on the periplasmic side of the plasma membrane. The first enzyme that primes the galactan chain with Ara is AftA (Alderwick et al. 2006). In *C. glutamicum*, the next enzyme that extends the α1,5 Ara chain of the core arabinan is Emb (Alderwick et al. 2005). The homologs in mycobacteria, EmbA and EmbB, are proposed to function redundantly in the core α1,5 arabinan synthesis (Alderwick et al. 2005; Escuyer et al. 2001). Notably, these enzymes, EmbB in particular, are the targets of the frontline drug ethambutol (Takayama and Kilburn 1989; Mikusova et al. 1995; Telenti et al. 1997; Lety et al. 1997; Belanger et al. 1996). The α1,3 branching is mediated by the arabinosyltransferase, AftC, which plays a role in both arabinogalactan and LAM biosynthesis in mycobacteria (Birch et al. 2008, 2010). AftD was initially proposed as α1,3 branching enzyme in mycobacteria (Skovierova et al. 2009), but a more recent study suggests that it is a processive α1,5 arabinosyltransferase, which extends the α1,3 branching Ara primed by AftC (Alderwick et al. 2018). Finally, AftB adds the terminal β1,2 capping (Seidel et al. 2007). When *aftB* is deleted, mycoloylation sites on arabinogalactan are severely reduced (Bou Raad et al. 2010). Nevertheless, the *aftB* deletion mutant can still produce an outer membrane, although the stability of the outer membrane becomes significantly compromised.

Once arabinan is synthesized onto the galactan polymer, the arabinogalactan complex is attached to the 6-OH of MurNAc residues in the peptidoglycan glycan chain (Fig. 13.3c). The enzymes that mediate this reaction are the LytR-CpsA-Psr (LCP) phosphotransferases, which are variably termed CpsA1/CpsA2, LcpA/LcpB, or Lcp1/CpsA by different groups (Wang et al. 2015; Harrison et al. 2016; Baumgart et al. 2016; Grzegorzewicz et al. 2016). CpsA1 is widely conserved and appears to play a primary role in arabinogalactan anchoring. In contrast, CpsA2 is not found in fast-growing species of *Mycobacterium* and is implicated in processes associated with the host-pathogen interaction (Koster et al. 2017).

Outer Membrane

The outer membrane (OM) , also known as the mycomembrane, is a mycolic acid-rich pseudo-bilayer of lipids. It has 7 nm thickness as determined by cryo-electron microscopy in mycobacteria (Zuber et al. 2008; Hoffmann et al. 2008; Sani et al. 2010). In a proposed model, the inner leaflet of the OM is composed primarily of mycolic acids, which are covalently attached by an ester linkage to the non-reducing end of arabinan. The outer leaflet of the OM is comprised of diverse lipid species. Some of these extractable outer membrane lipids are conserved throughout mycolata. For instance, the OM of both *Corynebacterium* and *Mycobacterium* likely consist of trehalose di(coryno)mycolates, free mycolic acids and fatty acids (Bansal-Mutalik and Nikaido 2011, 2014). In contrast, other lipids such as glycopeptidolipids (GPLs), phthiocerol dimycocerosates (PDIMs) and phenolic glycolipids (PGLs) are specific to certain species within mycolata. Some of the major outer membrane lipids are highlighted below.

Mycolic Acids

Mycolic acids are long α-alkyl β-hydroxy fatty acids with the meromycolic acid carbon backbone ranging from C18 to C76 and the alkyl side chain ranging from C24 to C26. Comprehensive surveys of the mycolata from the 1980s demonstrated that the length of the mycolic acids varies significantly among species (Collins et al. 1982; Goodfellow et al. 1982). Additional modifications such as cyclopropane rings, double bonds, and methylations further add diversity to the structures of mycolic acids (Marrakchi et al. 2014; Minnikin et al. 2015; Quemard 2016). Because of the structural variety, mycolic acid structures have been used for taxonomic purposes. For instance, *Corynebacterium* produces some of the shortest mycolic acids, ranging from C22 to C36 (Collins et al. 1982; Welby-Gieusse et al. 1970). In contrast, recent studies revealed that *Segniliparus rotundus* produces the longest mycolic acid (C100), which is perhaps the longest fatty acyl chain currently known (Hong et al. 2012; Laneelle et al. 2013). *Mycobacterium* produces relatively long mycolic acids, typically ranging from C60 to C90 (Barry et al. 1998). However, *Hoyosella altamirensis* and *Hoyosella subflava*, two recently discovered environmental cocci that belong to the *Mycobacteriaceae* family and are closely related to the *Mycobacterium* genus, produce relatively short mycolic acids, ranging from C30 to C36 (Laneelle et al. 2012).

In mycobacteria, mycolic acid biosynthesis is initiated by type I and type II fatty acid synthases (FAS-I and FAS-II) (Fig. 13.4). FAS-I is a large multifunctional enzyme that utilizes acetyl-CoA and malonyl-CoA to create short chain fatty acyl-CoAs (C16–C18 and C24–C26) (Brindley et al. 1969). The longer of the two distinct pools of the products are then carboxylated, and the carboxyacyl-CoA serves as the donor of the alkyl side chain for the mycolic acid synthesis.

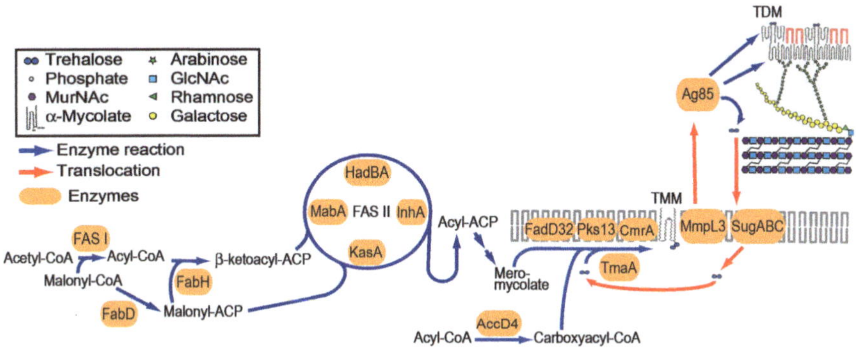

Fig. 13.4 Biosynthesis of mycolic acids. The chain length of mycolic acids can vary dramatically among mycolata bacteria. Further modifications of mycolic acids can also vary among species. For example, cyclopropane, methoxy, keto, and hydroxy modifications are found in *M. tuberculosis*, while epoxy and methyl modifications are found in *M. smegmatis*. Representative structures are shown. Biosynthesis of trehalose monomycolate (TMM) and trehalose dimycolate (TDM) is coordinated in conjunction with the biosynthesis of mycolic acid and its attachments to the arabinan layer

Next, β-ketoacyl-acyl carrier protein (ACP) synthase III (FabH) condenses malonyl-ACP and acyl-CoA produced by the FAS-I enzyme, producing β-ketoacyl-ACP. The FAS-II elongates the fatty acyl chain of β-ketoacyl-ACP to produce fully elongated acyl-ACPs (Odriozola et al. 1977; Bloch 1977; Slayden and Barry 2002). As mentioned above, *Corynebacterium* produces the shortest mycolic acids, and being consistent with this observation, *C. glutamicum* lacks the FAS-II elongation system. Instead, its genome encodes two FAS-I genes (*fasA* and *fasB*) with FasA playing the dominant role (Radmacher et al. 2005). In contrast to FAS-I being a single polypeptide carrying multiple catalytic domains, the FAS-II system is composed of several separate enzymes. First, MabA, the β-ketoacyl-ACP reductase, reduces the β-keto moiety of the β-ketoacyl-ACP. Second, dimeric β-hydroxyacyl-ACP dehydratases, HadBA/HadBC, dehydrate the product of MabA reaction, β-hydroxyacyl-ACP. Third, the resultant enoyl-ACP is reduced by *trans*-2-enoyl-ACP reductase, InhA. Finally, the fully saturated acyl-ACP is elongated by the β-ketoacyl-ACP synthases, KasA or KasB, using malonyl-ACP (Marrakchi et al. 2014; Duan et al. 2014).

Why are there two different heterodimers of β-hydroxyacyl-ACP dehydratases? HadB is proposed as the catalytic component (Biswas et al. 2015), implying that HadA and HadC play non-catalytic roles within the heterodimers HadBA and HadBC. Interestingly, HadC is mutated in an avirulent strain of *M. tuberculosis* (Lee et al. 2008; Zheng et al. 2008), and targeted gene disruptions in *M. tuberculosis* and *M. smegmatis* demonstrated changes in mycolic acid profile, and attenuation of virulence in the case of *M. tuberculosis* (Slama et al. 2016; Jamet et al. 2015). It is proposed that HadBA is the dehydratase in the early stage of meromycolic acyl chain elongation and HadBC mediates the late elongation steps. Furthermore, another recent study in *M. smegmatis* revealed an additional dehydratase, termed

HadD, which is involved in α- and epoxy-mycolic acid biosynthesis (Lefebvre et al. 2018). HadD is conserved in *Mycobacterium* genus, including *M. leprae*, but is absent in other genera of mycolata, suggesting that it is involved in specific steps found in mycobacteria.

Acyl-ACPs, produced by the FAS II, are modified by cyclopropane synthases and methyltransferases, resulting in meromycolates (Crellin et al. 2013; Marrakchi et al. 2014; Minnikin et al. 2015). Meromycolates are then activated to meromycoloyl-AMP by the fatty acyl-AMP ligase FadD32. Next, meromycoloyl-AMP is loaded onto Pks13, which condenses the meromycoloyl-AMP with carboxyacyl-CoA, and covalently links the resulting α-alkyl β-ketoacyl chain to the C-terminal ACP domain of Pks13 (Leger et al. 2009; Gavalda et al. 2009). Pks13 has an acyltransferase activity, which transfers the α-alkyl β-ketoacyl chain onto a trehalose, releasing the mono-α-alkyl β-ketoacyl trehalose (Gavalda et al. 2014). The released product is then reduced by CmrA to produce trehalose monomycolate (TMM) (Lea-Smith et al. 2007; Bhatt et al. 2008). In *C. glutamicum*, the TMM equivalent, namely trehalose monocorynomycolate (TMCM), is transiently acetylated by the acetyl transferase TmaA, and this acetylation is critical for the translocation of TMCM across the plasma membrane (Yamaryo-Botte et al. 2014). Acetylated TMCM is transported through the plasma membrane by the transporter MmpL3 (Grzegorzewicz et al. 2012; Varela et al. 2012; Xu et al. 2017; Li et al. 2016). TmaA is conserved in mycobacteria, suggesting that TmaA-mediated acetylation is a conserved mechanism of licensing mature TMM/TMCM for transport. Finally, Ag85 transfers mycolic acid from TMM to either another TMM molecule to produce TDM, or to the arabinan layer of the cell wall to create mycoloyl arabinogalactan peptidoglycan cell wall (Belisle et al. 1997; Backus et al. 2014) (Fig. 13.4).

Trehalolipids

Trehalolipids are bio-surfactants, which are important for bacteria to emulsify and utilize hydrophobic molecules. *Rhodococcus* species are prominent trehalolipid producers, and its production is induced when *Rhodococcus* is grown in the presence of hydrophobic molecules such as alkanes (Yakimov et al. 1999; Lang and Philp 1998). The capability of *Rhodococcus* species to produce trehalolipid surfactants attracts considerable interest in industrial applications, especially in oil recovery and oil spill treatment (Pacheco et al. 2010; Liu and Liu 2011). Structurally diverse variants of trehalolipids, which *Rhodococcus* can produce, include: mycoloylated trehaloses, such as TMM, TDM, and trehalose trimycolates (Niescher et al. 2006), and various forms of acylated trehalose, in which the acyl chains are generally shorter straight chain fatty acids (Singer et al. 1990; Philp et al. 2002; Uchida et al. 1989; Tuleva et al. 2008; Tokumoto et al. 2009; White et al. 2013; Espuny et al. 1995). Other mycolata bacteria, such as *Nocardia farcinica* and several species of *Tsukamurella*, are also known to produce tetraacyl or diacyl trehalose, respectively (Christova et al. 2015; Pasciak et al. 2010; Kugler et al. 2014; Vollbrecht et al. 1998). However, tre-

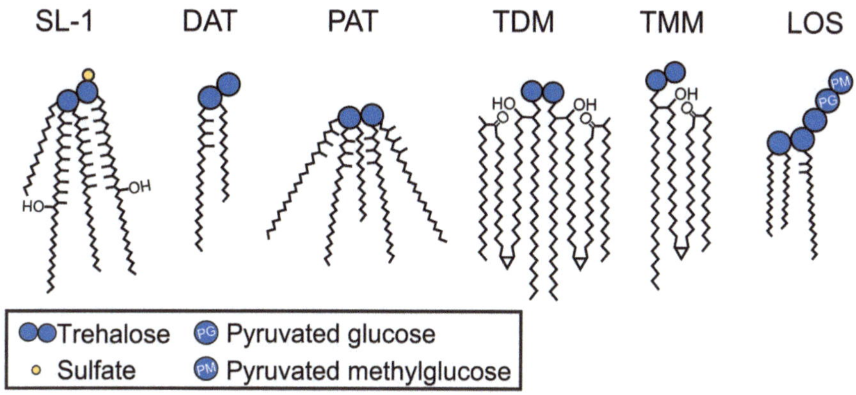

Fig. 13.5 Structures of trehalose-containing lipids. SL-1, a sulfolipid species; DAT, diacyltrehalose; PAT, pentaacyltrehalose; TMM, trehalose monomycolate; TDM, trehalose dimycolate; LOS, lipooligosaccharide. Fatty acid structures vary. Ester linkages and some unsaturated bonds are abbreviated. See text for additional structural variations

halolipids are not restricted to mycolata. *M. luteus* and *Arthrobacter* species produce trehalose tetraester (Tuleva et al. 2009; Passeri et al. 1991). These bacteria belong to the Micrococcales order, and exhibit a more typical Gram-positive cell wall without the outer membrane, suggesting that the presence of the outer membrane is not a prerequisite of producing trehalolipids. Rather, we wonder if the production of trehalolipid surfactants allowed the evolution of the outer membrane in mycolata.

Biosynthesis of trehalolipids has been studied more extensively in pathogenic *Mycobacterium* species than in other mycolata bacteria. In addition to TMM and TDM, bacteria that belong to the *M. tuberculosis* complex produce diacyl, triacyl and pentaacyl trehalose (DAT, TAT, and PAT) (Fig. 13.5). A unique feature of these mycobacterial trehalolipids is the extensive methyl branching of fatty acyl moieties. Mycolipenic acid is one such fatty acid, which is tri-methylated with one unsaturated bond. Mycolipenoyl modification of trehalose is only found in the *M. tuberculosis* complex such as *M. tuberculosis*, *M. bovis*, and *Mycobacterium africanum*. Other *Mycobacterium* species, such as *Mycobacterium fortuitum*, do produce acyl trehaloses, but the acyl chains are not multi-methylated (Ariza et al. 1994; Lopez-Marin et al. 1994). The polyketide synthase Msl3 synthesizes mycolipenic and mycosanoic acids, and the activation and loading of the fatty acid substrate are mediated by the fatty acid ligase FadD21 (Dubey et al. 2002; Rousseau et al. 2003; Belardinelli et al. 2014). Similarly, another polyketide synthase Msl5 produces a minor monomethyl branched unsaturated C16–C20 fatty acid found in acyl trehaloses (Dubey et al. 2003). Once fatty acids are made, PapA3 mediates the acyltransferase reactions using trehalose as the acceptor (Hatzios et al. 2009). It has been proposed that PapA3 can successively transfer fatty acyl groups to the 2- and 3-positions of trehalose at least in vitro. The acyltransferase Chp2 then mediates the last three acylation events to produce PAT (Belardinelli et al. 2014; Touchette et al. 2014). MmpL10 is the proposed plasma membrane flippase for the translocation of acyl trehalose species (Belardinelli et al. 2014; Touchette et al. 2014).

Sulfolipids are trehalolipids that are sulfonated at the 2-position of trehalose (Fig. 13.5) (Middlebrook et al. 1959; Goren 1970). The most abundant species is a tetra-acylated species known as SL-1, in which three acyl groups at the 6, 3′ and 6′-positions are hepta- or octa-methyl phthioceranic acids or hydroxyphthioceranic acids and one acyl group at the 2′-position is either palmitic or stearic acid. The biosynthesis starts with the sulfotransferase Sft0, a widely conserved protein in mycolata, which transfers sulfate from 3′-phosphoadenosine-5′-phosphosulfate to trehalose (Mougous et al. 2004). The second step is mediated by the acyltransferase PapA2, which transfers a straight chain fatty acid, palmitate or stearate, from its CoA donor substrate to the 2′-position of trehalose 2-sulfate, producing the monoacyl intermediate termed SL659 (Kumar et al. 2007). The third step is another acyltransferase reaction, in which PapA1 transfers (hydroxy) phthioceranoyl group from the polyketide synthase Pks2 to the 3′-position of trehalose residue of SL659 forming the diacyl intermediate SL1278 (Kumar et al. 2007; Sirakova et al. 2001; Bhatt et al. 2007). FadD23 is the proposed fatty acyl AMP ligase involved in Pks2-mediated (hydroxy)phthioceranoyl biosynthesis (Gokhale et al. 2007). Similar to the role of Chp2 for PAT synthesis, the acyltransferase Chp1 adds the remaining acyl groups at the 6- and 6′-positions of trehalose (Seeliger et al. 2011). The product, SL-1, is translocated across the plasma membrane, and MmpL8 and Sap are implicated in this process (Seeliger et al. 2011; Domenech et al. 2004; Converse et al. 2003). Precise functions of sulfolipids remain unknown, but these lipids are implicated in host-pathogen interactions and the establishment of infection (Angala et al. 2014; Schelle and Bertozzi 2006; Daffe et al. 2014).

Lipooligosaccharides (LOSs) are another type of trehalolipids found in *Mycobacterium* species, including *M. smegmatis, Mycobacterium kansasii, M. marium,* and *Mycobacterium canettii* (Hunter et al. 1983; Saadat and Ballou 1983; Daffe et al. 1991). LOSs play critical roles in colony morphology, biofilm formation, motility, as well as immune modulation during host infection (Alibaud et al. 2014; Rombouts et al. 2009; Ren et al. 2007; Sarkar et al. 2011; van der Woude et al. 2012). The core structure of LOS is similar to other acyl trehaloses in that trehalose is modified with either straight chain or branched chain fatty acids. The feature of LOS that distinguishes it from other trehalolipids is the additional glycan modification of the acyl trehalose core. The glycan structures vary among different *Mycobacterium* species. For example, *M. smegmatis* LOS consists primarily of Glc*p* (Fig. 13.5), whereas *M. kansasii, Mycobacterium gastri,* and *Mycobacterium marinum* produce several different LOS species containing Rha*p*, xylopyranose and *N*-acyl kanosamine in addition to Glc*p* (Saadat and Ballou 1983; Rombouts et al. 2009, 2010, 2011; Gilleron et al. 1993; Hunter et al. 1984). The biosynthesis of LOSs is not fully understood. Similar to other trehalolipid biosynthesis, the core acyl trehalose synthesis requires specific polyketide synthases such as Pks5 and Pks5.1, fatty acyl-AMP ligases such as FadD25, and acyltransferases such as PapA4 and PapA3 (Rombouts et al. 2011; Etienne et al. 2009). Glycosyltransferases presumably transfer glycans to the acyl trehalose core, but only a few genes have been experimentally validated (Ren et al. 2007; Sarkar et al. 2011; Burguiere et al. 2005; Chen et al. 2015; Nataraj et al. 2015).

Mannolipids

As discussed above, PIMs, LM, and LAM are at least partially present in the plasma membrane. Nevertheless, substantial evidence also suggests that these molecules are anchored to the outer membrane of mycolata bacteria as well. First, the majority of LM/LAM was accessible to surface biotinylation in *M. bovis* BCG (Pitarque et al. 2008). Second, only residual amounts of LM/LAM remain in the spheroplast of *M. smegmatis* (Dhiman et al. 2011). These data both suggest that the majority of LM/LAM could be in the outer membrane. Furthermore, LAM from pathogenic mycobacteria is capped with α1,2 Man residues, which are an epitope recognized by host lectins (Ishikawa et al. 2017; Kallenius et al. 2016; Turner and Torrelles 2018). Although host cell receptors can function to detect cell fragments rather than intact cells, the diverse repertoire of host immune molecules to identify these molecules may be more consistent with the concept that these molecules are exposed on the surface of mycobacterial cells. Finally, as detailed in the *Capsules and Extracellular Polysaccharides* section, mannan and arabinomannan are components of mycobacterial capsule. While the biosynthetic relationship between LM/LAM and mannan/arabinomannan is not established, there must be a mechanism to transport either lipidated or delipidated glycans across the cell wall and outer membrane. There have been studies suggesting that the lipoprotein LprG plays a role in this process (Drage et al. 2010; Alonso et al. 2017). However, there is also compelling evidence suggesting that LprG is involved in triacylglycerol trafficking, and acts in a more complex way by physically interacting with two other lipoproteins, LppK and LppI, as well as Ag85A (Touchette et al. 2017; Martinot et al. 2016).

Glycopeptidolipids

Glycopeptidolipids (GPLs) are found in the nontuberculous *Mycobacterium* species, such as *M. smegmatis, M. avium, Mycobacterium intracellulare*, and *Mycobacterium abscessus* and are important virulence factors for the pathogenic species (Schorey and Sweet 2008; Gutierrez et al. 2018; Mukherjee and Chatterji 2012). The defects in GPL biosynthesis results in compromised cell envelope integrity and abnormalities in growth, biofilm formation and sliding motility among others, suggesting their critical roles as a component of the outer membrane (Recht et al. 2000; Recht and Kolter 2001; Zanfardino et al. 2016). GPLs have a common core structure consisting of three parts: a mono-unsaturated 3-hydroxy/methoxy C26–C34 acyl chain, a D-phenylalanyl-D-*allo*-threonyl-D-alanyl-L-alaninol tetrapeptide, and two carbohydrate modifications, a 6-deoxy-α-L-talose linked to the D-*allo*-threonine residue and an α-L-Rha linked to L-alaninol (Fig. 13.6a). GPLs can be further glycosylated, and these glycan residues are additionally methylated to give rise to serotype-specific GPLs (Chatterjee and Khoo 2001). During the biosynthesis of GPLs, mono-unsaturated 3-hydroxy/methoxy acyl chain is synthesized by the actions of the acyl-CoA dehydrogenase FadE5, the polyketide synthase Pks, and *O*-methyltransferase Fmt (Jeevarajah et al. 2002; Sonden et al. 2005; Jeevarajah et al. 2004). Fmt is proposed to convert 3-hydroxy to 3 methoxy during the acyl chain synthesis. Mps1 and Mps2 are the nonribosomal peptide synthases that produce the tetrapeptide core (Sonden et al. 2005; Billman-Jacobe et al. 1999). PapA3, an acyltransferase involved in acyl trehalose biosynthesis in *M. tuberculosis*, is proposed to transfer the acyl chain bound to Pks to the tetrapeptide in nontuberculous mycobacteria (Ripoll et al. 2007). Gtf1 and Gtf2 are the 6-deoxytalosyltransferases and rhamnosyltransferase involved in the synthesis of the core glycan residues, respectively (Miyamoto et al. 2006). Additional rhamnosyltransferases, fucosyltransferases and glucosyltransferases, as well as glycan *O*-methyltransferases and *O*-acetyltransferase, produce serotype-specific GPLs (Recht and Kolter 2001; Jeevarajah et al. 2004; Miyamoto et al. 2006; Patterson et al. 2000; Miyamoto et al. 2007, 2008, 2010; Maslow et al. 2003; Eckstein et al. 1998; Naka et al. 2011; Fujiwara et al. 2007; Nakata et al. 2008). Once synthesized, GPLs are transported to the outer membrane by the actions of MmpL4a/b, Gap, and MmpS4 (Sonden et al. 2005; Medjahed and Reyrat 2009; Nessar et al. 2011; Bernut et al. 2016) (Fig. 13.6b).

GPLs are specific to certain species of *Mycobacterium*, but structurally different types of GPLs have been reported from other bacteria in the Corynebacteriales order. *Gordonia hydrophobica* produces a mono-glucosylated *N*-acyl tridecapeptide, in which the beta-hydroxy residue of the fatty acid is interlinked to the C-terminus of the peptide chain, forming a cyclic lactone ring (Moormann et al. 1997). Related molecules are also found in *Rhodococcus erythropolis* (Koronelli 1988).

Mycocerosyl Lipids

Phthiocerol dimycocerosates (PDIMs) are waxy lipids, in which a phthiocerol (a long (C33-C41) carbon chain β-diol) is esterified with two poly-methylated fatty acids called mycocerosic acids (Fig. 13.6a). PDIMs are found in pathogenic *Mycobacterium*, such as *M. tuberculosis*, *M. bovis*, and *M. leprae* (Daffe et al. 2014; Daffe and Laneelle 1988). Instead of mycocerosic acids, some species such as *Mycobacterium ulcerans* and *M. marinum* utilize phthioceranic acids, in which the chirality of the methyl branches are L configurations instead of D configurations found in mycocerosic acids. PDIMs are essential for *M. tuberculosis* to establish infection in animal models (Goren et al. 1974; Cox et al. 1999; Camacho et al. 1999), and the molecular mechanisms governing the host-pathogen interaction are actively investigated (Arbues et al. 2014). The PDIM synthesis is proposed to take place in four distinct stages (Trivedi et al. 2005) (Fig. 13.6c). First, FadD26, a fatty acyl-AMP ligase, activates long-chain fatty acids to fatty acyl-AMP and transfers the acyl moiety

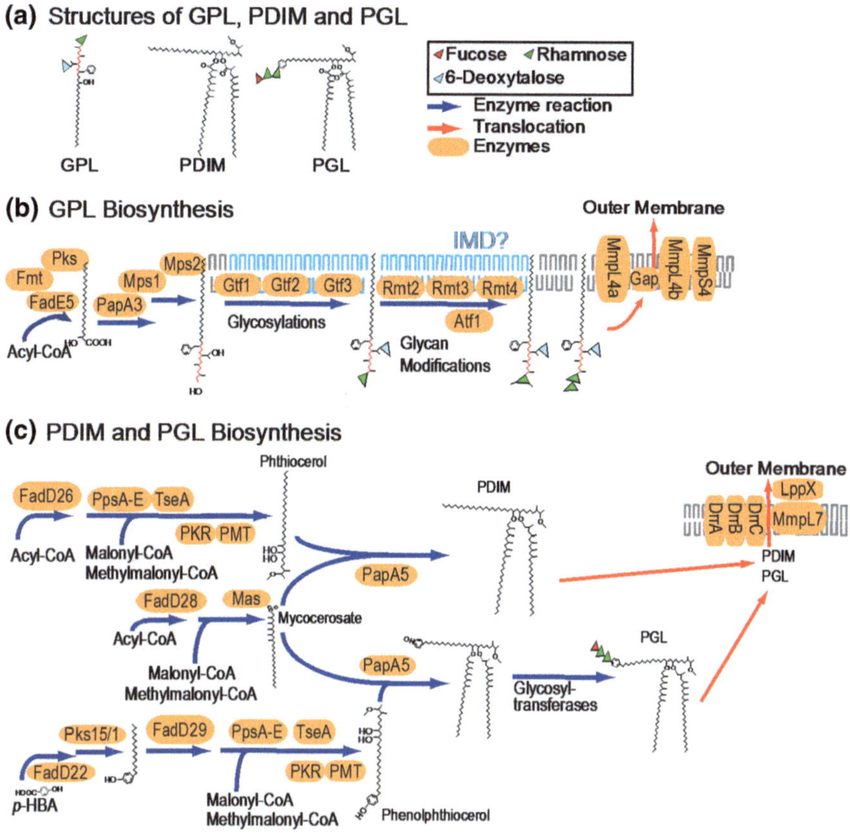

Fig. 13.6 Biosynthesis of glycopeptidolipids and mycocerosyl lipids. **a** Structures of glycopeptidolipids (GPLs) , phthiocerol dimycocerosates (PDIMs) and phenolic glycolipids (PGLs) . The structure of the core (nonspecific) GPL is shown. GPLs are modified by additional glycosylation, methylation and acetylation that result in a variety of serotypes, most well characterized in *M. avium*. The red line indicates the peptide backbone. Significant variations are also found in PDIMs and PGLs, and representative structures are shown. Fucose and rhamnose in PGLs are often methylated (not shown). **b** Biosynthesis of GPL. Gtf, glycosyltransferases; Rmt, *O*-methyltransferases; Atf, *O*-acetyltransferase. The IMD association of Gtf1-3 and Rmt2-4 is suggested from the proteomic analyses (Hayashi et al. 2016). **c** Biosynthesis of PDIMs and PGLs. Acyl-CoA is synthesized from FAS I. PKR, phthiodiolone ketoreductase; PMT, phthiotriol methyltransferase; *p*-HBA, *p*-hydroxybenzoic acid. PGLs are likely exported through the same machinery as that for PDIMs. Some structural details are abbreviated for simplicity

to the polyketide synthase PpsA for the synthesis of phthiocerol moiety (Trivedi et al. 2004, 2005; Camacho et al. 2001; Azad et al. 1997). PpsA and the next enzyme PpsB lack the dehydratase and the enoylreductase domains, allowing the formation of the β-diol structure of phthiocerol. Second, additional polyketide synthases PpsC, PpsD and PpsE continue the chain extension of phthiocerol using malonyl-CoA or methyl malonyl-CoA as the hydrocarbon donor. TesA is a thioesterase, which interacts with PpsE (Rao and Ranganathan 2004), and is proposed to release the phthiodiolone product upon completion of the synthesis (Waddell et al. 2005; Chavadi et al. 2011). Structural variations are introduced by phthiodiolone ketoreductases and phthiotriol methyltransferases (Pérez et al. 2004a; Onwueme et al. 2005; Simeone et al. 2007) to produce mature phthiocerol. Third, FadD28 activates and transfer a fatty acid onto the mycocerosic acid synthase Mas, which elongates the fatty acyl substrate with methyl malonyl-CoA to produce poly-methylated mycoserosic acids (Cox et al. 1999; Trivedi et al. 2005; Simeone et al. 2010; Rainwater and Kolattukudy 1985; Azad et al. 1996; Mathur and Kolattukudy 1992; Rainwater and Kolattukudy 1983; Fitzmaurice and Kolattukudy 1997, 1998). In contrast to modular polyketide synthases PpsA-E, Mas is an iterative polyketide synthase, which can extend the methyl branched carbon chain by multiple rounds of the elongation reaction. Finally, mycocerosic acids are transferred onto the diol of phthiocerol by the acyl transferase PapA5 (Trivedi et al. 2005; Onwueme et al. 2004). PapA5 is proposed to directly transfer the mycoserosic acid attached to Mas to the hydroxyl groups of phthiocerol.

Once PDIMs are synthesized, they are transported to the outer membrane. MmpL7 is a transporter of the RND permease superfamily and DrrABC are homologous to ABC transporters. Both of these putative transporters are proposed to function as PDIM transporters (Cox et al. 1999; Camacho et al. 2001; Waddell et al. 2005; Choudhuri et al. 2002). MmpL7 has been suggested to interact with PpsE, the last enzyme of the modular phthiocerol biosynthesis (Jain and Cox 2005), which may play a role in efficient transport of newly synthesized PDIMs across the plasma membrane. LppX is a β-barrel protein proposed to translocate PDIM across the outer membrane (Sulzenbacher et al. 2006).

PGLs are structurally related to PDIMs. PGLs harbor a phenolphthiocerol instead of a phthiocerol, and the phenolic residue is further modified by carbohydrates (Fig. 13.6a). The synthesis of the phenolphthiocerol moiety is similar to that of phthiocerol. However, before being loaded onto PpsA, a fatty acid must first be modified to *p*-hydroxyphenylalkanoate. The production of *p*-hydroxyphenylalkanoate is achieved by two enzymes. First, the fatty acyl-AMP ligase FadD22, which is specifically involved in PGL biosynthesis, activates *p*-hydroxybenzoic acid to form *p*-hydroxybenzoyl-AMP and transfers the *p*-hydroxybenzoyl moiety onto Pks15/1 (Simeone et al. 2010; Ferreras et al. 2008). Second, Pks15/1 extends the acyl chain by using 8–9 molecules of malonyl-CoA as the carbon donor. The product, *p*-hydroxyphenylalkanoate, is released from Pks15/1 and loaded onto PpsA by another fatty acyl-AMP ligase, FadD29 (Simeone et al. 2010). Once loaded onto PpsA, the biosynthesis of PGLs is the same as that of PDIMs except that the final product of the PapA5 reaction is further glycosylated to become PGLs (Pérez et al. 2004b). PGLs are involved in many aspects of host-pathogen interaction, including macrophage recruitment in zebrafish infection model of *M. marinum*, the escape of *M. marinum* from macrophages, inhibition of pro-inflammatory cytokines and Th1 response by *M. tuberculosis*, and nerve damage in leprosy patients (Cambier et al. 2014, 2017; Madigan et al. 2017; Reed et al. 2004; Ordway et al. 2007).

Capsules and Exopolysaccharides

Bacterial capsules surround the cell envelope and play a part in a variety of processes including biofilm development, protection from the environment, and pathogenesis. Capsules are typically constructed of secreted polysaccharides that adhere to the cell surface, though other macromolecules may also be present within this matrix. Released polysaccharides that do not form an adherent glycocalyx are called exopolysaccharides. Though they are not capsular under observed conditions, exopolysaccharides may share common functions and biosynthetic pathways to capsules. Actinobacterial capsules and exopolysaccharides are varied in structure and function, though some commonality between taxa is evident.

Mycobacterial capsules have attracted far more research interest than those of other Actinobacteria. The mycobacterial capsule is a loose, non-covalently attached layer with a thickness ranging from negligible to 40 nm in vitro (Sani et al. 2010). In pathogenic mycobacteria, capsules can be seen in vivo by electron microscopy as a 50–100 nm thick electron transparent zone (ETZ) surrounding the phagocytized bacilli between the envelope of the bacteria and the host material (Chapman et al. 1959). This ETZ remains stable in the phagosome during infection, but degrades quickly from phagocytized dead cells, implying mycobacteria actively maintain this structure within macrophages (Frehel et al. 1986). Electron-dense material, assumed to be host-derived, can be seen excluded to the outer boundary of the ETZ within these phagosomes, indicating that the capsule layer prevents the diffusion of host macromolecules (Daffe and Etienne 1999). Furthermore, the capsule layer of *M. bovis* BCG plays roles in binding human monocyte-derived macrophages and dampening cytokine response (Sani et al. 2010). While the function of capsules is evident in pathogenesis, a less extensive but similar structure is also observed by electron microscopy in non-pathogenic species (Daffe and Draper 1998). For example, non-pathogenic *M. smegmatis* has a thinner capsule than the high capsule producing pathogens *M. marinum, M. bovis*, and *M. tuberculosis* (Sani et al. 2010). In addition to thickness, mycobacterial capsules can differ in composition. Though all are composed of glycans and protein, the ratio of glycan/protein comprising the capsule can vary between species. Capsular material derived from the slow-growing *M. gastri* and *M. kansasii* contains up to 95% carbohydrate, while capsules of fast-growing *M. phlei* and *M. smegmatis* are highly proteinaceous. *M. avium* has a more balanced mix of protein and carbohydrate within their capsules (Lemassu et al. 1996).

Mycobacterial capsular glycans consist primarily of three types of neutral polysaccharides: α-glucan, arabinomannan, and mannan. The α-glucan is the most abundant species and consists of α1,4-D-Glc polymer with extensive α1,6-D-Glc branching

(Lemassu and Daffe 1994; Ortalo-Magne et al. 1995). These extracellular α-glucans are >100 kDa in size, which is 1,000 times smaller than structurally related cytoplasmic α-glucans (Lemassu and Daffe 1994). The α-glucans are recognized by host receptors, complement receptor 3 and DC-SIGN, and play a role in survival within the host (Geurtsen et al. 2009; Stokes et al. 2004; Sambou et al. 2008; Cywes et al. 1997). The structure of arabinomannan and mannan are identical or near-identical to that of the carbohydrate moieties of LAM and LM, which are described above (Lemassu and Daffe 1994; Ortalo-Magne et al. 1995; Maes et al. 2007). The arabinomannan has an approximate mass of 13 kDa, while the mannan has an approximate mass of 4 kDa. Due to the structural similarities, arabinomannan and mannan are presumed to be derived from LAM and LM, respectively (Maes et al. 2007). Because of its structural similarity to LAM, it is reasonable to speculate that capsular arabinomannan shares functionality with LAM, such as the ability to bind to macrophages. Although over 90% of capsule glycans are neutral, there are small amounts of phosphorylated species of mannan and arabinomannan bearing a negative charge in the mycobacterial capsule (Maes et al. 2007).

In addition to glycans and glycolipids, many proteins are embedded within the mycobacterial capsular matrix (Sani et al. 2010). Many of these capsular proteins are transported to the capsule via secretion systems such as the type VII secretion system ESX-1. Among the various substrates of ESX-1 are T cell antigens that promote the escape of engulfed *M. tuberculosis* from phagosome into the cytosol (Sani et al. 2010; van der Wel et al. 2007). Capsular proteins in some *Mycobacterium* species appear to be cytoplasmic proteins, as they lack secretion signals (Daffe and Etienne 1999). The mechanism for their transport to the capsule thus seems to be independent of the general secretory pathway, but this mechanism remains unknown. The ESX-5 secretion system also appears to support capsular maintenance through transport of key capsular proteins such as PPE10, without which the capsule has altered composition and physical/morphological properties (Ates et al. 2016). Capsule defects caused by ESX-5 deficiency result in reduced pathogenicity in a zebrafish model of tuberculosis, once again implicating the capsule in pathogenesis.

Other mycolata organisms also produce capsules and extracellular polysaccharides. *Corynebacterium*, one of the closest genera to *Mycobacterium*, possesses an outer layer similar to the mycobacterial capsule. This 35–40 nm thick carbohydrate-rich outer layer is composed mostly of neutral polymers of Ara (10–20%), Man (20–35%), and Glc (50–70%) (Puech et al. 2001). Lectins with specificity to GlcNAc, *N*-acetyl-D-galactosamine, D-Gal, and sialic acid bind to the corynebacterial surface possibly implicating these carbohydrates as additional capsule constituents (Mattos-Guaraldi et al. 1999). Glucan is the major polysaccharide, and comes in two apparent masses, 110 and 1.7 kDa, in *Corynebacterium xerosis*. Arabinomannan size distribution is also bimodal with a 13 kDa and a 1 kDa species (Puech et al. 2001). The smaller glucans and arabinomannans are notable, being absent in mycobacterial capsules. Proteins are generally minor components of corynebacterial capsules, accounting for less than 10% of the dry weight of the capsular material. However, some strains of *C. glutamicum* have an outermost paracrystalline S-layer composed almost exclusively of one protein, PS2, which appears to associate with the outer membrane through hydrophobic interactions (Chami et al. 1997; Peyret et al. 1993). The gene encoding PS2 is located on a chromosomal island, suggesting that a horizontal gene transfer event led to the development of the proteinaceous surface layer in these *C. glutamicum* strains (Hansmeier et al. 2006).

Rhodococcus equi expresses an immunogenic and antigenically varied 50–100 nm thick polysaccharide-rich capsule layer, which confers a mucoid appearance when grown on nutrient agar (Sydor et al. 2008). The antigenic capsular polysaccharide of one of the 27 identified *R. equi* serovars is made up of equal amounts of D-Glc, D-Gal, D-glucuronic acid, 4-*O*-(1-carboxyethyl)-D-Man, and pyruvic acid (Severn and Richards 1992). FbpA, a homolog of the mycobacterial mycoloyltransferase Ag85, is important for maintaining the integrity of this capsule in *Rhodococcus*, implying an importance for the mycolic acid layer in capsule stability (Sydor et al. 2008). However, capsule deficiency by the disruption of this gene does not impair pathogenicity, implying that the capsule is not a major virulence factor in *Rhodococcus* (Sydor et al. 2008).

Another mycolata genus, *Gordonia*, like mycobacteria, forms biofilms that are held together by a matrix of bacterially derived macromolecules, and therefore highly suggestive of capsule and exopolysaccharide production (Linos et al. 2000). *Gordonia polyisoprenivorans* and another *Gordonia* strain Y-102 produce exopolysaccharides (Kondo et al. 2000). The acidic polysaccharide from Y-102, termed gordonan, has a molecular weight of about 5,000 kDa and a repeating [-3-4-*O*-(1-carboxyethyl)-Man*p*-β1,4-D-GlcA*p*-β1,4-D-Glc*p*-β1-] trisaccharide structure. A nearly identical exopolysaccharide with GlcA*p* in α configuration is reported in *Gordonia rubripertincta* (formerly *Mycobacterium lacticolum*) (Kochetkov et al. 1979). Notably, the antigenic capsular polysaccharides of *R. equi* referred to above also share similar sugar composition, namely 4-*O*-(1-carboxyethyl)-D-Man, D-Glc, and D-glucuronic acid residues (Severn and Richards 1992).

Exopolysaccharides have also been described in Actinobacteria outside of the mycolata, and some of these may form capsular polysaccharide surface layers. While there is no direct evidence for capsular structures, *Streptomyces* species do secrete exopolysaccharides into their environment (Selim et al. 2018; Wang et al. 2003). One example is ebosin, composed of Gal, Ara, Man, fucose, xylose, Rha, galacturonic acid, and Glc (Wang et al. 2003). The biosynthesis of ebosin remains sketchy, but carbohydrates are proposed to be built on a lipid-linked precursor (Wang et al. 2003). *Bifidobacterium* is also a notable producer of exopolysaccharides. Exopolysaccharides produced by this genera are thought to facilitate numerous beneficial interactions for the host including immune modulation, host protection, and antagonizing pathogens (Castro-Bravo et al. 2018; Hidalgo-Cantabrana et al. 2014). The soil bacterium *Brevibacterium otitidis* also produces exopolysaccharides, which are ~127 kDa in size, containing Ara, Man, Glc, and mannouronic acid (Asker and Shawky 2010). While this is not an exhaustive review, their ubiquity and structural/compositional variation suggest that capsules have many critical functions for Actinobacteria beyond their traditionally defined role in pathogenesis. These extracellular materials likely provide a convenient yet non-essential and modifiable matrix to modulate interactions with diverse environments.

Conclusions and Outlook

While there have been leaps and bounds in understanding the structure and biosynthesis of the actinobacterial cell envelope, there are still numerous questions remaining. With the medical importance of *Mycobacterium* species, the cell envelope biosynthesis in this genus will continue to be an important focus of future research. Given the successful uses of ethambutol and isoniazid as frontline anti-TB drugs, it would not be surprising to find many more drug targets from cell envelope biosynthetic pathways. Indeed, the mycolic acid transporter MmpL3 is emerging as a promising target for chemotherapy against nontuberculous mycobacteria diseases (Viljoen et al. 2017; Li et al. 2018; Kozikowski et al. 2017). It will also continue to be exciting to discover many more lipids and glycans from Actinobacteria. One recent prominent example is a discovery in *M. abscessus* of a glycosyl diacylated nonadecyl diol alcohol, which is transported by MmpL8 ortholog in this bacterium (Dubois et al. 2018). We also have a limited understanding of spatiotemporal regulation of cell envelope biosynthesis. At the transcriptional and post-transcriptional levels, cells must have mechanisms to sense and respond to environmental changes. Polar restricted growth of many rod-shaped bacteria within the Actinobacteria class implies tight spatial subcellular regulation as well. Finally, how mycolata build this unique diderm cell envelope continues to be enigmatic. Such a complex macromolecular assembly seems like a highly demanding task, and yet mycolata are highly successful bacteria found in diverse environmental niches. An insight that both mycobacterial arabinogalactan and Gram-positive wall teichoic acids are similarly linked to the peptidoglycan makes us wonder how mycolata evolved this unique outer membrane. Continued research on the diverse repertoire of Actinobacteria will bring an in-depth understanding of how these amazing bacteria evolved their cell envelope and succeeded in their own way.

Acknowledgements The authors thank Julia Puffal, Corelle Rokicki and James Brenner for discussions. The recent work in our laboratory was supported by the American Lung Association, Pittsfield Anti-Tuberculosis Association, and NIH (AI140259).

References

Abou-Zeid C, Voiland A, Michel G, Cocito C (1982) Structure of the wall polysaccharide isolated from a group of corynebacteria. Eur J Biochem 128:363–370

Albesa-Jove D, Svetlikova Z, Tersa M et al (2016) Structural basis for selective recognition of acyl chains by the membrane-associated acyltransferase PatA. Nat Commun 7:10906. https://doi.org/10.1038/ncomms10906

Alderwick LJ, Radmacher E, Seidel M et al (2005) Deletion of Cg-*emb* in Corynebacterianeae leads to a novel truncated cell wall arabinogalactan, whereas inactivation of Cg-*ubiA* results in an arabinan-deficient mutant with a cell wall galactan core. J Biol Chem 280:32362–32371. https://doi.org/10.1074/jbc.M506339200

Alderwick LJ, Seidel M, Sahm H et al (2006) Identification of a novel arabinofuranosyltransferase (AftA) involved in cell wall arabinan biosynthesis in *Mycobacterium tuberculosis*. J Biol Chem 281:15653–15661. https://doi.org/10.1074/jbc.M600045200

Alderwick LJ, Dover LG, Veerapen N et al (2008) Expression, purification and characterisation of soluble GlfT and the identification of a novel galactofuranosyltransferase Rv3782 involved in priming GlfT-mediated galactan polymerisation in *Mycobacterium tuberculosis*. Protein Expr Purif 58:332–341. https://doi.org/10.1016/j.pep.2007.11.012

Alderwick LJ, Lloyd GS, Lloyd AJ et al (2011) Biochemical characterization of the *Mycobacterium tuberculosis* phosphoribosyl-1-pyrophosphate synthetase. Glycobiology 21:410–425. https://doi.org/10.1093/glycob/cwq173

Alderwick LJ, Birch HL, Krumbach K et al (2018) AftD functions as an alpha1→5 arabinofuranosyltransferase involved in the biosynthesis of the mycobacterial cell wall core. Cell Surf 1:2–14. https://doi.org/10.1016/j.tcsw.2017.10.001

Alibaud L, Pawelczyk J, Gannoun-Zaki L et al (2014) Increased phagocytosis of *Mycobacterium marinum* mutants defective in lipooligosaccharide production: a structure-activity relationship study. J Biol Chem 289:215–228. https://doi.org/10.1074/jbc.M113.525550

Alonso H, Parra J, Malaga W et al (2017) Protein *O*-mannosylation deficiency increases LprG-associated lipoarabinomannan release by *Mycobacterium tuberculosis* and enhances the TLR2-associated inflammatory response. Sci Rep 7:7913. https://doi.org/10.1038/s41598-017-08489-7

Alsteens D, Verbelen C, Dague E et al (2008) Organization of the mycobacterial cell wall: a nanoscale view. Pflugers Arch 456:117–125. https://doi.org/10.1007/s00424-007-0386-0

Angala SK, Belardinelli JM, Huc-Claustre E et al (2014) The cell envelope glycoconjugates of *Mycobacterium tuberculosis*. Crit Rev Biochem Mol Biol 49:361–399. https://doi.org/10.3109/10409238.2014.925420

Arbues A, Lugo-Villarino G, Neyrolles O et al (2014) Playing hide-and-seek with host macrophages through the use of mycobacterial cell envelope phthiocerol dimycocerosates and phenolic glycolipids. Front Cell Infect Microbiol 4:173. https://doi.org/10.3389/fcimb.2014.00173

Ariza MA, Martin-Luengo F, Valero-Guillen PL (1994) A family of diacyltrehaloses isolated from *Mycobacterium fortuitum*. Microbiology 140:1989–1994. https://doi.org/10.1099/13500872-140-8-1989

Asker MMS, Shawky BT (2010) Structural characterization and antioxidant activity of an extracellular polysaccharide isolated from *Brevibacterium otitidis* BTS 44. Food Chem 123:315–320. https://doi.org/10.1016/j.foodchem.2010.04.037

Ates LS, van der Woude AD, Bestebroer J et al (2016) The ESX-5 system of pathogenic mycobacteria is involved in capsule integrity and virulence through its substrate PPE10. PLoS Pathog 12:e1005696-26. https://doi.org/10.1371/journal.ppat.1005696

Azad AK, Sirakova TD, Rogers LM, Kolattukudy PE (1996) Targeted replacement of the mycocerosic acid synthase gene in *Mycobacterium bovis* BCG produces a mutant that lacks mycosides. Proc Natl Acad Sci USA 93:4787–4792

Azad AK, Sirakova TD, Fernandes ND, Kolattukudy PE (1997) Gene knockout reveals a novel gene cluster for the synthesis of a class of cell wall lipids unique to pathogenic mycobacteria. J Biol Chem 272:16741–16745

Backus KM, Dolan MA, Barry CS et al (2014) The three *Mycobacterium tuberculosis* antigen 85 isoforms have unique substrates and activities determined by non-active site regions. J Biol Chem 289:25041–25053. https://doi.org/10.1074/jbc.M114.581579

Ballou CE, Vilkas E, Lederer E (1963) Structural studies on the myo-inositol phospholipids of *Mycobacterium tuberculosis* (var. *bovis*, strain BCG). J Biol Chem 238:69–76

Bansal A, Kar D, Murugan RA et al (2015) A putative low-molecular-mass penicillin-binding protein (PBP) of *Mycobacterium smegmatis* exhibits prominent physiological characteristics of DD-carboxypeptidase and beta-lactamase. Microbiology 161:1081–1091. https://doi.org/10.1099/mic.0.000074

Bansal-Mutalik R, Nikaido H (2011) Quantitative lipid composition of cell envelopes of *Corynebacterium glutamicum* elucidated through reverse micelle extraction. Proc Natl Acad Sci USA 108:15360–15365. https://doi.org/10.1073/pnas.1112572108

Bansal-Mutalik R, Nikaido H (2014) Mycobacterial outer membrane is a lipid bilayer and the inner membrane is unusually rich in diacyl phosphatidylinositol dimannosides. Proc Natl Acad Sci USA 111:4958–4963. https://doi.org/10.1073/pnas.1403078111

Baranowski C, Welsh MA, Sham L-T, et al (2018) Maturing *Mycobacterium smegmatis* peptidoglycan requires non-canonical crosslinks to maintain shape. eLife. https://doi.org/10.7554/elife.37516

Barry CE, Lee RE, Mdluli K et al (1998) Mycolic acids: structure, biosynthesis and physiological functions. Prog Lipid Res 37:143–179

Batrakov SG, Bergelson LD (1978) Lipids of the streptomycetes structural investigation and biological interrelation: a review. Chem Phys Lipids 21:1–29

Baumgart M, Luder K, Grover S et al (2013) IpsA, a novel LacI-type regulator, is required for inositol-derived lipid formation in corynebacteria and mycobacteria. BMC Biol 11:122. https://doi.org/10.1186/1741-7007-11-122

Baumgart M, Schubert K, Bramkamp M, Frunzke J (2016) Impact of LytR-CpsA-Psr proteins on cell wall biosynthesis in *Corynebacterium glutamicum*. J Bacteriol 198:3045–3059. https://doi.org/10.1128/JB.00406-16

Belanger AE, Besra GS, Ford ME et al (1996) The *embAB* genes of *Mycobacterium avium* encode an arabinosyl transferase involved in cell wall arabinan biosynthesis that is the target for the antimycobacterial drug ethambutol. Proc Natl Acad Sci USA 93:11919–11924

Belanova M, Dianiskova P, Brennan PJ et al (2008) Galactosyl transferases in mycobacterial cell wall synthesis. J Bacteriol 190:1141–1145. https://doi.org/10.1128/JB.01326-07

Belardinelli JM, Larrouy-Maumus G, Jones V et al (2014) Biosynthesis and translocation of unsulfated acyltrehaloses in *Mycobacterium tuberculosis*. J Biol Chem 289:27952–27965. https://doi.org/10.1074/jbc.M114.581199

Belisle JT, Vissa VD, Sievert T et al (1997) Role of the major antigen of *Mycobacterium tuberculosis* in cell wall biogenesis. Science 276:1420–1422

Bentley SD, Chater KF, Cerdeno-Tarraga A-M et al (2002) Complete genome sequence of the model actinomycete *Streptomyces coelicolor* A3(2). Nature 417:141–147. https://doi.org/10.1038/417141a

Bernut A, Viljoen A, Dupont C et al (2016) Insights into the smooth-to-rough transitioning in *Mycobacterium bolletii* unravels a functional Tyr residue conserved in all mycobacterial MmpL family members. Mol Microbiol 99:866–883. https://doi.org/10.1111/mmi.13283

Bertram R, Schlicht M, Mahr K et al (2004) *In silico* and transcriptional analysis of carbohydrate uptake systems of *Streptomyces coelicolor* A3(2). J Bacteriol 186:1362–1373. https://doi.org/10.1128/JB.186.5.1362-1373.2004

Bhamidi S, Scherman MS, Rithner CD et al (2008) The identification and location of succinyl residues and the characterization of the interior arabinan region allow for a model of the complete primary structure of *Mycobacterium tuberculosis* mycolyl arabinogalactan. J Biol Chem 283:12992–13000. https://doi.org/10.1074/jbc.M800222200

Bhamidi S, Scherman MS, Jones V et al (2011) Detailed structural and quantitative analysis reveals the spatial organization of the cell walls of *in vivo* grown *Mycobacterium leprae* and *in vitro* grown *Mycobacterium tuberculosis*. J Biol Chem 286:23168–23177. https://doi.org/10.1074/jbc.M110.210534

Bhatt K, Gurcha SS, Bhatt A et al (2007) Two polyketide-synthase-associated acyltransferases are required for sulfolipid biosynthesis in *Mycobacterium tuberculosis*. Microbiology 153:513–520. https://doi.org/10.1099/mic.0.2006/003103-0

Bhatt A, Brown AK, Singh A et al (2008) Loss of a mycobacterial gene encoding a reductase leads to an altered cell wall containing beta-oxo-mycolic acid analogs and accumulation of ketones. Chem Biol 15:930–939. https://doi.org/10.1016/j.chembiol.2008.07.007

Billman-Jacobe H, McConville MJ, Haites RE et al (1999) Identification of a peptide synthetase involved in the biosynthesis of glycopeptidolipids of *Mycobacterium smegmatis*. Mol Microbiol 33:1244–1253

Birch HL, Alderwick LJ, Bhatt A et al (2008) Biosynthesis of mycobacterial arabinogalactan: identification of a novel alpha(1→3) arabinofuranosyltransferase. Mol Microbiol 69:1191–1206. https://doi.org/10.1111/j.1365-2958.2008.06354.x

Birch HL, Alderwick LJ, Appelmelk BJ et al (2010) A truncated lipoglycan from mycobacteria with altered immunological properties. Proc Natl Acad Sci USA 107:2634–2639. https://doi.org/10.1073/pnas.0915082107

Biswas R, Dutta A, Dutta D et al (2015) Crystal structure of dehydratase component HadAB complex of mycobacterial FAS-II pathway. Biochem Biophys Res Commun 458:369–374. https://doi.org/10.1016/j.bbrc.2015.01.119

Bloch K (1977) Control mechanisms for fatty acid synthesis in *Mycobacterium smegmatis*. Adv Enzymol Relat Areas Mol Biol 45:1–84

Bou Raad R, Meniche X, de Sousa-d'Auria C et al (2010) A deficiency in arabinogalactan biosynthesis affects *Corynebacterium glutamicum* mycolate outer membrane stability. J Bacteriol 192:2691–2700. https://doi.org/10.1128/JB.00009-10

Boutte CC, Baer CE, Papavinasasundaram K, et al (2016) A cytoplasmic peptidoglycan amidase homologue controls mycobacterial cell wall synthesis. eLife 5:a021113. https://doi.org/10.7554/elife.14590

Brammer Basta LA, Ghosh A, Pan Y et al (2015) Loss of a functionally and structurally distinct LD-transpeptidase, $LdtM_{t5}$, compromises cell wall integrity in *Mycobacterium tuberculosis*. J Biol Chem 290:25670–25685. https://doi.org/10.1074/jbc.M115.660753

Brennan PJ, Lehane DP (1971) The phospholipids of corynebacteria. Lipids 6:401–409

Brindley DN, Matsumura S, Bloch K (1969) *Mycobacterium phlei* fatty acid synthetase—a bacterial multienzyme complex. Nature 224:666–669. https://doi.org/10.1038/224666a0

Burguiere A, Hitchen PG, Dover LG et al (2005) LosA, a key glycosyltransferase involved in the biosynthesis of a novel family of glycosylated acyltrehalose lipooligosaccharides from *Mycobacterium marinum*. J Biol Chem 280:42124–42133. https://doi.org/10.1074/jbc.M507500200

Camacho LR, Ensergueix D, Perez E et al (1999) Identification of a virulence gene cluster of *Mycobacterium tuberculosis* by signature-tagged transposon mutagenesis. Mol Microbiol 34:257–267

Camacho LR, Constant P, Raynaud C et al (2001) Analysis of the phthiocerol dimycocerosate locus of *Mycobacterium tuberculosis*. Evidence that this lipid is involved in the cell wall permeability barrier. J Biol Chem 276:19845–19854. https://doi.org/10.1074/jbc.M100662200

Cambier CJ, Takaki KK, Larson RP et al (2014) Mycobacteria manipulate macrophage recruitment through coordinated use of membrane lipids. Nature 505:218–222. https://doi.org/10.1038/nature12799

Cambier CJ, O'Leary SM, O'Sullivan MP et al (2017) Phenolic glycolipid facilitates mycobacterial escape from microbicidal tissue-resident macrophages. Immunity 47:552–565.e4. https://doi.org/10.1016/j.immuni.2017.08.003

Cashmore TJ, Klatt S, Yamaryo-Botte Y et al (2017) Identification of a membrane protein required for lipomannan maturation and lipoarabinomannan synthesis in Corynebacterineae. J Biol Chem 292:4976–4986. https://doi.org/10.1074/jbc.M116.772202

Castro-Bravo N, Wells JM, Margolles A, Ruas-Madiedo P (2018) Interactions of surface exopolysaccharides from *Bifidobacterium* and *Lactobacillus* within the intestinal environment. Front Microbiol 9:2426. https://doi.org/10.3389/fmicb.2018.02426

Celler K, Koning RI, Willemse J et al (2016) Cross-membranes orchestrate compartmentalization and morphogenesis in *Streptomyces*. Nat Commun 7:1–8. https://doi.org/10.1038/ncomms11836

Chami M, Bayan N, Peyret JL et al (1997) The S-layer protein of *Corynebacterium glutamicum* is anchored to the cell wall by its C-terminal hydrophobic domain. Mol Microbiol 23:483–492

Chao MC, Kieser KJ, Minami S et al (2013) Protein complexes and proteolytic activation of the cell wall hydrolase RipA regulate septal resolution in mycobacteria. PLoS Pathog 9:e1003197. https://doi.org/10.1371/journal.ppat.1003197.s011

Chapman GB, Hanks JH, Wallace JH (1959) An electron microscope study of the disposition and fine structure of *Mycobacterium lepraemurium* in mouse spleen. J Bacteriol 77:205–211

Chatterjee D, Khoo KH (2001) The surface glycopeptidolipids of mycobacteria: structures and biological properties. Cell Mol Life Sci 58:2018–2042. https://doi.org/10.1007/PL00000834

Chaudhary DK, Kim J (2018) *Rhodococcus olei* sp. nov., with the ability to degrade petroleum oil, isolated from oil-contaminated soil. Int J Syst Evol Microbiol 68:1749–1756. https://doi.org/10.1099/ijsem.0.002750

Chauhan A, Lofton H, Maloney E et al (2006) Interference of *Mycobacterium tuberculosis* cell division by Rv2719c, a cell wall hydrolase. Mol Microbiol 62:132–147. https://doi.org/10.1111/j.1365-2958.2006.05333.x

Chavadi SS, Edupuganti UR, Vergnolle O et al (2011) Inactivation of *tesA* reduces cell wall lipid production and increases drug susceptibility in mycobacteria. J Biol Chem 286:24616–24625. https://doi.org/10.1074/jbc.M111.247601

Chen Y-Y, Yang F-L, Wu S-H et al (2015) *Mycobacterium marinum mmar_2318* and *mmar_2319* are responsible for lipooligosaccharide biosynthesis and virulence toward *Dictyostelium*. Front Microbiol 6:1458. https://doi.org/10.3389/fmicb.2015.01458

Choudhuri BS, Bhakta S, Barik R et al (2002) Overexpression and functional characterization of an ABC (ATP-binding cassette) transporter encoded by the genes *drrA* and *drrB* of *Mycobacterium tuberculosis*. Biochem J 367:279–285. https://doi.org/10.1042/BJ20020615

Christova N, Lang S, Wray V et al (2015) Production, structural elucidation, and *in vitro* antitumor activity of trehalose lipid biosurfactant from *Nocardia farcinica* strain. J Microbiol Biotechnol 25:439–447. https://doi.org/10.4014/jmb.1406.06025

Collins MD, Goodfellow M, Minnikin DE (1982) Fatty acid composition of some mycolic acid-containing coryneform bacteria. J Gen Microbiol 128:2503–2509. https://doi.org/10.1099/00221287-128-11-2503

Converse SE, Mougous JD, Leavell MD et al (2003) MmpL8 is required for sulfolipid-1 biosynthesis and *Mycobacterium tuberculosis* virulence. Proc Natl Acad Sci USA 100:6121–6126. https://doi.org/10.1073/pnas.1030024100

Cot M, Ray A, Gilleron M et al (2011) Lipoteichoic acid in *Streptomyces hygroscopicus*: structural model and immunomodulatory activities. PLoS ONE 6:e26316. https://doi.org/10.1371/journal.pone.0026316

Cox JS, Chen B, McNeil M, Jacobs WR (1999) Complex lipid determines tissue-specific replication of *Mycobacterium tuberculosis* in mice. Nature 402:79–83. https://doi.org/10.1038/47042

Crellin PK, Kovacevic S, Martin KL et al (2008) Mutations in *pimE* restore lipoarabinomannan synthesis and growth in a *Mycobacterium smegmatis lpqW* mutant. J Bacteriol 190:3690–3699. https://doi.org/10.1128/JB.00200-08

Crellin PK, Luo C-Y, Morita YS (2013) Metabolism of plasma membrane lipids in mycobacteria and corynebacteria. In: Baez RV (ed) Lipid Metabolism. InTech, London, UK, pp 119–148

Cywes C, Hoppe HC, Daffe M, Ehlers MR (1997) Nonopsonic binding of *Mycobacterium tuberculosis* to complement receptor type 3 is mediated by capsular polysaccharides and is strain dependent. Infect Immun 65:4258–4266

Daffe M, Draper P (1998) The envelope layers of mycobacteria with reference to their pathogenicity. Adv Microb Physiol 39:131–203

Daffe M, Etienne G (1999) The capsule of *Mycobacterium tuberculosis* and its implications for pathogenicity. Tuber Lung Dis 79:153–169

Daffe M, Laneelle MA (1988) Distribution of phthiocerol diester, phenolic mycosides and related compounds in mycobacteria. J Gen Microbiol 134:2049–2055. https://doi.org/10.1099/00221287-134-7-2049

Daffe M, Brennan PJ, McNeil M (1990) Predominant structural features of the cell wall arabinogalactan of *Mycobacterium tuberculosis* as revealed through characterization of oligoglycosyl alditol fragments by gas chromatography/mass spectrometry and by ^{1}H and ^{13}C NMR analyses. J Biol Chem 265:6734–6743

Daffe M, McNeil M, Brennan PJ (1991) Novel type-specific lipooligosaccharides from *Mycobacterium tuberculosis*. Biochemistry 30:378–388

Daffe M, McNeil M, Brennan PJ (1993) Major structural features of the cell wall arabinogalactans of *Mycobacterium*, *Rhodococcus*, and *Nocardia* spp. Carbohydr Res 249:383–398

Daffe M, Crick DC, Jackson M (2014) Genetics of capsular polysaccharides and cell envelope (glyco)lipids. Microbiol Spectr 2:MGM2-0021-2013. https://doi.org/10.1128/microbiolspec.mgm2-0021-2013

Daniel RA, Errington J (2003) Control of cell morphogenesis in bacteria: two distinct ways to make a rod-shaped cell. Cell 113:767–776. https://doi.org/10.1016/S0092-8674(03)00421-5

Dhiman RK, Dinadayala P, Ryan GJ et al (2011) Lipoarabinomannan localization and abundance during growth of *Mycobacterium smegmatis*. J Bacteriol 193:5802–5809. https://doi.org/10.1128/JB.05299-11

Dianiskova P, Kordulakova J, Skovierova H et al (2011) Investigation of ABC transporter from mycobacterial arabinogalactan biosynthetic cluster. Gen Physiol Biophys 30:239–250. https://doi.org/10.4149/gpb_2011_03_239

Domenech P, Reed MB, Dowd CS et al (2004) The role of MmpL8 in sulfatide biogenesis and virulence of *Mycobacterium tuberculosis*. J Biol Chem 279:21257–21265. https://doi.org/10.1074/jbc.M400324200

Donovan C, Bramkamp M (2014) Cell division in Corynebacterineae. Front Microbiol 5:132. https://doi.org/10.3389/fmicb.2014.00132

Donovan C, Sieger B, Krämer R, Bramkamp M (2012) A synthetic *Escherichia coli* system identifies a conserved origin tethering factor in Actinobacteria. Mol Microbiol 84:105–116. https://doi.org/10.1111/j.1365-2958.2012.08011.x

Drage MG, Tsai H-C, Pecora ND et al (2010) *Mycobacterium tuberculosis* lipoprotein LprG (Rv1411c) binds triacylated glycolipid agonists of Toll-like receptor 2. Nat Struct Mol Biol 17:1088–1095. https://doi.org/10.1038/nsmb.1869

Draper P, Kandler O, Darbre A (1987) Peptidoglycan and arabinogalactan of *Mycobacterium leprae*. J Gen Microbiol 133:1187–1194. https://doi.org/10.1099/00221287-133-5-1187

Draper P, Khoo KH, Chatterjee D et al (1997) Galactosamine in walls of slow-growing mycobacteria. Biochem J 327(Pt 2):519–525

Duan X, Xiang X, Xie J (2014) Crucial components of mycobacterium type II fatty acid biosynthesis (Fas-II) and their inhibitors. FEMS Microbiol Lett 360:87–99. https://doi.org/10.1111/1574-6968.12597

Dubey VS, Sirakova TD, Kolattukudy PE (2002) Disruption of *msl3* abolishes the synthesis of mycolipanoic and mycolipenic acids required for polyacyltrehalose synthesis in *Mycobacterium tuberculosis* H37Rv and causes cell aggregation. Mol Microbiol 45:1451–1459

Dubey VS, Sirakova TD, Cynamon MH, Kolattukudy PE (2003) Biochemical function of *msl5* (*pks8* plus *pks17*) in *Mycobacterium tuberculosis* H37Rv: biosynthesis of monomethyl branched unsaturated fatty acids. J Bacteriol 185:4620–4625. https://doi.org/10.1128/JB.185.15.4620-4625.2003

Dubois V, Viljoen A, Laencina L et al (2018) $MmpL8_{MAB}$ controls *Mycobacterium abscessus* virulence and production of a previously unknown glycolipid family. Proc Natl Acad Sci USA 115:E10147–E10156. https://doi.org/10.1073/pnas.1812984115

Eagen WJ, Baumoel LR, Osman SH, et al (2018) Deletion of PimE mannosyltransferase results in increased copper sensitivity in *Mycobacterium smegmatis*. FEMS Microbiol Lett 365:fny025. https://doi.org/10.1093/femsle/fny025

Ealand C, Rimal B, Chang J et al (2018) Resuscitation-promoting factors are required for *Mycobacterium smegmatis* biofilm formation. Appl Environ Microbiol 84:643. https://doi.org/10.1128/AEM.00687-18

Eckstein TM, Silbaq FS, Chatterjee D et al (1998) Identification and recombinant expression of a *Mycobacterium avium* rhamnosyltransferase gene (*rtfA*) involved in glycopeptidolipid biosynthesis. J Bacteriol 180:5567–5573

Escuyer VE, Lety MA, Torrelles JB et al (2001) The role of the *embA* and *embB* gene products in the biosynthesis of the terminal hexaarabinofuranosyl motif of *Mycobacterium smegmatis* arabinogalactan. J Biol Chem 276:48854–48862. https://doi.org/10.1074/jbc.M102272200

Espuny MJ, Egjido S, Mercade ME, Manresa A (1995) Characterization of trehalose tetraester produced by a waste lube oil degrader *Rhodococcus* sp. 51T7. Toxicol Environ Chem 48:83–88. https://doi.org/10.1080/02772249509358154

Etienne G, Malaga W, Laval F et al (2009) Identification of the polyketide synthase involved in the biosynthesis of the surface-exposed lipooligosaccharides in mycobacteria. J Bacteriol 191:2613–2621. https://doi.org/10.1128/JB.01235-08

Ferreras JA, Stirrett KL, Lu X et al (2008) Mycobacterial phenolic glycolipid virulence factor biosynthesis: mechanism and small-molecule inhibition of polyketide chain initiation. Chem Biol 15:51–61. https://doi.org/10.1016/j.chembiol.2007.11.010

Fitzmaurice AM, Kolattukudy PE (1997) Open reading frame 3, which is adjacent to the mycocerosic acid synthase gene, is expressed as an acyl coenzyme A synthase in *Mycobacterium bovis* BCG. J Bacteriol 179:2608–2615

Fitzmaurice AM, Kolattukudy PE (1998) An acyl-CoA synthase (*acoas*) gene adjacent to the mycocerosic acid synthase (*mas*) locus is necessary for mycocerosyl lipid synthesis in *Mycobacterium tuberculosis* var. *bovis* BCG. J Biol Chem 273:8033–8039

Fiuza M, Canova MJ, Patin D et al (2008) The MurC ligase essential for peptidoglycan biosynthesis is regulated by the serine/threonine protein kinase PknA in *Corynebacterium glutamicum*. J Biol Chem 283:36553–36563. https://doi.org/10.1074/jbc.M807175200

Flärdh K (2003) Essential role of DivIVA in polar growth and morphogenesis in *Streptomyces coelicolor* A3(2). Mol Microbiol 49:1523–1536

Flärdh K, Richards DM, Hempel AM et al (2012) Regulation of apical growth and hyphal branching in *Streptomyces*. Curr Opin Microbiol 15:737–743. https://doi.org/10.1016/j.mib.2012.10.012

Frehel C, Ryter A, Rastogi N, David H (1986) The electron-transparent zone in phagocytized *Mycobacterium avium* and other mycobacteria: formation, persistence and role in bacterial survival. Ann Inst Pasteur Microbiol 137B:239–257

Fujiwara N, Nakata N, Maeda S et al (2007) Structural characterization of a specific glycopeptidolipid containing a novel *N*-acyl-deoxy sugar from *Mycobacterium intracellulare* serotype 7 and genetic analysis of its glycosylation pathway. J Bacteriol 189:1099–1108. https://doi.org/10.1128/JB.01471-06

Fukuda T, Matsumura T, Ato M, et al (2013) Critical roles for lipomannan and lipoarabinomannan in cell wall integrity of mycobacteria and pathogenesis of tuberculosis. mBio 4:e00472–12. https://doi.org/10.1128/mbio.00472-12

Gao B, Gupta RS (2012) Phylogenetic framework and molecular signatures for the main clades of the phylum Actinobacteria. Microbiol Mol Biol Rev 76:66–112. https://doi.org/10.1128/MMBR.05011-11

Garcia-Heredia A, Pohane AA, Melzer ES, et al (2018) Peptidoglycan precursor synthesis along the sidewall of pole-growing mycobacteria. eLife 7:100. https://doi.org/10.7554/elife.37243

Gavalda S, Leger M, Van-der-Rest B et al (2009) The Pks13/FadD32 crosstalk for the biosynthesis of mycolic acids in *Mycobacterium tuberculosis*. J Biol Chem 284:19255–19264. https://doi.org/10.1074/jbc.M109.006940

Gavalda S, Bardou F, Laval F et al (2014) The polyketide synthase Pks13 catalyzes a novel mechanism of lipid transfer in mycobacteria. Chem Biol 21:1660–1669. https://doi.org/10.1016/j.chembiol.2014.10.011

Geurtsen J, Chedammi S, Mesters J et al (2009) Identification of mycobacterial alpha-glucan as a novel ligand for DC-SIGN: involvement of mycobacterial capsular polysaccharides in host immune modulation. J Immunol 183:5221–5231. https://doi.org/10.4049/jimmunol.0900768

Gibson KJC, Eggeling L, Maughan WN et al (2003) Disruption of Cg-Ppm1, a polyprenyl monophosphomannose synthase, and the generation of lipoglycan-less mutants in *Corynebacterium glutamicum*. J Biol Chem 278:40842–40850. https://doi.org/10.1074/jbc.M307988200

Gibson KJC, Gilleron M, Constant P et al (2005) A lipomannan variant with strong TLR-2-dependent pro-inflammatory activity in *Saccharothrix aerocolonigenes*. J Biol Chem 280:28347–28356. https://doi.org/10.1074/jbc.M505498200

Gilby AR, Few AV, McQuillen K (1958) The chemical composition of the protoplast membrane of *Micrococcus lysodeikticus*. Biochim Biophys Acta 29:21–29

Gilleron M, Vercauteren J, Puzo G (1993) Lipooligosaccharidic antigen containing a novel C4-branched 3,6-dideoxy-alpha-hexopyranose typifies *Mycobacterium gastri*. J Biol Chem 268:3168–3179

Gilleron M, Garton NJ, Nigou J et al (2005) Characterization of a truncated lipoarabinomannan from the Actinomycete *Turicella otitidis*. J Bacteriol 187:854–861. https://doi.org/10.1128/JB.187.3.854-861.2005

Gokhale RS, Saxena P, Chopra T, Mohanty D (2007) Versatile polyketide enzymatic machinery for the biosynthesis of complex mycobacterial lipids. Nat Prod Rep 24:267–277. https://doi.org/10.1039/b616817p

Goodfellow M, Weaver CR, Minnikin DE (1982) Numerical classification of some rhodococci, corynebacteria and related organisms. J Gen Microbiol 128:731–745. https://doi.org/10.1099/00221287-128-4-731

Goren MB (1970) Sulfolipid I of *Mycobacterium tuberculosis*, strain H37Rv. II. Structural studies. Biochim Biophys Acta 210:127–138

Goren MB, Brokl O, Schaefer WB (1974) Lipids of putative relevance to virulence in *Mycobacterium tuberculosis*: phthiocerol dimycocerosate and the attenuation indicator lipid. Infect Immun 9:150–158

Goude R, Amin AG, Chatterjee D, Parish T (2008) The critical role of *embC* in *Mycobacterium tuberculosis*. J Bacteriol 190:4335–4341. https://doi.org/10.1128/JB.01825-07

Grover S, Alderwick LJ, Mishra AK et al (2014) Benzothiazinones mediate killing of Corynebacterineae by blocking decaprenyl phosphate recycling involved in cell wall biosynthesis. J Biol Chem 289:6177–6187. https://doi.org/10.1074/jbc.M113.522623

Grzegorzewicz AE, Pham H, Gundi VAKB, et al (2012) Inhibition of mycolic acid transport across the *Mycobacterium tuberculosis* plasma membrane. Nat Chem Biol 1–8. https://doi.org/10.1038/nchembio.794

Grzegorzewicz AE, de Sousa-d'Auria C, McNeil MR et al (2016) Assembling of the *Mycobacterium tuberculosis* cell wall core. J Biol Chem 291:18867–18879. https://doi.org/10.1074/jbc.M116.739227

Guerin ME, Kaur D, Somashekar BS et al (2009) New insights into the early steps of phosphatidylinositol mannoside biosynthesis in mycobacteria: PimB' is an essential enzyme of *Mycobacterium smegmatis*. J Biol Chem 284:25687–25696. https://doi.org/10.1074/jbc.M109.030593

Gupta R, Lavollay M, Mainardi J-L et al (2010) The *Mycobacterium tuberculosis* protein LdtMt2 is a nonclassical transpeptidase required for virulence and resistance to amoxicillin. Nat Med 16:466–469. https://doi.org/10.1038/nm.2120

Gurcha SS, Baulard AR, Kremer L et al (2002) Ppm1, a novel polyprenol monophosphomannose synthase from *Mycobacterium tuberculosis*. Biochem J 365:441–450

Gutierrez AV, Viljoen A, Ghigo E et al (2018) Glycopeptidolipids, a double-edged sword of the *Mycobacterium abscessus* complex. Front Microbiol 9:1145. https://doi.org/10.3389/fmicb.2018.01145

Haites RE, Morita YS, McConville MJ, Billman-Jacobe H (2005) Function of phosphatidylinositol in mycobacteria. J Biol Chem 280:10981–10987. https://doi.org/10.1074/jbc.M413443200

Hamada M, Iino T, Tamura T et al (2009) *Serinibacter salmoneus* gen. nov., sp. nov., an actinobacterium isolated from the intestinal tract of a fish, and emended descriptions of the families *Beutenbergiaceae* and *Bogoriellaceae*. Int J Syst Evol Microbiol 59:2809–2814. https://doi.org/10.1099/ijs.0.011106-0

Hansmeier N, Albersmeier A, Tauch A et al (2006) The surface (S)-layer gene cspB of *Corynebacterium glutamicum* is transcriptionally activated by a LuxR-type regulator and located on a 6 kb genomic island absent from the type strain ATCC 13032. Microbiology 152:923–935. https://doi.org/10.1099/mic.0.28673-0

Harrison J, Lloyd G, Joe M, et al (2016) Lcp1 is a phosphotransferase responsible for ligating arabinogalactan to peptidoglycan in *Mycobacterium tuberculosis*. mBio. https://doi.org/10.1128/mbio.00972-16

Hatzios SK, Schelle MW, Holsclaw CM et al (2009) PapA3 is an acyltransferase required for polyacyltrehalose biosynthesis in *Mycobacterium tuberculosis*. J Biol Chem 284:12745–12751. https://doi.org/10.1074/jbc.M809088200

Hayashi JM, Luo C-Y, Mayfield JA et al (2016) Spatially distinct and metabolically active membrane domain in mycobacteria. Proc Natl Acad Sci USA 113:5400–5405. https://doi.org/10.1073/pnas.1525165113

Hayashi JM, Richardson K, Melzer ES, et al (2018) Stress-induced reorganization of the mycobacterial membrane domain. mBio 9:e01823–17. https://doi.org/10.1128/mbio.01823-17

Hempel AM, Wang S-B, Letek M et al (2008) Assemblies of DivIVA mark sites for hyphal branching and can establish new zones of cell wall growth in *Streptomyces coelicolor*. J Bacteriol 190:7579–7583. https://doi.org/10.1128/JB.00839-08

Hett EC, Chao MC, Steyn AJ et al (2007) A partner for the resuscitation-promoting factors of *Mycobacterium tuberculosis*. Mol Microbiol 66:658–668. https://doi.org/10.1111/j.1365-2958.2007.05945.x

Hett EC, Chao MC, Deng LL, Rubin EJ (2008) A mycobacterial enzyme essential for cell division synergizes with resuscitation-promoting factor. PLoS Pathog 4:e1000001. https://doi.org/10.1371/journal.ppat.1000001

Hett EC, Chao MC, Rubin EJ (2010) Interaction and modulation of two antagonistic cell wall enzymes of mycobacteria. PLoS Pathog 6:e1001020. https://doi.org/10.1371/journal.ppat.1001020.s002

Hidalgo-Cantabrana C, Sanchez B, Milani C et al (2014) Genomic overview and biological functions of exopolysaccharide biosynthesis in *Bifidobacterium* spp. Appl Environ Microbiol 80:9–18. https://doi.org/10.1128/AEM.02977-13

Hodgson DA (2000) Primary metabolism and its control in streptomycetes: a most unusual group of bacteria. Adv Microb Physiol 42:47–238

Hoffmann C, Leis A, Niederweis M et al (2008) Disclosure of the mycobacterial outer membrane: cryo-electron tomography and vitreous sections reveal the lipid bilayer structure. Proc Natl Acad Sci USA 105:3963–3967. https://doi.org/10.1073/pnas.0709530105

Hoischen C, Gura K, Luge C, Gumpert J (1997) Lipid and fatty acid composition of cytoplasmic membranes from *Streptomyces hygroscopicus* and its stable protoplast-type L form. J Bacteriol 179:3430–3436

Hong S, Cheng T-Y, Layre E et al (2012) Ultralong C100 mycolic acids support the assignment of *Segniliparus* as a new bacterial genus. PLoS ONE 7:e39017. https://doi.org/10.1371/journal.pone.0039017

Howlett R, Anttonen K, Read N, Smith MCM (2018) Disruption of the GDP-mannose synthesis pathway in *Streptomyces coelicolor* results in antibiotic hyper-susceptible phenotypes. Microbiology 164:614–624. https://doi.org/10.1099/mic.0.000636

Hunter SW, Murphy RC, Clay K et al (1983) Trehalose-containing lipooligosaccharides. A new class of species-specific antigens from *Mycobacterium*. J Biol Chem 258:10481–10487

Hunter SW, Fujiwara T, Murphy RC, Brennan PJ (1984) *N*-acylkansosamine. A novel *N*-acylamino sugar from the trehalose-containing lipooligosaccharide antigens of *Mycobacterium kansasii*. J Biol Chem 259:9729–9734

Hunter SW, Gaylord H, Brennan PJ (1986) Structure and antigenicity of the phosphorylated lipopolysaccharide antigens from the leprosy and tubercle bacilli. J Biol Chem 261:12345–12351

Ishikawa E, Mori D, Yamasaki S (2017) Recognition of mycobacterial lipids by immune receptors. Trends Immunol 38:66–76. https://doi.org/10.1016/j.it.2016.10.009

Jackson M, Crick DC, Brennan PJ (2000) Phosphatidylinositol is an essential phospholipid of mycobacteria. J Biol Chem 275:30092–30099. https://doi.org/10.1074/jbc.M004658200

Jain M, Cox JS (2005) Interaction between polyketide synthase and transporter suggests coupled synthesis and export of virulence lipid in *M. tuberculosis*. PLoS Pathog 1:e2. https://doi.org/10.1371/journal.ppat.0010002

Jamet S, Slama N, Domingues J et al (2015) The non-essential mycolic acid biosynthesis genes *hadA* and *hadC* contribute to the physiology and fitness of *Mycobacterium smegmatis*. PLoS ONE 10:e0145883. https://doi.org/10.1371/journal.pone.0145883

Jankute M, Cox JAG, Harrison J, Besra GS (2015) Assembly of the mycobacterial cell wall. Annu Rev Microbiol 69:405–423. https://doi.org/10.1146/annurev-micro-091014-104121

Jankute M, Alderwick LJ, Noack S et al (2017) Disruption of mycobacterial AftB results in complete loss of terminal β(1→2) arabinofuranose residues of lipoarabinomannan. ACS Chem Biol 12:183–190. https://doi.org/10.1021/acschembio.6b00898

Jankute M, Alderwick LJ, Moorey AR et al (2018) The singular *Corynebacterium glutamicum* Emb arabinofuranosyltransferase polymerises the alpha(1→5) arabinan backbone in the early stages of cell wall arabinan biosynthesis. Cell Surf 2:38–53. https://doi.org/10.1016/j.tcsw.2018.06.003

Jeevarajah D, Patterson JH, McConville MJ, Billman-Jacobe H (2002) Modification of glycopeptidolipids by an *O*-methyltransferase of *Mycobacterium smegmatis*. Microbiology 148:3079–3087. https://doi.org/10.1099/00221287-148-10-3079

Jeevarajah D, Patterson JH, Taig E et al (2004) Methylation of GPLs in *Mycobacterium smegmatis* and *Mycobacterium avium*. J Bacteriol 186:6792–6799. https://doi.org/10.1128/JB.186.20.6792-6799.2004

Jiang T, He L, Zhan Y, et al (2011) The effect of MSMEG_6402 gene disruption on the cell wall structure of *Mycobacterium smegmatis*. Microb Pathog 1–5. https://doi.org/10.1016/j.micpath.2011.04.005

Jin Y, Xin Y, Zhang W, Ma Y (2010) *Mycobacterium tuberculosis* Rv1302 and *Mycobacterium smegmatis* MSMEG_4947 have WecA function and MSMEG_4947 is required for the growth of *M. smegmatis*. FEMS Microbiol Lett 310:54–61. https://doi.org/10.1111/j.1574-6968.2010.02045.x

Jyothikumar V, Klanbut K, Tiong J et al (2012) Cardiolipin synthase is required for *Streptomyces coelicolor* morphogenesis. Mol Microbiol 84:181–197. https://doi.org/10.1111/j.1365-2958.2012.08018.x

Kallenius G, Correia-Neves M, Buteme H et al (2016) Lipoarabinomannan, and its related glycolipids, induce divergent and opposing immune responses to *Mycobacterium tuberculosis* depending on structural diversity and experimental variations. Tuberculosis 96:120–130. https://doi.org/10.1016/j.tube.2015.09.005

Kang CM, Nyayapathy S, Lee JY et al (2008) Wag31, a homologue of the cell division protein DivIVA, regulates growth, morphology and polar cell wall synthesis in mycobacteria. Microbiology 154:725–735. https://doi.org/10.1099/mic.0.2007/014076-0

Kato K, Strominger JL, Kotani S (1968) Structure of the cell wall of *Corynebacterium diphtheriae*. I. Mechanism of hydrolysis by the L-3 enzyme and the structure of the peptide. Biochemistry 7:2762–2773

Kaur D, Berg S, Dinadayala P et al (2006) Biosynthesis of mycobacterial lipoarabinomannan: role of a branching mannosyltransferase. Proc Natl Acad Sci USA 103:13664–13669. https://doi.org/10.1073/pnas.0603049103

Kaur D, McNeil MR, Khoo K-H et al (2007) New insights into the biosynthesis of mycobacterial lipomannan arising from deletion of a conserved gene. J Biol Chem 282:27133–27140. https://doi.org/10.1074/jbc.M703389200

Kaur D, Obregon-Henao A, Pham H et al (2008) Lipoarabinomannan of *Mycobacterium*: mannose capping by a multifunctional terminal mannosyltransferase. Proc Natl Acad Sci USA 105:17973–17977. https://doi.org/10.1073/pnas.0807761105

Kaur D, Angala SK, Wu SW et al (2014) A single arabinan chain is attached to the phosphatidylinositol mannosyl core of the major immunomodulatory mycobacterial cell envelope glycoconjugate, lipoarabinomannan. J Biol Chem 289:30249–30256. https://doi.org/10.1074/jbc.M114.599415
Kawanami J, Kimura A, Otsuka H (1968) Siolipin A: a new lipoamino acid ester isolated from *Streptomyces sioyaensis*. Biochim Biophys Acta 152:808–810
Kieser KJ, Baranowski C, Chao MC et al (2015a) Peptidoglycan synthesis in *Mycobacterium tuberculosis* is organized into networks with varying drug susceptibility. Proc Natl Acad Sci USA 112:13087–13092. https://doi.org/10.1073/pnas.1514135112
Kieser KJ, Boutte CC, Kester JC et al (2015b) Phosphorylation of the peptidoglycan synthase PonA1 governs the rate of polar elongation in mycobacteria. PLoS Pathog 11:e1005010. https://doi.org/10.1371/journal.ppat.1005010
Kimura A, Kawanami J, Otsuka H (1967) Lipids of *Streptomyces sioyaensis*. J Biochem 62:384–385
Klatt S, Brammananth R, O'Callaghan S et al (2018) Identification of novel lipid modifications and intermembrane dynamics in *Corynebacterium glutamicum* using high-resolution mass spectrometry. J Lipid Res 59:1190–1204. https://doi.org/10.1194/jlr.M082784
Koch D, Schleifer KH, Kandler O (1970) The amino acid sequence of the serine and aspartic acid containing mureins of *Bifidobacterium bifidum* Orla Jensen. Z Naturforsch B 25:1294–1301
Kochetkov NK, Sviridov AF, Arifkhodzhaev KA et al (1979) The structure of the extracellular polysaccharide from *Mycobacterium lacticolum* strain 121. Carbohydr Res 71:193–203. https://doi.org/10.1016/S0008-6215(00)86070-X
Kondo T, Yamamoto D, Yokota A et al (2000) Gordonan, an acidic polysaccharide with cell aggregation-inducing activity in insect BM-N4 cells, produced by *Gordonia* sp. Biosci Biotechnol Biochem 64:2388–2394. https://doi.org/10.1271/bbb.64.2388
Kordulakova J, Gilleron M, Mikusova K et al (2002) Definition of the first mannosylation step in phosphatidylinositol mannoside synthesis. PimA is essential for growth of mycobacteria. J Biol Chem 277:31335–31344. https://doi.org/10.1074/jbc.M204060200
Kordulakova J, Gilleron M, Puzo G et al (2003) Identification of the required acyltransferase step in the biosynthesis of the phosphatidylinositol mannosides of *Mycobacterium* species. J Biol Chem 278:36285–36295. https://doi.org/10.1074/jbc.M303639200
Koronelli TV (1988) Investigation of the lipids of saprophytic mycobacteria in the U.S.S.R. J Chromatogr 440:479–486
Koster S, Upadhyay S, Chandra P et al (2017) *Mycobacterium tuberculosis* is protected from NADPH oxidase and LC3-associated phagocytosis by the LCP protein CpsA. Proc Natl Acad Sci USA 114:E8711–E8720. https://doi.org/10.1073/pnas.1707792114
Kovacevic S, Anderson D, Morita YS et al (2006) Identification of a novel protein with a role in lipoarabinomannan biosynthesis in mycobacteria. J Biol Chem 281:9011–9017. https://doi.org/10.1074/jbc.M511709200
Kozikowski AP, Onajole OK, Stec J et al (2017) Targeting mycolic acid transport by indole-2-carboxamides for the treatment of *Mycobacterium abscessus* infections. J Med Chem 60:5876–5888. https://doi.org/10.1021/acs.jmedchem.7b00582
Kugler JH, Muhle-Goll C, Kuhl B et al (2014) Trehalose lipid biosurfactants produced by the actinomycetes *Tsukamurella spumae* and *T. pseudospumae*. Appl Microbiol Biotechnol 98:8905–8915. https://doi.org/10.1007/s00253-014-5972-4
Kumar P, Schelle MW, Jain M et al (2007) PapA1 and PapA2 are acyltransferases essential for the biosynthesis of the *Mycobacterium tuberculosis* virulence factor sulfolipid-1. Proc Natl Acad Sci USA 104:11221–11226. https://doi.org/10.1073/pnas.0611649104
Kumar P, Arora K, Lloyd JR et al (2012) Meropenem inhibits D,D-carboxypeptidase activity in *Mycobacterium tuberculosis*. Mol Microbiol 86:367–381. https://doi.org/10.1111/j.1365-2958.2012.08199.x
Laneelle MA, Prome D, Laneelle G, Prome JC (1990) Ornithine lipid of *Mycobacterium tuberculosis*: its distribution in some slow- and fast-growing mycobacteria. J Gen Microbiol 136:773–778. https://doi.org/10.1099/00221287-136-4-773

Laneelle M-A, Launay A, Spina L et al (2012) A novel mycolic acid species defines two novel genera of the Actinobacteria, *Hoyosella* and *Amycolicicoccus*. Microbiology 158:843–855. https://doi.org/10.1099/mic.0.055509-0

Laneelle M-A, Eynard N, Spina L et al (2013) Structural elucidation and genomic scrutiny of the C60–C100 mycolic acids of *Segniliparus rotundus*. Microbiology 159:191–203. https://doi.org/10.1099/mic.0.063479-0

Lang S, Philp JC (1998) Surface-active lipids in rhodococci. Antonie Van Leeuwenhoek 74:59–70

Larrouy-Maumus G, Skovierova H, Dhouib R et al (2012) A small multidrug resistance-like transporter involved in the arabinosylation of arabinogalactan and lipoarabinomannan in mycobacteria. J Biol Chem 287:39933–39941. https://doi.org/10.1074/jbc.M112.400986

Lavollay M, Arthur M, Fourgeaud M et al (2008) The peptidoglycan of stationary-phase *Mycobacterium tuberculosis* predominantly contains cross-links generated by L,D-transpeptidation. J Bacteriol 190:4360–4366. https://doi.org/10.1128/JB.00239-08

Lavollay M, Arthur M, Fourgeaud M et al (2009) The beta-lactam-sensitive D,D-carboxypeptidase activity of Pbp4 controls the L,D and D,D transpeptidation pathways in *Corynebacterium jeikeium*. Mol Microbiol 74:650–661. https://doi.org/10.1111/j.1365-2958.2009.06887.x

Lavollay M, Fourgeaud M, Herrmann J-L et al (2011) The peptidoglycan of *Mycobacterium abscessus* is predominantly cross-linked by L,D-transpeptidases. J Bacteriol 193:778–782. https://doi.org/10.1128/JB.00606-10

Lea-Smith DJ, Pyke JS, Tull D et al (2007) The reductase that catalyzes mycolic motif synthesis is required for efficient attachment of mycolic acids to arabinogalactan. J Biol Chem 282:11000–11008. https://doi.org/10.1074/jbc.M608686200

Lea-Smith DJ, Martin KL, Pyke JS et al (2008) Analysis of a new mannosyltransferase required for the synthesis of phosphatidylinositol mannosides and lipoarabinomannan reveals two lipomannan pools in Corynebacterineae. J Biol Chem 283:6773–6782. https://doi.org/10.1074/jbc.M707139200

Lechevalier MP, De Bievre C, Lechevalier H (1977) Chemotaxonomy of aerobic Actinomycetes: phospholipid composition. Biochem Syst Ecol 5:249–260. https://doi.org/10.1016/0305-1978(77)90021-7

Lee YC, Ballou CE (1964) Structural studies on the myo-inositol mannosides from the glycolipids of *Mycobacterium tuberculosis* and *Mycobacterium phlei*. J Biol Chem 239:1316–1327

Lee RE, Mikusova K, Brennan PJ, Besra GS (1995) Synthesis of the arabinose donor beta-D-arabinofuranosyl-1-monophosphoryldecaprenol, development of a basic arabinosyl-transferase assay, and identification of ethambutol as an arabinosyl transferase inhibitor. J Am Chem Soc 117:11829–11832. https://doi.org/10.1021/ja00153a002

Lee RE, Brennan PJ, Besra GS (1997) Mycobacterial arabinan biosynthesis: the use of synthetic arabinoside acceptors in the development of an arabinosyl transfer assay. Glycobiology 7:1121–1128

Lee A, Wu S-W, Scherman MS et al (2006) Sequencing of oligoarabinosyl units released from mycobacterial arabinogalactan by endogenous arabinanase: identification of distinctive and novel structural motifs. Biochemistry 45:15817–15828. https://doi.org/10.1021/bi060688d

Lee JS, Krause R, Schreiber J et al (2008) Mutation in the transcriptional regulator PhoP contributes to avirulence of *Mycobacterium tuberculosis* H37Ra strain. Cell Host Microbe 3:97–103. https://doi.org/10.1016/j.chom.2008.01.002

Lefebvre C, Boulon R, Ducoux M et al (2018) HadD, a novel fatty acid synthase type II protein, is essential for alpha- and epoxy-mycolic acid biosynthesis and mycobacterial fitness. Sci Rep 8:6034. https://doi.org/10.1038/s41598-018-24380-5

Leger M, Gavalda S, Guillet V et al (2009) The dual function of the *Mycobacterium tuberculosis* FadD32 required for mycolic acid biosynthesis. Chem Biol 16:510–519. https://doi.org/10.1016/j.chembiol.2009.03.012

Lemassu A, Daffe M (1994) Structural features of the exocellular polysaccharides of *Mycobacterium tuberculosis*. Biochem J 297:351–357

Lemassu A, Ortalo-Magne A, Bardou F et al (1996) Extracellular and surface-exposed polysaccharides of non-tuberculous mycobacteria. Microbiology 142:1513–1520. https://doi.org/10.1099/13500872-142-6-1513
Lennarz WJ, Talamo B (1966) The chemical characterization and enzymatic synthesis of mannolipids in *Micrococcus lysodeikticus*. J Biol Chem 241:2707–2719
Lerat S, Forest M, Lauzier A et al (2012) Potato suberin induces differentiation and secondary metabolism in the genus *Streptomyces*. Microbes Environ 27:36–42. https://doi.org/10.1264/jsme2.ME11282
Lerouge P, Lebas MH, Agapakis-Causse C, Prome JC (1988) Isolation and structural characterization of a new non-phosphorylated lipoamino acid from *Mycobacterium phlei*. Chem Phys Lipids 49:161–166
Letek M, Ordonez E, Vaquera J et al (2008) DivIVA is required for polar growth in the MreB-lacking rod-shaped actinomycete *Corynebacterium glutamicum*. J Bacteriol 190:3283–3292. https://doi.org/10.1128/JB.01934-07
Lety MA, Nair S, Berche P, Escuyer V (1997) A single point mutation in the *embB* gene is responsible for resistance to ethambutol in *Mycobacterium smegmatis*. Antimicrob Agents Chemother 41:2629–2633
Leyh-Bouille M, Bonaly R, Ghuysen JM et al (1970) LL-diaminopimelic acid containing peptidoglycans in walls of *Streptomyces* sp. and of *Clostridium perfringens* (type A). Biochemistry 9:2944–2952
Li W, Xin Y, McNeil MR, Ma Y (2006) *rmlB* and *rmlC* genes are essential for growth of mycobacteria. Biochem Biophys Res Commun 342:170–178. https://doi.org/10.1016/j.bbrc.2006.01.130
Li W, Obregon-Henao A, Wallach JB et al (2016) Therapeutic potential of the *Mycobacterium tuberculosis* mycolic acid transporter, MmpL3. Antimicrob Agents Chemother 60:5198–5207. https://doi.org/10.1128/AAC.00826-16
Li W, Yazidi A, Pandya AN et al (2018) MmpL3 as a target for the treatment of drug-resistant nontuberculous mycobacterial infections. Front Microbiol 9:1547. https://doi.org/10.3389/fmicb.2018.01547
Linos A, Berekaa MM, Reichelt R et al (2000) Biodegradation of cis-1,4-polyisoprene rubbers by distinct actinomycetes: microbial strategies and detailed surface analysis. Appl Environ Microbiol 66:1639–1645
Liu C-W, Liu H-S (2011) *Rhodococcus erythropolis* strain NTU-1 efficiently degrades and traps diesel and crude oil in batch and fed-batch bioreactors. Process Biochem 46:202–209. https://doi.org/10.1016/j.procbio.2010.08.008
Logsdon MM, Aldridge BB (2018) Stable regulation of cell cycle events in mycobacteria: insights from inherently heterogeneous bacterial populations. Front Microbiol 9:514. https://doi.org/10.3389/fmicb.2018.00514
Lopez-Marin LM, Gautier N, Laneelle MA et al (1994) Structures of the glycopeptidolipid antigens of *Mycobacterium abscessus* and *Mycobacterium chelonae* and possible chemical basis of the serological cross-reactions in the *Mycobacterium fortuitum* complex. Microbiology 140:1109–1118
Ma Y, Mills JA, Belisle JT et al (1997) Determination of the pathway for rhamnose biosynthesis in mycobacteria: cloning, sequencing and expression of the *Mycobacterium tuberculosis* gene encoding alpha-D-glucose-1-phosphate thymidylyltransferase. Microbiology 143:937–945. https://doi.org/10.1099/00221287-143-3-937
Ma Y, Stern RJ, Scherman MS et al (2001) Drug targeting *Mycobacterium tuberculosis* cell wall synthesis: genetics of dTDP-rhamnose synthetic enzymes and development of a microtiter plate-based screen for inhibitors of conversion of dTDP-glucose to dTDP-rhamnose. Antimicrob Agents Chemother 45:1407–1416. https://doi.org/10.1128/AAC.45.5.1407-1416.2001
Ma Y, Pan F, McNeil M (2002) Formation of dTDP-rhamnose is essential for growth of mycobacteria. J Bacteriol 184:3392–3395. https://doi.org/10.1128/JB.184.12.3392-3395.2002

Machowski EE, Senzani S, Ealand C, Kana BD (2014) Comparative genomics for mycobacterial peptidoglycan remodelling enzymes reveals extensive genetic multiplicity. BMC Microbiol 14:75. https://doi.org/10.1186/1471-2180-14-75

Madigan CA, Cambier CJ, Kelly-Scumpia KM et al (2017) A macrophage response to *Mycobacterium leprae* phenolic glycolipid initiates nerve damage in leprosy. Cell 170:973–985.e10. https://doi.org/10.1016/j.cell.2017.07.030

Maes E, Coddeville B, Kremer L, Guerardel Y (2007) Polysaccharide structural variability in mycobacteria: identification and characterization of phosphorylated mannan and arabinomannan. Glycoconj J 24:439–448. https://doi.org/10.1007/s10719-007-9036-1

Mainardi J-L, Villet R, Bugg TD et al (2008) Evolution of peptidoglycan biosynthesis under the selective pressure of antibiotics in Gram-positive bacteria. FEMS Microbiol Rev 32:386–408. https://doi.org/10.1111/j.1574-6976.2007.00097.x

Makarov V, Manina G, Mikusova K et al (2009) Benzothiazinones kill *Mycobacterium tuberculosis* by blocking arabinan synthesis. Science 324:801–804. https://doi.org/10.1126/science.1171583

Maloney E, Stankowska D, Zhang J et al (2009) The two-domain LysX protein of *Mycobacterium tuberculosis* is required for production of lysinylated phosphatidylglycerol and resistance to cationic antimicrobial peptides. PLoS Pathog 5:e1000534. https://doi.org/10.1371/journal.ppat.1000534.g009

Maloney E, Lun S, Stankowska D et al (2011) Alterations in phospholipid catabolism in *Mycobacterium tuberculosis* lysX mutant. Front Microbiol 2:19. https://doi.org/10.3389/fmicb.2011.00019

Marrakchi H, Laneelle M-A, Daffe M (2014) Mycolic acids: structures, biosynthesis, and beyond. Chem Biol 21:67–85. https://doi.org/10.1016/j.chembiol.2013.11.011

Marshall CG, Wright GD (1998) DdlN from vancomycin-producing *Amycolatopsis orientalis* C329.2 is a VanA homologue with D-alanyl-D-lactate ligase activity. J Bacteriol 180:5792–5795

Martinot AJ, Farrow M, Bai L et al (2016) Mycobacterial metabolic syndrome: LprG and Rv1410 regulate triacylglyceride levels, growth rate and virulence in *Mycobacterium tuberculosis*. PLoS Pathog 12:e1005351. https://doi.org/10.1371/journal.ppat.1005351.s013

Maslow JN, Irani VR, Lee S-H et al (2003) Biosynthetic specificity of the rhamnosyltransferase gene of *Mycobacterium avium* serovar 2 as determined by allelic exchange mutagenesis. Microbiology 149:3193–3202. https://doi.org/10.1099/mic.0.26565-0

Mathur M, Kolattukudy PE (1992) Molecular cloning and sequencing of the gene for mycocerosic acid synthase, a novel fatty acid elongating multifunctional enzyme, from *Mycobacterium tuberculosis* var. *bovis Bacillus Calmette-Guerin*. J Biol Chem 267:19388–19395

Mathur AK, Murthy PS, Saharia GS, Venkitasubramanian TA (1976) Studies on cardiolipin biosynthesis in *Mycobacterium smegmatis*. Can J Microbiol 22:354–358

Mattos-Guaraldi AL, Cappelli EA, Previato JO et al (1999) Characterization of surface saccharides in two *Corynebacterium diphtheriae* strains. FEMS Microbiol Lett 170:159–166

McNeil M, Daffe M, Brennan PJ (1990) Evidence for the nature of the link between the arabinogalactan and peptidoglycan of mycobacterial cell walls. J Biol Chem 265:18200–18206

Medjahed H, Reyrat J-M (2009) Construction of *Mycobacterium abscessus* defined glycopeptidolipid mutants: comparison of genetic tools. Appl Environ Microbiol 75:1331–1338. https://doi.org/10.1128/AEM.01914-08

Melzer ES, Sein CE, Chambers JJ, Siegrist MS (2018) DivIVA concentrates mycobacterial cell envelope assembly for initiation and stabilization of polar growth. Cytoskeleton (Hoboken) 75:498–507. https://doi.org/10.1002/cm.21490

Meniche X, de Sousa-d'Auria C, Van-der-Rest B et al (2008) Partial redundancy in the synthesis of the D-arabinose incorporated in the cell wall arabinan of Corynebacterineae. Microbiology 154:2315–2326. https://doi.org/10.1099/mic.0.2008/016378-0

Meniche X, Otten R, Siegrist MS et al (2014) Subpolar addition of new cell wall is directed by DivIVA in mycobacteria. Proc Natl Acad Sci USA 111:E3243–E3251. https://doi.org/10.1073/pnas.1402158111

Middlebrook G, Coleman CM, Schaefer WB (1959) Sulfolipid from virulent tubercle bacilli. Proc Natl Acad Sci USA 45:1801–1804

Mikusova K, Slayden RA, Besra GS, Brennan PJ (1995) Biogenesis of the mycobacterial cell wall and the site of action of ethambutol. Antimicrob Agents Chemother 39:2484–2489

Mikusova K, Huang H, Yagi T et al (2005) Decaprenylphosphoryl arabinofuranose, the donor of the D-arabinofuranosyl residues of mycobacterial arabinan, is formed via a two-step epimerization of decaprenylphosphoryl ribose. J Bacteriol 187:8020–8025. https://doi.org/10.1128/JB.187.23.8020-8025.2005

Mikusova K, Belanova M, Kordulakova J et al (2006) Identification of a novel galactosyl transferase involved in biosynthesis of the mycobacterial cell wall. J Bacteriol 188:6592–6598

Mills JA, Motichka K, Jucker M et al (2004) Inactivation of the mycobacterial rhamnosyltransferase, which is needed for the formation of the arabinogalactan-peptidoglycan linker, leads to irreversible loss of viability. J Biol Chem 279:43540–43546. https://doi.org/10.1074/jbc.M407782200

Minnikin DE, Patel PV, Alshamaony L, Goodfellow M (1977) Polar lipid-composition in classification of *Nocardia* and related bacteria. Int J Syst Bacteriol 27:104–117. https://doi.org/10.1099/00207713-27-2-104

Minnikin DE, Lee OY-C, Wu HHT et al (2015) Pathophysiological implications of cell envelope structure in *Mycobacterium tuberculosis* and related taxa. In: Ribon W (ed) Tuberculosis—expanding knowledge. InTech, London, UK, pp 145–175

Mishra AK, Alderwick LJ, Rittmann D et al (2007) Identification of an alpha(1→6) mannopyranosyltransferase (MptA), involved in *Corynebacterium glutamicum* lipomanann biosynthesis, and identification of its orthologue in *Mycobacterium tuberculosis*. Mol Microbiol 65:1503–1517. https://doi.org/10.1111/j.1365-2958.2007.05884.x

Mishra AK, Alderwick LJ, Rittmann D et al (2008a) Identification of a novel alpha(1→6) mannopyranosyltransferase MptB from *Corynebacterium glutamicum* by deletion of a conserved gene, NCgl1505, affords a lipomannan- and lipoarabinomannan-deficient mutant. Mol Microbiol 68:1595–1613. https://doi.org/10.1111/j.1365-2958.2008.06265.x

Mishra AK, Klein C, Gurcha SS et al (2008b) Structural characterization and functional properties of a novel lipomannan variant isolated from a *Corynebacterium glutamicum* pimB' mutant. Antonie Van Leeuwenhoek 94:277–287. https://doi.org/10.1007/s10482-008-9243-1

Miyamoto Y, Mukai T, Nakata N et al (2006) Identification and characterization of the genes involved in glycosylation pathways of mycobacterial glycopeptidolipid biosynthesis. J Bacteriol 188:86–95. https://doi.org/10.1128/JB.188.1.86-95.2006

Miyamoto Y, Mukai T, Maeda Y et al (2007) Characterization of the fucosylation pathway in the biosynthesis of glycopeptidolipids from *Mycobacterium avium* complex. J Bacteriol 189:5515–5522. https://doi.org/10.1128/JB.00344-07

Miyamoto Y, Mukai T, Maeda Y et al (2008) The *Mycobacterium avium* complex *gtfTB* gene encodes a glucosyltransferase required for the biosynthesis of serovar 8-specific glycopeptidolipid. J Bacteriol 190:7918–7924. https://doi.org/10.1128/JB.00911-08

Miyamoto Y, Mukai T, Naka T et al (2010) Novel rhamnosyltransferase involved in biosynthesis of serovar 4-specific glycopeptidolipid from *Mycobacterium avium* complex. J Bacteriol 192:5700–5708. https://doi.org/10.1128/JB.00554-10

Moormann M, Zahringer U, Moll H et al (1997) A new glycosylated lipopeptide incorporated into the cell wall of a smooth variant of *Gordona hydrophobica*. J Biol Chem 272:10729–10738

Morita YS, Velasquez R, Taig E et al (2005) Compartmentalization of lipid biosynthesis in mycobacteria. J Biol Chem 280:21645–21652. https://doi.org/10.1074/jbc.M414181200

Morita YS, Sena CBC, Waller RF et al (2006) PimE is a polyprenol-phosphate-mannose-dependent mannosyltransferase that transfers the fifth mannose of phosphatidylinositol mannoside in mycobacteria. J Biol Chem 281:25143–25155. https://doi.org/10.1074/jbc.M604214200

Morita YS, Fukuda T, Sena CBC et al (2011) Inositol lipid metabolism in mycobacteria: biosynthesis and regulatory mechanisms. Biochim Biophys Acta 1810:630–641. https://doi.org/10.1016/j.bbagen.2011.03.017

Mougous JD, Petzold CJ, Senaratne RH et al (2004) Identification, function and structure of the mycobacterial sulfotransferase that initiates sulfolipid-1 biosynthesis. Nat Struct Mol Biol 11:721–729. https://doi.org/10.1038/nsmb802

Movahedzadeh F, Smith DA, Norman RA et al (2004) The *Mycobacterium tuberculosis ino1* gene is essential for growth and virulence. Mol Microbiol 51:1003–1014

Mukherjee R, Chatterji D (2012) Glycopeptidolipids: immuno-modulators in greasy mycobacterial cell envelope. IUBMB Life 64:215–225. https://doi.org/10.1002/iub.602

Naka T, Nakata N, Maeda S et al (2011) Structure and host recognition of serotype 13 glycopeptidolipid from *Mycobacterium intracellulare*. J Bacteriol 193:5766–5774. https://doi.org/10.1128/JB.05412-11

Nakata N, Fujiwara N, Naka T et al (2008) Identification and characterization of two novel methyltransferase genes that determine the serotype 12-specific structure of glycopeptidolipids of *Mycobacterium intracellulare*. J Bacteriol 190:1064–1071. https://doi.org/10.1128/JB.01370-07

Nampoothiri KM, Hoischen C, Bathe B et al (2002) Expression of genes of lipid synthesis and altered lipid composition modulates L-glutamate efflux of *Corynebacterium glutamicum*. Appl Microbiol Biotechnol 58:89–96. https://doi.org/10.1007/s00253-001-0861-z

Nataraj V, Pang P-C, Haslam SM et al (2015) *MKAN27435* is required for the biosynthesis of higher subclasses of lipooligosaccharides in *Mycobacterium kansasii*. PLoS ONE 10:e0122804. https://doi.org/10.1371/journal.pone.0122804

Naumova IB, Kuznetsov VD, Kudrina KS, Bezzubenkova AP (1980) The occurrence of teichoic acids in streptomycetes. Arch Microbiol 126:71–75

Nessar R, Reyrat J-M, Davidson LB, Byrd TF (2011) Deletion of the *mmpL4b* gene in the *Mycobacterium abscessus* glycopeptidolipid biosynthetic pathway results in loss of surface colonization capability, but enhanced ability to replicate in human macrophages and stimulate their innate immune response. Microbiology 157:1187–1195. https://doi.org/10.1099/mic.0.046557-0

Nguyen TM, Kim J (2015) *Streptomyces gilvifuscus* sp. nov., an actinomycete that produces antibacterial compounds isolated from soil. Int J Syst Evol Microbiol 65:3493–3500. https://doi.org/10.1099/ijsem.0.000447

Nguyen L, Scherr N, Gatfield J et al (2007) Antigen 84, an effector of pleiomorphism in *Mycobacterium smegmatis*. J Bacteriol 189:7896–7910. https://doi.org/10.1128/JB.00726-07

Niederweis M (2008) Nutrient acquisition by mycobacteria. Microbiology 154:679–692. https://doi.org/10.1099/mic.0.2007/012872-0

Niescher S, Wray V, Lang S et al (2006) Identification and structural characterisation of novel trehalose dinocardiomycolates from *n*-alkane-grown *Rhodococcus opacus* 1CP. Appl Microbiol Biotechnol 70:605–611. https://doi.org/10.1007/s00253-005-0113-8

Noda M, Kawahara Y, Ichikawa A et al (2004) Self-protection mechanism in D-cycloserine-producing *Streptomyces lavendulae*. Gene cloning, characterization, and kinetics of its alanine racemase and D-alanyl-D-alanine ligase, which are target enzymes of D-cycloserine. J Biol Chem 279:46143–46152. https://doi.org/10.1074/jbc.M404603200

Normand P, Lapierre P, Tisa LS et al (2007) Genome characteristics of facultatively symbiotic *Frankia* sp. strains reflect host range and host plant biogeography. Genome Res 17:7–15. https://doi.org/10.1101/gr.5798407

Odriozola JM, Ramos JA, Bloch K (1977) Fatty acid synthetase activity in *Mycobacterium smegmatis*. Characterization of the acyl carrier protein-dependent elongating system. Biochim Biophys Acta 488:207–217

Onwueme KC, Ferreras JA, Buglino J et al (2004) Mycobacterial polyketide-associated proteins are acyltransferases: proof of principle with *Mycobacterium tuberculosis* PapA5. Proc Natl Acad Sci USA 101:4608–4613. https://doi.org/10.1073/pnas.0306928101

Onwueme KC, Vos CJ, Zurita J et al (2005) Identification of phthiodiolone ketoreductase, an enzyme required for production of mycobacterial diacyl phthiocerol virulence factors. J Bacteriol 187:4760–4766. https://doi.org/10.1128/JB.187.14.4760-4766.2005

Ordway D, Henao-Tamayo M, Harton M et al (2007) The hypervirulent *Mycobacterium tuberculosis* strain HN878 induces a potent TH1 response followed by rapid down-regulation. J Immunol 179:522–531

Ortalo-Magne A, Dupont MA, Lemassu A et al (1995) Molecular composition of the outermost capsular material of the tubercle bacillus. Microbiology 141:1609–1620. https://doi.org/10.1099/13500872-141-7-1609

Pacheco GJ, Ciapina EMP, Gomes E de B, Junior NP (2010) Biosurfactant production by *Rhodococcus erythropolis* and its application to oil removal. Braz J Microbiol 41:685–693. https://doi.org/10.1590/S1517-83822010000300019

Pakkiri LS, Waechter CJ (2005) Dimannosyldiacylglycerol serves as a lipid anchor precursor in the assembly of the membrane-associated lipomannan in *Micrococcus luteus*. Glycobiology 15:291–302. https://doi.org/10.1093/glycob/cwi003

Pakkiri LS, Wolucka BA, Lubert EJ, Waechter CJ (2004) Structural and topological studies on the lipid-mediated assembly of a membrane-associated lipomannan in *Micrococcus luteus*. Glycobiology 14:73–81. https://doi.org/10.1093/glycob/cwh012

Pan F, Jackson M, Ma Y, McNeil M (2001) Cell wall core galactofuran synthesis is essential for growth of mycobacteria. J Bacteriol 183:3991–3998. https://doi.org/10.1128/JB.183.13.3991-3998.2001

Pardeshi P, Rao KK, Balaji PV (2017) Rv3634c from *Mycobacterium tuberculosis* H37Rv encodes an enzyme with UDP-Gal/Glc and UDP-GalNAc 4-epimerase activities. PLoS ONE 12:e0175193. https://doi.org/10.1371/journal.pone.0175193

Pasciak M, Kaczynski Z, Lindner B et al (2010) Immunochemical studies of trehalose-containing major glycolipid from *Tsukamurella pulmonis*. Carbohydr Res 345:1570–1574. https://doi.org/10.1016/j.carres.2010.04.026

Passeri A, Lang S, Wagner F, Wray V (1991) Marine biosurfactants, II. Production and characterization of an anionic trehalose tetraester from the marine bacterium Arthrobacter sp. EK 1. Z Naturforsch, C: J Biosci 46:204–209

Patterson JH, McConville MJ, Haites RE et al (2000) Identification of a methyltransferase from *Mycobacterium smegmatis* involved in glycopeptidolipid synthesis. J Biol Chem 275:24900–24906. https://doi.org/10.1074/jbc.M000147200

Peng W, Zou L, Bhamidi S et al (2012) The galactosamine residue in mycobacterial arabinogalactan is alpha-linked. J Org Chem 77:9826–9832. https://doi.org/10.1021/jo301393s

Pérez E, Constant P, Laval F et al (2004a) Molecular dissection of the role of two methyltransferases in the biosynthesis of phenolglycolipids and phthiocerol dimycoserosate in the *Mycobacterium tuberculosis* complex. J Biol Chem 279:42584–42592. https://doi.org/10.1074/jbc.M406134200

Pérez E, Constant P, Lemassu A et al (2004b) Characterization of three glycosyltransferases involved in the biosynthesis of the phenolic glycolipid antigens from the *Mycobacterium tuberculosis* complex. J Biol Chem 279:42574–42583. https://doi.org/10.1074/jbc.M406246200

Perkins HR (1971) Homoserine and diaminobutyric acid in the mucopeptide-precursor-nucleotides and cell walls of some plant-pathogenic corynebacteria. Biochem J 121:417–423

Perkins HR, Cummins CS (1964) Chemical structure of bacterial cell walls. Ornithine and 2,4-diaminobutyric acid as components of the cell walls of plant pathogenic corynebacteria. Nature 201:1105–1107

Petit JF, Adam A, Wietzerbin-Falszpan J et al (1969) Chemical structure of the cell wall of *Mycobacterium smegmatis*. I. Isolation and partial characterization of the peptidoglycan. Biochem Biophys Res Commun 35:478–485

Petrickova K (2003) Eukaryotic-type protein kinases in *Streptomyces coelicolor*: variations on a common theme. Microbiology 149:1609–1621. https://doi.org/10.1099/mic.0.26275-0

Peyret JL, Bayan N, Joliff G et al (1993) Characterization of the *cspB* gene encoding PS2, an ordered surface-layer protein in *Corynebacterium glutamicum*. Mol Microbiol 9:97–109

Philp JC, Kuyukina MS, Ivshina IB et al (2002) Alkanotrophic *Rhodococcus ruber* as a biosurfactant producer. Appl Microbiol Biotechnol 59:318–324. https://doi.org/10.1007/s00253-002-1018-4

Pitarque S, Larrouy-Maumus G, Payre B et al (2008) The immunomodulatory lipoglycans, lipoarabinomannan and lipomannan, are exposed at the mycobacterial cell surface. Tuberculosis (Edinb) 88:560–565. https://doi.org/10.1016/j.tube.2008.04.002

Powell DA, Duckworth M, Baddiley J (1975) A membrane-associated lipomannan in micrococci. Biochem J 151:387–397

Puech V, Chami M, Lemassu A et al (2001) Structure of the cell envelope of corynebacteria: importance of the non-covalently bound lipids in the formation of the cell wall permeability barrier and fracture plane. Microbiology 147:1365–1382. https://doi.org/10.1099/00221287-147-5-1365

Puffal J, Garcia-Heredia A, Rahlwes KC, et al (2018) Spatial control of cell envelope biosynthesis in mycobacteria. Pathog Dis 76:fty027. https://doi.org/10.1093/femspd/fty027

Qu H, Xin Y, Dong X, Ma Y (2007) An *rmlA* gene encoding D-glucose-1-phosphate thymidylyltransferase is essential for mycobacterial growth. FEMS Microbiol Lett 275:237–243. https://doi.org/10.1111/j.1574-6968.2007.00890.x

Quemard A (2016) New insights into the mycolate-containing compound biosynthesis and transport in mycobacteria. Trends Microbiol 24:725–738. https://doi.org/10.1016/j.tim.2016.04.009

Radmacher E, Alderwick LJ, Besra GS et al (2005) Two functional FAS-I type fatty acid synthases in *Corynebacterium glutamicum*. Microbiology 151:2421–2427. https://doi.org/10.1099/mic.0.28012-0

Rahlwes KC, Ha SA, Motooka D et al (2017) The cell envelope-associated phospholipid-binding protein LmeA is required for mannan polymerization in mycobacteria. J Biol Chem 292:17407–17417. https://doi.org/10.1074/jbc.M117.804377

Rahman O, Pfitzenmaier M, Pester O et al (2009) Macroamphiphilic components of thermophilic actinomycetes: identification of lipoteichoic acid in *Thermobifida fusca*. J Bacteriol 191:152–160. https://doi.org/10.1128/JB.01105-08

Rainczuk AK, Yamaryo-Botte Y, Brammananth R et al (2012) The lipoprotein LpqW is essential for the mannosylation of periplasmic glycolipids in corynebacteria. J Biol Chem 287:42726–42738. https://doi.org/10.1074/jbc.M112.373415

Rainwater DL, Kolattukudy PE (1983) Synthesis of mycocerosic acids from methylmalonyl coenzyme A by cell-free extracts of *Mycobacterium tuberculosis* var. *bovis* BCG. J Biol Chem 258:2979–2985

Rainwater DL, Kolattukudy PE (1985) Fatty acid biosynthesis in *Mycobacterium tuberculosis* var. *bovis Bacillus Calmette-Guerin*. Purification and characterization of a novel fatty acid synthase, mycocerosic acid synthase, which elongates *n*-fatty acyl-CoA with methylmalonyl-CoA. J Biol Chem 260:616–623

Ramos A, Honrubia MP, Valbuena N et al (2003) Involvement of DivIVA in the morphology of the rod-shaped actinomycete *Brevibacterium lactofermentum*. Microbiology 149:3531–3542. https://doi.org/10.1099/mic.0.26653-0

Rana AK, Singh A, Gurcha SS et al (2012) Ppm1-encoded polyprenyl monophosphomannose synthase activity is essential for lipoglycan synthesis and survival in mycobacteria. PLoS ONE 7:e48211. https://doi.org/10.1371/journal.pone.0048211

Rao A, Ranganathan A (2004) Interaction studies on proteins encoded by the phthiocerol dimycocerosate locus of *Mycobacterium tuberculosis*. Mol Genet Genomics 272:571–579. https://doi.org/10.1007/s00438-004-1088-3

Recht J, Kolter R (2001) Glycopeptidolipid acetylation affects sliding motility and biofilm formation in *Mycobacterium smegmatis*. J Bacteriol 183:5718–5724. https://doi.org/10.1128/JB.183.19.5718-5724.2001

Recht J, Martinez A, Torello S, Kolter R (2000) Genetic analysis of sliding motility in *Mycobacterium smegmatis*. J Bacteriol 182:4348–4351

Reed MB, Domenech P, Manca C et al (2004) A glycolipid of hypervirulent tuberculosis strains that inhibits the innate immune response. Nature 431:84–87. https://doi.org/10.1038/nature02837

Ren H, Dover LG, Islam ST et al (2007) Identification of the lipooligosaccharide biosynthetic gene cluster from *Mycobacterium marinum*. Mol Microbiol 63:1345–1359. https://doi.org/10.1111/j.1365-2958.2007.05603.x

Ripoll F, Deshayes C, Pasek S, et al (2007) Genomics of glycopeptidolipid biosynthesis in *Mycobacterium abscessus* and *M. chelonae*. BMC Genomics 8:114. https://doi.org/10.1186/1471-2164-8-114

Rojas ER, Billings G, Odermatt PD et al (2018) The outer membrane is an essential load-bearing element in Gram-negative bacteria. Nature 559:617–621. https://doi.org/10.1038/s41586-018-0344-3

Rombouts Y, Burguiere A, Maes E et al (2009) *Mycobacterium marinum* lipooligosaccharides are unique caryophyllose-containing cell wall glycolipids that inhibit tumor necrosis factor-alpha secretion in macrophages. J Biol Chem 284:20975–20988. https://doi.org/10.1074/jbc.M109.011429

Rombouts Y, Elass E, Biot C et al (2010) Structural analysis of an unusual bioactive *N*-acylated lipo-oligosaccharide LOS-IV in *Mycobacterium marinum*. J Am Chem Soc 132:16073–16084. https://doi.org/10.1021/ja105807s

Rombouts Y, Alibaud L, Carrere-Kremer S et al (2011) Fatty acyl chains of *Mycobacterium marinum* lipooligosaccharides: structure, localization and acylation by PapA4 (MMAR_2343) protein. J Biol Chem 286:33678–33688. https://doi.org/10.1074/jbc.M111.273920

Rousseau C, Neyrolles O, Bordat Y et al (2003) Deficiency in mycolipenate- and mycosanoate-derived acyltrehaloses enhances early interactions of *Mycobacterium tuberculosis* with host cells. Cell Microbiol 5:405–415

Saadat S, Ballou CE (1983) Pyruvylated glycolipids from *Mycobacterium smegmatis*. Structures of two oligosaccharide components. J Biol Chem 258:1813–1818

Sambou T, Dinadayala P, Stadthagen G et al (2008) Capsular glucan and intracellular glycogen of *Mycobacterium tuberculosis*: biosynthesis and impact on the persistence in mice. Mol Microbiol 70:762–774. https://doi.org/10.1111/mmi.2008.70.issue-3

Sanders AN, Wright LF, Pavelka MS (2014) Genetic characterization of mycobacterial L,D-transpeptidases. Microbiology 160:1795–1806. https://doi.org/10.1099/mic.0.078980-0

Sandoval-Calderon M, Geiger O, Guan Z et al (2009) A eukaryote-like cardiolipin synthase is present in *Streptomyces coelicolor* and in most Actinobacteria. J Biol Chem 284:17383–17390. https://doi.org/10.1074/jbc.M109.006072

Sandoval-Calderon M, Nguyen DD, Kapono CA et al (2015) Plasticity of *Streptomyces coelicolor* membrane composition under different growth conditions and during development. Front Microbiol 6:1465. https://doi.org/10.3389/fmicb.2015.01465

Sani M, Houben ENG, Geurtsen J et al (2010) Direct visualization by cryo-EM of the mycobacterial capsular layer: a labile structure containing Esx-1-secreted proteins. PLoS Pathog 6:e1000794. https://doi.org/10.1371/journal.ppat.1000794.t001

Sarkar D, Sidhu M, Singh A et al (2011) Identification of a glycosyltransferase from *Mycobacterium marinum* involved in addition of a caryophyllose moiety in lipooligosaccharides. J Bacteriol 193:2336–2340. https://doi.org/10.1128/JB.00065-11

Schelle MW, Bertozzi CR (2006) Sulfate metabolism in mycobacteria. ChemBioChem 7:1516–1524. https://doi.org/10.1002/cbic.200600224

Scher M, Lennarz WJ (1969) Studies on the biosynthesis of mannan in *Micrococcus lysodeikticus*. I. Characterization of mannan-^{14}C formed enzymatically from mannosyl-1-phosphoryl-undecaprenol. J Biol Chem 244:2777–2789

Scherman M, Weston A, Duncan K et al (1995) Biosynthetic origin of mycobacterial cell wall arabinosyl residues. J Bacteriol 177:7125–7130

Schleifer KH, Kandler O (1972) Peptidoglycan types of bacterial cell walls and their taxonomic implications. Bacteriol Rev 36:407–477

Schoonmaker MK, Bishai WR, Lamichhane G (2014) Nonclassical transpeptidases of *Mycobacterium tuberculosis* alter cell size, morphology, the cytosolic matrix, protein localization, virulence, and resistance to beta-lactams. J Bacteriol 196:1394–1402. https://doi.org/10.1128/JB.01396-13

Schorey JS, Sweet L (2008) The mycobacterial glycopeptidolipids: structure, function, and their role in pathogenesis. Glycobiology 18:832–841. https://doi.org/10.1093/glycob/cwn076

Seeliger JC, Holsclaw CM, Schelle MW, et al (2011) Elucidation and chemical modulation of sulfolipid-1 biosynthesis in *Mycobacterium tuberculosis.* J Biol Chem. https://doi.org/10.1074/jbc.m111.315473

Seidel M, Alderwick LJ, Birch HL et al (2007) Identification of a novel arabinofuranosyltransferase AftB involved in a terminal step of cell wall arabinan biosynthesis in Corynebacterianeae, such as *Corynebacterium glutamicum* and *Mycobacterium tuberculosis*. J Biol Chem 282:14729–14740. https://doi.org/10.1074/jbc.M700271200

Selim MS, Amer SK, Mohamed SS et al (2018) Production and characterisation of exopolysaccharide from *Streptomyces carpaticus* isolated from marine sediments in Egypt and its effect on breast and colon cell lines. J Genet Eng Biotechnol 16:23–28. https://doi.org/10.1016/j.jgeb.2017.10.014

Sena CBC, Fukuda T, Miyanagi K et al (2010) Controlled expression of branch-forming mannosyltransferase is critical for mycobacterial lipoarabinomannan biosynthesis. J Biol Chem 285:13326–13336. https://doi.org/10.1074/jbc.M109.077297

Senzani S, Li D, Bhaskar A et al (2017) An amidase_3 domain-containing *N*-acetylmuramyl-L-alanine amidase is required for mycobacterial cell division. Sci Rep 7:1140. https://doi.org/10.1038/s41598-017-01184-7

Severn WB, Richards JC (1992) The acidic specific capsular polysaccharide of *Rhodococcus equi* serotype 3. Structural elucidation and stereochemical analysis of the lactate ether and pyruvate acetal substituents. Can J Chem 70:2664–2676. https://doi.org/10.1139/v92-336

Shi L, Berg S, Lee A et al (2006) The carboxy terminus of EmbC from *Mycobacterium smegmatis* mediates chain length extension of the arabinan in lipoarabinomannan. J Biol Chem 281:19512–19526. https://doi.org/10.1074/jbc.M513846200

Simeone R, Constant P, Malaga W et al (2007) Molecular dissection of the biosynthetic relationship between phthiocerol and phthiodiolone dimycocerosates and their critical role in the virulence and permeability of *Mycobacterium tuberculosis*. FEBS J 274:1957–1969. https://doi.org/10.1111/j.1742-4658.2007.05740.x

Simeone R, Leger M, Constant P et al (2010) Delineation of the roles of FadD22, FadD26 and FadD29 in the biosynthesis of phthiocerol dimycocerosates and related compounds in *Mycobacterium tuberculosis*. FEBS J 277:2715–2725. https://doi.org/10.1111/j.1742-464X.2010.07688.x

Singer ME, Finnerty WR, Tunelid A (1990) Physical and chemical properties of a biosurfactant synthesized by *Rhodococcus species* H13-A. Can J Microbiol 36:746–750. https://doi.org/10.1139/m90-128

Sirakova TD, Thirumala AK, Dubey VS et al (2001) The *Mycobacterium tuberculosis pks2* gene encodes the synthase for the hepta- and octamethyl-branched fatty acids required for sulfolipid synthesis. J Biol Chem 276:16833–16839. https://doi.org/10.1074/jbc.M011468200

Skovierova H, Larrouy-Maumus G, Zhang J et al (2009) AftD, a novel essential arabinofuranosyltransferase from mycobacteria. Glycobiology 19:1235–1247. https://doi.org/10.1093/glycob/cwp116

Slama N, Jamet S, Frigui W et al (2016) The changes in mycolic acid structures caused by *hadC* mutation have a dramatic effect on the virulence of *Mycobacterium tuberculosis*. Mol Microbiol 99:794–807. https://doi.org/10.1111/mmi.13266

Slayden RA, Barry CE (2002) The role of KasA and KasB in the biosynthesis of meromycolic acids and isoniazid resistance in *Mycobacterium tuberculosis*. Tuberculosis 82:149–160

Sonden B, Kocincova D, Deshayes C et al (2005) Gap, a mycobacterial specific integral membrane protein, is required for glycolipid transport to the cell surface. Mol Microbiol 58:426–440. https://doi.org/10.1111/j.1365-2958.2005.04847.x

Stern RJ, Lee TY, Lee TJ et al (1999) Conversion of dTDP-4-keto-6-deoxyglucose to free dTDP-4-keto-rhamnose by the *rmlC* gene products of *Escherichia coli* and *Mycobacterium tuberculosis*. Microbiology 145:663–671. https://doi.org/10.1099/13500872-145-3-663

Stokes RW, Norris-Jones R, Brooks DE et al (2004) The glycan-rich outer layer of the cell wall of *Mycobacterium tuberculosis* acts as an antiphagocytic capsule limiting the association of the bacterium with macrophages. Infect Immun 72:5676–5686. https://doi.org/10.1128/IAI.72.10.5676-5686.2004

Sulzenbacher G, Canaan S, Bordat Y et al (2006) LppX is a lipoprotein required for the translocation of phthiocerol dimycocerosates to the surface of *Mycobacterium tuberculosis*. EMBO J 25:1436–1444. https://doi.org/10.1038/sj.emboj.7601048

Sydor T, von Bargen K, Becken U et al (2008) A mycolyl transferase mutant of *Rhodococcus equi* lacking capsule integrity is fully virulent. Vet Microbiol 128:327–341. https://doi.org/10.1016/j.vetmic.2007.10.020

Szczepina MG, Zheng RB, Completo GC et al (2009) STD-NMR studies suggest that two acceptor substrates for GlfT2, a bifunctional galactofuranosyltransferase required for the biosynthesis of *Mycobacterium tuberculosis* arabinogalactan, compete for the same binding site. ChemBioChem 10:2052–2059. https://doi.org/10.1002/cbic.200900202

Takayama K, Kilburn JO (1989) Inhibition of synthesis of arabinogalactan by ethambutol in *Mycobacterium smegmatis*. Antimicrob Agents Chemother 33:1493–1499

Tatituri RVV, Illarionov PA, Dover LG et al (2007) Inactivation of *Corynebacterium glutamicum* NCgl0452 and the role of MgtA in the biosynthesis of a novel mannosylated glycolipid involved in lipomannan biosynthesis. J Biol Chem 282:4561–4572. https://doi.org/10.1074/jbc.M608695200

Telenti A, Philipp WJ, Sreevatsan S et al (1997) The *emb* operon, a gene cluster of *Mycobacterium tuberculosis* involved in resistance to ethambutol. Nat Med 3:567–570

Thakur M, Chakraborti PK (2008) Ability of PknA, a mycobacterial eukaryotic-type serine/threonine kinase, to transphosphorylate MurD, a ligase involved in the process of peptidoglycan biosynthesis. Biochem J 415:27–33. https://doi.org/10.1042/BJ20080234

Thanky NR, Young DB, Robertson BD (2007) Unusual features of the cell cycle in mycobacteria: polar-restricted growth and the snapping-model of cell division. Tuberculosis 87:231–236. https://doi.org/10.1016/j.tube.2006.10.004

Tokumoto Y, Nomura N, Uchiyama H et al (2009) Structural characterization and surface-active properties of a succinoyl trehalose lipid produced by *Rhodococcus* sp. SD-74. J Oleo Sci 58:97–102

Touchette MH, Holsclaw CM, Previti ML et al (2014) The *rv1184c* locus encodes Chp2, an acyltransferase in *Mycobacterium tuberculosis* polyacyltrehalose lipid biosynthesis. J Bacteriol 197:201–210. https://doi.org/10.1128/JB.02015-14

Touchette MH, Van Vlack ER, Bai L et al (2017) A screen for protein-protein interactions in live mycobacteria reveals a functional link between the virulence-associated lipid transporter lprg and the mycolyltransferase Antigen 85A. ACS Infect Dis 3:336–348. https://doi.org/10.1021/acsinfecdis.6b00179

Trivedi OA, Arora P, Sridharan V et al (2004) Enzymic activation and transfer of fatty acids as acyl-adenylates in mycobacteria. Nature 428:441–445. https://doi.org/10.1038/nature02384

Trivedi OA, Arora P, Vats A et al (2005) Dissecting the mechanism and assembly of a complex virulence mycobacterial lipid. Mol Cell 17:631–643. https://doi.org/10.1016/j.molcel.2005.02.009

Tuleva B, Christova N, Cohen R et al (2008) Production and structural elucidation of trehalose tetraesters (biosurfactants) from a novel alkanothrophic *Rhodococcus wratislaviensis* strain. J Appl Microbiol 104:1703–1710. https://doi.org/10.1111/j.1365-2672.2007.03680.x

Tuleva B, Christova N, Cohen R et al (2009) Isolation and characterization of trehalose tetraester biosurfactants from a soil strain *Micrococcus luteus* BN56. Process Biochem 44:135–141. https://doi.org/10.1016/j.procbio.2008.09.016

Tul'skaya EM, Shashkov AS, Streshinskaya GM et al (2011) Teichuronic and teichulosonic acids of actinomycetes. Biochemistry (Moscow) 76:736–744. https://doi.org/10.1134/S0006297911070030

Turner J, Torrelles JB (2018) Mannose-capped lipoarabinomannan in *Mycobacterium tuberculosis* pathogenesis. Pathog Dis 76:S1130. https://doi.org/10.1093/femspd/fty026

Uchida Y, Tsuchiya R, Chino M et al (1989) Extracellular accumulation of mono- and di-succinoyl trehalose lipids by a strain of *Rhodococcus erythropolis* grown on *n*-alkanes. Agric Biol Chem 53:757–763. https://doi.org/10.1271/bbb1961.53.757

Udou T, Ogawa M, Mizuguchi Y (1983) An improved method for the preparation of mycobacterial spheroplasts and the mechanism involved in the reversion to bacillary form: electron microscopic and physiological study. Can J Microbiol 29:60–68

Vadrevu IS, Lofton H, Sarva K et al (2011) ChiZ levels modulate cell division process in mycobacteria. Tuberculosis (Edinb) 91(Suppl 1):S128–S135. https://doi.org/10.1016/j.tube.2011.10.022

van der Wel N, Hava D, Houben D et al (2007) *M. tuberculosis* and *M. leprae* translocate from the phagolysosome to the cytosol in myeloid cells. Cell 129:1287–1298. https://doi.org/10.1016/j.cell.2007.05.059

van der Woude AD, Sarkar D, Bhatt A et al (2012) Unexpected link between lipooligosaccharide biosynthesis and surface protein release in *Mycobacterium marinum*. J Biol Chem 287:20417–20429. https://doi.org/10.1074/jbc.M111.336461

van Straaten KE, Kuttiyatveetil JRA, Sevrain CM et al (2015) Structural basis of ligand binding to UDP-galactopyranose mutase from *Mycobacterium tuberculosis* using substrate and tetrafluorinated substrate analogues. J Am Chem Soc 137:1230–1244. https://doi.org/10.1021/ja511204p

Varela C, Rittmann D, Singh A et al (2012) MmpL genes are associated with mycolic acid metabolism in mycobacteria and corynebacteria. Chem Biol 19:498–506. https://doi.org/10.1016/j.chembiol.2012.03.006

Veerkamp JH (1971) The structure of the cell wall peptidoglycan of *Bifidobacterium bifidum* var. *pennsylvanicus*. Arch Biochem Biophys 143:204–211

Vences-Guzman MA, Geiger O, Sohlenkamp C (2012) Ornithine lipids and their structural modifications: from A to E and beyond. FEMS Microbiol Lett 335:1–10. https://doi.org/10.1111/j.1574-6968.2012.02623.x

Verma V, Qazi GN, Parshad R, Chopra CL (1989) A fast spheroplast formation procedure in some 2,5-diketo-D-gluconate- and 2-keto-L-gulonate-producing bacteria. Biotechniques 7:449–452

Viljoen A, Herrmann J-L, Onajole OK et al (2017) Controlling extra- and intramacrophagic *Mycobacterium abscessus* by targeting mycolic acid transport. Front Cell Infect Microbiol 7:388. https://doi.org/10.3389/fcimb.2017.00388

Vollbrecht E, Heckmann R, Wray V et al (1998) Production and structure elucidation of di- and oligosaccharide lipids (biosurfactants) from *Tsukamurella* sp. nov. Appl Microbiol Biotechnol 50:530–537

Vollmer W, Blanot D, de Pedro MA (2008) Peptidoglycan structure and architecture. FEMS Microbiol Rev 32:149–167. https://doi.org/10.1111/j.1574-6976.2007.00094.x

von Wintzingerode F, Gobel UB, Siddiqui RA et al (2001) *Salana multivorans* gen. nov., sp. nov., a novel actinobacterium isolated from an anaerobic bioreactor and capable of selenate reduction. Int J Syst Evol Microbiol 51:1653–1661. https://doi.org/10.1099/00207713-51-5-1653

Waddell SJ, Chung GA, Gibson KJC et al (2005) Inactivation of polyketide synthase and related genes results in the loss of complex lipids in *Mycobacterium tuberculosis* H37Rv. Lett Appl Microbiol 40:201–206. https://doi.org/10.1111/j.1472-765X.2005.01659.x

Wang L-Y, Li S-T, Li Y (2003) Identification and characterization of a new exopolysaccharide biosynthesis gene cluster from *Streptomyces*. FEMS Microbiol Lett 220:21–27. https://doi.org/10.1016/S0378-1097(03)00044-2

Wang Q, Zhu L, Jones V et al (2015) CpsA, a LytR-CpsA-Psr family protein in *Mycobacterium marinum*, is required for cell wall integrity and virulence. Infect Immun 83:2844–2854. https://doi.org/10.1128/IAI.03081-14

Wehmeier S, Varghese AS, Gurcha SS et al (2009) Glycosylation of the phosphate binding protein, PstS, in *Streptomyces coelicolor* by a pathway that resembles protein *O*-mannosylation in eukaryotes. Mol Microbiol 71:421–433. https://doi.org/10.1111/j.1365-2958.2008.06536.x

Weidenmaier C, Peschel A (2008) Teichoic acids and related cell-wall glycopolymers in Gram-positive physiology and host interactions. Nat Rev Microbiol 6:276–287. https://doi.org/10.1038/nrmicro1861

Welby-Gieusse M, Laneelle MA, Asselineau J (1970) Structure of the corynomycolic acids of *Corynebacterium hofmanii* and their biogenetic implication. Eur J Biochem 13:164–167

Wesener DA, Levengood MR, Kiessling LL (2017) Comparing galactan biosynthesis in *Mycobacterium tuberculosis* and *Corynebacterium diphtheriae*. J Biol Chem 292:2944–2955. https://doi.org/10.1074/jbc.M116.759340

Weston A, Stern RJ, Lee RE et al (1997) Biosynthetic origin of mycobacterial cell wall galactofuranosyl residues. Tuber Lung Dis 78:123–131

Wheatley RW, Zheng RB, Richards MR et al (2012) Tetrameric structure of the GlfT2 galactofuranosyltransferase reveals a scaffold for the assembly of mycobacterial arabinogalactan. J Biol Chem 287:28132–28143. https://doi.org/10.1074/jbc.M112.347484

White DA, Hird LC, Ali ST (2013) Production and characterization of a trehalolipid biosurfactant produced by the novel marine bacterium *Rhodococcus* sp., strain PML026. J Appl Microbiol 115:744–755. https://doi.org/10.1111/jam.12287

Wietzerbin J, Das BC, Petit JF et al (1974) Occurrence of D-alanyl-(D)-meso-diaminopimelic acid and meso-diaminopimelyl-meso-diaminopimelic acid interpeptide linkages in the peptidoglycan of mycobacteria. Biochemistry 13:3471–3476

Wolucka BA, de Hoffmann E (1995) The presence of beta-D-ribosyl-1-monophosphodecaprenol in mycobacteria. J Biol Chem 270:20151–20155

Wolucka BA, McNeil MR, de Hoffmann E et al (1994) Recognition of the lipid intermediate for arabinogalactan/arabinomannan biosynthesis and its relation to the mode of action of ethambutol on mycobacteria. J Biol Chem 269:23328–23335

World Health Organization (2018) Global tuberculosis report 2018, pp 1–277

Xin Y, Lee RE, Scherman MS et al (1997) Characterization of the in vitro synthesized arabinan of mycobacterial cell walls. Biochim Biophys Acta 1335:231–234

Xu Z, Meshcheryakov VA, Poce G, Chng S-S (2017) MmpL3 is the flippase for mycolic acids in mycobacteria. Proc Natl Acad Sci USA 114:7993–7998. https://doi.org/10.1073/pnas.1700062114

Yague P, Willemse J, Koning RI et al (2016) Subcompartmentalization by cross-membranes during early growth of *Streptomyces* hyphae. Nat Commun 7:12467. https://doi.org/10.1038/ncomms12467

Yakimov MM, Giuliano L, Bruni V et al (1999) Characterization of antarctic hydrocarbon-degrading bacteria capable of producing bioemulsifiers. New Microbiol 22:249–256

Yamaryo-Botte Y, Rainczuk AK, Lea-Smith DJ, et al (2014) Acetylation of trehalose mycolates is required for efficient MmpL-mediated membrane transport in Corynebacterineae. ACS Chem Biol 141209130005000. https://doi.org/10.1021/cb5007689

Yano I, Furukawa Y, Kusunose M (1969) Phospholipids of *Nocardia coeliaca*. J Bacteriol 98:124–130

Zanfardino A, Migliardi A, D'Alonzo D et al (2016) Inactivation of MSMEG_0412 gene drastically affects surface related properties of *Mycobacterium smegmatis*. BMC Microbiol 16:267. https://doi.org/10.1186/s12866-016-0888-z

Zhang N, Torrelles JB, McNeil MR et al (2003) The Emb proteins of mycobacteria direct arabinosylation of lipoarabinomannan and arabinogalactan via an N-terminal recognition region and a C-terminal synthetic region. Mol Microbiol 50:69–76

Zheng H, Lu L, Wang B et al (2008) Genetic basis of virulence attenuation revealed by comparative genomic analysis of *Mycobacterium tuberculosis* strain H37Ra versus H37Rv. PLoS ONE 3:e2375. https://doi.org/10.1371/journal.pone.0002375

Zhou X, Halladin DK, Theriot JA (2016) Fast mechanically driven daughter cell separation is widespread in Actinobacteria. mBio 7:e00952–16-6. https://doi.org/10.1128/mbio.00952-16

Zhou X, Rodriguez-Rivera FP, Lim HC et al (2019) Sequential assembly of the septal cell envelope prior to V snapping in *Corynebacterium glutamicum*. Nat Chem Biol 2:a000414. https://doi.org/10.1038/s41589-018-0206-1

Zuber B, Chami M, Houssin C et al (2008) Direct visualization of the outer membrane of mycobacteria and corynebacteria in their native state. J Bacteriol 190:5672–5680. https://doi.org/10.1128/JB.01919-07

Zuneda MC, Guillenea JJ, Dominguez JB et al (1984) Lipid composition and protoplast-forming capacity of *Streptomyces antibioticus*. Lipids 19:223–228

Chapter 14
Archaeal Cell Walls

Andreas Klingl, Carolin Pickl and Jennifer Flechsler

Abstract The cell wall of archaea, as of any other prokaryote, is surrounding the cell outside the cytoplasmic membrane and is mediating the interaction with the environment. In this regard, it can be involved in cell shape maintenance, protection against virus, heat, acidity or alkalinity. Throughout the formation of pore like structures, it can resemble a micro sieve and thereby enable or disable transport processes. In some cases, cell wall components can make up more than 10% of the whole cellular protein. So far, a great variety of different cell envelope structures and compounds have be found and described in detail. From all archaeal cell walls described so far, the most common structure is the S-layer. Other archaeal cell wall structures are pseudomurein, methanochondroitin, glutaminylglycan, sulfated heteropolysaccharides and protein sheaths and they are sometimes associated with additional proteins and protein complexes like the STABLE protease or the bindosome. Recent advances in electron microscopy also illustrated the presence of an outer(most) cellular membrane within several archaeal groups, comparable to the Gram-negative cell wall within bacteria. Each new cell wall structure that can be investigated in detail and that can be assigned with a specific function helps us to understand, how the earliest cells on earth might have looked like.

Keywords S-layer · Pseudomurein · Methanochondroitin · Glutaminylglycan · Halomucin · Outer(most) cellular membrane

Introduction to Archaeal Cell Envelopes

Archaea represent the third major lineage or "domain" next to the bacteria and eukaryotes (Woese et al. 1990), and recent studies even suggest, that the eukaryotes might as well be a side branch of the archaea (e.g. Spang et al. 2015). The S-layer was suggested to be the predominant and most ancient archaeal cell envelope structure

A. Klingl (✉) · C. Pickl · J. Flechsler
Plant Development and Electron Microscopy, Department of Biology I, Biocenter LMU Munich, Großhaderner Str. 2-4, 82152 Planegg-Martinsried, Germany
e-mail: andreas.klingl@biologie.uni-muenchen.de

A. Kuhn (ed.), *Bacterial Cell Walls and Membranes*, Subcellular Biochemistry 92, https://doi.org/10.1007/978-3-030-18768-2_14

within the archaea since their discovery and first descriptions of archaeal cell walls (Figs. 14.1a and 14.2a; Albers and Meyer 2011). Especially within those organisms that are described as so-called extremophiles, the S-layer, a surface protein forming paracrystalline 2D crystal lattices with p1-, p2-, p3-, p4- and p6-symmetry, in the majority of cases seems to be an absolute necessity for growth and viability in general (Sleytr et al. 2007; Pum and Sleytr 2014). For example, the Sulfolobales thrive optimally at temperatures between 65 and 75 °C and a pH around 2. This means, they are able to grow in hot sulfuric acid and the only protection barrier between outside and the cytoplasm is the S-layer with a characteristic p3-symmetry. Also, the current record holder for the highest growth temperature at 113 °C, *Pyrolobus fumarii* (Blöchl et al. 1997), is surrounded by an S-layer with, in this case, p4-symmetry. In general, it is thought that the environmental conditions like high temperature, anoxia or low pH in which these extremophilic microorganisms can grow resemble the situation on earth when the first living cells developed. Therefore, the organisms that inhabit those biotopes today might look quite similar to the first living cells. At the moment, it is still an accepted theory, that the S-layer was the first archaeal cell wall that developed (Albers and Meyer 2011). However, in recent years, an increasing number of archaeal cell envelopes have been described differently from S-layers, pseudomurein, methanochondroitin, and protein sheaths (Table 14.1). These distinct outermost layers include single and double membranes, polysaccharides, and mucous layers (Albers and Meyer 2011). Remarkably, two microorganisms not exhibiting an S-layer or any other solid cell wall rank among the extremophiles, and they seem to have adapted to their extreme environments without the support of a protecting S-layer: *Ferroplasma acidiphilum* (Golyshina and Timmis 2005) and *Thermoplasma spp.* (Darland et al. 1970). In the following section, we elucidate archaea with different cell envelopes, and their living conditions.

Archaea with One Membrane

Despite living under extreme conditions (e.g. low pH values of 1–2 and at 60 °C), *Ferroplasma acidiphilum* (Golyshina and Timmis 2005) and *Thermoplasma spp.* (Darland et al. 1970) do not have a cell wall at all (Figs. 14.1b and 14.2b). Being bounded by only a single membrane, these organisms exhibit a pleomorphic cell shape. Cells of *Thermoplasma acidophilum* were proposed to preserve their intactness via mannose-rich glycoproteins and lipoglycans anchored in the plasma membrane, forming a protective glycocalix at the cell surface (Yang and Haug 1979; Smith 1984; Langworthy 1972; Albers and Meyer 2011). Membranes of *Ferroplasma acidamanus* were reported to contain tetraether lipids forming stable monolayer membranes instead of bilayer membranes, enabling the cells to live in acidic environments (Gulik et al. 1988; Macalady et al. 2004). The stability of these monolayer membranes is mainly caused by their higher resistance to hydrolysis. Additionally, effective proton pumps and acid stable membrane proteins might also play an important role in this regard.

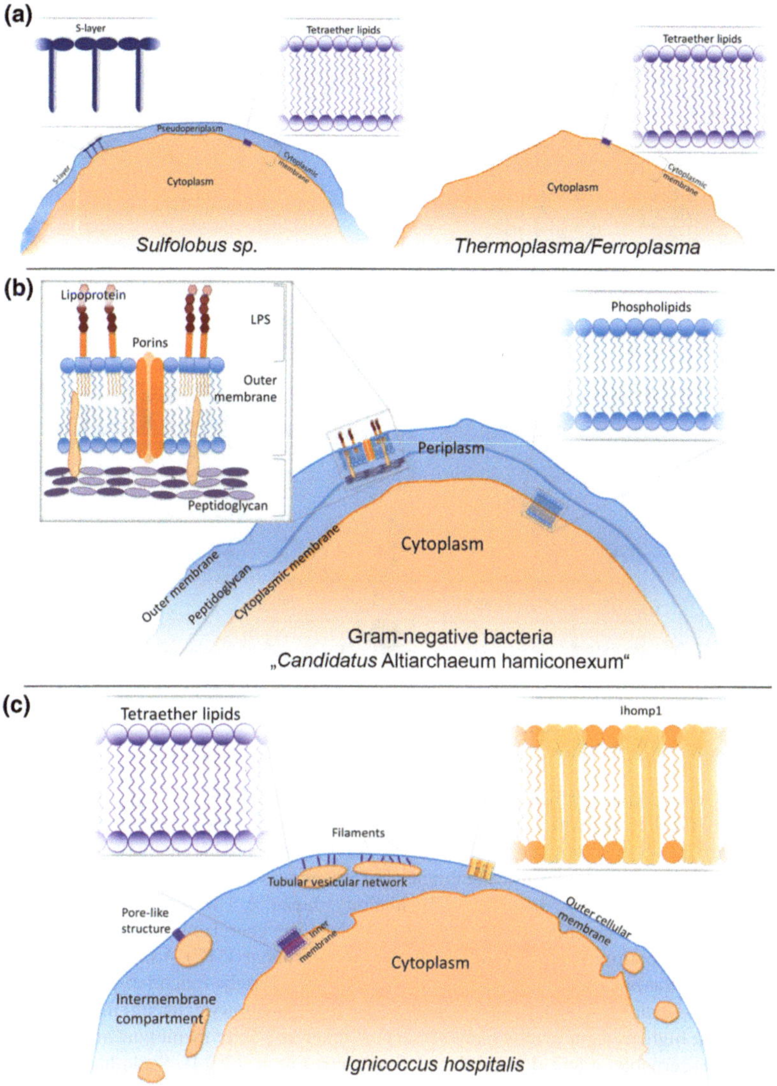

Fig. 14.1 Schematic drawing of different cell wall types. **a** As it could be shown for all *Sulfolobus* species so far, the majority of archaea described so far exhibit an S-layer as the only cell wall compound outside of the cytoplasmic membrane. In hyperthermophiles, the membrane often consists of tetraether lipids and less often diether lipids. **b** No cell wall at all could be found for *Thermoplasma* and *Ferroplasma.* The cytoplasm is surrounded by the cytoplasmic membrane alone, which is consisting of tetraether lipids. **c** A Gram-negative cell wall architecture with cytoplasmic membrane, periplasm containing the peptidoglycan and an outer membrane with lipopolysaccharide (LPS) from inside to outside is shown. It was thought to be restricted to bacteria but recent findings concerning "*Candidatus* Altiarchaeum hamiconexum" points to quite similar cell wall architecture in this archaeon (Probst et al. 2014). **d** *Ignicoccus hospitalis* represents a special case with an energized outer cellular membrane and a large intermembrane compartment. The inner membrane mainly consists of tetraether lipids. The major protein found in the outer membrane is Ihomp1

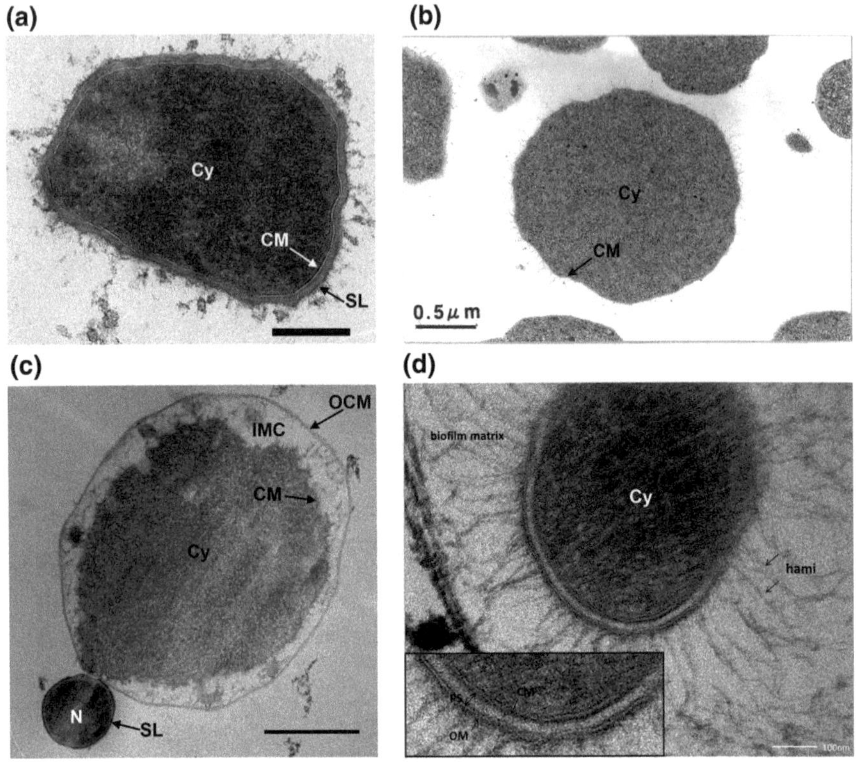

Fig. 14.2 Transmission electron micrographs of selected archaeal cell wall types. **a** The *Sulfolobus metallicus* cell reveals the typical S-layer, which is superimposing the cytoplasmic membrane and thereby forming a pseudoperiplasm. **b** *Thermoplasma acidophilum* does not have a cell wall and is surrounded by a single cytoplasmic membrane. The image was kindly provided by Tairo Oshima, Professor Emeritus of Tokyo Institute of Technology, Japan. **c** The special case of *Ignicoccus hospitalis*, depicting a cytoplasmic membrane (CM) and the energized outer cellular membrane (OCM), which are both enclosing the intermembrane compartment (IMC). A smaller *Nanoarchaeum equitans* cell at the bottom, which has an S-layer, is attached to the much larger *Ignicoccus* cell. **d** The image of "*Candidatus* Altiarchaeum hamiconexum" (SM1 euryarchaeon) illustrates a double membrane. At higher magnification (small inset) the inner and outer membrane and the periplasm are clearly discernable. **c** and **d** Reproduced from Perras et al. (2014). Scale bars: 100 nm (**d**), 200 nm (**a**) and 500 nm (**b**, **c**). CM, cytoplasmic membrane; Cy, Cytoplasm; IMC intermembrane compartment; N, *Nanoarchaeum equitans*; OCM, outer cellular membrane; SL, S-layer

Archaea with Double Membranes

Microorganisms exhibiting double membranes as outermost structures can be found in different phyla of the domain of the archaea. Interestingly, the double-membraned archaea described so far in the literature, all have in common the close interaction with other microorganisms (Klingl 2014; Perras et al. 2014). The architecture of double-membraned archaea is reminiscent of the cell envelope of Gram-negative

Table 14.1 Overview of the cell wall organization of archaea showing their major components and representative genera and species

Cell wall organization	Component	Organism
	SL CM	*Methanococcus*, *Halobacterium*, *Pyrodictium*, *Sulfolobus*, *Thermoproteus*
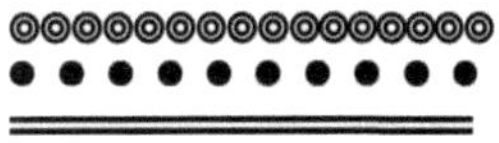	PS SL CM	*Methanospirillum*
	GC CM + LP	*Thermoplasma*
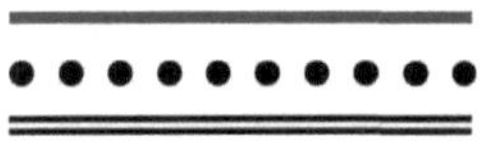	MC SL CM	*Methanosarcina*
	PM, HP, GG CM	*Methanobacterium*, *Methanosphaera*, *Methanobrevibacter*, *Halococcus*, *Natronococcus*
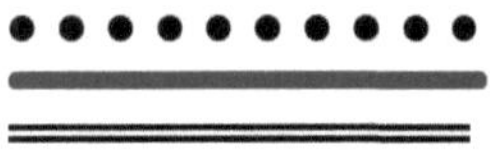	SL PM CM	*Methanothermus*, *Methanopyrus*
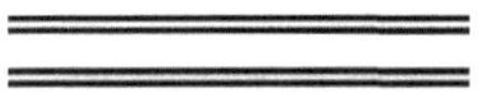	OCM CM	*Ignicoccus*, SM1 Euryarchaeon, ARMAN cells

CM cytoplasmic membrane, *GC* glycocalyx, *GG* glutaminylglycan, *HP* heteropolysaccharide, *LP* lipoglycans, *MC* methanochondroitin, *OCM* outermost cellular membrane or outer membrane, *PM* pseudomurein, *PS* protein sheath, *SL* S-layer. Based on Klingl (2014) and König (2001)

bacteria (Fig. 1c; Silhavy et al. 2010; Klingl 2014). The Gram-negative cell envelope briefly consists of three layers, namely the outer membrane, the peptidoglycan layer and the inner membrane (Silhavy et al. 2010). The outer membrane is a bilayer consisting of an inner leaflet containing phospholipids and an outer leaflet containing glycolipids (e.g. lipopolysaccharides) (Kamio and Nikaido 1976; Silhavy et al. 2010). In addition, the outer membrane contains two main classes of outer membrane proteins, which are mainly β-barrel proteins and lipoproteins (Silhavy et al. 2010). In contrast, the inner membrane forms a phospholipid bilayer (Silhavy et al. 2010). The periplasm is an aqueous compartment, which is enclosed by the inner- and the outer membrane (Mitchell 1961; Silhavy et al. 2010). In Gram-negative bacteria, the periplasmic space occupies approximately 10% of the total cell volume, and it contains a thin peptidoglycan layer, and soluble proteins such as hydrolytic enzymes. These are important for initial substrate degradation, binding proteins that are involved in transportation of substances, and receptors for chemotaxis (Ruiz et al. 2006; Klingl 2014). Although very similar at the first glance, archaeal double membranes are different from the Gram-negative cell envelope in regard to biochemical composition, and function of the different compartments. In archaea the space between inner and outer layer is referred to as pseudoperiplasmic space, filled with pseudo-periplasm. Since more and more archaea with double membranes have been described in the last years, it is unclear whether this cell wall anatomy is a general feature of many archaea, or it is an exception within the archaea (Perras et al. 2014).

The best-known example of double membrane archaea is the Crenarchaeon *Ignicoccus hospitalis*, a hyperthermophile growing optimally at 90 °C (Figs. 14.1d and 14.2c; Paper et al. 2007). It was isolated from a hydrothermal submarine vent system at the Kolbensey Ridge, situated north of Iceland (Paper et al. 2007). In the original sample, very tiny cocci were also found, directly attached to the cell surface of *Ignicoccus hospitalis*. These tiny cocci are referred to as *Nanoarchaeum equitans* (Paper et al. 2007). *N. equitans* is not able to grow in pure culture and is strictly dependent on its host *I. hospitalis* (Jahn et al. 2008). The attachment of *N. equitans* usually occurs in those areas on the surface of *I. hospitalis*, where the distance between the inner membrane (IM) and the outermost cellular membrane (OCM) has a minimum (Heimerl et al. 2017). In these areas the distance is just 20–50 nm compared to 200–500 nm on the opposite side of the host cell (Heimerl et al. 2017) *I. hospitalis* is a strictly anaerobic microorganism that uses elemental sulfur as a substrate. As boundaries, it features two membranes, namely an inner and an outer cellular membrane (Näther and Rachel 2004; Huber et al. 2012). The inner membrane consists of archaeol and caldarchaeol forming a tetraether monolayer, and the outer membrane consists of archaeol (Rachel et al. 2002; Burghardt et al. 2008). The most abundant protein of the outer membrane is Ihomp1 that forms oligomeric pore complexes (Burghardt et al. 2007). The pores formed by Ihomp1 have an inner diameter of about 2 nm and they may be involved in the interaction with *N. equitans* but no specific transport function could be demonstrated so far. A striking feature of *Ignicoccus* is the fact that the A_1A_O ATP synthase and the H_2:sulfur oxidoreductase are located in the outer cellular membrane, not at the inner membrane, leading to an energized outer membrane (Küper et al. 2010). Similar to Gram-negative bacteria, *Ignicoccus* exhibits a pseudoperiplasmic space between the two outer layers, which is, in this special case, termed inter-membrane compartment. The inter-membrane compartment occupies up to 40% of the total cell volume, and according to recent studies contains a complex tubular vesicular endomembrane system originating from the cytoplasm (Heimerl et al. 2017). The inner membrane system and the outer membrane system seem to interact via cylindrical shaped structures of an outer diameter of 100 nm and with a length of 50–100 nm underneath the outer cellular membrane (Heimerl et al. 2017). In addition, a filamentous matrix that putatively serves as cytoskeleton was described to be accommodated in the intermembrane compartment (Heimerl et al. 2017). Closely associated with the outermost cellular membrane (OCM), the acetyl-CoA-synthetase, catalyzing the first step of the carbon fixation pathway in *Ignicoccus hospitalis*, was the first enzyme to be located in the intermembrane compartment (Mayer et al. 2012). Given the ultrastructural data and the location of the archaeal ATP synthase, a structural and functional compartmentalization of the cell becomes apparent (Küper et al. 2010). DNA and ribosomes are located in the cytoplasm, while energy is generated at the outer cellular membrane (Küper et al. 2010; Huber et al. 2012). From these findings, it remains plausible that more metabolic processes might take place in the inter-membrane compartment. In addition to *I. hospitalis*, two further *Ignicoccus* species (*I. islandicus* and *I. pacificus*) and one other *Ignicoccus* isolate with unknown species have been described, so far, exhibiting the same ultrastructure and potentially the same structural and functional compartmentalized character (Huber et al. 2000, 2012).

Further archaea featuring a double membrane are the ARMAN archaea (archaeal Richmond Mine acidophilic nanoorganisms). ARMAN archaea are ultrasmall and low abundance microorganisms that thrive in acidic environments at a pH around 1.5 (Comolli et al. 2009). With cell volumes of 0.009–0.04 μm^3, they are close to the theoretical lower limit for life (Comolli et al. 2009). As *I. hospitalis*, some ARMAN archaea directly attach to representatives of the Thermoplasmatales via their double membranes (Comolli and Banfield 2014). 3D cryotomography analysis of biofilms containing ARMAN archaea revealed an inner and an outer membrane, but no S-layer (Comolli and Banfield 2014). Dark contrasted structures of 5–8 nm in diameter were detected in the periplasmic space, connected with the cytoplasm via the inner membrane (Comolli and Banfield 2014). In accordance with the predominant metabolisms of microorganisms in these environments, these structures could most probably be cytochromes (Comolli and Banfield 2014). Remarkably, an interspecies connection between ARMAN archaea and members of the Thermoplasmatales was observed to be mediated by "synapse-like" structures between the cytoplasmic spaces (Comolli and Banfield 2014).

The yet uncultivated archaeum "*Candidatus* Altiarchaeum hamiconnexum" (also known as SM1 Euryarchaeum) inhabits low temperature biotopes (Moissl et al. 2005; Perras et al. 2015). The archaeum grows in a biofilm, either as pure biofilm or in community with a bacterial sulfur reducer (Perras et al. 2014). Regarding ultrastructure, the SM1 Euryarchaeum is of coccoid shape with an average cell diameter of 0.6 μm (Probst et al. 2014). The cells also possess an inner and an outer membrane (5–6 nm thickness each) with a periplasmic space of 25 nm in between them (Fig. 14.2d; Henneberger et al. 2006; Perras et al. 2014). Neither particles, nor elongated structures were found in that compartment (Perras et al. 2014). Another remarkable ultrastructural feature of these archaea is their cell surface appendages that seem to span through both membranes (Perras et al. 2014). These pili-like structures are called "hami" (latin: anchor) and consist of a central and a distal part (Moissl et al. 2005). The central part contains barbwire-like prickles, and the distal part terminates in hook-shaped structure. Hundreds of these cell appendages were detected per cell, enabling the archaeum to adhere to surfaces, or to hami of other cells (Perras et al. 2014). Interestingly, the amino acid sequence of the major hamus subunit protein did not show similarities to any known proteins involved in cell appendage formation. Instead, it showed similarities to archaeal S-layer proteins (Perras et al. 2015).

Another representative having a double membrane is *Methanomassiliicoccus luminyensis*. The organism was isolated from human faeces, and is coccoid shaped with an average cell diameter of 850 nm. *M. luminyensis* is an obligatory anaerobic, and mesophilic microorganism. In addition to a cytoplasmic membrane, electron microscopy revealed an electron dense layer, most likely representing an outer membrane. A thick transparent layer was described, possibly outlining the periplasm (Dridi et al. 2012; Klingl 2014). Similar results with a cell wall showing two membranes could also be obtained for two other methanogenic isolates from termite guts, provisionally named "*Candidatus* Methanoplasma termitum" (Lang et al. 2015).

S-Layer

With the exception of archaea that are just surrounded by a cytoplasmic membrane without an additional cell wall (Figs. 14.1b and 14.2b), the surface layer or S-layer represents the simplest and most frequently found cell wall variation (Figs. 14.1a and 14.2a; Rodrigues-Oliveira et al. 2017).

The earliest studies dealing with S-layer proteins described these structures in *Halobacterium salinarum* and *Haloferax volcanii* (Houwink 1956; Mescher and Strominger 1976a,b; Lechner and Sumper 1987; Sumper et al. 1990; Sumper and Wieland 1995) or *Halococcus* (Brown and Cho 1970). In addition, hey could also be shown in methanogens like *Methanosarcina* (Kandler and Hippe 1977), the *Methanococcus* species *Methanococcus vannielii* and *Methanococcus thermolithotrophicus* (Koval and Jarrell 1987; Nußer and König 1987) and in *Methanothermus fervidus* (Kandler and König 1993; Kärcher et al. 1993). Interestingly, the first described glycoprotein in prokaryotes was the glycosylated S-layer protein of *H. salinarum* (Mescher and Strominger 1976a, b). Several other studies supported the fact that S-layer proteins are usually highly glycosylated with an abundance of sometimes more than 30% of the total protein mass caused by the glycan side chains (Kandler and König 1998; König et al. 2007; Veith et al. 2009; Albers and Meyer 2011; Klingl 2014). It is thereby thought that the high sugar content prevents the proteins from degradation and increases their stability in halophilic and otherwise extremophilic environments like high temperature and acidity (Fig. 14.3c, d; Yurist-Doutsch et al. 2008; Jarrell et al. 2014).

The S-layer protein automatically arranges in a 2-dimensional and regularly arranged pseudocrystal on the cell surface in an entropically driven process (Sleytr et al. 1988; Eichler 2003; Sleytr et al. 2007). It is the most abundant cell wall variant of all archaea described so far, especially within the Crenarchaea, but can also be found in various bacterial species including cyanobacteria (Rachel et al. 1997). This S-layer can be composed of one major (glyco)protein (large subunit), which also includes an amino acid stretch that forms the anchor in the membrane or pseudoperiplasmic murein. Sometimes, the S-layer is formed by two proteins where the anchor is represented the second smaller protein, the small subunit (Veith et al. 2009). The major S-layer protein, which can range in protein mass from 40 to 325 kDa, is forming a 2D pseudo-crystalline array with a distinct symmetry on the cell surface (Kandler and König 1985; Beveridge and Graham 1991; Baumeister and Lembcke 1992; Messner and Sleytr 1992; Kandler and König 1993; Sumper and Wieland 1995; König et al. 2007; Veith et al. 2009; Albers and Meyer 2011; Klingl et al. 2013; Klingl 2014). Together with the stalk-like anchoring structure, the S-layer array is superimposing

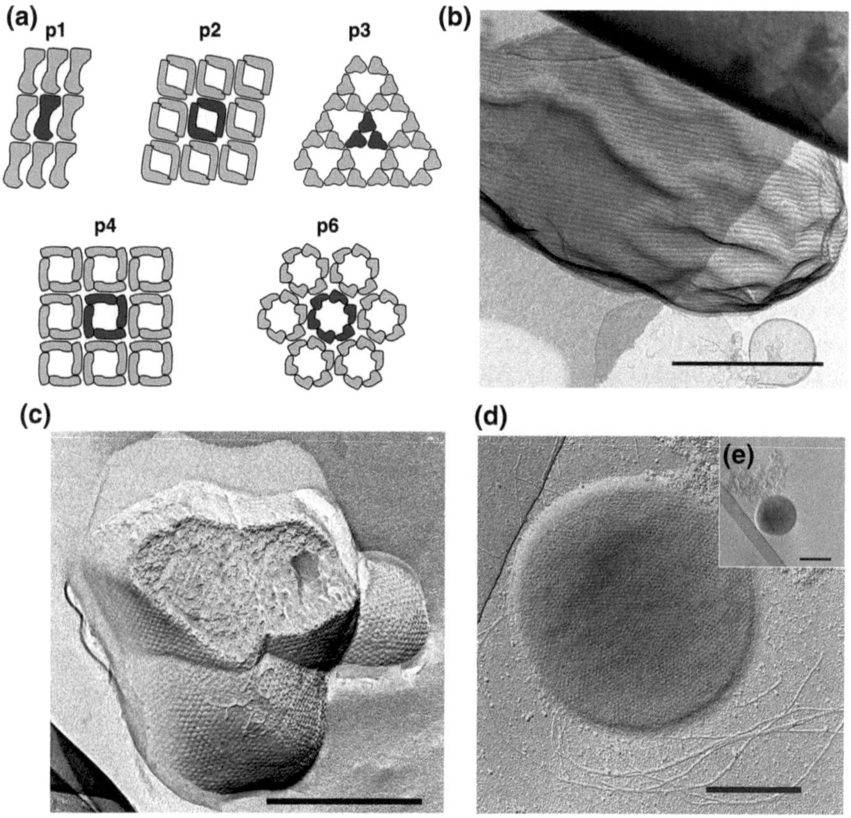

Fig. 14.3 Overview about S-layer and symmetry types. **a** The scheme illustrates the number of protein subunits that constitute the central crystal unit of an S-layer. There are either one, two, three, four or six, which lead to p1-, p2-, p3-, p4- or p6-symmetry, respectively. Reproduced from Klingl (2014). **b** The stripe-like p2-symmetry S-layer of the rod-shaped Gram-negative bacterium *Acidithiobacillus sp.* HV2/2 is shown. Reproduced from Klingl (2007). **c** Several freeze-fractured and freeze-etched *Nitrososphaera viennensis* cells lying close to each other shown. Two cells are cross-sectioned and the cytoplasm is exposed. This replica technique clearly depicts the regular S-layer pattern. *N. viennensis* is the only non-*Sulfolobus* archaeon described so far that has a p3-symmetry S-layer (Stieglmeier et al. 2014). **d** The S-layer of the coccoid *Ignisphaera aggregans*, a hyperthermophile growing optimally at 92 to 95 °C, has p6-symmetry. The cell also shows several cell appendages, most likely representing archaeal flagella. **e** Overview of the cell shown in (d). Scale bars: 500 nm (**b**–**d**) and 1000 nm (**e**)

the cytoplasmic membrane and thereby forming a quasi-periplasmic space between cytoplasmic membrane and S-layer in contrast to (pseudo)periplasm, which is the space between inner and outer membrane. This quasi-periplasm can be in extreme cases up to 70 nm wide as shown for *Staphylothermus marinus* (Peters et al. 1995, 1996). In the latter case, the long stalks that are anchoring the S-layer are called tetrabrachion.

The central crystal unit of S-layer arrays consists of two, three, four or six subunits, which leads to p2-, p3-, p4- or p6-symmetry, respectively (Fig. 14.3a; Klingl 2014). The lattice constants of the 2D crystals range between 11 and 30 nm (Messner and Sleytr 1992; König et al. 2007). The symmetry and center-to-center spacing were sometimes found to be genus specific, as it could be shown for the Sulfolobales (König et al. 2007; Klingl et al. 2011). All representatives that were described so far had glycosylated S-layers with p3-symmetry and lattice constants around 21 nm (König et al. 2007; Veith et al. 2009). There is just one documented case of an archaeon outside the Sulfolobales that is covered with a p3-symmetry S-layer: *Nitrososphaera viennensis* (Fig. 14.3c), which belongs to the recently described Thaumarchaeota (Stieglmeier et al. 2014). Something similar could be shown for p2-symmetry within the bacterial genus *Acidithiobacillus* (Fig. 14.3b; Klingl et al. 2011).

A unique character within many *Thermococcus* species is the fact, that many of them are not just surrounded by one S-layer but by a double S-layer (Miroshnichenko et al. 1998; Kostyukova et al. 1999; Atomi et al. 2004; Gorlas et al. 2014), e.g. *Thermococcus stetteri* shows two 5 nm thick layers of glycoproteins atop of each other, separated by a translucent 10–12 nm space (Gongadze et al. 1993; Rodrigues-Oliveira et al. 2017).

The high stability of the proteins against degradation like hydrolysis under different harsh environmental conditions directly leads to possible functions of S-layer proteins like the protection from the environment (Engelhardt and Peters 1998). For some crenarchaea like the Sulfolobales, the S-layer together with the cytoplasmic membrane has to withstand pH 2 and temperatures of 60–80 °C (Veith et al. 2009). This temperature and acid stability might be the result of a high content of charged amino acids and ionic interactions between cell wall moieties (Haney et al. 1999). As a prime example of stability, the S-layer proteins of *Thermoproteus tenax* and *Thermofilum pendens* have to be mentioned. The S-layer sacculus of these two organisms can be boiled at 100 °C for 30 min in 2% SDS (sodium dodecyl sulfate) without any hint of degradation (König and Stetter 1986; Wildhaber and Baumeister 1987; König et al. 2007). A clear counterexample to the high stability S-layer proteins of crenarchaea are the surface layers that can be found in most euryarchaea, e.g. *Archaeoglobus fulgidus*, which are labile and tend to disintegrate and are therefore difficult to isolate (König et al. 2007). In this regard, the euryarchaeon *Picrophilus* is an exception with a relatively stable S-layer protein, although the stability might just be a side effect of its resistance to acidity.

Further functions of S-layer proteins are the sustainment of cell integrity in alternating osmotic conditions (osmoprotection) as well as maintaining or at least supporting cell shape (Engelhardt 2007a, b; Klingl et al. 2011). Other potential functions of S-layers could be that they play a role as molecular sieves or that they are generating an additional cellular compartment similar to the periplasm in Gram-negative bacteria. The space between the cytoplasmic membrane and the umbrella-like envelope formed by the S-layer is therefore named quasiperiplasm. There is some evidence that the S-layer could be involved in cell attachment but it is more likely that this ability has to be attributed to cell appendages like flagella or pili like it could be shown for *Acidithiobacillus* (Klingl et al. 2011). Taking into account in which extreme environments like *Sulfolobus* species are thriving, it can even be speculated that the S-layer represents a natural protection layer against virus, prohibiting the attachment or entrance of such potentially dangerous particles.

Regarding the large amount of information on S-layer proteins of *Methanococcus* species (Akça et al. 2002), mesophilic and thermophilic archaea in particular (Claus et al. 2002) and S-layers in general (Claus et al. 2001, 2005; König et al. 2007; Albers and Meyer 2011), it is surprising that we are still lacking a clear proof for a distinct function of S-layer proteins. But the fact that S-layer proteins can make up more than 10% of the whole cellular protein is a strong indication that they are there for a certain reason.

Finally it should be mentioned, that cell appendages like hami might have more in common with S-layers than originally thought, although not crystallizing in a regular 2D lattice (Perras et al. 2015). This automatic arrangement in regular lattices is one of the major characteristics that defines a protein as S-layer. Otherwise, it is just a simple amorphous protein layer or sheath. Some obvious parallels exist between the S-layer stalks of *Staphylothermus marinus* and the hami of "*Candidatus* Altiarchaeum hamiconexum" (SM1 euryarchaeon) but the inability of the latter structure to form 2D crystals just corroborate these differentiation problems.

Cell Wall Structures of Methanogenic Archaea

Methanogens, which belong to the euryarchaeal kingdom, are divided into 5 orders: Methanobacteriales, Methanococcales, Methanomicrobiales, Methanosarcinales and Methanopyrales (Ferry and Kastead 2007). They represent a very diverse group within the archaeal domain, showing a great variety in phylogeny, physiology and morphology. This variety is also reflected in the structure and composition of the methanogenetic cell wall, including pseudomurein (Fig. 14.4a, c) and specific S-Layers as well as methanochondroitin (Fig. 14.4b) or proteinaceous sheaths (König et al. 2010).

Pseudomurein

The cell wall of most bacterial species contains the peptidoglycan murein, a polymer consisting of sugars and amino acids, forming a netlike layer outside the cell (Schleifer and Kandler 1972). Due to structural similarities like the construction of

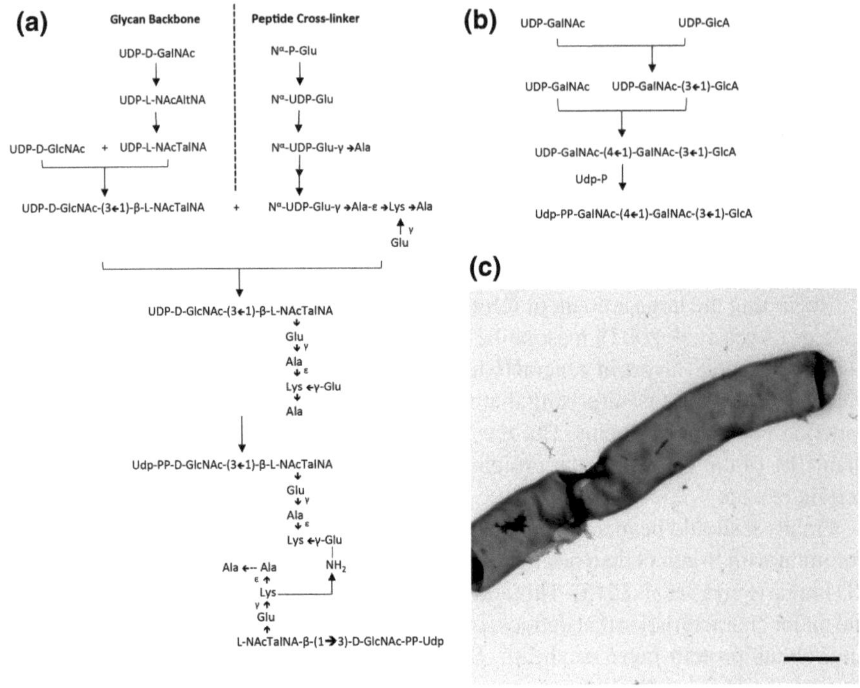

Fig. 14.4 **a** Proposed biosynthesis pathway of the archaeal peptidoglycan homologue pseudomurein according to König et al. (1994). It depicts the stepwise formation of the glycan backbone and the peptide cross-linker and ends up with the end product pseudomurein (bottom panel **a**). **b** Model of the biosynthesis pathway of the methanogenic cell wall compound methanochondroitin (bottom, panel **b**) via its building blocks UDP-GalNAc and UDP-GlcA (König et al. 1994). **c** The rod shaped pseudomurein sacculus of *Methanothermobacter thermautotrophicus*. Due to its high stability against detergents and harsh mechanical treatment, it remains intact after complete degradation of other cell wall components. Udp, undecaprenyl. Scale bar: 500 nm (**c**)

approximately 50% glycan and 50% peptide, the cell wall of Methanobacteriales and Methanopyrales was named pseudomurein (Fig. 14.3a; Kandler and König 1978). This term may be distracting, although there are similarities between these two structures. Actually, a common origin of murein and pseudomurein is unlikely, as the datasets of all pseudomurein expressing methanogens showed no indication for homologue genes or pathways that would indicate this (Hartmann and König 1990). Together with the fact that both show differences in cell wall chemistry, a convergent evolution of both polymers seems more likely (König et al. 2010; Steenbakkers et al. 2006).

Pseudomurein is one of the largest known biomolecules found in nature, forming huge molecules that enfold the whole cell with a thickness of approximately 15–20 nm (Kandler and König 1993). It is responsible for the maintenance of the cell shape and may also have protective functions because of its high stability.

Pseudomurein is built up from chains of disaccharides that are cross-linked via peptide subunits. In general, the glycan strand is built up by alternating units of β(1→3)-linked N-acetyl-D-glucosamine and N-acetyl-L-talosaminuronic acid. The glycan strands their selves are, in turn, cross-linked by peptide subunits that are composed of L-amino acids (in most cases glutamic acid, alanine, lysine) exclusively. The lack of D-amino acids, the use of N-acetyl-L-talosaminuronic acid instead of N-acetyl-muramic acid, the β(1→4)-linkage to N-acetylglucosamine and the increased occurrence of α- and ε-peptide bonds represent the main differences to the bacterial murein (Leps et al. 1984a,b; Kandler and König 1978, König and Kandler 1979; König et al. 1983). As the linkage between the two glycan components is β(1→3) there is the possibility of a replacement of N-acetyl-glucosamine by N-acetyl-galactosamine.

The first step in the biosynthesis of the glycan moiety is the epimerization and oxidation of UDP-N-acetyl-D-galactosamine to UDP-N-acetyl-L-talosaminuronic acid via UDP-N-acetyl-D-altrosaminuroic acid as an intermediate. In a next step UDP-N-acetyl-D-glucosamine is linked to UDP-N-acetyl-L-talosaminuronic acid via β(1→3)-glycosidic linkage, forming a disaccharide as repeating unit of the glycan strand. Simultaneously, an UDP-activated pentapeptide is formed which derives from N^{α}-P-glutamic acid that is converted to N^{α}-UDP-glutamic acid (Fig. 14.3a). After this, the amino acids alanine, lysine and glutamate are linked to N^{α}-UDP-glutamic acid at the expense of ATP before the pentapeptide is linked to the disaccharide (Hartmann et al. 1990). The resulting molecule is transferred to undecaprenylphosphate and added to the growing glycan strand.

It is an interesting fact that, to date, no archaeal or bacterial enzyme is known that cleaves the β(1→3) glycosidic bond of pseudomurein, whereas lots of bacterial cell wall hydrolases are able to break the β(1→4) glycosidic bond of murein. There are just two archaeal enzymes known that are able to split the peptide linkages of pseudomurein, the pseudomurein isopeptidases PeiW and PeiP, cleaving the isopeptide bond between L-lysine and L-alanine (Kiener et al. 1987; Luo et al. 2002; Visweswaran et al. 2010).

Methanochondroitin

Methanochondroitin is exclusively found in aggregated cells of *Methanosarcina* spp. where the methanochondroitin layer represents the shape maintaining structure (Kreisl and Kandler 1986). By increasing osmolarity, concentrations of divalent cations like Ca^{2+} or Mg^{2+} or during nutrient deficiencies it is possible to completely disaggregate the cells, with *Methanosarcina barkeri* showing for example a 20-fold decrease of glucuronic acid, a major component of the methanochondroitin layer (Sowers et al. 1993; Boone and Mah 1987). However, methanochondroitin is not the only cell wall structure in *Methanosarcina*, as some species also possess an additional S-Layer between their cytoplasmic membrane and the methanochondroitin layer (Francoleon et al. 2009; Arbing et al. 2012).

Structurally, methanochondroitin is built from proteoglycans, with the trimer [→)β-D-GlcA-(1→3)-β-D-GalNAc-(1→4)-β-D-GalNAc-(1→]$_n$ as the fundamental glycan structure (Fig. 14.3b). This structure is arranged to a fibrillar polymer and builds up an either compact or loose matrix (Kreisl and Kandler 1986). Because of its similarity to the eukaryotic chondroitin sulfate, a main component of cartilage-tissue, this structure was named methanochondroitin, although the structures differ in two main features compared to chondroitin: the absence of sulphate residues in methanochondroitin and a different molar ratio of GalNAc to GlcA (Levene and La Forge 1913). Similar structures can also be found within the bacteria, for example the cell wall component teichuronic acid [D-GlcA-(1→3)-GalNAc-(1→4)-]$_{23–25}$ in *Bacillus licheniformis* (Hughes and Thurman 1970).

In the first step, UDP-GalNAc and UDP-GlcA form UDP-N-acetylchondrosine that forms with an additional unit of UDP-GalNAc the repeating unit of methanochondroitin. This is linked to undecaprenol, a lipid carrier transferring the trisaccharide trough the cytoplasmic membrane where the polymerization of the growing glycan chain takes place (Hartmann and König 1991).

Proteinaceous Sheaths

One of the most striking cell wall structures within the methanogens are the proteinaceous sheaths found in *Methanospirillum hungatei* and *Methanosaeta concilii*. Both organisms form long filamentous chains, consisting of single cells arranged one after another (Zeikus and Bowen 1975; Beveridge et al. 1986). Each individual cell is surrounded by an inner cell wall, which is also involved in septal formation. An amorphous granular layer builds the inner cell wall in *M. concilii*, whereas in *M. hungatei* this inner cell wall was identified as an S-Layer showing a p6-symmetry (Firtel et al. 1993). Two unglycosylated proteins, a 110 kDa and a 114 kDa protein, form this S-Layer. There is another thin layer covering the plasma membrane, which has no influence on cell shape, as cells lose their rod-shape character when the outer sheath was removed (Beveridge et al. 1986; Sprott et al. 1979).

Besides the inner cell wall, *M. hungatei* and *M. concilii* possess an additional outermost sheath, arranged in a paracrystalline manner that surrounds the whole arrangement of cells. This outermost layer is not a coherent structure and therefore not a classical cell wall in a strict sense, as it is interrupted by so called spacer plugs that isolate the individual cells from each other. It was shown that the sheaths are built by stacked hoop-like structures, which can be separated under reducing conditions (Sprott et al. 1986; Patel et al. 1986).

The outermost sheath of *M. hungatei* and *M. concilii* can be isolated by classical cell wall isolation techniques and display a high stability against detergents, common proteases and chaotropic agents. In *M. hungatei*, the isolated sheath is built by amino acids and neutral sugars, whereas *M. concilii* showed a more complex amino acid pattern with glucose and mannose as the predominate sugar components (Sprott and McKellar 1980; Kandler and König 1978). Southam and Beveridge expanded the knowledge on the sheath composition of *M. hungatei* by the discovery of two different polypeptide groups: one that is soluble in sodium dodecyl sulfate (SDS)-β-mercaptoethanol (β-ME)-EDTA and includes polypeptides with masses from 10 to 40 kDa (building 74% of sheath's mass), and a phenol-soluble group representing approximately 19% of the sheath's mass. Therefore, they concluded a trilaminar structure of the sheath with phenol-soluble polypeptides that are surrounded by SDS-β-ME-EDTA soluble polypeptides. In this structure, the phenol-soluble polypeptides should convey rigidity to cells whereas the SDS-β-ME-EDTA soluble ones are necessary for the sheaths resilience (Southam and Beveridge 1991, 1992).

The outermost sheaths serve as a micro sieve and are responsible for the rod-shaped morphology. Moreover, they provide protection against protozoan grazing and the cellular turgor pressure while allowing the release of products like methane or the uptake of substances like H_2, CO_2 or acetate (Walsby 1994).

In 2015, Dueholm et al. could demonstrate that the sheath of *Methanosaeta thermophila* is morphologically similar to those of *M. hungatei* and *M. concilii*. They identified MspA (~60 kDa) as the major sheath protein. It was the first description of an archaeal with an amyloidogenic character and could therefore explain the enormous resistance of the sheath coincidentally with elastic properties that allow diffusible substrates to penetrate.

Two layered cell walls: *Methanothermus* and *Methanopyrus*
In transmission electron micrographs of the hyperthermophilic methanogens, *Methanothermus fervidus* and *Methanopyrus kandleri*, a double-layered cell envelope is visible (König et al. 2007). In *M. fervidus*, it consists of a pseudomurein layer covered by an S-layer with 6-fold symmetry (König et al. 2007). The arrangement in *M. kandleri* is quite similar although the outer layer does not reveal regular lattice pattern and is therefore not described as a typical S-layer (König et al. 2007).

Further Types of Cell Envelopes and Connected Elements

The cell envelope of *Halococcus morrhuae* is an inflexible, electron dense layer of 50–60 nm in width (Kocur et al. 1972). Biochemical investigations showed the cell wall to be composed of a highly sulfated heteropolysaccharide including glucose, mannose, galactose, glucuronic acid, galacturonic acid, glucosamine, and galosamunronic acid (Steber and Schleifer 1975; Schleifer et al. 1982; König et al. 2010). In addition, this heteropolysaccharide contains N-acetylated amino sugars (Schleifer et al. 1982; König et al. 2010). Furthermore, the cell wall was proposed to contain glycine that bridges amino groups of glucosamines and the carboxyl group of uronic or gulosaminouronic acid residues in the glycan strands (Steber and Schleifer 1979; König et al. 2010). A similar cell wall composition, but with different molar ratios of the cell wall components, was described for *Halococcus salifodinae*, another halophilic Archaeum isolated from a Permian rock salt deposit in Austria (Legat et al. 2013).

Natronococcus occultus is a haloalkalophilic organism growing optimally at pH values of 8.5–10 and in hypersaline environments containing 3.5 M salt (Tindall et al. 1984; König et al. 2010). In accordance with its extreme environment, the archaeum possesses a very unique cell envelope among archaea composed of a glutaminylglycan polymer (König et al. 2010; Albers and Meyer 2011). Similar cell wall polymers can be found among the *Bacillus* species *Sporosarcina halophile* and *Planococcus halophilus* (Niemetz et al. 1997; Klingl 2014). In contrast to bacteria, the archaeal version of the polymer is glycosylated, consisting of about 60 monomers with the glutamic acid residues linked via a ϒ-carboxylic group (König et al. 2007; Klingl 2014). Furthermore, the cell wall polymer of *Natronococcus occultus*, which has no equivalent in bacteria, comprises L-glutamate, N-acetyl-D-glucosamine, N-acetyl-D-galactosamine, D-galacturonic acid, D-glucuronic acid, and D-glucose in a ratio of 5:7:1:8:0.5:0.3 (Niemetz et al. 1997; König et al. 2010).

Haloquadratum walsbyi, a member of the Halobacteriales exhibits one (strain C23), or two (strain HBSQ001) S-layers, depending on the strain (Burns et al. 2007). The organism inhabits extremely salty environments. According to genomic data, the cells were predicted to develop a poly-ϒ-glutamate capsule (Bolhuis et al. 2006; Albers and Meyer 2011). A specific characteristic of H. walsbyi is the highly unusual cell shape: almost perfect quadrate flattened cells (Bolhuis et al. 2006). Another feature is the secretion of halomucin, a very large protein (927 kDa), which is highly similar to mammalian mucin (Hollingsworth and Swanson 2004; König et al. 2010). It was suggested to surround the cells as an aqueous shield and to protect the cells against their extremely salty environmental conditions (König et al. 2010). Furthermore, it might play a role in defense against halophages (Zenke et al. 2015). Recently, it was shown that halomucin is in fact only loosely associated with the cells, instead of totally surrounding the cells as a closed outermost layer (Zenke et al. 2015). Despite this loose association, it was proposed that halomucin could still fulfill its protecting functions (Zenke et al. 2015).

In addition, it should be mentioned that cell walls like the S-layer can be in contact and interact with other proteins. One such example is the STABLE protease of *Staphylothermus marinus*, which is associated with the long stalks of the S-layer (Peters et al. 1995). The so-called bindosome was described as a component of the cell envelope of *Sulfolobus solfataricus* was described, representing a major complex, which is involved in sugar binding and uptake in the thermoacidophilic archaeon and apparently interacts with the S-layer of *S. solfataricus* (Zolghadr et al. 2011). It is likely that further extracellular enzymes and metabolism-relevant proteins that are associated with the S-layer or another cell wall component (Matuschek et al. 1994) and will be found and described in detail in the future.

Conclusion

This overview of archaeal cell wall types shows its great variety and sometimes also complexity. It often seems to be specifically adapted to the respective environment like the cell walls of halophilic, thermophilic or acidophilic archaea. The high stability of the respective cell wall compound under the usual growth condition of these archaea seems to be a prerequisite for their adaptation. Future experiments will therefore deal with the question, how the cell wall of the first living cells on earth might have looked like. This will not only deliver an answer, which archaeal cell wall architecture developed first but it will also give us a better understanding of the origin of life in general.

Acknowledgements We wish to thank Dr. Michaela Stieglmeier, Prof. Dr. Christa Schleper, Prof. Dr. Reinhard Rachel and Prof. Dr. Tairo Oshima for their scientific input and Marion Debus, Cornelia Niemann and Jennifer Grünert for technical assistance.

References

Akça E, Claus H, Schultz N, Karbach G, Schlott B, Debaerdemaeker T, Declercq J-P, König H (2002) Genes and derived amino acid sequences of S-layer proteins from mesophilic, thermophilic and extremely thermophilic methanococci. Extremophiles 6:351–358. https://doi.org/10.1007/s00792-001-0264-1

Albers S-V, Meyer BH (2011) The archaeal cell envelope. Nat Rev Microbiol 9:414–426

Arbing MA, Chan S, Shin A, Phan T, Ahn CJ, Rohlin L, Gunsalus RP (2012) Structure of the surface layer of the methanogenic archaeon *Methanosarcina acetivorans*. PNAS 109:11812–11817

Atomi H, Fukui T, Kanai T, Morikawa M, Imanaka T (2004) Description of *Thermococcus kodakaraensis* sp. nov., a well studied hyperthermophilic archaeon previously reported as *Pyrococcus sp.* KOD1. Archaea 1:263–267. https://doi.org/10.1155/2004/204953

Baumeister W, Lembcke G (1992) Structural features of archaebacterial cell envelopes. J Bioenerg Biomembr 24:567–575

Beveridge TJ, Patel GB, Harris BJ, Sprott GD (1986) The ultrastructure of *Methanothrix concilii*, a mesophilic aceticlastic methanogen. Can J Microbiol 32:703–710

Beveridge TJ, Graham L (1991) Surface layers of bacteria. Microbiol Mol Biol Rev 55:684–705

Blöchl E, Rachel R, Burggraf S, Hafenbradl D, Jannasch HW, Stetter KO (1997) *Pyrolobus fumarii*, gen. and sp-nov., represents a novel group of archaea extending the upper temperature limit for life to 113 degrees C. Extremophiles 1(1):14–21

Bolhuis H, Palm P, Wende A, Falb M, Rampp M, Rodriguez-Valera F, Pfeiffer F, Oesterhelt D (2006) The genome of the square archaeon *Haloquadratum walsbyi*: life at the limits of water activity. BMC Genom 7:169

Boone DR, Mah RA (1987) Effects of calcium, magnesium, pH, and extent of growth on the morphology of *Methanosarcina barkeri* S-6. Appl Environ Microbiol 53:1699–1700

Brown AD, Cho KJ (1970) The cell walls of extremely halophilic cocci. Gram-positive bacteria lacking muramic acid. J Gen Microbiol 62:267–270. https://doi.org/10.1099/00221287-62-2-267

Burghardt T, Näther DJ, Junglas B, Huber H, Rachel R (2007) The dominating outer membrane protein of the hyperthermophilic archaeum *Ignicoccus hospitalis*: a novel pore-forming complex. Mol Microbiol 63:166–176

Burghardt T, Saller M, Gürster S, Müller D, Meyer C, Jahn U, Hochmuth E, Deutzmann R, Siedler F, Babinger P, Wirth R, Huber H, Rachel R (2008) Insight into the proteome of the hyperthermophilic

crenarchaeon *Ignicoccus hospitalis*: the major cytosolic and membrane proteins. Arch Microbiol 190:379–394
Burns DG, Janssen PH, Itoh T, Kamekura M, Li Z, Jensen G, Rodriguez-Valera F, Bolhuis H, Dyall-Smith ML (2007) *Haloquadratum walsbyi* gen. nov., sp. nov., the square haloarchaeon of Walsby, isolated from saltern crystallizers in Australia and Spain. IJSEM 57:387–392
Claus H, Akça E, Schultz N, Karbach G, Schlott B, Debaerdemaeker T, Declercq J-P, König H (2001) Surface (glyco-)proteins: primary structure and crystallization under microgravity conditions. In: Proceedings of the first European workshop on exo-/astro-biology, ESA SP-496, Frascati, pp 806–809
Claus H, Akça E, Debaerdemaeker T, Evrard D, Declercq JP, König H (2002) Primary structure of selected archaeal mesophilic and extremely thermophilic outer surface layer proteins. Syst Appl Microbiol 25:3–12
Claus H, Akça E, Debaerdemaeker T, Evrard D, Declercq JP, Robin Harris J, Schlott B, König H (2005) Molecular organization of selected prokaryotic S-layer proteins. Can J Microbiol 51:731–743
Comolli LR, Baker BJ, Downing KH, Siegerist CE, Banfield JF (2009) Three-dimensional analysis of the structure and ecology of a novel, ultra-small archaeon. ISME J 3:159–167
Comolli LR, Banfield JF (2014) Inter-species interconnections in acid mine drainage microbial communities. Front Microbiol 5:367
Darland G, Brock TD, Samsonoff W, Conti SF (1970) A thermophilic, acidophilic mycoplasma isolated from a coal refuse pile. Science 170:1416–1418
Dridi B, Fardeau M-L, Ollivier B, Raoult D, Drancourt M (2012) *Methanomassiliicoccus luminyensis* gen. nov., sp. nov., a methanogenic archaeon isolated from human faeces. IJSEM 62:1902–1907
Dueholm MS, Larsen P, Finster K, Stenvang MR, Christiansen G, Vad BS, Bøggild A, Otzen DE, Nielsen PH (2015) The tubular sheaths encasing *Methanosaeta thermophila* filaments are functional amyloids. J Biol Chem 290(33):20590–20600. https://doi.org/10.1074/jbc.M115.654780
Eichler J (2003) Facing extremes: archaeal surface-layer (glyco)proteins. Microbiology 149:3347–3351
Engelhardt H, Peters J (1998) Structural research on surface layers: a focus on stability, surface layer homology, and surface layer-cell wall interactions. J Struct Biol 124:276–302
Engelhardt H (2007a) Are S-layers exoskeletons? The basic function of protein surface layers revisited. J Struct Biol 160:115–124
Engelhardt H (2007b) Mechanism of osmoprotection by archaeal S-layers: a theoretical study. J Struct Biol 160:190–199
Ferry JG, Kastead KA (2007) Methanogenesis. In: Cavicchioli R (ed) Archaea—molecular and cell biology. ASM Press, Washington, DC, pp 288–314
Firtel M, Southam G, Harauz G, Beveridge TJ (1993) Characterization of the cell wall of the sheathed methanogen *Methanospirillum hungatei* Gp1 as an S-layer. J Bacteriol 175:7550–7560
Francoleon DR, Boontheung P, Yang Y, Kim U, Ytterberg AJ, Denny PA, Denny PC, Loo JA, Gunsalus RP, Ogorzalek Loo RR (2009) S-layer surface-accessible and Concanavalin A binding proteins of *Methanosarcina acetivorans* and *Methanosarcina mazei*. J Proteome Res 8:1972–1982
Golyshina OV, Timmis KN (2005) *Ferroplasma* and relatives, recently discovered cell wall-lacking archaea making a living in extremely acid, heavy metal-rich environments. Environ Microbiol 7:1277–1288
Gongadze GM, Kostyukova AS, Miroshnichenko ML, Bonch-Osmolovskaya EA (1993) Regular proteinaceous layers of *Thermococcus stetteri* cell envelope. Curr Biol 27:5–9
Gorlas A, Croce O, Oberto J, Gauliard E, Forterre P, Marguet E (2014) *Thermococcus nautili* sp. nov., a hyperthermophilic archaeon isolated from a hydrothermal deep-sea vent. Int J Syst Evol Microbiol 64:1802–1810. https://doi.org/10.1099/ijs.0.060376-0
Gulik A, Luzzati V, DeRosa M, Gambacorta A (1988) Tetraether lipid components from a thermoacidophilic archaebacterium. Chemical structure and physical polymorphism. J Mol Microbiol 201:429–435

Haney PJ, Badger JH, Buldak GL, Reich CI, Woese CR, Olsen GJ (1999) Thermal adaption analyzed by comparison of protein sequences from mesophilic and extremely thermophilic *Methanococcus* species. Proc Natl Acad Sci USA 36:3578–3583

Hartmann E, König H (1990) Comparison of the biosynthesis of the methanobacterial pseudomurein and the eubacterial murein. Naturwissenschaften 77:472–475. https://doi.org/10.1007/BF01135923

Hartmann E, König H, Kandler O, Hammes W (1990) Isolation of nucleotide activated amino acid and peptide precursors of the pseudomurein of *Methanobacterium thermoautotrophicum*. FEMS Microbiol Lett 69:271–276

Hartmann E, König H (1991) Nucleotide-activated oligosaccharides are intermediates of the cell wall polysaccharide of *Methanosarcina barkeri*. Biol Chem. Hoppe-Sleyer 372:971–974

Heimerl T, Flechsler J, Pickl C, Heinz V, Salecker B, Zweck J, Wanner G, Geimer S, Samson RY, Bell SD, Huber H, Wirth R, Wurch L, Podar M, Rachel R (2017) A complex endomembrane system in the archaeon *Ignicoccus hospitalis* tapped by *Nanoarchaeum equitans*. Front Microbiol 8:1072

Henneberger R, Moissl C, Amann T, Rudolph C, Huber R (2006) New insights into the lifestyle of the cold-loving SM1 euryarchaeon: natural growth as a monospecies biofilm in the subsurface. Appl Environ Microbiol 72:192–199

Hollingsworth MA, Swanson BJ (2004) Mucins in cancer: protection and control of the cell surface. Nat Rev Cancer 4:45–60

Houwink AL (1956) Flagella, gas vacuole sand cell-wall structure in *Halobacterium halobium*; an electron microscope study. J Gen Microbiol 15:146–150. https://doi.org/10.1099/00221287-15-1-146

Huber H, Burggraf S, Mayer T, Wyschkony I, Rachel R, Stetter KO (2000) *Ignicoccus* gen. nov., a novel genus of hyperthermophilic, chemolithoautotrophic archaea, represented by two new species, *Ignicoccus islandicus* sp. nov. and *Ignicoccus pacificus* sp. nov. IJSEM 50(6):2093–2100

Huber H, Küper U, Daxer S, Rachel R (2012) The unusual cell biology of the hyperthermophilic crenarchaeon *Ignicoccus hospitalis*. Antonie Van Leeuwenhoek 102:203–219

Hughes RC, Thurman PF (1970) Some structural features of the teichuronic acid of *Bacillus licheniformis* N.C.T.C. 6346 cell walls. Biochem J 117:441–449

Jahn U, Gallenberger M, Paper W, Junglas B, Eisenreich W, Stetter KO, Rachel R, Huber H (2008) *Nanoarchaeum equitans* and *Ignicoccus hospitalis*: new insights into a unique, intimate association of two archaea. J Bacteriol 190:1743–1750

Jarrell KF, Ding Y, Meyer BH, Albers SV, Kaminski L, Eichler J (2014) N-linked glycosylation in Archaea: a structural, functional, and genetic analysis. Microbiol Mol Biol Rev 78(2):304–41. https://doi.org/10.1128/MMBR.00052-13

Kamio Y, Nikaido H (1976) Outer membrane of *Salmonella typhimurium*: accessibility of phospholipid head groups to phospholipase c and cyanogen bromide activated dextran in the external medium. Biochemistry 15:2561–2570

Kandler O, Hippe H (1977) Lack of peptidoglycan in the cell walls of *Methanosarcina barkeri*. Arch Microbiol 113:57–60. https://doi.org/10.1007/BF00428580

Kandler O, König H (1978) Chemical composition of the peptidoglycan-free cell walls of methanogenic bacteria. Arch Microbiol 118:141–152

Kandler O, König H (1985) Cell envelopes of archaebacteria. In: Woese CR, Wolfe RS (eds) The bacteria. A treatise on structure and function. Archaebacteria, vol VIII. Academic Press, New York, pp 413–457

Kandler O, König H (1993) Cell envelopes of archaea: structure and chemistry. In: Kates M et al (eds) The biochemistry of archaea (Archaebacteria). Elsevier Science Publ. B.V., pp 223–259

Kandler O, König H (1998) Cell wall polymers in archaea (Archaebacteria). Cell Mol Life Sci 54(4):305–308

Kärcher U, Schröder H, Haslinger E, Allmeier G, Schreiner R, Wieland F, Haselbeck H, König H (1993) Primary structure of the heterosaccharide of the surface glycoprotein of *Methanothermus fervidus*. J Biol Chem 268:26821–26826

Kiener A, König H, Winter J, Leisinger T (1987) Purification and use of *Methanobacterium wolfeii* pseudomurein endopeptidase for lysis of *Methanobacterium thermoautotrophicum*. J Bacteriol 169:1010–1016

Klingl A (2007) Ultrastrukturelle Aspekte Pyrit-oxidierender Mikroorganismen. Diploma thesis, Institute for Microbiology, University of Regensburg

Klingl A (2014) S-layer and cytoplasmic membrane—exceptions from the typical archaeal cell wall with a focus on double membranes. Front Microbiol 5:624

Klingl A, Flechsler J, Heimerl T, Rachel R (2013) Archaeal cells. In: eLS. John, Chichester. https://doi.org/10.1002/9780470015902.a0000383.pub2

Klingl A, Moissl-Eichinger C, Wanner G, Zweck J, Huber H, Thomm M, Rachel R (2011) Analysis of the surface proteins of Acidithiobacillus ferrooxidans strain SP5/1 and the new, pyrite-oxidizing Acidithiobacillus isolate HV2/2, and their possible involvement in pyrite oxidation. Arch Microbiol 193(12):867–82. https://doi.org/10.1007/s00203-011-0720-y

Kocur M, Smid B, Martinec T (1972) The fine structure of extreme halophilic cocci. Microbios 5:101–107

König H, Kandler O (1979) The amino acid sequence of the peptide moiety of the pseudomurein from *Methanobacterium thermoautotrophicum*. Arch Microbiol 121:271–275

König H, Kandler O, Jensen M, Rietschel ET (1983) The primary structure of the glycan moiety of the pseudomurein from *Methanobacterium thermoautotrophicum*. Hoppe-Sleyers Z Physiol Chem 364:627–636

König H, StetterK O (1986) Studies on archaebacterial S-layers. Syst Appl Microbiol 7:300–309

König H, Hartmann E, Kärcher U (1994) Pathways and principles of the biosynthesis of methanobacterial cell wall polymers. Syst Appl Microbiol 16:510–517

König H (2001) Archaeal cell walls. eLS. https://doi.org/10.1038/npg.els.0000384

König H, Rachel R, Claus H (2007) Proteinaceous surface layers of archaea: ultrastructure and biochemistry. In: Cavicchioli R (ed) Archaea: molecular and cell biology

König H, Claus H, Varma A (eds) (2010) Cell envelopes of methanogens. Procaryotic cell wall compounds. Springer, Berlin, pp 231–251

Kostyukova AS, Gongadze GM, Polosina YY, Bonch-Osmolovskaya EA, Miroshnichenko ML, Chernyh NA, Obraztsova MV, Svetlichny VA, Messner P, Sleytr UB, L'Haridon S, Jeanthon C, Prieur D (1999) Investigation of structure and antigenic capacities of Thermococcales cell envelopes and reclassification of "*Caldococcus litoralis*" Z-1301 as *Thermococcus litoralis* Z-1301. Extremophiles 3:239–245. https://doi.org/10.1007/s007920050122

Koval S, Jarrell K (1987) Ultrastructure and biochemistry of the cell wall of *Methanococcus voltae*. J Bacteriol 169:1298–1306

Kreisl P, Kandler O (1986) Chemical structure of the cell wall polymer of *Methanosarcina*. Syst Appl Microbiol 7:293–299. https://doi.org/10.1016/S0723-2020(86)80022-4

Küper U, Meyer C, Müller V, Rachel R, Huber H (2010) Energized outer membrane and spatial separation of metabolic processes in the hyperthermophilic archaeon *Ignicoccus hospitalis*. PNAS 107:3152–3156

Lang K, Schuldes J, Klingl A, Daniel R, Brune A (2015) Comparative genome analysis of "*Candidatus* Methanoplasma termitum" indicates a new mode of energy metabolism in the seventh order of methanogens. Appl Environ Microbiol 81(4):1338–1352. https://doi.org/10.1128/AEM.03389-14

Langworthy TA, Smith PF, Mayberry WR (1972) Lipids of *Thermoplasma acidophilum*. J Bacteriol 112:1193–1200

Lechner J, Sumper M (1987) The primary structure of a prokaryotic glycoprotein. Cloning and sequencing of the cell surface glycoprotein gene of halobacteria. J Biol Chem 262:9724–9729

Legat A, Denner EBM, Dornmayr-Pfaffenhuemer M, Pfeiffe P, Knopf B, Claus H, Gruber C, König H, Wanner G, Stan-Lotter H (2013) Properties of *Halococcus salifodinae*, an isolate from permian rock salt deposits, compared with halococci from surface waters. Life 3:244–259

Leps B, Barnickel G, Bradaczek H (1984a) Structural studies on the bacterial cell wall peptidoglycan pseudomurein. I. Conformational energy calculations on the glycan strands in C1 conformation and comparison with murein. J Theor Biol 107:85–114
Leps B, Labischinski H, Barnickel G, Bradaczek H, Giesbrecht P (1984b) A new proposal for the primary and secondary structure of the glycan moiety of pseudomurein. Conformational energy calculations on the glycan strand with talosaminuronic acid in 1C conformation and comparison with murein. Eur J Biochem 144:279–286
Levene PA, La Forge FB (1913) On chondroitin sulphuric acid. J Biol Chem 15:69–79
Luo Y, Pfister P, Leisinger T, Wasserfallen A (2002) Pseudomurein endoisopeptidase PeiW and PeiP, two moderately related members of a novel family of proteases produced in *Methanothermobacter* strains. FEMS Microbiol Lett 208:47–51
Macalady JL, Vestling MM, Baumler D, Boekelheide N, Kaspar CW, Banfield JF (2004) Tetraether-linked membrane monolayers in *Ferroplasma spp.*: a key to survival in acid. Extremophiles 8:411–419
Matuschek M, Burchhardt G, Sahm K, Bahl H (1994) Pullulanase of *Thermoanaerobacterium thermosulphurigenes* EM1 (*Clostridium thermosulphurogenes*): molecular analysis of the gene, composite structure of the enzyme, and a common model for its attachment to the cell surface. J Bacteriol 176:3295–3302
Mayer F, Küper U, Meyer C, Daxer S, Müller V, Rachel R, Huber H (2012) AMP-forming acetyl coenzyme A synthetase in the outermost membrane of the hyperthermophilic crenarchaeon *Ignicoccus hospitalis*. J Bacteriol 194:1572–1581
Mescher MF, Strominger JL (1976) Structural (shape-maintaining) role of the cell surface glycoprotein of *Halobacterium salinarium*. Proc Natl Acad Sci USA 73:2687–2691
Mescher MF, Strominger JL (1976b) Purification and characterisation of a prokaryotic glycoprotein from the cell envelope of *Halobacterium salinarium*. J Biol Chem 251:2005–2014
Messner P, Sleytr UB (1992) Crystalline bacterial cell-surface layers. Adv Microb Physiol 33:213–274
Miroshnichenko ML, Gongadze GM, Rainey FA, Kostyukova AS, Lysenko AM, Chernyh NA, Bonch-Osmolovskaya EA (1998) *Thermococcus gorgonarius* sp. nov. and *Thermococcus pacificus* sp. nov.: heterotrophic extremely thermophilic archaea from New Zealand submarine hot vents. Int J Syst Evol Microbiol 48:23–29. https://doi.org/10.1099/00207713-48-1-23
Mitchell P (1961) Approaches to the analysis of specific membrane transport. In: Goodwin TW, Lindberg O (ed) Biological structure and function. Academic Press, New York, pp 581–603
Moissl C, Rachel R, Briegel A, Engelhardt H, Huber R (2005) The unique structure of archaeal 'hami', highly complex cell appendages with nano-grappling hooks. Mol Microbiol 56:361–370
Niemetz R, Kärcher U, Kandler O, Tindall BJ, König H (1997) The cell wall polymer of the extremely halophilic archaeon *Natronococcus occultus*. Eur J Biochem 249:905–911
Näther DJ, Rachel R (2004) The outer membrane of the hyperthermophilic archaeon *Ignicoccus*: dynamics, ultrastructure and composition. Biochem Soc Trans 32:199–203
Nußer E, König H (1987) S-layer studies on three species of *Methanococcus* living at different temperatures. Can J Microbiol 33:256–261. https://doi.org/10.1139/m87-043
Paper W, Jahn U, Hohn MJ, Kronner M, Näther DJ, Burghardt T, Rachel R, Stetter KO, Huber H (2007) *Ignicoccus hospitalis* sp. nov., the host of '*Nanoarchaeum equitans*'. IJSEM 57:803–808
Patel GB, Sprott GD, Humphrey RW, Beveridge TJ (1986) Comparative analyses of the sheath structures of *Methanothrix concilii* GP6 and *Methanospirillum hungatei* strains GP1 and JF1. Can J Microbiol 32:623–631
Perras AK, Wanner G, Klingl A, Mora M, Auerbach AK, Heinz V, Probst AJ, Huber H, Rachel R, Meck S, Moissl-Eichinger C (2014) Grappling archaea: ultrastructural analyses of an uncultivated, cold-loving archaeon, and its biofilm. Front Microbiol 5:397
Perras AK, Daum B, Ziegler C, Takahashi LK, Ahmed M, Wanner G, Klingl A, Leitinger G, Kolb-Lenz D, Gribaldo S, Auerbach A, Mora M, Probst AJ, Bellack A, Moissl-Eichinger C (2015) S-layers at second glance? Altiarchaeal grappling hooks (hami) resemble archaeal s-layer proteins in structure and sequence. Front Microbiol 6:543

Peters J, Nitsch M, Kuhlmorgen B, Golbik R, Lupas A, Kellermann J, Engelhardt H, Pfander JP, Müller S, Goldie K, Engel A, Stetter KO, Baumeister W (1995) Tetrabrachion: a filamentous archaebacterial surface protein assembly of unusual structure and extreme stability. J Mol Biol 245:385–401
Peters J, Baumeister W, Lupas A (1996) Hyperthermostable surface layer protein tetrabrachion from the archaebacterium *Staphylothermus marinus*: evidence for the presence of a right-handed coiled coil derived from the primary structure. J Mol Biol 257:1031–1041
Probst AJ, Weinmaier T, Raymann K, Perras A, Emerson JB, Rattei T, Wanner G, Klingl A, Berg IA, Yoshinaga M, Viehweger B, Hinrichs K-U, Thomas BC, Meck S, Auerbach AK, Heise M, Schintlmeister A, Schmid M, Wagner M, Gribaldo S, Banfield JF, Moissl-Eichinger C (2014) Biology of a widespread uncultivated archaeon that contributes to carbon fixation in the subsurface. Nat Commun 5:5497
Pum D, Sleytr UB (2014) Reassembly of S-layer proteins. Nanotechnology 25:312001. https://doi.org/10.1088/0957-4484/25/31/312001
Rachel R, Pum D, Šmarda J, Šmajs D, Komrska J, Krzyzánek V, Rieger G, Stetter KO (1997) II. Fine structure of S-layers. FEMS Microbiol Rev 20:13–23
Rachel R, Wyschkony I, Riehl S, Huber H (2002) The ultrastructure of *Ignicoccus*: evidence for a novel outer membrane and for intracellular vesicle budding in an archaeon. Archaea 1:9–18
Rodrigues-Oliveira T, Belmok A, Vasconcellos D, Schuster B, Kyaw CM (2017) Archaeal S-layers: overview and current state of the art. Front Microbiol 8:2597. https://doi.org/10.3389/fmicb.2017.02597
Ruiz N, Kahne D, Silhavy TJ (2006) Advances in understanding bacterial outer-membrane biogenesis. Nat Rev Microbiol 4:57–66
Schleifer KH, Kandler O (1972) Peptidoglycan types of bacterial cell walls and their taxonomic implications. Bacteriol Rev 36:407–477
Schleifer KH, Steber J, Mayer H (1982) Chemical composition and structure of the cell wall of *Halococcus morrhuae*. Zentralbl Bakteriol Mikrobiol Hyg Ser C3:171–178
Silhavy TJ, Kahne D, Walker S (2010) The bacterial cell envelope. Cold Spring Harb Perspect Biol 2:a000414
Sleytr UB, Messner P, Pum D, Sára M (1988) Crystalline bacterial cell surface layers. In: Proceedings of EMBO workshop on crystalline bacterial surface layers. Springer Verlag, Berlin, Heidelberg
Sleytr UB, Huber C, Ilk N, Pum D, Schuster B, Egelseer EM (2007) S-layers as a tool kit for nanobiotechnical applications. FEMS Microbiol Lett 267:131–144
Smith PF (1984) Lipoglycans from mycoplasmas. Crit Rev Microbiol 11:157–186
Sowers KR, Boone JE, Gunsalus RP (1993) Disaggregation of *Methanosarcina* spp. and growth as single cells at elevated osmolarity. Appl Environ Microbiol 59:3832–3839
Southam G, Beveridge TJ (1991) Immunochemical analysis of the sheath of the archaeobacterium *Methanospirillum hungatei* strain GP1. J Bacteriol 173:6213–6222
Southam G, Beveridge TJ (1992) Characterization of novel, phenol-soluble polypeptides which confer rigidity to the sheath of *Methanospirillum hungatei* GP1. J Bacteriol 174:935–946
Spang A, Saw JH, Jørgensen SL, Zaremba-Niedzwiedzka K, Martijn J, Lind AE, van Eijk R, Schleper C, Guy L, Ettema TJG (2015) Complex archaea that bridge the gap between prokaryotes and eukaryotes. Nature 521:173–179. https://doi.org/10.1038/nature14447
Sprott GD, Colvin JR, McKellar RC (1979) Spheroplasts of *Methanospirillum hungatii* formed upon treatment with dithiothreitol. Can J Microbiol 25:730–738
Sprott GD, McKellar RC (1980) Composition and properties of the cell wall of *Methanospirillum hungatei*. Can J Microbiol 26:115–120
Sprott GD, Beveridge TJ, Patel GB, Ferrante G (1986) Sheath disassembly in *Methanospirillum hungatei* strain GP1. Can J Microbiol 32:847–854
Steber J, Schleifer KH (1975) *Halococcus morrhuae*: a sulfated heteropolysaccharide as the structural component of the bacterial cell wall. Arch Microbiol 105:173–177
Steber J, Schleifer KH (1979) N-glycyl-glucosamine: a novel constituent in the cell wall of *Halococcus morrhuae*. Arch Microbiol 123:209–212

Steenbakkers PJM, Geerts WJ, Ayman-Oz NA, Keltjens JT (2006) Identification of pseudomurein cell wall binding domains. Mol Microbiol 62:1618–1630

Stieglmeier M, Klingl A, Rittmann S, Alves RE, Melcher M, Leisch N, Schleper C (2014) *Nitrososphaera viennensis* sp. nov., an aerobic and mesophilic ammonia oxidizing archaeon from soil and member of the novel archaeal phylum Thaumarchaeota. Int J Syst Evol Microbiol 64(8):2738–2752. https://doi.org/10.1099/ijs.0.063172-0

Sumper M, Berg E, Mengele R, Strobel L (1990) Primary structure and glycosylation of the S-layer protein of *Haloferax volcanii*. J Bacteriol 172:7111–7118

Sumper M, Wieland FT (1995) Bacterial glycoproteins. In: Montreuil J, Vliegenthart JFG, Schachter H (eds) Glycoproteins. Elsevier, Amsterdam, pp 455–473

Tindall BJ, Ross HMN, Grant WD (1984) *Natronobacterium gen. nov.* and *Natronococcus gen. nov.*—two genera of halalkaliphilc archaebacteria. Syst Appl Microbiol 5:41–57

Veith A, Klingl A, Zolghadr B, Lauber K, Mentele R, Lottspeich F, Rachel R, Alber S-V, Kletzin A (2009) *Acidianus*, *Sulfolobus* and *Metallosphaera* surface layers: structure, composition and gene expression. Mol Microbiol 73:58–72

Visweswaran GR, Dijkstra BW, Kok J (2010) Two major archaeal pseudomurein endoisopeptidases: PeiW and PeiP. Archaea 2010:480492. https://doi.org/10.1155/2010/480492

Walsby AE (1994) Gas vesicles. Microbiol Rev 58:94–144

Yang LL, Haug A (1979) Structure of membrane lipids and physico-biochemical properties of the plasma membrane from *Thermoplasma acidophilum*, adapted to growth at 37 degrees C. Biochem Biophys Acta 573:308–320

Wildhaber I, Baumeister W (1987) The cell envelope of *Thermoproteus tenax*: three-dimensional structure of the surface layer and its role in shape maintenance. EMBO J 6:1475–1480

Woese CR, Kandler O, Wheelis ML (1990) Towards a natural system of organisms: proposal for the domains Archaea, Bacteria, and Eucarya. Proc Natl Acad Sci USA 87:4576–4579. https://doi.org/10.1073/pnas.87.12.4576

Yurist-Doutsch S, Chaban B, VanDyke DJ, Jarrell KF, Eichler J (2008) Sweet to the extreme: protein glycosylation in Archaea. Mol Microbiol 68(5):1079–84. https://doi.org/10.1111/j.1365-2958.2008.06224.x

Zeikus JG, Bowen VG (1975) Fine structure of *Methanospirillum hungati*. J Bacteriol 121:373–380

Zenke R, von Gronau S, Bolhuis H, Gruska M, Pfeiffer F, Oesterhelt D (2015) Fluorescence microscopy visualization of halomucin, a secreted 927 kDa protein surrounding *Haloquadratum walsbyi* cells. Front Microbiol 6:249

Zolghadr B, Klingl A, Rachel R, Driessen AJM, Albers S-V (2011) The bindosome is a structural component of the *Sulfolobus solfataricus* cell envelope. Extremophiles 15:235–244

Index

A. Kuhn (ed.), *Bacterial Cell Walls and Membranes*, Subcellular Biochemistry 92, https://doi.org/10.1007/978-3-030-18768-2

FSC
www.fsc.org
MIX
Papier aus verantwortungsvollen Quellen
Paper from responsible sources
FSC® C105338

If you have any concerns about our products,
you can contact us on
ProductSafety@springernature.com

In case Publisher is established outside the EU,
the EU authorized representative is:
Springer Nature Customer Service Center GmbH
Europaplatz 3, 69115 Heidelberg, Germany

Printed by Libri Plureos GmbH
in Hamburg, Germany